K. Hsieh

Monolithic Microwave
Integrated Circuits

OTHER IEEE PRESS BOOKS

Kalman Filtering: Theory and Application, *Edited by H. W. Sorenson*
Spectrum Management and Engineering, *Edited by F. Matos*
Digital VLSI Systems, *Edited by M. T. Elmasry*
Introduction to Magnetic Recording, *Edited by R. M. White*
Insights Into Personal Computers, *Edited by A. Gupta and H. D. Toong*
Television Technology Today, *Edited by T. S. Rzeszewski*
The Space Station: An Idea Whose Time Has Come, *By T. R. Simpson*
Marketing Technical Ideas and Products Successfully! *Edited by L. K. Moore and D. L. Plung*
The Making of a Profession: A Century of Electrical Engineering in America, *By A. M. McMahon*
Power Transistors: Device Design and Applications, *Edited by B. J. Baliga and D. Y. Chen*
VLSI: Technology and Design, *Edited by O. G. Folberth and W. D. Grobman*
General and Industrial Management, *By H. Fayol; revised by I. Gray*
A Century of Honors, *An IEEE Centennial Directory*
MOS Switched-Capacitor Filters: Analysis and Design, *Edited by G. S. Moschytz*
Distributed Computing: Concepts and Implementations, *Edited by P. L. McEntire, J. G. O'Reilly, and R. E. Larson*
Engineers and Electrons, *By J. D. Ryder and D. G. Fink*
Land-Mobile Communications Engineering, *Edited by D. Bodson, G. F. McClure, and S. R. McConoughey*
Frequency Stability: Fundamentals and Measurement, *Edited by V. F. Kroupa*
Electronic Displays, *Edited by H. I. Refioglu*
Spread-Spectrum Communications, *Edited by C. E. Cook, F. W. Ellersick, L. B. Milstein, and D. L. Schilling*
Color Television, *Edited by T. Rzeszewski*
Advanced Microprocessors, *Edited by A. Gupta and H. D. Toong*
Biological Effects of Electromagnetic Radiation, *Edited by J. M. Osepchuk*
Engineering Contributions to Biophysical Electrocardiography, *Edited by T. C. Pilkington and R. Plonsey*
The World of Large Scale Systems, *Edited by J. D. Palmer and R. Saeks*
Electronic Switching: Digital Central Systems of the World, *Edited by A. E. Joel, Jr.*
A Guide for Writing Better Technical Papers, *Edited by C. Harkins and D. L. Plung*
Low-Noise Microwave Transistors and Amplifiers, *Edited by H. Fukui*
Digital MOS Integrated Circuits, *Edited by M. I. Elmasry*
Geometric Theory of Diffraction, *Edited by R. C. Hansen*
Modern Active Filter Design, *Edited by R. Schaumann, M. A. Soderstrand, and K. B. Laker*
Adjustable Speed AC Drive Systems, *Edited by B. K. Bose*
Optical Fiber Technology, II, *Edited by C. K. Kao*
Protective Relaying for Power Systems, *Edited by S. H. Horowitz*
Analog MOS Integrated Circuits, *Edited by P. R. Gray, D. A. Hodges, and R. W. Broderson*
Interference Analysis of Communication Systems, *Edited by P. Stavroulakis*
Integrated Injection Logic, *Edited by J. E. Smith*
Sensory Aids for the Hearing Impaired, *Edited by H. Levitt, J. M. Pickett, and R. A. Houde*
Data Conversion Integrated Circuits, *Edited by D. J. Dooley*
Semiconductor Injection Lasers, *Edited by J. K. Butler*
Satellite Communications, *Edited by H. L. Van Trees*
Frequency-Response Methods in Control Systems, *Edited by A. G. J. MacFarlane*
Programs for Digital Signal Processing, *Edited by the Digital Signal Processing Committee, IEEE*
Automatic Speech & Speaker Recognition, *Edited by N. R. Dixon and T. B. Martin*
Speech Analysis, *Edited by R. W. Schafer and J. D. Markel*
The Engineer in Transition to Management, *By I. Gray*
Multidimensional Systems: Theory & Applications, *Edited by N. K. Bose*
Analog Integrated Circuits, *Edited by A. B. Grebene*
Integrated-Circuit Operational Amplifiers, *Edited by R. G. Meyer*
Modern Spectrum Analysis, *Edited by D. G. Childers*
Digital Image Processing for Remote Sensing, *Edited by R. Bernstein*
Reflector Antennas, *Edited by W. Love*
Phase-Locked Loops & Their Application, *Edited by W. C. Lindsey and M. K. Simon*
Digital Signal Computers and Processors, *Edited by A. C. Salazar*
Systems Engineering: Methodology and Applications, *Edited by A. P. Sage*
Modern Crystal and Mechanical Filters, *Edited by D. F. Sheahan and R. A. Johnson*
Electrical Noise: Fundamentals and Sources, *Edited by M. S. Gupta*
Computer Methods in Image Analysis, *Edited by J. K. Aggarwal, R. O. Duda, and A. Rosenfeld*

Monolithic Microwave Integrated Circuits

Edited by
Robert A. Pucel
Consulting Scientist
Research Division, Raytheon Company

A volume in the IEEE PRESS Selected Reprint
Series, prepared under the sponsorship of the
IEEE Microwave Theory and Techniques Society.

The Institute of Electrical and Electronics Engineers, Inc., New York

IEEE PRESS
1985 Editorial Board

M. E. Van Valkenburg, *Editor in Chief*

M. G. Morgan, *Editor, Selected Reprint Series*

Glen Wade, *Editor, Special Issue Series*

J. M. Aein	Thelma Estrin	J. O. Limb
J. K. Aggarwal	L. H. Fink	R. W. Lucky
James Aylor	S. K. Ghandhi	E. A. Marcatili
J. E. Brittain	Irwin Gray	J. S. Meditch
R. W. Brodersen	H. A. Haus	W. R. Perkins
B. D. Carroll	E. W. Herold	A. C. Schell
R. F. Cotellessa	R. C. Jaeger	Herbert Sherman
M. S. Dresselhaus		D. L. Vines

W. R. Crone, *Managing Editor*

Hans P. Leander, *Technical Editor*

Teresa Abiuso, *Administrative Assistant*

Emily Gross, *Associate Editor*

Copyright © 1985 by
THE INSTITUTE OF ELECTRICAL AND ELECTRONICS ENGINEERS, INC.
345 East 47th Street, New York, NY 10017-2394
All rights reserved.

PRINTED IN THE UNITED STATES OF AMERICA

IEEE Order Number: PC01867

Library of Congress Cataloging in Publication Data
Main entry under title:

Monolithic microwave integrated circuits.

(IEEE Press selected reprint series)
Includes indexes.
1. Microwave integrated circuits—Addresses, essays, lectures.   2. Gallium arsenide semiconductors—Addresses, essays, lectures.   I. Pucel, Robert A. II. Series.
TK7876.M65   1985   621.381'32   85-11796
ISBN 0-87942-192-4

Contents

Preface ix

Introduction 1

Part I: Design Considerations 11
Design Considerations for Monolithic Microwave Circuits, *R. A. Pucel* (*IEEE Transactions on Microwave Theory and Techniques*, June 1981) 13
Hybrid vs. Monolithic Microwave Circuits: A Matter of Cost, *R. S. Pengelly* (*Microwave System News*, January 1983) 35
GaAs MMICs: Manufacturing Trends and Issues, *A. Christou* (*IEEE Gallium Arsenide Integrated Circuit Symposium Technical Digest*, 1984) 56

Part II: Materials and Processing Considerations 61
High-Purity Semi-Insulating GaAs Material for Monolithic Microwave Integrated Circuits, *H. M. Hobgood, G. W. Eldridge, D. L. Barrett, and R. N. Thomas* (*IEEE Transactions on Electron Devices*, February 1981) 63
Growth of High-Purity Semi-Insulating Bulk GaAs for Integrated-Circuit Applications, *R. D. Fairman, R. T. Chen, J. R. Oliver, and D. R. Ch'en* (*IEEE Transactions on Electron Devices*, February 1981) 73
Growth and Characterization of Dislocation-Free GaAs Mixed Crystals for IC Substrates, *K. Tada, S. Murai, S. Akai, and T. Suzuki* (*IEEE Gallium Arsenide Integrated Circuit Symposium Technical Digest*, 1984) 79
The Manufacturability of GaAs Integrated Circuits, *B. M. Welch and Y. Shen* (*IEEE Gallium Arsenide Integrated Circuit Symposium Technical Digest*, 1982) 83
A Manufacturing Process for Analog and Digital Gallium Arsenide Integrated Circuits, *R. L. Van Tuyl, V. Kumar, D. C. D'Avanzo, T. W. Taylor, V. E. Peterson, D. P. Hornbuckle, R. A. Fisher, and D. B. Estreich* (*IEEE Transactions on Microwave Theory and Techniques*, July 1982) 87
Transient Capless Annealing of Ion-Implanted GaAs, *R. C. Clarke and G. W. Eldridge* (*IEEE Transactions on Electron Devices*, August 1984) 94
Proton Isolation for GaAs Integrated Circuits, *D. C. D'Avanzo* (*IEEE Transactions on Microwave Theory and Techniques*, July 1982) 100
Properties of Surface Passivation Dielectrics for GaAs Integrated Circuits, *H. Hasegawa, T. Sawada, and T. Kitagawa* (*IEEE Gallium Arsenide Integrated Circuit Symposium Technical Digest*, 1983) 108
A Tantalum-Based Process for MMIC On-Chip Thin-Film Components, *M. Durschlag and J. L. Vorhaus* (*IEEE Gallium Arsenide Integrated Circuit Symposium Technical Digest*, 1982) 112
Tantalum Oxide Capacitance for GaAs Monolithic Integrated Circuits, *M. E. Elta, A. Chu, L. J. Mahoney, R. T. Cerretani, and W. E. Courtney* (*IEEE Electron Device Letters*, May 1982) 116
GaAs Microwave Devices and Circuits with Submicron Electron-Beam Defined Features, *W. R. Wisseman, H. M. Macksey, G. E. Brehm, and P. Saunier* (*Proceedings of the IEEE*, May 1983) 119
Yield Considerations for Ion-Implanted GaAs MMIC's, *A. Gupta, W. C. Petersen, and D. R. Decker* (*IEEE Transactions on Electron Devices*, January 1983) 128
Backgating in GaAs MESFET's, *C. Kocot and C. A. Stolte* (*IEEE Transactions on Microwave Theory and Techniques*, July 1982) 132
Effect of Substrate Conduction and Backgating on the Performance of GaAs Integrated Circuits, *C. P. Lee, R. Vahrenkamp, S. J. Lee, Y. D. Shen, and B. M. Welch* (*IEEE Gallium Arsenide Integrated Circuit Symposium Technical Digest*, 1982) 138
Radiation Effects in GaAs Integrated Circuits: A Comparison with Silicon, *M. Simons* (*IEEE Gallium Arsenide Integrated Circuit Symposium Technical Digest*, 1983) 142
Gamma Ray Radiation Effects on MMIC's Elements, *K. Aono, O. Ishihara, K. Nishitani, M. Nakatani, K. Fujikawa, M. Ohtani, and T. Odaka* (*IEEE Gallium Arsenide Integrated Circuit Symposium Technical Digest*, 1984) 147

Part III: Monolithic Circuit Applications 151
    Section III-A: Low-Noise Amplifiers and Other Receiver Circuits 152
GaAs Monolithic MIC's for Direct Broadcast Satellite Receivers, *S. Hori, K. Kamei, K. Shibata, M. Tatematsu, K. Mishima, and S. Okano* (*IEEE Transactions on Microwave Theory and Techniques*, December 1983) 153
12-GHz-Band Low-Noise GaAs Monolithic Amplifiers, *T. Sugiura, H. Itoh, T. Tsuji, and K. Honjo* (*IEEE Transactions on Microwave Theory and Techniques*, December 1983) 160
Monolithic Circuits for 12 GHz Direct Broadcasting Satellite Reception, *C. Kermarrec, P. Harrop, C. Tsironis, and J. Faguet* (*IEEE Microwave and Millimeter-Wave Monolithic Circuits Symposium Digest of Papers*, 1982) 166

A Low Cost GaAs Monolithic LNA for TVRO Applications, S. Moghe, T. Andrade, and H. Sun (*IEEE Gallium Arsenide Integrated Circuit Symposium Technical Digest*, 1984) 172

12 GHz-Band GaAs Dual-Gate MESFET Monolithic Mixers, T. Sugiura, K. Honjo, and T. Tsuji (*IEEE Gallium Arsenide Integrated Circuit Symposium Technical Digest*, 1983) 175

An 8-18 GHz Monolithic Two-Stage Low Noise Amplifier, L. C. T. Liu, D. W. Maki, C. Storment, M. Sokolich, and W. Klatskin (*IEEE Microwave and Millimeter-Wave Monolithic Circuits Symposium Digest of Papers*, 1984) 179

Single and Dual Stage Monolithic Low Noise Amplifiers, L. C. Liu, D. W. Maki, M. Feng, and M. Siracusa (*IEEE Gallium Arsenide Integrated Circuit Symposium Technical Digest*, 1982) 182

10 GHz Monolithic GaAs Low Noise Amplifier with Common-Gate Input, R. E. Lehmann, G. E. Brehm, D. J. Seymour, and G. H. Westphal (*IEEE Gallium Arsenide Integrated Circuit Symposium Technical Digest*, 1982) 186

Monolithic Microwave Gallium Arsenide FET Oscillators, J. S. Joshi, J. R. Cockrill, and J. A. Turner (*IEEE Transasctions on Electron Devices*, Feburary 1981) 190

Monolithic Voltage Controlled Oscillator for X- and Ku-Bands, B. N. Scott and G. E. Brehm (*IEEE Transactions on Microwave Theory and Techniques*, December 1982) 195

A Monolithic GaAs IC for Heterodyne Generation of RF Signals, R. L. Van Tuyl (*IEEE Transactions on Electron Devices*, February 1981) 201

Section III-B: Power Amplifiers 206

Design, Fabrication, and Characterization of Monolithic Microwave GaAs Power FET Amplifiers, H. Q. Tserng, H. M. Macksey, and S. R. Nelson (*IEEE Transactions on Electron Devices*, February 1981) 207

Monolithic Broadband Power Amplifier at X-Band, A. Platzker, M. S. Durschlag, and J. Vorhaus (*IEEE Microwave and Millimeter-Wave Monolithic Circuits Symposium Digest of Papers*, 1983) 215

Wideband 3W Amplifier Employing Cluster Matching, R. G. Freitag, J. E. Degenford, D. C. Boire, M. C. Driver, R. A. Wickstrom, and C. D. Chang (*IEEE Microwave and Millimeter-Wave Monolithic Circuits Symposium Digest of Papers*, 1983) 218

A 2-8 GHz 2 Watt Monolithic Amplifier, J. Dormail, Y. Tajima, R. Mozzi, M. Durschlag, A. M. Morris, and S. A. McOwen (*IEEE Gallium Arsenide Integrated Circuit Symposium Technical Digest*, 1983) 223

A GaAs Monolithic 6-18 GHz Medium Power Amplifier, C. D. Palmer, P. Saunier, and R. E. Williams (*IEEE Microwave and Millimeter-Wave Monolithic Circuits Symposium Digest of Papers*, 1984) 227

Section III-C: Broad-Band Amplifiers 230

Multi-Octave Performance of Single-Ended Microwave Solid-State Amplifiers, K. B. Niclas (*IEEE Transactions on Microwave Theory and Techniques*, August 1984) 231

A Monolithic GaAs 1-13-GHz Traveling-Wave Amplifier, Y. Ayasli, R. L. Mozzi, J. L. Vorhaus, L. D. Reynolds, and R. A. Pucel (*IEEE Transactions on Microwave Theory and Techniques*, July 1982) 244

2 to 30 GHz Monolithic Distributed Amplifier, J. M. Schellenberg, H. Yamasaki, and P. G. Asher (*IEEE Gallium Arsenide Integrated Circuit Symposium Technical Digest*, 1984) 249

A 2-18-GHz Monolithic Distributed Amplifier Using Dual-Gate GaAs FET's, W. Kennan, T. Andrade, and C. C. Huang (*IEEE Transactions on Microwave Theory and Techniques*, December 1984) 252

2-20-GHz GaAs Traveling-Wave Power Amplifier, Y. Ayasli, L. D. Reynolds, R. L. Mozzi, and L. K. Hanes (*IEEE Transactions on Microwave Theory and Techniques*, March 1984) 257

On Noise in Distributed Amplifiers at Microwave Frequencies, K. B. Niclas and B. A. Tucker (*IEEE Transactions on Microwave Theory and Techniques*, August 1983) 262

X, Ku-Band GaAs Monolithic Amplifier, Y. Tajima, T. Tsukii, E. Tong, R. Mozzi, L. Hanes, and B. Wrona (*IEEE MTT-S International Microwave Symposium Digest*, 1982) 270

The Matched Feedback Amplifier: Ultrawide-Band Microwave Amplification with GaAs MESFET's, K. B. Niclas, W. T. Wilser, R. B. Gold, and W. R. Hitchens (*IEEE Transactions on Microwave Theory and Techniques*, April 1980) 273

A Monolithic GaAs DC to 2-GHz Feedback Amplifier, W. C. Petersen, A. Gupta, and D. R. Decker (*IEEE Techniques on Electron Devices*, January 1983) 283

A 2.2 dB NF 30-1700 MHz Feedback Amplifier, M. Nishiuma, S. Katsu, S. Nambu, M. Hagio, and G. Kano (*IEEE International Solid-State Circuits Conference Digest of Technical Papers*, 1983) 286

Broadband Monolithic Low-Noise Feedback Amplifiers, P. N. Rigby, J. R. Suffolk, and R. S. Pengelly (*IEEE MTT-S International Microwave Symposium Digest*, 1983) 288

A Monolithic Multi-Stage 6-18 GHz Feedback Amplifier, A. M. Pavio, S. D. McCarter, and P. Saunier (*IEEE Microwave and Millimeter-Wave Circuits Symposium*, 1984) 293

GaAs FET Ultraband-Band Amplifiers for Gbit/s Data Rate Systems, K. Honjo and Y. Takayama (*IEEE Transactions on Microwave Theory and Techniques*, July 1981) 297

Monolithic GaAs Direct-Coupled Amplifiers, D. P. Hornbuckle and R. L. Van Tuyl (*IEEE Transactions on Electron Devices*, Feburary 1981) 305

A Monolithic Direct-Coupled GaAs IC Amplifier with 12-GHz Bandwidth, S. B. Moghe, H.-J. Sun, T. Andrade, C. C. Huang, and R. Goyal (*IEEE Transactions on Microwave Theory and Techniques*, December 1984) 313

A Monolithic Wide-Band GaAs IC Amplifier, D. B. Estreich (*IEEE Journal of Solid-State Circuits,* December 1982) 319
Inductively Coupled Push-Pull Amplifiers for Low Cost Monolithic Microwave ICs, S. A. Jamison, A. Podell, M. Helix, P. Ng, C. Chao, G. E. Webber, and R. Lokken (*IEEE Gallium Arsenide Integrated Circuit Symposium Technical Digest,* 1982) 327
Wideband Monolithic Cascadable Feedback Amplifiers Using Silicon Bipolar Technology, J. Kukielka and C. Snapp (*IEEE Microwave and Millimeter-Wave Circuits Symposium,* 1982) 330
Bipolar Monolithic Amplifiers for a Gigabit Optical Repeater, M. Ohara, Y. Akazawa, N. Ishihara, and S. Konaka (*IEEE Journal of Solid-State Circuits,* August 1984) 332

Section III-D: Transmit/Receive Modules 338
GaAs Monolithic Microwave Circuits for Phased-Array Applications, R. S. Pengelly (*IEE Proceedings,* August 1980) 339
Microwave Switching with GaAs FETs, Y. Ayasli (*Microwave Journal Magazine,* November 1982) 350
A Monolithic Single-Chip X-Band Four-Bit Phase Shifter, Y. Ayasli, A. Platzker, J. Vorhaus, and L. D. Reynolds, Jr. (*IEEE Transactions on Microwave Theory and Techniques,* December 1982) 358
An Analog X-Band Phase Shifter, D. E. Dawson, A. L. Conti, S. H. Lee, G. F. Shade, and L. E. Dickens (*IEEE Microwave and Millimeter-Wave Monolithic Circuits Symposium Digest of Papers,* 1984) 363
A Multi-Chip GaAs Monolithic Transmit/Receive Module for X-Band, R. A. Pucel, Y. Ayasli, D. Wandrei, J. L. Vorhaus, S. Temple, R. Waterman, A. Platzker, and C. Cavicchio (*IEEE MTT-S International Microwave Symposium Digest,* 1984) 372

Section III-E: Millimeter-Wave Circuits 376
Receiver Technology for the Millimeter and Submillimeter Wave Regions, B. J. Clifton (*Integrated Optics and Millimeter and Microwave Integrated Circuits, Proceedings SPIE,* 1981) 377
GaAs Integrated Microwave Circuits, E. W. Mehal and R. W. Wacker (*IEEE Transactions on Microwave Theory and Techniques,* July 1968) 386
A 31-GHz Monolithic GaAs Mixer/Preamplifier Circuit for Receiver Applications, A. Chu, W. E. Courtney, and R. W. Sudbury (*IEEE Transactions on Electron Devices,* February 1981) 390
A 69 GHz Monolithic FET Oscillator, D. W. Maki, J. M. Schellenberg, H. Yamasaki, and L. C. T. Liu (*IEEE Microwave and Millimeter-Wave Monolithic Circuits Symposium Digest of Papers,* 1984) 396
Ka-Band Monolithic GaAs Balanced Mixers, C. Chao, A. Contolatis, S. A. Jamison, and P. E. Bauhahn (*IEEE Transactions on Microwave Theory and Techniques,* January 1983) 401
94 GHz Planar GaAs Monolithic Balanced Mixer, P. Bauhahn, T. Contolatis, J. Abrokwah, C. Chao, and C. Seashore (*IEEE Microwave and Millimeter-Wave Monolithic Circuits Symposium Digest of Papers,* 1984) 406
A W-Band Monolithic GaAs Crossbar Mixer, L. T. Yuan (*IEEE Microwave and Millimeter-Wave Monolithic Circuits Symposium Digest of Papers,* 1984) 410

Section III-F: Special Components and Circuits 413
GaAs Monolithic Lange and Wilkinson Couplers, R. C. Waterman, Jr., W. Fabian, R. A. Pucel, Y. Tajima, and J. L. Vorhaus (*IEEE Transactions on Electron Devices,* February 1981) 414
A Microwave Phase and Gain Controller with Segmented-Dual-Gate MESFETs in GaAs MMICs, Y. C. Hwang, Y. K. Chen, R. J. Naster, and D. Temme (*IEEE Microwave and Millimeter-Wave Monolithic Circuits Symposium Digest of Papers,* 1984) 419
GaAs Monolithic Wideband (2-18 GHz) Variable Attenuators, Y. Tajima, T. Tsukii, R. Mozzi, E. Tong, L. Hanes, and B. Wrona (*IEEE MTT-S International Microwave Symposium Digest,* 1982) 424
Design of Interdigitated Capacitors and Their Application to Gallium Arsenide Monolithic Filters, R. Esfandiari, D. W. Maki, and M. Siracusa (*IEEE Transactions on Microwave Theory and Techniques,* January 1983) 427

Part IV: CAD, Measurement, and Packaging Techniques 435
Computer-Aided Design of Hybrid and Monolithic Microwave Integrated Circuits—State of the Art, Problems and Trends, R. H. Jansen (*13th European Microwave Conference Proceedings,* 1983) 437
Computer-Aided Design for the 1980's, L. Besser, C. Holmes, M. Ball, M. Medley, and S. March (*IEEE MTT-S International Microwave Symposium Digest,* 1981) 449
GaAs FET Large-Signal Model and Its Application to Circuit Designs, Y. Yajima, B. Wrona, and K. Mishima (*IEEE Transactions on Electron Devices,* February 1981) 452
Large-Signal GaAs FET Amplifier CAD Program, A. Platzker and Y. Tajima (*IEEE MTT-S International Microwave Symposium Digest,* 1982) 457
Accurate Models for Microstrip Computer-Aided Design, E. Hammerstad and O. Jensen (*IEEE MTT-S International Microwave Symposium Digest,* 1980) 460
Design of Microwave GaAs MESFET's for Broad-Band Low-Noise Amplifiers, H. Fukui (*IEEE Transactions on Microwave Theory and Techniques,* July 1979) 463
Addendum to "Design of Microwave GaAs MESFET's for Broad-Band Low-Noise Amplifiers" (*IEEE Transactions on Microwave Theory and Techniques,* October 1981) 470

Accurate Coupling Predictions and Assessments in MMIC Networks, *H. J. Finlay, J. A. Jenkins, R. S. Pengelly, and J. Cockrill* (*IEEE Gallium Arsenide Integrated Circuit Symposium Technical Digest*, 1983) 471

The Design and Calibration of a Universal MMIC Test Fixture, *J. A. Benet* (*IEEE Microwave and Millimeter-Wave Monolithic Circuits Symposium Digest of Papers*, 1982) 475

Calibration Methods for Microwave Wafer Probing, *E. W. Strid and K. R. Gleason* (*IEEE Microwave and Millimeter-Wave Monolithic Circuits Symposium Digest of Papers*, 1984) 481

WAFFLELINE—A Packaging Technique for Monolithic Microwave Integrated Circuits, *D. E. Heckaman, J. A. Frisco, J. B. Schappacher, and D. A. Koopman* (*IEEE Gallium Arsenide Integrated Circuit Symposium Technical Digest*, 1984) ... 486

A Low Cost Multiport Microwave Package for GaAs ICs, *D. A. Rowe, B. Y. Lao, R. E. Dietterle, and M. A. Moacanin* (*IEEE Gallium Arsenide Integrated Circuit Symposium Technical Digest*, 1984) 490

Author Index 493

Subject Index 495

Editor's Biography 501

Preface

THIS reprint volume is a selective compilation of papers on monolithic microwave integrated circuits (MMIC's) and related topics. It is intended to serve as a reference volume for workers already engaged in this field, such as materials, device, processing, and microwave engineers and technicians. Those contemplating entering this exciting field or wishing to learn more about it, for example, research planners and graduate students in physics, chemistry, and microwave engineering, should find this book useful. We hope that it also may play a role in the formulation of courses at universities.

The papers in this volume aim to provide an historical basis for the emergence of the MMIC field, and a basic understanding of the developments in the various scientific and engineering disciplines which have helped shape the rapid progress made within the last seven years.

It is my belief that the influence of MMIC's on the design of future microwave systems and on consumer applications of microwave circuits will be so profound that we are not yet able to fully appreciate it. From a modest beginning only six years ago with circuits designed for the S–X bands, the range of applicability of GaAs MMIC's has been extended to frequencies as low as 50 MHz and as high as 100 GHz!

It goes without saying that this volume would not have been possible without the contributions of the many research and development workers in the field who, through their creativity and tireless efforts and unbounded enthusiasm, have made the MMIC field what it is today. I wish to single out, in particular, my colleague and supervisor, R. W. Bierig, who through his vision and encouragement in the early days of development has exerted a major influence on the rapid emergence of the MMIC field.

It is my pleasure to acknowledge the assistance of my co-workers at Raytheon who helped me in the final selection from the many excellent papers which I would like to have included in the reprint volume. To the extent that we have failed in this mission, the responsibility rests with me.

It is with sincere gratitude that I express my appreciation to the members of the Administrative Committee of the IEEE Microwave Theory and Techniques Society, in particular, R. H. Knerr, for the confidence shown in me by presenting the opportunity and honor of compiling this volume. The cheerful cooperation and support given me by the members of the IEEE PRESS Editorial Board and the staff of the IEEE PRESS, in particular W. R. Crone and H. P. Leander, in the preparation of this reprint volume is gratefully acknowledged.

Special thanks are due H. Ellowitz of the *Microwave Journal* and C. Braun of the *Microwave System News* for their kind permission to reprint articles from their journals and for their preparation of "camera ready" copies.

I am also deeply indebted to the members of the Publication Department at the Research Division of Raytheon for their enthusiastic support of this project, and their professional assistance in the preparation of this volume.

ROBERT A. PUCEL
Editor

Introduction

RECENT years have witnessed the emergence of a new and exciting microwave technology which is expected to exert a profound influence on future microwave system designs in the military, commercial, and consumer markets. This new approach, monolithic microwave integrated circuits (MMIC's), is the subject of this reprint volume.

During the last two decades, remarkable advances have been made in the integration of microwave circuits ($f > 1$ GHz). Much of the progress during this period has mirrored the advances in the development of solid-state (semiconductor) microwave devices—first with silicon bipolar technology, and more recently, with gallium arsenide (GaAs) field-effect transistor (FET) technology. However, this progress in microwave integrated circuit (MIC) design could not have developed as it did without concommitant advances in the technology of planar circuits.

It is appropriate, before we proceed further, to define precisely what we mean by "microwave integrated circuits." We mean circuits which are formed in part, or in totality, on a planar dielectric surface by one or more deposition schemes. If, on the one hand, some or all of the component parts of the circuit are deposited on the dielectric surface, *except the active elements*, that is, the semiconductor diodes, triodes, etc., then the circuit approach is called the "hybrid" approach. This is the approach that has been with us for nearly 20 years. In this case, the substrate is generally a low-loss material, such as alumina, which serves two purposes 1) as a surface onto which parts are mounted and 2) as a wave propagating medium.

On the other hand, if all of the circuit components are deposited, *including active elements*, then the approach is called the "monolithic" approach. To be precise:

> the monolithic approach is an approach wherein all active and passive circuit elements and interconnections are formed *in situ* on or within a semi-insulating semiconductor substrate by a combination of deposition schemes such as epitaxy, ion implantation, sputtering, and evaporation.

Active elements include, for example, diodes and transistors used as signal amplifiers, modulators, and switches, to name a few applications. The full implication of this definition should be understood, since it underlies the reasons for designing and fabricating MMIC's. The reasons are embedded in the following promising attributes of the monolithic approach:

1) potentially low-cost circuits through batch processing;
2) improved reliability and reproducibility through minimization of wire bonds and discrete components;
3) small size and weight;
4) circuit design flexibility and multi-function performance on a chip.

Note that the monolithic approach requires the semi-insulating substrate to include a layer of semiconducting material into which the active devices are formed. Most of the papers contained in this volume concern themselves with MMIC's based on gallium arsenide (GaAs) substrates, with a few dealing with silicon (Si) substrates.[1] Other semiconductors, such as InP, have been considered for MMIC applications, but there is little progress to report at this time.

It should be evident to the reader that the salient feature of the monolithic approach is the elimination of all wire-bonding of components within the circuit itself. Needless to say, this approach, therefore, has the potential for low-cost batch production of high reliability circuits of extremely small-size and weight. Table I summarizes the features of the hybrid and monolithic approaches.

MMIC's: A Brief History

The concept of MMIC's is not new. Its origin can be traced to 1964 to a government-funded program based on silicon technology, which had as its objective a transmit-receive (T/R) module for an aircraft phased-array antenna. Unfortunately, the results were disappointing because the semi-insulating properties of the silicon substrate could not be maintained through the high-temperature diffusion and oxidation cycles. Thus, very lossy substrates resulted; these were unacceptable for microwave circuitry [1]. Because of this and other difficulties the attempt to form a monolithic circuit based on a semiconductor substrate lay dormant until Mehal and Wacker [3] revived the approach by using semi-insulating gallium arsenide as the base material and Schottky barrier diodes and Gunn devices as the active elements to fabricate a 94 GHz receiver front end. Because of the high-temperature properties of GaAs and the absence of high-temperature process steps, the problems of silicon were not present. However, it was not until workers at Plessey Co. applied this approach to an X-band amplifier, based on the Schottky-gate field-effect transistor, or MESFET (metal-semiconductor field-effect transistor), as the key active element that the present intense activity began [4].

What brought on this revival? First, the rapid development of GaAs material technology, namely, epitaxy and ion implantation, and the speedy evolution of the GaAs FET based on the

TABLE I
FEATURES OF HYBRID AND MONOLITHIC APPROACHES

Feature	Hybrid	Monolithic
Type of substrate	Insulator	Semiconductor
Passive components	Discrete/deposited	Deposited
Active components	Discrete	Deposited
Interconnects	Deposited and wire-bonded	Deposited
Batch processed?	No	Yes
Labor intensive?	Yes	No

---

[1] The silicon approach usually employs junction isolation in lieu of a semi-insulating Si substrate in contemporary MMIC's.

TABLE II
SOME PROPERTIES OF SEMICONDUCTORS AND INSULATORS

Property	GaAs	Silicon	Semi-insulating GaAs	Semi-insulating Silicon	Sapphire	Alumina
Dielectric Constant	12.9	11.7	12.9	11.7	11.6 (C-axis)	9.7
Density (gm/cc)	5.32	2.33	5.32	2.33	3.98	3.89
Thermal Cond. (watts/cm-°K)	0.46	1.45	0.46	1.45	0.46	0.37
Resistivity (ohm-cm)	---	---	$10^7 - 10^9$	$10^3 - 10^5$	$>10^{14}$	$10^{11} - 10^{14}$
Elec. Mobility (cm^2/v-sec.)	4300*	700*	---	---	---	---
Sat. Elec. Vel. (cm/sec.)	1.3×10^7	9×10^6	---	---	---	---

* At $10^{17}/cm^3$ doping

metal Schottky gate during the last decade led to high-frequency device performance previously unattained. Second, resolution of many troublesome device reliability problems made FET's more attractive for systems applications. Third, recognition on the part of circuit designers of the excellent microwave properties of semi-insulating GaAs (approaching that of alumina) removed the major objection associated with silicon. Fourth, hybrid circuits were becoming very complex and labor intensive because of the prolific use of wire bonds, and hence too costly. Fifth, clearly defined and discernable systems applications for MMIC's began to emerge. Sixth, the development of sophisticated computer-aided design (CAD) software allowed one to optimize the design of a circuit before building it, an essential requirement with the monolithic approach. Thus, it was the confluence of all of these factors, and others, which stimulated the accelerated development of GaAs MMIC's during the last seven years.

SILICON OR GALLIUM ARSENIDE?

It is ironic that this revival of MMIC's based on GaAs technology has, in turn, re-stimulated the interest in silicon MMIC's. The first attempts with silicon were based on the silicon-on-sapphire (SOS) approach [2]. The use of a sapphire substrate eliminated the earlier problems with thermal conversion of the silicon substrates. However, these circuits, based on MESFET devices, achieved only limited success because of the marginal microwave performance of the FET's. The SOS approach has not been pursued.

A more fruitful approach has followed along more traditional lines of bipolar technology. Here, semi-insulating substrates are not used. However, because of the frequency limitations of silicon bipolars, the MMIC's are relegated to operation below C-band. Several papers dealing with this approach are included in this volume.

Table II lists some of the pertinent physical and electrical properties of GaAs and silicon (n-type) in their conducting and semiconducting states, as well as that of sapphire and alumina. It is evident from this table that as a high-resistivity substrate, semi-insulating GaAs, sapphire, and alumina are for all practical purposes, comparable. Also evident is the fact that the carrier mobility of gallium arsenide is over six times that of silicon. For this reason and others, GaAs MESFET's are operable at higher frequencies and powers than silicon MESFET's of equivalent dimensions. For example, silicon MESFET's based on 1 μ gate technology will be limited in operation to upper S-band (4 GHz) at best, whereas GaAs MESFET's operate well at X-band and higher. The availability of large GaAs wafers, approaching 3.5 inches in diameter [6] obtained by the Czochralski method, will allow high circuit counts per wafer, and hence reduced costs. The early success of direct-coupled FET analog circuitry [7], which is leading to high component density at S-band, will also help. Finally, the proven success of gigahertz high-speed GaAs logic circuitry will allow, for the first time, complete integration of logic and analog microwave circuitry. This opens up the feasibility of high-speed signal processing on a chip.

We maintain because of these and other reasons, that before this decade is over, it will be GaAs monolithic integrated circuits that will exert the greatest influence on the way semiconductor device circuitry is used in future microwave systems. Before proceeding to a more detailed discussion of the GaAs approach to MMIC's, we shall step back for a moment to make a brief comparison of this approach with that of the more common hybrid approach.

COMPARISON OF THE HYBRID AND MONOLITHIC APPROACHES

Some insight can be gained into the relative characteristics of MMIC's based on GaAs and the more conventional hybrid integrated circuit (HIC) approach by consideration of the technologies underlying them.[2]

---

[2] We prefer the acronym HIC, rather than the more common MIC, to denote hybrid integrated circuits because it is more descriptive of the approach. The reader should keep this equivalence in mind as he refers to the papers in this volume.

MMIC technology depends on the exploitation of the high-resistivity properties of semi-insulating (SI) semiconductor substrates for passive components and the good microwave device properties of the semiconducting layers grown or deposited on this SI substrate. The active devices most commonly used in MMIC's are the MESFET and the Schottky-barrier diode in the case of GaAs, and the bipolar transistor in the case of silicon.

HIC technology, in contrast, depends on the insulating properties of the substrate primarily as a low-loss microwave medium which supports planar transmission line interconnect patterns as well as passive components bonded onto these interconnects or deposited *in situ*. In other words, the sole function of the substrate is that of a "holder" for all of the passive and active components. In principle, any low-loss dielectric exhibiting good adherence to metal conductors and available with a sufficiently smooth surface to allow good line definition and high yield of deposited thin-film components, such as capacitors and resistors, will suffice. Good thermal conductivity is also important in high-power applications. Note that semi-insulating GaAs, in the case of MMIC's, also satisfies these requirements.

We mentioned earlier that a salient feature of the MMIC approach is the potential that batch processing will lead to low cost, high performance, small size, and reproducibility—as is the case with silicon integrated circuits at lower frequencies. However, the microwave industry to date has not been one demanding high-volume production to capitalize on the advantages of batch processing. Indeed, for the demand that exists, a tremendous skill in HIC's has been established to meet the market requirements. In a sense, then, the MMIC approach must create new markets where volume production will be required, or where the limitations of hybrid technology simply do not allow performance and/or cost objectives to be met. This subject is dealt with in more detail by Pengelly in this volume.

Examples of potential mass market developments are in

1) direct broadcast satellite-to-TV receiver applications (DBS);
2) air/space-borne electronically steerable antennas;
3) high-speed (microwave) signal processing;
4) low-cost missile decoys;
5) collision avoidance radar for automobiles.

Millions of identical, low-cost circuits will be required for the first and last applications. In the second, third, and fourth applications, small size, light weight, and reliability, as well as low cost will be of paramount importance.

The digital monolithic approach in silicon has lead to the establishment of a series of standardized functional circuit chips. This is also a desirable objective for MMIC's. Indeed, many people believe that this development could be of greater significance than the establishment of a mass market, especially for system applications in the military. However, because of the greater diversity of microwave circuit functions as compared to digital functions, the extent to which a "standard parts" philosophy will develop remains to be seen. In any case, to be economical, such a development must be predicated on a large volume usage.

Examples of applications where technological objectives are difficult to meet by the hybrid approach are

1) multi-octave bandwidth amplifiers
2) millimeter circuits.

In both of these cases, the limitations imposed by bond-wire inductances and other parasitics and by discrete elements preclude wide-band and/or high-frequency performance. We shall cite in papers to follow examples of monolithic circuits which simply cannot be duplicated in performance by HIC's.

In fairness to the HIC approach, at the present state of development of MMIC's, there are certain components which are difficult to realize with the same performance as their hybrid counterparts. An example is the narrow-band (high-Q) filter. Because of the higher transmission line losses associated with the thinner GaAs substrates, coupled with the higher parasitic capacitances to the ground plane, lower-Q performance can be expected from MMIC's. This low-Q limitation is a problem in some circuit applications such as low-noise, stable VCO's. The lack of a high-Q resonating element in MMIC format is an outstanding, unsolved problem.

There are other drawbacks associated with the monolithic approach which must be recognized. These are principally the following:

1) device/chip area ratio is small;
2) circuit tuning (tweaking) is impractical;
3) trouble-shooting (debugging) is difficult;
4) undesired RF coupling (crosstalk) may be a problem;
5) integration of high-power sources (IMPATT's) is difficult;
6) thermal heat sinking is poor;
7) tight control on material quality and fabrication tolerances is required.

The first item refers to the fact that only a small fraction, typically 3-5 percent of the chip, is occupied by active devices, hence the high processing cost and lower yield associated with active devices unavoidably applies to the larger area occupied by the circuitry. A corollary of this is that the lower yield processes of device fabrication dominate the overall chip yield. Although this problem diminishes as the chip size becomes smaller, that is, for higher frequencies, it is absent in the hybrid approach since the circuit and device technologies are separated.

The second and third items are related and can be considered together. The small chip sizes characteristic of the monolithic approach make it virtually impossible to tune ("tweak") and troubleshoot circuits. Indeed, to want to do so would violate one of the precepts of this approach, namely, to reduce costs by minimizing all labor-intensive steps. What then can be done?

First, it is necessary to minimize the need for tweaking. This can be done by adopting a design philosophy which leads to circuits that are insensitive to manufacturing tolerances in the active devices and in the physical dimensions of the circuit components. This will be a difficult compromise to accept on the part of the circuit designer who seeks the ultimate in performance from each active device by circuit optimization. However, here computer-aided design techniques come to the rescue. Not only do CAD techniques play a decisive role in monolithic circuit design, but they can also be used to assess the effect of tolerances on circuit performance during the design phase—and rather easily. CAD programs for doing this reside on many in-house computer systems and are also available commercially.

Second, the customer must accept the fact that tolerances will be larger in MMIC's than in HIC's, and be willing to compromise by stating his system requirements in terms of a realistic "window" of acceptable performance.

The use of CAD also helps alleviate the problem of troubleshooting a working circuit. Until microwave probes suitable for monolithic circuits become practical [5], troubleshooting must be based on terminal RF measurements of the circuit, usually the input and output ports. If a certain component is suspected of being faulty, it is a simple matter of building this defect into the CAD circuit file and comparing the resultant calculated circuit response with that measured. This can be done for a series of suspected faults, and convergence to the true fault can be achieved rather expeditiously.

The potential problem of undesirable RF coupling within the circuit is real, because of the small chip sizes involved. In practice, line spacings of the order of two substrate thicknesses are sufficient. This proximity "rule" plays a major role in determining the chip area and, hence, the chip cost. This restriction on circuit packing density, somewhat unique to MMIC's, can be alleviated measurably if direct-coupled circuitry is used, that is, if no distributed or lumped componentry is involved. We shall see examples of this approach in the papers that follow.

Turning to the fifth item, though both low-noise and power FET circuitry can be integrated easily on the same chip, when very high powers, more precisely, power densities are involved, the monolithic approach may face some fundamental limitations. These limitations are associated with the need for special means of removing heat from the device. A case in point is the diamond heatsink used with millimeter-wave IMPATT diodes. Although it would be desirable to integrate avalanche diode sources in monolithic circuits for millimeter-wave applications, the high-power densities involved cannot be handled by heat transfer through the chip. This is not a problem with power FET's but, of course, FET's cannot deliver the powers available from IMPATT's. Integration of high-power sources in monolithic circuits is a problem that is now being addressed.

Even for FET power amplifiers, tradeoffs must be made between good thermal performance and good RF design. For example, to minimize the thermal resistance through the substrate, one must use a wafer as thin as practical. However, a thin wafer increases the skin losses of microstrip lines, and hence the attenuation within the circuit. Furthermore, since heat-sinking requires metallization of the bottomside of the chip, additional parasitic capacitance to ground is introduced and must be taken into account in the design.

Last, but undoubtedly the most important limitation at present is the low wafer chip yield associated with variations in substrate and active layer quality, such as purity, defect density, and traps, and lack of reproducibility in device performance attributable to wafer processing. Indeed, MMIC's will not fulfill their ultimate objective of cost effectiveness until these processing and material problems are brought under control.

In summary, despite these limitations, the flexibility of circuit realizations based on the MMIC approach is beyond doubt. The importance of MMIC development is that systems applications based on a large number of identical components, for instance, space-borne phased-array radars, requiring light-weight and reliable low-cost transmit/receive modules, may finally become cost effective. One might consider this type of application as the microwave analog of the computer (which spurred the growth of the silicon digital monolithic circuit market), since both require a large number of identical circuits.

Maximum cost effectiveness, as well as improved reliability, derives in part from the fact that wire bonding is eliminated in MMIC's, at least within the chip area itself, and is relegated to less critical and fewer locations at the periphery of the chip. Wire bonds have always been a serious factor in reliability and reproducibility. Furthermore, wire bonding, being labor-intensive, is not an insignificant factor in the cost of the circuit.

Small size and volume, and their corollary, light weight, are intrinsic properties of the MMIC approach. Small size allows batch processing of hundred of circuits per wafer. Since the essence of batch processing is that the cost of fabrication is determined by the cost of processing the entire wafer, it follows that the processing cost per chip is proportional to the area of the chip. Thus, the higher the circuit count per wafer, the lower the circuit cost.

The small circuit size intrinsic to the MMIC approach will enable circuit integration on a chip level, ranging from the lowest degree of complexity such as an oscillator, mixer, or an amplifier, to a next higher "functional block" level, for example, a receiver front end or a phase shifter. A still higher level of circuit complexity, such as a transmit-receive module, may or may not be integrated on a single chip. It will depend on the yield that can be achieved.

The elimination of wire bonding and the integration of active components within a printed circuit remove many of the undesired parasitics limiting the broad-band performance of circuits employing packaged discrete devices. The MMIC approach certainly eases the difficulty of attaining multi-octave performance. Furthermore, such broad-banding approaches as distributed amplifiers, heretofore shunned as being too wasteful of active elements, are now feasible, especially since the unavoidable parasitics associated with the active devices can be incorporated into the propagating circuit and rendered less harmful. Indeed, the performance obtained with broad-band MMIC's is unexcelled as will be shown in the papers included in this volume.

Finally, MMIC's will co-exist with HIC's and not displace them. In the near term, some systems will likely consist of monolithic chips incorporated into a hybrid format. In the long term, however, MMIC's will gravitate to the high-volume markets which, as yet, have not developed, or to applications where technical performance can only be met with this approach. In the more conventional markets, MMIC's will have to compete with HIC's, principally, on the basis of cost. Here it is unlikely that MMIC's can succeed because the cost advantages of batch fabrication will be outweighed by initial design and development costs associated with this approach. Table

TABLE III
THE POSITIVE AND NEGATIVE ATTRIBUTES OF THE HYBRID (HIC) AND
MONOLITHIC (MMIC) APPROACHES

ATTRIBUTE	HICs	MMICs
• BATCH PROCESSING	−	+
— LABOR INTENSIVE	−	+
— LOW COST/CIRCUIT	−	+
— ON-CHIP WIRE BONDS	−	+
— POTENTIAL RELIABILITY	−	+
• SMALL SIZE/WEIGHT	−	+
• CIRCUIT DESIGN FLEXIBILITY	−	+
• BANDWIDTH/FREQUENCY RANGE	−	+
• CONTROLLED PARASITICS	−	+
• SUBSTRATE COST	+	−
• ON-CHIP RF TESTING	+	−
— "TWEAKABILITY"	+	−
— REPAIRABILITY	+	−
• VARIETY OF ACTIVE ELEMENTS	+	−
• WAFER HANDLING	+	−
• HEAT SINKING	+	−
• CIRCUIT-Q-FACTORS	+	−
• MS IMPEDANCE RANGE	+	−
• CAPITAL EQUIPMENT COSTS	+	−

III summarizes the positive and negative attributes of the MMIC and HIC approaches.

Let us turn to a detailed discussion of the GaAs approach.

THE GALLIUM ARSENIDE APPROACH

A cornerstone of the monolithic approach will be the development of a highly reproducible device technology. This in turn will depend, in part, on the control of the starting substrate material and the active (semiconducting) layer(s).

Two general techniques are available for forming this layer on GaAs substrates, namely, chemical vapor deposition (CVD) and ion implantation. Examples of the CVD approach are vapor phase epitaxy (VPE), molecular beam epitaxy (MBE), and metal-organic chemical vapor deposition (MOCVD). Of the two approaches, the CVD approach is more widely used and developed. In this approach a doped single crystal semiconducting layer is deposited on a semi-insulating GaAs substrate, usually with an intervening high-resistivity epitaxial "buffer" layer to screen out diffusion of impurities from the substrate during the active layer growth. With ion implantation, dopant atoms usually are implanted directly into the surface of a semi-insulating GaAs substrate. This procedure requires a higher state of purity of the substrate.

VPE, the most commonly used CVD approach in MMIC technology, does not have the control or flexibility associated with implantation. With implantation, more uniform conducting layers are possible over a larger area—more uniform in doping level as well as in thickness. Furthermore, with implantation, selective doping is easy, allowing for higher levels of integration. By this we mean that formation of different doping profiles in different parts of the wafer can be achieved readily, whereas with epitaxy it is difficult. The potential device reproducibility achievable with implantation is a definite advantage for it.

It should be noted, however, that as MMIC's enter the millimeter bands, where device designs require thinner active layers with more complex and finely defined impurity profiles, the MBE and MOCVD approaches will become increasingly more important. However, the lack of a selective doping capability will remain.

It should be added that implantation can be used in conjunction with epitaxy. One such application is the isolation implant, where oxygen, for example, is implanted in the unused portions of the epitaxial layer to convert it to a high-resistivity region suitable for supporting microwave circuitry such as microstrip lines. By this means a truly planar surface is maintained, since little or no mesa etching is required to remove the undesired regions of epitaxial layer. This also eliminates yield problems associated with metallization patterns extending over mesa steps. It is likely that, once substrate purity reaches the necessary level for ion implantation (as it is approaching with unintentionally doped Czochralski-pulled crystals), ion implantation will supplant epitaxy as the preferred method for monolithic circuits.

The processing technology used for FET fabrication is also applicable to the monolithic circuit elements. The high degree of dimensional definition associated with FET photolithography is more than adequate for the circuit elements. Presently, the most critical phases of MMIC development are material growth, qualification, and circuit fabrication. For example, material handling (wafer breakage) is a major cost factor that can be eliminated by capitalization of automatic processing equipment. However, this will happen only if a large market is developed on which such capitalization can be amortized. These are major problems impacting on circuit yield and cost which must be solved. To underline the seriousness of these topics we have included a series of papers in this volume which address the problems of materials, technology, and yield.

The size advantages of GaAs MMIC's will be lost if proper packaging techniques are not used. Little effort has been devoted to this problem to date. Perhaps the efficient techniques adopted for low-frequency and digital circuitry can be modified for microwave applications. Much thought must be devoted to this very important problem. Several papers covering this topic are contained in this volume.

A SYNOPSIS OF THE MONOLITHIC CIRCUIT DESIGN PROCESS

The evolution of an MMIC design from its inception to its final realization as a measurable circuit is a complex one. In order that the reader may better understand the tasks involved and appreciate the papers contained in the reprint volume which address these tasks we shall review briefly the steps involved as graphically illustrated in Fig. 1.

Starting at the bottom of this "stepping stone," we note that a set of design goals is dictated usually by some system requirement or other application. From this starting point, in general terms, a circuit approach or topology evolves. The circuit topology and system requirements dictate the types of active devices to be used, for example, single or dual-gate FET's, low-noise or power devices, FET's and/or diodes, and so forth.

Next, if these devices are not available, they must be fabricated in a monolithic format. This is followed by dc-probing of the wafer for selection of acceptable specimens, acceptable

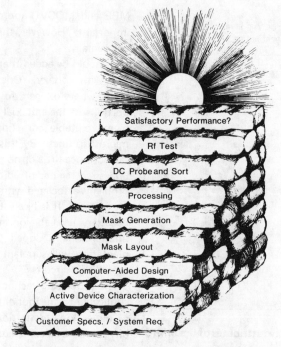

Fig. 1. The steps in the development of a monolithic circuit.

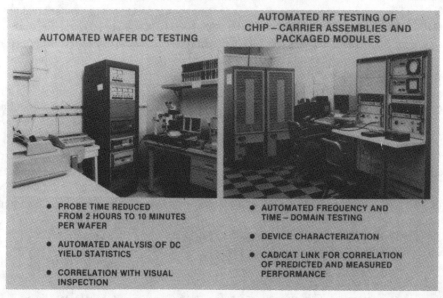

Fig. 2. Computer-aided testing of monolithic microwave circuits and modules.

in the sense that the devices satisfy certain dc criteria, such as IDSS, breakdown voltage, etc., and fall within some specified tolerance limits. The devices chosen on this basis are then prepared for RF characterization in carefully designed jigs which exhibit minimum parasitics. These jigs have been characterized earlier for their microwave performance (scattering parameters).

Usually the dc probe and microwave measurements are very extensive to enable one to obtain data at various operating voltages and to establish a measure of the spread in microwave properties amongst the selected devices. This latter data will be useful later when tolerance studies are made on the completed circuit. The sheer bulk of data that is usually desired dictates that automated measurement techniques be employed, such as illustrated in Fig. 2.

The "raw" microwave data cannot be used for further analysis until the jig properties (parasitics) have been removed. This process, often called "de-embedding," is a very critical and difficult step in that the accuracy of the monolithic circuit is dependent entirely on how carefully the device parameters are extracted from the measured data. There are many methods of de-embedding, but they all depend on the principles described in Fig. 3. Basically, the technique is based on the fact that for a linear system, the device parameters can be extracted from the measured two-port parameters by a linear matrix

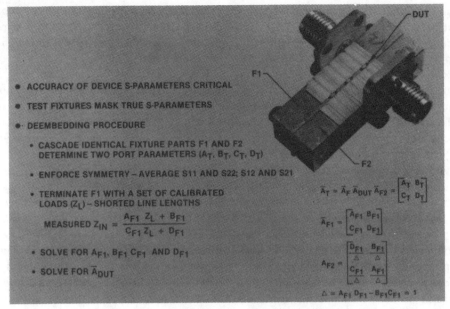

Fig. 3. De-embedding—a key characterization tool.

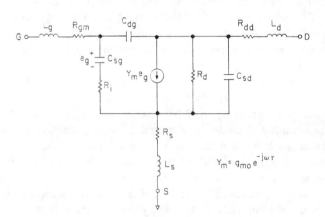

Fig. 4. Equivalent circuit of an FET.

Fig. 5. Computer-aided mask design.

transformation, given the two-port S-parameters representing the input and output transitions of the measurement jig. Often, because of measurement idiosynchrasies, certain data points are "suspect" and must be "smoothed" by some mathematical technique.

It is evident that the need for much of the de-embedding effort is self-inflicted because microwave measurements on devices and circuits are not taken on the wafer itself. This, however, requires the development of sophisticated microwave probes which will permit automatic step and repeat measurements over an entire wafer. Indeed, as wafer yields increase and chip sizes diminish, a large fraction of the chip cost will be determined by the cost of the microwave measurements if automation is not forthcoming. It is imperative, therefore, that automated microwave probing equipment be developed as quickly as possible. Fortunately, some efforts along these lines have been started [5].

Once the above tasks have been performed, it is often convenient to represent the de-embedded data by an equivalent circuit of the device, the topology of the circuit being rooted in the underlying physics of the device. Such an equivalent circuit obtained for an FET is illustrated in Fig. 4. This procedure, often called circuit modeling, involves obtaining numerical values for the equivalent circuit parameters which best fit the data in some "error" sense. This fitting is a feature usually found on CAD programs.

A model is useful in several ways: 1) it allows extrapolation and interpolation of the device performance to frequencies differing from the measurement conditions and 2) it is useful for tolerance studies and for "debugging" and troubleshooting.

The next major step in the building block shown in Fig. 1 is the CAD design phase. Here the "polished" device data is incorporated into the circuit topology file to allow analysis and optimization studies to be performed on the circuit design under consideration and for tolerance studies later on. The CAD step is a most essential one in that it allows extreme accuracy of design based on the starting data, a requirement for success because "tweaking" is out of the question and because the "brassboarding" stage is omitted normally.

After the circuit design is completed, it is committed to mask layout. Since much of the layout process is cut-and-try, or based on prior experience, it is convenient to use a computerized mask layout system such as the one illustrated in Fig. 5. Shown on the screen in a multi-colored format, representing various mask layers, is a layout for a two-stage low-noise FET amplifier. The operator, who works "side-by-side" with the circuit designer, manipulates the various components of the circuit into place with his "magic wand." Often, frequently used components are stored in data form in "memory" and are called into the mask layout as needed.

Once the circuit layout has been completed to the satisfaction of both the designer and the technologist, the numerical coordinate data describing the topologies of the various mask layers are recorded on magnetic tape. This tape is turned over to the mask fabrication specialists who convert the numerical data to actual photomasks with automated pattern generators designed for this purpose. The completed photomasks are delivered to the technology laboratory for fabrication of the monolithic circuit.

Following fabrication, the circuits are dc-probed on the wafer and sorted as was done for the single FET's. The difference, of course, is that much more probing per chip is involved. Finally, the last step, Fig. 1, following the dicing of the wafer into chips, is the selection of the chips for RF measurement which have passed dc-probe testing. These chips are carefully mounted on carriers and inserted in special jigs for RF testing. If the measured circuit performance is satisfactory, the design is complete and only packaging of the mounted chips remains. Usually, however, several design iterations are necessary before the desired circuit performance is obtained.

The design of a large-signal device, such as a power stage, is somewhat more complicated. The major difference from the above procedure is the inclusion of a large-signal analysis following the small-signal CAD design and circuit optimization. (Most CAD software can only be used for linear or small-signal design). Software for large-signal analyses is usually written at the individual laboratories since it is not available commercially.

The large-signal study consists of subjecting the circuit design, which was based on small-signal criteria, to an analysis of operation wherein the active devices are given some voltage- or current-dependent nonlinearities based on measurements or theory. Often this analysis indicates that certain circuit component values must be changed slightly to produce a good compromise between the small- and large-signal response. The small and large-signal analyses often are used in an interative fashion to facilitate convergence to the desired response. Several papers in this reprint volume address this task.

It should be obvious to the reader by this time that the design and fabrication of a microwave monolithic circuit requires a sequence of many complex and critical steps. There are no short cuts!

CONCLUSIONS AND SUMMARY

Monolithic microwave circuits based on gallium arsenide technology have become a practical reality. Owing its origin to early experiments based on silicon bipolar technology, the gallium arsenide approach, except for some scattered results in the sixties, emerged as a serious development only within the last seven years.

The factors most responsible for this rapid growth can be traced to 1) the development of the Schottky-barrier FET, 2) the excellent microwave properties of semi-insulating GaAs substrates, 3) the perfection of GaAs epitaxy and ion implantation, 4) the establishment of GaAs crystal pulling facilities capable of large-diameter crystal growth, 5) the emergence of potential mass markets for MMIC's, and 6) the development of powerful computer-aided design techniques.

Despite the important role played by CAD techniques to date, the available computer design programs have serious limitations which dictate the need for a new generation of CAD software. Specifically, more efficient algorithms are necessary to reduce CPU costs, and to allow inclusion of tolerance data *during* the optimization cycle, rather than as an afterthought in a *post* design analysis. Furthermore, software must be designed to encompass nonlinear as well as linear microwave elements. More "user-friendliness" would also be welcomed.

The promising attributes of monolithic technology to cut fabrication costs, to improve reliability and reproducibility, and to reduce size and weight will overcome many of the shortcomings of the hybrid approach.

Based on cost considerations, the potential markets of MMIC's will be for the most part systems requiring large quantities of circuits of the same type. Because of this, and because of the large capital expenditures required of an organization to become a viable contender for these markets, it is most likely that the major efforts in MMIC's will reside in the systems houses themselves, at least for the time being. If, however, the trend is towards the development of inventories of standardized chips, as some expect, or if the consumer markets develop, and here cost will be the prime objective, then this picture may change. It is too early to say.

We hope by this brief introduction that we have given the reader a better understanding of the developments that have been taking place in this very important and exciting field, and hence, a better appreciation of the papers contained in this volume.

EDITORIAL COMMENTS

This volume is divided into eight categories comprising a total of 82 papers. In some cases, a paper could have been included in several categories.

We have introduced each category with a short introduction. Where possible, we have tried to select a lead article which would serve as a suitable introduction to each topic.

The choice of papers has been a difficult one. We have attemped, when possible, to select papers which would serve a tutorial role as well as a description of work performed. We have purposely avoided compiling a "catalog" of papers in the field. It has been our intention to strike a balance among the papers covering all facets of the MMIC field.

We have tried our best to conform to the page count and journal guidelines prescribed to us. Thus we have emphasized in our selection papers from IEEE publications. We realize all too well that many good papers, especially those dealing with materials technology, have been excluded. Nevertheless, we believe that the present reprint volume more than adequately represents the exciting developments in the MMIC field.

REFERENCES

[1] T. M. Hyltin, "Microstrip transmission on semiconductor substrates," *IEEE Trans. Microwave Theory Tech.*, vol. MTT-13, pp. 777-781, Nov. 1965.

[2] D. Laighton, J. Sasonoff, and J. Selin, "Silicon-on-sapphire (SOS) monolithic transceiver module components for L- and S-band," in *Government Microcircuit Applications Conf. (GOMAC), Dig. Papers*, vol. 8, 1980, pp. 299-302.

[3] E. Mehal and R. W. Wacker, "GaAs integrated microwave circuits," *IEEE Trans. Microwave Theory Tech.*, vol. MTT-16, pp. 451-454, July 1968.

[4] R. S. Pengelly and J. A. Turner, "Monolithic broadband GaAs FET amplifiers," *Electron. Lett.*, vol. 12, pp. 251-252, May 13, 1976.

[5] E. Strid and K. Reed, "A microstrip probe for microwave measurements on GaAs FET and IC wafers," in *GaAs IC Symp. Research Abstracts*, paper no. 31, Oct. 1980.

[6] R. N. Thomas, "Advances in bulk silicon and gallium arsenide materials technology" in *IEDM Tech. Dig.*, 1980, pp. 118-119.

[7] R. van Tuyl, "A monolithic GaAs FET RF signal generation chip," in *ISSCC Dig. Tech. Papers*, 1980, pp. 118-119.

Part I
Design Considerations

THE intimate relationship between the fabrication technology of MMIC's and the design configurations permitted by this technology is a unique feature of the MMIC approach. The MMIC designer, unlike his hybrid counterpart, is not completely free to choose circuit topologies at will. Indeed, if his efforts are to lead to a cost-effective design, he must possess a working knowledge of the capabilities and constraints of the technology he has at his disposal.

It is a fact that, in some instances, the MMIC approach may *not* be the cost-effective approach, whereas the more traditional hybrid technique may be the appropriate one. The following three papers address these very important design issues.

Design Considerations for Monolithic Microwave Circuits

ROBERT A. PUCEL, FELLOW, IEEE

MTT National Lecture Invited Paper

Abstract — Monolithic microwave integrated circuits based on silicon-on-sapphire (SOS) and gallium arsenide technologies are being considered seriously as viable candidates for satellite communication systems, airborne radar, and other applications. The low-loss properties of sapphire and semi-insulating GaAs substrates, combined with the excellent microwave performance of metal-semiconductor FET's (MESFET's), allows, for the first time, a truly monolithic approach to microwave integrated circuits. By monolithic we mean an approach wherein all passive and active circuit elements and interconnections are formed into the bulk, or onto the surface of the substrate by some deposition scheme, such as epitaxy, ion implantation, sputtering, evaporation, and other methods.

The importance of this development is that microwave applications such as airborne phased-array systems based on a large number of identical circuits and requiring small physical volume and/or light weight, may, finally, become cost effective.

The paper covers in some detail the design considerations that must be applied to monolithic microwave circuits in general, and to gallium arsenide circuits in particular. The important role being played by computer-aided design techniques is stressed. Numerous examples of monolithic circuits and components which illustrate the design principles are described. These provide a cross section of the world-wide effort in this field. A glimpse into the future prospects of monolithic microwave circuits is made.

I. INTRODUCTION

THE LAST two to three years have witnessed an intensive revival in the field of analog monolithic microwave integrated circuits (MMIC's), that is, microwave circuits deposited on a semiconductor substrate, or an insulating substrate with a semiconductor layer over it. In this paper, we shall address the design and technology considerations of monolithic microwave integrated circuits as well as the potential applications of these circuits to microwave systems, such as satellite communications and phased-array radar, as well as instrumentation.

It is important that the reader understand what we mean by the term "monolithic" circuit. By monolithic, we mean an approach wherein all active and passive circuit elements or components and interconnections are formed into the bulk, or onto the surface, of a semi-insulating substrate by some deposition scheme such as epitaxy, ion implantation, sputtering, evaporation, diffusion, or a combination of these processes and others.

It is essential that the full implication of this definition be understood, since it strikes at the very core of why one would want to design and fabricate a microwave monolithic circuit. The reasons are embedded in the following promising attributes of the monolithic approach:

1) low cost;
2) improved reliability and reproducibility;
3) small size and weight;
4) multioctave (broad-band) performance; and
5) circuit design flexibility and multifunction performance on a chip.

The importance of this development is that systems applications based on a large number of identical components, for instance, space-borne phase-array radars planned for the future which require lightweight and reliable, low-cost transmit–receive modules, may finally become cost effective. One might consider this type of application as the microwave system analog of the computer (which spurred the growth of the silicon digital monolithic circuit market), since both require a large number of identical circuits.

Maximum cost effectiveness, as well as improved reliability, derives in part from the fact that wire bonding is eliminated in MMIC's, at least within the chip itself, and is relegated to less critical and fewer locations at the periphery of the chip. Wire bonds have always been a serious factor in reliability and reproducibility. Furthermore, wire bonding, being labor intensive, is not an insignificant factor in the cost of a circuit.

Small size and volume, and their corollary, light weight, are intrinsic properties of the monolithic approach. Small size allows batch processing of hundreds of circuits per wafer of substrate. Since the essence of batch processing is that the cost of fabrication is determined by the cost of processing the entire wafer, it follows that the processing cost per chip is proportional to the area of the chip. Thus, the higher the circuit count per wafer, the lower the circuit cost.

The elimination of wire bonding and the embedding of active components within a printed circuit eliminate many of the undesired parasitics which limit the broad-band performance of circuits employing packaged discrete devices. The monolithic approach will certainly ease the difficulty of attaining multioctave performance. Furthermore, such broad-banding approaches as distributed

Manuscript received January 16, 1981. This work summarizes the lecture given by the National Lecturer throughout the United States, Canada, and Europe, during 1980–1981.

The author is with the Research Division, Raytheon Company, Waltham, MA 02254.

amplifier stages, heretofore shunned as too wasteful of active elements, will now become feasible, because a cost penalty will not accrue from the prolific use of low-gain stages, and because the unavoidable parasitics associated with the active devices will be incorporated in the propagating circuit and rendered less harmful.

The small circuit size intrinsic to the monolithic approach will enable circuit integration on a chip level, ranging from the lowest degree of complexity such as an oscillator, mixer, or amplifier, to a next higher "functional block" level, for example a receiver front end or a phase shifter. A still higher level of circuit complexity, for example, a transmit–receive module, will be integrated, most likely in multichip form.

So far we have discussed only the virtues of the monolithic approach. Now let us consider some of its disadvantages and problem areas. These are principally the following:

1) unfavorable device/chip area ratio;
2) circuit tuning (tweaking) impractical;
3) trouble-shooting (debugging) difficult;
4) suppression of undesired RF coupling (crosstalk), a possible problem; and
5) difficulty of integrating high power sources (IMPATT's)

The first item refers to the fact that only a small fraction of the chip area is occupied by devices, hence the high processing cost and lower yield associated with active device fabrication is unavoidably applied to the larger area occupied by the circuitry. A corollary of this is that the lower yield processes of device fabrication dominate the overall chip yield. Although these problems diminish as the chip size becomes smaller, that is, for higher frequencies, they are absent in the hybrid approach where the circuit and device technologies are separated.

The second and third items are related and can be considered together. The small chip sizes characteristic of the monolithic approach make it virtually impossible to tune ("tweak") and troubleshoot circuits. Indeed, to want to do so would violate one of the precepts of this approach, namely, to reduce costs by minimizing all labor-intensive steps. What then can be done about these very real problems?

First, it is necessary to minimize the need for tweaking. This can be done by adopting a design philosophy which leads to circuits that are insensitive to manufacturing tolerances in the active devices and physical dimensions of the circuit components. This will be a difficult compromise to accept on the part of the circuit designer, who expects the ultimate in performance from each active device by circuit tuning. However, here computer-aided design (CAD) techniques come to the rescue. Not only will CAD techniques play a major, if not mandatory, role in monolithic circuit design, they will also be used to assess the effect of tolerances on circuit performance during the design phase—and rather easily. CAD program for doing this

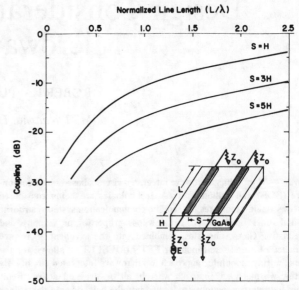

Fig. 1. Calculated coupling between adjacent parallel microstrip lines as a function of spacing and frequency.

reside on many internal computer systems and are also available commercially [4].

The use of CAD also helps alleviate the problem of troubleshooting a working circuit. Until microwave probes suitable for monolithic circuits become practical [19], troubleshooting must be based on terminal RF measurements of the circuit, usually the input and output ports. If a certain component is suspected of being faulty, it is a simple matter of building this defect into the CAD data file and comparing the resultant calculated circuit response with that measured. This can be done for a series of suspected faults, and convergence to the true fault can be achieved rather expeditiously.

The potential problem of undesirable RF coupling within the circuit is real because of the small chip sizes involved. To illustrate this point, Fig. 1 is a theoretical calculation of the coupling between two parallel microstrip lines on a GaAs substrate, one of which is excited by a generator. Both lines are matched at either end. Shown is the fraction of power coupled from the excited line to the adjacent line as a function of line length and line spacing. It is obvious that the coupling can become unacceptably high for long line lengths approaching a wavelength or more. Even for short lines, of the order of a quarter-wavelength or less, a feedback problem may exist if, say, a high-gain amplifier exists in one of the lines. In practice, line spacings of the order of three substrate thicknesses or more ($S > 3H$) have been found adequate in most cases. This proximity "rule" plays a major role in determining the chip area and, hence, the chip cost. This restriction on circuit packing density, somewhat unique to MMIC's, can be alleviated measurably if direct-coupled circuitry is used, that is, if no distributed or lumped componentry is involved. We shall see examples of this approach later.

Turning to the fifth item, though both low-noise and power FET circuitry can easily be integrated on the same chip, where very high powers, more precisely, power densities are involved, the monolithic approach may face some fundamental limitations. These limitations are associated with the need for special means of removing heat from the device. A case in point is the diamond heatsink used with millimeter-wave IMPATT diodes. Though it would be desirable to integrate avalanche diode sources in monolithic circuits for millimeter wave applications, the high-power densities involved cannot be handled by heat transfer through the chip. This is not a problem with power FET's, but of course, FET's cannot deliver the powers available from IMPATT's. Integration of high power sources in monolithic circuits is a problem that, as yet, has not been addressed.

Even for FET power amplifiers, tradeoffs must be made between good thermal performance and good RF design. For example, to minimize the thermal resistance through the substrate, it is desirable to use as thin a wafer as practical. However, a thin wafer increases the circuit skin effect losses, and hence the attenuation. Furthermore, since heat-sinking requires metallization of the chip bottomside, additional parasitic capacitance to ground is introduced and corrections must be made to planar inductors to account for "image" currents in the ground plane.

Despite these limitations on power, it is possible that with on-chip power combining techniques applied to FET's which are thermally isolated from each other [17], power outputs of the order of 10-W CW or so may be realizable from a single chip at the lower microwave frequencies, that is, at X-band.

II. MMIC's — A Brief History

The concept of MMIC's is not new. Its origin goes back to 1964 to a government-funded program based on silicon technology, which had as its objective a transmit–receive module for an aircraft phased-array antenna. Unfortunately, the results were disappointing because of the inability of semi-insulating silicon to maintain its semi-insulating properties through the high-temperature diffusion processes. Thus, very lossy substrates resulted, which were unacceptable for microwave circuitry [12]. Because of these and other difficulties the attempt to form a monolithic circuit based on a semiconductor substrate lay dormant till 1968 when Mehal and Wacker [15] revived the approach by using semi-insulating gallium arsenide (GaAs) as the base material and Schottky barrier diodes and Gunn devices to fabricate a 94-GHz receiver front end. However, it was not until Plessey applied this approach to an X-band amplifier, based on the Schottky-gate field-effect transistor, or MESFET (MEtal-Semiconductor Field-Effect Transistor), as the key active element that the present intense activity began [16].

What brought on this revival? First, the rapid development of GaAs material technology, namely, epitaxy and ion implantation, and the speedy evolution of the GaAs FET based on the metal Schottky gate during the last decade led to high-frequency semiconductor device performance previously unattained. A few examples are high-efficiency and high-power amplifier performance through Ku-band, low-noise amplifiers, variable-gain dual-gate amplifiers, and FET mixers with gain. The dual-gate FET will play a major role in MMIC's because of its versatility as a linear amplifier whose gain can be controlled either digitally or in analog fashion. With dual-gate FET's, multi-port electronically switchable RF gain channels are feasible. Second, resolution of many troublesome device reliability problems made FET's more attractive for systems applications. Third, recognition of the excellent microwave properties of semi-insulting GaAs (approaching that of alumina), removed the major objection of silicon. Fourth, hybrid circuits were becoming very complex and labor intensive because of the prolific use of wire bonds, and hence too costly. Fifth, the emergence of clearly defined and discernible systems applications for MMIC's became more apparent. Thus it was the confluence of all of these factors, and others, which stimulated the development of GaAs MMIC's within the last five years.

TABLE I
SOME PROPERTIES OF SEMICONDUCTORS AND INSULATORS

Property	GaAs	Silicon	Semi-insulating GaAs	Semi-insulating Silicon	Sapphire	Alumina
Dielectric Constant	12.9	11.7	12.9	11.7	11.6 (C-axis)	9.7
Density (gm/cc)	5.32	2.33	5.32	2.33	3.98	3.89
Thermal Cond. (watts/cm-°K)	0.46	1.45	0.46	1.45	0.46	0.37
Resistivity (ohm-cm)	---	---	$10^7 - 10^9$	$10^3 - 10^5$	$>10^{14}$	$10^{11} - 10^{14}$
Elec. Mobility (cm^2/v-sec.)	4300*	700*	---	---	---	---
Sat. Elec. Vel. (cm/sec.)	1.3×10^7	9×10^6	---	---	---	---

* At $10^{17}/cm^3$ doping

III. Silicon or Gallium Arsenide?

It is ironic that this revival based on GaAs technology has, in turn, restimulated the interest in silicon MMIC's—but now based on the silicon-on-sapphire (SOS) approach [13]. There are understandable reasons for this. First, the use of sapphire as a substrate eliminates the losses associated with semi-insulating silicon mentioned earlier. Second, silicon technology is an extremely well developed technology—much more so than GaAs. Third, the availability of the simpler MESFET technology, developed in GaAs, could now be used in place of the more complex bipolar technology, which, however, was still available should it be needed. Nevertheless, gallium arsenide has the "edge" for reasons to be discussed next.

Table I lists some of the pertinent physical and electrical properties of GaAs and silicon (n-type) in their insulating and semiconducting states, as well as that of sapphire and alumina. As is evident from this table, as a high-resistivity substrate, semi-insulating GaAs, sapphire, and alumina are, for all practical purposes, comparable. Also evident is that the carrier mobility of gallium arsenide is over six times that of silicon. For this reason and others, GaAs MESFET's are operable at higher frequencies and powers than silicon MESFET's of equivalent dimensions. For example, silicon MESFET's based on 1-μm gate technology will be limited in operation to upper S-band at best, whereas GaAs MESFET's operate well at X-band and higher. Therefore, it is highly likely that the performance of 1-μm gate silicon MESFET's will be matched, and perhaps exceeded by that of 2-μm gate GaAs MESFET's at S-band. The near-future availability of much larger GaAs wafers, approaching 3.5 in in diameter [20], obtained by the Czochralski method, will overcome the size limitations imposed by the present 1-in wafers grown by the Bridgman method. The early success of direct-coupled FET analog circuitry [11], [21], which leads to high component density at S-band, will also help overcome wafer size limitations in GaAs. Finally, the proven success of gigahertz high-speed GaAs logic circuitry will allow, for the first time, complete integration of logic and analog microwave circuitry. This opens up the feasibility of high-speed signal processing on a chip.

We do not wish to imply that MMIC work based on SOS technology should be diminished; however, we believe its major role will be found in the range below 2 GHz, for example, in IF circuitry and other applications. In light of this conclusion, we shall direct the following discussion to the GaAs approach. However, much of what we shall say, as will be obvious to the reader, will also apply, with minor changes, to the SOS approach or to other approaches which may emerge in the future. Nevertheless, we maintain that before this decade is over, it will be GaAs monolithic integrated circuits that will exert the greatest influence on the way solid-state device circuitry is used in microwave systems.

IV. THE GALLIUM ARSENIDE APPROACH

A cornerstone of the monolithic approach will be the availability of a highly reproducible device technology. This in turn is related, in part, to the control of the starting material, especially the active (semiconducting) layer.

Two general techniques are available for forming this layer on GaAs substrates, namely, epitaxy and ion implantation. Of the two approaches, the former at present is more widely used and developed. In this approach a doped single crystal semiconducting layer is deposited on a semi-insulating GaAs substrate, usually with an intervening high-resistivity epitaxial "buffer" layer to screen out diffusion of impurities from the substrate during the active layer growth. With ion implantation, dopant atoms are implanted directly into the surface of a semi-insulating GaAs substrate. This procedure requires a higher state of purity of the substrate—a problem at present.

Expitaxy does not have the control or flexibility associated with implantation. With implantation, more uniform conducting layers are possible over a large area—more uniform in doping level as well as in thickness. Furthermore, with implantation, selective doping is easy, that is, formation of different doping profiles in different parts of the wafer is easy to achieve, whereas with epitaxy it is difficult. The potential device reproducibility achievable with implantation is a definite advantage for it.

It should be added that implantation can also be used in conjunction with epitaxy. One such application is the isolation implant, wherein oxygen is implanted in the unused portions of the epitaxial layer to produce a high-resistivity region within the epitaxial layer onto which microwave circuitry may be situated. Thus a truly planar surface is maintained, since no mesa etching is required to remove the undesired regions of epitaxial layer. This also eliminates yield problems associated with metallization patterns extending over mesa steps.

It is likely that, once substrate purity reaches the necessary level for ion implantation (as it is approaching with unintentionally doped Czochralski-pulled crystals), ion implantation will supplant epitaxy as the preferred method for monolithic circuits.

The processing technology used for FET fabrication is also applicable to the monolithic circuit elements. The high degree of dimensional definition associated with FET photolithography is more than adequate for the circuit elements.

V. GENERAL DESIGN CONSIDERATIONS

We turn now to a discussion of the design considerations for MMIC's.

A. Constraints on Chip Size

Present substrate sizes corresponding to that of GaAs boules are approximately 1 in in diameter, though larger boules approaching 3 in in diameter are now being grown by the Czochralski method. Given the expected limits on substrate size, it is instructive to estimate the circuit count/wafer achievable as a function of frequency, since the processing cost per circuit is inversely proportional to this density.

We assume that the maximum linear dimension per circuit will fall between $\lambda_g/10$ and $\lambda_g/4$, where λ_g is the wavelength of the propagation mode (microstrip-coplanar, etc.) in GaAs. The lower limit takes into account the approximate maximum size of lumped elements; the upper limit, the typical maximum size of distributed elements. It seems reasonable to assume that in the vicinity of 10 to 20 GHz some distributed elements of the order of $\lambda_g/4$ (for example, hybrid and branch line couplers) will be used. Therefore, above this frequency range, linear circuit dimensions of the order of $\lambda_g/4$ will be the rule. Let us choose 16 GHz as the demarcation frequency. We then postulate a "linear" admixture of lumped- and distributed-element weighting so that we obtain $\lambda_g/10$ at 1 GHz and $\lambda_g/4$ at 16 GHz as the probable linear dimension of a circuit function "chip."

Fig. 2 is a plot of the approximate density of these circuits as a function of frequency for two sizes of wafer. (The 2-in square wafer corresponds to a 3-in diameter

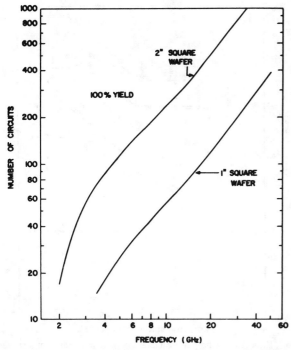

Fig. 2. Estimated number of circuits per wafer taking dicing and edge waste into account.

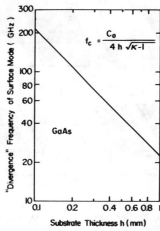

Fig. 3. Frequency of onset of lowest order TE surface wave on a GaAs substrate as a function of substrate thickness.

wafer.) The circuit density estimates take into account edge and cutting waste, but not "proximity effects," wafer yield, and other factors which will reduce these numbers.

A yield factor is associated with each fabrication step, the overall yield being the product of the individual yield factors. Thus, since active devices generally require the most processing steps (about 40 for an FET), the overall yield is determined by the device processing technology. The "proximity effect," that is, the RF coupling problem mentioned earlier, will put stringent limitations on how closely packed the signal lines may be, and hence how much circuitry can be compressed into the chip area, which is fixed by wavelength or lumped-element dimensions as just described.

The circuit count estimate must be modified for very low microwave frequencies (below C-band) if active components such as FET's are used to simulate resistors and capacitors and if inductors are dispensed with because tuning is not necessary. In this so-called direct-coupled design, packing densities approaching those normally associated with digital circuitry is possible [21], that is, much higher than that indicated in Fig. 2. However, it must be cautioned that this circuit approach is not suitable for all monolithic applications, for example, high-efficiency power amplifiers or low-noise circuits. The reason is that the use of active (FET) devices as resistive elements in the gate and drain circuits introduces high dc power dissipation and mismatch, as well as additional noise [11].

It is appropriate at this time to point out that the size advantages of GaAs MMIC's will be lost if proper packaging techniques are not used. Perhaps the efficient techniques adopted for low-frequency and digital circuitry can be suitably modified for microwave applications. Much thought must be devoted to this very important problem.

B. Constraints on Wafer Thickness

So far we have discussed requirements on the substrate area. There also are constraints imposed on the substrate thickness. Some of these are:

1) volume of material used;
2) fragility of wafer;
3) thermal resistance;
4) propagation losses;
5) higher order mode propagation;
6) impedance-linewidth considerations; and
7) thickness tolerance versus impedance tolerance.

Obviously, to keep material costs down one wishes to use as thin a substrate as can be handled without compromising the fragility. Thermal considerations also require the thinnest wafer possible. On the other hand, a thin wafer emphasizes the effect of the ground plane. For example, propagation losses increase inversely with substrate thickness in the case of microstrip. Furthermore, the Q-factor and inductance of thin-film inductors decrease with decreasing substrate thickness. In contrast, undesired higher-order surface mode excitation is inhibited for thinner substrates.

Fig. 3 is a graph of the frequency denoting the onset of the lowest order (TE) surface mode as a function of substrate thickness. For example, for a substrate thickness of 0.1 mm (4 mils) the "safe" operating frequency range is below 200 GHz. It appears that, for presently contemplated circuit applications, surface mode propagation is not a limiting factor in the choice of substrate thickness. The linewidth dimensions for a given impedance level of some propagation modes, such as microstrip, are proportional to substrate thickness. Therefore, thicker substrates alleviate the effect of thickness and linewidth tolerances.

The point being made here is that the choice of substrate thickness is a tradeoff of the factors listed above, being strongly dependent on the frequency of operation and the

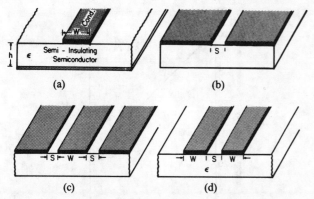

Fig. 4. Four candidate propagation modes for monolithic circuits. (a) Microstrip (MS). (b) Slot line (SL). (c) Coplanar waveguide. (d) Coplanar strips (CS).

power dissipation of the circuit. It is true that perhaps the most important of the factors is the thermal consideration. We believe that in the frequency range up to 30 GHz a substrate thickness of the order of 0.1 mm to 0.15 mm is appropriate for power amplifier circuits, with thicknesses up to 0.6 mm tolerable for low-noise amplifiers and similar circuits, provided a satisfactory means of dicing the thicker wafers can be found.

C. Propagation Modes

At microwave frequencies, the interconnections between elements on a high dielectric constant substrate such as GaAs, where considerable wavelength reduction occurs, must be treated as waveguiding structures. On a planar substrate, four basic modes of propagation are available, as illustrated in Fig. 4. The first mode (Fig. 4(a)) is microstrip (MS), which requires a bottomside ground metallizaton. Its "inverse," slot line (SL), is shown in Fig. 4(b). The third mode is the coplanar waveguide (CPW) shown in Fig. 4(c); it consists of a central "hot" conductor separated by a slot from two adjacent ground planes. Its "inverse," the coplanar stripline (CS), is illustrated in Fig. 4(d); here, one of the two conductors is a ground plane. Both the coplanar waveguide and coplanar strips are generally considered to be on infinitely thick substrates. Of course, this condition cannot be met. We shall see the implication of this later.

Of the four modes, only the slot line is not TEM-like. For this reason, and because it uses valuable "topside" area, we do not expect slot line to be a viable candidate for monolithic circuits, except possibly in special cases.

The principal losses of microstrip and the coplanar modes consist of ohmic losses. Since the coplanar structures are, in essence, "edge-coupled" devices, with high concentration of charge and current near the strip edges, the losses tend to be somewhat higher than for microstrip, as verified by experiment [5].

The lack of a ground plane on the topside surface of the microstrip structure is a considerable disadvantage when shunt element connections to the hot conductor are required. However, as we shall see later, there are ways to overcome this disadvantage. Table II summarizes, in a

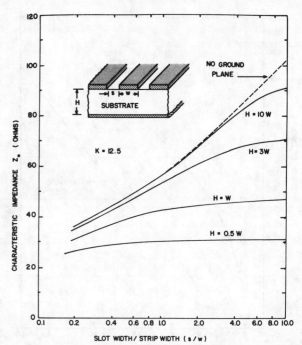

Fig. 5. Effect of ground plane on characteristic impedance of a coplanar waveguide.

TABLE II
QUALITATIVE COMPARISON OF PROPAGATION MODES

	MICROSTRIP	COPLANAR WAVEGUIDE	COPLANAR STRIPS	SLOT LINE
Attenuation Loss	low	medium	medium	high
Dispersion	low	medium	medium	high
Impedance Range (ohms)	10-100	25-125*	40-250*	high
Connect Shunt Elements	diff.	easy	easy	easy
Connect Series Elements	easy	easy	easy	diff.

*Infinitely thick substrate

qualitative way, the features of the four modes of propagation illustrated in Fig. 4.

The impedance range achievable with CPW and CS is somewhat greater than for MS, particularly at the higher end of the impedance scale, provided an infinitely thick substrate is assumed for CPW and CS. This range is reduced considerably when practical substrate thicknesses are used and the bottomside of the chip is metallized. Fig. 5 shows how the high impedance end of the scale is lowered when substrates of the order of 0.1–0.25 mm thick are mounted on a metal base (for heat-sinking purposes). The considerable reduction in Z_0 makes the design of monolithic circuitry with CPW nearly as dependent on substrate thickness as with MS, at least at the high end of the impedance scale.

Microstrip has its own unique restriction on the realizable impedance range. This is dictated by technology considerations. The limitation stems from the fact that for MS

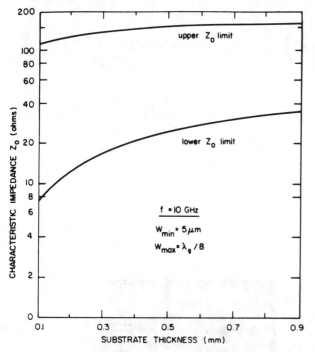

Fig. 6. Range of characteristic impedance of microstrip on GaAs substrate as a function of substrate thickness.

the characteristic impedance Z_0 is a function of the ratio W/H (see Fig. 4). The highest achievable impedance is determined by the smallest linewidth W that can be realized with acceptable integrity over a long length, say, a quarter of a wavelength. Our experience is that a minimum linewidth of 5 µm is reasonable. With this restriction, and an additional limit imposed on the maximum linewidth to be well below a quarter-wavelength, say, one-eighth-wavelength, the realizable range of characteristic impedance as a function of substrate thickness and frequency is constrained within the range indicated by Fig. 6.

It is evident that the usable impedance range for a 0.1-mm thick substrate is approximately 10–100 Ω, and somewhat less for thicker substrates and higher frequencies. For higher frequencies, the lower curve moves "up." This limited impedance range is a severe restriction in the design of matching networks, a problem not faced in the hybrid approach.

Weighing all of these factors, we believe that of the four candidate modes, MS and CPW are the most suitable for GaAs monolithic circuits, with preference toward MS. Indeed, there will be instances where both modes may be used on the same chip to achieve some special advantage. The transition from one mode to the other is trivial. Most of the examples of MMIC's to be described later are based on MS.

Fig. 7(a) is a graph of the wavelength of a 50-Ω MS line on GaAs as a function of frequency, with dispersion neglected. The wavelength of CPW is similar. Fig. 7(b) illustrates the attenuation of MS as a function of characteristic impedance and substrate thickness at 10 GHz. The loss in decibels per centimeter increases as the square root of frequency. The loss per wavelength, on the other hand, decreases as the square root of frequency. Note the inverse dependence of loss on substrate thickness.

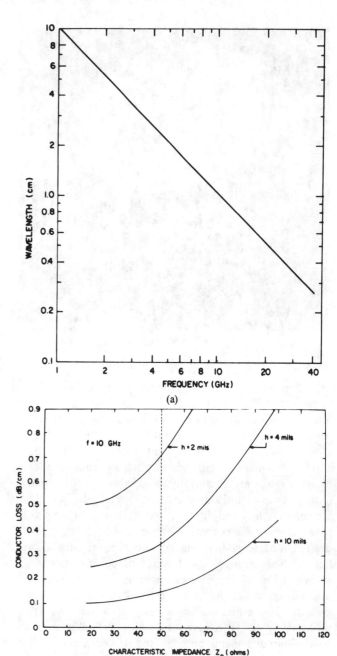

Fig. 7. (a) Wavelength as a function of frequency for microstrip on a GaAs substrate for $h = 0.1$ mm. (b) Conductor loss of microstrip on a GaAs substrate as a function of substrate thickness and characteristic impedance for $f = 10$ GHz.

D. Low Inductance Grounds and Crossovers

Microstrip and coplanar waveguide are adequate for interconnections that do not require conductor crossovers or that are not to contact the bottomside ground metallization. Often, however, such connections are needed. In particular, with MS, which does not have any topside ground planes, some means of achieving a low-inductance ground is essential.

Fig. 8. 50-μm diameter "via" hole etched in a GaAs wafer.

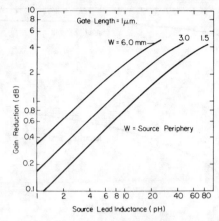

Fig. 9. Calculated gain reduction of a GaAs power FET as a function of source lead inductance.

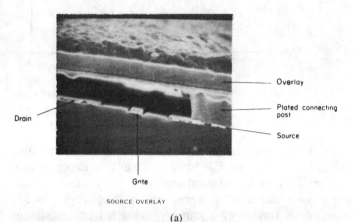

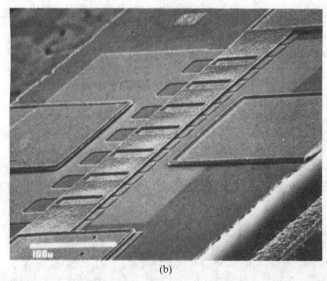

Fig. 10. (a) SEM microphotograph of a segment of source overlay (airbridge) of a power FET showing gate and drain contacts. (b) Top view of a GaAs power FET showing an air-bridge overlay connecting all source pads.

Two general methods of grounding are available: 1) the "wrap-around" ground; and 2) the "via" hole ground. The former requires a topside metallization pattern near the periphery of the chip which can be connected to the chip ground. The "via" hole technique, on the other hand, allows placement of grounds through the substrate where desired. Holes are chemically milled through the substrate until the top metallization pattern is reached. These holes are subsequently metallized at the same time as the ground plane to provide continuity between this plane and the desired topside pad. A microphotograph of such a "via" hole etched through a test wafer of 50-μm thickness is shown in Fig. 8. The hole diameter, in this case, is only 50 μm, much smaller than those normally used in monolithic circuits. The estimated inductance of a via hole is approximately 40–60 pH/mm of substrate thickness. Examples of circuits using both grounding techniques will be described later.

Low inductance grounds are especially important in source leads of power FET's. An inductance in the source lead manifests itself as resistive loss in the gate circuit, and hence a reduction in power gain. To illustrate this, Fig. 9 is a graph of the calculated gain reduction as a function of source lead inductance for an unconditionally stable power FET, corresponding to a power output of 1, 2, and 4 W ($W = 1.5$, 3.0, and 6.0 mm).

The second interconnect problem arises when it is necessary to connect the individual cells of a power FET without resorting to wire bonds. A requirement is that these interconnects also have a low inductance. Here the so-called "air-bridge" crossover is useful. This crossover consists of a deposited strap which crosses over one or more conductors with an air gap in between for low capacitive coupling.

Fig. 10(a) is a cross-sectional view of a source crossover which interconnects two adjacent source pads of a power FET. The air gap is approximately 4 μm. Clearly shown is the 1-μm gate and the larger drain pad underneath the crossover. Fig. 10(b) is a closeup, angular view of a power FET which employs an airbridge (overlay) interconnect bus.

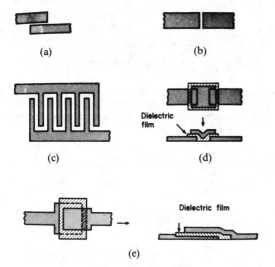

Fig. 11. Some planar capacitor designs. (a) Broadside coupled. (b) End coupled. (c) Interdigitated. (d) End-coupled overlay. (e) Overlay.

It is evident that the airbridge technology allows one to interconnect all cells without recourse to wire-bonding and therefore nicely satisfies the criterion for a monolithic circuit. Airbridge interconnects, of course, are also useful for microstrip and other crossovers. A good example is a planar spiral inductor, which requires a contact to the inner terminal.

E. Thin-Film Components

A flexible monolithic circuit design philosophy must include both lumped elements (dimensions <0.1 wavelength) and distributed elements, that is, elements composed of sections of transmission line. Lumped elements, R's, C's, and L's, are also useful for the RF circuitry, and in some cases mandatory, as for example, in thin-film resistive terminations for couplers. Lumped thin-film capacitors are absolutely essential for bias bypass applications, because of the large capacitance values required. Planar inductors can be extremely useful for matching purposes, especially at the lower end of the microwave band where stub inductors are very large, physically.

The choice of lumped or distributed elements depends on the frequency of operation. Lumped elements are suitable through X-band up to, perhaps, 20 GHz. It is likely, however, that beyond this frequency range, distributed elements will be preferred. It is difficult to realize a truly lumped element, even at the lower frequencies, because of the parasitics to ground associated with thin substrates. In this section we shall review the design principles of planar lumped elements.

1) Planar Capacitors: There are a variety of planar capacitors suitable for monolithic circuits—those achieved with a single metallization scheme, and those using a two-level metallization technology in conjunction with dielectric films. Some possible geometries for planar capacitors are shown in Fig. 11. The first three, which use no dielectric film and depend on electrostatic coupling via the substrate, generally are suitable for applications where low values of capacitance are required (less than 1.0 pF) for instance, high-impedance matching circuits. The last three geometries, the so-called overlay structures which use dielectric films, are suitable for low-impedance (power) circuitry and bypass and blocking applications. Capacitance values as high as 10 to 30 pF are achievable in small areas.

Two sources of loss are prevalent in planar capacitors, conductor losses in the metallization, and dielectric losses of the films, if used. Since the first three schemes illustrated in Fig. 11 are edge-coupled capacitors, high charge and current concentrations near the edges tend to limit the Q-factors. At X-band, typical Q-factors measured to date have been in the range of 50, despite the fact that no dielectric losses are present. The last three geometries distribute the current more uniformly throughout the metal plates because of the intervening film. However, even here, Q-factors only in the range of 50–100 are typical at X-band (10 GHz) because of dielectric film losses. Let us turn now to a more detailed analysis of the overlay structures depicted in Fig. 11, in particular the structure in Fig. 11(e).

First, we review briefly some general requirements of dielectric films for the overlay geometry. Some properties of dielectric films of importance are 1) dielectric constant, 2) capacitance/area, 3) microwave losses, 4) breakdown field, 5) temperature coefficient, 6) film integrity (pinhole density, stability over time), and 7) method and temperature of deposition. This last requirement is obviously important, because the technology used for film deposition must be compatible with the technology used for the active device (FET). Dielectric films which easily satisfy this criterion are SiO_x and Si_3N_4.

Some useful figures of merit for dielectric films are the capacitance-breakdown voltage product

$$F_{cV} = \left(\frac{C}{A}\right) V_b \tag{1a}$$

$$= \kappa \epsilon_0 E_b \tag{1b}$$

$$\cong (8-30) \times 10^3 \text{ pF} \cdot \text{V/mm}^2$$

and the capacitance-dielectric Q-factor product

$$F_{cq} = \left(\frac{C}{A}\right) Q_d \tag{1c}$$

$$= \frac{(C/A)}{\tan \delta_d} \tag{1d}$$

where C/A is the capacitance per unit area, V_b is the breakdown voltage, E_b is the corresponding breakdown field, κ is the dielectric constant, and $\tan \delta_d$ is the dielectric loss tangent. Breakdown fields of the order of 1–2 MV/cm are typical of good dielectric films. Dielectric constants are in the order of 4–20. Loss tangents can range from 10^{-1} to 10^{-3}. It is desirable to have as high figures of merit as possible. Table III is a list of candidate films and their properties.

We return, now, to the overlay structure of Fig. 11(e). A closeup perspective view is shown in Fig. 12. Taking into

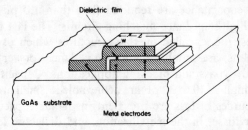

Fig. 12. Perspective of an overlay thin-film capacitor.

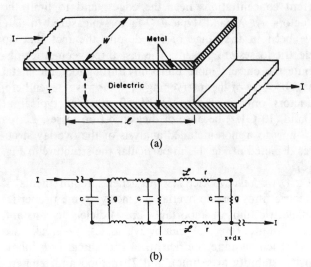

Fig. 13. Diagrams relevant to analysis of impedance of a thin-film capacitor. (a) Thin-film capacitor. (b) Circuit model.

TABLE III
PROPERTIES OF SOME CANDIDATE DIELECTRIC FILMS

DIELECTRIC	κ	TCC (ppm/°C)	C/A* (pF/mm²)	(C/A)·Q_d	(C/A)·V_b	COMMENTS
SiO	4.5–6.8	100–500	300	low	medium	Evaporated
SiO$_2$	4–5	50	200	medium	medium	Evaporated, CVD, or Sputtered
Si$_3$N$_4$	6–7	25–35	300	high	high	Sputtered or CVD
Ta$_2$O$_5$	20–25	0–200	1100	medium	high	Sputtered and Anodized
Al$_2$O$_3$	6–9	300–500	400	high	high	CVD, anodic oxidation, sputtering
Schottky-Barrier Junction	12.9	--	550	very low	high	Evaporated Metal on GaAs
Polyimide	3–4.5	−500	35	high	--	Spun and Cured Organic Film

*Film thickness assumed = 2000 Å, except for polyimide, 10,000 Å.

account the longitudinal current paths in the metal contacts, one may analyze this device as a lossy transmission line as indicated in Fig. 13. In Fig. 13(b), $\mathcal{L}$ and r represent the inductance and resistance per unit length of the metal plates, and c and g denote the capacitance and conductance per unit length of the dielectric film. The relation between g and c is determined by the loss tangent, $g = \omega c \tan \delta_d$. The series resistance in the plates is determined by the skin resistance if the metal thickness exceeds the skin depth, or the bulk metal resistance if the reverse is true. Usually, the bottom metal layer is evaporated only, and hence is about 0.5 μm thick, which may be less than the skin depth. The top metal is normally built up to a thickness of several micrometers or more by plating.

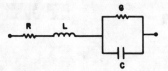

Fig. 14. Equivalent circuit of a thin-film capacitor. $R = 2/3\, rl$. $C = cl$. $G = \omega c \tan \delta$. $L = \mathcal{L} l$. r = resistance/length (electrodes). c = capacitance/length. $\mathcal{L}$ = inductance/length (electrodes). $\tan \delta$ = loss tangent of dielectric film.

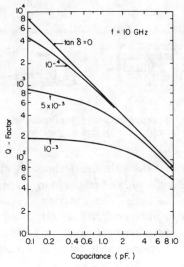

Fig. 15. Quality factor of a square thin-film capacitor as a function of capacitance and dielectric loss tangent for $f = 10$ GHz.

For a well-designed capacitor, the longitudinal and transverse dimensions are small compared with a wavelength in the dielectric film. In this case, a good approximation to the capacitor is the equivalent circuit shown in Fig. 14. When the skin loss condition prevails, the Q-factor corresponding to these losses is given by the expression

$$Q_c = \frac{3}{2\omega R_s (C/A) l^2} \quad (2)$$

where R_s is the surface skin resistivity and l is the electrode length (see Fig. 12). Note the strong dependence on electrode length. This arises because of the longitudinal current path in each electrode. Note that if one electrode, say the bottom plate, is very thin, Q_c is decreased.

The dielectric Q-factor is $Q_d = 1/\tan \delta$, and the total Q-factor is given by the relation $Q^{-1} = Q_d^{-1} + Q_c^{-1}$. Fig. 15 is a graph of the calculated Q-factor as a function of capacitance for various loss trangents. Note that for a 1-pF capacitor, and no dielectric losses, the predicted Q-factor is approximately 800! Yet, experimentally, values more like one-tenth of this are obtained, suggesting that dielectric films are extremely lossy—much more so than their bulk counterparts. No satisfactory explanation for this observation has yet been advanced.

2) Planar Inductors: Planar inductors for monolithic circuits can be realized in a number of configurations, all achieved with a single-layer metallization scheme. Fig. 16 illustrates various geometries that can be used for thin-film inductors. Aside from the high-impedance line section, all

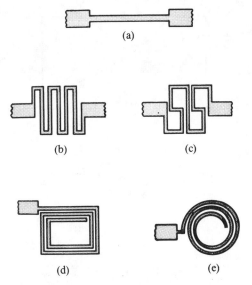

Fig. 16. Some planar inductor configurations. (a) High-impedance line section. (b) Meander line. (c) S-line. (d) Square spiral. (e) Circular spiral.

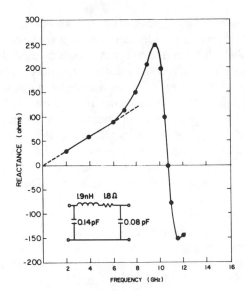

MEASURED REACTANCE OF A TEN-SEGMENT
SQUARE-SPIRAL GROUNDED INDUCTOR
ON A 0.1 MM THICK SI-GAAs SUBSTRATE

Fig. 18. Measured reactance of a ten-segment square spiral inductor on a 0.1-mm thick GaAs substrate (equivalent circuit shown in inset).

of the structures depend on mutual coupling between the various line segments to achieve a high inductance in a small area. In any multisegment design, one must insure that the total line length is a small fraction of a wavelength, otherwise the conductor cannot be treated as "lumped." Unfortunately, this latter condition is not often satisfied. Fig. 17 is a SEM photograph of a multisegment square–spiral inductor. Note the crossover connections.

When thin substrates are used, corrections must be made to the calculated inductance to account for the ground plane. These corrections are always in the direction to reduce the inductance, and are typically in the range of 15 percent, though for large-area inductors, the reduction can be as high as 30 percent.

Typical inductance values for monolithic circuits fall in the range from 0.5 to 10 nH. The higher values are difficult to achieve in strictly lumped form because of intersegment fringing capacitance. A more serious problem is that of shunt capacitance to ground, especially in the case of the microstrip format. This capacitance to ground can become important enough to require its inclusion in determining the performance of the inductor.

An illustration of the serious effect of capacitance to ground is demonstrated by the data of Fig. 18. This is a graph of the measured reactance of a ten-segment square spiral inductor as a function of frequency. The inductor is approximately 0.4 mm square, consisting of segments 1 mil wide, separated by 1 mil (see Fig. 17). The inductance, as designed, was nominally 1.9 nH. Note that above 10 GHz the reactance becomes capacitive! The equivalent circuit, as deduced from two-port S-parameters, is shown in the inset. The substrate thickness was 0.1 mm.

Of course, the inductor is usable, provided all of the parasitics indicated in Fig. 18 are taken into account.

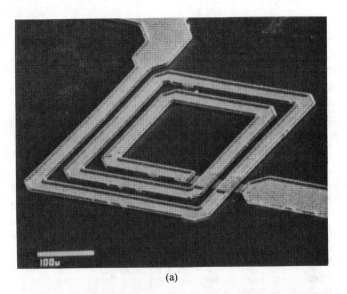

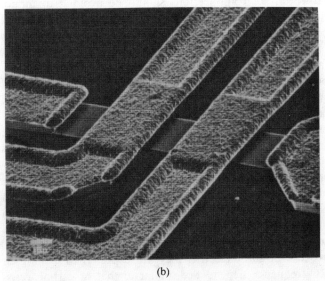

Fig. 17. SEM photographs of a thin-film square spiral inductor showing air-bridge crossovers.

Unfortunately, these parasitics are not known in advance. Thus, in a computer-aided approach, corrections to the circuit in which the inductor appears must be made in later iterations. This can become a costly procedure. It is often more sensible to use an inductive transmission line segment whose electrical behavior is known in advance.

Some of the skin losses in the inductor reside in the ground plane (assuming a metallized bottom side) and increase as the ground plane approaches the film inductor (not unlike shielding losses). However, the dependence on substrate thickness is mild, since most of the losses reside in the film turns, because of their small cross section.

In practice, inductor Q-factors of the order of 50 are observed at X-band, with higher values at higher frequencies. There appears to be no way to improve the Q-factor significantly, because of the highly unfavorable ratio of metal surface area to dielectric volume.

Somewhat higher Q-factors are achievable with microstrip resonant stub sections. These are more properly considered as distributed inductors, or more correctly, as distributed resonant elements. Three sources of loss are important here, skin losses, dielectric losses, and radiation losses. For microstrip stubs, the skin losses are those associated with microstrip, as are the dielectric losses. Skin losses vary inversely with the substrate thickness, and increase as the line impedance increases. Assuming negligible dielectric losses, one may show that the conductor Q-factor for a quarter-wave open circuit stub is given by

$$Q_c = \frac{27.3}{(\alpha \lambda g)} \quad (3)$$

where $(\alpha \lambda g)$ is the loss in the line section in decibels per wavelength. Since $(\alpha \lambda g)$ decreases as $f^{-1/2}$, Q_c increases as the square root of frequency, as for thin-film inductors. On the other hand, radiation losses from the open circuit end vary as [8]

$$Q_r = \frac{R}{(fh)^2} \quad (4)$$

where h is the substrate thickness and R is a function of w/h and the dielectric constant of the substrate. (The radiation factor R is considerably larger for a quarter-wave stub grounded at its far end.) Note that the radiation Q decreases as the square of the frequency and the substrate thickness h. Thus any attempt to increase the conductor Q-factor by increasing the frequency and substrate thickness is eventually overcompensated by the decrease in radiation Q. Fig. 19 illustrates this fact for practical substrate thicknesses. Thus, above X-band, open-circuit stub resonators are dominated by radiation losses, unless the substrate is less than 0.25 mm thick. This radiation also can cause coupling to adjacent circuits. A way to overcome both problems is to use a ring resonator.

The choice then as to whether reactive lumped elements or distributed elements should be used must be considered for each individual application. If high-Q narrow-band

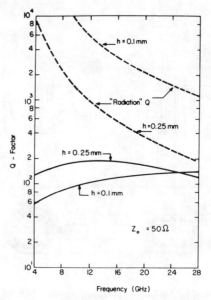

Fig. 19. Quality factor of a quarter-wave microstrip resonator on a GaAs substrate.

circuits are to be realized, distributed elements are recommended, provided space is available. On the other hand, broad-band circuits are probably easier to design with lumped elements, though even here synthesis techniques based on transmission line stubs are now available. Some circuits are more readily designed with distributed elements. Examples are four-port couplers and power combiners/dividers.

3) Planar Loads: Planar loads are essential for terminating such components as hybrid couplers, power combiners and splitters, and the like. Some factors to be considered in the design of such loads are: 1) the sheet resistivity available; 2) thermal stability or temperature coefficient of the resistive material; 3) the thermal resistance of the load; and 4) the frequency bandwidth. Other applications of planar resistors are bias voltage dividers and dropping resistors. However, such applications should be avoided in monolithic circuits, where power conservation is usually an objective.

Planar resistors can be realized in a variety of forms but fall into three categories: 1) semiconductor films; 2) deposited metal films; and 3) cermets. Resistors based on semiconductors can be fabricated by forming an isolated land of conducting epitaxial film on the substrate, for example, by mesa etching or by isolation implant of the surrounding conducting film. Another way is by implanting a high-resistivity region within the semi-insulating substrate. Metal film resistors are formed by evaporating a metal layer over the substrate and forming the desired pattern by photolithography. These techniques are illustrated in Fig. 20. Cermet resistors are formed from films consisting of a mixture of metal and a dielectric. However, because of the dielectric, they are expected to exhibit an RC frequency dependence similar to that of carbon resistors, which may be a problem in the microwave band.

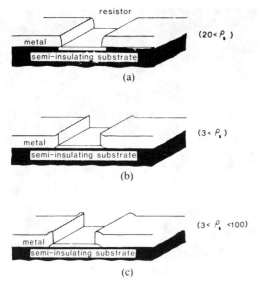

Fig. 20. Examples of planar resistor designs. (a) Implanted resistor. (b) Mesa resistor. (c) Deposited resistor.

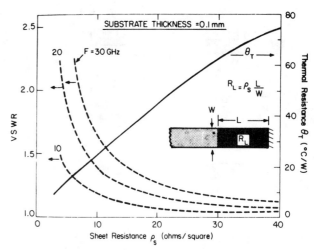

Fig. 21. Thermal resistance and VSWR of a planar resistor as a function of sheet resistance and frequency.

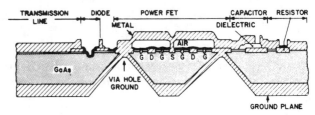

Fig. 22. Composite sketch illustrating technologies used in monolithic circuits.

TABLE IV
PROPERTIES OF SOME RESISTIVE FILMS

MATERIAL	RESISTIVITY ($\mu\Omega$-cm)	TCR (ppm/°C)	METHOD OF DEPOSITION	STABILITY	COMMENTS
Cr	13 (BULK)	+3000 (BULK)	EVAPORATED SPUTTERED	G-E	EXCELLENT ADHERENCE TO GaAs
Ti	55-135	+2500	EVAPORATED SPUTTERED	G-E	EXCELLENT ADHERENCE TO GaAs
Ta	180-220	-100 TO +500	SPUTTERED	E	CAN BE ANODIZED
Ni Cr	60-600	200	EVAP. (300°C) SPUTTERED	G-E	STABILIZED BY SLOW ANNEAL AT 300°C
TaN	280	-180 TO -300	REACTIVELY SPUTTERED	G	CANNOT BE ANODIZED
Ta$_2$N	300	-50 TO -110	REACTIVELY SPUTTERED	E	CAN BE ANODIZED
BULK GaAs	3-100 ohms/sq.	+3000	EPITAXY OR IMPLANTATION	E	NONLINEAR AT HIGH CURRENT DENSITIES

Metal films are preferred over semiconducting films because the latter exhibit a nonlinear behavior at high dc current densities and a rather strong temperature dependence—as some metal films do. Not all metal films are suitable for monolithic circuits, since their technology must be compatible with that of GaAs. Table IV lists some candidate metal films along with GaAs.

A problem common to all planar resistors used as microwave loads is the parasitic capacitance attributable to the underlying dielectric region and the distributed inductance of the film, which makes such resistors exhibit a frequency dependence at high frequencies. If the substrate bottomside is metallized, one may determine the frequency dependence by treating the load as a very lossy microstrip line.

For low thermal resistance, one should keep the area of the film as large as possible. To minimize discontinuity effects in width, the width of the resistive film load should not differ markedly from the width of the line feeding it. This means that the resistive element should be as long as possible to minimize thermal resistance. This length is specified by the sheet resistivity of the film and is given by the formula

$$l = \frac{wR}{\rho_s} \quad (5)$$

where w is the width of the film, R the desired load resistance, and ρ_s the sheet resistance of the film.

If one increases the length of the load (by decreasing the sheet resistivity) to achieve a low thermal resistance, one may get into trouble because the load may begin to exhibit the behavior of a transmission line (albeit a very lossy one) rather than a lumped resistor. Fig. 21 shows how the VSWR increases dramatically at low values of ρ_s because the length of the load becomes too large. Also shown is the thermal resistance. Clearly, a tradeoff is necessary between VSWR and thermal resistance.

All of the technologies we have discussed above are conveniently summarized in the cross-sectional view of a hypothetical monolithic circuit shown in Fig. 22.

4) Transmission Line Junction Effects: The many junctions and bends required of transmission lines in monolithic circuits to achieve close packing introduce unwanted parasitic inductances and capacitances. Fig. 23 illustrates some of the circuit representations of these junctions. Since such discontinuities cannot be avoided, but only minimized, the frequency dependencies must be taken into account, especially when the frequency is above X-band. It is particularly important to include junction effects in any broad-band design, that is, octave bandwidths. Unfortunately, though much work has been done on this topic, the results are not generally in a form useful for the circuit

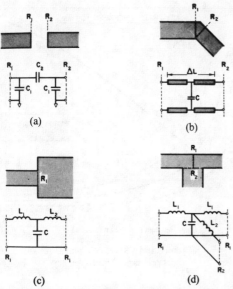

Fig. 23. Some microstrip discontinuities and their equivalent circuits. (a) Gap. (b) Bend. (c) Width discontinuity. (d) Tee junction.

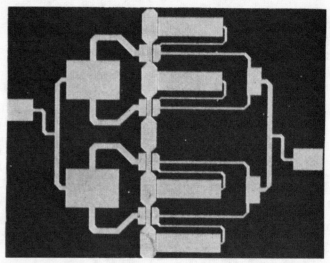

Fig. 24. Monolithic GaAs four-FET X-band power combiner. Chip size is 4.8×6.3×0.1 mm. (Raytheon Company.)

designer. As a consequence, computer-aided design programs do not incorporate corrections for junctions at present.

VI. EXAMPLES OF MONOLITHIC CIRCUITS

We shall present examples of some practical monolithic circuits which demonstrate the design principles discussed above. These circuits are representative of the research being conducted at laboratories around the world.

Fig. 24 is a photograph of a GaAs chip containing a single-stage four-FET power combiner designed at Raytheon (Research Division) [17]. This amplifier, an X-band microwave circuit, was the first to dispense with wire bonds on the chip by use of "via" holes for grounding the source pads. Built on a chip 4.8×6.3×0.1 mm in size, and using a microstrip format with on-chip matching to a 50-Ω system, the circuit exhibited a 5-dB small-signal gain at 9.5 GHz and a saturated CW power output of 2.1 W at 3.3-dB gain (see Fig. 25). Bias was supplied through bias tees via the RF terminals. Although large by present standards, the chip area could be reduced by 30 percent if the capacitive stubs were replaced by thin-film capacitors, which were not available at the time.

An extension of this technology to a two-stage X-band power amplifier also designed at this laboratory [22] is shown in Fig. 26. In this circuit, thin-film capacitors, based on SiO or Si_3N_4 technology, were incorporated on the chip for RF blocking and bias applications. Another innovation, clearly evident in the future, is the use of extended integral (grown) beam leads, an offshoot of the airbridge technology. The beam leads allow off-chip bonding of the RF and dc supply connections to the chip, thus avoiding damage to the chip. The amplifier, built on a 2.5×3.2 ×0.1-mm chip, exhibited a saturated CW power output of 550 mW and 8.5-dB gain at 9.5 GHz and a small-signal gain of 10 dB.

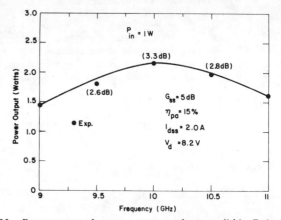

Fig. 25. Power output–frequency response for monolithic GaAs four-FET power combiner.

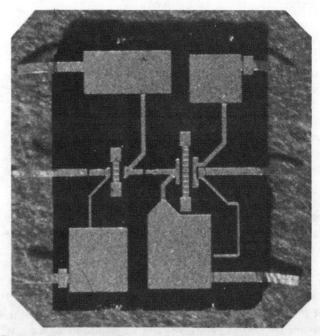

Fig. 26. Two-stage GaAs monolithic X-band amplifier. Chip size is 2.5×3.2×0.1 mm. (Raytheon Company.)

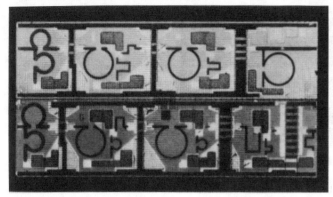

Fig. 27. Three- and four-stage GaAs monolithic X-band power amplifiers. Circuit sizes are 1.0×4.0×0.1 mm (Courtesy, W. Wisseman, Texas Instruments, Inc.)

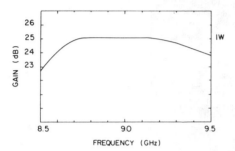

Fig. 28. Measured power gain–frequency response of four-stage amplifier of Fig. 27. (Courtesy, W. Wisseman, Texas Instruments, Inc.)

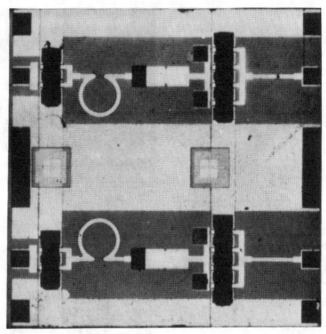

Fig. 29. Two-stage GaAs monolithic X-band push–pull amplifier. Chip size is 2.0×2.0×0.1 mm. (Courtesy, W. Wisseman, Texas Instruments, Inc.)

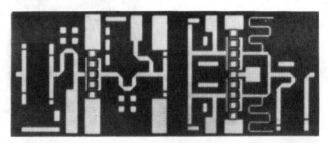

Fig. 30. Two-stage 5.7–11-GHz GaAs monolithic power amplifier. Chip size is 2.0×4.75×0.1 mm. (Courtesy, J. Oakes, Westinghouse.)

Turning to results obtained at other laboratories, Fig. 27 represents some of the research at Texas Instruments [18]. Shown is a chip containing side-by-side X-band amplifiers: the top, a three-stage FET amplifier; the bottom, a four-stage amplifier. Each chip is 1×4×0.1 mm in size. Both designs are based on a lumped-element approach which uses spiral inductors, clearly evident in the photographs, and thin-film capacitors of the end-coupled variety (Fig. 11(d)). Grounding is achieved by means of a metallized peripheral strip, and bias connections are made by wirebonds to pads on the chip. The three-stage amplifier delivers 400 mW at 23-dB gain and the four-stage delivers 1 W at 27-dB gain and 15–17-percent power-added efficiency in the 8.8 to 9.2-GHz range (see Fig. 28).

Another circuit reported by this laboratory [18] is the push–pull amplifier shown in Fig. 29. Each channel is a two-stage power amplifier, again based on the lumped-element approach, situated on a 2.0×2.0×0.1-mm chip. Although not monolithic in the strict sense of the word because inductive wire bonds interconnect the two channels, the design is unique in that a "virtual" ground is achieved by connection of the corresponding source pads of the adjacent channels; thus the need for a low inductance ground for the sources is avoided. Over 12-dB gain was obtained at 9.0 GHz with a combined CW power output of 1.4 W. All three amplifiers interface with a 50-Ω system.

An octave bandwidth GaAs amplifier designed at Westinghouse (R. and D. Center) is shown in Fig. 30. This circuit, similar to the one reported by Degenford et al. [7], consists of 1200-μm and 2400-μm periphery power FET's in cascade formed by selective ion implantation into a semi-insulating substrate. Built on a 2.0×4.75×0.1-mm chip, the circuit is based on a microstrip format with via holes, and makes liberal use of interdigitated capacitors. Source pads are grounded individually with vias. The amplifier produces a power output of 28±0.7 dBm at a gain of 6±0.7 dB across the 5.7 to 11-GHz band.

Another monolithic wideband amplifier is the 4–8-GHz eight-stage GaAs circuit reported by TRW [3] shown in Fig. 31. The design, based on the lumped-element approach (spiral inductors and SiO_2 thin-film capacitors) uses a coplanar feed at the input and output 50-Ω ports, with coplanar ground planes extending the full length of the 2.5×5.0-mm chip.

A departure from the GaAs approach is the SOS three-stage L-band amplifier built at Raytheon (Equipment Division) [13] (Fig. 32). This circuit, occupying a chip 7.5×7.5×0.46 mm in size, delivers 200-mW CW output at 20-dB gain at 1.3 GHz. The circuit uses spiral inductors. Dielec-

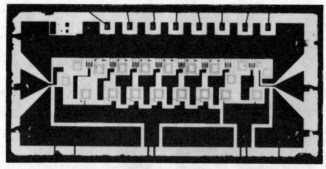

Fig. 31. Eight-stage 4–8-GHz GaAs monolithic amplifier. Chip size is 2.5×5.0 mm. (Courtesy, A. Benavides, T.R.W., Inc.)

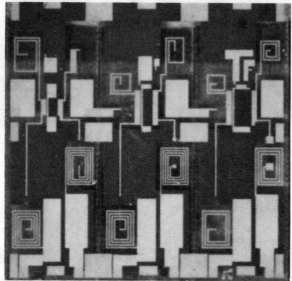

Fig. 32. Three-stage L-band silicon-on-sapphire amplifier. Chip size is 7.5×7.5×0.46 mm. (Courtesy, D. Laighton, Raytheon Company.)

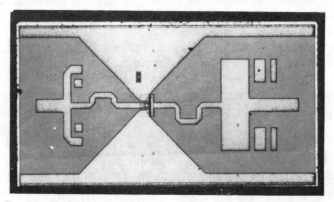

Fig. 33. Single-stage 20-GHz GaAs monolithic low-noise amplifier. Chip size is 2.75×1.95×0.15 mm. (Courtesy, A. Higashisaka, Nippon Electric Company.)

tric (SiO_2) films are used for capacitors and conductor crossovers.

So far we have described power amplifiers only. The first monolithic low-noise amplifier was reported by NEC (Central Research Laboratories) [10] (Fig. 33). This is a one-stage circuit on a 2.75×1.95×0.15-mm GaAs chip. The matching circuits use microstrip lines and stubs to interface with a 50-Ω system through bias tees. Large topside pads are used for the source RF grounds. The circuit, using a 0.5-μm gate, exhibited a noise figure of 6.2 dB and an associated gain of 7.5 dB in the 20.5–22.2-GHz band.

Most of the circuits we have described so far are based on the lumped-element or the microstrip approach or on a

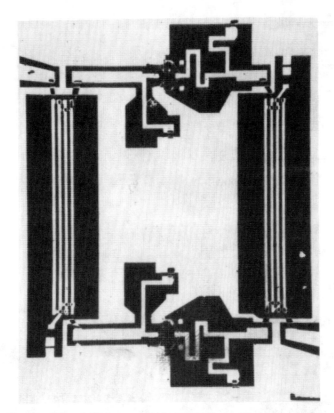

Fig. 34. X-band GaAs monolithic balanced amplifier using coplanar coupler. Chip size is 4.0×4.0 mm. (Courtesy, E. M. Bastida, CISE SpA.)

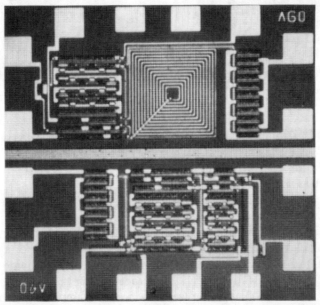

Fig. 35. Multistage direct-coupled GaAs monolithic amplifiers. Circuit sizes are 300×650 μm. (Courtesy, D. Hornbuckle, Hewlett Packard.)

combination of the two. Fig. 34 is a photograph of an X-band circuit using coplanar waveguides. This is a balanced amplifier reported by CISE SpA [2] built on a 4.0×4.0-mm GaAs chip, which uses two 90°, 3-dB broadband couplers. The couplers employ CPW rather than MS to obviate the need for micron-line spacings which are necessary with MS couplers. Lumped inductors and thin film (SiO_2) capacitors are used for RF matching and bypass. The circuit utilizes 0.8-μm gate MESFET's and has demonstrated a gain slightly below 10 dB between 8.5 and 11 GHz.

The next circuits, Fig. 35 represent a complete departure from the design philosophy considered so far. Shown is a photograph of two wide-band (0–4.5 GHz) amplifiers de-

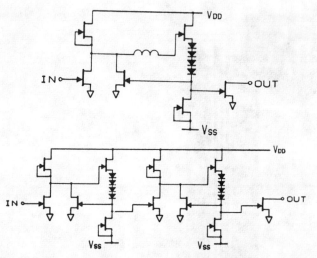

Fig. 36. Circuit schematics for direct-coupled amplifiers shown in Fig. 35. (Courtesy, D. Hornbuckle, Hewlett Packard.)

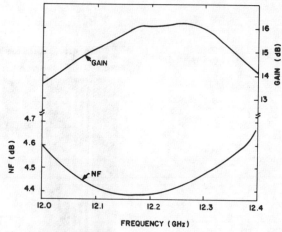

Fig. 38. Performance curves for receiver front end shown in Fig. 37. (Courtesy, P. Harrop, LEP.)

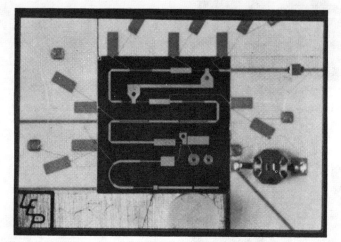

Fig. 37. 12-GHz GaAs monolithic receiver front end. Chip size is 1.0×1.0 cm. (Courtesy, P. Harrop, LEP.)

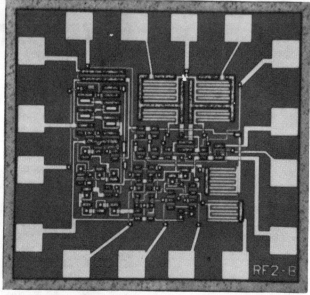

Fig. 39. Direct-coupled GaAs monolithic FET RF signal generation chip. Chip size is 600×650 μm. (Courtesy, R. Van Tuyl, Hewlett Packard.)

signed at Hewlett Packard [11]. What is unique about these circuits is the fact that, except for the spiral inductor, MESFET's are used throughout as active devices and as replacements for resistors and capacitors. The elimination of lumped elements, in conjunction with a direct-coupled circuit approach, allows a very high circuit packing density. Fig. 36 illustrates the circuit complexity achieved in each 0.3×0.65-mm area. Both amplifiers exhibited a gain in excess of 10 dB over the band.

Up until now we have described circuits which earlier we referred to as the lowest level of complexity. The next series of circuits represent integration on a functional block level. The first circuit (Fig. 37) is an integrated receiver front end on a GaAs chip intended for 12-GHz operation. This was reported by LEP [9]. The circuit, deposited on a large 1-cm square chip of GaAs, consists of a two-stage low-noise 12-GHz MESFET amplifier, an 11-GHz FET oscillator, and a dual-gate FET mixer. The matching circuits use microstrip lines and quarter-wave dc blocks. The oscillator is stabilized by an off-chip dielectric resonator. Bias circuits are included on the surrounding alumina substrate. Preliminary results are summarized in Fig. 38. The circuit is intended for a potential consumer market for domestic satellite-to-home TV reception planned for Europe.

Another example of the functional block approach is the monolithic GaAs FET RF signal generation chip (Fig. 39) designed at Hewlett Packard [21]. An extremely high degree of integration was achieved by use of the direct-coupled approach described earlier. Contained within the 0.65× 0.65-mm chip is the circuit shown in the schematic of Fig. 40. The local oscillator is resonated by an off-chip inductor which is tuned over the 2.1–2.5-GHz range by an on-chip Schottky barrier junction capacitor. The circuit is intended for an instrument application.

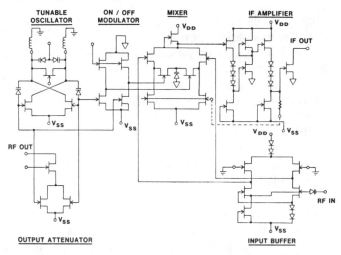

Fig. 40. Schematic for direct-coupled signal generation chip shown in Fig. 39. (Courtesy, R. Van Tuyl, Hewlett Packard.)

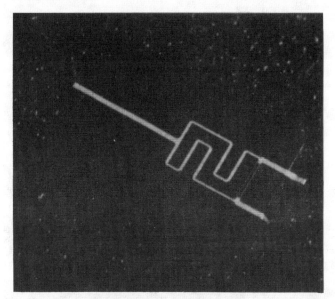

Fig. 42. GaAs monolithic X-band Wilkinson combiner/divider. Chip size is $1.5 \times 2.5 \times 0.1$ mm. (Raytheon Company.)

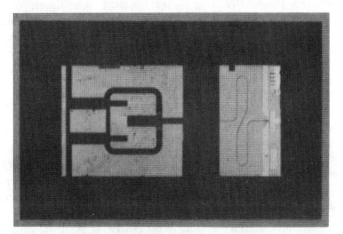

Fig. 41. GaAs monolithic mixer/IF circuit for millimeter-wave receiver applications. Chip size is $2.7 \times 5.3 \times 0.18$ mm. (Courtesy, R. Sudbury, Lincoln Laboratories.)

Our final functional block circuit is the monolithic balanced Schottky-barrier diode mixer/IF FET preamplifier chip illustrated in Fig. 41. This MS circuit, reported by Lincoln Laboratories [6], is built on a $2.7 \times 5.3 \times 0.18$-mm GaAs chip in MS format. The circuit operates between a 31-GHz signal source and a 2-GHz IF output. An external oscillator signal is injected through one of the coupler ports. The circuit exhibits an overall gain of 4 dB and a single-sideband noise figure of 11.5 dB.

We now turn to some special passive components fabricated in monolithic form. The first is a Wilkinson combiner/divider reported by Raytheon [23] shown in Fig. 42. Built on a $1.5 \times 2.5 \times 0.1$-mm chip, the circuit uses a thin-film titanium balancing resistor and was designed to operate at a center frequency of 9.5 GHz. Note the extended beam leads. As an illustration of the extremely good electrical balance that one can achieve with the high-resolution photolithography intrinsic to the monolithic approach, we show in Fig. 43 a graph of the power division and phase balance measured for the two 3-dB ports.

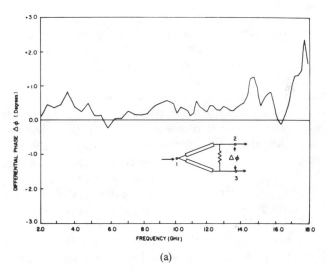

(a)

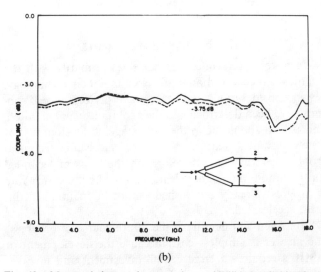

(b)

Fig. 43. Measured phase and power balance of Wilkinson divider shown in Fig. 42.

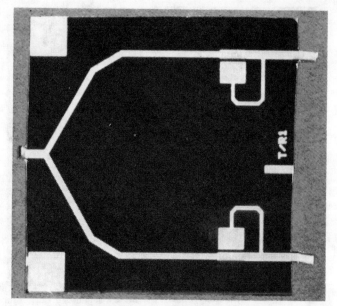

Fig. 44. GaAs monolithic *X*-band transmit/receive switch. Chip size is $3.0 \times 3.0 \times 0.1$ mm. (Raytheon Company.)

Another component designed at this laboratory is the all-FET T/R switch shown in Fig. 44 [1]. This switch, intended for phased-array applications at *X*-band, requires no dc hold power in either state. Built on a $3.0 \times 3.0 \times 0.1$-mm chip, the switch exhibits an isolation in excess of 33 dB between the transmitter and receiver ports in the 7–13-GHz range, and an insertion loss as low as 0.5 dB within this band. An alternative approach, also using FET's, was reported by McLevige *et al.* [14]. Both approaches utilize the change in source–drain resistance with gate bias.

The examples we have shown, though not exhaustive, are representative of the work reported so far (December 1980) and are intended to give the reader a good perspective of the advances made in the field during the last two years. Needless to say, the next several years will see the emergence of a still higher level of circuit integration in this rapidly developing field.

VII. Future Developments

We have so far concerned ourselves primarily with the technical aspects of monolithic circuits—their technology, design considerations, and microwave performance. Problems have been described and their solutions demonstrated. This is as it should be in the early stage of development of a new technical venture. No major unsolvable technical problems are evident; therefore, on the basis of technical considerations alone, there is no reason why the steady rate of progress already established cannot be maintained, indeed, accelerated.

What then will determine the future course of progress? The answer is simple—cost! Because the development of MMIC's requires a large capital investment and involves time-consuming and expensive processing technology plus a sophisticated testing procedure, the future development of this field will rest squarely on the as yet unproven expectations of reductions in cost and, to a lesser extent, improvements in reliability and reproducibility accruing from the monolithic approach.

The matter of cost reductions, in turn, rests on the answers to two questions.

1) Will the many complex technology steps required of MMIC's lend themselves to a high-yield production process?

2) Will a mass market develop in the microwave system area—a mass market necessary to capitalize on the high-volume low-cost attributes of batch processing?

Both of these requirements were eventually satisfied for silicon technology. Will this happen for gallium arsenide microwave technology? Time will tell. Since the silicon development was helped along by a vast domestic market (radios, TV's, etc., and more important, the commercial computer) and military markets, what are the expected large-volume markets for MMIC's?

Two potentially large markets appear to be developing, one military, the other consumer. In the military area, one such market includes electronically scanned radar systems, especially airborne and space-borne systems being planned for the future. For it is in the phased-array antenna, which may require modules as high as 10^5 in number, that we find a microwave system analog of the computer, which gave impetus to the growth of the silicon IC market. The antenna module requirements have already spurred developments of such module subsystems as transmitters, low-noise receivers, phase shifters, and transmit–receive (T/R) switches, some examples of which were described earlier. Here, along with cost, important design performance criteria will be reliability and small weight and size.

Another military application is in ECM systems, which require low-cost high-gain broad-band amplifiers. The difficult technical problems and projected high manufacturing cost associated with the hybrid integrated approach to this task have in essence mandated the use of monolithic circuits. Finally, the possibility of merging high-speed GaAs digital and microwave circuitry on the same chip may encourage use of such circuitry in signal processing at the RF level.

Turning to the nonmilitary markets, one potentially large outlet may be receiver front ends for the direct satellite-to-home-TV consumer market. Numerous such systems are being planned, for example, in Europe. We have described earlier one circuit intended for this market.

A third potential market, though much smaller in size, is instrumentation. Here cost and possibly circuit packing density are most important. Several examples of circuits earmarked for this application have been described.

We have not said much of the millimeter-wave spectrum. It is perhaps premature to do so, as this field itself is in its early stage of development. Here monolithic applications might develop, more for technical reasons than for economic reasons, because of the important role played by undesired packaging parasitics associated with discrete devices at these high frequencies. It is not unlikely that here too, as at lower frequencies, military applications may spur initial development. Now we turn to the question of costs.

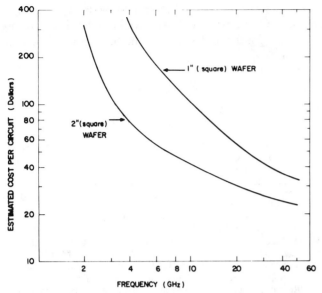

Fig. 45. Zero-order estimate of circuit cost as a function of frequency and wafer size.

Cost being the crucial item that it is, what are the factors contributing to it? They are the following:

1) materials;
2) materials processing;
3) circuit/device technology;
4) circuit assembly and packaging; and
5) testing (dc and RF).

These items, as is evident, do not include the important but nonrecurring costs such as capital investment, engineering, and mask design. In the materials category we include cost of substrate qualification, epitaxial growth and/or ion implantation, and profile evaluation, among others. Testing includes both dc and RF testing at the wafer probe level as well as circuit performance evaluation at the jig or package level.

Can a dollar figure be attached to these costs? At this stage, no! It is hazardous, at best, to attempt an accurate cost analysis based on laboratory experience, for large volume production, because ultimate module costs will be directly dependent on circuit yield in a manufacturing environment.

It is helpful, nevertheless, to attempt at least a "zeroth" order estimate of potential circuit costs, not so much to obtain an absolute level of cost, but to pinpoint the high cost items in the list above. To do this we have estimated the number of available circuits per wafer as a function of circuit operating frequency. This estimate was shown in Fig. 2. In the context of our present discussion, a circuit is equivalent to one submodule, for example, a transmitter stage or two, a phase shifter, etc. Using this estimate, we have determined the cost per circuit as a function of operating frequency. This data is shown in Fig. 45.

The cost estimates were derived by assumption of a 50-percent processing yield, independent of frequency. The base cost includes material cost, fully loaded labor cost, and circuit qualification at the dc and RF level. We feel that the cost estimates shown are useful guidelines but they should not be considered accurate in any absolute sense. For instance, depending on circuit complexity, current laboratory yields at X-band range from near zero to 20 percent. The development of a 50-percent yield fabrication process technology, deemed necessary, requires much additional experience and substantial simplification of monolithic circuit fabrication techniques.

Adjustments may be necessary at either end of the frequency scale. For example, in the range below 3–4 GHz, a drastic cost reduction may ensue, at least for some circuit applications, if the direct-coupled approach can be used. At the other end of the scale, above, say, 10–12 GHz, the cost figures should be elevated. The reason is that, because of the necessity of submicron gate technology, the lower throughput of the ultrahigh resolution electron beam (EB) lithography will increase costs substantially. Here what will be needed is optimization of the processing technology by appropriate merging of the EB lithography for the active devices and the higher throughput photolithography which is more than adequate for the circuit elements. This problem has not yet been addressed.

On the basis of our cost analysis, certain definite conclusions can be reached about the expected relative cost of the several items listed above. First, the two materials factors, under large production lots (>100 K parts) will contribute a negligible amount to the total cost—of the order of 5 percent or less. Second, next to wafer processing, the cost of packaging and microwave testing will be the largest cost factor. Indeed, because these latter costs will be fairly independent of the frequency band, and because of the decreasing processing cost per circuit with increasing frequency and wafer size, it is expected that packaging and testing will be the dominant cost factor at the higher frequencies (perhaps above 10 GHz).

It seems evident that, in light of this conclusion, the reduction of assembly and testing costs will be of paramount importance and must be addressed rather soon. Not only must as many functions as possible be integrated on one chip, consistent with high yield, but RF testing of chips and monolithic circuits and modules must be automated, just as dc tests have been. This will be very difficult because RF probes small enough for chip use are still in the laboratory stage, and their extension to performance tests on an entire circuit are nonexistent.

VIII. Conclusions and Summary

Monolithic microwave circuits based on gallium-arsenide technology have finally become a practical reality. Owing its origin to early experiments based on silicon bipolar technology, the gallium-arsenide approach, except for some scattered results in the sixties, emerged as a serious development only within the last three years.

The factors most responsible for this rapid growth can be traced to: 1) the development of the Schottky-barrier field-effect transistor; 2) the excellent microwave properties of semi-insulating GaAs as a low-loss substrate; 3) the perfection of GaAs epitaxy and ion implantation; 4) the establishment of GaAs crystal pulling facilities capable of large-diameter crystal growth; and 5) the emergence of potential systems applications for monolithic microwave circuits.

We have attempted to demonstrate in this paper some of the many design considerations and tradeoffs that must be made to optimize the performance of GaAs monolithic microwave circuits. Attention has been focussed, primarily, on the nondevice aspects of monolithic circuit design.

Despite the small physical size of the circuitry, interconnections between components often must be treated as wave-propagating structures because of the high dielectric constant of GaAs, which reduces the wavelength within the substrate. Both coplanar waveguide and microstrip lines, as well as combinations of both, are appropriate for monolithic circuits.

A typical circuit design may use both distributed and lumped-element components. Lumped elements, it was shown, are not truly lumped, because of built-in parasitics arising from the dielectric substrate. These must be taken into account at X-band and higher frequencies. A major drawback of thin-film inductors and capacitors is the limited Q-factor achieved to date. Much has yet to be learned about loss reduction in thin dielectric films.

We have shown that MMIC's are realized rather easily. Via hole grounding and source airbridge interconnections are eminently suited for them. Computer-aided design techniques are a "must" to reduce the number of iterations necessary.

Many examples of monolithic circuits have been shown which demonstrate the design principles described. These circuits, representing a world-wide cross section of the efforts in this field, have emerged within the last two to three years, and demonstrate the variety of circuit applications amenable to the monolithic approach. The promising attributes of the monolithic technology to cut fabrication costs, improve reliability and reproducibility, and reduce size and weight will overcome many of the shortcomings of the hybrid approach.

We have argued that, based on the cost considerations, the potential markets for MMIC's will be for the most part systems requiring large quantities of circuits of the same type. Because of this, and because of the large capital expenditures required of an organization to become a viable contender for these markets, it is most likely that the major efforts in MMIC's will eventually reside in the systems houses themselves.

Acknowledgment

The progress reported in this paper represents the cumulative effort of many people, too numerous to mention individually. However, the author wishes to express his deep appreciation to his associates at Raytheon, whose work is described here, and to his colleagues at many other laboratories who so graciously gave him permission to use their photographs and latest results to help describe their research.

References

[1] Y. Ayasli, R. A. Pucel, J. L. Vorhaus, and W. Fabian, "A monolithic X-band single-pole, double-throw bidirectional GaAs FET Switch," in *GaAs IC Symp. Res. Abstracts*, 1980, paper no. 21.

[2] E. M. Bastida, G. P. Donzelli, and N. Fanelli, "An X-band monolithic GaAs balanced amplifier," in *GaAs IC Symp. Res. Abstracts*, 1980, paper no. 25.

[3] A. Benavides, D. E. Romeo, T. S. Lin, and K. P. Waller, "GaAs monolithic microwave multistage preamplifier," in *GaAs IC Symp. Res. Abstracts*, 1980, paper no. 26.

[4] L. Besser, "Synthesize amplifiers exactly," *Microwave Syst. News*, pp. 28–40, Oct. 1979.

[5] D. Ch'en and D. R. Decker, "MMIC's the next generation of microwave components," *Microwave J.*, pp. 67–78, May 1980.

[6] A. Chu, W. E. Courtney, L. J. Mahoney, G. A. Lincoln, W. Macropoulos, R. W. Sudbury, and W. T. Lindley, "GaAs monolithic circuit for millimeter-wave receiver application," in *ISSCC Dig. Tech. Pap.*, pp. 144–145, 1980.

[7] J. E. Degenford, R. G. Freitas, D. C. Boire, and M. Cohn, "Design considerations for wideband monolithic power amplifiers," in *GaAs IC Symp. Res. Abstracts*, 1980, paper no. 22.

[8] E. Denlinger, "Losses of microstrip lines," *IEEE Trans. Microwave Theory Tech.*, vol. MTT-28, pp. 513–522, June 1980.

[9] P. Harrop, P. Lesarte, and A. Collet, "GaAs integrated all-front-end at 12 GHz," in *GaAs IC Symp. Res. Abstracts*, 1980, paper no. 28.

[10] A. Higashisaka, in *1980 IEEE MTT-S Workshop Monolithic Microwave Analog IC's*.

[11] D. Hornbuckle, "GaAs IC direct-coupled amplifiers," in *1980 IEEE MTT-S Int. Microwave Symp. Dig.*, pp. 387–388.

[12] T. M. Hyltin, "Microstrip transmission on semiconductor substrates," *IEEE Trans. Microwave Theory Tech.*, vol. MTT-13, pp. 777–781, Nov. 1965.

[13] D. Laighton, J. Sasonoff, and J. Selin, "Silicon-on-sapphire (SOS) monolithic transceiver module components for L- and S-band," in *Government Microcircuit Applications Conf. Dig. Pap.*, vol. 8, 1980, pp. 299–302.

[14] W. V. McLevige and V. Sokolov, "A monolithic microwave switch using parallel-resonated GaAs FET's," in *GaAs IC Symp. Res. Abstracts*, 1980, paper no. 20.

[15] E. Mehal and R. W. Wacker, "GaAs integrated microwave circuits," *IEEE Trans. Microwave Theory Tech.* vol. MTT-16, pp. 451–454, July 1968.

[16] R. S. Pengelly and J. A. Turner, "Monolithic broadband GaAs FET amplifiers," *Electron. Lett.* vol. 12, pp. 251–252, May 13, 1976.

[17] R. A. Pucel, J. L. Vorhaus, P. Ng, and W. Fabian, "A monolithic GaAs X-band power amplifier," in *IEDM Tech. Dig.* 1979, pp. 266–268.

[18] V. Sokolov and R. E. Williams, "Development of GaAs monolithic power amplifiers," *IEEE Trans. Electron Devices*, vol. ED-27, pp. 1164–1171, June 1980.

[19] E. Strid and K. Reed, "A microstrip probe for microwave measurements on GaAs FET and IC wafers," in *GaAs IC Symp. Res. Abstracts*, 1980, paper no. 31.

[20] R. N. Thomas, "Advances in bulk silicon and gallium arsenide materials technology," in *IEDM Tech. Dig.*, 1980, pp. 13–17.

[21] R. Van Tuyl, "A monolithic GaAs FET RF signal generation chip," in *ISSCC Dig. Tech. Pap.* 1980, pp. 118–119.

[22] J. L. Vorhaus, R. A. Pucel, Y. Tajima, and W. Fabian, "A two-stage all-monolithic X-band power amplifier," in *ISSCC Dig. Tech. Pap.*, pp. 74–75, 1980.

[23] R. C. Waterman, W. Fabian, R. A. Pucel, Y. Tajima, and J. L. Vorhaus, "GaAs monolithic Lange and Wilkinson couplers," *GaAs IC Symp. Res. Abstracts*, 1980, paper no. 30.

GaAs MMICs promise low cost, high performance, small size and reproducibility. However, hybrid circuits still provide an optimum solution for the particular microwave component needed.

Hybrid Vs. Monolithic Microwave Circuits: A Matter of Cost

By **Raymond S. Pengelly**
Plessey Research Ltd.

During the last few years readers of both learned microwave journals and popular magazines have become aware of a significant sector of microwave research and development—the GaAs monolithic microwave integrated circuit (MMIC). The cleverness of the technology and circuit design of these MMICs has impressed many members of the microwave community who are used to "conventional" hybrid MIC or waveguide components and systems. Recently the development of MMIC technology and circuit design, the latter helped much by CAM and CAD, has demonstrated some remarkable results up to about 35 GHz. Perhaps the most interesting aspect of MMIC results to date has been the way in which the GaAs MESFET has been used to perform a wide variety of functions. Some of these functions include: amplification, frequency conversion, oscillation, phase shifting, switching, and so forth.[1,2,3,4,5] Significantly, in many cases the FET has performed better than its predecessors in these particular circuit areas.

Although the promise of batch-processed GaAs MMICs is one of low cost, high performance, small size and reproducibility—like the Si IC that goes before it—the microwave industry is one where large number requirements are not often met. A tremendous skill in hybrid MIC assembly has already been built up worldwide and hybrid MICs can provide an optimum solution for the particular microwave component needed. Quite rightly, therefore, there are many members of the microwave industry who treat MMICs with much skepticism, pointing out the disadvantages of that particular medium. While MMIC designers have been maintaining a strong pace, devotees of hybrid MICs have been continually improving their ability to produce more sophisticated, smaller and lower-cost components. Reviewing the relative advantages and the state-of-the-art of the two methods of producing microwave components leads to several conclusions about the prospects of the two competing technologies.

Some Basic MMIC and Hybrid Techniques

To gain some insight into the relative characteristics of monolithic microwave circuits based on GaAs and the more conventional hybrid MIC circuits, we must first consider the fabrication of the two types. Obviously since there are a large number of hybrid techniques varying from finline through suspended substrate to microstrip and slotline, we will consider here lumped element and microstrip distributed circuits on substrate materials such as alumina, sapphire and beryllia.

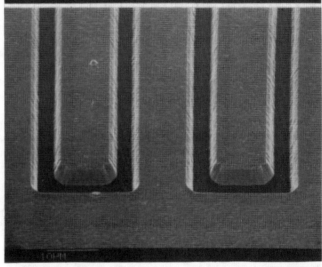

1. Details of parts of an MMIC, showing excellent definition of interdigitated capacitors and spiral inductors. A dielectric layer was used to separate metal layers.

Raymond S. Pengelly is group leader for GaAs MICs at Plessey Research (Caswell) Ltd., Caswell, Towcester, Northants, England. The author gratefully acknowledges the cooperation of various workers in the field in preparing this article, including E. Belohoubek and F. Sechi, of RCA; R. Pucel of Raytheon Research; K. Niclas of Watkins-Johnson; J. Degenford of Westinghouse; P. Harrop of LEP; D. Maki of Hughes Aircraft; and E.M. Bastida of CISE, Italy. The author is also indebted to his co-workers at Plessey Research, including P.D. Cooper, J. Suffolk, T. Bambridge, P. Rigby, A. Hughes, S. Greenhalgh, C. Suckling, J. Cockrill, R. Butlin, and C. Oxley. Parts of this paper include work carried out with the support of Procurement Executive, Ministry of Defence, sponsored by DCVD. The work done in this paper has been performed in part under the sponsorship and technical direction of the International Telecommunications Satellite Organization (INTELSAT). Any views expressed are not necessarily those of INTELSAT.

Reprinted with permission from *Microwave System News*, Jan. 1983.

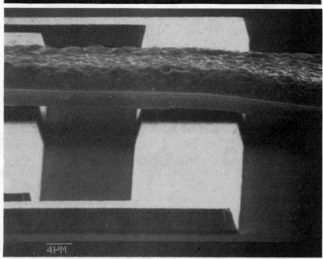

2. Air bridge connection between the center of a square spiral inductor and a gold track. For many applications, low-inductance, low-fringing capacitance interconnections are required between components, and an air bridge connection is the ideal way to solve the problem.

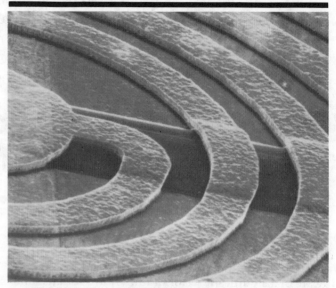

3. Alternatives to the air bridge are connections on top of thick (2 to 3 μm) polyimide, silicon nitride, or narrow underpasses. Here is such an example, applied to a circular spiral inductor.

MMIC Technology. GaAs ICs depend on exploiting the high-resistivity properties of the semi-insulating (SI) substrate material for passive components and the high mobility properties of the semiconducting layer that is either epitaxially grown onto or implanted into the SI substrate. The active device most commonly used in MMIC is the MESFET, while Schottky barrier diodes are only used as DC level shifting devices or at millimeter-wave frequencies as mixers. Recently varactor diodes have also been incorporated into MMICs for tuning purposes.[6]

A typical MMIC process consists of implanting donors or growing epitaxial layers for the active devices. The former gives improved uniformity of characteristics. Si^{28} is usually used for the implant species. Capacitors are formed using either interdigital or MIM (metal-insulator-metal) structures while resistors are formed using thin films of metals or cermets. A dielectric layer such as silicon nitride or polyimide is used to separate metal layers. In order to produce thick metalization to either carry current or reduce skin-depth losses, two methods are commonly employed—plated gold or sputtered gold, which is defined using ion beam milling. Figure 1, for example, shows details of parts of an MMIC where excellent definition of interdigitated capacitors and spiral inductors can be seen. Metal thickness is 3 μm.

For many applications low-inductance and low-fringing capacitance interconnection are required between components. For example, Figure 2 shows a so-called "air bridge connection" between the center of a square spiral inductor and a gold track. Usually air bridge connections are somewhat shorter than the example in Figure 2, thus giving better mechanical strength. Alternatives to the air bridge are connections on top of thick polyimide (2 to 3 μm thick), silicon nitride or narrow underpasses.

An example of the latter as applied to a circular spiral inductor is shown in Figure 3. Of particular importance in GaAs ICs operating at microwave frequencies is an effective and low-inductance method of connecting components (particularly common source connected FETs) to RF ground. For simple monolithic circuits where the

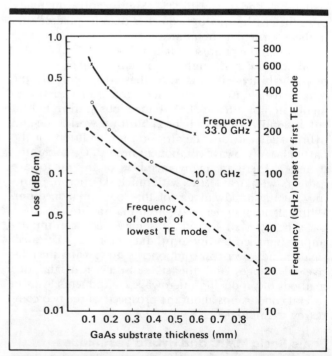

4. This relationship between microstrip loss and GaAs substrate thickness shows that if the GaAs substrate is made thin to relieve the task of producing small vias, then the microstrip loss will increase considerably at the higher frequencies.

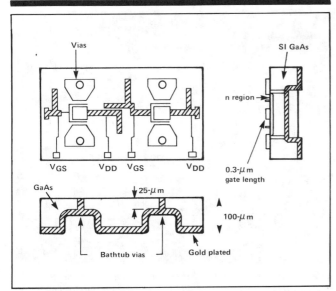

5. *For frequencies where 50-μm thick GaAs is undesirable, thicker (150 to 200 μm) substrates are used and the wafers thinned in the immediate vicinity of the FETs. The wafers are then plated up on their reverse sides and lapped down to form a flat back surface for subsequent chip mounting.*

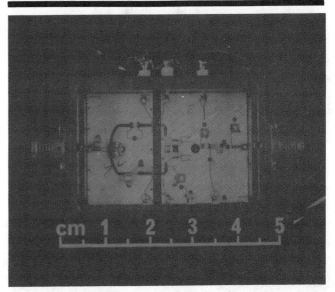

6. *This experimental hybrid lumped-element vector modulator was used to test out a design before producing MIC masks. Common source FETs are used in an active power divider.*

FETs can be located near the chip edges, fine gold mesh has been used to connect the circuit ground areas to the package ground. As the number of FETs increases per chip or the layout of the chip prohibits edge grounding, then related through holes, or "vias," etched through the GaAs in convenient positions are used. This via technology is, however, not necessarily straightforward since chemical etching produces an unsymmetrical hole particularly when substrate thicknesses much above a few tens of microns are employed.

Figure 4 shows the relationship between microstrip loss and GaAs substrate thickness, for example. Thus, if the GaAs substrate is made thin to relieve the task of producing small vias, the microstrip loss will increase particularly at the higher frequencies. Equally, in order to enable effective heat-sinking of power FET monolithic amplifiers, thin GaAs substrates are required because of the poor thermal impedance of the material. Thin substrates in this case are, however, undesirable since high-Q circuit components are particularly needed to transform the low-power FET impedances to 50 Ohms without excessive loss factors being involved.[7] In monolithic power amplifiers, the MESFETs are made to occupy sufficient GaAs area with FET source connections produced using thick air bridges and with several vias per device. Hence, with substrate thicknesses of about 50 μm, total thermal resistance is kept to a value whereby maximum channel temperatures do not exceed a safe limit (about 125°C). For frequencies where 50-μm thick GaAs is undesirable from the viewpoint of RF losses, thicker substrates (150 to 200 μm) are used and the wafers are then thinned only in the immediate vicinity of the FETs. The wafers are then plated up on their reverse sides and lapped down to form a flat back surface for subsequent chip mounting. Figure 5 shows a schematic of this "bathtub" via technique.

From the foregoing, it may be appreciated that in a complex MMIC there are many stages of fabrication. Let us say that the typical number of masks required for an MMIC circuit is 15. Thus, in order to produce cost-effective circuits using the MMIC method, the yield of such fabrication stages has to be high. At the moment, average component yields are about 90 percent with some large gate-width FETs having yields considerably lower. MMICs that are in the development stage at the present time can have about 100 components per chip (made up of FETs, MIM capacitors, resistors, vias, air bridges and so forth.) With a 90-percent mean component yield the MMIC technologist would have to make more than 37,000 circuits before finding one that was fully functional. Increasing that average component yield to 95 percent results in one in 169 working. To achieve an overall yield of, say, 25 percent, a mean component yield of 98.65 percent is required. This demonstrates that cost advantages to be gained from batch production of MMICs will only become a reality if there is the right balance between circuit complexity and the state of production technology at any one time. Otherwise it may be more appropriate to use a more "labor intensive" hybrid technique where production yield can be increased by reworking the microwave circuit. Reworking of GaAs ICs is virtually impossible.

Hybrid IC Technology. It is undoubtedly true that there are many circuit functions now available with GaAs MMICs that would have been impossible to produce using conventional substrate-based hybrid technology. This is particularly true of circuits requiring many small gate-width FETs[8,9] and in circuits where discrete

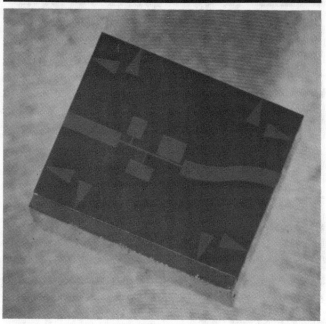

7. *This Ku-band filter uses fine* line geometries on quartz. It was ion-etched with a 23-GHz cutoff frequency. The narrow lines and gaps are 17-μm wide.

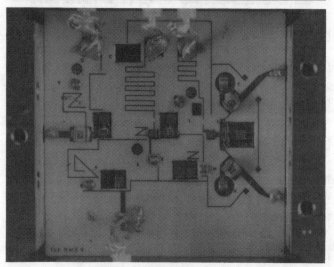

8. *One of the advantages of hybrid* techniques is that the circuit is far more flexible than any MMIC since bonding in extra capacitor pads and changing wire bond lengths or introducing different high-impedance microstrip lengths can be used to tune the circuit.

FETs were not designed suitably for their intended use. In many cases, however, it is possible to use hybrid ICs to demonstrate the concept in a future MMIC design. For example, Figure 6 shows an experimental vector modulator using lumped/distributed components on alumina substrates where common gate and common source FETs are used in an active power divider and FETs are also used in π arrangements for attenuators. This hybrid circuit, which was not designed to be particularly compact, measures 38 × 25 mm. The monolithic version measures 4.5 × 3 mm and will be described later as part of a T/R module.

In recent years there has been much work in producing a new generation of hybrid MICs having improved characteristics. Considerable advances have been made in new media such as finline,[10] suspended substrates[11] and microstrip/slotline.[12] Some of the work, however, has been aimed specifically at using lumped elements in order to produce circuits which are comparable in size to MMICs. In many amplifier circuits, for example, T networks consisting of series inductors and shunt capacitors are used as the matching components. This topology is often chosen because it can be conveniently realized using high-impedance transmission lines or bond wires and MIM capacitors which in many cases are simply metalized areas above high dielectric constant substrates.[13] As in MMICs self-biasing techniques are used for active devices. Complete amplifier matching circuits can be produced on 0.25-mm thick alumina substrates using conventional photolithographic masks and photoresist stages to define the thin-film gold circuits. Recently, ion milling has been used to replace chemical etching to produce high-resolution geometries on quartz and polished alumina substrates. Figure 7 shows, a Ku-band filter which uses fine line geometries on quartz.

Capacitors having typical values of 0.5 pF at X-band can be conveniently produced on 0.25-mm thick alumina substrates using metalized 1.27-mm squares. Since the wavelength at 10 GHz on alumina is about 12 mm, these capacitors are about λ/10. Usually the active devices are soldered or epoxied to grounded ledges between the substrates as shown in Figure 6. But they may also be mounted on small grounded carriers which are attached to the ground plane on the reverse side of the substrate as shown in Figure 8. One of the advantages of such a hybrid technique is that the circuit is far more flexible than an MMIC since bonding in extra capacitor pads and changing wire bond lengths or introducing different high-impedance microstrip lengths can be used to tune the circuit. In practice, trimming of circuits can usually be achieved by adjusting only one or two circuit elements. In order to produce reliable circuits the chip devices are mounted onto carriers made of a material—such as Kovar, certain copper alloys, or molybdenum—which has a thermal expansion coefficient close to the chip material (Si or GaAs). In cases where miniature hybrid circuits use chip capacitors, bond wires, and so forth, the microwave design engineer depends on the experience of operators in assembling his designs. To reduce such dependence to a minimum, therefore allowing greater reproducibility of performance, the design of the hybrid circuit should reduce the number of variables influenced by the circuit assembler. To this end, particularly for circuits which involve more than one or two active devices, wire bonding needs to be minimized and substrates processed on a batch basis to reduce costs. Standard hybrid circuits usually use packaged devices particularly if bare chip-bonding facilities are not available to the circuit designer. Hybrid circuits composing a

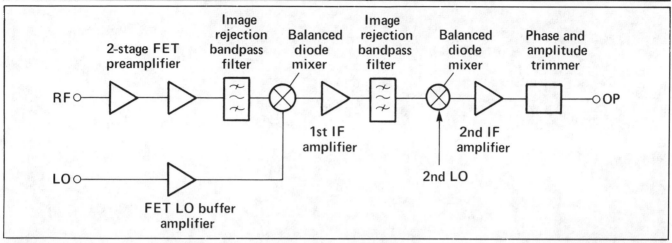

9. Hybrid microwave circuits often consist of complete subsystems in one housing. The double superhet receiver shown might consist of balanced FET amplifiers followed by a diode mixer fabricated on 0.635-mm thick alumina substrates mounted in copper alloy carriers which in turn are mounted on demountable "trays" containing waveguide-to-microstrip taper transitions.

subsystem or "supercomponent" will invariably consist of separate circuit substrates for each function especially when the circuits are being fabricated on substrate materials such as alumina or quartz. The reasons for this are complex and include the following:
- Cost effectiveness
- Testing
- Repairability and maintainability
- Yield
- The need for below waveguide cutoff sections to prevent spurious moding

In the earlier days of hybrid MIC manufacture, the lack of accurate data for active devices and inadequate knowledge of dispersion and discontinuity effects on microstrip, for example, made the "cut-and-try" method of design a popular technique. With the advent of programmable desktop computers and automatic network analyzer systems, very accurate characterization of components and structures over wide frequency ranges has led to the adoption of a more systematic approach to design aided by sophisticated computer programs such as SUPERCOMPACT. This has resulted in only final adjustments being made to circuits during production to correct for the inevitable tolerances in both device and circuit characteristics.

Hybrid microwave circuits today often consist of complete subsystems in one housing. For example, the double superhet receiver shown schematically in Figure 9 might consist of balanced FET amplifiers followed by a diode mixer fabricated on 0.635-mm thick alumina substrates mounted in copper alloy carriers which in turn are mounted on demountable "trays" containing waveguide-to-microstrip taper transitions. These trays allow the complete assembly to be easily repaired if any of the components in the front end fail by replacing the suspect unit. The first IF stage and second mixer might be built using 6010 Duroid with hermetically packaged hybrid mixers and amplifiers obtained from specialist vendors. The mechanical and electrical approaches to this particular design would be aimed at making the characterization of each of the subassemblies as easy as possible prior to final packaging.

Considerable engineering skill in terms of both mechanical and electrical knowledge is required to design such a production unit. Even more skill is needed if the unit must be low cost with good performance. In hybrid subsystems it is more often than not the mechanical aspects of the design which determine the reliability of the product. In an effort to improve on cost, size and reliability, the miniature hybrid module concept mentioned earlier has recently undergone further development. For example, Sechi and co-workers at RCA have been developing miniature, lightweight high-performance GaAs power amplifiers.[14] Like MMICs the miniature beryllium-oxide circuits (MBCs) use a more sophisticated technology than standard microstrip-based hybrids.

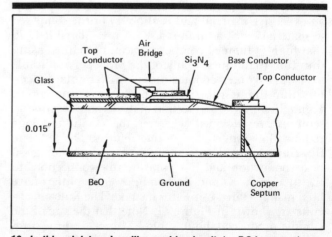

10. In this miniature beryllium-oxide circuit, by RCA, ground septums are provided wherever the circuit topology requires low-inductance ground returns. The FETs are flip-chip mounted onto the substrate to provide good heat conduction and low-parasitic interconnections to the RF circuit.

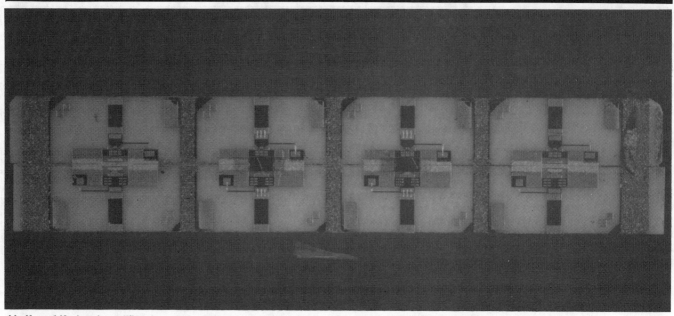

11. X- and Ku-band amplifiers have been built using beryllia circuits. This Ku-band amplifier is quite competitive with GaAs MMICs since it is about 5 mm x 5 mm. This technique is particularly attractive for power applications at a relatively simple circuit complexity level.

The technique adopted is the outcome of attempting to find a solution to several requirements:
- A substrate material which can be batch-processed
- Good RF grounds having low resistance and low inductance
- A substrate material offering high thermal conductivity
- A technique which uses thin-film deposition of high-Q lumped inductors and capacitors
- A medium which enables active devices to be mounted onto the circuit without wire bonds
- A medium offering small size, low weight and potential for low cost

Based on six points, GaAs is a good candidate apart from its thermal properties and cost. GaAs is at present an expensive material and is also very brittle compared to materials such as alumina. RCA uses glazed BeO so that high-Q lumped components can be defined using thin-film techniques. Many of the processes applicable to MMICs have been adopted such as Si_3N_4 overlay capacitors, thin-film resistors and air bridge interconnections. Figure 10 shows a cross-section of the technique. Ground septums are provided wherever the circuit topology requires low-inductance ground returns. The FETs are flip-chip mounted onto the substrate to provide good heat conduction and low-parasitic interconnections to the RF circuit. X- and Ku-band power amplifiers have been manufactured using this method. The Ku-band circuits are shown in Figure 11. Note that the size of the circuits produced (about 5 mm × 4 mm) is quite competitive with GaAs MMICs. Such a technique is particularly attractive for power applications at a relatively simple circuit complexity level. Although multiple-grounded areas can be envisaged using such a construction, the reliability of flip mounting a large number of small FETs in a complex circuit, for example, would be a limiting factor.

Having had a brief look at some MMIC and hybrid circuits, some of the items which distinguish the advantages of the two technologies will be discussed:
- Performance, bandwidth and flexibility
- Power capabilities
- Range of active devices available
- Reproducibility

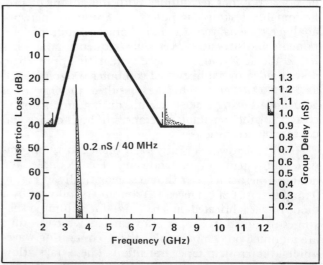

12. Filters are difficult to realize at the present stage of MMIC circuit design confidence and technology. The requirement for a monolithic bandpass filter with a rejection of 40 dB at 2.45 GHz, 4.9 GHz, 7.35 GHz, and 30 dB at 12.25 GHz is a demanding one. Bandpass frequency should be 3.4 to 4.625 GHz. A distributed design on GaAs would be large and costly at these frequencies.

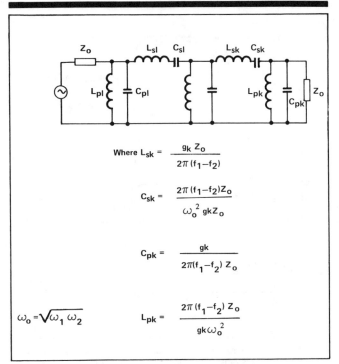

13. A lumped element of a 0.1-dB Chebyschev five-section filter prototype, showing the low-pass bandpass transformation.

- Ease of fabrication, size and handling
- Testing
- Cost
- Reliability and repairability

Performance, Bandwidth and Flexibility

Some of the earliest examples of MMICs demonstrated that such circuits could produce good performance over large bandwidths particularly in the area of small-signal amplifiers.[15] The basic techniques of wideband passive matching have been applied to GaAs FETs successfully by a number of workers exploiting the fact that one can place components electrically closer to the FET in an MMIC than in a hybrid circuit.[16, 17, 18] Both single-ended and balanced MMIC amplifiers, for example, have been fabricated covering octave and greater bandwidths. However, only a small percent of the GaAs area actually

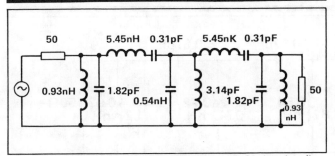

14. Element values of a lumped-element 0.1-dB Chebyschev five-section filter prototype, using frequency scaling.

contains active devices. Thus it can be argued the designer is using an expensive substrate material which is actually inferior to much cheaper alumina. The hybrid circuit devotees' argument is even stronger when broadband hybrid amplifier performance is noted.[19, 20] Some workers have to adopted a "half-way house" position with very simple MMICs containing two FETs and interstage networks (the so-called SUPERFET of Watkins-Johnson, for example) so that the advantages of the MMIC medium can be gained. The FETs, however, still occupy nearly 50 percent of the GaAs area. The SUPERFET, for example, has been used to produce multistage 7- to 13-GHz amplifiers having 36 ±0.5-dB gain within a total packaged volume of only 16.5 cm³ including voltage regulator and so forth. In circuit designs involving FETs used in other than common source connection, MMICs have distinct performance advantages over hybrid circuits. Although simple hybrid circuits containing common gate and source follower FETs have been made, they use chip capacitors, resistors and wire bonding techniques which lead to high cost and skilled labor needs. Several examples of both feedback and actively matched MMIC amplifiers have been developed giving good performance over decade bandwidths.[21, 22, 23] Level shifting and DC coupling have made wide bandwidth video and pulse amplifiers a reality.[24] The production of such circuits in hybrid form is difficult to imagine. Because the active devices and passive circuits can be placed closely together in an MMIC, and because the surface quality of GaAs is excellent compared to some substrate materials such as alumina, circuit techniques which cannot be considered as normal hybrid practice have become popular with GaAs IC designers. Combined with novel uses of FETs in active splitters and switches, for example, a range of MMICs have been demonstrated each performing an attractive function, for example, phase shifters,[4, 25] low- and high-power switches,[5, 25] and VOC.[6]

Where large numbers of FETs and diodes of small dimensions are used,[27] the attractions—first apparent to Si IC designers many years ago—of GaAs MICs are obvious. These circuits, currently at least, fall into the frequency range below about 6 GHz. Examples of such circuits are logarithmic amplifiers, operational amplifiers, phase detectors, and fast AD converters.[28, 29] While there are many cases where hybrid circuits can be made more cost effective than MMICs, particularly where low numbers are involved, this may only be applicable up to about 20 GHz. Above that frequency the interface between active devices and matching circuits or other components becomes dependent on the accurate knowledge of, or minimization of, parasitics. In this case MMICs offer a well-controlled environment for MESFETs and diodes where via-hole or wrap-around grounding together with the excellent definition of components leads to an optimum solution. Although circuit adjustment is not easily available in an MMIC environment (unlike a waveguide housing, for example) the medium allows much more accurate circuit design from the start. Since a quarter wavelength at, say 30 GHz on GaAs is only about 0.88 mm, the relative percentage of GaAs

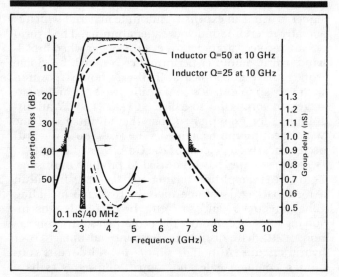

15. *Effect of* finite inductor Q on bandpass filter characteristics.

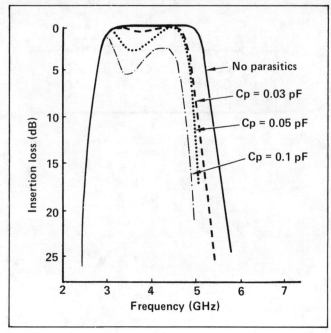

16. *Effect of Inductor* parasitic capacitances on five-secion Chebyschev bandpass filter insertion loss vs. frequency.

area used by passive and active devices is much greater than at lower frequencies. Provided that feedback and grounding problems can be minimized and microstrip modes in coplanar structures can be suppressed by incorporating effective grounds,[30] the prospects for GaAs MMICs in the millimeter-wave region are excellent.

At the current stage of MMIC circuit design confidence and technology yield, there are certain components which are difficult to realize with the same performance as their hybrid counterparts. Filters come into this category where the higher transmission-line loss of microstrip on GaAs (even if the latter is several hundred microns thick) or the low-Q and parasitics of lumped components makes filter design, at least, not so straightforward as in higher Q media.

An example will clarify this statement. Consider the requirement for a monolithic bandpass filter having a rejection of 40 dB at 2.45 GHz, 7.35 GHz and 30 dB at 12.25 GHz (Fig. 12). The bandpass frequency should be 3.4 to 4.625 GHz. A distributed design on GaAs would be rather large at these frequencies and, therefore, costly.

A lumped-element 0.1-dB Chebyschev five-section filter prototype is shown in Figure 13. Using frequency scaling the element values are as shown in Figure 14.

The response of this filter is shown in Figure 15. Note that the rejection is only about 35 dB at 2.45 GHz, but group delay over the passband is satisfactory. If inductor Qs of 50 at 10 GHz are incorporated into the filter model the passband insertion loss shown in Figure 15 results (i.e., 3 dB). If the inductor Q were only 25 at 10 GHz the insertion loss would be about 5 to 6 dB.

The realization of such a filter on GaAs using lumped elements requires inductance values of about 5 nH. Spiral inductors on GaAs have shunt parasitics which produce the results in Figure 16 where three different values of shunt capacitance are shown. The sensitivity to small values of capacitance is due to the inductors being at high-impedance points within the filter.

The use of a direct-coupled filter design such as that shown in Figure 17 together with impedance inverters restricts inductance values to about 1 nH. These can be implemented as high-impedance transmission lines on GaAs. Figure 18 shows how the higher the impedance of

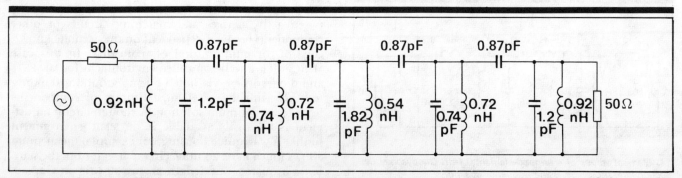

17. *Results of a direct-coupled* bandpass filter using capacitive impedance inverters.

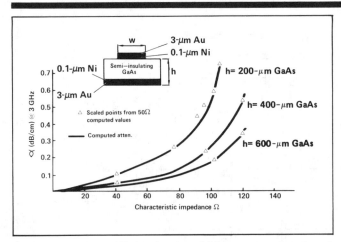

18. Microstrip loss of GaAs of various thicknesses as a function of line impedance.

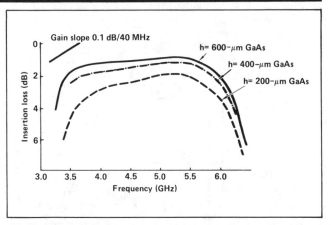

19. The effect of GaAs substrate thickness on the insertion loss of a direct-coupled bandpass filter using high-impedance short-circuit stubs.

the lines the more sensitive to substrate thickness the loss factor of these lines. This can be directly translated into the bandpass insertion loss characteristic shown in Figure 19. The conclusion is that for reasonable filter performance on GaAs the substrate thickness must be kept to about 400 to 500 μm, and that the filter prototype must be chosen carefully to avoid using high-value components with inconveniently large parasitics. In many filter cases, therefore, it may be concluded that the MMIC approach is not optimum and that miniature hybrid filter designs should be adopted, such as the dielectrically resonated designs that are available.[31]

In summary, the flexibility of circuit realizations using MMICs is beyond doubt and in many cases the performances achievable are impressive. There are certain cases, however, where the superior Q factors of more conventional microwave circuit techniques are advantageous. In addition, in many cases there is the need to interface the MMIC through a simple microstrip circuit to waveguide or other coaxial components.

Power Capabilities

The power handling of any medium depends on several interrelated factors:
- The inherent power limitations of the active devices themselves
- The thermal conductivity of the active device semiconductor material and that of the material to which it is attached
- The current handling properties of the metalization associated with the medium
- The frequency of operation with relation to device size

As far as the MESFET is concerned the inherent power limitations of the device are due to voltage breakdown between gate and drain which is caused by avalanche breakdown and high fields in the active channel-substrate interface.[32] This limitation is equally valid for power FETs in MMIC or hybrid environments. Of significance, though, is that there is no RF power generating two-terminal device presently available which is compatible with the MMIC medium. This is because such devices as the IMPATT and Gunn diode depend for their operation on being bulk "vertical" devices—a geometry which is incompatible with a basically planar structure. Although planar TEDs have been developed these are not used to generate RF power. As a consequence of gate yield, device terminal impedances and physical size discrete power FETs with much more than 10-mm total gate width are not common.[33, 34] Rather, power combining techniques are used to generate higher powers with several discrete power FETs either using direct combining methods[35] or by using prematching and indirect combining.[36]

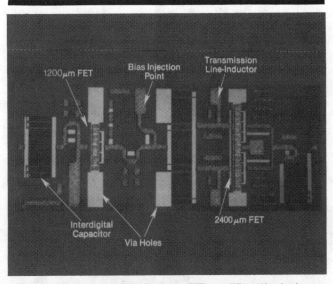

20. In this broadband MMIC power FET amplifier, Westinghouse divided the interstage matching network into two halves feeding the power MESFETs.

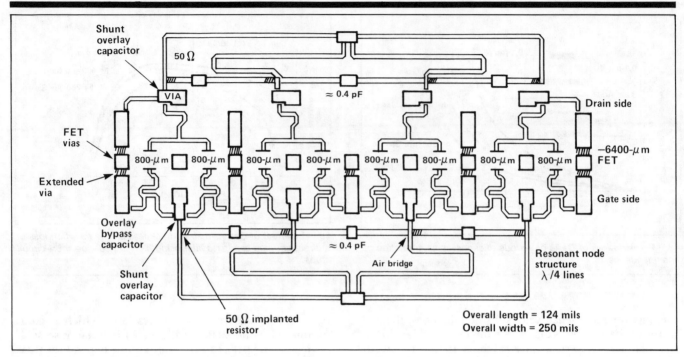

21. Cell cluster matching approach for a Westinghouse monolithic three-Watt 8- to 12-GHz FET amplifier.

In a monolithic IC environment the direct combining of several large gate-width FETs can cause unique problems since the overall "width" of the FETs becomes incompatible with the requirements of the matching networks. This can be seen in the case of the broadband power amplifier of Figure 20 where Westinghouse has divided the interstage matching network into two halves feeding the power MESFETs.[37] This philosophy can be further adapted for large gate-width devices as shown in Figure 21.

Attempting to match gate-width FET devices becomes increasingly difficult as frequency increases, particularly over large relative bandwidths. In order to ease the problem and ensure that all parts of the FET are fed in-phase, the FET can be divided into several "cells" of say one-quarter the total gate width and matched to a convenient impedance. Then the four individual and matched cells are combined via Wilkinson power combiners to produce a 50-Ohm match. Intercell interactions due to the non-perfect matching of individual parts of the FET can be reduced by exploiting the isolation of the Wilkinson combiners. Somewhat similar concepts have also been adopted by Hughes Aircraft[7] in their hybrid power FET amplifiers where output powers of up to 2.3 Watts at 15 GHz with 4-dB associated gain and 25-percent overall efficiency operating in Class A have been reported.

One of the dominant factors in determining how much power can be generated and, therefore, dissipated by virtue of the finite DC to RF efficiency of the device, is the operating channel temperature and the associated derating of the device when operated above ambient conditions. In the case of discrete power FETs the heat generated in the channels of the devices is usually dissipated by one of several methods all of which allow the device to be mounted in a hybrid circuit. These methods of heat sinking are:

- Through the substrate
- By flip-chipping the device
- By air bridging the sources
- By introducing vias in the sources

Table 1 shows the calculated rise in the channel temperature of a power FET whose total gate width is 12 mm as a function of GaAs substrate thickness when the unit gate width is 600 μm. This device has air bridged source connections and vias through the GaAs at the ends and middle of the structure. The DC power dissipation per unit area shown in the Table, and thus temperature rise in the channel, can be made markedly different by changing the overall size of the FET device. For

Table 1. Channel temperature rises in a GaAs FET having 12-mm total gate width* vs. substrate thickness and DC power density.

Substrate Thickness (μm)	Maximum Channel Temperature Rise (°C)	DC Power Density (W/mm²)
50	59	18.8
100	115	18.8
100	57	9.4
200	69	6.3
200	103	9.4
200	180	18.8

*FET assumed to have 20 × 600-μm wide parallel gates with 4-μm thick metalization.

example the approximate area of GaAs encompassed by a device with an 18.8-W/mm² DC power dissipation is 0.5 mm², while the area of a device fabricated on 100-μm thick GaAs having a dissipation of 9.4 W/mm² is 1 mm². The "safe" current density of both film resistors and gold metalization on MMICs has also to be considered. For example, if the gold metalization has a current density of 10^6 A/cm² with track widths of 60- and 4-μm metal thickness, the current handling is 2.4 amps. If the FET has 35-percent power added efficiency then the maximum output power with a typical power gain of 8 dB at X-band (assuming a drain voltage of 8 volts) will be limited to 8 W. To increase the amount of power available, therefore, RF power combining techniques or increasing the number of DC current paths are required. Thin-film resistor current density limitations may further reduce output power capabilities.

In an MMIC environment flip-chipping of the entire monolithic power circuit is impractical since the parasitics introduced between the inverted chip and the ground plane would be unacceptably high, particularly at high frequencies. Thus for MMICs a compromise has to be made between substrate thickness, the area of GaAS consumed by the power FETs, and the resultant Qs of the matching components employed.

In hybrid power FET circuits, the thermal rise in the FET channels can be minimized by correct thermal device design and using an optimum method for heat dissipation. Substrate thinning to only 5 or 10 microns with an integral heat sink is a popular means of providing a low thermal resistance to the circuit carrier (heat sink).

For the above reasons it is suggested that the power limitations of FET-based MMICs at the present state-of-the-art is about 10-W CW up to X-band and 1 W at 20 GHz, the latter being determined by device performance and matching losses rather than thermal considerations.

It is apparent that hybrid modules containing a number of power MESFETs can produce powers in excess of 25-W CW at C-band[40] and combining techniques have produced powers close to 10 W in the 20-GHz band.[41] Such techniques will be used to combine the powers of MMIC power amplifiers in the future.

Range of Active Devices Available

There are many applications where low-noise and power GaAs FETs have already earned a strong-hold and have replaced silicon diodes, bipolar transistors, and TWTAs. Because the GaAs MMIC is essentially a planar medium, devices such as IMPATTs and PIN diodes are incompatible with the technology used to produce the MESFET and Schottky diode commonly used. In fact, attempts have been made to produce bipolar transistors using GaAs as the base material, but performance to date is disappointing in comparison with the MESFET.[42, 43]

The areas where IMPATTs and PIN diodes cannot be ousted by the MMIC are in high-power and very high frequency, millimeter-wave applications. Planar Gunn diodes have only been successfully used as logic elements[44] but little success has been achieved with the devices being used as microwave sources or amplifiers.

A Solution to the Disadvantages of MMICs?

At the present state-of-the-art of GaAs MMICs there are three areas where this technology has disadvantages over hybrid MICs:
- Unfavorable device to chip area ratio
- Impractical circuit tuning
- Difficulties in trouble-shooting circuits which do not perform correctly

Considerable work is in progress to ease these three difficulties. CISE in Italy have recently produced a novel technique termed the "microhybrid" which combines passive components built-in GaAs substrates (although substrates of different materials could be used) together with flip-chip mounted FETs which are thermocompression bonded to the substrate. This technique, it is claimed, is particularly suitable for initial testing of circuit designs without the need for expensive IC mask sets and aids in tuning circuit response by actual FET positional changes. The illustration shows a single-stage low-noise preamplifier operating at X-band which forms part of a DBS front end. ■

Where close-to-carrier noise is of importance the 1/f noise contribution of the FET is, at least at the moment, considerably worse than the bipolar Si device. Such 1/f noise is also of concern in DC-coupled amplifiers, low-IF mixers, upconverters and multipliers.

In terms of device flexibility, the hybrid approach to microwave circuit production has obvious advantages. The circuit media depends on the device being used—waveguide for IMPATTs, E-plane for low-noise millimeter-wave mixers, stripline for varactor diode phase shifters, ridged-waveguide for high-power diode limiters, to mention just a few. It is highly unlikely that such circuits will be produced as MMICs in the future. However, there are signs that considerable improvements can still be made in circuit design. For example, the use of Class-B power FET amplifiers, particularly for phased-array transmitters, is gaining momentum[45] where power-added efficiency improvements of 50 percent over Class-A operation are presently available. The use of distributed or traveling-wave FET devices or amplifiers is also seen as a method by which wide bandwidths and higher-output powers may be achieved.[46]

Reproducibility

Reproducibility of performance from one circuit to the next is one advantage of the MMIC. Provided the initial design is correct, the need for any adjustment following processing is unnecessary. This argument is only partially true. As has been shown in the literature,[47,48] MMICs are just as likely to variations in performance as any other microwave circuit. Variations in gate length, channel depth and doping concentration can vary the gain, noise and intermodulation performance of MESFETs alone. Reproducibility of MMICs results mainly from the excellent definition and repeatability of passive components and choosing an optimum circuit of the function needed. For example, the use of active matching together with active biasing can produce amplifiers with a high degree of tolerance to FET variations and temperature. Reproducibility of the MMIC is also determined by the ability to keep the GaAs substrate thickness and quality of metalization constant from batch-to-batch. To compensate for the inevitable small changes in performance, particularly from batch-to-batch, the user either must accept the variations in his specification to keep costs low or some means of simple adjustment must be included in the IC design. In some cases this can be produced by allowing external adjustment of gate-to-source MESFET voltage.

Because of the overall physical size of GaAs monolithic circuits and miniature hybrid circuits, the phase repeatability of performance is often superior to conventional hybrid versions of the circuit since errors introduced by wire bond length variations, discrete chip placement, and so forth, are removed. In order to minimize the amount of external tuning required after the installation of MMICs into hybrid subsystems, the technologist is very aware of the need to keep close control over all process steps in IC manufacture. Where matching of IC performance is important, RF-on-wafer testing of all circuits prior to packaging will be an important method of obtaining RF performance information. Several sys-

Table 2. Comparison of hybrid and monolithic high-gain amplifiers

Parameter	Details
Hybrid	
Frequency	8–18 GHz
Noise figure	4.5 dB typical
Gain	40 ± 2 dB
No. of wire bonds	400
No. of FETs	16
No. of substrates	16
No. of carriers	8
No. of chip resistors (DC and RF)	32
No. of chip capacitors	40
Size	Approx. 64 x 8 x 4 mm (unpackaged)
Monolithic *(using via technology and alumina Lange couplers)*	
Frequency	7.5–18.5 GHz
Noise figure	5.2 dB typical
Gain	57 ±1.5 dB
No. of wire bonds	14
No. FETS	16
No. of GaAs chips	5
No. of carriers	1
Size	19.5 x 9.5 x 5 mm (packaged)

22. Assembly detail of a typical broadband GaAs FET low-noise amplifier.

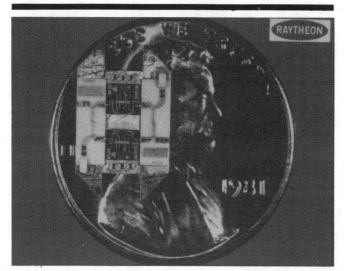

23. *Example of a monolithic two-stage 8- to 12-GHz amplifier, by Raytheon Research, balanced by the use of alumina Lange couplers.*

tems are already well established to allow sufficient RF measurements on each chip prior to dicing to be performed.[49]

Improvements have also been made in hybrid circuit repeatability in the last few years with the increased use of new production techniques such as automatic and semi-automatic bonders, chip placement, laser scribing and adjustment.

Ease of Fabrication, Size and Handling

GaAs MMICs undoubtedly require a large capital investment in equipment, both in development and production. If we discount the costs involved with LEC ingot growth and ion implantation equipment and the electron beam lithography or photolithography facilities associated with discrete FET manufacture, there is still a considerable facility required to produce ICs, both from the processing viewpoint and the computer-aided design, testing and packaging standpoints. The difference between using hybrid circuits with bare chip or packaged devices and using MMICs is that in the latter case many companies in the late 1970s involved themselves in the design and manufacture of the ICs rather than depend on future supplies from commercial vendors. More systems companies who have no involvement in MMIC development, have now begun to realize the potential advantages of such circuits and are now actively pursuing commercial orders or cooperative programs with MMIC houses. Equally they realize the enormous expense involved in setting up their own GaAs IC development production facility.

In order to produce a GaAs MMIC, which even today is relatively simple in its level of complexity, a complicated process routine is required compared to a hybrid circuit. In many cases, it is not cost effective to produce an MMIC version of a hybrid design unless the quantities involved are high (i.e., in thousands). It is instructive, however, to take a simple example of the relative complexity of a typical complete microwave product—the FET amplifier.

Figure 22 shows a typical Plessey eight-stage hybrid broadband amplifier covering 8 to 18 GHz (shown as a series of interconnected subcarriers prior to insertion in its final package). Figure 23 shows a typical MMIC two-stage broadband amplifier from Raytheon covering the same frequency range.[16] In this latter circuit the MMIC is contained between two coventional Lange couplers on alumina substrates, but these couplers could equally be included on the GaAs.[50, 51] Table 2 attempts to show the difference in the "statistics" of the two approaches particularly in the areas of wire bonding, assembly of substrates onto carriers, and so forth—i.e., the labor intensive aspects of the two approaches. To reduce discrete component count with hybrid MICs, attempts are being made to include tuning capacitors and so forth, on the substrates as MIM structures. However, where the MMIC wins is in its ability to totally confine all components including bias networks to the GaAs chip, thus dramatically reducing the time taken to "build up" modules.

At present, because the MMIC is a small "supercomponent" compared with previous technologies, the packaging of the chips, the associated DC voltage regulators and control circuitry, which may also be needed, tend to dominate the overall size of the subsystem module. Also, because the use of RF-on-wafer testing is only just becoming a reality the DC acceptable MMICs have to be premounted onto subcarriers, RF tested and then inserted into the final housing. The latter is particularly true where many chips containing a number of FETs are involved. Although the overall size of the subsystem may be quite small, the amount of labor involved in assembly is still large.

Figure 25 shows, for example, a transceiver module being developed by Plessey, for 2 to 4 GHz consisting of a number of pretested MMICs each contained on a subcarrier. The main chip components consist of:
- A low-noise preamplifier for receive
- A phase shifter on transmit providing gain
- A driver and power amplifier providing a 1-W CW output power
- SPDT and power T/R switches

Each of the chips is contained on a small subcarrier with small interface microstrip substrates allowing the chips to be tested prior to insertion in the final assembly and also allowing DC bias, and so forth, to be conveniently interfaced to the chips. In many applications local power supply regulation is required and thus bare chip Si IC regulators are chosen to complement the size of the GaAs ICs. These are contained on the reverse side of the package. The limiting factor as far as overall size is concerned will be determined by the packaging technique used, particularly as more functions are designed within single chips.

Referring to the transceiver shown in Figure 24, on transmit, the reference signal is directed via the SPDT switch at the bottom of the unit into the active phase shifter which consists of an analog vector modulator together with 90- and 180-degree phase shifters. The analog vector modulator is a circuit which divides the signal into two equal amplitude paths, delays each by ±/45 degrees, attenuates these two signals and recom-

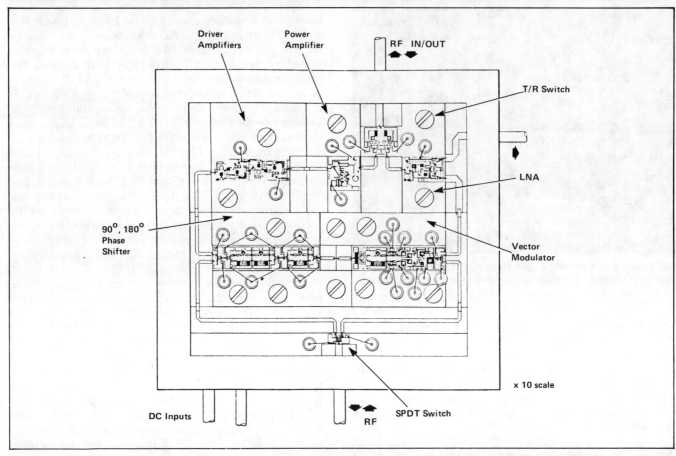

24. Representation of an S-band transmit/receive module using GaAs MMICs mounted onto subcarriers for initial RF testing before insertion into the final housing. Assembly of the complete T/R module is still complex but could not be considered without the use of MMICs.

bines them to produce any vector in a 0- to 90-degree quadrant.[4] The phase shifting elements are high-pass/low-pass structures allowing broadband operation due to the use of lumped elements. Well-known microwave components, such as Wilkinson power dividers and branched line hybrids, can be produced as lumped-element circuits.

The 90- and 180-degree phase shifters use SPDT switches and LC phase shifters. Following phase encoding the signal is passed through a driver and power amplifier stage providing about 18 dB gain. Together with the small signal amplifier prior to the driver stage a total gain of more than 29 dB is provided since the phase shifter also has a small amount of gain. Since the phase shifter is active and, therefore, non-reciprocal two SPDT switches are needed on either side of it to redirect the signal on receiver. On receiver the T/R switch is also changed over to provide a low-loss path to the low-noise preamplifier. each subcarrier containing the GaAs chips is screwed into the package with glass-isolated DC feedthroughs providing bias and control signals to the chips. The complete transceiver module measures about 3.2 × 2.9 cm excluding coaxial connectors.

Dramatic reductions in overall packaging can be seen by comparing the outline of two superhet receivers in

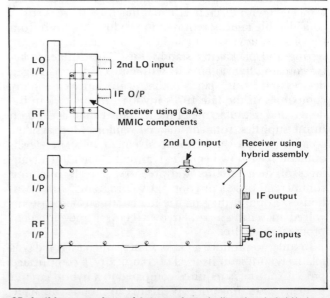

25. In this comparison of two receivers built using hybrid/microstrip/waveguide components and monolithic components, the overall size of the latter type is dominated by the waveguide housing. Size can be estimated by OSM connectors.

48

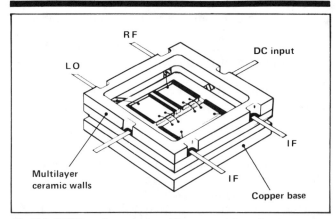

26. *This MMIC package uses multilayer* ceramic walls on a copper base. Although package performance may not be broadband, it is lower in cost than alternative techniques.

27. *An MMIC contained in an all-metal package* with 50-Ohm glass-to-metal feedthroughs. This package is broader band than that shown in Figure 26, but is probably costlier even in large quantities.

Figure 25. One is designed using waveguide/microstrip/hybrid techniques while the other uses a waveguide/MMIC technique. This example shows the potential problems involved in interfacing very small RF circuits to more conventional low-media such as waveguide components particularly at X-band and below. Will a range of standard microwave packages become available for MMICs? This question is prompted by the range of standard packages used in Si ICs.

The constraints facing the microwave IC package designer are the ones of acceptable microwave performance over, if possible, wide bandwidths, small size, good thermal characteristics where needed, hermeticity and low cost. To produce the smallest package size either the MMIC alone or the MMIC and a low level of Si bare chip IC voltage regulation and control can be included in the package. The MICPAC[53] used by Texas Instruments is one example of an attempt to produce a low-cost package using a multilayer ceramic technique. This type of package depends for its performance on reducing low-Q resonances occurring between the microstrip lines and the metal sealing ring (in the case of a ceramic lid) or the metal lid (Fig. 26). Thus it is particularly suitable for modest bandwidth operation. It has good thermal characteristics since the GaAs chips are soldered directly to the OFHC base. An alternative package using 5-Ohm glass-to-metal seals being developed at Plessey is shown in Figure 27. Although this package is expensive in small quantities it has superior frequency performance and since it is made entirely of metal it is more "well-behaved" when inserted into hybrid circuits.

In many cases the system application dictates the packaging concept to be used since invariably the package will contain GaAs MMICs, other microwave components such as circulators and DC and logic circuits (in a phased-array radar T/R module, for example).

In recent years there has been a marked trend in certain areas of the microwave industry to produce very small components for high-density packaging with one outline capable of containing many circuit options. Two examples of this are the OMNIPAC range of GaAs FET amplifiers from Omni Spectra and the MINPAC range of microwave circuits from Watkins-Johnson. This latter range of products uses a glass-to-metal feedthrough system.

The likelihood is that the multilayer ceramic package will become a standard for low-cost MMIC products, while the glass-to-metal seal package will be the hi-rel, high-performance option for wideband, military requirements. This latter package will gain popularity particularly if miniature hybrid circuits continue to use the same package concept.

As with discrete devices the bare chip MMIC will be used by those companies who have the necessary facilities for installing these chips into a high-yield subsystem.

Testing

As far as MMICs are concerned, one of the overall factors affecting the ultimate low cost of such parts is the degree to which yields can be improved over those presently being achieved.[54] Assuming that realistic yields for each processing stage and component are reasonable (i.e., over 95 percent in most cases) the overall yield of chips from 50- or 75-mm diameter wafers after DC probing can be expected to be about 30 percent with complexity envisaged for ICs in the next two to three years. In order to ensure that these DC tested chips have a high probability of being fully functional at RF, some form of "RF-on-wafer" testing is essential for all except the most simple circuits where a higher yield is to be expected anyway. Such RF-on-wafer testing identifies chips which perform to an RF specification and, therefore, packaging of out-of-specification chips is removed thereby decreasing cost. In a subsystem requirement where perhaps six to 10 chips with today's MMIC complexity are being used (e.g., 10 FETs, 16 capacitors, 77 air bridges, 26 via holes, etc.[55]) it is essential to ensure a high probability that the completely assembled module will perform satisfactorily. At present, most MMIC workers mount DC tested chips onto subcarriers, insert these subcarriers into test jigs and assess the RF performances. These tested circuits are then inserted into the final housing. To increase overall packaged device yield, it is necessary to develop a testing system for MMICs in close collaboration with the packaging concept. Thus, if chips-on-subcarrier testing[56] is unacceptable, bare chip-on-wafer testing is required.[49]

In hybrid circuits, both conventional and miniature, the active devices are usually DC probed and RF assessed (either 100 percent or on a representative basis) before insertion into the circuit. In most cases to date, testing of discrete transistors is only fully achieved on packaged devices. Companies have avoided developing transistor-on-wafer RF test facilities as there is a high probability that, if samples from a batch are satisfacotry, all other visually acceptable devices will also perform satisfactorily. MMICs are distinctly different since they can contain a large number of active and passive components.

Cost

The cost of any microwave circuit can be related directly to five main factors shown in Figure 28. Currently MMICs are generally expensive in development mainly because various engineering and technology activities have not been optimized. These include the IC process itself, mask making, CAD and testing. Hybrid circuits tend to be cheaper in their initial development but usually undergo an expensive preproduction exercise more often than not to achieve reliability in and reproducibility of performance.

The cost effectiveness of MMICs over hybrids can be conveniently expressed as an equation:

$$\text{Cost ratio} = \frac{(\text{Monolithic IC Cost} + \text{Testing} + \text{Packaging})}{(\text{Hybrid Circuit Cost} + \text{Device cost} + \text{Mounting} + \text{Testing} + \text{Packaging})}$$

MMIC yield is directly related to FET and other active device yield, passive component yield and the number of process steps.

Knowing the yield of each part of either the MMIC or hybrid approach, it is possible to calculate two important break points. First, the level of integration of MMICs to give an acceptable yield can be defined (say 40 to 50 percent). Second, those circuits which are cheaper to produce as hybrids can be defined. This is directly related to the level of maturity of the MMIC and hybrid technologies.

On the assumption that acceptable IC yields are available and that semi-automatic testing and packaging facilities are available it has been estimated that MMICs will cost, on average, about $25 to $35 (£15-£20) in 1,000 off quantities in a few year's time. Obviously this cost is directly related to IC yield, GaAs material costs, and so forth, and to the number of circuits involved.

Just as with the well-established Si IC technologies of the 1970s and lower-frequency circuits, it appears that the GaAs IC will be able to make significant inroads into the cost reduction of microwave circuits particularly where high-volume requirements are concerned. This is because batch processing of the total microwave circuit becomes a small part of the cost of the units. In the case of phased-array radars where the number of elements can vary from a few hundred to many thousands, GaAs monolithic circuits become attractive in making such systems feasible.

To estimate the cost of a typical monolithic circuit it is necessary to access accurately the overall yield of individual chips throughout the various production stages. In order to do this consider three examples of typical chips using different technological approaches.

■ The realization of a low-noise J-band amplifier using electron-beam exposed FETs with 0.3-μm gate lengths.

■ The realization of a low-noise single-ended mixer using dual-gate FETs with 0.5-μm gate lengths.

■ The realization of an IF amplifier with an AGC facility using 1-μm gate length FETs.

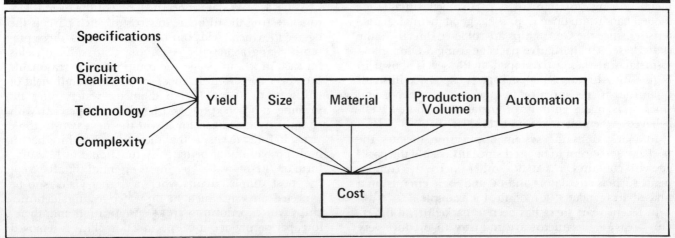

28. The cost of any microwave circuit can be related directly to the five main factors shown above. MMICs are generally expensive in development because various engineering and technology activities have not been optimized.

Table 3. Relative percentage yields of GaAs ICs on vapor-phase epitaxial and ion-implanted material

Circuit Description	Small Area VPE	5 cm VPE	7.5 cm VPE	5 cm Ion-Implanted	7.5 cm Ion-Implanted
0.3-μm EB preamplifier	n	2.5n	2.8n	4.2n	4.7n
0.5-μm photolithography mixer	n	2.5n	2.8n	4.2n	4.7n
1-μm photolithography IF amplifier	2n	3n	3.3n	4.3n	4.9n

Table 3 gives a comparison of the expected yields of the three circuits based on use of present-day processing techniques. Many relatively simple monolithic circuits already realized in the United States and Europe have used small-area vapor-phase epitaxial (VPE) material. Present yields are dominated by four factors.

- Edge of wafer defects which account for 35-percent loss on typical VPE wafer areas
- Gate metalization faults
- Variations in DC parameters over the slice caused by active-layer thickness variations and material defects; surface defects usually account for a small percentage of failures
- Yield of large value Si_3N_4 overlay capacitors

As indicated in Table 3, the use of 50-mm diameter GaAs wafers greatly increased the yield while going to ion-implanted material gives an even greater yield. On 50-mm material edge, defects account for 12 percent of failures.

To accurately assess the cost of a particular monolithically based module, consider the example of combining the three previously mentioned circuits into an overall receiver front end. For circuits presently being considered this involves the chips being put into a microwave package much as shown in Figure 27, together with DC regulation circuits, temperature sensing circuits, and so on. Thus the overall cost of a module can be subdivided into the following:

- Cost of basic materials including GaAs
- Cost of packaging—microwave chip package and overall module
- Cost of processing GaAs MMICs
- Cost of DC regulators, etc. (using the lowest cost techniques available, e.g., thick film or IC)
- Assembly cost

- Testing cost (including individual chip testing)

As may be appreciated from Table 4 for 50- or 75-mm ion-implanted GaAs wafers, the ultimate cost of each module is dictated by the assembly and testing stages. Thus it becomes apparent that, in order to realize the lowest cost, assembly must be reduced by eventually integrating as many functions as possible onto a chip and must be made as simple as possible to the point where automatic procedures can be adopted. This is also equally applicable to testing where both DC and RF testing must be achieved on an automatic basis.

This is an area where much concentration will be needed in the future to enable the lowest production costs to be achieved for the higher-volume applications.

The cost of a circuit consists of more than the cost of the component parts—to consider IC costs in isolation from the systems they go into is wrong. However it has already been indicated that the cost of components based on GaAs ICs is made up of the cost of making the chip, packaging and testing it, taking account of the yield at each stage. The cost of design and product engineering has to be recovered as well.

The cost of producing the chips depends on the cost of the process in man-hours, materials, capital depreciation and yield. The cost of packaging is related to the number of chips in that package so that the more circuits to be put onto the same chip by increasing packing density, the lower the package cost per function.

The cost of testing depends on the test time, the cost and depreciation of the test equipment, man-hours and yield. This applies to both wafer probing and final testing. One major advantage of GaAs ICs is that they will require the minimum of adjustment and select on test procedures. The cost of assembly depends on package type, cost, labor, yield, and so forth.

Table 4. Percentage costs at each stage towards total cost

Cost breakdown for complete monolithic receiver front end for 500 off

Process	Small Area VPE Slice	5 cm Ion-Implanted	5 cm Ion-Implanted (2,000 off)
Preamplifier manufacture	34%	3%	5%
Mixer manufacture	22%	2%	3%
IF manufacture	12%	2%	2%
Packages	5%	13%	13%
Assembly	18%	60%	50%
Testing	9%	20%	27%

Assumptions: (i) Final number of working slices to start slices is 1 in 3
(ii) Assembly of chips into packages gives a 50% yield with 80% tested yield.

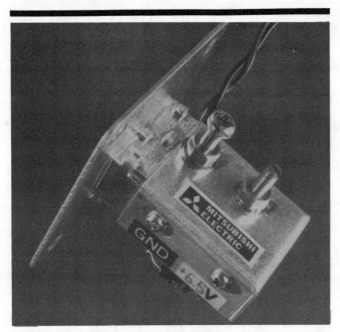

29. Two views of a low-cost GaAs FET dielectrically resonated waveguide oscillator by Mitsubishi. Note the construction of the waveguide housing and ceramic mounting.

A yield loss at any stage is significant but it is desirable to achieve the best possible yield towards the end of the process. Testing cost does not depend heavily on the process technology used. Slice costs are expected to increase substantially as a result of going from 1-μm to sub-0.5-μm geometries because the processes used are different in terms of mask making, yield and higher capital costs involved. The use of electron beam (EB) technology also produces a relatively lower throughput. Taking into account slice sizes (7.5-cm ion-implanted wafers) and packing density going from 1 μm to 0.25 μm, FET devices will probably increase the chip cost on "typical" chip complexities by a factor of at least two or three times. Photolithography costs are reflected in the cost of equipment—a factor of 5:1 between conventional UV and electron beam. The advantages to be gained on going to sub-0.5-μm geometries on GaAs ICs are very uncertain both for analog and digital circuits because of the high capital cost of EB equipment and the slow rate of throughput. This situation could be transformed by technical advances in EB machines and/or high-speed resists which could increase work rate and reduce capital cost per slice. Thus it is seen that the use of 0.5 μm or greater FET geometries combined with large-area GaAs slices, simple packaging techniques and

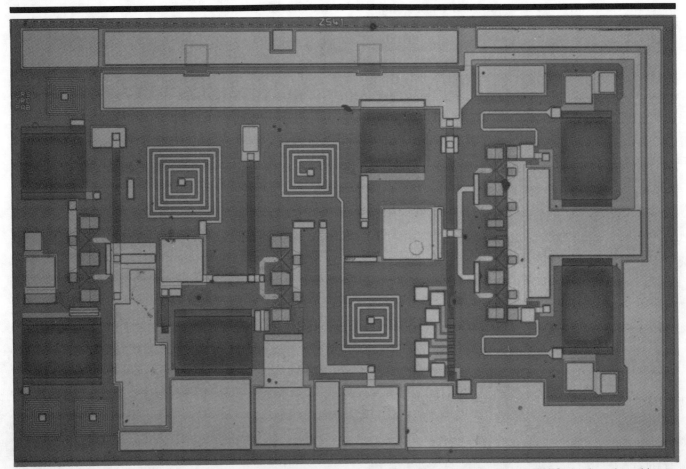

30. There are many stages of fabrication for a complex MMIC. The typical number of masks required for an MMIC circuit such as this one may be as many as 15. Thus, in order to produce cost-effective circuits using the MMIC method, the yield of such fabrication stages has to be high.

multifunction chips will signficantly reduce the overall cost to a point where the use of GaAs ICs in systems can become a realistic concept.

In order for hybrid circuits to compete in cost, concepts such as those adopted by RCA,[14] Raytheon[13] or Fujitsu[57] need to be developed further. In many respects, of course, the technology required for miniature hybrid circuits is close to that needed for GaAs MMICs, the significant differences being in terms of substrate materials and device mounting. As active devices in particular become more reproducible (an outcome of MMIC technology) automatic equipment can be used to produce high-frequency circuits akin to the popular (and low cost) thick-film hybrids available up to 2 GHz or so at the present time from such companies as Watkins-Johnson, Alpha, and Avantek.

Reliability and Repairability

MMICs and conventional hybrid microwave integrated circuits are almost extremes in terms of potential reliability and the ease of maintenance. On the one hand a well-qualified IC process can be highly reliable especially in the area where hybrid circuits have difficulties—the mechanic aspects of bringing together a mix of technologies—chip device bonding, thin-film resistors, metalization processes, wire bonding, and so forth. On the other hand the hybrid circuit producer can rework his circuits relatively easily if failure occurs. Indeed most hermetically packaged circuits have a lid/package interface designed to be reworked several times in the event of circuit failure. One of the most powerful arguments for GaAs MMICs is that, if costs become very low, then replacement of the complete component becomes possible. In practice this is the only course of action anyway regardless of cost. Thus, a vital part of the preproduction developments in hand for GaAs MMICs at the moment is ensuring that the reliability of the process is as high as economically possible and therefore the MTBF of each MMIC is long. Because most MMIC processes are complex with FETs, thin-film resistors, MIM capacitors, vias, air bridges, and so forth, the reliability of each part of the process and the interaction between the various parts of the process have to be monitored carefully. In this respect most companies involved in MMICs are at an early stage. Life tests on individual components such as resistors and capacitors are underway to establish failure criteria and the necessary information to allow the accurate calculation of MTBFs for customer requirements. Reliability of MMICs is directly related to the number of components per chip which in turn is directly related to yield.

In hybrid circuits the overall reliability is dominated usually by the active devices. However, in a very complex circuit, interconnections between circuits and devices can also have a substantial effect on reliability.

At the present stage of MMIC development, together with the past track record and current development of hybrid MICs, it is clear that both technologies have relative advantages. In the near future it is likely that MMICs will be used in large volume including requirements such as phased array (naval, air- and satellite-borne). For this to happen it is necessary to ensure that the design techniques and technologies in the laboratory today are effectively transferred to production.

Since there is no one technology that can solve all problems, as we have seen in this article, it is certain that MMICs will provide only part of the microwave industry's needs.

As has been seen, GaAs MMICs do not always provide the best medium for the specification required. MMICs are difficult to integrate directly to certain types of transmission media, e.g., waveguide, although some ingenious solutions have or are being developed particularly at millimeter waves.[58]

The microwave loss of thin GaAs substrates (which are required to keep thermal impedance within reasonable limits) is higher than desirable particularly at the higher frequencies.

At the moment certain components such as filters and non-reciprocal devices are difficult to produce. Although research into IC-compatible components is at an early stage,[59] there are good reasons for questioning whether there is any need for changing from the more conventional methods.

Because the number of high-volume requirements in the microwave industry is few (phased-array radars, navigation receivers, satellite TVROs) the concept that is most likely to be effective is the MMIC/hybrid approach where a relatively small number of MMIC "building blocks" are used in a variety of applications. By producing 50-Ohm compatible MMIC modules the engineer can then use various MIC mediums to interconnect the MMICs with other devices. This concept also allows the integration of more conventional components such as miniature filters and non-reciprocal devices with the MMICs. The complexity of the MMIC modules can be increased as the yield and reliability of the devices in production is improved and established. The analogy here is with first-generation Si ICs being interconnected on glass-fiber PCBs. Today, as Si IC technology has matured, complete boards can be turned into ICs using gate array techniques, for example. Whether GaAs MMICs ever get to the point of being commercially viable for every application regardless of numbers (within limits) depends to a large extent on the amount of effort and funding attributed to military requirements.

In current commercial microwave applications such as TVRO equipment[60] and microwave radio relays, emphasis has been placed on using a low-cost technology such as fiber board MICs, tin-plated extruded housings (Fig. 29) etc., with a build level which requires semiskilled labor or semi-automatic equipment. This tends to be the exact opposite to most military MICs where alumina- or quartz-based microstrip circuits, for example, are soldered or epoxied onto milled subcarriers and contained in finely machined enclosures with the build being done by experienced labor and final tuning achieved with skilled technicians. With the introduction of miniature hybrid ICs involving such techniques as flip-chipping, and deposited thin-film capacitors, batch processing concepts have to be adopted with miniature packaging techniques. Thus, many needs of the hybrid MIC and

GaAs MMIC manufacturer are common to the extent that interfacing the two techniques should be eased. Thus, in a transmit/receive module at X-band, it is likely that MMICs would be used for all circuit functions except where superior thermal dissipation is needed for the driver and output stages. These may well be produced using GaAs FETs on BeO substrates.

The future for both GaAs MMICs and hybrid MICs therefore looks very promising. Well-established MIC production lines will not be displaced easily and both the GaAs MMIC and the newer miniature hybrid MICs have yet to prove themselves in terms of cost, reliability and yield. These aspects are being addressed and thus the next few years will indeed be exciting as both MMICs and new-concept MICs are introduced into subsystems. ■

References

1. H. Yamasaki, "GaAs FET Technology. A Viable Approach to Millimeterwaves," *Microwave Journal*, Vol. 25, No. 6, June 1982, pp. 93-105.
2. R. Van Tuyl, "A Monolithic GaAs FET RF Signal Generation Chip," *IEEE Trans. on Electron Devices*, Vol. ED-28, No. 2, Feb. 1981, pp. 166-170.
3. J.S. Joshi, J.R. Cockrill, and J.A. Turner, "Monolithic Microwave Gallium Arsenide FET Oscillators," *IEEE Trans. on Electron Devices*, Vol. ED-28, No. 2, Feb. 1981, pp. 158-162.
4. C.W. Suckling, R.S. Pengelly, and J.R. Cockrill, "S-Band Phase Shifter Using Monolithic GaAs Circuits," *1982 Digest of Technical Papers*, ISSCC, San Francisco, pp. 134-135.
5. Y. Ayasli, R.A. Pucel, J.L. Vorhaus, and W. Fabian, "A Monolithic X-Band Single-Pole Double-Throw Bidirectional GaAs FET Switch," *IEEE Integrated Circuits Symposium*, No. 1980, Las Vegas, U.S., Paper No. 21.
6. B.N. Scott and G.E. Brehm, "Monolithic Voltage Controlled Oscillator for X- and Ku-Bands," *1982 IEEE MTT-S Digest*, Dallas, U.S., pp. 482-485.
7. J.M. Schellenberg and H. Yamasaki, "A New Approach to FET Power Amplifiers," *Microwave Journal*, Vol. 25, No. 3, March 1972, pp. 51-66.
8. I. Shahriary, T.S. Lin, and K. Weller, "A Practical Wideband GaAs Phase Detector," *IEEE Microwave and Millimeter Wave Monolithic Integrated Circuit Symposium*, Dallas, June 1982, pp. 47-49.
9. D.P. Hornbuckle and R.L. Van Tuyl, "Monolithic GaAs Direct-Coupled Amplifiers," *IEEE Trans. Electron Devices*, Vol. ED-28, No. 2, pp. 175-182, Feb. 1981.
10. R.N. Bates and M.D. Coleman, "Millimeter Wave Finline Balanced Mixers," *Proc. 9th European Microwave Conference*, 1979, pp. 721-735.
11. I.C. Hunter and J.D. Hunter, "Varactor Tuned Microwave Filters," *1982 IEEE MTT-S Intern. Microwave Symposium Digest*, Dallas, U.S., pp. 339-401.
12. L. W. Chua and M.A.G. Upton, "The Crossbow Modulator—A Microwave PSK Switch," *Colloquium on Microwave Integrated Circuit Developments IEE*, Savoy Place, London, January 1982, Digest No. 1982/9, pp. 13/1-13/4.
13. S.B. Moghe, "Quasi-Lumped Element Impedance Matching Networks for Wideband Miniature GaAs FET Amplifiers," *Microwave Journal* Vol. 24, No. II, November 1981, pp. 71-75.
14. F.N. Sechi, R. Brown, H. Johnson, E. Belohoubek, E. Mykietyn, and M. Oz, "Miniature Beryllia Circuits—A New Technology for Microwave Power Amplifiers," *RCA Review*—to be published.
15. R.S. Pengelly and J.A. Turner, "Monolithic Broadband GaAs FET Amplifiers," *Electronic Letters*, Vol. 12, No. 10, May 1976, pp. 251-252.
16. Y. Tajima, T. Tsukii, E. Tong, R. Mozzi, L. Hanes, and B. Wrona, "X-, Ku-Band GaAs Monolithic Amplifier," *1982 IEEE MTT-S Int. Microwave Symposium Digest*, pp. 476-478.
17. E.W. Strid and K.R. Gleason, "A DC-12 GHz Monolithic GaAs FET Distributed Amplifier," *IEEE Trans. on MTT* July 1982, Vol. MTT-30, No. 7, pp. 969-976.
18. J.E. Degenford, D.C. Boire, R.G. Freitag, and Cohn, "A Study of Optimal Matching Circuit Topologies for Broadband Monolithic Power Amplifiers," *1981 IEEE MTT-S Intern. Microwave Symposium Digest*, Los Angeles, pp. 351-353.
19. S. Yamamura, et al., "4-8 GHz Miniature GaAs FET Amplifier," *1979 MTT-S International Microwave Symposium Digest*, pp. 335-337.
20. D.P. Hornbuckle, "A 2 to 6.2 GHz, 300 mW GaAs MESFET Amplifier," *1978 IEEE MTT-S International Microwave Symposium Digest*, Ottawa, pp. 288-290.
21. H.P. Weidlich, J.A. Archer, E. Pettenpaul, F.A. Pety, and J. Huber, "A GaAs Monolithic Broadband Amplifier," *ISSCC Digest of Technical Papers*, Feb. 1981, pp. 192-193.
22. W.C. Petersen, D.R. Decker, A.K. Gupta, J. Dully, D.R. Ch'en, "A Monolithic GaAs 0.1 to 10 GHz Amplifier," *1981 IEEE MTT-S Intern. Microwave Symposium Digest*, Los Angeles, pp. 354-355.
23. R.S. Pengelly, J.R. Suffolk, "GaAs Monolithic Circuit Possibilities," *Proceedings of the 1981 Internepcon Conference*, Brighton, England.
24. S. Hori, K. Kamei, et al., "Direct Coupled GaAs Monolithic IC Amplifiers," *IEEE 1982 Microwave and Millimeter-Wave Monolithic Circuits Symposium Digest*, Dallas, pp. 16-19.
25. J.L. Vorhaus, R.A. Pucel, and Y. Tajima, "Monolithic Dual-Gate GaAs FET Digital Phase Shifter," *GaAs IC Symposium Research Abstracts*, 1981, paper 35.
26. Y. Ayasli, R. Mozzi, L. Hanes, and L.O. Reynolds, "An X-Band 10 Watt Monolithic Transmit-Receive GaAs FET Switch," *IEEE 1982 Microwave and Millimeter-Wave Monolithic Circuits Symposium Digest*, Dallas, pp. 42-44.
27. S.I. Long, B.M. Welch, et al., "High Speed GaAs Integrated Circuits," *Proc. IEEE*, Vol. 70, January 1982, pp. 35-45.
28. K. Weller, N. Duttathriya, and T. Lin, "Monolithic GaAs X-Band Balanced Dual-Gate FET BPSK Modulator," *1982 GaAs IC Symposium*, New Orleans, paper 18.
29. L.C. Upadhyayula and W.R. Curtice, "Design and Fabrication of GaAs Analog-to-Digital ICs," *1982 IEEE Microwave and Millimeter-Wave Monolithic Circuits Symposium Digest*, Dallas, pp. 54-56.
30. R.S. Pengelly, M.G. Stubbs, J.R. Cockrill, and J.A. Turner, "GaAs FET Microwave Amplifiers in X and Ku-Band Using Casaded Chips—Design and Performance," *1979 GaAs IC Symposium Proceedings*.
31. L. Accatino and A. Angelucci, "A Dielectric Resonator Filter as Low Loss Delay Element for 14 GHz On-Board 4 Phase DCPSK Demodulation," *IEEE Microwave Symposium Digest*, Los Angeles, June 1981, pp. 405-407.
32. S. Tiwari, D.W. Woodard, and L.F. Eastman, "Domain Formation in MESFETs—Effect of Device Structure and Materials Parameters," *Proceedings of the 7th Biennial Cornell Electrical Engineering Conference*, Cornell University, New York, 1979, pp. 237-248.
33. H.M. Macksey, H.Q. Tserng, and G.H. Westphal, "S-Band GaAs Power FET," *1982 IEEE MTT-S International Microwave Symposium Digest*, pp. 150-152.
34. A. Higashisaka, K. Hanjo, Y. Tahayama, and F. Hasegawa, "A 6 GHz, 25W GaAs MESFET With an Experimentally Optimised Pattern," *1980 IEEE MTT-S International Microwave Symposium*, May 1980, pp. 9-11.
35. J.M. Schellenberg and H. Yamasaki, "An FET Chip-Level Cell Combiner," *ISSCC Digest of Technical papers*, Feb. 1981, pp. 76-77.
36. Y. Mitsui, M. Kobiki, M. Wataze, K. Segawa, et al., "10 GHz 10W Internally Matched Flip-Chip GaAs Power FETs," *1980 IEEE Int. Microwave Symp. Digest*, May 1980, pp. 6-8.
37. J.E. Degenford, R.G. Freitag, D.C. Boire, and M. Cohn, "Broadband Monolithic MIC Ppower Amplifier Development," *Microwave Journal*, Vol. 25, No. 3, March 1982, pp. 89-96.
38. Y. Mitsui, M. Otsubo, T. Ishii, S. Mitsui, and K. Shirahata, "Flip-Chip Mounted GaAs Power FET With Improved Performance in X-and Ku-Band," *Proceedings of the 1979 European Microwave Conference*, Brighton, England, pp. 272-276.
39. L.A. D'Asaro, J.V. Di Lorenzo, and H. Fukui, "Improved Performance of GaAs Microwave Field Effect Transistors With Via-

Connections Through the Substrate," *Int. Electron Devices Meeting Tech. Digest*, December 1977, pp. 370-371.
40. N. Fukuden, F. Ogata, et al., "A 4.5 GHz 40 Watt GaAs FET Amplifier," *1982 IEEE MTT-S International Microwave Symposium Digest*, pp. 66-68.
41. J. Gael, et al., "A 1 Watt GaAs Power Amplifier for the NASA 30/20 GHz Communication System," *1982 IEEE MTT-S International Microwave Symposium Digest*, pp. 225-227.
42. H. Beneking and L.M. Su, "GaAlAs/GaAs Heterojunction Microwave Bipolar Transistor," *Electronics Letters*, 16 April 1981, Vol. 17, No. 8, pp. 301-302.
43. O. Ankri and A. Scavennec, "Design and Evaluation of a Planar GaAlAs-GaAs Bipolar Transistor," *Electronics Letters*, 3rd Jan. 1980, Vol. 16, No. 1, pp. 41-42.
44. D.C. Claxton, T.D. Leisher, and T.G. Mills, "TED BPSK Modulator/Demodulator Integrated Circuit Development," *Final Report, January 1978, TRW Defense and Space Systems*, Contract No. M00014-76-C-0570 ONR.
45. M. Cohn, J.E. Degenford, and R.G. Freitag, "Class B Operation of Microwave FETs for Array Module Applications," *1982 IEEE MTT-S International Microwave Symposium Digest*, pp. 169-171.
46. Y. Ayasli, R.L. Mozzi, J.L. Vorhaus, L.D. Reynolds, and R.A. Pucel, "A Monolithic GaAs 1-13 GHz Traveling-Wave Amplifier," *IEEE Trans. MTT*, July 1982, Vol. MTT-30, No. 7, pp. 976-982.
47. R.S. Pengelly, J.R. Suffolk, J.R. Cockrill, and J.A. Turner, "A Comparison Between Actively and Passively Matched S-Band GaAs Monolithic FET Amplifiers," *1981 IEEE MTT-S International Microwave Symposium Digest*, pp. 367-369.
48. J.E. Degenford, M. Cohn, R.R. Frietag, and D.C. Boire, "Processing Tolerance and Trim Considerations in Monolithic FET Amplifiers," *1980 IEEE ISSCC Digest*, San Francisco, pp. 120-121.
49. E. Strid and K. Gleason, "A Microstrip Probe for Microwave Measurements on GaAs FET and IC Wafers," presented at *1980 GaAs IC Symp. paper 31*, 1980.
50. R.C. Waterman, et al., "GaAs Monolithic Lange and Wilkinson Couplers," *IEEE Trans. Electron Devices, Vol. ED-28, No. 2*, Feb. 1981, pp. 212-216.
51. M. Kumar, S.N. Subbarao, R.T. Menna, and H.C. Huang, "Monolithic GaAs Interdigitated 90° Hybrids with 50 and 25 Ohm Impedances," *IEEE 1982 Microwave and Millimeter-Wave Monolithic Circuits Symposium Digest of Papers*, pp. 50-53.
52. R.S. Pengelly, "Monolithic GaAs ICs Tackle Analog Tasks," *Microwaves*, Vol. 18, No. 7, 1979, pp. 56-65.
53. C. Orr, "A Cumulative Spot Transmissive Lens Antenna," *1982 IEEE MTT-S International Microwave Symposium Digest*, pp. 188-189.
54. A. Gupta, W.C. Petersen, D.R. Decker, "Yield Considerations for Ion Implanted GaAs MMICs," *IEEE 1982 Microwave and Millimeter-Wave Monolithic Circuits Symposium, Digest of Papers*, pp. 31-35.
55. Y. Ayasli, A. Platzker, J. Vorhaus, and L.D. Reynolds, "A Monolithic X-Band Four-Bit Phase Shifter," *1982 IEEE MTT-S Int. Microwave Symp. Digest*, pp. 486-488.
56. J.A. Benet, "The Design and Calibration of a Universal MMIC Test Fixture," *IEEE 1982 Microwave and Millimeter-Wave Monolithic Circuits Symposium, Digest of Papers*, pp. 36-41.
57. H. Yokouchi, H. Kurematsu, K. Ogawa, and H. Ashida, "4 GHz 3 Watts FET Amplifier for Digital Transmission," *1978 IEEE MTT-S International Microwave Symposium Digest*, Ottawa, pp. 276-277.
58. A. Contolatis, C. Chao, S. Januson, and C. Butter, "Ka-Band Monolithic GaAs Balanced Mixers," *IEEE 1982 Microwave and Millimeter-Wave Monolithic Circuits Symposium Digest*, pp. 28-30.
59. S. Talisa, and D. Bolle, "Performance Characteristics of Magneto-Plasmon Based Submillimeter Wave Nonreciprocal Devices," *1981 IEEE MTT-S International Microwave Symposium Digest*, pp. 287-289.
60. J.R. Forrest, "Assessing Antennas for Small Satcom Terminals," *MSN*, Vol. 11, No. 10, October 1981, pp. 77-99.

GaAs MMICs: MANUFACTURING TRENDS AND ISSUES

A. Christou

Naval Research Laboratory
Washington, D.C. 20375

ABSTRACT

The critical issues impacting the manufacturability of GaAs monolithic microwave integrated circuits are reviewed. The manufacturing problems encompass those of material quality, processing, and circuit design. In reality, these problems are closely related. Good substrates are a necessity for a reproducible process. The success of a particular circuit design approach is very much dependent on the limitations imposed by materials, processing capability and chip architecture. The manufacturing technology for fabricating GaAs monolithic microwave integrated circuits has improved over the past two years. The refinements now being made are directed primarily at enhancement of yield, realization of performance, and increase in circuit complexity. The purpose of this review paper is to (1) address the issues affecting the manufacture of GaAs MMICs and (2) present the industrial status (through an industrial data base) in addressing such issues as yield, throughput, design rules, chip architecture, reliability, design for yield and manufacturability, substrate qualification, choice of processing technology and current status of process related models and sensitivity analysis.

INTRODUCTION

During the time period of 1980-1984 a great deal of progress has been made in the advancement of basic GaAs integrated circuit manufacturing technology. A well defined technology for manufacturing integrated circuits with a variety of different device profiles within the circuit, and with the provision for two level metallization for passive element formation and interconnections, has been established.

The problems involved in the new technology encompass those in material quality, processing techniques, and the approach to circuit design. In reality, the problems in these areas are closely related. Good starting materials are a necessity before a reproducible process can be established. It is often difficult to separate problems in materials from those in processing because the success of a particular processing technique is dependent on the quality of the material used. Likewise, the success of a particular circuit design approach is very much dependent on the limitations imposed by the materials and processing capability.

For the purpose of this paper, the problem areas have been catagorized in this manner, but the interdependency should be kept in mind. The status of GaAs analog integrated circuits are reviewed with particular emphasis on the solution of manufacturing problems as envisioned by industry.

MATERIALS FOR LINEAR INTEGRATED CIRCUITS

In order to obtain improved uniformity and reproducibility using ion implantation for channel formation many laboratories engaged in the development of GaAs ICs initiated internal efforts to produce improved quality GaAs material. The major effort has been devoted to the establishment of the liquid encapsulated Czochralski (LEC) technique for bulk crystal growth. This technique provides for the growth of large diameter boules of semi-insulating GaAs without the introduction of chromium.

The LEC material growth technique has proven to be of great benefit in the development of GaAs ICs. Although the crystal defect density for LEC material is higher than for HB, this has not had a noticable impact on circuit performance. Good quality LEC material is now available in quantity from several commercial material vendors, which largely removes the requirement for each developer of GaAs ICs to establish their own source for starting materials. The quality of starting material, which only a short time ago seemed to be the biggest obstacle preventing establishment of a viable GaAs analog IC manufacturing process, is now not considered as serious a problem by industrial manufacturers.

The material status for manufacturing GaAs analog ICs indicates that three basic problems must still be addressed:

(1) Basic materials analysis to better understand the nature of the etch pits in LEC material.

(2) Investigation of the influence of material quality on backgating effects observed between adjacent devices in an IC.

(3) An effort on the part of material suppliers to establish a better data base on the reproducibility and uniformity of material supplied in large quantities for IC manufacturing.

Much of the materials work reported to date[1-3] involves the success of ion implantation and annealing in the IC process. The success of direct ion implantation is dependent on the method used for capping and annealing. Direct ion implantation has been accomplished in both LEC and Horizontal Bridgeman material.

Substrate Wafers

In comparing the physical properties of HB and LEC GaAs substrates two major factors are addressed: etch pit density and size. LEC substrates are presently available in 2 inch and 3 inch diameter wafers, whereas only 2 inch cored round HB material is presently available. HB substrates are characterized by an etch pit density 10-20 times lower than LEC material. The most significant issue with regard to substrate defects is the correlation of defect density with yield and/or reliability of a monolithic microwave integrated circuit. This issue is only being addressed by two U.S. manufacturers.[4] Initial investigations[5] have demonstrated a strong correlation between FET DC performance and non-uniformities of the LEC material. Table I compares the electrical characteristics of HB and LEC material.

Table I

Electrical Characteristics of GaAs Substrates

Electrical Properties	HB	LEC
Resistivity	$>10^7$-Ω-cm	$>10^7$-Ω-cm
Thermal Stability	Excellent	Variable
Impurities	Silicon Chromium	Si, Cr, B, Mn
Compensation Technique	Chromium	EL2

The LEC process uses control of stoichiometry to provide compensation of residual impurities. The advantage is that no additional impurities have been added, however, this process is difficult to control, reproducibility suffers and is very vendor dependent. The variability of LEC material leads to substantially lower yield than one normally experiences with HB wafers.

Epitaxial Layers

Direct ion implantation into GaAs is the preferred technique for creating the active GaAs layer. The versatility of ion implantation for producing high resistivity or high conductivity regions in GaAs wafers allows for the production of planar MMICs. Three inch wafers can be implanted with ease with typical throughputs of 200 wafers per hour.

A variety of approaches have been explored for annealing implanted GaAs, including some which do not involve the application of a capping film. The merits of the various capped and capless annealing techniques have not been fully documented in terms of uniformity, reproducibility, and surface morphology. Further information is needed on the surface physics associated with the application of a thin capping film, the pros and cons associated with having a thin film on the GaAs throughout the selective ion implant and anneal step, and the influence of the cap material quality (density, purity, etc.) on the process. A relatively new technique, known as flash annealing, has been developed which deserves further study. The anneal is performed at a much higher temperature for a much shorter period of time than a conventional furnace anneal. This approach promises to provide for a higher resolution channel profile, which should lead to improved device performance within an IC. A systematic evaluation of all promising techniques is required in order to establish a viable GaAs IC manufacturing process.

The importance of good quality materials and a uniform and reproducible method for forming the active device profiles within a linear microwave IC cannot be overemphasized. tTe circuit is operating in the linear region of the device characteristic, and therefore, the detailed behavior of the linear characteristic is critical. This is determined, in large part, by the semiconductor doping profiles. Unlike hybrid microwave integrated circuits, which are typically designed with some tuning possible to compensate for active device variations, the monolithic IC is more likely to be fixed tuned. This places a greater reliance on the inherent uniformity of the devices. The degree of success in establishing a GaAs monolithic microwave IC technology for real applications is ultimately dependent on how uniform and reproducible the active device profiles can be defined.

Both VPE and MBE growth technologies have been proven for producing active layers for power FETs. VPE processes have long been associated with high throughput of GaAs epitaxial layers. The ability to produce ultra-pure layers by VPE has been proven as has the ability to produce high performance microwave circuits on this material. Thus, the high level of production experience with this active layer and its established performance make this technique highly attractive from a cost point of view. Therefore VPE closely rivals ion implantation in U.S. industry as the preferred active layer formation technique.

In addressing the MBE (molecular beam epitaxy) technique, one must also consider cost, throughput and yield. Initial cost is high and the limited throughput tends to raise the cost of the active material deposited by this technique. But the potential for greatly enhanced circuit performance exists. The increased materials uniformity, quality and control[6,7] should result in a significant return on investment due to increased yields that can be achieved for MMICs. As is the case for VPE material, first an undoped buffer layer is deposited, followed by an intentionally silicon doped GaAs active layer. Silicon doping is readily achieved in the range from 10^{15} to 2×10^{18} cm^{-3} with room temperature electron mobilities of approximately 4000-5000 cm^2/v-sec at 10^{17} cm^{-3}.

GAAS MMIC PROCESSING

The processing technology for fabrication of GaAs monolithic microwave integrated circuits has improved dramatically over the past three years. The refinements now being made are directed primarily at enhancement of yield, extension of performance, and increase in circuit complexity. The GaAs IC process is actually relatively simple compared to the silicon IC fabrication procedures. Many of the techniques used to fabricate an IC on GaAs have been adapted from well established techniques used in the silicon IC industry. However, applying the techniques to the new material has not always been straight forward. Each processing step in a GaAs IC process must be optimized for application to GaAs.

FET Structure Building Block

The primary objective of industry has been the establishment of a planar process for MMIC's similar to that used for digital ICs. However, techniques which involve significant etching of GaAs, for mesa isolation or channel recessing, cannot be avoided. Selective ion implantation for the channel is limited in its ability to realize an optimum FET structure.

Discrete device manufacturers routinely use a deep channel recess in order to minimize parasitics and optimize channel pinchoff and transconductance characteristics. The concern in using this approach for MMIC manufacturing is that the additional variable introduced will degrade the uniformity of the FET characteristics across a wafer. However, dry etching techniques are presently used to produce the uniformity required for MMIC FET channel recessing. The effect of these techniques on the GaAs etched surface needs further study as does the uniformity of the process.

A totally different approach to improving FET RF performance, which might prove more uniform, is to use self-aligned n^+ implantation after gate formation. This technique has been introduced by workers in enhancement mode FET digital ICs. The device requirements for linear ICs are different and will no doubt require further device refinement. A significant aspect of the approach is the use of a refractory metal for the Schottky gate which can withstand the high anneal temperature following the n^+ implantation. This type of barrier is well suited to an MMIC process which requires high temperature depositions following the gate deposition. An optimized self-aligned process for linear circuits should result in RF performance comparable to the deeply recessed channel structure. In addition, both yield and RF performance will be enhanced.

Lithography

The production proven contact photolithography is the industrial baseline. However, the trend initiated in 1983-1984 is the hybrid approach of e-beam direct write and direct-step-on-wafer projection lithography. The most common problems experienced with contact lithography are photoresist liftoff and mask related losses. The former is controlled by a develop inspection step after each exposure. Mask related losses are controlled by using only featureless wafers after epitaxial growth. In addition to contact lithography, the hybrid technique looks promising for high throughput, low cost MMICs. The specific approach taken in industry has been e-beam direct writing of submicron features, such as transistor gates, while the rest of the lithography is carried out on optical projection aligners.

The e-beam direct-write technique for submicron feature definition is powerful, providing uniform gate definition of $0.5\mu m \pm 10\%$ and is not limited by the small depth of field offered by conventional lithography. Yield of the $0.5\mu m$ features is expected to be greater than 80% compared to 60% achievable by contact lithography.

Ohmic Contacts

Another important problem area in GaAs IC processing is that of ohmic contacts. The standard approach to forming an ohmic contact to GaAs involves the alloying of a AuGe eutectic on the GaAs surface which results in a heavily doped GaAs region under the contact. The problems associated with this technique include the control of surface morphology, the critical time/temperature control necessary to get good ohmic contacts, and the sensitivity of the contact to later high temperature processing. The standard contact technology has been satisfactory for discrete device fabrication because there are no subsequent high temperature steps required and the uniformity of the contact electrical behavior is not quite as critical.

For GaAs IC processing, it is desirable to have an ohmic contact process that can withstand at least moderately high temperatures without degradation. The contact should have excellent electrical properties and good surface morphology so the contacts can be made small and relatively densely packed compared to a discrete device. The requirement for withstanding high post-alloy temperature is imposed by the need to deposit a high quality dielectric material following device fabrication for the two level metal system.

This requirement has been met by utilizing the AuGeNi/Refractory/Au metal system. To improve device performance an n^+ epitaxial layer on top of the active layer is used by at least three manufacturers. The improved ohmic contact resistance expected should enhance device performance, but is a tradeoff against increased processing complexity and a higher mesa (non-planar structure). Thus, the effect of surface planarity versus lower contact resistance must be considered by an appropriate process-yield model.

Schottky Gate Formation

The baseline industrial metal system for the Schottky contacts is the Ti/Refractory/Au metal system because of its proven reliability. The refractory barrier is typically TiW, Pt, W, Mo or TiW-Ta[8]. The manufacturing technology calls for defining $0.5\mu m$ gate lines in a thin layer of photoresist. The $0.5\mu m$ features are subsequently covered with a sputtered thin film layer of the base metal such as Ti/Refractory. The wafer is patterned again, using the same mask. Alignment accuracy at this step is very critical and $\pm 0.25\mu m$ alignment is being reproduced routinely. Low resistance gates are the result of Au plating the structures to form the mush-room shaped cross-section. Gates defined using e-beam lithography are fabricated using multiple resist layers and a single exposure, eliminating the need for the alignment of two exposures. Before the gate metal is deposited, recessing of the gate region must be performed to achieve the proper value of saturation current.

Passive Element Processing

The quality of the passive element processing can significantly impact performance and yield. The problems present in the manufacturing of MMIC's include:

(1) Deposition of a low loss dielectric for MIM formation.

(2) Step coverage in non-planar structures.

(3) High resolution patterning in thick top metal.

(4) Wafer thinning for controlled impedance transmission structures.

(5) Provisions for via grounding through the substrate.

Dielectric deposition is restricted by the sensitivity of the ohmic and Schottky contacts to high temperature stress following their formation. The three dielectrics utilized are: Si_3N_4, Ta_2O_5 and SiO_2. Going to higher temperature results in better dielectric properties, but can only be done if the process also possesses a high temperature stable contact technology. Hence, improvements in this area are closely tied to work on contacts. Also important is the requirement for reducing the thickness of pinhole free deposition. If the dielectric is used to separate upper and lower level metals in crossovers, coverage by the dielectric over steps formed by the edge of the lower level metal is a processing problem which must be solved. The technique which has promise for high yield crossovers is the use of polymide to separate the metal levels. Further work is required to settle on an optimum approach for MMIC manufacturing.

For higher frequency circuits, two additional techniques for manufacturing MMICs have importance. These are controlled wafer thinning and etched via fabrication. The wafer must be thinned and metallized on the back to provide a microstrip ground. Substrate vias are now used universally by all manufacturers in order to provide for improved ground connections and for internal ground connections in complex circuits. Further development of via etching is required to establish better quality control. The baseline technology for vias is the use of a combination of wet chemistry and reactive ion etching to fabricate via holes through GaAs wafers. This combination gives the best yield for the reduced via hole size necessary for the high performance MMICs. Presently, most companies utilize an isotropic wet chemical etching process in order to fabricate 100µm thick wafers. This process is very high yield if the wafers are accurately thinned. Reactive ion etching (RIE), on the other hand, has proven to be a much more directional process and thus less sensitive to wafer thickness variations. The extreme directionality results in steep via hole side walls which are difficult to metallize. Therefore, the combination of wet and dry etching appears to be the optimum technique.

GAAS MMIC CHIP DESIGN

In principle, the design of a circuit for monolithic implementation uses the same tools and body of knowledge as that of hybrid MIC design. In practice, the design rules imposed by the monolithic format greatly influence the approach taken in design. Some of the general constraints imposed include:

(1) Space taken by the passive elements in the circuit must be kept to a minimum.

(2) The range of element values permitted is limited by the planar format, or the near planar format.

(3) The design must be tolerant of the expected variation in device parameters due to lack of tuning capability.

(4) Circuit complexity and chip partitioning are determined by yield objectives, what can and cannot be implemented monolithically, problems in biasing, and isolation requirements.

Device Modeling

Circuit design is presently being accomplished by first developing an experimental data base on the devices to be used through extensive discrete device characterization.[9,10] This data provides the basic circuit parameters of interest for the device including the variation of these parameters to be expected over a wafer and from wafer to wafer. With an adequate device data base as a starting point, most circuit designs can be carried out using available CAD tools. Circuit topologies can be chosen which maximize the RF yield based on this type of analysis. Circuit topologies are typically chosen which will maximize the RF yield based on this type of analysis.

The modeling of passive components for MMIC development is a much less formidable task than that of active elements, but some problems exist at the present time. The greatest problem has been experienced with the analysis and characterization of small inductive elements. Analytical models for a spiral inductor does not yield a value consistent with measured data for values less than 0.5nh. To a lessor extent, the accurate modelling of small capacitors is a problem. Although the analysis of an interdigital capacitor structure is relatively difficult compared to that of a parallel plate MIM capacitor, there is inherently better control over the value in processing since the value is not dependent on the characteristics of a deposited dielectric film. Currently, empirical data is used to design the interdigital structure. This is adequate for designs using a previously characterized interdigitated geometry. An improved analytical design approach would be useful to provide greater flexibility in design.

Computer Aided Design Tools

The needed aid to MMIC design available to industry include COMPACT, SPICE, AMPSYN. Although they are not satisfactory programs by themselves, they can be used together for building block development. AMPSYN is a program which can be used to synthesize a lossless network for matching a complex series impedance or parallel admittance to a characteristic transmission line real impedance. This has been used to generate matching network topology and starting values for more accurate analysis and optimization. COMPACT is used to analyze single-ended circuits including loss and other parasitics. It has optimization routines which are used to determine the best set of element values for a given topology. SPICE is used to analyse the small signal and transient behavior of a network defined by model interconnections. It has been used to analyze balanced circuits and to perform first order large signal analysis based on an approximate FET model.

More work is required to develop a user oriented program allowing for the analysis and optimization of complex circuits. The program should incorporate large signal analysis with the option of defining a device by large signal S-parameters or load-pull parameters.

CONCLUSIONS

The principal attraction of monolithic microwave integrated circuits on a single substrate is its potential for cost reduction in volume production. By integrating active devices and matching circuits on a single substrate, component assembly is simplified. The cost of MMIC is leveraged by batch processing, with a large number of potentially good devices being available for each wafer processed. The establishment of a manufacturing technology will eventually be measured by the bottom line cost of amplifiers for various system applications. In order to optimize the manufacturing technology, a thorough understanding of the cost elements of the IC is essential. The manufacturing technology problems addressed in this paper must be solved in order to meet the user's requirements for performance, cost and reliability. In the previous discussion, a number of areas for further work were suggested. Most of the required investigations are necessary either to improve yield and uniformity of the MMIC product or to enhance performance. At the present time, both areas are important: yield must be improved in order to realize the low cost potential; the performance of the MMIC chip must exceed that of its hybrid counterpart for it to be acceptable for use in most system applications.

References

1. R.C. Eden, B.M. Welch, R. Zucca and S.I. Long, IEEE Tranms. Electron Devices, ED-26, pp. 299, (1979).
2. R.C. Eden, B.M. Welch, and R. Zucca, IEEE J. Solid State Circuits, SC-13, pp. 419, (1978).
3. H. Kusakawa, K. Suyama, S. Okamura and M. Fukuta, Proc. 12th Conf. Solid-State Devices, Tokyo 1980, Japan, J. Appl. Phys., Suppl. 20-1, pp. 229, (1981).
4. R.D. Fairman, R.T. Chen, J.R. Oliver and D.R. Chen, IEEE Trans. Electron Devices, ED-28, pp. 135, (1981).
5. H.M. Hobgood, G.W. Eldridge, D.A. Barrett, and R.N. Thomas, IEEE Trasns. Electron Devices, ED-28, pp. 135 (1981).
6. J.C.M. Hwang, Appl. Phys. Lett. 42, 66 (1983).
7. J.C.M. Hwang, J. Electrochem. 130, 493 (1983).
8. B.M. Welch, Interfaces and Contascts, pp. 401, Mat. Res. Soc. Vol. 18 (1983).
9. H.Fukui, IEEE Trans. MTT, 27, pp. 643, (1979).
10. R. Esfandiari, IEEE Trans. MTT, 31, No. 1, (1983).

Part II
Materials and Processing Considerations

CRUCIAL to achievement of good circuit performance and high chip yield is the proper choice and qualification of the semi-insulating substrate.

Equally as important, and perhaps more so, are the various technologies which are essential to the fabrication of MMIC's. These include the technology of the active devices, i.e., FET's, diodes, bipolars, etc., as well as of the passive thin-film components, i.e., resistors, capacitors, inductors, and interconnections. These technologies encompass epitaxy and ion implantation for the active devices and some passive components, dielectric and resistive film growth, as well as isolation and passivation techniques.

These and related topics are addressed in this part of the volume.

High-Purity Semi-Insulating GaAs Material for Monolithic Microwave Integrated Circuits

H. M. HOBGOOD, GRAEME W. ELDRIDGE, DONOVAN L. BARRETT, AND R. NOEL THOMAS

Abstract — Liquid-Encapsulated Czochralski (LEC) growth of large-diameter bulk GaAs crystals from pyrolytic boron nitride (PBN) crucibles has been shown to yield high crystal purity, stable high resistivities, and predictable direct ion-implantation characteristics. Undoped ($\lesssim$low 10^{14} cm^{-3} chromium) and lightly Cr-doped (low 10^{15} cm^{-3} range) $\langle 100 \rangle$-GaAs crystals, synthesized and pulled from PBN crucibles contain residual shallow donor impurities typically in the mid 10^{14} cm^{-3}, exhibit bulk resistivities above 10^7 $\Omega \cdot$ cm, and maintain the high sheet resistances required for IC fabrication ($>10^6$ $\Omega/\square$) after implantation anneal. Direct ^{29}Si channel implants exhibit uniform (± 5 percent) and predictable LSS profiles, high donor activation (75 percent), and 4800- to 5000-cm^2/V $\cdot$ s mobility at the (1 to 1.5) $\times 10^{17}$ cm^{-3} peak doping utilized for power FET's.

It has also been established that LEC crystals can provide the large-area, round $\langle 100 \rangle$-substrates which will be required to realize a reproducible, low-breakage GaAs IC processing technology. The fabrication of 2-in-diameter GaAs substrates to tight dimensional tolerances, with $\langle 110 \rangle$-orientation flats and edge rounding has been demonstrated experimentally.

I. INTRODUCTION

THE ADVENT of monolithic GaAs IC's is expected to have a broad impact on microwave signal processing and power amplification with impressive advances in the performance and cost-effectiveness of future advanced systems for military radar and telecommunications being anticipated. The present development of monolithic microwave GaAs circuits is based on a selective direct ion implantation of semi-insulating substrates because of its greater flexibility compared with epitaxial techniques for planar device processing. However, direct-implantation technology imposes severe demands on the quality of the semi-insulating GaAs, and the unpredictable and often inferior properties of commercially available Bridgeman and gradient freeze substrates in the past has been a major limitation to at-

Manuscript received September 8, 1980; revised October 1, 1980. This research was partially supported by the Defense Advanced Research Projects Agency and monitored by the Office of Naval Research under Contract N00014-80C-0445.

The authors are with Westinghouse Research and Development Laboratories, Pittsburgh, PA 15235.

taining uniform and predictable device characteristics by implantation. These problems of substrate reproducibility are now well recognized in a symptomatic sense and are probably associated with excessive and variable concentrations of impurities, particularly silicon and chromium, and with the redistribution of these impurities during implantation and thermal processing. Monolithic GaAs circuits require substrates which 1) exhibit stable, high resistivities during thermal processing, to maintain both good electrical isolation and low parasitic capacitances associated with active elements; 2) contain very low total concentrations of ionized impurities so that the implanted FET channel mobility is not degraded; and 3) permit fabrication of devices of predictable characteristics so that active and passive elements can be matched in monolithic circuit designs.

Another important consideration is the need for uniformly round, large-area substrates. Broad acceptance of GaAs IC's by the systems community will occur only if a reliable GaAs IC manufacturing technology capable of yielding high-performance monolithic circuits at *reasonable costs* is realized. Unfortunately, the characteristic D-shaped slices of boat-grown GaAs material have been a serious deterrent to the achievement of this goal, since much of the standard semiconductor processing equipment developed for the silicon IC industry relies on uniformly round substrate slices. Improvements in the basic GaAs materials and its availability as round, large-area substrates are, therefore, key requirements if a high-yield, low-cost GaAs IC processing technology is to be realized in the near future.

To address these problems, the growth of high-quality large-diameter GaAs crystals by Liquid-Encapsulated Czochralski (LEC) and the fabrication technology for producing cassette-compatible large-area substrates from these circular cross-section crystals has been investigated. The growths were performed in a high-pressure LEC puller, where crystal purity can be preserved by *in situ* elemental compounding [1] and *reproducible* semi-insulating GaAs achieved by employing silicon-free pyrolytic boron nitride crucible techniques [2]-[5]. The growth of 2- and 3-in diameter, ⟨100⟩-oriented GaAs crystals, investigations of materials purity and semi-insulating properties, and their characterization for directly implanted power IC fabrication are reported here.

II. LEC Crystal Growth and Substrate Preparation

Liquid-Encapsulated Czochralski (LEC) growth was first demonstrated experimentally in 1962 by Metz *et al.* [6] for the growth of volatile PbTe crystals, and has since been developed by Mullin *et al.* [7] for several III-V crystals. In this Czochralski technique, the dissociation of the volatile As from the GaAs melt which is contained in a crucible is avoided by encapsulating the melt in an inert molten layer of boric oxide and pressurizing the chamber with a nonreactive gas, such as nitrogen or argon, to counterbalance the As dissociation pressure. *In situ* compound synthesis can be carried out from the elemental Ga and As components since the boric oxide melts before significant sublimation starts to take place (~450°C). Compound synthesis occurs rapidly and exothermally at about 820°C under a sufficient inert gas pressure (~60 atm) to prevent sublimation of the arsenic component. Crystal growth is

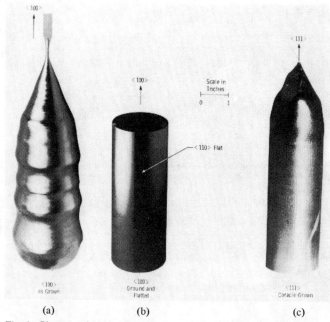

Fig. 1. Photograph illustrating: (a) as-grown, nominally 2-in-diameter semi-insulating, ⟨100⟩ GaAs crystal; (b) ⟨100⟩ GaAs crystal centerless ground to 1.975 ± 0.005-in diameter with ⟨110⟩ flat; (c) ⟨111⟩ GaAs crystal grown with $2 \pm \frac{1}{16}$-in diameter utilizing "coracle" method.

initiated from the stoichiometric melt by seeding and slowly pulling the crystal through the transparent boric oxide layer.

The Melbourn high-pressure LEC puller (which is manufactured by Metals Research Ltd., Cambridge, England, and is the outcome of developmental efforts at the Royal Radar and Signals Establishment) is currently being introduced by many laboratories for the growth of large bulk GaAs as well as GaP and InP. The puller consists of a resistance-heated 6-in-diameter crucible system capable of charges up to about 10 kg and can be operated at pressures up to 150 atm. The GaAs melt within the pressure vessel can be viewed by means of a closed-circuit TV system. A high-sensitivity weight cell continuously weighs the crystal during growth and provides a differential weight signal for manual diameter control. In addition, a unique diameter control technique which involves growing the crystal through a diameter-defining aperture made of silicon nitride, has been developed for ⟨111⟩-oriented growth. In this "coracle" technique, the defining aperture is fabricated from pressed silicon nitride which conveniently floats at the GaAs melt/B_2O_3 encapsulant interface.

In our studies, a reproducible growth methodology for preparing nominally 2- and 3-in-diameter, ⟨100⟩-oriented GaAs crystals free of major structural defects such as twin planes, lineage, inclusions, and precipitates has been successfully established over the course of about 40 experimental growths. A photograph of a typical 2-in nominal diameter, ⟨100⟩-semi-insulating GaAs crystal grown from a 3-kg charge is shown in Fig. 1(a). The use of very clean conditions during crucible loading and growth, vacuum baking of the boric oxide encapsulant to assure a water-free oxide, and a growth technique to produce crystals with sharp cone angles [8] during the initial stages of growth were found to be essential ingredients to achieving twin-free ⟨100⟩ growths. The growth of chromium-

Fig. 2. Photograph of high-purity semi-insulating GaAs wafers sawn on the ⟨100⟩ growth axis from a centerless ground crystal with diameter of 1.975 ± 0.005 in. Wafers have double ⟨110⟩ flats and rounded edges.

free semi-insulating GaAs crystals from melts synthesized *in situ* from 6/9's purity Ga and As charges and contained in a pyrolytic boron nitride crucible has been emphasized in these studies. However, GaAs crystals pulled from undoped and chromium doped melts contained in conventional fused silica crucibles have also been carried out for comparison of materials properties.

Limited experience has also been gained in coracle growths which yield significantly improved diameter uniformity (±1/16 in) as shown in Fig. 1(c). Unfortunately, its utilization is restricted to ⟨111⟩ growths because of severe twinning in ⟨100⟩-coracle growths [9]. However, device fabrication employing ion implantation or epitaxial techniques, is normally carried out on (100)-GaAs surfaces. The coracle technology, therefore, offers little advantage at the present time since the ⟨111⟩-oriented crystals would have to be sliced at a 54° angle and would thus yield elliptical (100) wafers. Our preliminary assessments of the structural quality of ⟨111⟩-oriented LEC crystals grown by the "coracle" method (as determined by X-ray topography and dislocation-etching techniques) indicate that such crystals are often characterized by excessive dislocation-generation and formation of microtwins near the crystal periphery while extensive activation of {111} ⟨110⟩ glide systems occurs in the interior of the crystal. The poor structural quality of the outer regions of coracle-grown ⟨111⟩ crystals often leads to extensive twinning as shown in Fig. 1(c) where a twin boundary can be seen cutting diagonally across the crystal approximately $\frac{2}{3}$ down the axis.

An alternative approach to achieving uniform, circular cross section (100)-oriented GaAs wafers is illustrated in Fig. 1(b), where a ⟨100⟩ ingot has been ground accurately to diameter with a (110) orientation flat by conventional grinding techniques. Surface work damage was removed by etching. Approximately 100 polished wafers of uniform diameter with a thickness of about 0.020 in have been obtained from such crystals. Fig. 2 shows a photograph of a batch of high-purity semi-insulating GaAs wafers sawn on the (100) orientation and fabricated to tight dimensional tolerances (1.975 ± 0.005-in diameter and

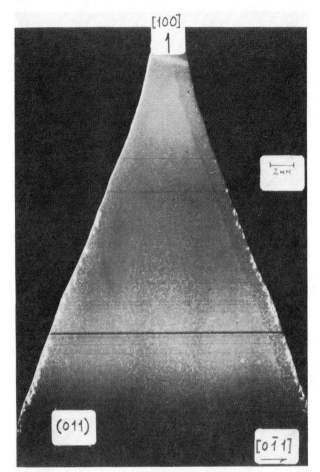

Fig. 3. Reflection X-ray topograph ($g = \langle 260 \rangle$) of a (011) longitudinal section cut from a ⟨100⟩ GaAs crystal grown with a cone angle of 27°. Crystal pulled from pyrolytic boron nitride crucible.

0.020 ± 0.001-in thickness), with ⟨110⟩ flats and edge rounded to minimize breakage during handling.

III. CHARACTERIZATION

A. Crystalline Perfection

Large-diameter (≤3-in) ⟨100⟩ LEC GaAs crystals are usually characterized by radially nonuniform dislocation distributions exhibiting maximum dislocation densities in the 10^4- to 10^5-cm^{-2} range. The dislocation-generation mechanism is thought to be due primarily to thermally induced stresses which, owing to the large convective heat transfer coefficient of the B_2O_3 encapsulating layer and the temperature difference between the crystal interior and the B_2O_3 ambient near the growth interface, can exceed the critical resolved shear stress for dislocation motion [10]. For crystal growths at diameters $\sim \frac{1}{2}$ in, the attendant thermal stresses are diminished, and the crystals can be grown entirely dislocation free [11]. For these small crystals, successful growth free of twins and dislocations depends upon growth conditions which yield proper stoichiometry [12], interface shape [13], [14], and crystal cone angle [8], [15].

Our investigations on improving the structural quality of large-diameter ⟨100⟩ GaAs crystals have concentrated on determining the optimum conditions for initiating a dislocation-free growth and eliminating the tendency toward twinning which has often frustrated ⟨100⟩ GaAs growth efforts in the past. Fig. 3 shows a reflection X-ray topograph of a longitudinal section of the seed-end cone of a ⟨100⟩ GaAs crystal cor-

TABLE I
HIGH-SENSITIVITY SECONDARY ION MASS SPECTROSCOPY ANALYSIS[†]
OF LEC SEMI-INSULATING GaAs CRYSTALS PULLED FROM
HIGH-PURITY PYROLITIC BORON NITRIDE CRUCIBLES

Crystal	C[*]	O[*]	Si	S	Se	Te	Cr	Mg	Mn	B	Doped Yes	Doped No
BN-1 (s)	2e15	1.5e16	4.8e14	5.7e14	2e13	1e12	<4e14	1e15	7e14	2.3e15		✓
BN-2 (s)	1e16	6e16	6e14	2e15	1e15	7e13	5.6e14	4e14	8e14	1.7e17		✓
BN-3 (s)	3e15	1.3e16	9.3e14	4.2e14	3e14	<5e12	4.3e14	2e15	1e15	5.4e17		✓
BN-4 (s)	-	-	2e15	8e14	2e14	1e14	1e16	2.5e15	2e15	1.5e17	Cr	
BN-6 (t)	2.9e15	9.8e15	5.4e14	5e14	9.4e13	<5e12	5.4e14	2.5e15	1.7e15	6.9e17		✓
BN-8 (s)	1.1e16	3.7e16	8.7e15	1.1e15	5e13	7e12	<4e14	8.5e14	<7e14	7.3e16	Si	
BN-10 (s)	2e15	1e16	8.6e14	1.5e15	7e14	7e12	<4e14	3.6e15	1.2e15	2e18		✓
BN-11 A (s)	4.3e15	1.8e16	6.4e14	1e15	8e13	6e12	6.4e15	3.2e15	1.4e15	1.9e17	Cr	
BN-11 B (s)	3e15	1.3e16	7.6e14	8.7e14	1e14	8e12	6.3e15	4.4e15	1.6e15	1.9e17	Cr	
BN-12 (s)	2e16	7e16	1e15	2e15	5e14	6e13	3.2e15	3e14	8e14	1.8e17	Cr	
BN-13 (s)	-	-	<3e14	4e15	2e14	1e13	2e15	5e14	7e14	7e16	Cr	
BN-13 (t)	-	-	<3e14	3e15	5e14	1e13	4e15	8e14	7e14	2e17	Cr	
BN-14 (s)	-	-	4e14	8e14	9e14	8e12	6e14	1e15	8e14	3e17		✓
BN-14 (t)	-	-	7e14	1e15	1e15	1e13	1e15	9e14	1e15	7e17		✓
BN-15 (s)	-	-	<3e14	1e15	1e14	9e12	3e15	7e14	8e14	7e16	Cr	
BN-15 (t)	-	-	<3e14	2e15	5e14	2e13	5e15	1e15	8e14	2e17	Cr	
Detection Limit			3e14	3e14	5e12	5e12	4e14	3e14	7e14	8e12		

[†] Sims analysis courtesy Charles Evans & Associates, San Mateo CA
(s) Seed end sample
(t) Tang end sample
[*] Detection limits for C, O, Fe (< mid 10^{15} cm^{-3}) not well defined

TABLE II
HIGH-SENSITIVITY SECONDARY ION MASS SPECTROSCOPY ANALYSIS[†]
OF GaAs CRYSTALS GROWN BY LEC AND BOAT-GROWTH METHODS

Crystal	C	O	Si	S	Se	Te	Cr	Mg	Mn	B	Doped Yes	Doped No
LEC/PBN (a)	5e15	2e16	5e14	1e15	4e14	1e13	4e14	7e14	8e14	5e17		✓
LEC/QTZ (b)	6e15	-	1e15	2e15	3e14	4e13	2e16	2e14	8e14	1e15	Cr	
	8e15	4e16	7e14	2e15	3e13	4e13	5e14	3e14	7e14	1e15		✓
	7e15	4e16	9e14	9e14	7e13	3e13	5e14	3e14	6e14	3e15		✓
	5e16	4e16	8e14	1e15	2e14	5e13	1e16	5e14	7e14	2e15	Cr	
	8e16	3e16	2e15	2e15	4e13	2e13	1e16	5e14	8e14	1e15	Cr	
LEC/QTZ (c)	-	-	5e15	6e15	5e14	5e13	2e16	2e15	3e15	2e15	Cr	
	-	-	8e15	4e15	4e14	4e13	6e15	3e15	2e15	2e15	Cr	
	-	-	4e16	4e15	9e14	4e14	1e15	2e15	1e15	1e15		✓
	-	-	2e15	5e15	3e13	7e14	3e16	2e15	1e15	1e15	Cr	
Boat Grown (c)	-	-	8e15	3e15	3e14	2e13	2e16	2e15	1e15	1e14	Cr	
	-	-	6e15	6e15	3e14	2e13	4e16	1e15	1e15	6e14	Cr	
	-	-	2e16	8e15	2e13	4e15	9e16	2e15	1e15	7e14	Cr	
	-	-	5e15	3e15	5e13	5e13	2e16	1e15	2e15	1e15	Cr	

(a) Representative of the Values Shown in Table I
(b) Material Prepared In-House
(c) Material from Commercial Suppliers
[†] Sims Analysis Courtesy Charles Evans & Assoc., San Mateo, CA.

responding to a cone angle of 27° to the crystal axis. Dislocation-free growth is initiated using a Dash-type seeding [16]. As the cone diameter increases, the regions of highest dislocation density ($<10^3$ cm^{-2}) are confined to the crystal interior and a layer near the crystal surface corresponding to regions of maximum thermal stress; however, severe glide plane activation such as that typically observed in flat-top growths [5] has been eliminated. In addition, a steep cone angle like that shown in Fig. 3 has been found to reduce dramatically the tendency toward twinning which normally accompanies shouldering in flat-top growths of ⟨100⟩ GaAs crystals. A dislocation density maximum of $\sim 10^4$ cm^{-2} is observed in etched substrates corresponding to the full crystal diameter (2- and 3-in diameter).

B. Impurity Content and Electrical Characterization

Secondary ion mass spectrometry (SIMS) bulk analysis[1] of LEC GaAs material pulled from both quartz and pyrolytic boron nitride (PBN) crucibles at our laboratories, as well as commercially supplied large-area boat-grown substrates and LEC GaAs pulled from quartz containers have been carried out. A wide range of impurity species were examined. The SIMS data (Tables I and II) taken together with the resistivity data on corresponding substrates (Table III) show clearly the LEC

[1] The SIMS analyses were performed at Charles Evans and Associates, San Mateo, CA, using a Cameca 1MS-3F ion microanalyzer.

TABLE III
RESISTIVITY MEASUREMENTS ON SUBSTRATES[†] REPRESENTATIVE OF GaAs CRYSTALS PULLED FROM PYROLITIC BORON NITRIDE AND QUARTZ CRUCIBLES

Chromium Content (cm^{-3})	As-Grown Resistivity ($\Omega \cdot cm$)	As-Grown R_s ($\Omega/\square$)	Post-Implant Anneal R_s ($\Omega/\square$)
LEC/PBN			
Undoped (5e14)	7e7	1.2e9	1.4e7
	6e7	5.3e9	1.3e7
	4.8e7	4.3e8	9e7
5e15	2e8	3.1e9	4.1e8
	1.7e8	4.0e9	4.5e7
LEC/QTZ			
Undoped (5e14)	3e7 tang	7.7e8	$\ll 10^6$ [b]
	9e5	2e7	—
	2.5e3	6.1e4	—
	0.09	—	—
	0.35	—	—
5e16	5.8e8	1e10	1.4e9
	3.9e8	1.2e10	4e9
	5.7e8	1.2e10	7e8

[a] Si_3N_4 Capping and 860°C/15 m Annealing.
[b] FETs and monolithic circuits fabricated on these substrates found to exhibit excessive leakage.
[†] Samples are from near crystal shoulder unless otherwise specified

growths from PBN crucibles yield reproducibly high-purity GaAs substrates exhibiting consistently high, thermally stable semi-insulating behavior. Undoped GaAs crystals pulled from quartz crucibles often exhibit variable and anomalously low as-grown resistivities, while commercially supplied Cr-doped semi-insulating GaAs substrates are characterized by impurity contents which vary from supplier to supplier. The detailed SIMS data for crystals pulled from pyrolytic boron nitride crucibles are shown in Table I. Quantitative estimates of impurity concentrations were obtained by calibration against GaAs samples which had been implanted with known doses of specific impurities. Comparative SIMS results for LEC GaAs pulled from PBN and quartz crucibles and boat-grown substrate material are shown in Table II. For LEC GaAs prepared at our laboratories, residual silicon concentrations in the mid 10^{14}-cm^{-3} range are observed in GaAs/PBN samples and somewhat higher concentrations in crystals pulled from quartz crucibles. Commercially supplied LEC and boat-grown crystals grown in quartz containers exhibit silicon levels which range up to mid 10^{16} cm^{-3}, depending upon growth technique and the substrate supplier. The residual chromium content in undoped LEC GaAs crystals pulled from either PBN or fused silica crucibles is below the detection limit of the SIMS instrument to be in the low 10^{14}-cm^{-3} range. Analyses of LEC-grown crystals pulled from Cr-doped melts contained in quartz crucibles reveal that the Cr content (typically 2×10^{16} cm^{-3} at the seed end and approaching 10^{17} cm^{-3} at the tang end) is close to the anticipated doping level based on the amount of Cr dopant added to the melt and its reported segregation behavior [17]. Cr dopant levels of (2 to 9) $\times 10^{16}$ cm^{-3} were observed in material grown by horizontal gradient freeze or Bridgeman methods. The reduced concentration of shallow donor impurities in growths from PBN crucibles permits lower Cr doping levels to be utilized as shown in Table I (BN-11, 12, 13, 15) where typical Cr dopant concentrations range from 3×10^{15} cm^{-3} at the crystal seed end to 6×10^{15} cm^{-3} near the crystal tang end. The corresponding resistivity levels measured in lightly Cr doped ($<5 \times 10^{15}$ cm^{-3}) GaAs/PBN material (Table III) approach those typ-

ically observed in heavily Cr doped ($>1 \times 10^{16}$ cm^{-3}) GaAs pulled from fused silica containers. The SIMS studies also indicate that LEC growths from PBN crucibles generally result in high boron concentrations (10^{17}-cm^{-3} range) in the GaAs/PBN material versus low 10^{15}-cm^{-3} concentrations in quartz crucible growth. The results reported in Section IV suggest that boron in GaAs/PBN substrates does not contribute significantly to ionized impurity scattering and remains electrically neutral through implantation processing [18]. No significant differences in carbon ($\sim 10^{16}$ cm^{-3}), oxygen (10^{16} cm^{-3}), selenium (mid 10^{14} cm^{-3}), tellurium ($<10^{14}$ cm^{-3}), and iron (mid 10^{15} cm^{-3}) contents of different GaAs samples are revealed by these investigations; however, the results for carbon and oxygen must be viewed as tentative since detection limits for these elements are not as yet well defined for the SIMS technique.

Resistivity and thermal stability measurements have been carried out on LEC-grown GaAs substrates to determine the suitability of the substrates for ion implantation studies. Table III shows the results of resistivity measurements on substrates representative of GaAs crystals pulled from pyrolytic boron nitride crucibles and, for comparison, quartz crucibles. The substrates were taken from near the crystal shoulder unless indicated otherwise. Substrate resistivities in the mid 10^8-$\Omega \cdot cm$ range are observed in conventionally Cr-doped substrate material (containing mid 10^{16} cm^{-3} Cr concentrations) pulled from quartz crucibles compared to resistivities of mid 10^7 $\Omega \cdot cm$ in undoped GaAs/PBN crystals ($<5 \times 10^{14}$-cm^{-3} Cr content) and resistivities of 10^8 $\Omega \cdot cm$ observed in lightly Cr-doped GaAs/PBN crystals (mid 10^{15}-cm^{-3} Cr content). In contrast to other work [19], undoped GaAs crystals pulled in our laboratories from fused silica crucibles show lower resistivities and exhibit large seed to tang variations from crystal to crystal, ranging from 10^6 to 10^7 $\Omega \cdot cm$ in one to resistivities as low as 0.09-$\Omega \cdot cm$ range in another crystal, even though the crystals exhibit SIMS impurity concentrations roughly equivalent to those observed in GaAs pulled from pyrolytic boron nitride crucibles.

Thermal stability of substrates for ion implantation was assessed by means of resistivity measurements following an encapsulated anneal of the semi-insulating slice (prior to implantation) to determine whether any conducting surface layers formed as a result of thermal treatment. Samples for the unimplanted, encapsulated anneal test were prepared using the same plasma-enhanced silicon nitride capping process as is used for implantation (see Section IV). The encapsulated wafers were annealed at 860°C for 15 min in a forming gas atmosphere. Surface sheet resistances exceeding 10^6 $\Omega/\square$ are desired after annealing to ensure isolation of passive, as well as active elements in analog IC processing.

Typical sheet resistance data for semi-insulating undoped and lightly Cr-doped ($<5 \times 10^{15}$ cm^{-3}) GaAs/PBN substrates as well as undoped and conventionally Cr-doped ($>1 \times 10^{16}$ cm^{-3}) GaAs substrates before and after encapsulated anneals are shown in Table III. Some degradation in the sheet resistance was always observed as a result of the thermal treatment, but the leakage currents measured in the low-field measurements (10^3 V/cm) and the RF losses observed in monolithic circuits fabricated on these substrates were low. In particular, surface leakages of <30 μA at 15-V bias and high breakdown voltages of 25 V were measured in interdigitated capacitor structures (5-μm separation and 13.2-mm periphery) to test the performance under the high operating fields normally utilized in monolithic circuits [20]. Additional assessment of the substrate quality was provided by measurements of transport properties of n-doped layers formed by direct ion implantation into the semi-insulating substrates and is discussed in the following section.

IV. ION IMPLANTATION

A direct ion-implantation technology yielding uniform and reproducible doping characteristics across each substrate and from substrate to substrate is highly desired for GaAs IC processing. Our approach implicitly assumes that higher purity semi-insulating GaAs substrates are required to avoid some of the difficulties encountered in the past such as: spurious activation of residual impurities, redistribution phenomena, and interactions with the implanted species. Silicon implantation of undoped LEC GaAs/PBN has, therefore, been emphasized with data from Cr-doped LEC GaAs included for comparison. The characterization of implanted substrates involves diagnostic techniques which are modified to meet the particular requirements of monolithic power amplifier development. Knowledge of the undepleted net carrier concentration per unit area of implanted layers is needed since this quantity will determine both the full channel current and the gate-drain breakdown voltage of gate-recessed power FET's fabricated on the layer [21], [22]. Although output power per unit periphery may be only weakly dependent on this quantity, reproducible matching to the passive output tuning circuit should be critically dependent upon it. Attention is then focused on carrier concentration per unit volume since this affects matching to passive input, interstage tuning circuits, and the gain. The pinch-off voltage and its uniformity are not considered directly.

Wafers for implantation are nominally 2-in diameter, cut on the ⟨100⟩ crystal growth axis, and bromine-methanol front surface polished to a thickness of approximately 0.020 in. Partial wafers, but more than $1\frac{1}{4}$ in in greatest dimension, were used for almost all of the tests described here. A 900-Å Si_3N_4 front surface encapsulation is deposited immediately following cleaning (to remove bromine and carbon contamination) and etch-back. This sequence implies that implantation through the Si_3N_4 encapsulant is found to be necessary to ensure a clean Si_3N_4/GaAs interface. Plasma-enhanced reaction of SiH_4 and N_2 at 340°C is used to deposit the Si_3N_4 at a rate of 70 Å/min. Uniformity of Si_3N_4 thickness is ±3 percent while the refractive index is 1.96 ± 1 percent. Implantation is performed at ambient temperature and 7° off normal incidence using the 3-in-diameter cassette-load end station of a 400-kV Varion Extrion implanter. All of the implantations discussed here employed $^{29}Si^+$ generated from an SiF_4 source. The initial choice of Si as the primary implant species was made on the basis of achievable range [23], integrity of the implanted profile through annealing [24], and ability to activate ambient temperature implants [25]. The test implants described here were performed at either 100, 200, and 400 kV or at a pair of lower energies such as 125 and 275 kV to approximate a flat ^{29}Si concentration; recoil implantation from the encapsulant was

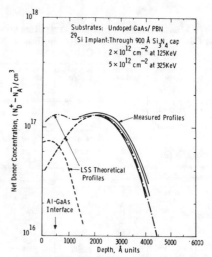

Fig. 4. Uniformity of implant profiles in undoped GaAs/PBN substrates measured by C–V profiling.

not considered in calculating doses. 2500 Å of 7-percent phosphosilicate glass was deposited on both surfaces prior to annealing. The implanted samples were annealed at 860°C for 15 min in flowing forming gas. Both the heating and cooling cycles were controlled.

A. Implant Profiles and Activation

Fig. 4 shows the net donor profiles achieved by $^{29}Si^+$ implantation into undoped GaAs/PBN substrates. The data obtained by surface Al Schottky-barrier diodes are representative of variations across each slice and slice-to-slice from approximately $\frac{2}{3}$ of one crystal. The superimposed theoretical curves are generated from tabulated joined half-Gaussian tables [23]. The standard deviations are modified by $\sigma_i^2 \to \sigma_i^2 + 3 \times 10^{-11}$ cm^2 which may reflect a diffusion process. The stopping power of the encapsulant is assumed equal to that of GaAs and recoil implantation from encapsulant is ignored. The theoretical curves have been scaled in concentration to reflect an activation efficiency of 75 percent.

A more detailed model of the form

$$n(z) = \eta \sum_{E_i} n_I(z, E_i) - N_{A0} \tag{1}$$

provides a better representation of the profiles that is also consistent with Hall data. Equation (1) assumes that the carrier concentration $n(z)$ can be represented by a constant activation efficiency η times the superposition of the implanted Si profiles $n_I(z, E_i)$ less a uniform concentration N_{A0} representing the density of levels that must be filled to raise the Fermi level from its position in the as-grown material to the neighborhood of the conduction band edge. Hall data for the undoped GaAs/PBN samples yield $\eta = 0.72$ and $N_{A0} = 1 \times 10^{16}/cm^3$, compared with Cr-doped GaAs substrates which yield $\eta \simeq 0.85$ and $N_{A0} \simeq 4 \times 10^{16}/cm^3$. It should be reiterated that these samples remain n-type semi-insulating through unimplanted, encapsulated anneals; N_{A0}^- may, in fact, be the density of deep donors ionized as a result of the acceptor concentration exceeding the residual shallow donor concentration.

Surface Hall-effect measurements yield a net donor concen-

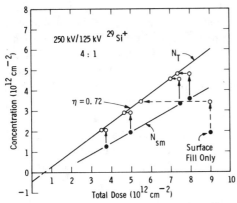

Fig. 5. Activation efficiency by Hall measurements of ^{29}Si implants in undoped GaAs/PBN substrates. The data as measured (N_{sm}) and after correction for surface depletion and implant deposition in nitride cap (N_t) are shown.

tration per unit area

$$Q_u/q = (N_D^+ - N_A^-)$$
$$= \left\{\int_{\lambda_d}^{\infty} \mu(z)n(z)\,dz\right\}^2 \bigg/ \int_{\lambda_d}^{\infty} \mu^2(z)n(z)\,dz \qquad (2)$$

where λ_d is the same surface depletion depth that determines undepleted charge Q_u in actual FET structures. The measured Q_u is potentially modified by depth-dependent Hall and drift mobilities, as well as the Hall factor itself. Both effects are neglected here. This approach is justified by the observation that the saturation current of FET structures prior to gate recess can be represented by

$$I = (1.18 \pm 0.1 \times 10^7 \text{ cm/s})Q_u \qquad (3)$$

(where I has units of amperes per centimeters gate width) independent of concentration level (0.6-$1.8 \times 10^{17}/\text{cm}^3$), type of semi-insulating substrate, activation efficiency achieved, and Hall mobility observed. The saturation velocity noted in (3) may imply a systematic underestimate of the actual Q_u, as a result of the Hall-effect measurement technique [21].

Fig. 5 illustrates the activation data achieved from an undoped GaAs/PBN crystal by Hall measurements. In this case, a fairly extensive series of samples was implanted at a fixed pair of energies and fixed surface/channel dose ratio to yield approximately flat Si profiles at different concentrations. The total dose required at this energy to achieve a specified undepleted concentration can be interpolated directly from the raw data. Corrections are required to obtain η and N_{A0}. An a priori correction for surface depletion can be achieved if it is assumed that the net donor concentration profile is flat between the Si_3N_4-GaAs interface and the range of maximum concentration (R_M) of the channel implant; the equivalent, uniform concentration implant depth ($R_M + \sqrt{(\pi/2)}\,\sigma_d$) is known; and the surface depletion barrier is known. For the undoped substrates, the profile is approximately flat; the equivalent depth can be approximated by $(11.4)\,E$ angstroms, where E is the channel implant energy in kiloelectronvolts, and the surface depletion barrier can be taken to equal 0.6 eV. The vertical arrows in Fig. 5 indicate these corrections to yield total net donor concentration cm^{-2}. The horizontal arrows correct the total implanted Si concentration for 40 percent deposition of the surface fill implant into the Si_3N_4 rather than the GaAs. The slope of the corrected data is the activation efficiency $\eta = 0.72$ while the intercept on the vertical axis scaled by the effective depth yields $N_{A0} = 0.9 \times 10^{16}$ cm^{-3}. Direct C-V measurement of these samples yields volume concentrations which agree within ±5 percent of those implied by the corrected Hall data and effective depth agreement ±3 percent. The significance of η and N_{A0} to the ion-implantation technology is that they can be employed in an inverse process to predict undepleted charge and net volume carrier concentration at other channel implant energies and down to volume concentrations as low as $1.2 \times 10^{16}/\text{cm}^3$. (Prediction at net carrier concentrations in excess of $2 \times 10^{17}/\text{cm}^3$ is complicated by a concentration dependent activation efficiency.) An even more significant tentative result is that the undoped LEC GaAs/PBN can be consistently grown to yield $\eta = 0.75 \pm 0.03$ and $N_{A0} = 1.0 \pm 0.3 \times 10^{16}/\text{cm}^3$.

Comparable data have been assembled for high Cr (0.6-1.2×10^{17} cm^{-3}) LEC GaAs crystals grown from fused silica crucibles and low Cr (3-6×10^{15} cm^{-3}) LEC GaAs grown from pyrolytic boron nitride. Analysis of the high Cr material is complicated by a depth-dependent activation efficiency which is presumably related to Cr pileup near the Si_3N_4-GaAs interface and by thickness variations in the conducting layer owing to compensating Cr in the substrate. For net donor concentrations less than 2×10^{17} cm^{-3} the channel implant activation efficiency $\eta = 0.90 \pm 0.05$ and $N_{A0} = (4$-$8) \times 10^{16}/\text{cm}^3$. The low Cr material can be analyzed by the technique described above to yield $\eta = 0.82 \pm 0.06$ and $N_{A0} = (1$-$2) \times 10^{16}/\text{cm}^3$, but the behavior at low implanted Si concentrations [$(0.2$-$6) \times 10^{16}/\text{cm}^3$] suggests a possible Si-Cr interaction.

B. Channel Mobility

The ambient temperature surface Hall mobility of FET channel layers ($n = 1.2 \times 10^{17}/\text{cm}^3$) implanted into undoped LEC GaAs/PBN is found to be 4900 ± 100 cm^2/V·s. Values of 4300-4600 cm^2/V·s can also be achieved in selected Cr-doped LEC GaAs crystals however. Analysis of this mobility over a broad range of net donor densities reveals significant differences in the behavior of these two types of material as is shown in Fig. 6. The data shown here correspond to the mobility average

$$\mu = \int_{\lambda_d}^{\infty} \mu^2(z)n(z)\,dz \bigg/ \int_{\lambda_d}^{\infty} \mu(z)n(z)\,dz \qquad (4)$$

where each point corresponds to a separate sample prepared to yield an approximately flat $n(z)$ profile. The data on undoped samples were obtained from three separate test implant runs and include data from two different crystals. At the lower end of the implanted Si concentration range (1.7×10^{16}-1.5×10^{18} cm^{-3}), the surface depletion depth is so large that an appreciable fraction of the measured carriers for those 100-, 200-, plus 400-kV implants is associated with the half-Gaussian lying beyond R_M. If the undepleted profile were exactly half-Gaussian and $\mu(z)$ could be approximated as a linear function of $n(z)$ only, μ should more properly be associated with $n(R_M)/\sqrt{2}$.

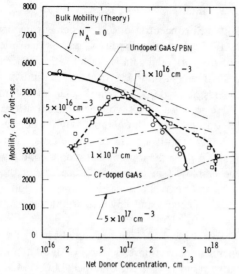

Fig. 6. Measured direct ^{29}Si implant mobility in semi-insulating undoped and Cr-doped GaAs compared with theoretical bulk mobility.

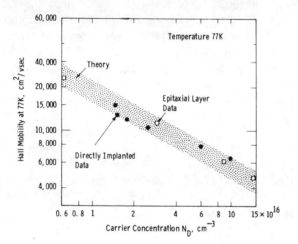

Fig. 7. Directly implanted channel mobility at 77 K in undoped GaAs substrates.

Since this ambiguity is significant only at the lowest concentrations where μ is a weak function of $n(R_M)$, all μ values are plotted at $n(R_M)$. Substrate and process control is demonstrated by implantations designed for $n(R_M) = 1.6 \times 10^{16}$ cm^{-3} reproducibly yielding $n(R_M) = 1.6 \times 10^{16} \pm 0.4 \times 10^{16}$ cm^{-3} with mobilities of 5300–5700 cm^2/V·s. The mobilities of these layers at 77 K are 12 000–14 500 cm^2/V·s (see Fig. 7). These results are believed to be significant improvements over previously reported work.

The experimental data shown in Fig. 6 are overlaid by theoretical bulk drift mobility data drawn from tables published by Walukiewicz et al. [26]. Although comparison of depth averaged Hall-mobility data with this theory may not be strictly valid, this approach is believed to provide a valuable engineering perspective. Fig. 6 shows that the observed channel mobilities in undoped GaAs/PBN substrates are consistent with the presence of a compensating acceptor density of about 1×10^{16} cm^{-3} in agreement with the value derived from activation measurements. At net donor concentrations greater than $(1.5-2) \times 10^{17}$ cm^{-3}, the mobility is observed to decrease rapidly implying an increase in the net acceptor density. The observed decrease in mobility strongly suggests that an increasing fraction of the implanted ^{29}Si resides on acceptor sites such as Si_{As}^- or $Si_{Ga}V_{Ga}^-$. Similar "amphoteric" behavior has been documented for both Group IV and Group VI impurities in high-purity VPE growth [27]. Detailed mobility analysis [28] indicates that, provided Si acceptor formation during implantation is taken into account, the total activation of electrically active $N_D + N_A$ centers (above the residual background acceptor density of about 1×10^{16} cm^{-3}) is always 100 percent. This result suggests the absence of inactive ^{29}Si due to processes such as interstitial or pair formation, spurious activation of other impurities or the generation of active native defects, and that the net donor activation efficiency is determined by thermodynamic considerations.

Hall mobility of Cr-doped GaAs/silica ($>5 \times 10^{16}$ cm^{-3} Cr) exhibits strikingly different characteristics. The data were obtained from three different crystals to ensure representative behavior. Net doping saturation which occurs at about 2×10^{18} cm^{-3} is displaced to higher levels with increasing Cr content. At low carrier concentration levels (10^{16}-cm^{-3} range), anomalously low mobility values are observed in implanted Cr-doped GaAs and the observed mobility appears to be strongly dependent on the Cr content. Reproducibility at carrier concentrations below about 5×10^{16} cm^{-3} is very poor. Preliminary measurements of implanted channels formed in light Cr-doped GaAs/PBN substrates ($\sim 5 \times 10^{15}$-cm^{-3} Cr content) indicate mobilities which lie midway between the undoped GaAs/PBN and the Cr-doped GaAs/silica curves shown in Fig. 6. A self-consistent analysis of the mobility and activation of Cr-doped GaAs leads to the conclusions that formation of Cr-Si complexes and Cr control of the formation of silicon acceptors are important phenomena [29].

V. Conclusions

Significant progress has been made towards developing large-diameter semi-insulating GaAs crystals of improved quality by LEC growth for direct-ion implantation. The intent has been to establish a source of reproducibly high-quality GaAs substrates for analog power IC development [30] by achieving 1) thermally stable, semi-insulating properties without resorting to intentional Cr doping (or at least to reduce the Cr content significantly) to avoid the serious redistribution problems associated with this impurity [31], [32], 2) uniform and predictable doping characteristics by direct ^{29}Si implantation, and 3) demonstrating that uniform, round cross-section slices suitable for low-cost IC processing can be fabricated from LEC crystals.

The use of *in situ* compounding, PBN crucible techniques, and very low moisture content B_2O_3 (assured by vacuum baking) appear to be the key ingredients to achieving *reproducible* semi-insulating ⟨100⟩ crystals of high purity and structural quality. Quantitative SIMS analyses indicate that residual silicon, Group VI impurities, and chromium can be maintained below the 10^{15}-cm^{-3} range in undoped GaAs crystals pulled from both PBN and quartz containers [19]. However, *consistent* high resistivities ($>10^7$ $\Omega \cdot$cm) and thermally annealed sheet resistances ($>10^6$ $\Omega/\square$), which yield acceptably low leakages and RF losses in analog circuits, are achieved from only GaAs/PBN grown crystals and not, in contrast to other studies [19], from GaAs crystals pulled from quartz crucibles.

Direct ion-implantation studies of undoped and lightly Cr doped (low 10^{15}-cm^{-3} Cr content) using ^{29}Si implants yield reproducible implant profiles, showing excellent agreement with LSS theoretical predictions. Directly implanted channels with the (1 to 1.5) $\times 10^{17}$-cm^{-3} peak donor concentrations required for X-band power FET structures yield electron mobilities between 4800 and 5000 cm^2/V·s. A compensating acceptor density of 1×10^{16} cm^{-3} or less is observed in undoped GaAs/PBN substrates compared with up to 10^{17}-cm^{-3} levels in conventional Cr-doped GaAs. Channel dopings down to 2×10^{16} cm^{-3} have been implanted in undoped GaAs/PBN substrates and mobilities of 5700 at 298 K and 14 500 cm^2/V·s at 77 K have been measured. Strong amphoteric doping behavior is observed at silicon implant concentrations approaching 10^{18} cm^{-3} and the total electrical activation, $N_D + N_A$, is 100 percent.

Monolithic analog amplifier circuits fabricated on high-purity semi-insulating LEC GaAs substrates (and reported in the companion paper appearing in this Special Issue [30]) have resulted in RF output powers approaching 1 W at associated 5-dB gain with octave bandwidth at X-band frequencies.

The materials preparative techniques commonly used in the silicon industry have been applied successfully to bulk LEC crystals to demonstrate the fabrication of uniformly round, large-area $\langle 100 \rangle$-GaAs wafers, which will be needed to realize a reliable, low-cost GaAs IC manufacturing technology.

Acknowledgment

The authors wish to thank Dr. H. C. Nathanson for his advice and encouragement in this work. They are indebted to Dr. J. G. Oakes, Dr. M. C. Driver, and Dr. M. Cohn (Westinghouse Advanced Technology Laboratories, Baltimore, MD) for their support and for providing the FET performance data quoted in this work. They also wish to thank Dr. T. T. Braggins for electrical and transport measurements, and Dr. W. J. Takei for his contribution in the X-ray topographic studies. The excellent technical assistance of L. L. Wesoloski and W. E. Bing in crystal growth, T. A. Brandis in substrate preparation, and P. Kost, D. J. Gustafson and J. C. Neidigh in ion implantation is gratefully acknowledged.

References

[1] D. Rumsby, presented at the IEEE Workshop on Compound Semiconductors for Microwave Materials and Devices, Atlanta, GA, 1979, unpublished.

[2] E. M. Swiggard, S. H. Lee, and F. W. Von Batchelder, "GaAs synthesized in pyrolytic boron nitride (PBN)," *Inst. Phys. Conf. Ser.*, no. 336, p. 23, 1977. See also R. L. Henry and E. M. Swiggard, "LEC growth of InP and GaAs using PBN crucibles," *Inst. Phys. Conf. Ser.*, no. 336, p. 28, 1977.

[3] T. R. AuCoin, R. L. Ross, M. J. Wade, and R. O. Savage, "Liquid encapsulated compounding and Czochralski growth of semi-insulating gallium arsenide," *Solid State Technol.*, vol. 22, no. 1, p. 59, 1979.

[4] R. N. Thomas, D. L. Barrett, G. W. Eldridge, and H. M. Hobgood, "Large diameter, undoped semi-insulating GaAs for high mobility direct ion implanted FET technology," in *Proc. Semi-Insulating III-V Materials Conf.* (Nottingham, England, Apr. 1980).

[5] R. N. Thomas, H. M. Hobgood, G. W. Eldridge, D. L. Barrett, and T. T. Braggins, "Growth and characterization of large diameter, undoped semi-insulating GaAs for direct ion implanted FET technology," *Solid-State Electron.*, in press.

[6] E. P. A. Metz, R. C. Miller, and R. Mazelsky, "A technique for pulling crystals of volatile materials," *J. Appl. Phys.*, vol. 33, p. 2016, 1962.

[7] J. B. Mullin, R. J. Heritage, C. H. Holliday, and B. W. Straughan, "Liquid encapsulated pulling at high pressures," *J. Cryst. Growth*, vol. 34, p. 281, 1968.

[8] W. A. Bonner, "Reproducible preparation of twin-free InP crystals using the LEC technique," *Mater. Res. Bull.*, vol. 15, p. 63, 1980.

[9] R. Ware and D. Rumsby, private communication.

[10] A. S. Jordan, R. Caruso, and A. R. von Neida, "A thermoelastic analysis of dislocation generation in pulled GaAs crystals," *Bell Syst. Tech. J.*, vol. 59, no. 4, p. 593, 1980.

[11] A. Steinemann and U. Zimmerli, "Dislocation-free gallium arsenide single crystals," in *Proc. Int. Cryst. Growth Conf.* (Boston, MA, 1966, p. 81.

[12] —, "Growth peculiarities of gallium arsenide single crystals," *Solid-State Electron.*, vol. 6, p. 597, 1963.

[13] J. C. Brice, "An analysis of factors affecting dislocation densities in pulled crystals of gallium arsenide," *J. Cryst. Growth*, vol. 7, p. 9, 1970.

[14] B. C. Grabmaier and J. C. Grabmaier, "Dislocation-free GaAs by the liquid encapsulation technique," *J. Cryst. Growth*, vol. 13, p. 635, 1972.

[15] P. J. Roksnoer, J. M. P. L. Huybregts, W. M. van de Wiggert, and A. J. R. deKock, "Growth of dislocation-free gallium phosphide crystals from a stoichiometric melt," *J. Cryst. Growth*, vol. 40, p. 6, 1977.

[16] W. C. Dash, "Growth of silicon crystals free of dislocation," *J. Appl. Phys.*, vol. 28, no. 8, p. 882, 1957.

[17] R. K. Willardson and W. P. Allred, in *Proc. 1966 Int. Symp. on Gallium Arsenide* (Reading, England). Surrey, England: Adlard and Sons, 1967, p. 35.

[18] T. J. Magee, R. Ormond, R. Blattner, C. Evans, Jr., and R. Sankaran, "Front surface control of Cr redistribution and formation of stable Cr depletion channels in GaAs," presented at the Workshop on Process Technology for Direct Ion Implantation in Semi-Insulating III-V Materials, Santa Cruz, CA, Aug. 1980, unpublished.

[19] R. D. Fairman and J. R. Oliver, "Growth and characterization of semi-insulating GaAs for use in ion implantation," in *Proc. Semi-Insulating III-V Materials Conf.* (Nottingham, England, Apr. 1980).

[20] J. G. Oakes, private communication.

[21] S. H. Wemple, W. C. Niehaus, H. M. Cox, J. V. DiLorenzo, and W. O. Schlosser, "Control of gate-drain avalanche in GaAs MESFET's," *IEEE Trans. Electron Devices*, vol. ED-27, pp. 1013-1018, 1980.

[22] W. R. Wisseman, G. E. Brehm, F. H. Doerbeck, W. R. Frensley, H. M. Macksey, J. W. Maxwell, H. Q. Tserng, and R. E. Williams, "GaAs power field effect transistors," Interim Tech. Rep. Dec. 1979, Contract F33615-78-C-0510.

[23] J. F. Gibbons, W. S. Johnson, and S. W. Mylroie, *Projected Range Statistics-Semiconductors and Related Materials*. New York: Wiley, 1975.

[24] C. O. Bozler, J. P. Donnelly, R. A. Murphy, R. W. Laton, R. W. Sudbury and W. T. Lindley, "High efficiency ion implanted lo-hi-lo GaAs IMPATT diodes," *Appl. Phys. Lett.*, vol. 29, pp. 123-125, 1976.

[25] J. P. Donnelly, W. P. Lindley, and C. E. Hurwitz, "Silicon- and selenium-ion implanted GaAs reproducibly annealed at temperatures up to 950°C," *Appl. Phys. Lett.*, vol. 27, pp. 41-43, 1975.

[26] W. Walukiewicz, J. Lagowski, L. Jastrebski, M. Lichtensteiger, and H. C. Gatos, "Electron mobility and free-carrier absorption in GaAs: Determination of the compensation ratio," *J. Appl. Phys.*, vol. 50, pp. 899-908, 1979.

[27] C. M. Wolfe and G. E. Stillman, "Self compensation of donors in high purity GaAs," *Appl. Phys. Lett.*, vol. 27, pp. 564-567, 1975.

[28] G. W. Eldridge, H. M. Hobgood, D. L. Barrett, T. T. Braggins, and R. N. Thomas, "Direct ion implantation studies of large diameter undoped GaAs prepared by LEC growth for monolithic X-band power FET circuits," presented at the 38th Annual Device Research Conf., Cornell Univ., June 1980, to be published.

[29] This is demonstrated by comparison of the data in Fig. 6 with B. T. Debney and P. R. Jay, "The influence of Cr on the mobility of electrons in GaAs FET's," *Solid-State Electron.*, vol. 23, pp. 773-781, 1980.

[30] M. C. Driver, S. K. Wang, J. X. Przybysz, V. L. Wrick, R. A. Wickstrom, E. S. Coleman, and J. G. Oakes, "Monolithic microwave amplifiers formed by ion implantation into LEC gallium arsenide substrates," this issue, pp. 191-196.

[31] A. M. Huber, G. Morillot, and N. T. Linh, "Chromium profiles in semi-insulating GaAs after annealing with a Si_3N_4 encapsulant," *Appl. Phys. Lett.*, vol. 34, p. 358, 1979.

[32] C. S. Evans, Jr., V. R. Deline, T. W. Sigmon, and A. Lidow, "Redistribution of Cr during annealing of ^{80}Se-implanted GaAs," *Appl. Phys. Lett.*, vol. 35, no. 3, p. 291, 1979.

Growth of High-Purity Semi-Insulating Bulk GaAs for Integrated-Circuit Applications

ROBERT D. FAIRMAN, R. T. CHEN, MEMBER, IEEE, JOHN R. OLIVER, AND DANIEL R. CH'EN, MEMBER, IEEE

Abstract—Growth of high-purity bulk semi-insulating GaAs by the Liquid-Encapsulated Czochralski (LEC) method has produced thermally stable, high-resistivity crystals suitable for use in direct ion implantation. Large round substrates have become available for integrated-circuit processing. The implanted wafers have excellent electrical uniformity (± 4 percent V_p) and have shown electron mobility as high as 4800 cm^2/V·s for Se implants with 1.7×10^{17} cm^{-3} peak doping. Careful control of background doping through *in situ* synthesis has produced GaAs with Si concentrations as low as 6×10^{14} cm^{-3} grown from SiO$_2$ crucibles. Detailed results of qualification tests for ion implantation in LEC GaAs will be discussed. Feasibility of successful high-speed GaAs large-scale integrated circuits using LEC substrates will be described.

I. Introduction

RENEWED INTEREST in semi-insulating GaAs bulk crystals is being propelled by the emergence of high-performance GaAs integrated-circuit (IC) technologies. The new devices include the digital IC [1], the monolithic microwave IC (MMIC) [2], and high-speed charge-coupled devices [3]. The device fabrication methods have largely adapted ion-implantation technology to achieve good circuit reproducibility and high doping uniformity. Direct ion implantation into semi-insulating (SI) GaAs substrates to attain the desired doping properties can be achieved provided that the SI GaAs has the following characteristics: 1) low residual background dopants for well-behaved high-mobility device active layers, and 2) adequate compensation for device thermal stability.

Commercially available semi-insulating GaAs has shown variable and nonreproducible results in ion implantation. Much effort has been expended in developing qualification methods to select crystals suitable for use in the Rockwell ion-implantation process. The resulting electrical yields for ion implantation by Bridgman-grown and gradient-freeze-grown crystals are generally low and unpredictable. The irregularly shaped wafers obtained from these growth techniques are also incompatible with standard semiconductor manufacturing equipment. GaAs boules grown in the ⟨100⟩ direction by liquid-encapsulated Czochralski (LEC) methods have shown considerable promise for use in device applications. This growth method also has the potential of large round wafers and high growth yields. However, greatly improved understanding of the basic materials properties and refined methods of LEC GaAs growths are needed before this technology can support the predicted substrate requirement of the advanced GaAs device in a production environment.

This paper describes research in the LEC growth of SI GaAs suitable for device fabrication using ion implantation. In the next section, the desired materials properties related to the active circuit elements are discussed. Crystal growth apparatus and experiment techniques used are described in Section III. LEC crystal characteristics and electrical properties are described in Section IV. Finally, the highlights of this work are summarized in Section V.

II. Semi-Insulating Materials Requirements for Direct Ion Implantation

Attainment of low-cost, high-performance IC devices is critically dependent upon the availability of large-area semi-insulating GaAs substrates capable of direct ion implantation. A wide variety of low- and high-dose implants are required to fabricate components such as low-noise and high-power MESFET's, Schottky diodes, varactors, switches, as well as passive elements for full circuit applications. Planar varactor structures with ion-implantated p-n junctions will involve diffusion-resistant Be implants. Successful ion implantation involves recognition of the following substrate requirements:

1) Reproducible, high-resistivity GaAs, thermally stable to permit fabrication of planar, isolated active regions without etching. ($R_s \geqslant 10^7$ $\Omega/\square$.)

2) Low background doping with respect to shallow donor and acceptor impurities.

3) Capability of producing abrupt implanted carrier profiles with high electrical activation and carrier mobility for the implanted species.

4) Sufficiently high crystalline quality to avoid defect-related impurity and processing problems.

5) Large and regular size slice shapes to permit high device yields and efficient device fabrication.

A set of material specifications satisfying the above criteria is defined as follows:

A. Resistivity

Following a thermal anneal at 850°C with 1100-Å Si$_3$N$_4$ cap the GaAs surface must not degrade to $<10^7$ $\Omega/\square$ and must retain n-type conduction.

B. Background Doping

Shallow donor and acceptor concentration must be $\leqslant 5 \times 10^{15}$ cm^{-3} by chemical analysis (SIMS) and have $\leqslant 10^{15}$ cm^{-3}

Manuscript received August 8, 1980; revised October 2, 1980.
The authors are with Rockwell International, Electronics Research Center, Thousand Oaks, CA 91360.

electrically active impurities. Compensation can be achieved with native deep centers such as EL2 [4] or by intentional doping with deep-level impurities such as chromium.

C. Ion Implant Tests (Se)

Hot, bare surface implants (200°C) with Se^+ must produce the following results:

1) abrupt carrier profile which compares favorably with LSS profiles;
2) high degree of impurity activation (>80 percent);
3) high degree of implanted dopant uniformity, <±10 percent in depletion voltage within implanted region;
4) exhibit a high electron and drift mobility ⩾90 percent of theoretical mobility (after implant) and show an increase in drift mobility at the active layer–substrate interface.

D. Crystalline Perfection

The crystal must be free from major crystalline defects such as stacking faults, inclusions, precipitates, twins, and low-angle boundaries. Dislocation densities $<10^5$ cm^{-2} are acceptable for ion implantation. For use in epitaxial growth defect densities $<10^4$ cm^{-2} are preferable.

E. Crystal Diameter and Ingot Size

Ion implantation and IC fabrication favors round slices as a convenient shape which is less prone to breakage and easily adapted to automated process equipment. High-quality LEC GaAs is presently available commercially in both 2- and 3-in diameters with ⟨100⟩ orientation. However, ⟨111⟩ oriented crystals are more easily grown than ⟨100⟩ and have a higher diameter tolerance at the 2-in diameter level. Historically, ⟨100⟩ slices were required strictly for use in epitaxy in reducing dopant incorporation and for breaking devices in the scribing operation.

Material requirements stated herein combine all the stringent specifications called out earlier for each separate device, now all combined in one material. Cost and overall reproducibility are important factors in both materials growth and final device fabrication. Yield factors for each separate element of a complex monolithic microwave circuit directly reflect material properties.

III. LIQUID-ENCAPSULATED CZOCHRALSKI (LEC) CRYSTAL GROWTH

Liquid-encapsulated Czochralski (LEC) techniques for growth of III-V crystals have been successfully used to grown GaP, GaAs, and InP. Pioneering work by Bass and Oliver [4] as well as Mullin *et al* [5] were instrumental in developing the technique and instrumentation for growth of III-V semiconductors. Later work by Weiner [6], Swiggard [7], and AuCoin [8] have been directed towards improving crystal purity by synthesis and growth from Al_2O_3 and PBN crucibles and establishing direct synthesis methods for growth of high-purity GaAs. Our work concerns the use of a new generation, high-pressure LEC crystal puller from Metals Research Ltd. The "Melbourn" puller is a resistance-heated crystal-growth system with a maximum

Fig. 1. State-of-the-art LEC GaAs ingots.

pull length of 24 in and a 6-in-diameter crucible with a capacity for charges up to 10 kg. Overpressures up to 150 atm can be attained. A unique closed-circuit TV system is used for viewing the melt and crystal during growth. A weighing system is employed to monitor the crystal weight and provide differential weight data to aid in diameter control. Programmed diameter control using a "coracle" technique involves crystal growth through a floating diameter-defining aperture made from Si_3N_4. Diameter control of ⟨111⟩ oriented crystals with the coracle method is very effective; however, ⟨100⟩ oriented crystals are grown manually without a coracle due to a higher propensity for twinning using the coracle method.

At present, a GaAs charge of 3 kg is used in each growth run; however, charges up to 10 kg are possible. Crystals weighing as much as 6 kg have been grown in the Melbourn system [9].

Numbers of 2- and 3-in diameter, ⟨100⟩ and ⟨111⟩, undoped and Cr-doped GaAs crystals free of major crystalline defects, including twins, inclusions, and precipitates, have been grown successfully from both SiO_2 pyrolytic boron nitride (PBN) crucibles. Two 2-in ⟨111⟩ and 3-in ⟨100⟩ LEC ingots grown using the automatic "coracle" and manual diameter control method, respectively, are shown in Fig. 1. Excellent crystal diameter control is clearly demonstrated for both growth methods.

IV. EXPERIMENTAL RESULTS

Thermal stability and active ion-implantation tests were carried out on LEC grown semi-insulating GaAs substrates to determine the qualification of the substrates for ion implantation. The criteria for these two basic tests have been described in Section II as a part of material requirements for semi-insulating GaAs for direct ion-implantation technology.

Table I summarizes results of the thermal-stability test along with the dopant and crucible used in crystal growth. All of the crystals show sheet resistance after the thermal anneal in the 10^9-Ω/□ range regardless of whether they are undoped or chromium doped, from the front or tail portion of the ingot,

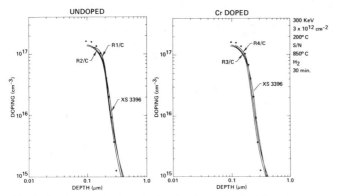

Fig. 2. Carrier profiles for selenium-implanted LEC GaAs substrates.

TABLE I
LEC GaAs—THERMAL STABILITY TEST

Sample No.	Sheet Resistance Ω/□ As Grown	Sheet Resistance Ω/□ After Anneal*	Crucible	Dopant
R1F	>5 × 10⁹	4.2 × 10⁹	SiO₂	None
R1T	>5 × 10⁹	4.2 × 10⁹	SiO₂	None
R2F	>5 × 10⁹	3.6 × 10⁹	SiO₂	None
R2T	>5 × 10⁹	5.0 × 10⁹	SiO₂	None
R3F	>5 × 10⁹	2.0 × 10¹⁰	SiO₂	Cr
R3T	>5 × 10⁹	4.2 × 10⁹	SiO₂	Cr
R4F	>5 × 10⁹	2.0 × 10¹⁰	SiO₂	Cr
R4T	>5 × 10⁹	1.5 × 10⁹	SiO₂	Cr
R5F	>5 × 10⁹	6.7 × 10⁹	SiO₂	Cr
R5T	>5 × 10⁹	5.0 × 10⁹	SiO₂	Cr
R7F	>5 × 10⁹	2.9 × 10⁹	PBN	None
R7T	>5 × 10⁹	1.0 × 10¹⁰	PBN	None

*1100Å Si₃N₄ cap, 850°C/30 min in H₂. F: front. T: tail.

TABLE II
CHEMICAL ANALYSIS OF HIGH-PURITY SEMI-INSULATING GaAs BY SECONDARY ION MASS SPECTROMETRY (SIMS)
($\times 10^{15}$ atoms·cm^{-3})

	Sample	Si	Cr	Mg	S	Mn	Fe	B
Undoped	R2/C F*	0.81	0.39	3.1	2.1	0.86	1.0	0.42
	R2/C T*	1.0	1.2	5.6	2.9	1.7	4.0	0.70
	R7/C F**	1.3	0.17	1.3	3.2	<0.5	2.1	0.62
	R7/C T**	2.4	0.19	3.2	6.3	0.82	2.3	3.4
Chromium Doped	R4/C F*	0.63	11	3.6	1.4	1.5	1.9	1.2
	R4/C T*	1.2	48	2.8	3.9	2.4	1.1	0.38
	R5/M F*	0.94	12	0.56	1.1	0.80	6.0	0.56
	R5/M T*	1.3	36	0.44	4.0	0.54	5.8	1.2
Bridgman	XS4033 F*	12	2.4	0.39	1.2	0.35	2.5	<0.1
	XS4033T*	4.9	20	0.27	4.2	<0.5	3.0	0.18
SIMS DETECTION LIMITS		(0.5)	(0.1)	(0.1)	(0.1)	(0.5)	(1.0)	(0.1)

*Ingot grown from SiO₂ crucible or boat.
**Ingot grown from PBN crucible.
F: front. T: tail. C: coracle. M: manual

or grown in PBN or SiO₂ crucible. These data demonstrate the excellent thermal stability for all LEC ingots, even for crystals R1–R5 which were grown from SiO₂ crucibles. The success indicated here is attributable to the high purity *in situ* synthesis technique employed in the growth of all these ingots.

Direct implantation of Se into chromium-doped and undoped LEC materials shows very high carrier mobilities of 4500–4800 cm²/V·s at 300 K from samples with peak doping concentration of 1.5×10^{17} cm^{-3}. Carrier profiles as shown in Fig. 2 are typical of the high activation achieved for Se implants in our LEC material and the excellent correlation with LSS range statistics over the entire carrier profile. No indication of straggling is apparent for either the undoped or chromium-doped ingots. Implantation profiles from both front and tail samples of each ingot show excellent agreement in the profile depth at 10^{16} cm^{-3}. Measurement of the depletion voltage distribution on a wafer from LEC ingot R5 have shown very high uniformity of depletion voltage (±4 percent) indicative of the highly uniform material quality existing in our large-diameter LEC GaAs wafers.

A. Impurity Analysis

As described previously, the availability of semi-insulating GaAs materials with low background impurity density is one of the critical factors required to insure the direct ion-implantation process with predictable reproducibility and overall doping uniformity. In the present study, assessment of residual impurities as well as deep centers in these materials was carried out by using highly sensitive secondary ion mass spectroscopy (SIMS), photo-induced current transient spectroscopy (PITS), and photo/dark conductivity measurements.

The SIMS measurement results on our undoped and chromium-doped LEC GaAs crystals pulled from both SiO₂ and PBN crucibles are shown in Table II, in which the result on one representative Bridgman crystal is included for purposes of comparison. Some common donor, acceptor, and deep-level impurities are observed in these materials. It is important to note the low silicon content observed in all our LEC crystals produced by *in situ* synthesis as compared to the Bridgman sample. Moreover, the low residual impurity content observed in all our LEC crystals whether pulled from SiO₂ or PBN crucibles demonstrates the high-purity growth capability of the Melbourn growth system.

Sulfur is a significant impurity in all GaAs samples. It is present as a common trace impurity in high-purity arsenic. Magnesium in the LEC materials is due to a contaminant in the Si₃N₄ coracle. Low chromium levels are observed in the undoped crystals grown from SiO₂ or PBN crucibles. The concentration of iron reported here is of some concern but it does not limit the use of these materials in present device applications. The low boron concentration of these crystals is especially notable compared to other workers [10] using similar equipment. Results of undoped, high-purity LEC GaAs grown from SiO₂ crucibles described here exceed any reported to date by other techniques. The encapsulant, B₂O₃, is effective in minimizing Si contamination arising from melt contact with the SiO₂ crucible. Other workers have commented on the reactive gettering of Si by B₂O₃ [5]–[8]. Other workers have used separate synthesis in conjunction with LEC growth and produced semi-insulating GaAs but not as thermally stable as that produced by the direct synthesis method.

Higher silicon concentrations have been detected consistently in Bridgman-grown crystals, demonstrating the contam-

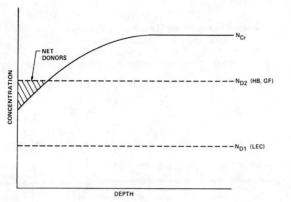

Fig. 3. Surface conversion mechanism following chromium redistribution.

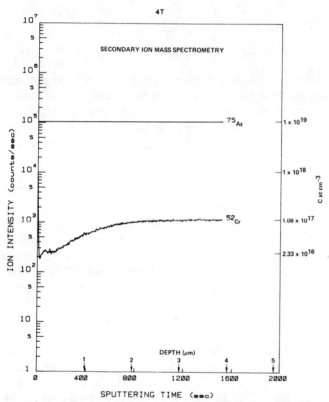

Fig. 4. Typical chromium redistribution profile in LEC GaAs.

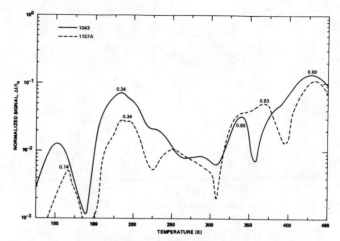

Fig. 5. Normalized photo-induced transient spectroscopy (PITS) spectra for chromium-doped VPE GaAs samples.

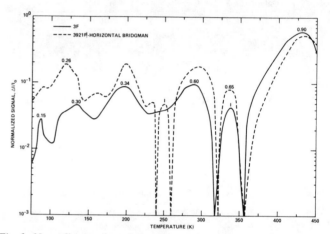

Fig. 6. Normalized photo-induced transient spectroscopy (PITS) spectra for bulk chromium doped semi-insulating LEC GaAs.

inating aspect of SiO_2 crucibles in the absence of B_2O_3. Higher Si concentrations are observed in the front sections of all Bridgman crystals as a result of the higher process temperatures in the initial portion of the crystal growth experiment [13].

The phenomenon of thermal conversion in Cr-doped GaAs during the post-implant anneal is well known. Recent work reported by Asbeck and other workers [11] has verified this conversion effect to be the Cr outdiffusion as first described by Tuck et al. [12]. Fig. 3 clearly demonstrates this thermal conversion mechanism which indicates how crystals with a high donor background become uncompensated following the thermal redistribution of chromium. A typical redistribution profile shown in Fig. 4 for our LEC samples illustrates effects similar to those described by Asbeck et al., but the background concentration of Si in the LEC samples was much lower than the Cr intercept and, consequently, the crystal remained semi-insulating.

B. Deep-Level Spectroscopy Measurements

Deep trap measurements are made by photo-induced transient spectroscopy (PITS) which involves the detection of the current decay due to the emission of trapped carriers after illumination by chopped bandgap light. The current decay is sampled at two points in time, with the difference logarithmically recorded as a function of temperature. Peaks will occur when the trap emission rate is equal to the sampling rate in the case of discrete levels, and hence by varying the sampling time constant and observing the peak shift in temperature, the trap energy and cross section can be computed. In the case of the measurements performed here, slightly greater than bandgap light (1.55 eV) and 5-μm separation ohmic surface contacts were used so that the measurements would be confined to the near-surface region.

PITS spectra for two epi-layers and two substrates are shown in Figs. 5 and 6, with the data plotted as the initial value of the current decay I, normalized by the photocurrent I_0 for a 3-ms sampling time constant. The VPE layers 1043 and 1157A

TABLE III
Trap Summary from PITS Measurements

T_m (°K)	E_t (eV)	σ (cm^2)	Identity	Sample
88	0.15	8×10^{-14} (n)	–	3F
117	0.14	1×10^{-16} (n)	–	1157A
126	0.26	2×10^{-12} (n)	–	3921F
149	0.30	7×10^{-14} (p)	HL12	3F, 3921F
170	0.34	6×10^{-13} (n)	–	1043, 1157A
190	0.34	4×10^{-14} (n)	EL6	All
245	0.51 (I)	1×10^{-12} (n)	EL4	3921F
280	0.60	1×10^{-12} (n)	EL3	3F
330	0.65 (I)	1×10^{-13} (n)	–	3F, 3921F, 1043
373	0.83	2×10^{-13} (p)	HL10	1157A
430	0.90	2×10^{-14} (p)	HL1	All

n: electron trap
p: hole trap
I: inversion

TABLE IV
Dark Conductivity Data

Sample	σ (ohm^{-1}-cm^{-1}) @ 300°K	E_A (eV)	E_F (eV)
3F	5.0×10^{-9}	0.72	0.72
3921F	2.0×10^{-8}	0.70	0.68
1043 VPE	1.3×10^{-8}	0.75	0.69
1157A VPE	1.2×10^{-8}	0.60	0.69

E_A and E_F are activation and Fermi energy, respectively (referenced to 0°K band gap).

in Fig. 5 are Cr doped, grown on Cr-doped substrates. Substrates 3F and 3921F in Fig. 6 are Cr-doped samples from the front of LEC and Bridgman ingots, respectively. A prominent feature of these spectra is the broad 0.90-eV peak associated with chromium acting as a hole trap. A second major peak occurs at T_m = 330 K as an inversion for the substrate samples and as a normal peak for sample 1043, with an activation energy of 0.65 eV. The hole trap at 0.83 eV, T_m = 373 K for 1157A is assigned to an (unidentified) deep recombination center. Another prominent feature of these spectra is the peak at 170–190 K due to two 0.34-eV levels. A summary of major levels measured in given in Table III. An interesting interaction between Cr and the recombination center is observed in the suppression of the 0.83-eV level upon increasing the Cr concentration in VPE materials. The 300 K dark conductivity varied by no more than a factor of 4 for the measured samples, with the LEC substrate 3F showing the lowest conductivity at $\sigma = 5 \times 10^{-9}$ $\Omega^{-1}\cdot$cm^{-1}. A summary of these data is shown in Table IV.

C. Crystalline Perfection

The role of dislocations in descrete and integrated GaAs devices remains undefined in spite of earlier attempts to relate defects with device yields. Only major crystalline defects such as inclusions, precipitates, and stacking faults have serious effects upon device operation and are sufficiently documented [14]–[16].

Recent efforts in the growth of 2- and 3-in-diameter crystals with new resistance-heated pullers (Metals Research Ltd.) have produced high-quality ⟨111⟩ and ⟨100⟩ ingots free of microstructured defects such as twins, inclusions, and precipitates. Many crystals such as these have been grown in our laboratory, using the high-pressure Melbourn puller, with defect densities between 10^4 and 10^5 cm^{-2}. Much of the high dislocation density arises from strain and is experienced in a region close to the outer periphery of the boule.

Our initial work in improving crystalline quality of large-diameter ingots follows earlier efforts in dislocation-free growth where many initial dislocations are "grown out" by necking experiments and controlled shoulder growth [16]–[19]. Results of carefully grown "shoulders" have demonstrated substantial reduction in dislocation density in GaP [19] and more recently in GaAs Czochralski growth [10]. Recent results in our laboratory have shown dislocation density reductions to 1.5×10^4 cm^{-2} in ⟨111⟩ crystal growth using the coracle method. Analysis of large-diameter slices with X-ray topography have shown evidence of strain-induced defects. Microstructural analysis of 3-in-diameter slices with Transmission Electron Microscopy have shown very clean sections free of dislocations and microprecipitates.

Although the crystalline perfection of LEC GaAs has not presently been developed to that of Bridgman crystals, many advantages of the high purity and excellent electrical uniformity make it preferrable to boat-grown GaAs. Substantial efforts in device fabrication will be necessary to make a clear definition concerning the role of dislocations and their related electrical activity in LEC crystals. However, initial device results from digital LSI circuit fabrication (5 × 5 multiplier) have shown reasonably high yields of complex circuits using LEC GaAs. Dislocation densities of 3×10^4 cm^{-2} were involved, and circuits with 260 gates (1300 active devices) have been fabricated and successfully operated with 190-ps speed and 134-fJ speed power product! In these devices, the probability of having a dislocation under a gate is >90 percent!

V. Conclusion

High-purity bulk semi-insulating GaAs crystals have been grown for use in direct ion implantation. The crystals show excellent qualification yields for use in monolithic microwave device fabrication. High qualification yields demonstrated by this material have been attributed to the low Si background doping obtained by the use of a new *in situ* synthesis technique. Undoped ingots grown from both SiO_2 and PBN crucibles have shown excellent thermal stability and abrupt carrier profiles for selenium-ion implantation. High electron mobilities in the range of 4500–4800 cm^2/V·s have been attained for selenium-ion-implanted layers with 1.7×10^{17} cm^{-3} peak doping. Feasibility of using LEC crystals for fabrication of high-density, high-speed logic devices has been demonstrated showing that the higher dislocation densities found in state-of-the-art LEC crystals (3×10^4 cm^{-2}) are not electrically active and do not adversly affect performance of a majority-carrier device.

Acknowledgment

The authors wish to thank Dr. P. Asbeck and Dr. R. Zucca for helpful discussion of this work, and R. Ware, R. Waldock of Metals Research, and J. Dreon for their assistance in crystal growth and characterization.

REFERENCES

[1] S. I. Long, F. S. Lee, R. Zucca, B. M. Welch, and R. C. Eden, "MSI high-speed low-power GaAs integrated circuits using Schottky diode FET logic," *IEEE Trans. Microwave Theory Tech.*, vol. MTT-28, no. 5, pp. 466-472, May 1980.
[2] D. R. Decker, A. K. Gupta, W. Peterson, and D. R. Ch'en, "A monolithic GaAs I.F. amplifier for integrated receiver applications," in *Tech. Dig. 1980 MTT-S Int. Microwaves Symp.*
[3] I. Deyhimy, J. S. Harris, R. C. Eden, and R. J. Anderson, "Reduced geometry GaAs CCD for high speed signal processing," in *Proc. 1979 Int. Conf. Solid State Devices* (Tokyo, 1979), *Japan. J. Appl. Phys.*, vol. 19, suppl. 19-1, pp. 169-272, 1980.
[4] S. J. Bass and P. E. Oliver, *J. Cryst. Growth*, vol. 3, p. 286, 1960.
[5] J. B. Mullin, R. J. Heritage, C. H. Holiday, and B. W. Stranghan, *J. Crystal Growth*, vol. 34, p. 281, 1968.
[6] M. E. Weiner, D. T. Lassota, and B. Schwartz, *J. Electrochem. Soc.*, vol. 118, p. 301, 1971.
[7] E. M. Swiggard and R. L. Henry, in *Gallium Arsenide and Related Compounds* (Inst. Phys. Conf. Ser., no. 33b), p. 28, 1976.
[8] T. R. Aucoin, R. L. Ross, M. J. Wade, R. D. Savage, *Solid-State Technol.*, vol. 22, p. 59, 1979.
[9] D. Rumsby, presented at the IEEE Workshop on Compound Semiconductors for Microwave Materials and Devices, Atlanta, GA, Feb. 1979.
[10] R. N. Thomas, H. M. Hobgood, D. L. Barret, and G. W. Eldridge, in *Proc. Semi-Insulating III-V Materials Conf.* (Nottingham, England, Apr. 14-16, 1980).
[11] P. Asbeck, J. Tandon, E. Babcock, B. Welch, C. Evans, and V. Deline, "Effects of Cr redistribution on device characteristics in ion-implanted GaAs IC's fabricated with semi-insulating GaAs" (abstract), *IEEE Trans. Electron Devices*, vol. ED-26, no. 11, p. 1853, Nov. 1979.
[12] B. Tuck, G. A. Adegoboyega, P. R. Jay, and M. J. Cardwell, in *Gallium Arsenide and Related Compounds* (Inst. Phys. Conf. Ser., no. 45), p. 114, 1978.
[13] J. B. McNeely, private communication.
[14] P. Petroff and R. L. Hartman, *Appl. Phys. Lett.*, vol. 23, p. 469, 1973.
[15] W. A. Brantley and D. A. Harrison, in *Proc. IEEE Reliability Physics Symp.*, p. 267, Apr. 1973.
[16] R. J. Roedel, A. R. Von Neida, R. Caruso, and L. R. Dawson, *J. Electrochem. Soc.*, vol. 126, p. 637, 1979.
[17] A. Steinemann and U. Zimmerli, in *Proc. Int. Cryst. Growth Conf.*, p. 81 (Boston, MA, 1966).
[18] B. C. Grabmaier and J. G. Grabmaier, *J. Crystal. Growth*, vol. 13/14, p. 653, 1972.
[19] P. J. Roksnoer, J. M. P. L. Huijbregts, W. M. Van De Wijgert, and A. J. R. Delsock, *J. Cryst. Growth*, vol. 40, p. 6, 1977.
[20] R. D. Fairman, F. J. Morin, and J. R. Oliver, in *GaAs and Related Compounds* (Inst. of Phys. Ser., no. 45), 1979.

GROWTH AND CHARACTERIZATION OF DISLOCATION-FREE GaAs MIXED CRYSTALS FOR IC SUBSTRATES

Kohji Tada, Shigeo Murai, Shin-ichi Akai, Takashi Suzuki

Sumitomo Electric Industries, Ltd.
1-3, Shimaya 1-chome, Konohana-ku, Osaka, 554 Japan

ABSTRACT

Semi-insulating GaAs single crystals mixed with In or InAs have been successfully grown dislocation-free (DF) by improving thermal and crystal growth conditions. The suitability of DF-GaAs for IC substrate was demonstrated by direct evaluation of the uniformity of FET arrays by ion implantation (I^2).
The average value of the threshold voltage of 625 FETs (gate length of 1μm and width of 5μm) of matrix arrays, each separated by 200μm, in 5mm squre region was -0.022V with standard deviation of 13mV which is much smaller than that in the conventional LEC crystals (50 ∼ 100mV).

INTRODUCTION

Recently, it has been found that, among the various defects, crystal defect of dislocations have the most detrimental effect upon the uniformity of threshold voltages (Vth) of FETs made by the I^2 process (1), (2). GaAs single crystals mixed with indium (In) or indium arsenide (InAs) have been successfully studied to reduce dislocations (3), (4). The problems which remain to be examined are the suitability of such doped/mixed crystals as a substrate for GaAs IC and the improvement of crystal technology to make the production possible.
In this paper we present our recent results on crystal growth of GaAs mixed with InAs which contains a DF area of more than 40% in a wafer of diameter 65mm and on a promising evaluation of the crystal's suitability as a substrate for GaAs IC.

CRYSTAL GROWTH

Adding or mixing large amount of In or InAs to GaAs heavily (0.1mol% ∼ 0.4mol%) is considered to be effective in suppressing the generation of microdefects which are responsible for the formation of dislocations.
Reduction of the temperature gradient (∇T) in the solid/liquid interface is necessary to prevent the generation and multiplication of dislocations resulting from localized thermal stresses.
Constitutional supercooling and segregation of solute (considered to be $Ga_{1-x}In_xAs$) the other hand, tend to occur during crystal growth if the temperature gradient is too small. Hence, it is quite important to optimize ∇T and growth rate (v) in order to have crystals with low dislocation density and without any segregation.
We have successfully grown DF single crystals of maximum diameter 65mm with <001> axis and a weight of 1,180g using a growth rate of 6mm/H. The value of $\partial T/\partial Z$ was estimated to be about 60°C/cm.
We achieved the most DF area in the part crystal where the fraction of solidified (g) was in the range 0.13 ∼ 0.60. A (001) wafer was cut and chemically etched by

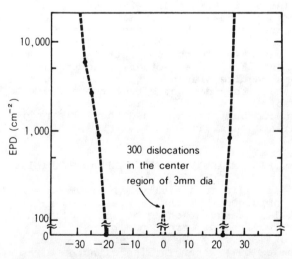

Fig. 1 Profile of EPD of DF-GaAs Wafer

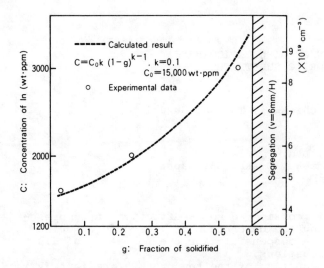

Fig. 2 Indium concentration in DF-GaAs ingot

KOH solution and the etch pit density (EPD) was measured. A profile of the EPD is shown in Fig. 1 for the wafer at g=0.4. We found a DF area of about 40% near the central region of the wafer, outside the central core having a diameter of about 3mm, which contains about 300 dislocations. The concentration of In calculated with a distribution coefficient (k) of 0.1 is shown in Fig. 2. The experimentally obtained values for the concentration of In nearly coincide with the calculated results.

For value of g greater than 0.6, segregations were found to occur. An x-ray topographic study showed relatively weak striations at intervals of a few hundred microns. These were perphaps due to local fluctuations of the In concentration. It was confirmed by chemical analysis that the variation in the concentration was well within ±2%. The effect of these striations requires further study, but in our crystal, they did not effect the uniformity of the intensity variation of CL (Cathode Luminescence) and the FET's properties, significantly.

SPECIFIC RESISTIVITY AND THERMAL STABILITY

The specific resistivity (ρ) of DF GaAs was obtained from the V-I relation in small rectangular chips of 1mm x 0.3mm x 8mm cut from different positions in the ingot (Fig. 3). We selected 8 ∿ 9 pieces from a (001) wafer in line with the direction <1$\bar{1}$0>, with the 1mm edges parallel to this axis and with a separation of 2mm between the chips. The specific resistivities of the chips were averaged to attain that of the wafer. The average ($\bar{\rho}$) and standard deviation are shown in the figure. The specific resistivity is $8 \times 10^7 \sim 10 \times 10^7 \Omega$cm and does not vary in the region of g=0.24 ∿ 0.56, which corresponds to an In concentration of $6 \times 10^{19} \sim 8 \times 10^{19}$ cm^{-3}.

The residual impurities of Si and S analysed by spark source mass spectroscopy were 4.3×10^{14} and 7.0×10^{14} cm^{-3} respectively in the front part of the ingot. [C]$_{AS}$, analysed by FTIR, was 6.0×10^{15} cm^{-3}, and [EL2] analysed by IR absorption, was 1.8×10^{16} cm^{-3}. We are not certain that the compensating mechanism at work in undoped GaAs crystals still works for the DF GaAs crystals but it seems [EL2] could compensate residual acceptor levels in the DF GaAs crystals.

Annealing of the crystal wafer of semi-insulating DF GaAs was done at the temperature of 800 ∿ 850°C for 20 minutes with Si$_3$N$_4$ cap (thickness of 1,160 Å) by P-CVD. After annealing, the Si$_3$N$_4$ cap was removed using HF at room temperature for 5 minutes. We measured the surface leakage current (i_L) between two probes in contact with the wafer separated by a distance of 1mm into which 1,000V was applied. At 800°C, i_L (before)=4.5µA and i_L (after) =4.2µA. Even at 850°C, i_L did not change within experimental error; i_L (before) = 4.9µA, i_L (after) = 3.9µA. We can therefore say that the annealing process does not induce significant thermal conversion.

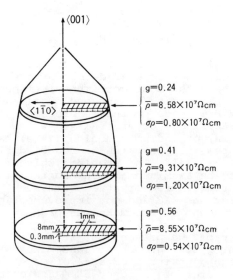

Fig. 3 Specific resistivity (ρ) of DF-GaAs crystal

HALL MOBILITY OF I^2 LAYER

An I^2 of $^{28}Si^+$ of 180kV and dose(ϕ) 1.5 x 10^{12} ~ 6 x 10^{12} cm^{-2} was examined to evaluate the electron mobility in the I^2 layer of DF GaAs. A (001) wafer from the front of the ingot was polished and the $^{28}Si^+$ implanted onto it. Chips of 2.6mm x 2.6mm were cut from the wafer and measured by van der Pauw method. In these experiments annealing was done by the following method; The implanted surface was placed face to face on a different GaAs wafer and together the wafers were placed in a quartz chamber filled with flowing N_2 gas and heated to 820°C for 20 min.
The sheet carrier density (N_s) is proportional to the dose, as expressed by $N_s = \eta\phi - N_{Res}$. (η: activation efficiency). The experimental value of η is about 66% with N_{Res}=5.6 x 10^{11}cm^{-2}, which is slightly lower than the 72 ~ 76% of undoped GaAs with N_{Res}=4.7 x 10^{11} ~ 7.0 x 10^{11}cm^{-2}. The surface Hall mobility decreases as ϕ increases as shown in Fig. 4. At ϕ = 1.5 x 10^{12}cm^{-2} or N_s = 5.5 x 10^{11}cm^{-2}, μ_H was 4854 cm^2/V·S, which is the averaged value over 10 points on the wafer. The standard deviation was 96cm^2/V·S, which is slightly higher than the value of 4680 cm^2/V·S for undoped LEC GaAs.

FET'S UNIFORMITY

The threshold voltages of FETs fabricated on DF-GaAs wafer were evaluated. A channel layer was formed by implantation of $^{28}Si^+$ with 180kV and 50kV for D-FET and E-FET respectively (Fig. 5). Each FET has a gate length of 1μm and a width of 5μm. The distance between the source and gate and between the gate and drain is 2μm, with the gate direction parallel to <110> crystal axis. The FETs are arranged into a matrix with separation of 200μm. Fig. 6 (a) shows mean threshold voltage ($\bar{V}$th) and standard deviation (σVth) for each cell, all of which contain 100 D-FETs in a square region of 2mm x 2mm. The total number of cells in the circular region of the wafer is 379. The data for conventional undoped GaAs is given in Fig. 6 (b), showing much more scattering than that of the DF-GaAs.
The data for the uniformity of E-FETs is shown in Fig. 7. The $\bar{V}$th of 625 E-FETs in a 5mm square region was -0.022V with σVth = 13mV, which is much smaller than that in the conventional LEC GaAs (50 ~ 100mV). The best value of σVth was 9.9mV with $\bar{V}$th = -0.199V, in a 2mm square region of 100 FETs.

CONCLUSIONS

Semi-insulating GaAs single crystals mixed with In or InAs were successfully grown with a DF area of about 40% in a wafer with a diameter of 65mm.
The suitability of this DF-GaAs single crystal as a substrate was demonstrated by direct evaluation of the uniformity of FET arrays made by ion implantation.
Localized nonuniformities of electrical properties of FETs, which have hitherto been influenced by dislocations were eliminated by using DF-GaAs as a substrate and the striking results of the uniformity of the FETs was obtained.

ACKNOWLEDGEMENT

The authors wish to express their thanks for the contribution of all members of the GaAs project at Sumitomo Electric Industries, Ltd. and to Dr. T. Nakahara for his continuous encouragement.

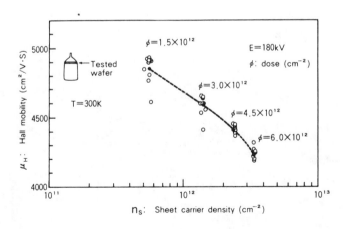

Fig. 4 Hall mobility and sheet carrier density

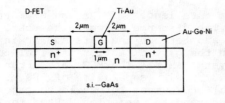
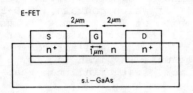

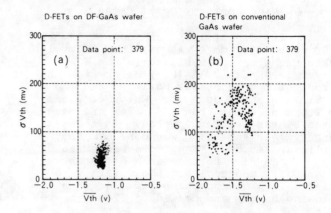

Fig. 6 $\overline{V}_{th}$ and σV_{th} for each cell which contain 100 FETs in 2mm x 2mm region D-FETs on DF-GaAs (a) and conventional LEC GaAs (b)

Fig. 5 Cross sectional structures of D-and E-FET and I^2 parameter

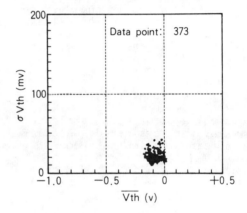

Fig. 7 E-FETs on DF-GaAs

REFERENCES

(1) S. Miyazawa, Y. Ishii, S. Ishida, and Y. Nanishi, Appl. Phys. Lett., 43, 853, 1983

(2) T. Takebe, M. Shimazu, A. Kawasaki, T. Kotani, R. Nakai, K. Kikuchi, S. Murai, K. Tada, S. Akai, and T. Suzuki, to be published in the proceeding of the 3rd International Conference on Semi-Insulating III-V Materials, Warm Spring, Oregon, USA, 1984

(3) M.G. Mil'vidsky, V.B. Osvensky, and S.S. Shifrin, J. Crystal Growth., 52, 396, 1981

(4) G. Jacob, M. Duseaux, J.P. Farges, M.M.B. Van Den Boom, and P.J. Roksnor, J. Crystal Growth., 61, 417, 1983

THE MANUFACTURABILITY OF GaAs INTEGRATED CIRCUITS*

Bryant M. Welch[†] and Yie-der Shen

Rockwell International/Microelectronics Research and Development Center
Thousand Oaks, CA

ABSTRACT

The establishment of a high yield GaAs integrated circuit manufacturing technology closely paralleling silicon processing and testing methodology has finally, after many years, become a reality. Technological barriers which had previously limited GaAs integrated circuits manufacturing activities have been largely circumvented through the availability of high quality GaAs 3 inch wafers, improved ion implantation techniques, direct step on wafer photolithography systems, and advanced dry replication processes. These key technological advances have been instrumental in the successful transition of an one inch wafer research and development activity into a 3 inch wafer pilot line capability at Rockwell. Device performance and statistical variations observed on 3 inch GaAs wafers are similar to those observed on one inch GaAs wafers. The successful demonstration of 3 inch GaAs integrated circuit processing provides a sound base from which a high yield GaAs manufacturing line can be established.

INTRODUCTION

Until very recently, GaAs technology was largely characterized by many diverse device process approaches, and unique process steps and custom equipment requirements having little or nothing in common with the world's semiconductor community. In the past, GaAs suffered with images of "black magic" process recipes and "dangerous" small black crumbly material, etc. Over the past several years this image has been rapidly disappearing as evidenced by the advanced state of development of microwave, analog and digital GaAs integrated circuits. In addition, a high level of interest and believability in GaAs technology has been demonstrated through the proliferation of papers on GaAs, new laboratories starting GaAs R&D activities and the establishment of new commercial GaAs ventures.

Observing the most recent progress and increased level of interest in GaAs leads us to question why GaAs technology, has, after so many years, finally matured as a semiconductor technology. While GaAs research and development over the years has played a strong role in improving the quality of the material and processes; this in itself is not the key reason why the technology has matured so fast. Two other factors stand out.

First, a major trend has occurred over the past few years where a commonality in planar "Si-like" fabrication approaches has dominated the GaAs community. This trend has provided a large "critical-mass" of technologists all pulling for the first time in the same direction. Second, and more important, the paths of Si and GaAs processing equipment and techniques have finally crossed due to the common requirements of dense, micron resolved geometries. The Japanese VLSI and U.S. VHSIC programs have provided a major thrust in the development of practical high yield processes and equipment compatible with 1 μm design rules. GaAs technology needed these tools ten years ago, but only recently have they become available due to the tremendous leverage of the Si semiconductor industry.

Any determination of whether GaAs ICs are ready for production cannot be generalized for all possible GaAs IC technology approaches. In this paper we will consider only the most mature GaAs technology based on depletion-mode MESFETs. While excellent progress is currently being made in more difficult technologies such as enhancement-mode E-MESFET, E-JFET, HEMT (TEGFET), etc., or advanced higher performance device structures using self-aligned gates (SAG), submicron gates, heterojunctions, etc., we do not consider these technologies production ready. The scope of depletion-mode MESFET technology includes monolithic microwave SSI/MSI and analog and digital MSI/LSI applications. The technology is characterized by the use of ion implanted material, planar device structures, 1 μm gate lengths, Schottky barrier level shifting and/or switching diodes, and MOM capacitors, as required. Buffered FET Logic (BFL), Schottky Diode FET Logic (SDFL), or similar circuit approaches in the medium to low power range are consistent with the following assessment.

GaAs MANUFACTURABILITY ASSESSMENT

A basic premise made in this analysis is that if GaAs ICs can be fabricated with reasonable yields using Si IC circuit manufacturing equipment methodologies, it should be considered manufacturable. An analysis of the key aspects of the technology will provide insight for this assessment.

A number of technical advances in GaAs semi-insulating material, ion implantation, and

[†]Presently employed: GigaBit Logic, Culver City, CA.

*This work supported, in part, by the Defense Advanced Res. Projects Agency of the Dept. of Defense & monitored by the Air Force Off. of Scien. Res. under Contract No. F49620-80-C-0087.

lithography techniques have provided a sound basis for the establishment of a practical GaAs IC manufacturing technology. Table I serves as a guide to the key technology areas that need to be assessed in order to determine production readiness. While only a few years ago quality GaAs material was difficult to obtain, today large LEC grown ingots yielding 3 inch diameter GaAs wafers of excellent semi-insulating quality are available. Currently, there are only three vendors supplying 3 inch GaAs wafers, however, with the strong trend toward using 3 inch compatible production processing equipment, an improved supply of material and a reduction in cost per wafer is anticipated in the near future. Figure 1 shows a 3 inch GaAs wafer fabricated at Rockwell.

Ion implantation is now universally accepted as the active layer fabrication approach for GaAs ICs. This has come about principally due to the improved quality of undoped Liquid Encapsulated Czochralski (LEC) semi-insulating material versus the Chrome Doped Horizontal Bridgman (HB) material used in the early development of GaAs ICs. Problems associated with active layer profile control due to Cr redistribution and non-reproducible caps have been eliminated through the use of improved substrates and dielectric deposition techniques. Today both Si_3N_4 and SiO_2 caps; deposited in commercial sputtering, plasma enhanced CVD or LPCVD systems, are used successfully. The majority of the GaAs laboratories routinely report FET channel active layer uniformities (≤ 2 inch wafers) in the 5-8% range for 1-2 volt threshold devices. Similar reports on 3 inch wafers are discussed later in this paper. This degree of material uniformity and control is quite adequate for GaAs depletion-mode logic approaches.

The main areas of concern are the availability of 3 inch wafers and wafer breakage. GaAs wafers do and always will break (cleave) easier than Si wafers. Optimized crystal growth techniques (less stress) and improved wafer preparation (edge rounding) are expected to minimize breakage. However, the main improvement in reducing breakage will come through improving and minimizing wafer process handling. The best solution is offered through the use of cassette to cassette process equipment. Experience with this approach at Rockwell using the CENSOR direct-step-on-wafer (DSW) system has proved to be excellent.

Another major barrier that had previously limited GaAs manufacturability was the very demanding lithography requirements (~ 1 μm gate lengths, etc.) of the technology. The development of 10X DSW projection photolithography systems have dramatically changed this situation. At Rockwell we are currently using a CENSOR DSW system for processing on GaAs 3 inch wafers and are experiencing outstanding results. Figure 2 is the photograph of a CENSOR SRA-100 DSW system. The automatic handling

ISSUE	COMMENTS	SATISFACTORY		
		YES	?	NO
MATERIAL (3" WAFERS)	HIGH QUALITY LEC ($\sim 10^8$ Ω·cm)	X		
AVAILABILITY	2" ($\sim$6-8 VENDORS), 3" (3 VENDORS)			X
COST	$\sim$$250/WAFER, <$150/WAFER PROJECTED	X		
BREAKAGE	BREAKAGE MINIMIZED WITH AUTOMATION		X	
ACTIVE LAYER	N-TYPE (Se, Si), P-TYPE (Be, Mg)	X		
IMPLANTATION	WELL CONTROLLED, 5-8% UNIFORMITY	X		
ENCAPSULATION	WELL DEVELOPED Si_3N_4 AND SiO_2 PROCESS	X		
PHOTOLITHOGRAPHY	AUTOMATED 10X DSW	X		
RESOLUTION	WELL CONTROLLED ≤ 1.0 μm	X		
OVERLAY	MEAN + 3σ ≤ 0.25 μm	X		
LITHOGRAPHY	FULL DRY PROCESS AVAILABLE	X		
PLASMA ETCHING	RAPIDLY MATURING TECHNOLOGY	X		
ENHANCED LIFT-OFF	HIGH YIELD PROCESS	X		
RELIABILITY	PROVEN ON DISCRETE DEVICES		X	
DC TESTING	AUTOMATED WELL DEVELOPED	X		
HIGH SPEED TESTING	NEEDS MUCH DEVELOPMENT			X
PACKAGING	NEEDS DEVELOPMENT			X

TABLE I - GaAs (1 μm) Manufacturing Status

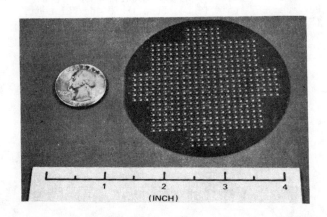

Figure 1 - Photograph of a processed 3 inch GaAs wafer

Figure 2 - Photograph of a CENSOR SRA-100 DSW System

of the GaAs wafers from cassette to cassette works well without any occurrence of breakage. This system has a 1 μm resolution specification which is extremely easy to attain on GaAs wafers. Data obtained over a thirteen wafer (416 fields) data base exhibited ≤1 μm resolved patterns on all fields. The overall distribution of resolution patterns was: 20% ≤1 μm; 51% ≤0.875 μm; and 29% ≤0.75 μm. Of equal importance and difficulty are the layer to layer alignment requirements. For a 1 μm GaAs technology with 0.5 μm design rules, an overlay accuracy of ± 0.25 μm is required. Statistical data for the CENSOR DSW alignment accuracy is presented in Figure 3. This data was obtained from ten randomly selected wafers out of twenty-five wafers sequentially processed at the rate of fifty-three wafers/hour (3 inch wafers; 32 fields/wafer). Verniers were measured in three positions x_{top}, x_{bottom} and y_{center} for each field (960 fields). The statistical results indicate a mean + 3σ alignment accuracy of less than 0.25 μm. Similar results were obtained on processed GaAs wafers demonstrating that the CENSOR DSW is a 1 μm compatible manufacturing tool.

Dry processing equipment, along with innovative processing techniques, are now providing high yield micron and submicron device feature replication. Previously many of us only had custom "in-house" designed and built R&D equipment to use in the development of GaAs ICs. Now many new process options exist due to the rapid emergence and availability of new plasma etching (PE), reactive ion etching (RIE), reactive ion beam etching (RIBE), and ion milling (IM) equipment. Processing on GaAs relies heavily on both dry processing and enhanced lift-off techniques for fineline pattern replication.[1] Traditionally, lift-off replication methods usually exhibited poor yield and were considered unreliable. The development of very high yield enhanced lift-off techniques has completely overcome the earlier limitations associated with lift-off. While lift-off techniques first manifested themselves in applications like 1 μm GaAs, today it is not uncommon for state-of-the-art 1 μm Si processes to use similar strategies.[2]

GaAs Schottky barrier, ohmic contact and interconnect technologies are already well developed at the MSI level. Metalizations are available on standard commercial semiconductor electron beam evaporation and magnetron sputtering systems. The reliability of the preferred Au based metal systems are well documented on discrete (e.g., low noise and power FETs) GaAs commercial devices. However, further work is needed to fully verify GaAs IC reliability, particularly at the MSI/LSI levels during the early stages of manufacturing.

Low frequency parametric testing on process monitor chips is routinely used on GaAs ICs at a number of development laboratories. Process monitoring at Rockwell has been instrumental in the control and reproducibility obtained in the process technology. These testing methods are all automated and are already consistent with manufacturing disciplines.[3] High speed testing, on the other hand, poses a much more formidable problem for GaAs ICs. Automated high speed testers in the ≥1 GHz range are not commercially available. The custom GaAs high speed test setups often required is a problem area which must be addressed for practical GaAs production. On the positive side, the VHSIC thrust is starting to place increased pressure on the need for automated high speed IC testing equipment.

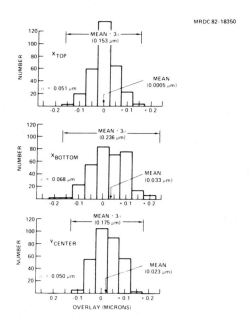

Figure 3 - Histograms of alignment data for a CENSOR DSW.

GaAs THREE INCH WAFER PROCESSING

Up to this point our discussion on the manufacturability of GaAs ICs has been based on the assessment of the various technology areas involved. Successful integration of all of the individual pieces of the technology is required in order to demonstrate GaAs IC production readiness. Perhaps the strongest evidence to date in support of GaAs manufacturability is the recent results obtained on 3 inch GaAs wafers processed at Rockwell.

The following describes the device results obtained on processing GaAs ICs on 3 inch wafers. The basic planar ion implanted processing steps have been previously described in detail.[1] Highlights of the latest fabrication methods incorporated into the process during the transition to 3 inch processing includes: the use of 3 inch diameter undoped LEC GaAs wafers; selenium (n⁻) and silicon (n⁺) implantation through a Si_3N_4 cap; the optional use of proton damage for isolation; use of a LPCVD SiO_2 overcap layer; use of a CENSOR automatic cassette to cassette DSW system on all mask levels; layer to layer registration to a single alignment mark layer; use of a dry process with no wet pre-ohmic contact etch or gate recess etch; and planetary fixturing on all ohmic, Schottky and interconnect metal depositions.

Initial 3 inch processing results closely resembled one inch processing with the exception of two key improvements. The quality and consistency of the lithography was clearly superior with the automatic DSW system and the larger wafers appear much cleaner than the small (edge dominated) one inch wafers. The material uniformity of the LEC wafers is excellent as evidenced by the histogram of depletion voltages in Figure 4, obtained from C-V profiles equally distributed across a 3 inch wafer. A standard deviation of 5.9% was obtained for these measurements. Devices fabricated on these layers also exhibited uniform characteristics.

PROCESS/DEVICE	PARAMETER	MEAN	REPRODU-CIBILITY	AVE UNIFORMITY (BEST WAFER)
IMPLANT (320 keV Se 2.5E12) (320 keV Si 4E12)	n^- (DEPLETION VOLTAGE) $n^- + n^+$ SHEET RESISTANCE (ρ_s)	1.167V 335 $\Omega/\square$	2.8% 1.8%	±6.66% (3.6%) ±2.2% (1.6%)
OHMIC CONTACT (AuGe/Ni)	CONTACT RESISTANCE (R_c) SPECIFIC CONTACT RESISTANCE (r_c)	0.137 $\Omega \cdot$ mm 6.03E-7 $\Omega \cdot cm^2$	11%	±24% (14.9%)
SCHOTTKY BARRIER (Ti/Pt/Au)	IDEALITY FACTOR (n_f)	1.071	4.6%	±14% (9.2%)
D-MESFET (1 μm Lg x 50 μm W)	PINCH-OFF VOLTAGE (V_p) SAT CURRENT (I_{dss})	1.056V 3.852 ma	7.5% 6.6%	±(14.78% (12%) ±14.35% (11.4%)
SCHOTTKY DIODE (1 μm x 1 μm)	SERIES RESISTANCE (R_s) FORWARD THRES VOLT (V_{th})	648 Ω 0.75V	4.8% 3.2% T	±15% (10.1%) ±6% (4.2%)

TABLE II - Summary of GaAs 3 inch wafer processing.

The data in Table II speaks for itself, however, the most impressive evidence of technology maturity is the statistical reproducibility exhibited between the five wafers. These five wafers were taken from two different wafer lots; three wafers from one lot and two wafers from another lot. Even though the wafers were not all processed simultaneously, minimal variations were observed between wafers. The n^- and n^+ implantation variations were <3%; the ohmic contact and Schottky barrier variations were 11% and 4.6%, respectively, and the key FET and diode parameters were under 7.5%. The data shown in this table is solid evidence in support of the present 3 inch fabrication capability and maturity of the GaAs D-MESFET technology.

CONCLUSION

The marketability and application of GaAs ICs was not part of the assessment presented in this paper. In terms of technology status, any reasonable assessment would conclude that a GaAs technology based on depletion-mode MESFETs has matured to the point of production readiness. Well documented technological progress in GaAs ICs, along with the availability of GaAs compatible Si process equipment and the recent demonstration of practical 3 inch GaAs wafer processing capabilities, solidly supports the manufacturability of GaAs MSI/LSI. These events are leading the way towards the current establishment of GaAs manufacturing activities.

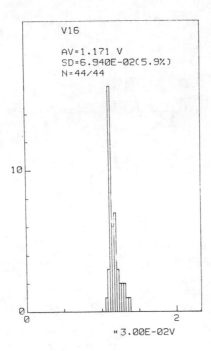

Figure 4 - Histogram of C-V profile depletion voltages for a 3 inch GaAs wafer.

Five 3 inch wafers were thoroughly d.c. characterized; a summary of these results is presented in Table II. Represented in this chart are the mean values of all of the important process and device parameters, including the variation observed between wafers (reproducibility) and the average uniformity (1σ values) and best wafer uniformity values. Beside the material and device uniformities, the individual ohmic contact and Schottky barrier processes also exhibit excellent figures of merit and control. For instance, the average ohmic contact resistance (r_c) and Schottky barrier ideality factor (n_f) measured was 0.137$\Omega \cdot$mm and n_f = 1.071, respectively. The summarized results in Table II indicate that no penalty was paid in device uniformities and performance from scaling up from 1 inch to 3 inch wafer processing.

1) B.M. Welch, Y.D. Shen, R. Zucca, R.C. Eden, & S.I. Long, "LSI Processing Technology for Planar GaAs ICs", IEEE Trans. on Elec. Dev. ED27,6,116,'80.

2) W.R. Hunter, L. Ephrath, W. Grobman, C.M. Osburn, B.L. Crowder, A. Cramer, & H. Luhn, "1 μm MOSFET VLSI Technology: A Single-Level Polysilicon Technology Using EBL", IEEE J. of Sol. St. Circuits, SC-14, 2, 275, 1979.

3) R. Zucca, B.M. Welch, C.P. Lee, R.C. Eden and S.I. Long, "Process Evaluation Test Structures For Planar GaAs ICs", IEEE Trans. on Elec. Dev. ED-27, 12, 2292, 1980.

A Manufacturing Process for Analog and Digital Gallium Arsenide Integrated Circuits

RORY L. VAN TUYL, SENIOR MEMBER, IEEE, VIRENDER KUMAR, DONALD C. D' AVANZO, MEMBER, IEEE, THOMAS W. TAYLOR, VAL E. PETERSON, MEMBER, IEEE, DERRY P. HORNBUCKLE, MEMBER, IEEE, ROBERT A. FISHER, and DONALD B. ESTREICH, MEMBER, IEEE

Abstract —A process for manufacturing small-to-medium scale GaAs integrated circuits is described. Integrated FET's, diodes, resistors, thin-film capacitors, and inductors are used for monolithic integration of digital and analog circuits. Direct implantation of Si into $>10^5$ $\Omega \cdot$cm resistivity substrates produces *n*-layers with ± 10-percent sheet resistance variation. A planar fabrication process featuring retained anneal cap (SiO_2), proton isolation, recessed Mo–Au gates, silicon nitride passivation, and a dual-level metal system with polyimide intermetal dielectric is described. Automated on-wafer testing at frequencies up to 4 GHz is introduced, and a calculator-controlled frequency domain test system described. Circuit yields for six different circuit designs are reported, and process defect densities are inferred.

I. INTRODUCTION

THE EARLIEST efforts at monolithic integration of gigahertz-bandwidth circuits in gallium arsenide [1] addressed the performance advantages of the technology, not the manufacturability. The subsequent application of ion implantation [2] and planar processing [3] did much to improve the GaAs integrated circuit (IC) fabrication technology. Now, as this technology moves into a more mature, applications-oriented phase, much attention is being paid to manufacturing techniques, both in fabrication and testing. This paper describes a manufacturing process for small-to-medium scale GaAs IC's of the type described in references [4]–[7] (amplifiers, counters, and RF subsystems).

The wafer fabrication portion of the process is presented, along with a description of components and process steps. Rationale is given for the incorporation of new or nonstandard process techniques. Automated on-wafer circuit testing is discussed, and a specially designed system for performing these tests at microwave frequencies is described. Finally, yield data for representative circuits fabricated and tested with this process are presented.

II. COMPONENTS

The manufacturing system described here is designed to fabricate small-to-medium scale gigahertz bandwidth analog and digital GaAs IC's. The principal attributes of this high-speed process are a low-capacitance substrate and interconnection scheme (dual-layer metal), and a variety of active, as well as passive, components. The components

Manuscript recieved January 18, 1982.
The authors are with Hewlett-Packard Company, Santa Rosa Technology Center, Santa Rosa, CA 95404.

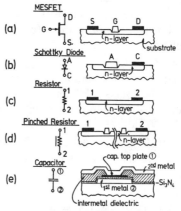

Fig. 1. Components of the GaAs IC process.

TABLE I
TYPICAL COMPONENT PARAMETERS

COMPONENT	PARAMETER	VALUE
FET	I_{DSS} (V_{DS}=3V)	200 mA/mm
	V_P (@5% I_{DSS})	-2.1 V
	g_m (V_{GS}=0V, V_{DS}=3V)	135 mS/mm
	g_m	100 mS/mm
	C_{gs} (V_{GS}=-1V, V_{DS}=3V)	1 pF/mm
	f_T	15 GHz
SCHOTTKY DIODE	$C_{JUNCTION}$	2 fF/μm^2
	J_{SAT}	1.6 x 10^{-14} A/μm^2
RESISTORS:		
STANDARD	R_{SHEET}	320 $\Omega/\square$
"PINCH"	R_{SHEET}	1400 $\Omega/\square$
THIN-FILM CAPACITORS	C	60,000 pF/cm^2
	$V_{BREAKDOWN}$	80 V
	$I_{LEAKAGE}$ (@15 V)	1 pA
SPIRAL INDUCTORS	L	1 nH - 20 nH
	Q (10 nH@1 GHz)	1

illustrated in Fig. 1 have their key electrical parameters listed in Table I.

Pinchoff voltage of the 1-μm gate-length FET is adjusted to -2.1 V by channel-thickness etching; a standard deviation of 0.31 V is maintained on a typical wafer. This pinchoff voltage choice represents a compromise between the lower pinchoff preferred for low-power digital circuits and the higher pinchoff voltage which can give lower distortion and higher output power in linear circuits. Since analog and digital circuits are often integrated together on a single chip, the use of a single-pinchoff-voltage FET tends to insure compatibility of signal amplitude between the two circuit types.

Schottky diodes, fabricated concurrently with the FET's, are used for RF detectors, switches, and level-shift elements [4], [5].

Resistors, made simply with the implanted *n*-layer, are used as loads and for bias adjustment, chiefly in linear circuits [7]. Where very high values of bias resistance are required, the "Pinch" resistor of Fig. 1(d) is useful, since it has over four times the average sheet resistance of the implanted resistor.

A particularly important component, and one which is not inherent to the GaAs FET process, is the thin-film metal-insulator-metal capacitor. The high capacitance per unit area (60 000 pF/cm^2) of this component makes capacitors of 1 pF to 20 pF practical, with parasitic capacitance-to-ground less than 0.1 percent. Such capacitors are especially useful for coupling and bypass applications within the chip [7].

The final component is the dual metal system itself which offers low resistance interconnects (0.37 $\Omega/\square$) and low capacitance crossovers (typ. < 4 fF). Spiral inductors, though not process components in themselves, can be designed with this dual-metal system. The low observed Q values limit inductors' usefulness to peaking applications [5] in the ≤4-GHz frequency range typical of circuits made with this process.

III. Wafer Fabrication

The first step in GaAs IC fabrication is the formation of an *n*-type layer by ion implantation. For this step to be successful, and for subsequent device electrical characteristics to be satisfactory, suitable GaAs starting material must be selected. The two most basic requirements for GaAs material electrical characteristics are:

1) that bulk resistivity be sufficiently high, even after ion implant anneal, to assure acceptably low leakage currents between circuit elements;
2) that residual impurities be low enough to insure uniform and repeatable implanted-layer sheet resistance and acceptably low back-depletion layer coupling between devices ("Backgating") [8].

In early semi-insulating GaAs [9] the high-resistivity condition was often met by addition of large amounts of Cr to the crystal growth melt. This often produced material which was unacceptable for ion implantation directly into the substrate. As a result, early ion-implanted GaAs IC's [2] used low-resistivity liquid-phase epitaxially grown (LPE) "Buffer" layers [10] as the implant medium, with satisfac-

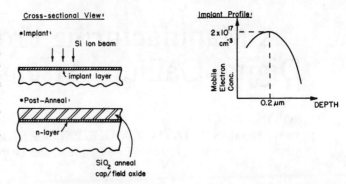

Fig. 2. Ion implantation step. Si ions are implanted at 230 keV to a dose of $6.25 \times 10^{12}/\text{cm}^2$. Mobile electron concentration is $2 \times 10^{17}/\text{cm}^3$ peak at 0.2 μm below surface.

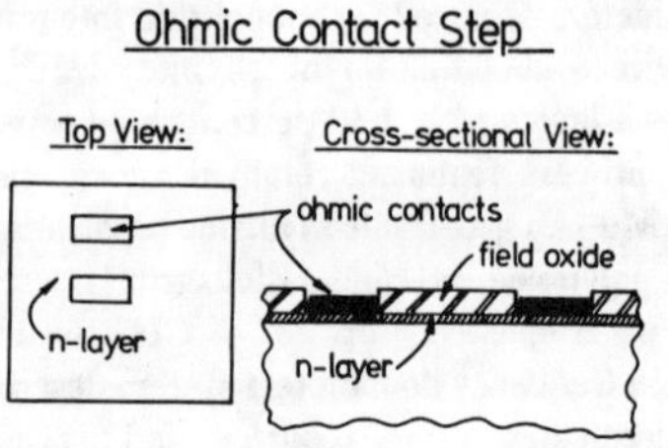

Fig. 3. Ohmic contact step. Au–Ge alloyed ohmic contacts are fabricated with the aid of field oxide lift assist.

tory results. However, cost-effective manufacturing of GaAs IC's dictates the use of direct-to-substrate ion-implantation. Suitable material is obtained by $\langle 111 \rangle$-growth of high-purity or lightly Cr-doped GaAs with the liquid-encapsulated Czochralski (LEC) method [11]. Because these ingots must be sliced along the $\langle 100 \rangle$ plane, irregularly shaped slices result. These slices are cleaved into rectangular (1.25–1.5-in) wafers for processing. The more desirable 2–3-in diameter round, $\langle 100 \rangle$-grown wafers, which are also suitable for direct-to-substrate ion implantation [12] have been evaluated, but not yet incorporated into the process described here. To assure that the high-resistivity condition is satisfied in all cases, each wafer is subjected to a nondestructive ac conductance test, and required to pass a specification equivalent to $\geq 10^5$-$\Omega \cdot$cm resistivity. Although material is routinely grown which far exceeds this resistivity specification, even the best ingots can exhibit conductive sections, necessitating a wafer-by-wafer screen.

Uncoated wafers are ion implanted with Si ions [13] at 230 keV and a dose of $6.25 \times 10^{12}/\text{cm}^2$, as shown in Fig. 2. The wafers are coated with chemical-vapor-deposited (CVD) SiO$_2$ 420 nm thick, and annealed in H$_2$ ambient at 850°C for 30 min. The resulting active doping profile (Fig. 2) is 200 nm deep at the peak concentration of $2 \times 10^{17}/\text{cm}^3$. Estimated electrical activation is 75–80 percent, and nominal sheet resistance is 320 $\Omega/\square$. Sheet resistance, as measured with a contactless RF conductivity meter [14], typically varies less than ±5 percent on a given wafer. A total sheet resistance variation for all wafers has been established at ±10 percent, with high percentage of acceptance.

Standard Au–Ge alloyed ohmic contacts are produced as shown in Fig. 3. A novel feature of this process step is the use of the anneal cap, which has been retained to serve as a process-assisting field oxide. Windows are chemically

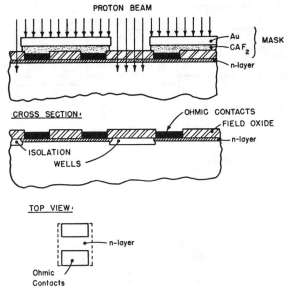

Fig. 4. Proton isolation step. Protons at 140 keV are implanted through field oxide, masked by Au-dielectric sandwich. Invisible isolation wells give superior isolation to selective implant process.

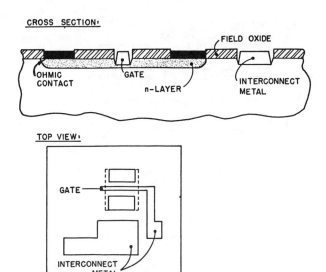

Fig. 5. Gate and first-level interconnect metal fabrication. Mo–Au gates are fabricated by contact photolithography, oxide and GaAs etching, and liftoff. Field oxide lift assist assures clean patterns.

etched in the field oxide where ohmic contacts are to appear. The undercutting of photoresist during this oxide etch produces a cantilevered photoresist overhang which insures trouble-free lifting of the deposited ohmic contact metal pattern.

The manifest advantages of planar isolation in GaAs IC fabrication have traditionally been realized with selective ion implantation [3]. The process presented here uses the slightly more complicated proton bombardment isolation, the details of which are described by D'Avanzo [8]. The principal advantages of proton isolation are:

1) improved dc leakage current isolation, especially between gate-level interconnect metal and n-regions, due to greatly decreased surface leakage current; and
2) reduced back-depletion layer coupling ("Backgating") between closely spaced active devices.

As shown in Fig. 4, protons are implanted through the field oxide at 140 keV with a dose of $5 \times 10^{14}/\text{cm}^2$. Chemical removal of the gold-dielectric mask leaves behind an essentially planar surface with proton-damaged isolation wells between devices. Leakage current between 3-μm-spaced n-islands is typically 10 nA at 16 V; leakage between gate-level interconnect metal and n-islands separated by 3 μm is 10 nA at 10 V. The "Backgating" threshold voltage is typically 7 V with undoped substrate material, effectively eliminating adjacent device backgating as a serious circuit design limitation.

The 1-μm FET gates and the first level of interconnect metal are simultaneously formed in the next process step (Fig. 5). Lithography at this step is by contact photomasking, as is the entire process. Submicrometer hard-surface masks define photoresist cuts which mask field oxide etch, gate channel etch, and metal lifting. Once again, the field oxide serves as an undercut support for overhanging photoresist, insuring a clean lift for the E-beam evaporated Mo–Au gate metal system. This metal system has been shown to produce Schottky barriers which are stable during subsequent (300°C) processing and accelerated device life tests. (A somewhat similar evaporated Mo–Au system proved not to be the critical lifetime-determining factor in GaAs FET reliability studies by Mizuishi et al. [15]).

At this point, test FET's are subjected to in-process dc and 1-MHz tests. The purpose of these tests is to reject wafers on which FET's do not meet process control specifications, thereby saving subsequent process costs. Wafers passing in-process electrical tests are next coated with a thin (100-nm) film of oxygen-free silicon nitride. This film is deposited in a parallel-plate reactor apparatus [16] at an RF power of 250 W (5 W/cm^2) onto a heated substrate (200°C). Although such films may not be stoichiometric, and contain substantial included-hydrogen, they are claimed to give increased device reliability with respect to long-term burnout, as compared to SiO$_2$ passivating films [17]. It has also been proposed that incorporated hydrogen in silicon nitride films plays an active role in increasing GaAs FET reliability by chemically combining with residual oxygen on the GaAs surface [17]. Plasma-enhanced CVD was chosen because:

1) it proceeds at temperatures well below the ohmic contact alloy temperature (430°C); and
2) it does not cause irreversible damage to n-type GaAs layers, as RF-sputtered films commonly do.

However, the high dielectric constant (6.9) of these silicon nitride films, and the tendency of thick depositions to crack, make them a poor choice for the dielectric which separates first and second metal layers, where minimum inter-layer capacitance is required.

The high dielectric constant of the silicon nitride passivation layer is exploited to make thin-film capacitors, as shown in Fig. 6. A titanium top plate is patterned by photoresist lifting over sections of first-level interconnect metal which serve as the lower capacitor plates. Nitride

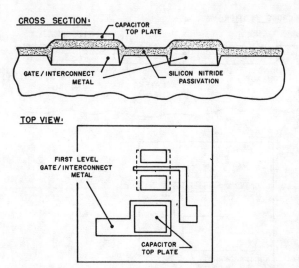

Fig. 6. Passivation and capacitor. Silicon nitride 100 nm thick is deposited by plasma CVD. Dielectric constant of 6.9 gives high (60 000 pF/cm^2) capacitance in areas where capacitor top plates are defined.

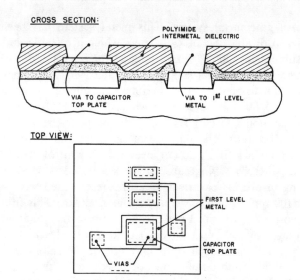

Fig. 7. Intermetal dielectric and via cut. Polyimide (Dupont PI-2555) is spun on and cured at 300°C to give smooth top surface, as shown. Vias are etched with dry ion-etch technique.

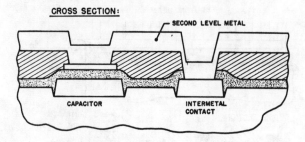

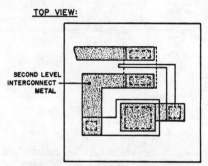

Fig. 8. Second level interconnect metal. Ti–Pt–Au metal is evaporated, then patterned with ion etching. Second metal interconnects ohmic metal, first layer metal, and capacitor top plates.

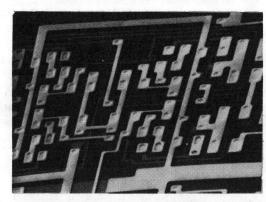

Fig. 9. Scanning electron micrograph of circuit section. First-level and second-level metal are shown, along with interconnecting vias. Wafer is later overcoated with polyimide as a scratch protector for second metal.

film quality and parameter control are such that 10-pF capacitors with active area of 1.7×10^{-4} cm^2 are produced with 90-percent yield and 5-percent standard deviation of capacitance value.

The intermetal dielectric is a spin-on polyimide film [18], [19]. This film produces an essentially flat top surface, regardless of underlying topology, and after cure at 300°C has dielectric constant, $\varepsilon_r = 3.8$, and loss tangent, $\tan \delta = 0.007$, measured at frequencies up to 2 GHz. Stress tests (15 lb/in^2 steam for 100 h, 150°C bake for 1700 h, 300°C bake for 300 h) have proven the durability of this polyimide layer. A combination of reactive ion etching (O$_2$ Plasma) and ion etching (Ar beam) is used to pattern "Via" openings in the polyimide and silicon nitride (Fig. 7). Where the capacitor top plate exists, it acts as an etch stop. Thus, the via openings to capacitors and first level metal are produced with one mask-and-etch operation.

Interconnection of components is accomplished with the second layer Ti–Pt–Au metal system (Fig. 8). Besides connecting first-level (gate) metal to ohmic contacts, the second metal provides low resistance (0.37 Ω/□) interconnects for high-current-carrying paths, and crossovers for signal-path wiring. Fig. 9 shows a scanning electron micrograph of the ion-etch patterned second metal on polyimide dielectric. Flatness of the polyimide over underlying features is evident.

IC fabrication is completed by coating the entire circuit with a second layer of polyimide film, whose function is to prevent handling damage to second-level metal. Bonding pads and scribe borders are opened with the via etch process.

Final wafer test consists of dc and 1-MHz audit of FET, diode, resistor, capacitor, and metallization test patterns. The purpose of these tests is wafer screening for process control. Functional testing of integrated circuits is described in the next section.

IV. CIRCUIT TESTING

Wafer testing of GaAs IC's in this manufacturing process is performed with an automated RF circuit test system (Figs. 10 and 11). There are two fundamental reasons for testing circuits this way:

1) high frequency IC's often do not function at dc or low frequencies (e.g., ac-coupled amplifiers); and
2) functional operation does not guarantee correct response over the desired range of frequencies and amplitudes.

Since rejection of chips after package assembly is expensive, an accurate on-wafer tester is essential to the economics of the process.

The high-frequency wafer test system, diagrammed in Fig. 10, consists of an RF signal source (0.01–8.4-GHz sweep signal generator), an RF detector Hz–22-GHz spectrum analyzer), dc power supplies, bias-voltage supplies, RF relays, and an automatic wafer prober. All these devices are calculator-controlled via an IEEE Std. 488 data bus. Circuit testing proceeds by applying voltages, measuring dc currents, then stimulating the circuit with an RF input, and monitoring its input and output signals at various frequencies and amplitudes.

The advantages of frequency-domain testing versus time domain testing are:

1) standard, programmable instruments can detect and analyze signals. Digital output of relatively few numbers can completely characterize complex waveforms, including distortion, modulation, and spurious responses;
2) bandwidth restrictions of real-time oscilloscopes and the triggering and aliasing problems associated with sampling oscilloscopes are avoided; and
3) large measurement range (>140 dB) facilitates measurements ranging from large-signal switching response to low-level thermal noise.

The disadvantages of this type of frequency domain testing are:

1) phase information can not be readily obtained; and
2) parameters of the time domain (e.g., risetime, delay time) cannot be measured directly.

As a result of these properties, the RF test system is very useful for characterization of RF circuits of all types, but of limited usefulness in characterizing and troubleshooting circuits whose response is essentially time domain. However, adequate tests can be performed on even these switching-type circuits as long as they can be stimulated to deliver a periodic output waveform with a recognizable "Signature". An example of such a test is given later.

The wafer-probing portion of this test system consists of a calculator-controlled automatic wafer stepper and a high-frequency probe card (Fig. 12). DC and RF signals alike are fed to and from the circuit under test on 50-Ω coaxial and microstrip lines. Contact to test wafers is made through a hole drilled in the sapphire probe card via

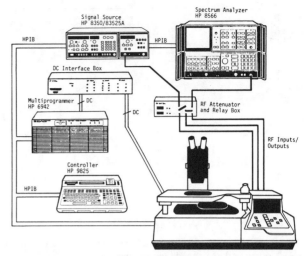

Fig. 10. Diagram of automated RF wafer test system. The calculator controls dc and RF instruments via an IEEE 488 bus. The basis of this system is the programmable spectrum analyzer which can measure linear and nonlinear responses of both digital and analog circuits.

Fig. 11. Photograph of automated RF wafer test system.

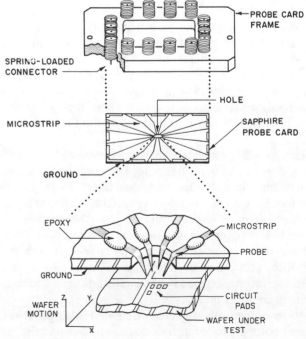

Fig. 12. High-frequency probe card. 50-Ω microstrip lines on a sapphire substrate convey signals to and from the wafer under test via Be–Cu probes protruding through a center hole. The wafer moves, probes are stationary.

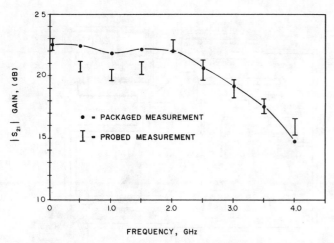

Fig. 13. Amplifier frequency response measurement; packaged and probed. (Bars represent measurement uncertainty).

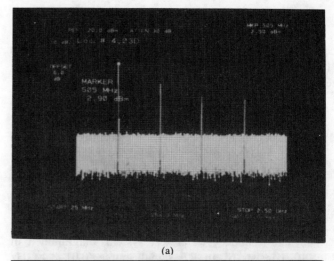

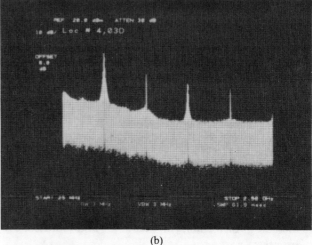

Fig. 14. Digital circuit frequency domain test example. A binary counter master–slave flip-flop is operating correctly in (a). The easily recognized spectrum displayed in (b) is for marginal, but incorrect, operation.

Be–Cu probes. Where ground is required, a probe is connected to the probe card bottom-side ground plane with a short gold ribbon bond. The probe inductance, 0.7 nH, can present problems, especially in signal ground paths. Also, the 50-Ω characteristic impedance of microstrip lines used as power supply leads can furnish unwanted signal feedback paths. For these reasons, it is often necessary to customize probe card designs (e.g., include power-supply bypass capacitors on the probe card) for particular circuits, and to consider the effects of probe parasitics when laying out chips. Very nearly, however, the same performance that can be obtained from a hybrid-substrate-mounted chip can be observed with the RF wafer probe.

An example of the difference between packaged circuit performance and wafer probe performance of an individual circuit is shown in the frequency response data of Fig. 13. Here, a particular 22-dB gain, 2.5-GHz bandwidth amplifier chip was measured with the probe card and in a microwave package. The measurement uncertainty varies from ±0.5 dB to ±1.0 dB, as indicated by the data bars. At low frequency and high frequency, the packaged performance and probed performance generally agree to within the probe system measurement uncertainty. Corrections for repeatable anomalies, such as the gain "dip" of probed data (0.5–1.5 GHz), can be introduced into the data analysis.

Examples of parameters commonly measured on amplifying circuits are: gain; frequency response; power saturation; harmonic distortion, and noise figure. In production test mode, such circuits can be checked at sufficient predetermined frequencies in 1 min or less. During this test, input bias is optimized, small-signal gain and noise figure are measured at five frequencies, and large-signal gain and distortion checked at five frequencies. The same system is used to perform extended characterization tests which require greater test time.

An example of the frequency domain test response of a circuit normally tested in time domain is shown in Fig. 14. Here, a binary counter, consisting of a master–slave flip-flop, is functionally tested with an input frequency of 1 GHz. Fig. 14(a) shows the signature spectrum expected from a properly functioning unit. Fig. 14(b) shows the response of this circuit with marginal input drive conditions. Often, such marginal operation is difficult to observe on sampling oscilloscopes. In contrast, frequency domain response is easily interpreted, even by an automatic test routine which does not rely on human interpretation. A test of a more complex digital counting circuit, with four different externally controlled modes of operation and a requirement for high-frequency input sensitivity measurement, is performed by this system in less than 3 min. Input bias for best sensitivity is determined, all functional modes tested, and counting sensitivity for nine frequencies is measured.

V. RESULTS

The automated RF test system has been used to collect functional yield data on a variety of circuits, both linear and digital. Data collected for six circuit types from 23 tested wafers is graphed in Fig. 15. Here, four different amplifier circuits (A1–A4) and two digital circuits (D1–D2) have their functional yield percentages (full-wafer, including edges) plotted against active circuit area (devices

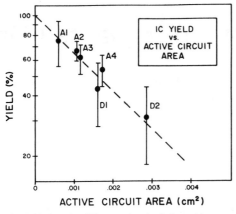

Fig. 15. IC yield for six different circuit designs (A = amplifier type, D = digital type). Average of individual wafer yields is plotted against active circuit area. Bars indicate one standard deviation.

and wiring). Functional yields ranging from 13 percent to 97 percent are included in the data. The inverse relation between yield and active area is evident, with an implied defect density of 440/cm^2 in a homogeneous Poisson distribution [20]. It has been inferred from visual inspection that a high correlation exists between metal-pattern defects of photolithographic origin and circuit defects, but that the majority of these defects are not associated with the 1-μm gate stripes. Therefore, it should not be assumed that this data represents any fundamental limit on GaAs IC technology in general, but rather that it reflects only the experience for the wafer population studied. In fact, the "Window Method" [20] of extrapolating yields was applied to a particularly high-yield wafer of this group. Analysis indicated a uniform Poisson distribution of 100 defects/cm^2, excluding edge defects, rising to a nonuniform density of 240/cm^2 when wafer edges were included. The conclusion drawn is that GaAs IC technology stands to benefit substantially in manufacturing yield by adopting larger wafers and cleaner lithographic procedures. However, present yields (Fig. 15) are wholly adequate for economic manufacturing of small-to-medium-scale GaAs IC's.

VI. Conclusion

We have presented a description of a manufacturing process for GaAs IC's which is intended for a low-volume high variety-of-circuits production environment. Key features include:

1) wafer selection on basis of substrate resistivity;
2) retained anneal cap field oxide for metal lifting assist;
3) proton-damage isolation for reduced backgating and increased isolation breakdown;
4) silicon nitride passivation used as a thin-film capacitor;
5) polyimide intermetal dielectric and scratch protection layers; and
6) on-wafer functional testing and characterization of circuits at microwave frequencies.

Circuit components (FET's, diodes, resistors, capacitors, inductors) were described.

In addition, an automated on-wafer IC tester was described which performs rapid frequency-domain functional tests and circuit characterization measurements at frequencies from 0.1 GHz to >4 GHz.

It is claimed that this manufacturing process offers a practical way to economically produce GaAs IC's in the low-microwave frequency range where we believe they will make substantial contributions to the design and manufacture of electronic instruments and systems.

Acknowledgment

The authors would like to recognize the contributions of C. Coxen, D. Schram, A. Fowler, and C. Hart to the development of this process. Also, the support of P. Wang and D. Gray for this project is gratefully acknowledged.

References

[1] R. L. Van Tuyl and C. A. Liechti, "High speed integrated logic with GaAs MESFET's," in *ISSCC Dig. Tech. Papers*, vol. 17, Feb. 1974, pp. 114–115.
[2] R. L. Van Tuyl and C. A. Liechti, "High-speed GaAs MSI," in *ISSCC Dig. Tech. Papers*, vol. 19, pp. 20–21.
[3] R. C. Eden, B. M. Welch, and R. Zucca, "Planar GaAs IC technology: Applications for digital LSI," *IEEE J. Solid-State Circuits*, vol. SC-13, Aug. 1978, pp. 419–426.
[4] R. L. Van Tuyl, C. A. Liechti, R. E. Lee, and E., "GaAs MESFET logic with 4-GHz clock rate," *IEEE J. Solid-State Circuits*, vol. SC-12, Oct. 1977, pp. 485–496.
[5] D. P. Hornbuckle and R. L. Van Tuyl, "Monolithic GaAs direct-coupled amplifiers," *IEEE Trans. Electron Devices*, vol. ED-28, Feb. 1981, pp. 175–182.
[6] R. L. Van Tuyl, "A monolithic GaAs IC for heterodyne generation of RF signals," *IEEE Trans. Electron Devices*, vol. ED-28, Feb. 1981, pp. 166–170.
[7] D. B. Estreich, "A wideband monolithic GaAs IC amplifier," in *ISSCC Dig. Tech. Papers*, Feb. 1982.
[8] D. D'Avanzo, "Proton isolation for GaAs integrated circuits," *IEEE Trans. Electron Devices*, vol. ED-30, July 1982.
[9] G. R. Cronin and R. W. Haisty, "The preparation of semi-insulating GaAs by chromium doping," *J. Electrochem. Soc.*, vol. 111, July 1964, pp. 874–877.
[10] C. Stolte, "Device quality n-type layers produced by ion implantation of Te and S into GaAs," in *IEDM Technical Digest*, Dec. 1975.
[11] W. Ford, G. Elliot, and R. C. Puttbach, "Liquid encapsulated synthesis and growth of GaAs at 2 Atm. pressure," in *Research Abstracts of the 29th IEEE GaAs IC Symp.*, Nov. 1980, paper 13.
[12] R. N. Thomas et al., "Growth and characterization of large diameter undoped semi-insulating GaAs for direct ion implanted FET technology," *Solid-State Electron.*, vol. 24, 1981, pp. 387–399.
[13] E. Stoneham, T. S. Tan, and J. Gladstone, "Fully implanted GaAs power FET's," in *Proc. Int. Elec. Dev. Meeting*, Dec. 1977, pp. 330–333.
[14] G. L. Miller et al., U.S. Patent, 4 000 458, Dec. 18, 1976.
[15] K. Mizuishi, H. Kurono, H. Sato, and H. Kodera, "Degradation mechanism of GaAs MESFET's," *IEEE Trans. Electron Devices*, vol. ED-26, July 1979, pp. 1008–1014.
[16] A. K. Sinha et al., "Reactive plasma deposited Si-N films for MOS-LSI passivation," *J. Electrochem. Soc.*, vol. 125, Apr. 1978, pp.
[17] S. H. Wemple et al., "Long-term and instantaneous burnout in GaAs power FET's: Mechanisms and solutions," *IEEE Trans. Electron Devices*, vol. ED-28, July 1981, pp. 834–840.
[18] Publication PC-1 "Pyralin* P.I. 2555," Dupont Co., Wilmington, Del.
[19] T. Herndon and R. L. Burke, "Intermetal polyimide insulation for VLSI," in *Proc. Kodak 179 Interface*, Oct. 25, 1979.
[20] R. M. Warner, "A note on IC-yield statistics," *Solid-State Electron.*, vol. 24, 1981, pp. 1045–1047.

Transient Capless Annealing of Ion-Implanted GaAs

R. CHRIS CLARKE AND GRAEME W. ELDRIDGE

Abstract—A method for high-temperature capless activation of implanted gallium arsenide has been developed that uses high-purity semi-insulating PBN LEC gallium arsenide [1] both as implant host and stabilizing medium. This capless technology, "transient capless annealing," has shown high activation of implanted dose (85 percent) with high uniformity (± 4.5 percent) and abrupt (500 Å/decade) carrier concentration depth profiles at 1.5×10^{17} cm^{-3} doping. Hall measurements taken from activated films show an electron mobility of 4500 cm^2/V·s at room temperature. S-band integrated circuits fabricated by transient capless annealing of discretely implanted ^{29}Si$^+$ delivered 20 dB of gain and 29 dB·m of output power between 3.0 and 3.6 GHz. The high-throughput batch-processing nature of transient capless annealing makes this process commercially attractive for high-yield integrated-circuit production in gallium arsenide.

I. Introduction

A COMPUTER search of the scientific journals [1] shows little published work on the activation of ion-implanted gallium arsenide prior to the mid-1960's. Fig. 1 shows a histogram of the number of scientific publications each year since then. Perhaps the most important reason for the rapid expansion in this field is the possibility of fabricating high-frequency monolithic integrated circuits using direct implantation into semi-insulating gallium arsenide substrates [2].

Direct ion implantation of doping species into semi-insulating wafers of gallium arsenide followed by thermal annealing and activation has been demonstrated as a viable technique in the preparation of several electronic devices, in particular field-effect transistors (FET's) [3], injection lasers [4], and monolithic integrated circuits [5], using a variety of capped and capless activation technologies.

For successful FET and integrated-circuit fabrication, the annealing and activation process should possess the following features:

1) high activation of implanted dose,
2) wide range of available activated dopant concentration,
3) uniform activation of dose across large-area wafers,
4) precise and predictable net donor density,
5) near theoretical electron mobility,
6) negligible surface compensation,
7) sharp and abrupt profile shape,
8) simultaneous multiple implant activation,
9) large-area wafer capacity (2-and 3-in diameter),
10) large-volume batch production,
11) flat mirror-like and damage-free wafer surfaces, and
12) low dielectric loss regions between discrete implants.

Manuscript received January 5, 1984; revised March 9, 1984.
The authors are with the Westinghouse Research and Development Center, Pittsburgh, PA 15235.

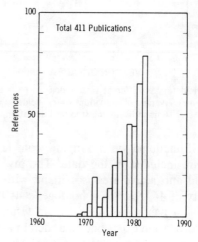

Fig. 1. Papers on annealing of ion-implanted GaAs.

This paper shows how "transient capless annealing" of implants into PBN LEC substrates can meet the requirements for FET and IC production.

II. Implant Activation by Thermal Annealing

The minimum temperature for implant activation in GaAs is between 750° and 800°C. This high temperature is necessary in order to anneal damage present in the gallium arsenide lattice caused by the penetration of high-energy ions, and to move the implanted ions to electrically active substitutional sites. Unprotected heating of an implanted GaAs wafer to the activation temperature suffers from two drawbacks. First, the vapor pressure of gallium arsenide is substantial in this temperature range [6], so that the gallium arsenide surface is eroded; secondly, the evaporation is not congruent, so that the arsenic and gallium losses are not equal. From the vapor pressure/temperature diagram for gallium arsenide shown in Fig. 2(a), it is clear that any temperature above the congruent evaporation temperature (650°C) that might be used to activate an implant will result in incongruent loss of the wafer surface. Fig. 3 is a micrograph of the surface of a gallium arsenide (100) wafer that was heated to 900°C in dry hydrogen. The result of incongruent wafer loss is evident as surface roughness and an arsenic vacancy concentration.

III. Transient Capless Annealing

This technology was evaluated using 2- and 3-in diameter (100) PBN LEC wafers implanted directly with ^{29}Si$^+$ at ambient temperature, using ion energies of 20–800 KeV at doses calculated by LSS theory, to achieve doping densities of 3×10^{16} cm^{-3} to 3×10^{19} cm^{-3}. An angle of 7° to the ion implant beam was employed to reduce channeling effects.

Reprinted from *IEEE Trans. Electron Devices*, vol. ED-31, pp. 1077–1082, Aug. 1984.

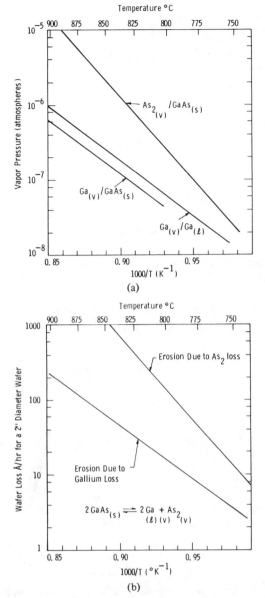

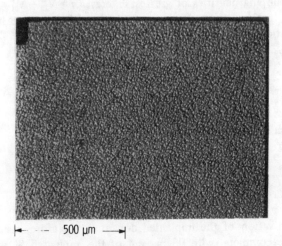

Fig. 2. Both a gallium and an arsenic pressure are present over gallium arsenide solid in the annealing temperature range. (a) The pressure of gallium and arsenic over GaAs against temperature; (b) erosion rate for a 2-in-diam gallium arsenide wafer is shown against temperature.

Fig. 3. A micrograph of a gallium arsenide wafer that was subjected to high-temperature erosion.

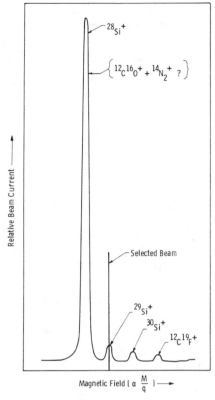

Fig. 4. Mass spectrographic resolution of ion beam (SiF_4) source.

A mass spectrum for the implant species available from an SiF_4 source is shown in Fig. 4. $^{28}Si^+$ is the most dominant silicon isotope, but has a mass conflict with $^{28}N_2^+$; $^{29}Si^+$ was, therefore, selected as the purest source of silicon ions.

Annealing of implanted wafers was performed in a silica tube containing a boat filled with crushed high-purity (PBN LEC) gallium arsenide solid. The slight decomposition of the crushed solid creates both a gallium and an arsenic pressure in the system greater than the decomposition pressure of the wafers by virtue of a temperature differential. Fig. 5 shows a schematic diagram of the annealing furnace and reaction chambers. The implanted wafers suffered no observable degradation of surface quality or of electrical behavior as a result of the high-temperature implant activation process.

The time-temperature curves for the significant regions of the annealing chamber are given in Fig. 6. Aside from the time needed for chamber flushing, the process time is less than 20 min and the time at activation temperature less than 3 min.

A. Electrical Assessment

Uniform activated implants were characterized for their differential activation, carrier mobility, and compensation ratio by the use of Hall data. Cloverleaf Van der Pauw samples were cut from the wafers and alloyed tin dots were used to make electrical contact. Information concerning peak electron density, profile shape, depth, and uniformity was obtained from surface patterns of aluminum dots evaluated with the use of a C-V profiler.

The activation process was temperature sensitive showing a region of peak differential activation. The activated doping

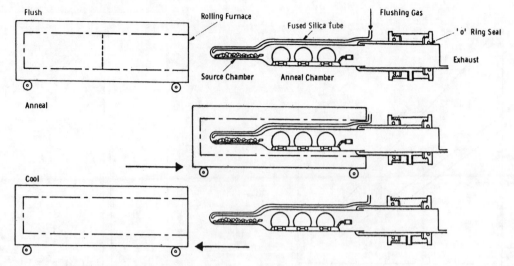

Fig. 5. Crushed PBN LEC solid is loaded into the source chamber and implanted wafers are loaded into the anneal chamber of the transient capless anneal tube.

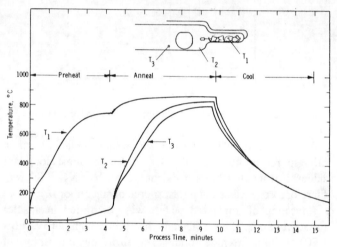

Fig. 6. Transient capless annealing cycle.

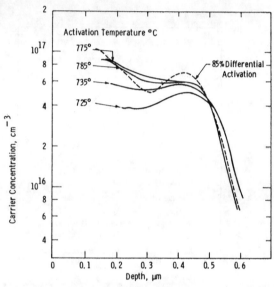

Fig. 7. The temperature dependence of activation is demonstrated here by the superimposed carrier concentration-depth profiles of similar implants, activated at different temperatures.

profiles of pieces of the same implanted wafer activated at different maximum temperatures is shown in Fig. 7. Below 725°C, no activation was observed. Activation then increased with temperature up to 775°C, but at 785°C an apparent fall in differential activation was seen, perhaps due to amphoteric shift of the silicon dopant. The peak differential activation temperature was 770°–780°C. This peak differential activation corresponds to 85-percent activation of the implanted dose.

In Fig. 8, the observed electrically active impurity density for transient capless anneals is plotted against the implanted ion density for ^{29}Si$^+$. The total ionized impurity density was deduced from room-temperature Hall data and theoretical compensation curves [7].

This analysis allows evaluation of both the ionized donor and ionized acceptor densities. Saturation of the net donor density and the total ionized center density at high implanted Si concentration may be explained as the exhaustion of the gallium vacancies required to achieve substitutional doping, e.g.,

$$Si + V_{Ga} \rightarrow [Si]_{Ga}. \qquad (1)$$

For doping densities of greatest interest ($n = N_D^+ - N_A^- \leqslant 2 \times 10^{17}$ cm^{-3}), the activation may be modeled by two linear equations

$$N_D^+ + N_A^- = [Si] + 1.5 \times 10^{16} \text{ cm}^{-3} \qquad (2)$$

and

$$N_D^+ - N_A^- = 0.86\,[Si] - 1.5 \times 10^{16} \text{ cm}^{-3} \qquad (3)$$

which can be solved to give

$$N_D^+ = 0.93\,[Si] \qquad (4)$$

and

$$N_A^- = 0.07\,[Si] + 1.5 \times 10^{16} \text{ cm}^{-3}. \qquad (5)$$

Activation of the implanted dopant is, therefore, assumed to be complete, with 93 percent of the implanted Si acting as donors and 7 percent as acceptors. The divergence of the

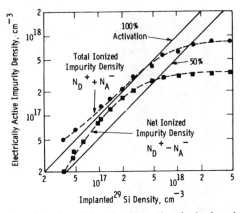

Fig. 8. A plot of implanted ion density against ionized purity density reveals information concerning differential activation, compensation ratio, and background impurity contributions.

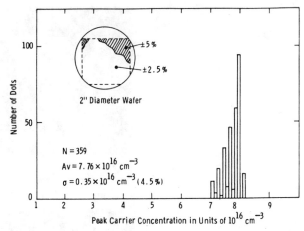

Fig. 10. Aluminum Schottky-barrier dots reveal the uniformity of transient capless-annealed profiles in PBN LEC gallium arsenide; of 369 dots, the average peak carrier concentration is 7.76×10^{16} cm^{-3} and the standard deviation is 0.35×10^{16} cm^{-3}.

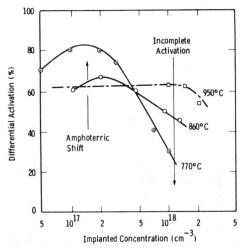

Fig. 9. Determination of the temperature for maximum differential activation.

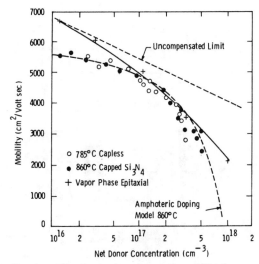

Fig. 11. Net donor density is plotted against mobility for capped and capless activations in PBN LEC GaAs wafers. Theoretical and epitaxial results are given for comparison.

logarithmic density curves at low concentrations is strong evidence for an inherent residual acceptor density of 1.5×10^{16} cm^{-3} in PBN LEC substrate wafers. This residual acceptor usefully preserves the semi-insulating nature of unimplanted regions of the wafer throughout the annealing and activation cycle.

Further information about this low background compensation is given by the room-temperature electron mobility curves of Fig. 11, where a comparison is made between Si$_3$N$_4$ capped [8], capless, and epitaxially produced films. Throughout the range 10^{16} cm^{-3} to 5×10^{17} cm^{-3}, the capped and capless data are equivalent, both showing the presence of a compensating impurity below 5×10^{16} cm^{-3} net donor concentration when compared to epitaxial and theoretical results. In the range of particular interest for FET's and IC's between 5×10^{16} cm^{-3} and 3×10^{17} cm^{-3}, the room-temperature electron mobility of transient capless-annealed implants compared favorably with observed data from epitaxially deposited films.

In the region of very high doping, greater than 5×10^{17} cm^{-3}, the differential activation of implanted dose is reduced. However, raising the activation temperature, as shown in Fig. 9, allows for greater activation of silicon, but at the expense of an amphoteric shift toward acceptor sites. In other words,

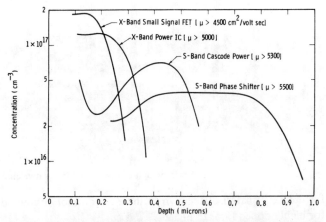

Fig. 12. Various electronic device profiles fabricated in PBN LEC gallium arsenide by implantation and transient capless activation.

more silicon is activated by increasing the activation temperature, but this increased activation is accompanied by an increase in silicon acceptors. This compensation effect at high doping levels is also found with group VI impurities [9] for capped and capless activated implants and to a lesser extent

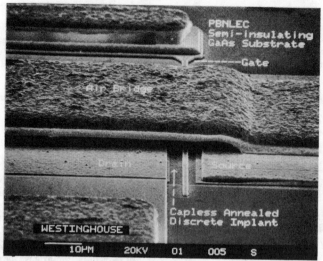

Fig. 13. A scanning electron micrograph of a transistor fabricated by direct ion implantation and capless annealing.

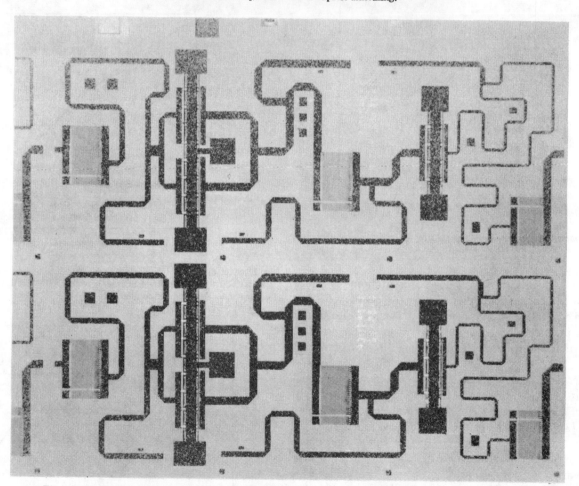

Fig. 14. A micrograph of the S-band integrated-circuit chip showing air bridges and interdigitated capacitors.

for epitaxial material as shown in the electron mobility curve for highly doped films in Fig. 11.

As indicated in Fig. 10 the activation of $^{29}\text{Si}^+$ in PBN LEC GaAs by transient capless annealing is most uniform. Aluminum dots were evaporated on a 2-in-diam capless-activated wafer and the peak carrier concentration was determined beneath each dot. Of 300 measurements, the average peak carrier concentration was 7.76×10^{16} cm^{-3} and the standard deviation was 4.5 percent.

IV. Electronic Device Structures

Transient capless annealing has been used to fabricate various profiles for electronic devices such as the X-band FET, X-band power IC, S-band cascode, power FET, and S-band phase shifter, which were each prepared by the simultaneous activation of multiple implants. Examples of the carrier concentration-depth profiles of such activated implants are shown in Fig. 12.

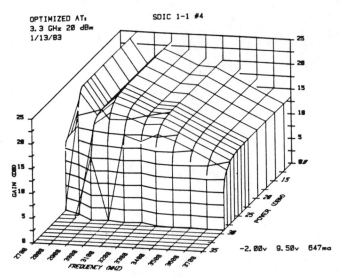

Fig. 15. A 3-D plot of gain power frequency for the S-band circuit of Fig. 14.

These data illustrate the flexibility of the implant and anneal approach.

V. FET's and Integrated Circuits

Selectively implanted FET channel layers ($N_D = 1.5 \times 10^{17}$ cm^{-3}, $\lambda = 3000$ Å) were formed by directly implanting ^{29}Si into undoped LEC semi-insulating GaAs substrates through a photoresist mask. Registration marks were then made by a self-aligned method and activation was accomplished by transient capless annealing. After implant characterization, IC fabrication was achieved using AuGe/Ni/Pt ohmics, 1-μm-long Ti/Pt/Au gates, plated Au circuits, air-bridge interconnects, and via-hole grounding through 4-mil-thick substrates. A scanning electron microscope picture of FET construction is detailed in Fig. 13.

A two-stage monolithic S-band GaAs power amplifier, designed as the driver in a phased-array T/R module, has been fabricated and tested [10]. The circuit contained interdigital capacitors and microstrips for low-loss impedance matching, dc voltage blocking, power splitting, and power combining (see Fig. 14).

The 1200-μm FET's demonstrated a dc transconductance of 120 mS at zero gate bias and a gate drain breakdown voltage near pinchoff in excess of 20 V for an I_{DSS} of 330 mA. The variation of I_{DSS} over a 2-in wafer was less than 6 percent.

During RF power testing, the two-stage amplifiers delivered 20 dB of gain and 29 dB·m of output power in the 3.0-to 3.6-GHz frequency range. A three-dimensional plot of gain, frequency, and power performance of an S-band driver chip is shown in Fig. 15. This clearly shows the viability of capless annealing for the fabrication of power monolithic circuits.

Acknowledgment

The authors would like to acknowledge the advice and encouragement of Dr. R. N. Thomas, Dr. M. Driver, and Dr. H. C. Nathanson, and the technical assistance of B. L. Bingham, P. Kost, R. Nye, W. Valek, W. A. Wickstrom, and S. Maystrovich. We also acknowledge the work of Dr. S. K. Wang, who directed the fabrication of the S-Band T/R circuits.

References

[1] DIALOG Information Services, Inc., 1982.
[2] H. M. Hobgood, G. W. Eldridge, D. L. Barrett, and R. N. Thomas, "High-purity semi-insulating GaAs material for monolithic microwave integrated circuits," *IEEE Trans. Electron Devices*, vol. ED-28, no. 2, p. 140, 1981.
[3] R. G. Hunsperger and N. Hirsch, "GaAs field effect transistors with ion-implanted channels," *Electron. Lett.*, vol. 9, p. 557, 1973.
[4] M. K. Barnoski, R. G. Hunsperger, and A. Lee, "Ion-implanted GaAs injection laser," *Appl. Phys. Lett.*, vol. 24, no. 12, p. 267, 1974.
[5] M. C. Driver, S. K. Wang, J. X. Przybysz, V. L. Wrick, R. A. Wickstrom, E. S. Coleman, and J. G. Oakes, "Monolithic microwave amplifiers formed by ion implantation into LEC gallium arsenide substrates," *IEEE Trans. Electron Devices*, vol. ED-28, no. 2, p. 141, 1981.
[6] C. T. Foxon, J. A. Harvey, and B. A. Joyce, "The evaporation of GaAs under equilibrium and nonequilibrium conditions using a modulated beam technique," *J. Phys. Chem. Solids*, vol. 34, p. 1693, 1973.
[7] W. Walukiewicz, J. Lagowski, L. Jastrebski, M. Lichtensteiger, and H. C. Gatos, "Electron mobility and free carrier absorption in GaAs: Determination of the compensation ratio," *J. Appl. Phys.*, vol. 50, p. 899, 1979.
[8] R. N. Thomas, H. M. Hobgood, G. W. Eldridge, D. L. Barrett, T. T. Braggins, L. B. Ta, and S. K. Wang, "High-purity LEC growth and direct ion implantation of GaAs for monolithic microwave circuits," in *Semiconductors and Semimetals*, R. K. Willardson and A. C. Beer, Eds. New York: Academic Press, 1984.
[9] C. M. Wolfe and G. E. Stillman, "Self compensation of donors in high purity gallium arsenide," *Appl. Phys. Lett.*, vol. 27, no. 10, pp. 564–567, 1975.
[10] R. C. Clarke, G. W. Eldridge, S. K. Wang, and W. Valek, "Transient capless annealing of ion-implanted PBN LEC GaAs for monolithic microwave integrated circuits," in *Proc. IEEE Microwave Millimeter Wave Monolithic Circuits Symp.*, 1983.

Proton Isolation for GaAs Integrated Circuits

DONALD C. D'AVANZO, MEMBER, IEEE

Abstract—Significant improvement in the electrical isolation of closely spaced GaAs integrated circuit (IC) devices has been achieved with proton implantation. Isolation voltages have been increased by a factor of four in comparison to a selective implant process. In addition, the tendency of negatively biased ohmic contacts to reduce the current flow in neighboring MESFET's (backgating) has been reduced by at least a factor of three. The GaAs IC compatible process includes implantation of protons through the SiO_2 field oxide and a three-layered dielectric–Au mask which is definable to 3-μm linewidths and is easily removed. High temperature storage tests have demonstrated that proton isolation, with lifetimes on the order of 10^5 h at 290°C, is not a lifetime limiting component in a GaAs IC process.

I. INTRODUCTION

THE SUCCESSFUL design and fabrication of medium- and large-scale GaAs integrated circuits (IC's) requires a high degree of electrical isolation between closely spaced active devices. Traditionally, mesa etched [1], [2] and selective implant [3] processes have been used to isolate GaAs IC's. These processes can result in significant current flow

Manuscript received January 18, 1982; revised February 12, 1982.
The author is with the Hewlett–Packard Company, Santa Rosa Technology Center, Santa Rosa, CA 95404.

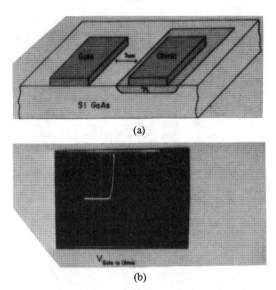

Fig. 1. (a) Cross section of a gate-to-ohmic isolation test pattern fabricated with a selective implant process. (b) Curve-tracer photograph of the current–voltage characteristic for the test pattern diagrammed in (a).

between isolated patterns separated by five microns or less. In this work, proton implantation is used to achieve a dramatic improvement in the electrical isolation of GaAs IC's.

A typical isolation characteristic for a selective implant process is displayed in Fig. 1. The test pattern consists of a gate metallization on semi-insulating GaAs separated by 3 μm from an ohmic-contacted n-type region. The measured current versus voltage characteristic is rectifying. Significant current flows when the gate metal is biased positively with respect to the ohmic contact, but negligible current flows with the reverse polarity. While the 5.0-V breakdown or isolation voltage is typical for selective implant and mesa isolated structures, lower isolation voltages, on the order of 2.0 V, are often observed. The variability in isolation voltage can be partially attributed to surface conditions, including chemical treatments and the presence or absence of dielectric passivation. Asymmetric characteristics are also observed when an ohmic metallization directly contacts the semi-insulating GaAs substrate. In both cases, high fields at the edges of the metal/semi-insulating GaAs interfaces are believed to be responsible for the relatively low breakdown voltages.

A second detrimental effect related to insufficient isolation is backgating. It occurs when a negatively-biased n-type region is in close proximity to an operating GaAs MESFET. As the negative bias on the ohmic contact is increased, the active layer-substrate depletion region beneath the MESFET widens, resulting in decreased current flow. In a typical mesa or selective implant wafer the drain current is reduced by 10 percent with only -3.0 V applied to a backgating ohmic contact separated from the MESFET by 5.0 μm.

The relatively low degree of electrical isolation in mesa and selective implant processes can lead to significant circuit design constraints, including an increase in minimum spacing rules, reduction in packing density, and severe design complications. Proton implantation constitutes a potentially superior isolation process. Implantation of protons into GaAs has been used successfully to isolate discrete diodes, lasers, and MESFET's [4]–[11]. Extensive characterization has indicated that proton damage is capable of producing high resistivity, thermally stable layers in both n and p type GaAs [4]–[9]. Recently proton [12], [13], as well as boron [14] and oxygen [15], ion-implantations have been applied to linear and digital monolithic IC's.

This paper describes a GaAs IC-compatible proton isolation process, with emphasis on the electrical (isolation and backgating) characteristics important to medium- and large-scale integration. Proton damage characterization and process development are presented in the second section. Measurement results including isolation, backgating, and reliability are described in the third section. The fourth section contains a summary of the work and important conclusions.

II. Process Development

The major objective in developing an IC-compatible proton isolation process is to eliminate isolation as a design consideration. The HP Santa Rosa GaAs IC process [16] is designed to fabricate small- and medium-scale gigahertz-bandwidth digital and analog IC's. For this type of technology, greater than 10-V isolation is required between patterns separated by the minimum spacing of 3.0 μm. To achieve this goal, a proton implantation mask must be designed which can be defined to a width of 3.0 μm while adequately protecting active areas from proton damage. In addition, the protons must be implanted through the SiO_2 anneal cap which is retained as a field oxide in order to optimize surface passivation and minimize process complexity [16]. The process features to be established are: the minimum implant energy required to achieve sufficient isolation; and the optimum dose for the specific active layer profile and temperature processing [16].

A. Proton Damage Characterization

Proton implantation into GaAs creates damage centers which effectively trap electrons and holes. These traps reduce the conductivity of an active layer by removing mobile electrons from the conduction band.

The damage profile has been characterized by measuring the mobile carrier concentration (with the CV technique) before and after partially compensating proton implantations. These measurements have been performed over a range of implant energies and doses. Typical results are plotted in Fig. 2. The initial electron profile is constant at 6×10^{16} cm^{-3} to a depth of 0.6 μm. Protons are implanted through the 4200-Å SiO_2 field oxide at an energy of 110 keV, which is large enough to penetrate the active layer. The electron profile is remeasured indicating the shape and magnitude of the as-implanted proton damage as a function of proton dose and energy. The results in Fig. 2

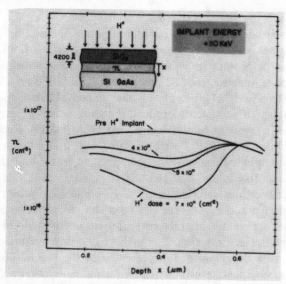

Fig. 2. The electron concentration n versus depth (measured by the CV technique) before and after partially compensating proton implantations. The protons are implanted through 4200 Å of SiO_2.

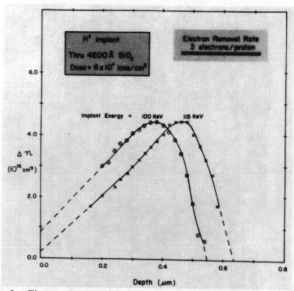

Fig. 3. Electron removal Δn versus depth for two implantation energies.

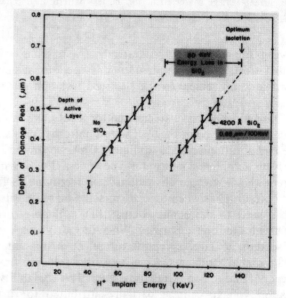

Fig. 4. Depth of as-implanted proton damage peak versus implantation energy with and without 4200-Å SiO_2 field oxide.

indicate that the damage (electron removal) concentration increases linearly with dose for the relatively low, partially compensating proton doses shown (4.0, 5.0, and 7.0×10^{11} cm^{-2}).

The concentration of electrons trapped or removed from conduction, Δn, can be computed as a function of depth by subtracting the electron profile after proton implantation from the initial electron profile. The results for a dose of 6.0×10^{11} protons/cm^{-2} and energies of 100 and 115 keV are plotted in Fig. 3. The protons are implanted through the field oxide. The total number of electrons removed from conduction can be estimated by integrating the electron removal profiles. Dividing the integral by the proton dose results in an electron removal rate of three electrons per proton. This implies that, at this relatively low dose, the average proton creates enough damage to trap three electrons. As will be discussed below, the electron removal rate may be somewhat lower for the high proton doses required for complete compensation.

The electron removal characteristics plotted in Fig. 3 indicate that the shape of the damage profile is relatively independent of implantation energy while the depth of the as-implanted damage profile increases with increasing energy. This effect is examined more closely in Fig. 4 where the depth of the damage peak is plotted as a function of energy for proton implantations with and without the SiO_2 field oxide. In both cases, the depth of the damage peak increases at a rate of 0.65 μm/100 keV. The offset between the damage peaks with and without the anneal cap indicates that protons lose approximately 50 keV in passing through the 4200-Å SiO_2 field oxide.

Current–voltage measurements demonstrate that the minimum energy required to achieve optimum isolation for the GaAs IC active layer profile [16] is 140 keV. Extrapolating the data in Fig. 4 implies that the depth of the damage peak is located at 0.62 μm for proton implantation through 4200 Å of SiO_2 at 140 keV. For the high doses required for complete compensation (between 1.0×10^{14} and 1.0×10^{15} protons/cm^2), significant damage extends beyond the depth of the damage peak. After proton implantation through the field oxide, with a dose of 5×10^{14} protons/cm^2 and energy of 140 keV, CV measurements indicate full compensation extends to approximately 0.9 μm. Since the 140-keV implant through the oxide corresponds to a 90-keV implant directly into GaAs, the measured depth agrees well with previously reported values [4], [8] between 0.85 and 1.0 μm/100 keV. The total damage depth is a factor of two larger than the depth of the GaAs IC active layer profile. For the same implantation conditions, the lateral spread of protons from a mask edge has been estimated (from measurements on narrow resistors) to be between 0.6 and 0.7 μm. This is consistent with experimental and theoretical values reported in the literature [7], [17].

The second implantation parameter to be considered is

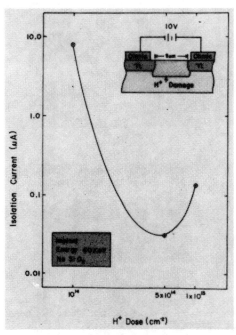

Fig. 5. Isolation current versus proton dose for two ohmic contacts separated by 5.0 μm of proton damaged area with 10 V applied.

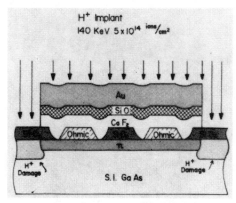

Fig. 6. Schematic cross section of a GaAs IC at the proton isolation step.

the proton dose. To determine the dose required to achieve sufficient isolation the current–voltage characteristics have been measured between two ohmic contacts separated by 5.0 μm of proton damaged area. The initial active layer is formed by a Si implantation and has a peak concentration of 2.0×10^{17} cm^{-3} at a depth of 0.2 μm [16]. The isolation current with 10 V applied is plotted as a function of dose in Fig. 5. While sufficient isolation is achieved between 1.0×10^{14} and 1.0×10^{15} protons/cm^2, the minimum isolation current (for the three doses included in this experiment) occurs at 5.0×10^{14} protons/cm^2. The increase in isolation current with increasing dose has been previously reported [6] and may be caused by defect-level banding [6] or enhanced hopping conduction mechanisms [9]. While the data presented in Fig. 5 are for unbaked samples with 60-keV proton implantations directly in GaAs, similar results are observed for 140-keV implantations through the SiO$_2$ field oxide, both before and after a four hour 300°C bake included in the GaAs IC process [16].

The apparent optimum dose is two orders of magnitude greater than what would be predicted from the low dose electron removal rate (Fig. 3). Several factors may contribute to the discrepancy. First, excess damage beyond that needed to trap all the conduction electrons can reduce the mean free path and as a result the electron mobility. Secondly, the maximum depth of damage increases with increasing dose, so that a larger portion of the substrate is converted to very high resistivity material. Finally, electron removal may not increase linearly with dose at high dose levels.

The optimum proton implantation parameters are established at 140 keV and 5.0×10^{14} cm^{-2}. The final process consideration is the development of a proton mask.

B. Proton Implantation Mask

The proton implantation mask must meet three requirements to be compatible with the GaAs IC process. First, it must stop protons implanted at 140 keV in order to effectively protect active areas. Second, the mask must be definable to the minimum linewidth of the IC process, 3.0 μm in this case. Finally, the mask must be easily removable with a process that is nondamaging to the exposed GaAs surface. Those objectives were achieved with a three-layer mask composed of CaF$_2$, SiO, and Au defined by a photoresist lift. Before describing the details of this process it will be helpful to briefly review the preceding fabrication steps [16].

The first step in the GaAs IC process is a direct, unmasked implantation of Si (6.25×10^{12} cm^{-2}, 230 keV) into high purity or lightly Cr-doped (less than 1.0×10^{15} cm^{-3}) substrates. This active layer implant is annealed after capping with 4200 Å of low temperature, CVD, SiO$_2$. The anneal cap is retained in field areas during subsequent processing. In the next step, ohmic contacts are deposited and alloyed.

The proton isolation mask, composed of the two dielectric layers and 1.3 μm of Au, is defined by a positive photoresist lift. A schematic cross section of the final structure is displayed in Fig. 6. Protons are then implanted at 140 keV to a dose of 5.0×10^{14} cm^{-2}. In the field areas, protons penetrate the SiO$_2$ and enter the GaAs with an effective approximate energy of 90 keV. The resulting damage creates a high resistivity layer to a depth of 0.9 μm. The Au–dielectric mask protects the active areas during the proton implantation. Finally, the mask is readily removed by etching the CaF$_2$ in dilute HCl. To date, this masking scheme has been used successfully in 20 production-prototype runs including over 100 wafers. The following sections present electrical measurement results for the proton isolation process.

III. RESULTS

A. Isolation

The gate-to-ohmic isolation characteristics, typical of the selective implant and the proton implant processes, are

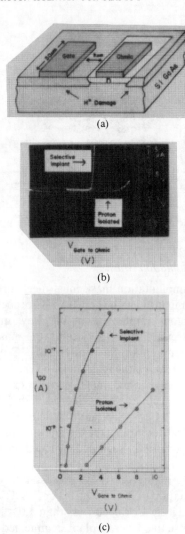

Fig. 7. (a) Cross section of a gate-to-ohmic isolation test pattern fabricated with a proton implantation process. (b) Curve-tracer photograph comparing the gate-to-ohmic isolation characteristics for typical selective implant and proton isolated processes. (c) Gate-to-ohmic isolation current I_{GO} versus voltage for typical selective implant and proton isolation processes.

compared in Fig. 7. The test pattern consists of a 50-μm-long gate metallization directly on high resistivity, proton implanted, GaAs separated by 3.0 μm from an ohmic-contacted n-type region. Both measurements are on fully processed wafers including a four hour bake at 300°C. The curve tracer photograph indicates that no significant current flows in the forward direction below 20 V for the proton isolated process. This value of isolation voltage is a factor of four greater than the value obtained with the selective implant process. Significant improvement is also realized at low current levels, as is demonstrated in Fig. 7(c). At typical IC operating voltages the isolation current is three orders of magnitude less for the proton isolated pattern. Even for a worst case IC operating voltage of 10 V, only 10 nA of isolation current flows for the proton isolated process.

A second test pattern is used to evaluate ohmic-to-ohmic isolation. It consists of two 50-μm-long ohmic-contacted n-type regions separated by a 3-μm-wide high resistivity

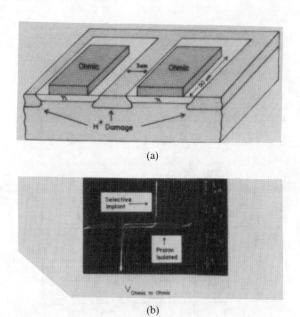

Fig. 8. (a) Cross section of an ohmic-to-ohmic isolation test pattern fabricated with a proton implantation process. (b) Curve-tracer photograph comparing typical ohmic-to-ohmic isolation characteristics for proton implant and selective implant processes.

proton implanted region, as shown in Fig. 8(a). For this test pattern the isolation characteristics, displayed in Fig. 8(b), are symmetric. Again, considerable improvement is obtained with proton isolation, in which case no significant current flows below 25 V. In contrast, large values of isolation current are observed at approximately 10 V for the selective implant process.

The preceding results indicate that significant improvement can be obtained with a proton isolation process. In fact, the results demonstrate that proton implantation eliminates isolation as a design constraint. The next section compares backgating effects for the two processes.

B. Backgating

Backgating occurs in a GaAs IC when a negatively biased ohmic-contacted n-type region is in close proximity to an operating MESFET. As the negative bias increases, the depletion region at the interface between the MESFET and the semi-insulating substrate widens, resulting in a reduction in drain current. Backgating is an extremely undesirable effect in an IC. When significant backgating occurs, FET dc characteristics depend not only on the internal device biases but also on the proximity of negative voltage lines. Normally straightforward design considerations, such as dc operating points and matching between devices, become layout dependent, leading to severe design complications and a relaxation of minimum design rules. The following results demonstrate that proton isolation helps avoid these problems by significantly reducing the susceptibility of GaAs IC's to the backgating effect.

The test pattern used to characterize backgating is schematically drawn in Fig. 9. The structure consists of a 100-μm-wide MESFET and a negatively biased, backgating, ohmic contact. To maximize backgating sensitivity, the

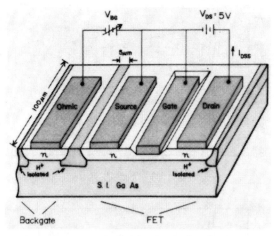

Fig. 9. Schematic cross section of the GaAs IC backgating test pattern.

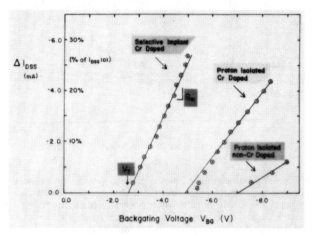

Fig. 10. Change in drain current ΔI_{DSS} versus backgating voltage V_{BG} measured on the test pattern diagrammed in Fig. 9. Typical characteristics for three groups are plotted: selective implant on lightly Cr-doped substrates, proton isolated on lightly Cr-doped, and proton isolated on non-Cr-doped.

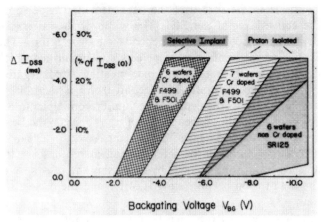

Fig. 11. Distribution in wafer averages of backgating threshold and transconductance for three groups of wafers: selective implant on lightly Cr-doped substrates, proton isolated on lightly Cr-doped, and proton isolated on non-Cr-doped. F499, F501, and SR125 are ingot identification numbers.

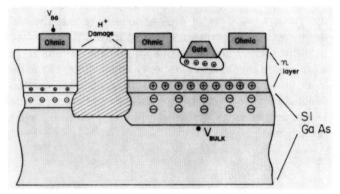

Fig. 12. Cross section of a GaAs MESFET in close proximity to a backgating ohmic contact.

ohmic-contacted n-type region runs the full width of the MESFET and is separated by 5.0 μm from the source. The MESFET is operated at $V_{DS} = 5.0$ V and $V_{GS} = 0.0$ V. For these bias conditions, the drain current and transconductance are nominally 20 mA and 12 mS, respectively [16]. Backgating is characterized by measuring the change in drain current ΔI_{DSS} as a function of backgating voltage V_{BG}.

Typical backgating characteristics for three groups of fully processed wafers (including a four hour bake at 300°C) are plotted in Fig. 10. In all cases, the change in drain current increases linearly with negatively increasing backgating voltage. The characteristics are well defined by a backgating threshold V_T below which no significant backgating occurs, and a backgating transconductance G_m equal to the slope of the line. For the selective implant process on lightly Cr-doped (less than 1.0×10^{15} cm^{-3}) substrates the threshold is approximately -2.5 V. At a typical IC operating voltage of -5.0 V, the current has decreased by an intolerable 30 percent. In contrast, with proton isolation on lightly Cr-doped material, no significant backgating occurs below 5.0 V. In addition, the backgating transconductance is halved. At a worst case IC operating voltage of -6.0 V, less than 10-percent reduction in drain current is observed. Further improvement is obtained with proton isolation on non-Cr-doped substrates. In this case, the backgating threshold is extended beyond the range of normal circuit operation to -7.0 V.

The backgating threshold and transconductance have been characterized on a large number of completed wafers. One hundred percent of the test patterns are measured and wafer averages are computed for both parameters. The spread in wafer averages for the three groups are graphically displayed in Fig. 11. For the selective implant process on lightly Cr-doped substrates, the threshold can be as low as -2.0 V, resulting in a 25-percent decrease in drain current at only -4.0 V. In contrast, the minimum thresholds observed for proton isolated wafers on lightly Cr-doped and on non-Cr-doped substrates are approximately -4.0 and -6.0 V, respectively.

The preceding results demonstrate that proton isolation significantly decreases the susceptibility of GaAs IC's to the backgating effect. The improvement is gained by proton implantation's ability to increase the resistivity of GaAs in the field area, between the backgating pad and the active MESFET. The structure is schematically diagrammed in Fig. 12. According to current understanding of backgating [18]–[20], the degradation in drain current results from the spread of the back depletion region into the active layer of the FET. The width of the interfacial depletion region primarily depends on two factors. The first is the concentration of defects and impurities which can trap electrons or emit holes creating fixed negative charge in the substrate. The second factor is the voltage developed across the depletion region. Proton implantation does not affect the concentration of traps beneath the FET. Instead, proton damage reduces the voltage developed across the interfacial depletion region in response to the applied backgating voltage, by increasing the resistivity of the field region between the MESFET and the backgating pad.

It has been observed that both isolation and backgating for the selective implant process are sensitive to surface conditions, including chemical treatment and dielectric passivation. The fact that mesa isolation is comparable to selective implant in isolation and backgating characteristics also implies that the surface plays a significant role. One of the prime advantages of proton isolation is that the high dose implantation creates a fairly uniform damage layer from the surface to a depth of approximately 1 μm. In this way, proton implantation not only compensates the active layer, but also passivates the surface. As a result, proton-isolated IC's are relatively insensitive to surface conditions. In the next section, thermal stability of proton damage is investigated.

C. Thermal Stability

Proton implantation compensates an active layer by creating lattice defects which trap mobile electrons and holes. It can be expected that the lattice damage will anneal out at a sufficiently high temperature. Previous work has demonstrated that high-resistivity proton-implanted layers are maintained for relatively short anneal times up to temperatures of 400°C for single implantations [4], [8] and 500°C for multiple implantations [7], [9], [21]. However, specific annealing characteristics are a strong function of the initial electron profile, and the dose and energy of the proton implantation [5], [6]. While high temperature annealing is important for evaluating stability during high temperature processing, the main consideration for GaAs IC's is the long-term stability of the damaged layer at worst case operating temperatures. High temperature storage tests have been performed to evaluate the long-term reliability of proton isolation. The results are discussed in the next few paragraphs.

To accelerate the aging process, completed wafers were stored at 290°C and 350°C in nitrogen ambients. At numerous intervals during the stress tests, the wafers were removed from the furnaces, cooled to room temperature, and measured. The isolation voltage V_{iso} is measured on the test pattern in Fig. 7(a) and is defined as the voltage applied between the gate and ohmic contact with 1.0 μA of leakage current. The gate metallization is biased positively to determine the worst case isolation. In addition, the backgating threshold was measured on the test pattern

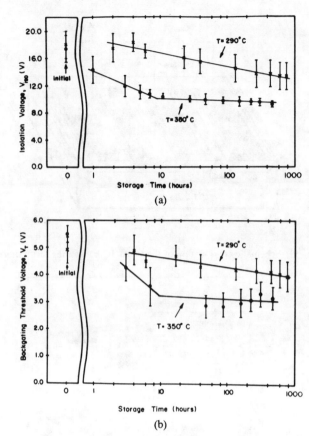

Fig. 13. Thermal stability of proton damage. (a) Isolation voltage V_{iso} versus storage time for two wafers, one stored at 290°C and the other stored at 350°C. The isolation voltage is measured at 1.0 μA on the test pattern diagrammed in Fig. 7(a). (b) Backgating threshold versus storage time for the two wafers included in (a). The backgating threshold is measured on the test pattern diagrammed in Fig. 9. The data points represent wafer averages (10 devices for V_{iso} and 35 devices for V_T) and the error bars indicate standard deviations.

diagrammed in Fig. 9. Both parameters are plotted as functions of high temperature storage time in Fig. 13(a) and (b). Two lightly Cr-doped wafers are included, one stored at 290°C and the other at 350°C. Each data point represents a wafer average, while the error bars indicate one standard deviation.

The results in Fig. 13(a) demonstrate a gradual logarithmic degradation in isolation voltage for the wafer stored at 290°C. Extrapolation of the data to longer storage times predicts a time to wafer-specification (8.5 V at 1.0 μA) exceeding 10^5 h. Similar behavior is observed for the backgating threshold voltage degradation at 290°C in Fig. 13(b). Again, a gradual logarithmic degradation rate dominates the aging characteristic with extrapolated failure times exceeding 10^5 h.

A more complex process is observed at 350°C. The isolation voltage in Fig. 13(a) decreases rapidly during the first 10 h of storage. However, the initial rapid degradation is then followed by an apparent saturation. The backgating threshold in Fig. 13(b) displays similar aging properties. The results imply that two or more annealing processes may be occurring simultaneously. A low activation energy process dominates at 290°C and during the initial stages at 350°C, while a second, higher activation energy process dominates for long storage times at 350°C. These observations are consistent with recent identification of from three to five electron traps in proton implanted material [22]–[24].

The complex annealing behavior prevents the derivation of a single activation energy from the measured data. However, the extrapolated isolation and backgating threshold lifetimes of greater than 10^5 h at 290°C compare favorably with lifetimes between 10^3 and 10^4 h reported in the literature for commercial GaAs MESFET's aged at approximately 290°C [25], [26]. The high temperature storage tests imply that proton isolation is not the life time limiting component of the GaAs IC process. This conclusion is supported by measured results on a group of devices aged at an ambient temperature of 175°C. After 1500 h of storage, negligible changes in isolation voltages and backgating thresholds were observed.

IV. SUMMARY

Proton implantation has been successfully employed in GaAs IC's to achieve dramatic improvements in isolation and backgating characteristics. In the IC, compatible process protons are implanted through the SiO_2 field oxide (the retained anneal cap) to simplify processing and to optimize surface passivation. An optimum proton dose and energy was determined from a systematic characterization of proton damage. The three-layer dielectric–Au mask is defined to 3-μm linewidths and is easily removed without damaging the exposed GaAs surface.

This process has achieved isolation voltages of 20 V between active regions separated by 3.0 μm, a factor of four improvement in comparison to a selective implant process. Backgating threshold voltages have been increased from -2.5 V for a selective implant process to -5.0 V for proton implantation on lightly Cr-doped substrates and -7.0 V for proton implantation on undoped substrates. These results indicate that proton isolation eliminates both isolation and backgating as design constraints.

High temperature storage tests have demonstrated that proton damage is thermally stable. Isolation lifetimes in excess of 10^5 h at 290°C have been observed. In addition, no measurable changes in isolation voltages or backgating thresholds were detected in devices stressed for 1500 h at 175°C. The high temperature storage tests imply that proton isolation does not limit the operating lifetime of a GaAs IC.

ACKNOWLEDGMENT

The author gratefully acknowledges the contributions of T. Taylor, V. Kumar, R. Van Tuyl, D. Estreich, C. C. Chang, and B. Hughes.

REFERENCES

[1] R. L. Van Tuyl and C. A. Liechti, "High-speed integrated logic with GaAs MESFET's," *IEEE J. Solid-State Circuits*, vol. SC-9, pp. 269–276, Oct. 1974.

[2] R. L. Van Tuyl, C. A. Liechti, R. E. Lee, and E. Gowen, "GaAs MESFET logic with 4-GHz clock rate," *IEEE J. Solid-State Circuits*, vol. SC-12, pp. 485–496, Oct. 1977.

[3] R. C. Eden, B. M. Welch, and R. Zucca, "Planar GaAs IC technology: Applications for digital LSI," *IEEE J. Solid-State Circuits*, vol. SC-13, pp. 419–426, Aug. 1978.

[4] A. G. Foyt, W. T. Lindley, C. M. Wolfe, and J. P. Donnelly, "Isolation of junction devices in GaAs using proton bombardment," *Solid-State Electron.*, vol. 12, pp. 209–214, 1969.

[5] J. C. Dyment, J. C. North, and L. A. D'Asaro, "Optical and electrical properties of proton-bombarded p-type GaAs," *J. Appl. Phys.*, vol. 44, pp. 207–213, Jan. 1973.

[6] B. R. Pruniaux, J. C. North, and G. L. Miller, "Compensation of n-type GaAs by proton bombardment," in *Proc. 2nd Inter. Conf. Ion Implantation in Semiconductors*, New York: Springer-Verlag, 1971, pp. 212–221.

[7] J. D. Speight, P. O'Sullivan, P. A. Leigh, N. McIntyre, K. Cooper, and S. O'Hara, "The isolation of GaAs microwave devices using proton bombardment," in *Proc. Inst. Phys. Conf.*, ser. no. 33a, pp. 275–286, 1977.

[8] T. Sakurai, Y. Bamba, and T. Furuya, "Effects of proton bombardment to n-type GaAs," *Fujitsu Sci. & Tech. J.*, pp. 71–80, June 1975.

[9] J. P. Donnelly and F. J. Leonberger, "Multiple-energy proton bombardment in n^+-GaAs," *Solid-State Electron.*, vol. 20, pp. 183–189, 1977.

[10] R. A. Murphy, W. T. Lindley, D. F. Peterson, A. G. Foyt, C. M. Wolfe, C. E. Hurwitz, and J. P. Donnelly, "Proton-guarded GaAs IMPATT diodes," in *Proc. Symp. on GaAs*, London: Institute of Physics, pp. 224–230, 1972.

[11] C. C. Chang, D. L. Lynch, M. D. Sohigian, G. F. Anderson, T. Schaffer, and G. I. Roberts, "A zero-bias GaAs millimeter wave integrated detector circuit," submitted to the *IEEE MTT-S Symp.*, 1982.

[12] D. D'Avanzo, "Proton isolation for GaAs integrated circuits," in *Research Abstracts of the IEEE GaAs IC Symp.*, paper 21, Oct. 1981.

[13] J. Mun, J. A. Phillips, and I. A. W. Vance, "Optimization of GaAs normally-off logic circuits," in *Research Abstracts of the IEEE GaAs IC Symp.*, paper 9, Oct. 1981.

[14] D. Boccon-Gibod, M. Gavant, M. Rocchi, and M. Cathelin, "A 3.5-GHz single-clocked binary frequency divider on GaAs," in *Research Abstracts of the IEEE GaAs IC Symp.*, paper 7, Nov. 1980.

[15] J. L. Vorhaus, W. Fabian, P. B. Ny, and Y. Tajima, "Dual-gate GaAs FET switches," *IEEE Trans. Electron Devices*, vol. ED-28, pp. 204–211, Feb. 1981.

[16] R. Van Tuyl, V. Kumar, D. D'Avanzo, T. Taylor, V. Peterson, D. Hornbuckle, R. Fisher, and D. Estreich, "A manufacturing process for analog and digital gallium arsenide integrated circuits," *IEEE Trans. Microwave Theory Tech.*, this issue, pp. 935–942.

[17] H. Matsumura and K. G. Stephens, "Electrical measurement of the lateral spread of the proton isolation layer in GaAs," *J. Appl. Phys.*, vol. 48, pp. 2779–2783, July 1977.

[18] P. L. Hower, W. W. Hooper, D. A. Tremere, W. Lehrer, and C. A. Bittmann, "The Schottky barrier gallium arsenide field-effect transistor," in *Proc. 1965 Int. Symp. Gallium Arsenide and Related Compounds*, pp. 187–195, 1968.

[19] T. Itok and H. Yanai, "Stability of performance and interfacial problems in GaAs MESFET's," *IEEE Trans. Electron Devices*, vol. ED-27, pp. 1037–1045, June 1980.

[20] C. Kocot and C. Stolte, "Backgating in GaAs MESFET's," *IEEE Trans. Microwave Theory Tech.*, this issue, pp. 963–968.

[21] K. Steeples, G. Dearnley, and A. M. Stoneham, "Hydrogen-ion bombardment of GaAs," *Appl. Phys. Lett.*, vol. 36, pp. 981–983, June 1980.

[22] S. S. Li, W. L. Wang, P. W. Lai, and R. T. Owen, "Deep-level defects and diffusion length measurements in low energy proton-irradiated GaAs," *Jour. Electronic Materials*, vol. 9, pp. 335–350, 1980.

[23] A. Nouailhat, G. Guillot, G. Vincent, and M. Baldy, "Analysis of defect states by transient capacitance methods in proton bombarded gallium arsenide at 300 K and 77 K," *Lecture Notes in Physics; New Developments in Semiconductor Physics*. New York: Springer-Verlag, 1980, pp. 107–115.

[24] Y. Yuba, K. Gamo, K. Murakami, and S. Namba, "Proton implantation damages in GaAs studied by capacitance transient spectroscopy," in *Inst. Phys. Conf.*, Ser. no. 59, pp. 329–334, 1981.

[25] I. Drukier and J. F. Silcox, "On the reliability of power GaAs FET's," in *17th Annual Proc. IEEE Reliability Physics Sym.*, pp. 150–155, 1979.

[26] K. Mizuishi, H. Kurono, H. Sato, and H. Kodera, "Degradation mechanism of GaAs MESFET's," *IEEE Trans. Electron Devices*, vol. ED-26, pp. 1008–1014, July, 1979.

PROPERTIES OF SURFACE PASSIVATION DIELECTRICS FOR GaAs INTEGRATED CIRCUITS

H. Hasegawa, T. Sawada and T. Kitagawa

Department of Electrical Engineering, Faculty of Engineering

Hokkaido University, Sapporo, 060 Japan

ABSTRACT

Electrical properties of surface passivation films for GaAs ICs were studied by measuring the surface I-V characteristics on semi-insulating wafers and the surface state parameters. Silicon dioxide and silicon nitride films were formed by plasma CVD processes. Characteristic activation energies of 0.3-0.5 eV were observed in both linear and superlinear regions of I-V curves and in electron emission from surface states, which suggests surface state conduction. It is shown that surface passivation becomes increasingly important in GaAs VLSIs.

INTRODUCTION

In spite of the recent rapid development of GaAs digital and analog integrated circuits, electrical properties of the dielectric layer on the semiconductor have neither been investigated in detail, nor been optimized. It is known that the insulator-semiconductor interface in compound semiconductors is characterized by high density of surface states, but their origin and their effects on device performance are not well understood at present(1).

This paper presents the results of a systematic study on the properties of two commonly used surface dielectrics, i.e., silicon dioxide and silicon nitride, both being prepared by plasma CVD processes. Surface I-V characteristics and interface characteristics were investigated in detail. It is shown that these characteristics are extremely sensitive to the insulator deposition process. Therefore, optimization of the surface passivation process becomes increasingly important as the circuit integration level is increased towards VLSI region.

EXPERIMENTAL

Insulator Deposition

Silicon dioxide and silicon nitride films were prepared by plasma CVD processes in $SiH_4 + O_2(+N_2)$ and $SiH_4 + N_2$ gas mixtures, respectively, at the total pressure of 0.4 - 1.0 Torr. Films were deposited onto both Cr-O doped Horizontal Bridgeman and undoped LEC semi-insulating GaAs wafers and also onto semi-conducting GaAs wafers. Wafers were kept at 300 °C during deposition. Figure 1 shows an example of the measured ellipsometric data on the deposited silicon nitride film. For convenience, following notations are used in this paper to indicate the wafer type and the surface condition:

O : silicon dioxide, N : silicon nitride
B : surface before deposition
D : surface covered by insulator
R : surface whose insulator is removed by etching
1 : wafer #1, Cr-O doped HB wafer
2 : wafer #2, Cr-O doped HB wafer
3 : wafer #3, undoped LEC wafer

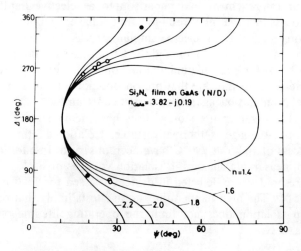

Fig.1 Ellipsometry of Nitride Film

Measurement Method

Surface I-V characteristics were measured on the surface passivated semi-insulating wafers, using the structure shown in Fig.2. AuGe/Ni ohmic contacts with different spacings were provided by the standard photolithographic process.

Interface electronic properties were investigated on MIS samples formed on semi-

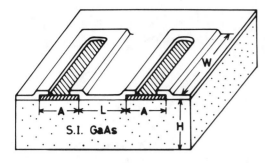

Fig.2 Sample for I-V Measurement

conducting wafers, using C-V, DLTS and PCTS(photocapacitance transient spectroscopy)(2) methods.

EXPERIMENTAL RESULTS

Surface I-V Characteristics

Figure 3 (a) and (b) show the measured I-V characteristics between two ohmic electrodes shown in Fig.2 for an oxide deposited (O/D) surface and a nitride removed (N/R) surface, respectively. Generally, the leakage current shows ohmic behavior up to a threshold voltage V_T, beyond which it first increases sharply in proportion to 6-8th power of the applied voltage for 4-5 decades, and then it further increases less steeply approximately in proportion to the square of the applied voltage. The general behavior is schematically shown in Fig.3(a).

The theoretical value of the resistance, R_b, in the ohmic region for the structure shown in Fig.2 is calculated by conformal mapping as

$$R_b = 4 \rho K(k)/K(k') \qquad (1)$$

where ρ = resistivity of semiconductor bulk
$K(k)$ = complete elliptic integral of the 1st kind
$k = ((p+1)^{-1}-(p-b)^{-1})/((p-1)^{-1}-(p-a)^{-1})$
$p = (2ab+a-b)/(a+b+(2(a-b)(a-1)(b+1))^{1/2}$
$a = \cosh(\pi L/2H), \quad b = \cosh(\pi(L+2A)/2H)$
$k' = (1-k^2)^{1/2}$

The calculated resistance R_b and the measured resistance, R_m, in the ohmic region are plotted vs. spacing L in Fig.4. The bulk resistivity of the substrates was carefully determined using sandwitch structures. As seen in Fig.4, R_m is extremely sensitive to the processing conditions, and can be 1-2 orders of magnitude smaller than R_b, although it can also be close to R_b under optimized conditions. Removal of the insulator films generally increases the leakage current. When R_m is much smaller than R_b, its dependence on L is different from that of R_b, being approximately proportional to L. This indicates formation of a surface ohmic conduction channel.

V_T is also sensitive to processing conditions, as shown in Fig.5. Under the same conditions, V_T was found to be proportional to L, which is different from the well-known L^2-dependence in the trap-filled voltage in the bulk space charge limited currents.

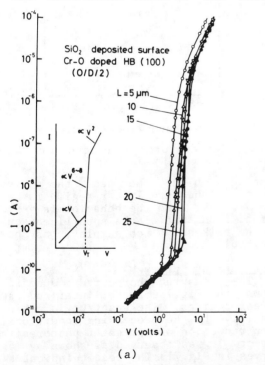

(a)

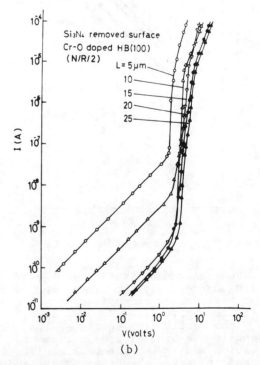

(b)

Fig.3 Measured Surface I-V Characteristics

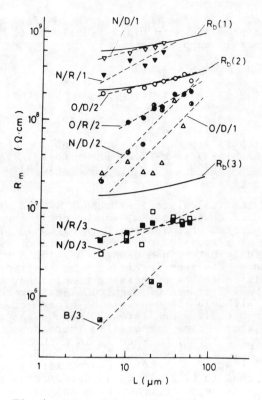

Fig.4 Dependence of R_b and R_m on L

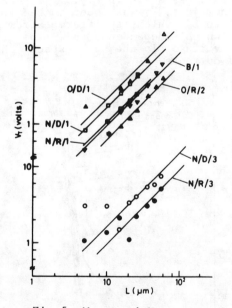

Fig.5 Measured V_T vs. L

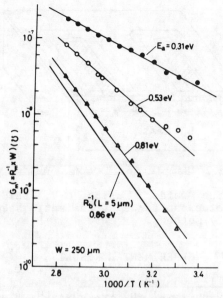

(a) Linear Region

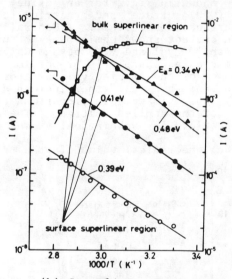

(b) Superlinear Region

Fig.6 Temperature Dependence of Leakage Current

The measured activation energy of the leakage current in the linear and superlinear regions is summarized in Fig.6(a) and (b), respectively. When R_m is close to R_b, the activation energy of the ohmic current is 0.8-0.85 eV, and is close to the activation energy of the bulk resistivity, whereas it takes substantially reduced values of 0.3-0.5eV, when R_m is much smaller than R_b. On the other hand, the activation energy in the superlinear region is always 0.3-0.5 eV, and the behavior is very different from that of the superlinear current in the bulk sandwitch samples, as shown in Fig.6(b).

Photocurrent Spectra

The measured photocurrent spectra are shown in Fig.7 for the linear and superlinear regions. The spectrum is again very sensitive to the processing conditions, and even reduction of current (photoresistance effect) was observed as shown by dashed curves in Fig.7. The result indicates that the conduction is dominated by surface effects.

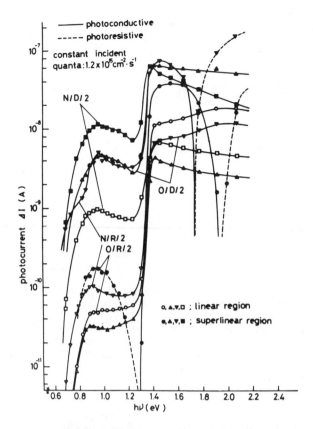

Fig.7 Photocurrent Spectra

Surface State Measurements

C-V, DLTS and PCTS measurements detected high density of surface states, which cause so-called surface Fermi level pinning. It was observed that nitride films tend to give a deeper pinning position from the conduction band edge than dioxide films, although the details of surface state distribution and the pinning position are sensitive to the processing conditions. Generally, the pinning position from the conduction band edge, as revealed by C-V and PCTS measurements, was within the range of 0.7-0.8 eV, but the DLTS activation energy for electron emission from the states at the pinning position was found to be anomalously low, being 0.3-0.5 eV.

DISCUSSION AND CONCLUSION

The results presented here strongly indicate that the surface I-V characteristics are governed by the properties of the semiconductor-insulator interface. Particularly, appearance of a common characteristic activation energy of 0.3-0.5 eV in the linear and superlinear regions of leakage currents and also in the surface state measurements, is suggestive. In a recent paper(3), surface leakage currents and frequency dispersion of transconductance in surface passivated GaAs MESFETs were correlated, where characteristic activation energies of 0.3-0.6 eV were also reported. Our tentative interpretation of the present results is that high-density of surface states form a surface conduction channel, and control the conduction. The linear dependence of V_T on L may be understood by filling of surface states, where the voltage sustained is reduced by the two-dimensional space charge reduction effect. If the charge density per area due to state filling is constant and equal to N_S, then "the surface state filling voltage" is calculated to be

$$V_T = 2 N_S L / \pi \varepsilon_0 (1 + \varepsilon_S) \qquad (2)$$

In an attempt to explain the anomalously small DLTS activation energy for emission from the pinning surface states, we proposed a surface disorder model for the origin of surface states in compound semiconductor MIS interfaces in 1979, whose details are explained in ref.(1). The observed sensitiveness of the characteristics to the processing conditions can be related to the degree of surface disorder induced during processing according to this model.

Since V_T could be less than 100 mV for L of several microns if the surface dielectric and its deposition process are not optimized, surface conduction can impose a serious limitation on the achievable packing density in GaAs VLSIs fabricated by the current technologies. The present results also predict fairly large and fluctuating subthreshold currents in low-power VLSI FET devices, which will affect the operation, yield and reliability of GaAs VLSIs. In addition, it is highly probable that the present surface state conduction channel is closely related to high 1/f noise in GaAs FET devices.

ACKNOWLEDGMENTS

The authors would like to thank Dr.H.Ohno for his helpful discussion. The present work is financially supported in part by a grant-in-aid for the special project research on " Nanometer Structure Electronics".

REFERENCES

(1) H. Hasegawa and T. Sawada, Thin Solid Films Vol.103, pp.119-140, 1983.
(2) H. Hasegawa and T. Sawada, J.Vac.Sci. Technol. Vol.21(2), pp.457-462, 1982.
(3) M. Ozeki, K. Kodama, M. Takikawa and A. Shibatomi, J.Vac.Sci.Technol. Vol. 21(2), pp.438-441, 1982.

A TANTALUM-BASED PROCESS FOR MMIC ON-CHIP THIN-FILM COMPONENTS

M. Durschlag and J. L. Vorhaus

Raytheon Company, Research Division
131 Spring Street
Lexington, Massachusetts 02173

ABSTRACT

A method is described for fabricating both large-value (> 10 pF), small-area capacitors on GaAs utilizing anodization of a Ta:N (20%) film and stable Ta film resistors during the same process providing a full complement of on-chip passive components. The technique has been employed to construct monolithic X-band amplifiers with 400 pF of on-chip capacitance.

INTRODUCTION

The versatility of plasma-deposited Si_3N_4 as a post-implant anneal cap (1), surface passivation layer (2), and dielectric material for the fabrication of thin-film capacitors (3) has led to its widespread utilization in GaAs IC technology. The relatively high capacitance-to-area ratios achievable with Si_3N_4 (120 pF/mm^2) yield small-value (< 4 pF) capacitors that require significantly less GaAs surface area than do capacitors fabricated using competitive technologies such as thin-film polyimide and metal interdigitated structures (C/A = 60 pF/mm^2 and 5 pF/mm^2, respectively).

However, many applications in monolithic microwave integrated circuits (MMICs) now require rather large-value capacitors (> 10 pF) in order to achieve sufficiently low rf impedance levels. For example, rf bypassing of the dc gate bias input of a power FET may require a lumped capacitance of 30-40 pF in order for the capacitor's impedance at X-band to be small compared to the input impedance of the FET. As the complexity and level of integration of MMICs increase, circuits with total on-chip capacitance requirements ranging well above 100 pF are being designed and built (4). If the conventional Si_3N_4 technology is used in fabricating such structures, capacitor areas can easily become the dominant feature in the circuit, consuming 40% or more of the valuable GaAs real estate. This in turn leads to a two-fold decrease in wafer yield. The circuit chip must be larger to accommodate the capacitors, which in turn results in a reduced chip count per wafer. At the same time, the yield of functioning circuits is reduced because of pinhole-induced shorts in the large-area dielectric films. The rf performance of these large structures is also suspect. It is questionable, for instance, if a 40 pF capacitor, which would be nearly one sixteenth of a wavelength on a side at X-band, can be considered a truly lumped element in its performance.

It has long been recognized that a higher dielectric constant film such as Ta_2O_5 (K = 21) would be advantageous in alleviating the above problems (5). In this paper we are reporting on a technique for fabricating large-value capacitors on GaAs MMICs which utilizes the anodization of reactively sputtered tantalum to form such a film. The method is both simple and compatible with standard GaAs processing.

The capacitance-to-area ratio (specific capacitance) achievable with this new technique is ten times typical values for Si_3N_4 capacitors and is comparable to the highest reported values for structures which use reactively sputtered Ta_2O_5 (6,7). In addition, the process is designed to utilize unanodized portions of the Ta film to serve as thin-film resistors, thus eliminating the need for separate resistor material deposition and patterning steps. Also, as explained below, Si_3N_4 films are retained as part of the process to serve as the dielectric layer for small-value (< 4 pF) precision thin-film capacitors and for passivation of the circuit if desired.

PROCESSING

The complete tantalum-based process is presented schematically in Fig. 1. First, the bottom electrodes are defined as part of the conventional source-drain ohmic metallization (Ni/Au-Ge/Au), after which the entire wafer is coated with approximately 5,000 Å of plasma-deposited Si_3N_4. This insulating layer serves a threefold purpose: it is utilized as the dielectric for small-value capacitors; it improves anodization yield by minimizing edge-crossing failures; and it can also be retained as a passivation layer if required. The nitride is next patterned in a freon plasma etch. This step opens holes through the nitride to the metal electrodes for the Ta_2O_5 capacitors and to the GaAs surface for the Ta resistors (Fig. 1a).

The next step is to reactively sputter a tantalum film over the entire surface. By maintaining a 1% partial pressure of N_2 in the argon sputtering gas, a 20 atomic percent nitrogen content is incorporated into the tantalum. This has been shown to lower the loss tangent and reduce the temperature

Reprinted from *IEEE Gallium Arsenide Integrated Circuit Symp. Tech. Dig.*, 1982, pp. 146-149.

coefficient of capacitance (TCC) of the anodized film to less than 200 ppm/°C (8,9). While the TCC is almost an order of magnitude greater than for Si_3N_4, it is nevertheless adequate for the typical circuit applications to which these large capacitors are put.

The nitrogen incorporation has another, even more important, effect. While its presence does decrease the film resistivity slightly compared to pure Ta, it dramatically reduces the temperature coefficient of resistivity (TCR)(8). This results in thin-film resistor material which not only has about a factor of two higher resistivity (at 150 $\mu\Omega$-cm) than the conventionally used titanium, but which also has a TCR more than an order of magnitude better (100 vs. 2500 ppm/°C) and is more chemically and thermally inert than Ti.

After the Ta film deposition, the wafer is masked with photoresist and selectively anodized, as shown in Fig. 1b. (Note that the anodized portion of the film creeps under the protective resist layer.) The initial 3000 Å thick film is typically anodized at 100 V to give a Ta_2O_5 film having a dielectric constant of 21 and a thickness of 1400 Å (10). This results in a specific capacitance of more than 1200 pF/mm^2. The resist is next stripped and a new layer patterned to define the actual lumped element structures. Unwanted material is then removed in a freon plasma (Fig. 1c). At this point the Si_3N_4 can be retained for passivation, if desired, by stopping the plasma etch when just the overlying Ta metal has been removed. Finally, counterelectrodes and contact pads are formed with plated gold using our standard MMIC process (11).

The use of anodization to form the Ta_2O_5 dielectric has several advantages over other methods such as sputtering or chemical vapor deposition. First of all, because it is a growth rather than a deposition technique, the films are somewhat less susceptible to defects and voids. This makes a thinner film more practical and thus further increases the specific capacitance. Secondly, the uniformity and reproducibility attainable with anodization are difficult to achieve in any other manner. Because the thickness is self-limiting, uniformity on a given wafer can be better than ± 2%. In addition, because the thickness is linearly dependent on the final anodization voltage -- a quantity which is generally reproducible to within 1 V without extraordinary measures -- run-to-run variations of less than 1% on a film anodized to 100 V can be readily achieved.

RESULTS AND CONCLUSIONS

Single capacitors as large as 600 pF have been fabricated using the above technique. Results on smaller structures indicate a loss tangent of 0.02 at 1 MHz with a Q of about 70 at that frequency. The breakdown field strength has been measured at between 3 and 6 × 10^6 V/cm. The yield of 40 pF capacitors has generally been good ($\sim$ 90%) with no special attempt made to adjust Ta deposition conditions to optimize yield. The uniformity has been excellent -- about ± 2% across a wafer.

Figure 2 is the schematic diagram of a GaAs MMIC four-stage power amplifier designed for fabrication using the Ta-based process technology. The amplifier consists of a 200 micron periphery FET biased for maximum gain ($V_g = 0$, $V_D = 6$ V) driving three additional FET stages (200, 600 and 1600 micron periphery) biased for maximum power ($I_{DS} = 1/2\ I_{DSS}$, $V_D = 10$ V). All input, output, and interstage matching circuitry is realized on-chip, as are the bias feeding networks. The nine bypass capacitors, which range in value from 20 to 30 pF, use Ta_2O_5 dielectric films (total circuit capacitance is about 200 pF). There are also thin-film Ta resistors with values between 100 and 300 Ω. Two of these are placed across the inputs of the 200 micron FETs to fix their input impedances in order to make stable matching easier across the 9-10 GHz circuit bandwidth. The third resistor is used in the gate bias feeding network to suppress potential low frequency ($\sim$ 10 MHz) oscillations. Both blocking and all three interstage coupling capacitors, which are all < 4 pF, are formed using Si_3N_4 films.

A composite drawing of the various mask layers for this circuit showing its layout is presented in Fig. 3. The chip is 2.65 × 6.3 mm. In the drawing the Ta capacitors are shown crosshatched and the resistors are shown as solid lines. It can be seen that the capacitors easily fit onto the 10 mil square pads required for our via-hole grounding technique (11). Thus, these bypass capacitors are no longer a significant factor in determining the overall area of the circuit chip. It is estimated that if Si_3N_4 capacitors had been used instead, the chip size would have been 20 to 30% larger. In addition, the yield of circuits in which all nine bypass capacitors would have been free of pinholes and thus free of electrical shorts is estimated at no better than a few percent if we had used our standard Si_3N_4 capacitor process. The actual circuit yield was about 25% -- an improvement due entirely to the use of small-area, large-value Ta_2O_5 capacitors for bypassing.

We have also designed and fabricated a two-stage high-power X-band amplifier which has eight 50 pF Ta_2O_5 bypass capacitors on chip. We believe that this total of 400 pF of on-chip capacitance is the largest yet achieved on a GaAs MMIC.

In summary, we have described a process technology based on a sputter deposited anodized Ta:N (20%) film which permits the fabrication of small-area high-value on-chip capacitors. The process is compatible with existing GaAs MMIC technology. Despite using no more masking/patterning processing steps than a simple Si_3N_4 capacitor/thin-film resistor process, this method allows for the simultaneous fabrication of two kinds of thin-film capacitors as well as thin-film Ta resistors while at the same time preserving the option of a retained Si_3N_4 layer for device passivation. The process results in a reduction of the total GaAs area consumed by on-chip capacitors leading to a significant overall yield improvement. Monolithic amplifiers with as much as 400 pF of on-chip capacitance have been successfully fabricated using this technique.

REFERENCES

(1) F. Eisen, C. Kirkpatrick, and P. Asbeck, "Implantation into GaAs," from GaAs FET Principles and Technology, ed. J. V. DiLorenzo and D. D. Khandelwal (Dedham, Massachusetts: Artech House, 1982), pp. 117-144.

(2) A. K. Sinha, H. J. Levinstein, T. E. Smith, G. Quintana, and S. E. Haszko, "Reactive Plasma Deposited Si-N Films for MOS-LSI Passivation," J. Electrochem. Soc., Vol. 125, pp. 601-608:April 1978.

(3) H. S. Veloric, J. Mitchell, Jr., G. E. Theriault, and L. A. Carr, Jr., "Capacitors for Microwave Applications," IEEE Trans. Parts, Hybrids, Packaging, Vol. PHP-12, pp. 83-89:June 1976.

(4) J. L. Vorhaus, R. A. Pucel, Y. Tajima, and W. Fabian, "A Two-Stage All Monolithic X-Band Power Amplifier," IEEE International Solid-State Circuits Conference Digest 24, pp. 74-75:1981.

(5) D. A. McLean, "Tantalum and Tantalum Compounds in Thin Film Microcircuitry," J. Electrochem. Soc. Jap., Vol. 34, pp. 1-11:1966.

(6) M. E. Elta, A. Chu, L. J. Mahoney, R. T. Cerretani, and W. E. Courtney, "Tantalum Oxide Capacitors for GaAs Monolithic Integrated Circuits," IEEE Electron Device Lett., Vol. EDL-3, pp. 127-129:May 1982.

(7) F. Vratny, "Tantalum Oxide Films Prepared by Oxygen Plasma Anodization and Reactive Sputtering," J. Am. Ceram. Soc., Vol. 50, pp. 283-287:June 1967.

(8) M. H. Rottersman, M. J. Bill, and D. Gerstenberg," Tantalum Film Capacitors with Improved AC Properties," IEEE Trans. Components, Hybrids, Man. Tech., Vol. CHMT-1, pp. 137-142:June 1978.

(9) P. K. Reddy and S. R. Jawalekar, "Dielectric Properties of Tantalum Oxynitride Films," Phys. Status Solidi, Vol. 54, pp. K63-K66:1979.

(10) R. W. Berry, P. M. Hall, and M. T. Harris, Thin-Film Technology, (New York:Van Nostrand, 1968), pp. 271-285.

(11) J. L. Vorhaus, R. A. Pucel, and Y. Tajima, "Monolithic Dual-Gate GaAs FET Digital Phase Shifter," IEEE Trans. Microwave Theory Tech. MTT-30, pp. 982-992:July 1982.

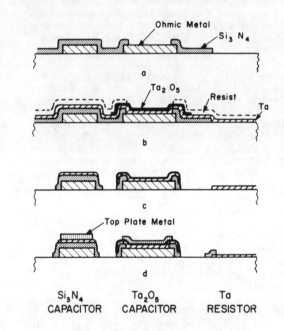

Figure 1. Schematic diagram of tantalum-based thin-film component fabrication process.

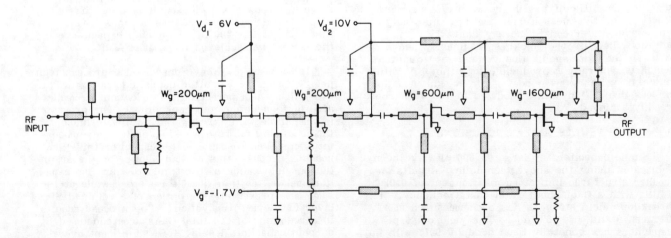

Figure 2. Four-stage X-band power amplifier schematic diagram.

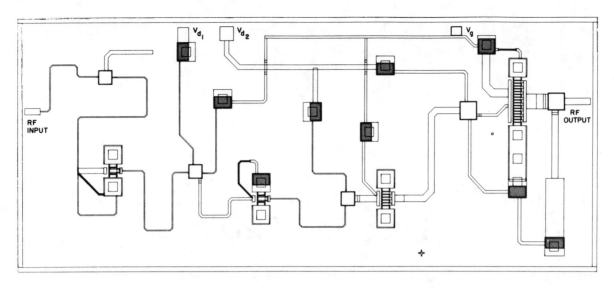

Figure 3. Four-stage X-band power amplifier circuit layout.

Tantalum Oxide Capacitors for GaAs Monolithic Integrated Circuits

M. E. ELTA, MEMBER, IEEE, A. CHU, L. J. MAHONEY, R. T. CERRETANI, AND W. E. COURTNEY

Abstract—The performance of a high-yield tantalum oxide capacitor for use in GaAs monolithic microwave integrated circuits is described. The integral metal-insulator-metal sandwich structure is reactively sputter-deposited at low temperatures, compatible with a photoresist lift-off process, on semi-insulating GaAs substrate. Dielectric constants of 20-25 were achieved in the capacitors fabricated. An initial application of this process as an interstage coupling capacitor for a two-stage preamplifier is given.

THERE HAS BEEN considerable interest in the past decade in monolithic microwave integrated circuits [1]. The integration of active devices such as GaAs MESFET's and mixers has received much attention [2]. However, these circuits also require the integration of active and passive circuit elements which in many cases are the most troublesome part of the monolithic integrated circuit.

The design of a small area capacitor of large capacitance can be very difficult. In order to reliably obtain such a capacitor, a reliable high dielectric constant material is required. This precludes the use of common dielectrics such as Si_3N_4, SiO_2, and polyimide, which have dielectric constants of less than eight. The two most common candidates for low temperature deposition are TiO_2 and Ta_2O_5, which can be formed thermally, anodically, or by reactive sputter deposition [3,4]. This article presents initial results for a reactive sputter deposited Ta_xO_y capacitor.

DESIGN PRINCIPLES

Monolithic microwave integrated circuits (MMIC) require extremely reliable large-value capacitors (50-200 pF) for bias and interstage coupling elements. The most important element in the capacitor is the fragile dielectric thin film which extends over a relatively large area. A low temperature sputter-deposited process which produces a "liftable" integral metal-insulator-metal (MIM) structure, results in the selective deposition of the layers only upon the areas where they are required, and does not expose the other areas of the device wafer to potentially hazardous removal operations. This feature permits the MIM structure to be deposited at the end of the fabrication sequence so that its integrity can be ensured.

The fabrication yield of the capacitor is very important. Sequential deposition of the capacitor films in a vacuum environment will greatly reduce defects, but it is also important to minimize the capacitor area to exclude mask and wafer defects. This requires that the capacitance per unit area be large or that a thin, large dielectric constant material be used. The dielectric layer cannot be made arbitrarily thin because a large breakdown voltage is also important. It is also crucial that the conduction through the dielectric thin film be minimized.

CAPACITOR FABRICATION

The integral MIM capacitors were formed using Ta and Au sputtering targets on a cooled wafer table (38°C) at a RF power level of 50W. The Au-Ta-Ta_xO_y-Ta-Au sandwich uses an Ar plasma for the metal layers and a Ar-O_2 plasma for the dielectric layer. The O_2 level is set to 15-20% of the Ar flow rate. No attempt has been made as yet to optimize the dielectric layer with respect to the O_2 flow rate. The thin Ta layers (100-250 Å) are used to promote adhesion between the gold and the dielectric layer, and is formed by sputter depositing the Ta a few minutes in the Ar plasma before the O_2 flow is switched on for the Ta_xO_y layer. Auger analysis has shown that the transition regions between the Ta and Ta_xO_y layers are less than 80 Å wide at the first interface and less than 200 Å wide at the second interface. The predeposition pressure is less than 2×10^{-7} torr.

The MIM sandwich is selectively deposited on the semi-insulating GaAs using lift-off with an AZ photoresist mask. The MIM layers are then patterned in a staircase structure as shown in Fig. 1. In this process, the top gold layer is wet-etched

---

Manuscript received February 9, 1982; revised March 10, 1982.
The authors are with the Lincoln Laboratory, Massachusetts Institute of Technology, Lexington, MA 02173.
This work was sponsored by the Department of the Army. The U.S. government assumes no responsibility for the information presented.

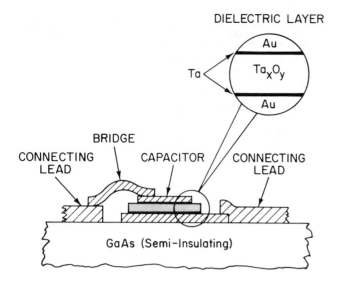

Fig. 1. Staircase capacitor structure with connecting leads for GaAs microwave integrated circuits.

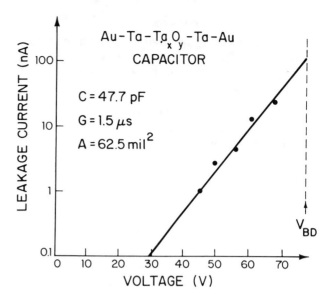

Fig. 2. Leakage current as a function of voltage.

and the Ta-Ta$_x$O$_y$-Ta layers are dry-etched using photoresist masks. The staircase structure eliminates premature edge breakdown and increases the breakdown voltage. Connecting leads and bridge structures can then be used to connect the capacitor to the other devices.

Experimental Results, Applications, and Discussion

Many MIM layers were grown to obtain useful capacitors in the 50-200 pF range. A typical test capacitor has an oxide thickness of 1500 Å with a capacitance of 50 pF. This structure yields a capacitance per unit area of 0.75 pF/mil^2, a conductance per unit area of 24 nS/mil^2, a relative dielectric constant of 20-25, a loss tangent of 0.03 at one megahertz, and a breakdown voltage of 80 volts. The conductance and loss tangent are larger than expected, and it is hoped that the loss tangent at one megahertz will be reduced to 0.01 in the near future. Burnout of the capacitor usually occurs at a small spot at the edge of the top contact. The capacitance and loss tangent have been measured up to 500 MHz with no degradation of quality; experiments are in progress to measure the microwave properties of the capacitor at higher frequencies. The capacitor will also withstand a moderate amount of heat treatment. A 300°C bake for 30 min in air produces no significant change, but a 460°C alloy cycle for 30 sec in flowing nitrogen will reduce C and G by a factor of 2-3, although the loss tangent remains the same. Figure 2 shows the leakage current as a function of voltage for a 47.7 pF capacitor. Notice that the current is a simple exponential function of voltage. No attempt has been made as yet to identify the conduction mechanism for the reactive-sputtered Ta$_x$O$_y$ layers as Mead has done for anodic Ta$_2$O$_5$ layers [5]. Future experiments will concentrate on minimizing the leakage current by studying the stochiometry of the layers produced as a function of deposition parameters.

Figure 3 shows an important demonstration of the capacitor process, a 130-140 pF interstage coupling capacitor for a two stage GaAs MESFET preamplifier. The capacitor layers are deposited and lifted after the source and drain of the MESFET are alloyed. The capacitor is then patterned into the staircase

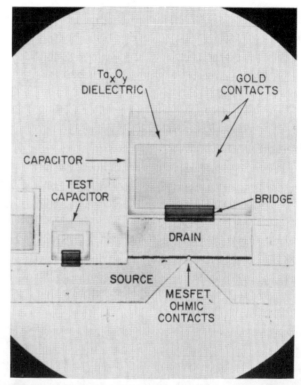

Fig. 3. GaAs integrated circuit showing MESFET ohmic contacts connected to coupling capacitor.

configuration and a polyimide bridge connection to the drain of the MESFET is fabricated. The MESFET gate pattern is then lifted and a high inductance lead is connected to the other capacitor terminal. The first two-stage preamplifier wafer processed had eight circuits with one capacitor failure because of a wafer defect. The mean capacitance was 135.8 pF with a range of values of 132.9-137.1 pF. There was no noticeable correlation to the capacitor location on the wafer. Additional two stage preamplifiers are being processed and designs which include integrating MESFET, coupling capacitors, and bias capacitors are being considered.

Acknowledgment

The authors gratefully acknowledge many helpful discussions with G. Lincoln and R. Sudbury. The authors would like to thank J. Lambert, J. Lawless, K. Molvar, and M. Pierce for technical assistance.

References

[1] R. A. Pucel, "Design considerations for monolithic microwave circuits," *IEEE Trans. Microwave Theory Tech*, MTT-29, pp. 513-534, June 1981.

[2] A. Chu, W. E. Courtney, and R. W. Sudbury, "A 31-GHz monolithic GaAs mixer/preamplifier circuit for receiver applications," *IEEE Trans. Elect. Devices*, ED-28, pp. 149-154, Feb. 1981.

[3] P. K. Reddy and S. R. Javalekar, "Improved properties of TaN-Ta_2O_5 N_x-Al capacitors," *Thin Solid Films*, vol. 64, pp. 71-76, 1979.

[4] T. Umezawa, S. Yazima, and K. Matsummoto, "The electrical properties of resistive and dielectric thin films prepared by reactive sputtering from a Tantalum-Titanium composite target," *Thin Solid Films*, vol. 52, pp. 69-75, 1978.

[5] C. A. Mead, "Electron transport mechanisms in thin insulating films," *Phys. Rev.*, vol. 128, Dec. 1, 1962.

GaAs Microwave Devices and Circuits with Submicron Electron-Beam Defined Features

WILLIAM R. WISSEMAN, SENIOR MEMBER, IEEE, H. MICHAEL MACKSEY, SENIOR MEMBER, IEEE, GAILON E. BREHM, SENIOR MEMBER, IEEE, AND PAUL SAUNIER, MEMBER, IEEE

Invited Paper

Abstract—This paper describes the fabrication and application of GaAs FET's, both as discrete microwave devices and as the key active components in monolithic microwave integrated circuits. The performance of these devices and circuits is discussed for frequencies ranging from 3 to 25 GHz. The crucial fabrication step is the formation of the submicron gate by electron-beam lithography.

I. INTRODUCTION

HISTORICALLY, microwave devices have been one of the primary driving forces of micron and submicron lithography. The initial work on solid-state microwave sources in the 1960's centered on the silicon bipolar transistor [1]. Since the high-frequency performance of solid-state devices is limited by transit-time effects, there was a need for small device geometries. With the advent of the Gunn diode in the middle 1960's and IMPATT diodes a few years later, two-terminal devices played a dominant role [2], [3]. In these devices, the carrier transit time is determined by material parameters rather than the lithographic definition of patterns on the semiconductor surface. Consequently, the rapid progress in the development of GaAs Gunn diodes and silicon, and later, GaAs IMPATT diodes that operated at microwave and millimeter-wave frequencies was not related to lithography improvements. Work also continued on microwave silicon bipolar transistors, and with the development of electron-beam lithography in the late 1960's, culminated in the demonstration of a 10-GHz transistor that delivered 1 W [4]. The effort to develop microwave bipolar transistors was a primary motivation for the development of electron-beam lithography at Texas Instruments.

Beginning in the early 1970's, the GaAs FET became the focus of microwave solid-state device development, and improved lithography was again critical [5]. The GaAs FET is important both as a low-noise amplifier of microwave signals and as a source of microwave power. It is being used in a variety of circuit applications in communications, radar, and electronic warfare. Unlike microwave Gunn and IMPATT diodes, GaAs FET's are planar devices that are fabricated on semi-insulating substrates. This makes it relatively easy to fabricate both active devices and passive impedance-matching elements on the same substrate, and has led to the development of a variety of GaAs monolithic microwave integrated circuits (MMIC's) over the past several years [6]. Complex

Manuscript received November 18, 1982. This work was supported in part by AFWAL, NRL, and NASA Lewis Research Center.
The authors are with Central Research Laboratories, Texas Instruments, Dallas, TX 75265.

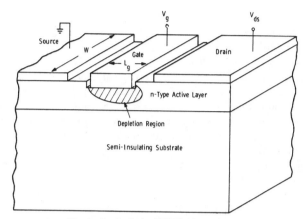

Fig. 1. Basic GaAs FET device structure.

MMIC's are now under development at a number of companies. They offer improved performance and/or reduced cost in comparison with conventional hybrid microwave circuits, and they are expected to result in a variety of systems applications. These include active-element phased-array radars, low-cost expendable jammers for electronic warfare, and receivers for direct satellite television broadcasting.

The development of GaAs FET's paralleled and benefitted from rapid advances in lithography, both optical and electron beam. While the primary motivation for most of these advances was the development of silicon IC's, much of the work on micron and submicron structures carried out to date has involved GaAs. The development of GaAs FET's at Texas Instruments will be described in this paper and their application as discrete devices in hybrid microwave circuits and as the key elements in MMIC's will be discussed. The need for submicron geometries and the means for fabricating structures with these geometries using electron-beam lithography will be covered.

II. REQUIREMENT FOR SMALL DEVICE GEOMETRIES

The basic structure of the GaAs FET is shown in Fig. 1. The structure is formed on a semi-insulating GaAs substrate with a resistivity of about $10^7 \; \Omega \cdot cm$. The conducting layer is formed by epitaxial growth or by direct ion implantation into the substrate. The doping of the conducting layer is n-type since electron mobilities in GaAs are more than an order of magnitude higher than hole mobilities. Current flow between the ohmic source and drain contacts is modulated by changing the voltage on the Schottky-barrier gate metal of length

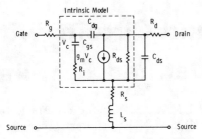

Fig. 2. Equivalent circuit of a GaAs FET.

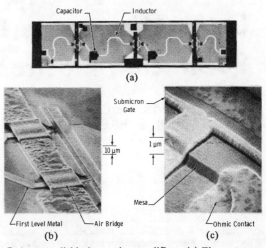

Fig. 3. GaAs monolithic low-noise amplifier. (a) Three-stage amplifier chip. (b) SEM of monolithic FET showing thick metal layers used for inductors and air bridges. (c) SEM showing submicron details of channel region.

L_g. The current handling capability of the FET is determined by the gate width W. Power FET's commonly have multiple-gate fingers in order to provide high current capability without the excessive gate resistance associated with a single-gate finger.

The velocity of electrons in GaAs depends strongly on the electric field. The velocity peaks at about 2×10^7 cm/s and saturates at about 8×10^6 cm/s at high fields. For gate lengths of 1 μm or less, the electron velocity is saturated during most of the transit period. Therefore, the transit time and maximum frequency of operation are determined by the ratio of the gate length to the saturated velocity. For microwave frequency operation, the transit time for the electrons to move through the region under the gate must be minimized. The most obvious way to do this is to reduce the gate length. However, in the limit of very-short gate lengths (<0.25 μm), it has been predicted that the electron transport will be ballistic in nature and that velocities several times the saturation velocity can be achieved [7]. The impact of ballistic transport on microwave FET's is uncertain at this point and is the subject of some controversy [8]. Ballistic transport does not apply to the devices considered in this paper since gate lengths are greater than 0.4 μm for the applications that are described.

In order to explicitly show the effect of gate length on device performance, the FET can be modeled by the lumped-element equivalent circuit shown in Fig. 2 [9], [10]. This equivalent circuit applies to both low-noise and power FET's operated at small-signal levels. The maximum available small-signal gain (MAG) calculated for this circuit is given by

$$\text{MAG} = \left(\frac{f_T}{f}\right)^2 \cdot \frac{1}{4g_{ds}(R_g + R_i + R_s + \pi f_T L_s) + 4\pi f_T C_{dg}(2R_g + R_i + R_s + 2\pi f_T L_s)} \quad (1)$$

and the cutoff frequency f_T is given by

$$f_T \approx \frac{g_m}{2\pi C_{gs}} \quad (2)$$

where

f operating frequency
g_{ds} drain conductance
R_g gate series resistance
R_i channel resistance between source and gate
R_s source series resistance
L_S common-source lead inductance
C_{dg} drain–gate capacitance
C_{gs} gate–source capacitance
g_m transconductance.

When component values determined from measurement of device S-parameters at microwave frequencies are inserted into these equations, the agreement with experimentally measured gain is very good at small-signal levels [11, pp. 13–60]. This approach must be modified to account for the performance of power FET's operated at large-signal levels [11, pp. 61–97]. For given device parameters, the MAG in (1) falls off 6 dB per octave increase in frequency (i.e., as $1/f^2$). The cutoff frequency varies approximately as $1/L_g$ since $C_{gs} \propto L_g$ so that short gate length structures are required to achieve usable gain at high frequencies.

Short gate lengths also are required for low noise figures. An approximate expression for the minimum noise figure in terms of device geometry and material parameters is given by

$$F_{\min} = 1 + K_1 L_g f \sqrt{g_m (R_g + R_S)} \quad (3)$$

where K_1 is a fitting factor [12]. The noise figure depends strongly on gate length, although the dependence is not quite linear because some of the device geometry and material parameters also depend on gate length. In summary, minimizing FET gate length maximizes the frequency of operation and the performance at any given frequency.

III. Device Fabrication

Fabrication of GaAs devices capable of operating at microwave frequencies involves the formation of submicron and micron features, as well as much larger structures, on and in the surface of a single-crystalline GaAs chip. A description of a typical process for fabrication of GaAs MMIC's is given in this section with special emphasis on the techniques used to form the submicron FET gates. The procedure used to fabricate discrete microwave GaAs FET's is a subset of the MMIC process.

Fig. 3 shows a three-stage low-noise amplifier that is representative of GaAs MMIC technology. The chip size is 1 mm × 4 mm. The overall layout and larger features of this circuit are shown in Fig. 3(a). Submicron features of the circuit are shown in the scanning electron micrographs (SEM) of Fig. 3(b) and (c). The arrows point out features that are formed during processing as described in the following. Several

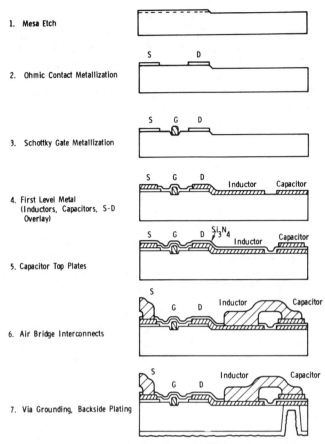

Fig. 4. GaAs MMIC process flow diagram.

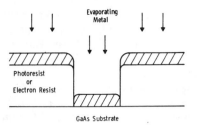

Fig. 5. Schematic diagram of liftoff process.

hundred MMIC's of this type or several thousand discrete devices are fabricated on a single wafer, and several wafers can be processed as a group so that low cost and uniformity of characteristics are obtained.

Two key processes have been used together in the development of submicron GaAs devices: electron-beam lithography for resist patterning and metal definition by resist liftoff. The combination of these two processes produces high-resolution lines with a high aspect ratio (the ratio of metal thickness to line width). Line widths of 0.5 µm with aspect ratios greater than 1 are achieved.

GaAs single-crystal wafers are used as substrates for fabrication of these devices and circuits. Preparation of these substrates involves the synthesis and growth from the purified elements gallium and arsenic of a single crystal of gallium arsenide. The boule is then sawed into wafers that are typically 50 mm in diameter and polished to a mirror finish free of mechanical damage or chemical contamination. The wafers are generally undoped or lightly doped with a compensating impurity such that they are semi-insulating (10^7 $\Omega \cdot$ cm or greater). High crystalline quality and surface finish are required so that submicron features can be formed in and on the substrate in large quantities with high yield. Fig. 4 is a flow diagram for the steps used in wafer processing.

A. Active Layer Formation and Mesa Etch

Using either ion implantation or epitaxial deposition, a continuous n-type layer is formed over the entire wafer surface. This layer is approximately 0.25 µm thick and is the first submicron feature of the circuit. Micron and submicron features will be patterned in the other two dimensions to form the active channels of the microwave FET's on the surface of this layer. To provide isolation, the n-type active layer is mesa etched to remove it from all areas of the wafer except where the active FET's or diodes will be formed. This is done using conventional contact photolithography to pattern a photoresist layer that protects those areas of the surface where the active layer is to remain during wet chemical etching. Alternative approaches for accomplishing this process step include selective ion implantation of the active area using a photoresist mask or ion bombardment of the area surrounding the active region to destroy the conductivity.

B. Source/Drain Ohmic Contact Metallization

An AuGe/Ni layer is evaporated and defined by photoresist liftoff. Liftoff is a technique used extensively to pattern metals on GaAs. The reasons for using liftoff rather than etching are that the GaAs surface is readily damaged by many of the chemicals commonly employed to etch metal patterns and that liftoff produces much steeper metal edges. Fig. 5 illustrates the liftoff technique. A cross section of a typical pattern defined in a photoresist or electron-resist layer is shown. The pattern is the negative image of the desired metal pattern. Metal atoms evaporated from a distant heated source condense on the substrate. If the resist sidewall is vertical or undercut, the metal deposited on the substrate is not connected to that deposited on the resist. When the resist is dissolved away, the unwanted metal is removed. A technique called dielectric-assisted liftoff can be used to improve liftoff of very thick metal layers. In this case, a dielectric layer such as silicon nitride is deposited prior to the resist. The dielectric layer is undercut during the etching process so that a resist overhang is formed. This procedure helps to prevent a connection between metal deposited on the GaAs substrate and that on the resist surface.

After deposition and liftoff of the source/drain metal layer, alloying at 430°C for 3 min forms ohmic contacts to the GaAs. The contact layers are 0.2 µm thick and they have minimum features of 3 µm. Edge smoothness of 0.25 µm is required since alignment of the gate metal to within 0.5 µm of the edge is sometimes required. This ohmic contact metal layer is also used to form alignment marks used by the electron-beam exposure system to precisely locate the critical submicron gate pattern.

C. Schottky-Gate Metallization

Electron-beam lithography is employed for the submicron gates. This technique allows high uniformity of submicron lines on large wafers. The wafer is divided into many fields having alignment marks at each corner. These alignment marks are deposited during the source/drain contact evaporation. A 2 mm × 2 mm maximum field size is usually used at Texas Instruments. The field size is limited by the maximum amount the electron beam can be deflected without distortion. The wafer is coated with an appropriate electron-sensitive material such as PMMA (polymethylmethacrylate) and loaded on a computer-controlled x–y stage in the vacuum chamber beneath the electron column. The controlling computer is programmed to step the stage from field to field within the array, align to each field, and direct the electron beam to expose the proper patterns. Alignment is accomplished by scanning across the alignment marks located at the four corners of the electron-beam field while detecting the backscattered electrons from the edges of these marks. Once these four marks are located, the computer corrects for x and y positional and rotational errors. It also corrects for magnification errors resulting from earlier operations, from variations in wafer temperature, or from the electron-beam machine itself. Special care must be taken in electron-beam patterning of GaAs to provide a ground connection to all patterns on the wafer surface to avoid charging effects caused by the semi-insulating substrate. After stepping, aligning, and exposing each field, the wafer is removed from the vacuum chamber and the pattern is developed. A tradeoff between the electron-beam exposure dose and the development time is made to optimize the resist contrast and the resist edge profile.

After electron-beam exposure and development, the gate area is recessed using a wet etch to achieve the desired device current, and Ti/Pt/Au is evaporated and defined by a liftoff process similar to that used for ohmic contacts. Fig. 3(b) shows the smooth edges and high aspect ratio that are achieved. This gate metal stripe is 0.6 µm thick and 0.5 µm wide.

D. First-Level Metal (Inductors, Capacitor Bottom Plates, Source/Drain Overlay)

Sequential layers of titanium and gold are deposited and patterned by liftoff to simultaneously form the inductors, the bottom plates of the metal–insulator–metal (MIM) capacitors, and to overlay the alloyed source and drain regions to reduce the spreading resistance. This metal can be almost 2 µm thick to reduce microwave losses in the various components. It is patterned using the dielectric-assisted liftoff procedure described previously.

E. Capacitor Top Plates

High-quality silicon nitride is deposited by a plasma-assisted reaction of silane and ammonia to form the capacitor dielectric. Ti/Au is again vacuum deposited and patterned by liftoff to form the capacitor top plates. For discrete devices, the silicon nitride is used only as a protective layer.

F. Air-Bridge Interconnects

Gold-plated air bridges are used to form low conductor loss, low parasitic capacitance interconnects between source regions of multiple-gate-finger FET's; this is a convenient way, as well, to connect capacitor top plates and first-level metal inductors without step coverage problems. A thick layer of photoresist supports the gold bridge during plating and another layer of resist confines the plating to the desired area. Approximately 5 µm of plated gold is used to form a strong low-loss structure. Elements of the circuit requiring additional conductivity are plated at the same time, as are the pads for wire bonding and dc probing.

G. Thinning, Backside Metallization, and Dicing

After completion of the preceding steps, topside wafer processing is complete and all the micron and submicron features of the device have been formed. The wafer must then be thinned to the proper thickness and metallized on the backside to form an RF ground plane and a strong metallurgical bond for packaging. In some cases, vias are etched through to the topside and backfilled with plated gold where grounding is required in the interior of the circuit. Finally, the individual MMIC chips are separated from each other by a high-speed saw or by a scribe and break technique.

IV. Discrete FET's

A number of discrete GaAs power FET's have been developed at Texas Instruments for use at frequencies from 3 to 25 GHz. The design guideline typically used is that at least 4-dB gain is required at the operating power level for a useful device. The small-signal gain is at least 4 dB higher. The gain drops off as $1/(\text{frequency})^2$ as shown in (1), so that special care must be taken in the design of FET's that are operated at the higher end of this frequency range. In order to obtain the necessary gain, the gate length is reduced in accordance with (1) and (2), but changes in device design are also required. Two of the most effective changes are to reduce the gate-finger widths [13] and the source lead inductance [10].

It has been shown that small devices can produce up to 1-W output power per millimeter gate width almost independent of frequency over most of the frequency range under discussion [15]. However, the maximum output power that has been obtained from GaAs FET's described in the literature decreases from about 30 W at 3 GHz to 1 W at 20 GHz. This decrease in output power with frequency is partially due to the inability to either match the impedance or uniformly feed large transistors as frequency increases, since the chip size is an increasing fraction of a wavelength. Also contributing to reduced output power at higher frequencies is the fact that the design changes to increase gain mentioned in the previous paragraph generally reduce the total gate width that can fit on a given size chip. In this section, three Texas Instruments devices are described in detail in order to illustrate the different design and lithography requirements for different frequency ranges. The devices are an X-band FET (8–12 GHz), a K-band FET (18–26 GHz), and an S-band FET (2–4 GHz). They are described in the order of their development. These devices can, of course, be operated at less than the design frequency at a higher gain level.

A. X-Band FET

The X-band FET has 4.8-mm total gate width in four cells. The large gate width is obtained by the conventional method of paralleling a number of gate fingers—each of the four cells has eight 150-µm fingers. Fig. 6 shows SEM photographs of a bonded device. The gate pad is located at the bottom of Fig. 6(a) and the drain pad at the top. The sources are interconnected by a plated-gold air bridge. In Fig. 6(b) it is seen that the sources are grounded by bond wires at the ends of the 2 mm × 0.5 mm × 0.1-mm-thick chip and also by wires going across the chip. If the wires going across the chip were left off, the source lead inductance would be excessive and the gain would be reduced by several decibels. The gain would

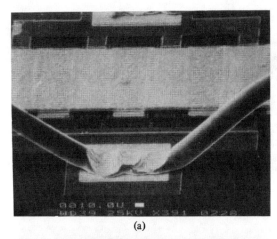

Fig. 6. SEM photographs of bonded X-band GaAs power FET. (a) Details of a single cell. (b) 4.8-mm gate-width chip.

also be degraded if the wires interconnecting the gate and drain pads were left off. These wires help equalize the amplitude and phase of the microwave signal at each cell.

The devices are fabricated by the process described in Section III with gates defined by electron-beam lithography. Each electron-beam field is 2 mm X 2 mm and contains three 4.8-mm gate-width devices plus several material and processing test patterns. The gate length is typically 0.75 μm and the gates are fabricated by the liftoff of 0.15 μm Ti/0.05 μm Pt/0.4 μm Au with about 0.7 μm of PMMA. The exposure time per field is about 10 s including about 3 s to move the stage between fields and acquire the alignment marks. The total exposure time for a 2.5 cm X 2.5 cm slice is about 25 min. This time could be reduced significantly by changing several photomasks slightly so that the electron-beam machine would not be used to write any part of the gate pad. This improvement has been made on devices developed more recently. Device gain is not significantly improved by reducing the gate length to 0.5 μm, because the increased gain due to the shorter gate length is canceled by gain reduction caused by increased signal attenuation along the more resistive 150-μm gate fingers. In order to take advantage of the higher gain accompanying 0.5-μm gates, it is necessary to reduce the finger width to 75 μm or less. This would cut the device total gate width, and hence the power capability in half. The present gate-gate spacing of the device in Fig. 6 is 30 μm across the drains and 55 μm across the sources. It is not desirable to reduce this

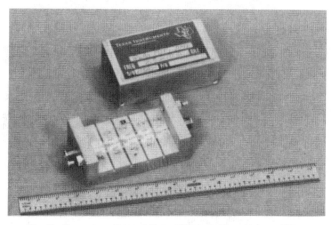

Fig. 7. Four-stage X-band GaAs FET hybrid power amplifier.

spacing significantly to increase device total gate width, because chip heating would become excessive.

The 4.8-mm gate-width devices typically have 2.5- to 3-W output power with 4- to 6-dB gain at 10 GHz, with the best devices having up to 3.9-W output power with 6-dB gain and 39-percent power-added efficiency. The highest gain devices can be operated into Ku-band, producing up to 2-W output power with 4-dB gain at 15 GHz. These devices were developed for use in the transmitter-amplifier of a phased-array radar module which also contains a low-noise receiver and a phase shifter. A large number of identical modules are required in a phased-array radar. The transmitters are typically capable of amplifying a transmitted signal with 25-dB gain and 2-W output power over a 1-GHz bandwidth at X-band. This gain requires four stages of amplification and the 4.8-mm device is used in the fourth stage. Fig. 7 is a photograph of a 2-W X-band amplifier with four stages of amplification. The input, output, and interstage impedances are matched by the microstrip circuits that take up most of the amplifier area in Fig. 7.

B. K-Band FET

The K-band FET has been designed especially for high-frequency (20-GHz) operation. The main feature of this device type is the single straight-line gate with two or more feeds (hence the name π-gate). Devices with 0.3-, 0.6-, 0.9-, 1.35-, and 1.8-mm gate widths have been developed at Texas Instruments.

Fig. 8 is a photograph of a 1.35-mm device. Nine gate pads connect the gate stripe; each gate pad feeds two individual 75-μm-wide gate fingers. This unit gate width is optimum for high-frequency operation and permits the use of 0.5-μm gate lengths to obtain higher gain. Since it is necessary to interconnect the gate pads in order to feed all gates equally, air bridges are needed to connect the individual sources to the grounding bar that extends the length of the chip.

At high frequencies, it is difficult to achieve adequate gain (>4 dB). Therefore, the attainment of small gate lengths is critical. By carefully controlling the exposure of the electron-beam resist and its development time, it is possible to define 0.4- to 0.5-μm-wide stripes and liftoff 0.6-μm-thick metal. For minimum distortion and maximum resolution, a reduced field size of 1.5 mm X 1.5 mm is used. It contains three 1.35-mm, two 0.6-mm, and one 0.3-mm gate-width devices

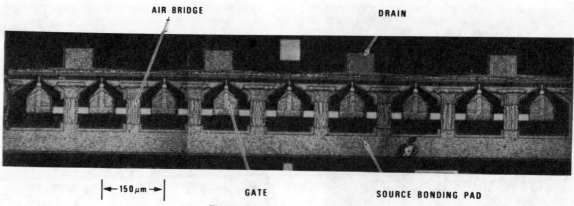

Fig. 8. K-band GaAs power FET.

plus several material and processing test patterns. Since the electron-beam machine is not used to write any part of the gate pads, the writing time per field is about 3 s.

The π-gate structure has several advantages over the conventional parallel-finger design. The parasitic reactances, which become increasingly important influences on performance as the frequency of operation increases, are smaller. Equivalent circuit calculations from the S-parameters demonstrate that the π-gate structure reduces the gate-to-drain capacitance by about 30 percent. The large-area source-grounding bar connected to ground by a low-inductance mesh or a sheet of solder insures a very low source lead inductance per unit gate width. Another advantage of this structure is that, with a large number of widely spaced gate and drain pads, the effective inductance of the bond wires used to contact the device can be made very low so that the FET input and output capacitances can be resonated at very high frequencies. Finally, the very symmetrical feeding of the gates improves the performance.

A 1.35-mm FET typically has 700- to 800-mW output power with 6-dB gain at 15 GHz. The available gain decreases as the frequency goes up: at 20 GHz the best devices have 675 mW with 5.8-dB gain; at 22 GHz, 355 mW with 4.5-dB gain; and at 23 GHz, 200 mW with 4.0-dB gain. Up to 200 mW at 25 GHz was obtained from a device operated as an oscillator [15].

These devices were used extensively in an amplifier developed for NASA operating in the 17.7–20.2-GHz communication band [16]. Fig. 9(a) is a photograph of the overall amplifier which combines 16 identical modules. Fig. 9(b) is a detailed picture of one module showing the six stages necessary to achieve the goal of 0.5 W with 30-dB gain. The first five stages use, respectively, 150-, 150-, 300-, 600-, and 1350-μm gate-width FET's. In the output stage, two 1350-μm devices are paralleled by using a Wilkinson combiner. The overall amplifier has a power of 8 W with 30-dB gain.

C. S-Band FET

The S-band FET was developed recently [17] with a goal of fabricating a device capable of producing 25-W pulsed output power with 6-dB gain across the 3- to 3.5-GHz band. These devices will be used in phased-array radar systems in a transmitter–amplifier application similar to that of the X-band device described earlier, but at a lower frequency and con-

(a)

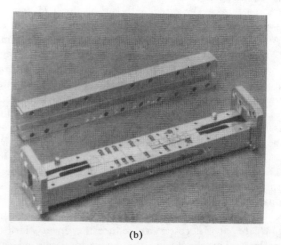

(b)

Fig. 9. K-band GaAs FET hybrid power amplifier. (a) 16-module amplifier having 8-W output power with 30-dB gain over the 17.7–20.2-GHz band. (b) Six-stage amplifier module having 30-dB gain and 0.5-W output power.

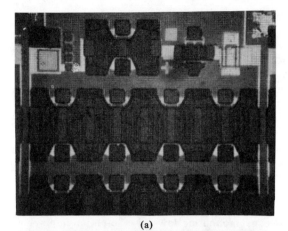

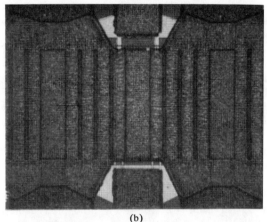

Fig. 10. S-band GaAs power FET slice during fabrication. (a) 19.2-mm gate-width device. (b) Details of a single cell.

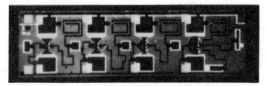

Fig. 11. Four-stage X-band GaAs FET monolithic power amplifier chip.

siderably higher power level. The 25-W device has 38.4-mm total gate width in eight cells. Two 19.2-mm gate-width chips are used instead of a single chip in order to increase yield and reduce thermal stresses during bonding. The chips are mounted adjacent to each other and treated as a single 38.4-mm gate-width device. In order to fit this large gate width on a chip, the gate-finger width is increased to 300 μm and the gate-to-gate spacing reduced to 20 μm. The chips are thinned to 0.05 mm to reduce the resulting high operating temperature. Fig. 10 shows photographs of a slice during processing. A 19.2-mm gate-width device is shown in the center of Fig. 10(a) and a single cell is shown in Fig. 10(b). The sources are interconnected by a plated-gold air bridge which also connects to the bonding areas along both sides of the chip.

The major problem with the K-band FET described previously is obtaining enough gain at that high a frequency. The gain requirement for the X-band device is not as severe, but the yield is more of a concern since large numbers of fairly large devices are required. With this S-band device, gain is not a problem due to the low frequency (see (1)). However, it is difficult to obtain high process yields with such a large device (38.4-mm total gate width distributed among 128 gate fingers). The use of longer gates is one way to increase yield. Because of the low operating frequency, adequate gain is obtained with a gate length of 1.5 μm. Although the 1.5-μm gates can be defined either with photolithography or electron-beam lithography, the latter has been found to give a con-

siderably higher yield. There is no mask–slice abrasion, mask runout, or variation in mask contact. A 2 mm $\times$ 2 mm electron-beam field contains two 19.2-mm gate-width chips plus two smaller gate-width devices and several test patterns. The exposure time is only about 10 s per field, since the photomasks are designed so that only the gate fingers are defined by the electron beam. Owing to the larger gate length, relatively thick gate metal can be lifted off, resulting in higher gain due to the reduced signal attenuation on the gate fingers. 0.7 μm of PMMA is used to lift off the 0.15 μm Ti/0.05 μm Pt/0.8 μm Au gates.

The best 38.4-mm gate-width devices fabricated to date have had 20-W CW output power with 7-dB gain and 34-percent power-added efficiency at 3 GHz. Pulsed devices have had more than 30-W output power at 3 GHz and 25 W across the 3.0- to 3.5-GHz band.

V. MONOLITHIC CIRCUITS

A number of MMIC's have been developed for a variety of purposes at Texas Instruments. These circuits are generally designed using well-characterized submicron-gate GaAs FET types in a chip layout that makes use of MIM capacitors and high-impedance microstrip lines as inductive elements. This semi-lumped-element approach minimizes the amount of chip area consumed by the passive circuit elements.

These element values are optimized using computer-aided design and are then precisely and repeatably realized during the MMIC fabrication process. On some circuits, plated-through vias are used to ground points within the circuit. The circuits described in this section are typical of power amplifiers, low-noise amplifiers, and oscillators being developed for phased-array radar, electronic warfare, and satellite communications applications from 7 to 20 GHz.

A. X-Band Power Amplifier

A four-stage power amplifier utilizing discrete GaAs FET's was shown in Fig. 7. A similar amplifier which has been integrated onto a single chip of GaAs is shown in Fig. 11 [18]. This amplifier is 5.75 mm $\times$ 1.63 mm $\times$ 0.15 mm, and GaAs FET's with gate widths of 0.3, 0.3, 0.6, and 1.5 mm are used in the first through fourth stages. Networks of inductors and MIM capacitors are used to provide 50-Ω input and output impedances and the appropriate impedances for interstage matching. The drain bias is brought to the FET's through the line running along the upper side of the chip and the gate bias through the line along the lower side. In Fig. 11, the FET's are the multifinger structures along the center of the chip and the capacitors are the large squares and rectangles. The eight very large pads are ~30-pF shunt capacitors to ground. The ground plane is brought to the capacitor bottom plates through etched vias to the back of the chip which also ground the FET sources. Only four bonds are required

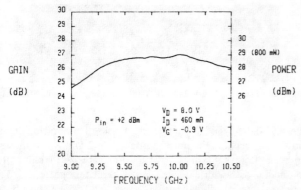

Fig. 12. Gain–frequency response of the power amplifier of Fig. 11.

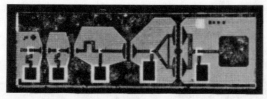

Fig. 13. Four-stage K-band GaAs FET monolithic power amplifier.

to operate this chip: drain bias, gate bias, RF input, and RF output.

The fabrication process was described in Section III. The gates are approximately 0.75 μm long and are defined by electron-beam lithography. The chip is divided into two electron-beam fields, one with the two 0.3-mm gate-width devices and one with the two larger devices. Fig. 12 shows the gain–frequency response of one of these amplifiers. This chip, with less than 10-mm^2 area, has the performance of a much larger hybrid amplifier employing discrete devices. The monolithic amplifier eliminates numerous bond wires and is potentially much more reliable than an equivalent hybrid amplifier.

A similar four-stage amplifier (without the on-chip bias network) has been developed that produces 1.25-W output power with 30-dB gain over the 7.25- to 7.75-GHz band [19]. Several hundred of these amplifiers have been fabricated for use in a satellite active-array antenna. The slice yields have averaged about 20 percent with some as high as 40 percent.

B. K-Band Power Amplifier

The amplifier discussed in this section makes full use of GaAs monolithic technology to replace the module of Fig. 9(b) with the single 1.4 mm × 4.4 mm chip shown in Fig. 13. The performance goal is 0.5-W output power with 20-dB gain across the 17.7–20.2-GHz bandwidth. This four-stage amplifier now under development at Texas Instruments uses 0.3-, 0.3-, and 0.6-mm FET's in the first through third stages. The output stage is a combination of two 0.75-mm FET's in parallel. The concept of the matching network is the same as for the X-band amplifier described in Section V-A, but the size of the inductors and capacitors are smaller since the circuit is designed for a higher operating frequency. Etched vias are used to ground the FET sources and the capacitor bottom plates.

As stated previously, the FET gate length is particularly critical to the performance of high-frequency FET's. In this case, 0.5-μm-long gates are needed to insure a sufficient gain, and gate-finger widths of 75 μm are required to minimize losses. The chip is divided into two electron-beam fields, one with the two 0.3-mm devices and one with the 0.6- and 1.5-mm devices.

The advantage of the monolithic approach over the conventional hybrid amplifier employing discrete FET's is particularly obvious if one considers the size difference between this chip and the module of Fig. 9(b). Also of prime consideration, if one were to use monolithic circuits in the 16-module amplifier of Fig. 9(a), would be the saving of most of the 700 bonding steps that are required to assemble and tune the 112 discrete FET's.

C. X-Band Low-Noise Amplifier

An X-band monolithic low-noise amplifier employing three 300-μm gate-width FET's in a three-stage common-source cascade was shown in Fig. 3 [20]. This chip is 4 mm × 1 mm × 0.15 mm in size, which is roughly one-fifth the length and width of an equivalent low-noise amplifier using conventional hybrid techniques. The input-matching circuit presents the optimum source impedance, or noise match, to the first FET of the amplifier, minimizing the noise figure. Succeeding stages incorporate matching circuits to maximize gain over the desired bandwidth. The output stage is matched to 50 Ω. Wire bonds are used only to connect the input and output terminals and to provide gate and drain bias to each FET.

The fabrication process for this amplifier chip was described in Section III. The metallurgical FET gate length (width of the gate stripe) is 0.5 μm. The gates are defined in 0.6-μm-thick Ti/Pt/Au using electron-beam lithography and metal liftoff. To satisfy the special requirements for low capacitance in low-noise devices, a thin layer of silicon nitride is used under the electron resist. This nitride layer is removed after electron-beam exposure and development by plasma etching in a reactor, which allows precise control of the undercut of the nitride layer beneath the resist overhang. This small undercut allows the edge of the gate recess, which is to be etched into the GaAs, to be moved back from the gate metal edge. The gate is then defined by the overhanging resist edge during the evaporation and liftoff process. By enlarging this excess gate recess to slightly more than the gate depletion depth, the depletion-layer capacitance is greatly reduced with only a minimal increase in parasitic resistance. Electron-beam exposure of this circuit, like most large MMIC's, must be accomplished using multiple fields. Actually, a pair of these chips, which are located side by side on the wafer, are exposed together in two electron-beam steps. The gates of the first two stages of the two amplifiers are exposed in one pass followed by the gates of the last stages. Since there is a relatively small amount of exposed area in each field, exposure times are minimal even for relatively insensitive electron resists.

This amplifier will have application in the receiver sections of phased-array radar and communications systems. Fig. 14 is a plot of typical gain and noise figure as a function of frequency. At 10.5 GHz, the noise figure is 3.1 dB and the gain is 26 dB.

D. Voltage-Controlled Oscillator

Fig. 15 shows an MMIC voltage-controlled oscillator (VCO) containing a 300-μm gate-width FET, two varactor diodes, and appropriate passive matching circuitry for achieving wide tuning bandwidths at X- and Ku-bands [21]. The looped trans-

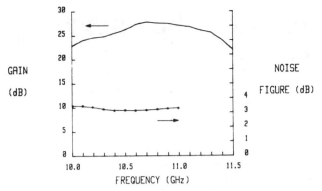

Fig. 14. Gain and noise figure as a function of frequency for the low-noise amplifier of Fig. 3.

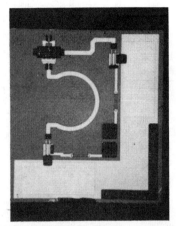

Fig. 15. Ku-band monolithic voltage-controlled oscillator chip.

mission line, which has approximately 1-nH inductance, resonates with the sum of the capacitances of the FET and the varactor-diode cathode.

The fabrication process used for these VCO chips is similar to that used for the monolithic power and low-noise amplifiers. The submicron (0.5-μm) gates are defined by electron-beam direct-slice writing as before. This chip is small enough that a complete circuit and surrounding test devices can be included in a single electron-beam field. The varactor-diode anodes, which are the same metallization as the FET gates but are 3 to 6 μm wide, are defined by electron beam for convenience and for precise control of their dimensions. Their size, and therefore their capacitance, can then be changed as needed from run to run by simply changing the electron-beam pattern-control program.

These monolithic VCO's have shown tuning bandwidths over 6 GHz in the 12- to 22-GHz frequency range. They will find application in electronic warfare systems. Other versions of this basic design can be used in microwave receivers with phase-lock-loop frequency control.

VI. Conclusion

We have shown why micron and submicron geometries are needed for microwave transistors and how electron-beam lithography is applied to the fabrication of submicron-gate GaAs FET's. Descriptions were given of discrete GaAs FET's and GaAs MMIC's that operate in the frequency range from 3 to 25 GHz. In the future it is expected that advances in techniques for achieving submicron geometries, such as electron-beam lithography, will continue to be applied first to microwave devices.

Acknowledgment

The authors wish to thank the members of the GaAs Microwave and Digital Technology Branch for their many contributions to the work described in this paper. They are especially grateful to K. Bradshaw and P. Tackett for their efforts in the area of electron-beam lithography.

References

[1] H. F. Cooke, "Microwave transistors: Theory and design," *Proc. IEEE*, vol. 59, pp. 1163-1181, Aug. 1971.

[2] J. B. Gunn, "Instabilities of current in III-V semiconductors," *IBM J. Res. Develop.*, vol. 8, pp. 141-159, Apr. 1964.

[3] R. L. Johnston, B. C. DeLoach, and B. G. Cohen, "A silicon diode microwave oscillator," *Bell Syst. Tech. J.*, vol. 44, pp. 369-372, Feb. 1965.

[4] H. T. Yuan, J. B. Kruger, and Y. S. Wu, "X-band silicon power transistor," in *1975 Int. Microwave Symp. Dig. Tech. Papers*, pp. 73-75.

[5] C. A. Liechti, "Microwave field-effect transistors—1976." *IEEE Trans. Microwave Theory Tech.*, vol. MTT-24, pp. 279-300, June 1976.

[6] R. A. Pucel, "Design considerations for monolithic microwave circuits," *IEEE Trans. Microwave Theory Tech.*, vol. MTT-29, pp. 513-534, June 1981.

[7] M. S. Shur and L. F. Eastman, "Ballistic transport in semiconductor at low temperatures for low-power high-speed logic," *IEEE Trans. Electron Devices*, vol. ED-26, pp. 1677-1683, Nov. 1979.

[8] J. Frey, "Ballistic transport in semiconductor devices," in *1980 IEDM Tech. Dig.*, pp. 613-617.

[9] P. Wolf, "Microwave properties of Schottky-barrier field-effect transistors," *IBM J. Res. Develop.*, vol. 14, pp. 125-141, Mar. 1970.

[10] M. Fukuta, K. Syama, H. Suzuki, and H. Ishikawa, "GaAs microwave power FET," *IEEE Trans. Electron Devices*, vol. ED-23, pp. 388-397, Apr. 1976.

[11] R. S. Pengelly, *Microwave Field-Effect Transistors—Theory, Design and Applications*. Chichester: Research Studies Press, 1982.

[12] H. Fukui, "Optimal noise figure of microwave GaAs MESFET's," *IEEE Trans. Electron Devices*, vol. ED-26, pp. 1032-1037, July 1979.

[13] W. R. Frensley and H. M. Macksey, "Effect of gate stripe width on the gain of GaAs MESFETs," in *1979 Cornell Electrical Engineering Conf. Proc.*, pp. 445-452.

[14] J. V. DiLorenzo and W. R. Wisseman, "GaAs power MESFET's: Design, fabrication, and performance," *IEEE Trans. Microwave Theory Tech.*, vol. MTT-27, pp. 367-378, May 1979.

[15] H. M. Macksey, H. Q. Tserng, and S. R. Nelson, "GaAs power FET for K-band operation," in *1981 ISSCC Dig. Tech. Papers*, pp. 70-71.

[16] R. C. Bennett, P. Saunier, R. P. Lindsley, C. H. Moore, and R. E. Lehmann, "20 GHz GaAs FET transmitter," presented at the Int. Conf. on Communications, Philadelphia, PA, June 1982.

[17] H. M. Macksey, H. Q. Tserng, and G. H. Westphal, "S-Band GaAs power FET," in *1982 Int. Microwave Symp. Dig. Tech. Papers*, pp. 150-152.

[18] H. Q. Tserng, H. M. Macksey, and S. R. Nelson, "A four stage monolithic X-band GaAs FET power amplifier with integrated bias network," in *1982 GaAs IC Symp. Tech. Dig.*, pp. 132-135.

[19] H. Q. Tserng and H. M. Macksey, "A monolithic GaAs power FET amplifier for satellite communications," in *1982 ISSCC Dig. Tech. Papers*, pp. 136-137.

[20] R. E. Lehmann, G. E. Brehm, and G. H. Westphal, "10 GHz 3-stage monolithic low-noise amplifier," in *1982 ISSCC Dig. Tech. Papers*, pp. 140-141.

[21] B. N. Scott and G. E. Brehm, "Monolithic voltage-controlled oscillator for X- and Ku- bands," in *1982 Int. Microwave Symp. Dig. Tech. Papers*, pp. 482-485.

Yield Considerations for Ion-Implanted GaAs MMIC's

ADITYA GUPTA, WENDELL C. PETERSEN, MEMBER, IEEE, AND D. R. DECKER, SENIOR MEMBER, IEEE

Abstract —An ion-implantation based process is described for fabricating GaAs monolithic microwave integrated circuits (MMIC's) incorporating active devices, RF circuitry, and bypass capacitors. Low ohmic contact resistance and good control of metal–insulator–metal (MIM) capacitance values is demonstrated and some factors affecting FET and capacitor yield are discussed. High dc yield of typical amplifier circuits is shown indicating that this process has the potential for achieving very high overall yields in a production environment. Good yield of functional MMIC modules with multicircuit complexity is projected.

I. INTRODUCTION

AN ION-IMPLANTATION based process has been developed for the fabrication of gallium arsenide monolithic microwave integrated circuits (MMIC's) incorporating active devices, RF circuitry, and all bypass capacitors. Multiple, localized ion implantation in semi-insulating GaAs substrates [1] is used for forming optimized active layers and n^+ contacts for low noise and power FET's, mixer diodes, etc. In addition to providing a planar structure, this approach allows the flexibility of accommodating different active layers on the same substrate as required for optimizing the performance of different devices in the circuit. Contact photolithography is used for all pattern steps. Silicon nitride deposited by plasma enhanced CVD is used as the dielectric in metal–insulator–metal (MIM) capacitors and as the insulator in a two level metallization process. Excellent uniformity and reproducibility of MIM capacitors has allowed their use for both RF tuning and bypassing. Except resistors, all microwave circuitry, air bridges, and beam leads are on the second metallization level which is electroplated to a thickness of 2–3 μm to minimize losses. Backside through-substrate via holes are etched where necessary. This paper presents an overview of the fabrication process and discusses yield limiting factors which have been investigated. The data presented here have been obtained on a process diagnostic test pattern [2] (Fig. 1) present on all our MMIC mask sets.

II. FABRICATION PROCESS

Fig. 2 is a schematic drawing of the various active and passive components comprising an MMIC. These include low noise and power MESFET's, Schottky-barrier diodes, thin film and bulk resistors, MIM capacitors for RF tuning and bypassing, transmission lines, air bridges, and backside

Manuscript received May 5, 1982.
The authors are with Rockwell International, Microelectronics Research and Development Center, Thousand Oaks, CA 91360.

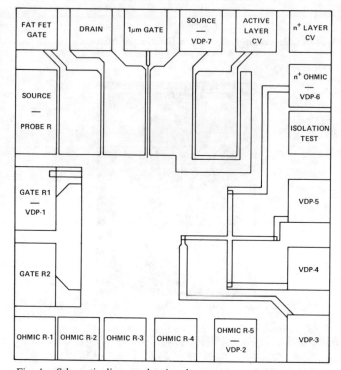

Fig. 1. Schematic diagram showing the process monitor test pattern.

through-substrate via holes. A pictorial representation of the fabrication procedure is given in Fig. 3. As shown there, fabrication of an MMIC begins with the synthesis of doping profiles for FET active layers, n^+ contacts and bulk resistors by localized Si^+ ion implantation in qualified semi-insulating GaAs substrates. Photoresist is used as the implantation mask. Substrate qualification consists of sampling the front and the tail of the ingot under consideration and checking the doping profile for a standard implant-cap-anneal cycle. Activation, pinch off voltage uniformity, and electron mobility are measured and compared with design specifications to determine the suitability of the ingot for the MMIC process. The isolation afforded by the SI substrate after undergoing an annealing cycle is also checked. A sheet resistance $\geq 10^7$ $\Omega/\square$ is required for passing this test. Fig. 4 shows the reproducibility of a 100-keV Si-implantation profile in different types of substrates, processed at different times. The minor profile variations observed arise as a result of differences in substrate background doping and compensation and may be corrected by slight adjustment of implant schedules based on qualification data. Additional data for a different im-

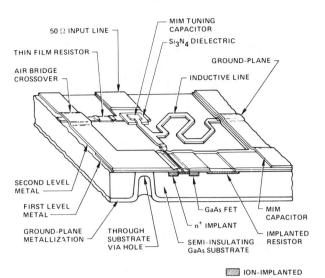

Fig. 2. Schematic drawing of an MMIC showing typical components needed.

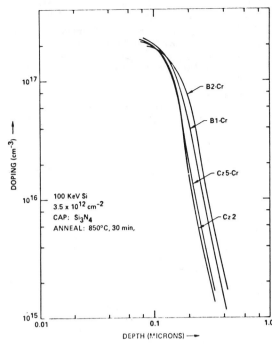

Fig. 4. Doping profiles in LEC and Bridgman semi-insulating GaAs substrates.

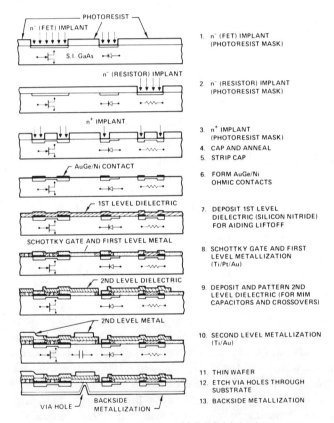

Fig. 3. Pictorial representation of MMIC fabrication process.

TABLE I
ACTIVE LAYER UNIFORMITY
(Implant: 5E12 cm^{-2}, 225 keV Si; 5E12 cm^{-2}, 40 keV Si)

ID	Substrate Type	$\langle I_{dss} \rangle$ mA Before Gate Recess	σ (%)
A	Undoped LEC	217.5	2.3
B	Undoped LEC	217.4	2.3
C	Undoped LEC	195.5	0.92
D	Cr-doped Bridgman	193.4	2.0
E	Cr-doped Bridgman	196.8	2.3
F	Cr-doped Bridgman	201.0	4.4

$\langle I_{dss} \rangle$ = 203.6 mA
σ_{IDSS} = 11.0 mA (5.4%)

plant schedule are provided in Table I where the average I_{dss} before gate recess of 200-μm wide FET's is given for various substrates. These wafers were also processed separately and the data show excellent uniformity and reproducibility of active layers made possible by direct ion implantation in SI GaAs.

Following active layer formation, ohmic contacts are defined by sequential evaporation of Au–Ge and Ni, liftoff, and alloying at 450°C. This metallization scheme results in low resistance contacts quite reliably as evidenced by the data in Table II. These data were obtained on randomly selected wafers at the completion of front-side processing and show that it is possible to maintain a low specific contact resistance ($\sim 1 \times 10^{-6}$ $\Omega \cdot$cm^2) through the 250°C silicon nitride deposition steps. The observed variation of specific contact resistance shown in Table II represents a combination of measurement error and actual resistance variation but contributes less than 10-percent variation to FET source resistance.

After contact metallization, the 1.0-μm gates are defined by contact photolithography, recess etch, Ti–Pt–Au evaporation, and liftoff. At present, gate yield is a significant circuit yield limiting factor. Some preliminary data are given in Table III where yields have been averaged over several wafers processed. The criterion "other process related defects" includes damage due to wafer handling, poor

TABLE II
Specific Contact Resistance Results
(After Completion of all MMIC Process Steps)

Date Measured	ID	$\langle R_c \rangle$ (10^{-6} Ω cm²)
10/80	SP-100	0.51
	SP-101	0.66
	SP-102	1.31
	R2C-IL5	0.51
	15-2	0.99
	16-1	0.49
	W7-1	1.6
	W7-2	1.7
Through	G19H-111	3.3
	W9-1	0.91
	W12-2	2.5
	W19-1	0.86
	60-1	3.1
	42-2	2.5
	40-1	0.56
	531	0.84
	71S421	0.30
	41-2	0.86
5/82	511	1.1

Metallization: AuGe/Ni

$\langle R_c \rangle = 1.3 \times 10^{-6}$ Ω cm²
$\sigma = 0.92 \times 10^{-6}$ Ω cm²

TABLE III
FET DC Yield
($\langle I_{dss} \rangle = 47.1$ mA; $\sigma = 5.5$ mA (11.6%))

Criterion	FET Yield for Different Gate Widths Gate Length = 1.0 μm		
	Width = 200 μm S-D gap = 3.8 μm	Width = 500 μm S-D gap = 4.8 μm	Width = 990 μm S-D gap = 4.8 μm
Broken Gate	0.90	0.94	0.82
Other Process Related Defects	0.92	0.98	0.95
Net Yield	0.83	0.92	0.78

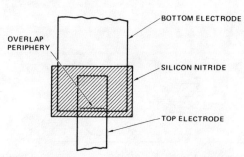

Fig. 5. Schematic diagram of a metal–insulator–metal (MIM) capacitor.

TABLE IV
Metal–Insulator–Metal Capacitors
(Insulator: 6000-Å Silicon Nitride Deposited by Plasma Enhanced CVD)

Approx. Measurement Date	ID Number	Average Cap. pF/mm²	σ (%)	
6/80	143	125	2.1	
	165	132	2.1	
	173	132	3.4	measured on in-process wafers
	174	128	1.9	
	175	131	2.4	
	R5M/1B	125	1.8	
	1265	135	3.2	
Through	1357-1	127	<0.3	
	1357-2	133	<0.3	
	1393	125	<0.3	
	1398	125	<0.3	measured on test chips
	1412	125	<0.3	
	1424	136	<0.3	
	1432	138	<0.3	
2/82	1439	139	<0.3	

Mean = 130 pF/mm²
σ = 3.9%
N = 15

source–drain definition, and shorts caused by metallization defects normally found with contact photolithography. The lower yield of 200-μm wide FET's as compared to the 500-μm wide FET's is probably due to the smaller source–drain gap of the 200-μm wide device. These data were obtained on ~10 cm² (half of a 2-in wafer) GaAs wafers. In order to maximize gate yield several precautions have to be taken. These include monitoring wafer flatness and ensuring that it is in the range of ±1 μm/in after capping and annealing, and using 0.090-in thick lithographic masks for minimum runout from mask bowing during contact printing.

Gate metal definition is followed by Ti/Au first level metallization which provides overlays for ohmic contacts and the lower electrodes of MIM capacitors. This first level metal pattern is defined by ion milling through a photoresist mask to remove unwanted areas so as to achieve rounded edges of the remaining metal which are necessary for good capacitor yield as discussed below. A 6000-Å layer of silicon nitride is deposited next using plasma enhanced chemical vapor deposition (PSN). This forms the dielectric for MIM capacitors and the crossover insulator in a two level metallization scheme. Finally, the second metal layer is defined by photolithography and gold electroplating to a thickness of 2 to 3 μm. It provides the top electrode of MIM capacitors, all interconnects, air bridges, and other microwave circuitry.

The uniformity, reproducibility, and dc yield of MIM capacitors have been studied. Data on the first two aspects is given in Table IV. These data, obtained on randomly selected wafers, span a period of 20 months and clearly indicate that by adequately monitoring the deposition process it is possible to have a tight control on the thickness and dielectric constant of the PSN. Such control has encouraged the use of MIM capacitors for RF tuning as well as bypassing. The first group of data in Table IV were obtained on actual wafers in process and the σ value reflects variations in both the PSN and the electroplated top electrode of the small (100 μm × 100 μm) test capacitors. Remaining data were obtained on test wafers where the top electrode was formed by liftoff. Negligible variation in capacitance values was observed in these cases.

DC yield of MIM capacitors was found to depend on both the area and the length of overlap periphery (Fig. 5) between the first and the second metallization levels. It was possible to get a good fit of measured yield data using multiple linear regression techniques to a linearized equation (valid for yields near unity) of the form

$$Y = 1 - \alpha A - \beta P$$

as shown in Table V where all the terms have also been defined. The last column of Table V shows the contribution to circuit yield of 5 typical (10 pF) bypass capacitors and clearly indicates the importance of this problem. The

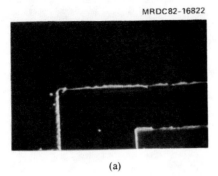

(a)

(b)

Fig. 6. (a) Sharp pattern edges obtained by direct liftoff of Ti(500 Å)/Au(4000 Å). (b) Smooth edges obtained by ion milling the metallization pattern.

TABLE V
DC YIELD OF MIM CAPACITORS
(Insulator: 6000-Å Silicon Nitride Deposited by Plasma Enhanced CVD)

$Y = 1 - \alpha A - \beta P$
Y = DC Yield
A = Capacitor Area (mm^2)
P = Overlap Periphery Between First and Second Metallization Levels (mm)

ID Number	α mm^{-2}	β mm^{-1}	RMS Prediction Error $\sqrt{\sum_i (y-y_i)^2}$	Yield of 10 pF Bypass Cap with Dimensions of $A = 0.077$ mm^2 $P = 0.5$ mm	Contribution to Circuit Yield Assuming 5 Bypass Capacitors
E	0.83	0.15	0.080	0.86	0.47
F	0.16	0.20	0.046	0.89	0.56
G	0.66	0.013	0.016	0.94	0.73

area dependence of capacitor yield is due to a pinhole density in the PSN. In practice, pinholes in the nitride are associated with debris on the wafer, metal splattering during first level metallization, etc., and can be reduced by controlling these factors. The overlap periphery dependence arises due to the sharp edges (as obtained by direct liftoff, Fig. 6(a)) which are not well covered by PSN and usually result in a short. Rounded edges as obtained by ion milling or special liftoff techniques [3] (Fig. 6(b)) are more reliably covered by PSN and cause fewer shorts. The improvement in capacitor yield attributable to improved processing of the first level metal may be seen from the dramatic reduction of β for wafer G which had ion milled first level metal compared to wafers E and F which were processed earlier and had first level metal defined by direct liftoff. The periphery problem can be effectively circumvented by using an airbridge to contact the top capacitor electrode but this approach may result in added process complexity (unless airbridges are being used elsewhere in the circuit) and constraints on circuit layout.

TABLE VI
DC CIRCUIT YIELD

	Buffer Amplifier	Driver Amplifier	Power Amplifier
Total Gate Periphery (mm)	0.2	1.0	1.98
Source-Drain Gap (μm)	3.8	4.8	4.8
Total MIM Capacitance (pF)	20	49.4	50.8
FET Yield (%)	81	86	59
Capacitor Yield (%)	97	88	76
Circuit Yield (%)	78	76	47

Preliminary data on dc probe plus visual (microscopic inspection) circuit yield of three different circuits are presented in Table VI. These data were obtained at the completion of front side processing and do not include attrition due to subsequent steps involving thinning, via hole etching, backside metallization and sawing. The data depicted in Table VI are commensurate with FET and capacitor yields presented earlier (with improved processing) and represent some of the highest yields observed using the above described MMIC fabrication process. These data indicate the potential for achieving high overall yields of functional MMIC modules with multicircuit complexity. For example, a module consisting of the three circuits described in Table VI would have an overall dc yield of about 28 percent before backside processing.

III. CONCLUSIONS

An ion-implantation based process has been developed for fabricating GaAs MMIC's incorporating active devices, RF circuitry, and bypass capacitors. Low ohmic contact resistance and good control of MIM capacitance values have been demonstrated which were achieved by careful monitoring of the associated fabrication processes. Yield limiting factors affecting MIM capacitors and FET's have been discussed and some preliminary data on overall circuit yields have been presented. High dc yield of typical amplifier circuits has been shown indicating that the above described MMIC fabrication process has the potential for achieving high overall yields in a production environment. Based on the data presented, it is expected that GaAs MMIC modules with multicircuit complexity can be fabricated with acceptable yield.

REFERENCES

[1] B. M. Welch, Y. D. Shen, R. Zucca, R. C. Eden, and S. I. Long, "LSI processing technology for planar GaAs integrated circuits," *IEEE Trans. Electron Devices*, vol. ED-27, no. 6, pp. 1116–1124, June 1980.

[2] A. A. Immorlica, Jr., D. R. Decker, and W. A. Hill, "A diagnostic pattern for GaAs FET material development and process monitoring," *IEEE Trans. Electron Devices*, vol. ED-27, no. 12, pp. 2285–2291, Dec. 1980.

[3] M. Hatzakis, B. J. Canavello, and J. M. Shaw, "Single step optical lift-off process," *IBM J. Res. Develop.*, vol. 24, no. 4, pp. 452–460, July 1980.

Backgating in GaAs MESFET's

CHRISTOPHER KOCOT AND CHARLES A. STOLTE, MEMBER, IEEE

Abstract — The phenomenon of backgating in GaAs depletion mode MESFET devices is investigated. The origin of this effect is electron trapping on the Cr^{2+} and EL(2) levels at the semi-insulating substrate-channel region interface. A model describing backgating, based on DLTS and spectral measurements, is presented. Calculations based on this model predict that closely compensated substrate material will minimize backgating. Preliminary experimental data support this prediction.

I. INTRODUCTION

THE characteristics of GaAs metal-semiconductor field effect transistors (MESFET's) depend strongly on the properties of the interface between the n-type active region and the semi-insulating substrate [1]. GaAs MESFET's can exhibit phenomena such as a drift in the drain current with time and a change in the drain current as a result of a change in the substrate bias. The decrease in the drain current when a negative voltage is applied to the substrate is termed backgating [2], [3]. Backgating is a detrimental effect in complex GaAs integrated circuits due to the interaction between closely spaced devices. This effect is caused by the relatively large capacitance of the substrate-active channel interface due to negative charge accumulated on deep traps in the interface region. The application of a bias to the substrate modulates this space charge region. This results in a change in the active channel region width and, therefore, a change in the drain current. In this paper, we will present the results of investigations into the physical nature of the deep traps responsible for backgating. Based on these investigations, we propose a model which explains backgating and demonstrate one solution, namely the use of closely compensated substrate material to minimize the back-side channel capacitance.

II. EXPERIMENTAL PROCEDURE

A. Substrate Material

GaAs MESFET's fabricated in four different types of substrate materials were investigated. The active region for the devices is produced by ion implantation into: 1) Cr-doped semi-insulating substrates (with Cr concentrations between 5×10^{15} and 1×10^{17} cm^{-3}); 2) high purity semi-insulating substrates (grown with no intentionally added dopants); 3) buffer layers on Cr-doped substrates; and 4) buffer layers on high purity substrates.

The substrate material, both Cr-doped and high purity, is grown by the two-atmosphere liquid encapsulated Czochralski (LEC) technique [4]. Chromium incorporates into the GaAs lattice on Ga sites and gives up three electrons to the bonds. The neutral state of Cr with respect to the lattice is Cr^{3+} with the electron configuration $3d^3$. Capture of electrons leads successively to the core states $3d^4$, Cr^{2+} (a singly, negatively charged acceptor), and $3d^5$, Cr^{1+} (a doubly, negatively charged acceptor). Chromium which is neutral, Cr^{3+}, is a double acceptor. The Cr^{2+} and Cr^{3+} levels are located 0.70 eV [5] and 0.45 eV [6] above the valence band, respectively, as shown in the energy level diagram of Fig. 1. There is uncertainty concerning the position of the Cr^{4+} and Cr^{1+} levels. According to the literature, the Cr^{4+} level is 0.15 eV [6] above the valence band and the Cr^{1+} level is degenerate with the conduction

Manuscript received January 4, 1982; revised February 1, 1982.
The authors are with Hewlett–Packard Laboratories, Palo Alto, CA 94304.

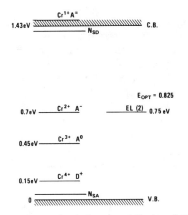

Fig. 1. Energy level diagram of Cr-doped GaAs.

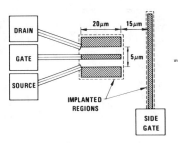

Fig. 2. The MESFET test structure used in the backgating, DLTS, and spectral response experiments.

band [7]. The Cr^{2+} and Cr^{3+} levels are the most important to the understanding of the behavior of Cr-doped semi-insulating GaAs.

The other commonly observed level in GaAs is the EL(2) donor level which is responsible for the high resistivity of nonintentionally doped, high-purity GaAs [8]. This level is believed to be due to an anti-site defect, Ga on an As site [9]. It is located 0.75 eV below the conduction band (thermal activation energy); the optical ionization energy is 0.82 eV [10]. The necessary conditions for the production of high-resistivity material is that the dopants satisfy the following relations:

$$\text{if } N_d > N_a \text{ then } (N_{da} - N_{dd}) > (N_d - N_a), [11] \quad (1)$$

or

$$\text{if } N_a > N_d \text{ then } (N_{dd} - N_{da}) > (N_a - N_d), [12] \quad (2)$$

where N_d and N_a are the concentrations of shallow donors and acceptors, respectively, and N_{dd} and N_{da} are the concentrations of deep donors, EL(2), and acceptors, Cr, respectively. The recent paper by Martin describes the compensation mechanisms in detail [13].

The high-purity buffer layers are grown on the Cr-doped or high-purity semi-insulating substrates by the liquid-phase epitaxy (LPE) technique. These layers are grown in a horizontal graphite slider system at 700°C. This epitaxial material, characterized using Hall measurements, is n-type with mobilities measured at room temperature of approximately 8000 cm^2 V^{-1} s^{-1} and mobilities measured at 77 K are greater than 120 000 cm^2 V^{-1} s^{-1}. The measured free-carrier concentration at room temperature is less than 1×10^{14} cm^{-3} [14]. The buffer layers used in this investigation are approximately 3 μm thick and are fully depleted by the surface and semi-insulator to buffer layer interface space charge regions.

The depletion mode MESFET test devices, shown in Fig. 2, are fabricated using selective region ion implantation. The MESFET channel regions are formed by localized ion implantation of 500-keV Se ions to a dose of 6×10^{12} cm^{-2}. A second localized Si implant at 500 keV to a dose of 1×10^{13} cm^{-2} is used in the ohmic contact regions to lower the sheet resistance. The implants are simultaneously annealed at 850°C for 15 min using a Si$_3$N$_4$ cap. The MESFET fabrication includes a recessed gate technology which reduces the modulation of the drain current due to changing depletion layer widths at the free surface during switching transients and also allows the adjustment of the gate cutoff voltage during the device processing [15]. The side-gate electrode is an ohmic contact to an implanted region. This region is isolated from the active region of the MESFET by the semi-insulating substrate or the fully depleted buffer layer.

III. EXPERIMENTAL RESULTS

The magnitude of the backgating effect is determined by measuring the change in the saturated drain current as a result of the application of a negative voltage to the side-gate electrode. In these experiments, the source–drain bias is 4 V, the gate is shorted to the source, and the side-gate bias is -4 V. Results obtained from these devices fabricated in wafers of the types listed above are presented in Fig. 3. These data represent the mean and one sigma variation of the drain-current change with the application of -4-V side-gate voltage for approximately 30 devices on a one-inch-square wafer. These data illustrate the wide spread in the magnitude of backgating on a single wafer and the large variation of the effect from wafer to wafer. In general, the effect is less for buffer layers on Cr-doped substrates than for high-purity substrates, with or without a buffer. The buffer layers used in this experiment had little influence on backgating. This is not understood since the buffer layer should decrease the capacitance between the substrate and the active layer and therefore decrease the magnitude of backgating. Experiments to investigate the effect of buffer layers are in progress. The sample on the Cr-doped substrate was fabricated in closely compensated material. This will be discussed in detail later.

The time dependence of the drain current following the application of a side-gate potential is shown in Fig. 4. There is a rapid decrease in current when the side-gate bias is applied followed by an additional slow decrease for Cr-doped substrates and a slow increase for high-purity substrates. The general form of the drain-current transients for buffer layers on the different types of substrates is the same as observed without the buffer layer. The magnitude of the drain-current change depends on the particular substrate as illustrated in Fig. 3. These current transients can be understood by considering the band diagram shown in Fig. 5. Negative charge is accumulated on deep levels on the substrate side of the substrate-channel region interface.

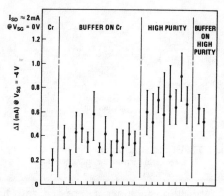

Fig. 3. Backgating results measured on devices fabricated in different substrate materials.

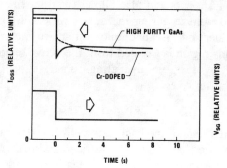

Fig. 4. Typical time dependence of the drain current for Cr-doped and high-purity substrates after the application of a side-gate voltage.

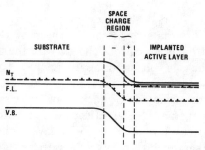

Fig. 5. The energy band diagram of the substrate-active region interface.

TABLE I
DEEP LEVELS DETERMINED BY DLTS MEASUREMENTS OF THE BACKGATING TRANSIENTS

Cr-DOPED			HIGH-PURITY		
LEVEL	ACTIVATION ENERGY (eV)	CAPTURE CROSS-SECTION (cm²)	LEVEL	ACTIVATION ENERGY (eV)	CAPTURE CROSS-SECTION (cm²)
Cr	0.781	7e−15	EL (2)	0.82	2.7e−13
HL	0.644	1.1e−13	EL	0.692	2.0e−13
HL	0.516	1.4e−13	HL	0.535	4.1e−15

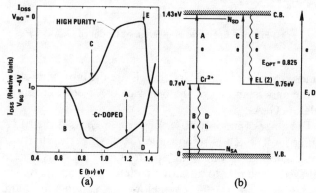

Fig. 6. (a) Spectral dependence of backgating for Cr-doped and high-purity substrates. (b) Energy level diagram showing the transitions responsible for the spectral response.

This negative charge is balanced by the positive space charge region in the channel region and produces the relatively large capacitance of the substrate-channel region interface. A negative potential applied to the substrate increases the width of the depletion region at the back side of the MESFET channel which results in the rapid decrease of the drain current. This change in the depletion layer width is followed by the emission of electrons or holes from the deep levels as equilibrium is reestablished. In the case of Cr-doped substrates, holes are emitted from the deep Cr level which results in a further decrease in the drain current because of the increase of the negative charge on the substrate side of the interface. In the high-purity substrates, electrons are emitted to the conduction band from the EL(2) level and the drain current increases to an equilibrium value.

These drain-current transients have been analyzed using an automated conductance DLTS system [16]. In this technique, the transients are analyzed at a number of temperatures to derive the activation energies and the capture cross sections of the levels responsible for the change in the drain current following the application of the side-gate voltage. The results obtained by this technique for Cr-doped and high-purity substrates are summarized in Table I. In Cr-doped substrates, the major DLTS peak is due to the emission of holes from the Cr level to the valence band. The major DLTS peak in high-purity substrates is due to the emission of electrons from the EL(2) level to the conduction band. The activation energies and the capture cross sections determined in this investigation agree with those reported by Martin [17].

The spectral dependence of backgating has been measured to obtain additional information regarding the levels responsible for the effect and to understand ambient light effects on the MESFET characteristics. Typical spectral dependence of the changes in the saturated drain current, measured with a side-gate voltage of −4 V, for Cr-doped and high-purity substrates are shown in Fig. 6(a). The transitions responsible for the spectral dependence of the drain current are indicated on the energy level diagram of Fig. 6(b). In the case of the Cr-doped substrates, the decrease of the current in the range of 0.65 eV to 1.0 eV is caused by the increase of backgating due to electron transitions from the valence band to the Cr^{3+} level. These transitions produce chromium in the Cr^{2+} charge state. The holes created in the valence band by this transition are swept away by the electric field in the space charge region at the substrate-channel region interface. This increase in the charge on the Cr^{2+} level increases depletion depth into the MESFET channel, as indicated in Fig. 5, and therefore the drain current decreases. The increase of the current at 1.0 eV is caused by electron transitions from the Cr^{2+} level to the conduction band which reduces the negative charge in the junction, decreases the depletion width, and increases the drain current. Since the Cr^{2+} level is close to the center of the band gap both transitions, valence band to the Cr^{2+} level and Cr^{2+} level to the conduction band, are possible. In the photon energy range from 0.65 eV to 1.0 eV, the optical cross section for electron transitions from the valence band to the Cr level σ_p^0 is greater than the optical cross section for electron transitions from the Cr level to the conduction band σ_n^0, and therefore the transition producing electrons in the Cr^{2+} state is dominant. In the range from 1.0 eV to 1.4 eV, $\sigma_n^0 > \sigma_p^0$, and therefore the transition of electrons from the Cr^{2+} level to the conduction band is dominant [10]. A steep increase in current is seen at the band-gap photon energy because electron-hole pairs are generated and the holes are trapped by the Cr^{2+} level while the electrons are swept away by the space charge field. This results in a decrease in the negative charge at the interface and an increase in the current.

The spectral dependence of backgating is quite different in the case of high-purity substrates. An increase of the drain current is observed with a threshold of 0.8 eV which agrees with the optical ionization energy of the EL(2) deep level [10]. These transitions decrease the concentration of the negative charge on the substrate side of the interface which produces a wider channel region and the increase in the drain current. The drain current decreases as the photon energy approaches the band-gap energy. In this region, electron-hole pairs are generated and the electrons are captured by the EL(2) level. This increase in the negative charge reduces the drain current as discussed above. It should be noted that there is almost no backgating effect when the high-purity substrate is illuminated with 1.2-eV light. In this condition, no change in the drain current is seen when a side-gate potential is applied.

IV. BACKGATING MODEL

A model which is consistent with the experimental results obtained on structures without buffer layers is postulated. The experiments described above indicate that backgating in Cr-doped substrates is caused by the accumulation of negative charge on the Cr^{2+} level (fast transient) and by the emission of holes from the Cr level (slow transient). For high-purity substrates, backgating is caused by the accumulation of negative charge on the EL(2) level (fast transient) and by the emission of electrons from the EL(2) level (slow transient). The net concentration of negative charge, at equilibrium, on the substrate side of the substrate-channel region interface N_{eff} depends on the occupancy of the deep traps in the bulk according to the following relationship:

where

N_{Cr}	total chromium concentration in the substrate,
$N_{EL(2)}$	total EL(2) level concentration in the substrate,
e_p^{Cr}	emission rate for holes from the Cr^{3+} level,
e_n^{Cr}	emission rate for electrons from the Cr^{2+} level,
$e_p^{EL(2)}$ and $e_n^{EL(2)}$	emission rates for holes and electrons from the EL(2) level, respectively,
$E_{Cr^{2+}}$ and $E_{EL(2)}$	thermal activation energies with respect to the valence band for the Cr^{2+} and EL(2) levels, respectively, and
E_f	energy of the Fermi level with respect to the valence band in the bulk of the substrate.

In this equation, the Fermi-function terms give the concentration of negative charge on the deep levels in the bulk. The emission-rate terms give the concentration of negative charge in the space charge region. The emission rates are derived from the capture cross sections given in the literature [17]. The difference between the emission-rate terms and the Fermi-function terms is the excess concentration of negative charge on the deep levels on the substrate side of the substrate-channel region interface.

Since the relative occupancy of the deep levels in the space charge region is fixed by the appropriate emission rates, the concentration of negative charge at the interface is controlled by the occupancy of the deep traps in the bulk of the substrate. For a low N_{eff}, the space charge region at the interface between the substrate and the active layer will be very diffuse and backgating will be minimal.

Calculations of the magnitude of backgating as a function of the substrate compensation using this model are discussed below. In these calculations, the change of the drain current ΔI for a -2-V and -4-V bias applied to the substrate side of the MESFET channel is determined. The relationship between the depletion layer width at the substrate-channel region interface and the excess charge N_{eff} in the space charge region is derived using the standard application of Poisson's equation and the neutrality condition which in this case is

$$\int_0^d N_D(x)\,dx = N_{eff} W \quad (4)$$

where d is the depletion width on the channel-region side ($N_D(x)$ is the charge distribution in this region) and W is the depletion width on the substrate side (N_{eff} is the constant negative charge in this region). Using these conditions, the general form for the voltage appearing across the

$$N_{eff} = N_{Cr}\{e_p^{Cr}/(e_p^{Cr}+e_n^{Cr}) - 1/[1+\exp[(E_{Cr^{2+}}-E_f)/kT]]\}$$
$$+ N_{EL(2)}\{e_p^{EL(2)}/[e_p^{EL(2)}+e_n^{EL(2)}] - 1/[1+\exp[(E_{EL(2)}-E_f)/kT]]\} \quad (3)$$

space charge region is derived to be

$$V = \frac{q}{\epsilon}\left\{ d\int_0^d N_D(x)\,dx - \int_0^d \left(\int_0^x N_D(x)\,dx\right) dx \right.$$
$$\left. + \frac{1}{2N_{\text{eff}}}\left(\int_0^d N_D(x)\,dx\right)^2 \right\} \quad (5)$$

where V is the sum of the built-in voltage and the applied backgate bias. In these calculations, it is assumed that the active region donor concentration can be represented by

$$N_D(x) = N_D[1 - \exp(-x/t)] \quad (6)$$

where N_D and t are determined by a fit to the free-carrier concentration profile used in the experimental devices. This approximation, made as a matter of computational convenience, is a good representation for the recessed gate devices investigated. Using (5) and (6), we obtain

$$V = N_D q/\epsilon \{ d^2/2 + t[d\exp(-d/t) + t\exp(-d/t) - t] + N_D/2N_{\text{eff}}[d - t[1 - \exp(d/t)]]^2 \}. \quad (7)$$

This relationship allows the determination of d for given values of N_D, t, N_{eff}, and V. The value of N_{eff} is given by (3) and the values of N_D and t are given by the fit to the experimental concentration profile in the channel region. Therefore, for different values of V, the depletion depth d in the back-channel region is determined.

The saturated drain current at zero gate bias is calculated using

$$I(d) = \int_d^T q v_s W N_D(x)\,dx \quad (8)$$

where v_s is the saturated velocity, W is the gate width, and T is the total channel region thickness minus the Schottky-gate depletion width. Using the assumed doping profile (6), the general expression for the drain current is

$$I_D = q v_s W N_D \{(T-d) + t[\exp(-T/t) - \exp(-d/t)]\}. \quad (9)$$

Finally, the results of these calculations are shown in Fig. 7 where the drain current with 0, -2-V, and -4-V backgate bias are plotted as a function of N_{eff}. The data shown in Fig. 7 were calculated using the following parameters: $N_D = 2.25 \times 10^{17}$ cm^{-3}, $t = 0.1$ μm, $T = 0.13$ μm, $W = 20$ μm, $v_s = 10^7$ cm s^{-1}, and $V_{bi} = 0.7$ V.

This figure graphically demonstrates the role of the substrate compensation on the drain current for zero, -2-V, and -4-V backgate bias. For example, if N_{eff} equals 1×10^{16} cm^{-3}, the application of a -2-V backgate bias reduces the drain current by 0.88 mA, 22 percent, and a -4-V backgate bias reduces the drain current by 1.55 mA, 39 percent. If N_{eff} equals 1×10^{15} cm^{-3}, the decrease in the drain current is 0.35 mA, 9 percent, and 0.55 mA, 14 percent, at -2-V and -4-V backgate bias, respectively. The shallow- and deep-level concentrations corresponding to these values of N_{eff} can be determined from (3). For example, if the EL(2) concentration is 1×10^{16} cm^{-3} and the difference in the shallow-level concentration $(N_d - N_a)$

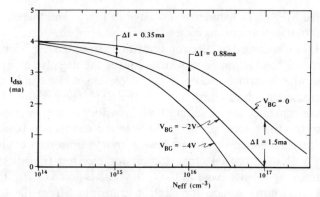

Fig. 7. The calculated dependence of the drain current I_{dss} on the degree of substrate compensation N_{eff} for different backgate bias voltages V_{BG}.

is 5×10^{15} cm^{-3} then a Cr concentration of 1.8×10^{16} cm^{-3} yields N_{eff} equal to 1×10^{16} cm^{-3} and a Cr concentration of 6×10^{15} cm^{-3} yields N_{eff} equal to 1×10^{15} cm^{-3}. Therefore, using these calculations, the magnitude of backgating can be determined if the properties of the substrate and the active region doping concentration profile are known. In general, the backgating effect is minimized if N_{eff} is small.

This backgating model has been evaluated in a preliminary experiment in which MESFET test devices were fabricated in low Cr-doped material, approximately 5×10^{15} cm^{-3} Cr concentration. The measured backgating characteristics, represented by the first data point, Cr (shown in Fig. 3), are in qualitative agreement with the model and demonstrate reduced backgating for closely compensated material. The calculations described above assume that the entire backgate voltage is applied at the substrate-channel region interface. A model for the voltage communication between the side-gate and the back-channel region will be developed to permit a more quantitative correlation between the calculated and experimental results. Experiments using proton bombardment isolation to reduce the voltage communication between the side-gate and the back-channel region have demonstrated a decrease in the magnitude of backgating for a given side-gate bias [19].

The model predicts that the use of an n^+ layer, separated from the active region by a buffer layer, would eliminate backgating. In this structure, the excess negative charge N_{eff} in the substrate would produce a narrow space charge region entirely in the n^+ region. Therefore, the space charge will not reach the channel region and no change in the channel width would be produced by a back-gate bias. This postulate, based on the backgating model, will be evaluated experimentally.

V. Conclusions

The results of this experimental investigation indicate that the phenomenon of backgating is the result of an accumulation of excess charge at the substrate-channel region interface. This charge resides on deep traps in the substrate material, either on Cr and EL(2) levels in Cr-doped substrates or on EL(2) levels in high-purity semi-insulating substrates. The deep levels responsible for backgating were determined from spectral response and DLTS measurements. Calculations based on this model predict that the magnitude of backgating is dependent on the degree of compensation in the substrate material. Closely compensated substrates have less backgating than substrates with a large excess of deep traps which are unoccupied in the bulk. Preliminary experimental results on lightly Cr-doped material support this model.

Other solutions to the backgate problem include the use of thick buffer layers or n^+ layers as described above. These solutions assure that the space charge region at the semi-insulating to n layer interface does not extend to the MESFET channel region. These solutions and their effect on integrated circuit performance are under investigation.

Acknowledgment

The authors thank the members of the Hewlett–Packard Laboratory who contributed to the work reported. The DLTS measurements were made in conjunction with D. Mars. The encouragement of and the stimulating discussions with R. Archer, C. Bittmann, and C. Liechti contributed to the results reported.

References

[1] P. L. Hower, W. W. Hooper, D. A. Tremere, W. Lehrer, and C. A. Bittmann, "The Schottky barrier galium arsenide field-effect transistor," in *Proc. 1968 Inf. Symp. Gallium Arsenide and Related Compounds*, pp. 187–195.
[2] T. Itoh and H. Yanai, "Stability and performance and interfacial problems in GaAs MESFET's," *IEEE Trans. Electron Devices*, vol. ED-27, no. 6, pp. 1037–1045, 1980.
[3] M. Tanimoto, K. Suzuki, T. Itoh, H. Yanai, L. M. F. Kaufmann, W. Nievendick, and K. Heime, "Anomalous phenomena of current-voltage characteristics observed in Gunn-effect digital devices under dc bias conditions," *Electron. Commun. Japan*, vol. 60-C, no. 11, pp. 102–110, 1977.
[4] W. M. Ford and T. L. Larsen, "LEC growth of large GaAs single crystals," in *Proc. Electro Chem. Soc.*, vol. 75-1, p. 517, 1975.
[5] H. R. Szawelska and J. W. Allen, "Photocapacitance measurements of the two acceptor levels of chromium in GaAs," *J. Phys. C.*, vol. 12, pp. 3359–3367, 1979.
[6] U. Kaufmann and J. Schneider, "Chromium as a hole trap in GaP and GaAs," *Appl. Phys. Lett.*, vol. 36, pp. 747–748, 1980.
[7] A. H. Hennel, W. Szuszkiewicz, G. Martinez, B. Clerjoud, A. M. Huber, G. Mouillot, and P. Merenda, "Activation of Cr^{1+} $(3d^5)$ level in GaAs:Cr induced by hydrostatic pressure," in *Proc. Semi-insulating III–V Materials, Nottingham, 1980*, G. J. Rees, Ed., 1980, pp. 228–232.
[8] G. M. Martin, A. Mitonneau, and A. Mircea, "Electron traps in bulk and epitaxial GaAs crystals," *Electron. Lett.*, vol. 13, pp. 191–193, 1977.
[9] J. Lagowski, H. C. Gatos, J. M. Parsey, K. Wada, M. Kaminska, and W. Walukiewicz, "Origin of the 0.82-eV electron trap in GaAs and its annihilation by shallow donors," *J. Appl. Phys.*, to be published.
[10] G. M. Martin, G. Jacob, G. Poilblaud, A. Goltzene, and C. Schwab, "Identification and analysis of near-infrared absorption bands in undoped and Cr-doped semi-insulating GaAs crystals," in *11th Int. Conf. on Defects and Radiation Effects in Semiconductors*, (Oiso, Tokyo), Sept. 1980.
[11] P. F. Linquist, "A model relating electrical properties and impurity concentrations in semi-insulating GaAs," *J. Appl. Phys.*, vol. 48, pp. 1262–1267, 1977.
[12] E. M. Swiggard, S. H. Lee, and F. W. Batchelder, "Electrical properties of PBN-LEC GaAs crystals," in *Inst. of Phys. Conf. Ser.* no. 45, pp. 125–133, 1979.
[13] G. M. Martin, J. P. Forges, G. Jacob, and J. P. Hollars, "Compensation mechanisms in GaAs," *J. Appl. Phys.*, vol. 51, pp. 2840–2852, 1980.
[14] C. A. Stolte, "The influence of substrate properties on the electrical characteristics of ion implanted GaAs," in *Proc. Semi-insulating III–V Materials, Nottingham, 1980*, G. J. Rees, Ed., 1980, pp. 93–99.
[15] C. A. Liechti, C. A. Stolte, M. Namjoo, and R. Joly, "GaAs Schottky-gate field effect transistor medium scale integration," Wright–Patterson AFB, Final Report AFAL-TR-81-1082, OH, Jan. 1981.
[16] E. E. Wagner, D. Hiller, and D. Mars, "Fast digital apparatus for capacitance transients analysis," *Rev. Sci. Instrum.*, vol. 51, pp. 1205–1211, 1980.
[17] G. M. Martin, "Key electrical parameters in semi-insulating materials; the methods to determine them in GaAs" in *Proc. Semi-insulating III–V Materials, Nottingham 1980*, R. J. Rees, Ed., 1980, pp. 13–28.
[18] A. Mitonnou, A. Mircea, G. M. Martin, and D. Pons, "Electron and hole capture cross sections at deep centers in Gallium Arsenide," *Rev. Phys. Appl.*, vol. 14, pp. 853–861, 1979.
[19] D. D'Avanzo, "Proton isolation for GaAs integrated circuits," this issue, pp. 955–963.

EFFECT OF SUBSTRATE CONDUCTION AND BACKGATING ON THE PERFORMANCE OF GaAs INTEGRATED CIRCUITS

C. P. Lee, R. Vahrenkamp, S. J. Lee, Y. D. Shen and B. M. Welch

Rockwell International/Microelectronics Research and Development Center
1049 Camino dos Rios, Thousand Oaks, CA 91360 (805) 498-4545

ABSTRACT

The influence of substrate conduction on the backgating effect and the performance of GaAs integrated circuits is presented. A direct correlation between substrate leakage current and backgating has been found and a model which links these two phenomena is described. Proton implantation has been found to be effective in increasing the threshold for substrate conduction and the onset of backgating. Circuit performance is also improved due to less backgating when proton isolation is used. Under light illumination, substrate conduction and backgating effect are greatly enhanced, and circuit performance is degraded with increasing negative voltage supply.

INTRODUCTION

As GaAs integrated circuits become more and more complex and densely packed severe demands are placed on semi-insulating substrates to provide adequate device isolation. Although high resistivity ($>10^8$ Ω-cm) substrate material is commonly available, we have found that substantial substrate leakage can occur when the substrates are subject to high electrical field where substrate conduction is controlled by carrier injection and/or some instability phenomenon.[1] This leakage current, although very small compared to the circuit operating current, could be enough to cause cross talk between different circuit elements and degrade the circuit performance.

In the past, the crosstalk phenomenon in GaAs integrated circuits has been attributed to the backgating effect,[2,3] which is a result of the modulation of the FET's current through the channel-substrate interface by the substrate potential. However, the understanding of the backgating effect has been mostly limited to the material aspects of the problem, such as: the deep level identification in the substrate and the properties of the interface junctions.[4,5] Only recently this problem was tied to the substrate leakage, and a much clearer picture of backgating was achieved.[1,6] In this paper, we will first present a model which describes the relationship between substrate conduction and backgating; based on this model we will explain the results of proton isolation and light sensitivity effect. Circuit performance at different substrate conditions will be described and explained.

MODEL

The cross talk between two devices fabricated on a semi-insulating substrate can be illustrated in Fig. 1. A FET and another device (shown as an ohmic contact on top of a n-type region) are isolated from each other by the substrate. The FET's channel current is influenced by the negative voltage (V_{BG}) applied to the other device through

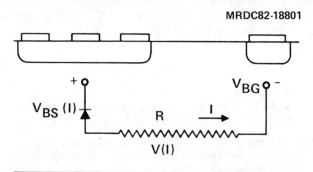

Fig. 1 The equivalent circuit of two devices fabricated on a semi-insulating substrate. The channel-substrate interface is represented by a diode and the substrate is represented by a nonlinear resistor.

the coupling of the substrate, a nonlinear resistor (R), and the channel-substrate interface junction, which can be represented by a diode.[7] If the substrate material were an ideal insulator, i.e., $R \to \infty$ or $I \to 0$, the FET would not be able to feel the existence of the other device and no backgating would occur. However, in reality, semi-insulating GaAs is not an ideal insulator; current does flow through the substrate when voltages are applied to the devices. Since the voltage drop (V_{BS}) across the interface diode is a monotonically increasing function of the reverse leakage current, a higher leakage current would cause a higher voltage drop and backgating effect would occur.

The relationship between substrate conduction and backgating has been studied using the test structure shown in Fig. 2. The substrate current which flows from the backgate electrode to the FET was measured as a function of the backgating voltage. Figure 3 shows the substrate I-V relations measured with the FET biased at three different

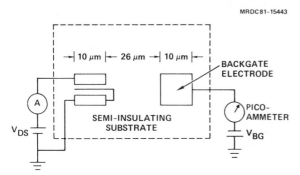

Fig. 2 The test structure and the test set-up for the backgating and the substrate conduction measurements.

drain voltages. At $V_{DS} = 0$, where the FET was just used as an anode, the I-V curve is linear at low voltages, then rises steeply at $V_{BG} = 5V$ and enters a high-order power law regime. This I-V behavior, similar to what has been observed on many high resistivity materials, has been explained by the trap-filled-limited carrier injection model.[1,8] However, in GaAs, other instability phenomena at high fields such as space charge domain formation and impact ionization can also contribute to such I-V relation. At $V_{DS} = 1.5$ V and 2.5 V the FET's backgating characteristics (I_{DSS} vs V_{BG}) were also measured. These characteristics, along with the substrate I-V relations, are shown in Fig. 3b and 3c. A strong correlation between backgating and substrate conduction is clearly seen in the figures. The threshold voltage for the sudden increase in the substrate current is exactly the same as the onset voltage for backgating. At low backgating voltages, because of the low free carrier concentration in the high resistive semi-insulating substrate, the substrate current is too small to maintain a substantial reverse bias voltage drop across the channel-substrate interface junction. Therefore, no backgating effect is observed and the devices are essentially isolated from one another. But when the voltage applied between devices exceeds the threshold value, a much higher substrate current flows between devices and through the channel-substrate junction of the FET, causing the junction reverse bias voltage to increase and the backgating effect to take place.

PROTON ISOLATION

Recently, D'Avanzo reported that proton bombardment in semi-insulating substrates can

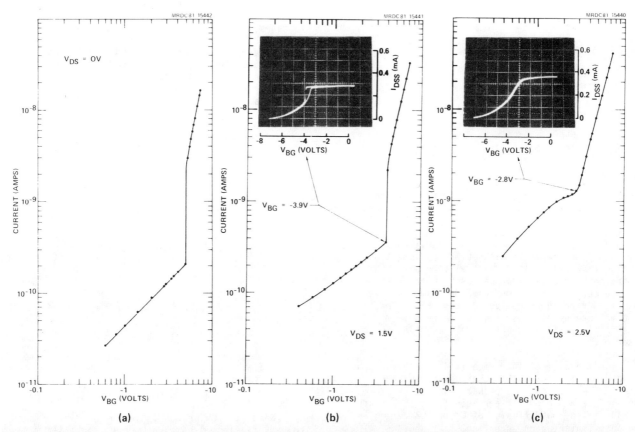

Fig. 3 The current-voltage relation of the substrate conduction between the FET and the backgate electrode when (a) $V_{DS} = 0$ V, (b) $V_{DS} = 1.5$ V, and (c) $V_{DS} = 2.5$ V. The backgating characteristics, I_{DSS} vs V_{BG}, at $V_{DS} = 1.5$ V and $V_{DS} = 2.5$ V are shown by the photographs in (b) and (c), respectively. The thresholds in backgating and substrate conduction are indicated by arrows.

increase device isolation and reduce the backgating effect.[9] This improvement was attributed to the increase of resistivity caused by proton damage. In recent experiments conducted at our laboratory, proton bombardment was also shown to greatly reduce the backgating effect. It was found, however, that the improvement in backgating was not due to an increase in resistivity but to an increase in the threshold for substrate conduction. Figure 4 shows the backgating characteristics (measured with the test structure shown in Fig. 2 at $V_{DS} = 2$ V) and the substrate I-V relations for samples with and without proton bombardment. The proton bombardment was carried out at 150 keV implant energy to a dose of 4×10^{14} cm^{-2}. Wtihout proton bombardment, the

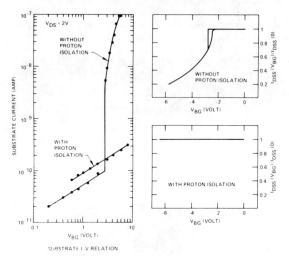

Fig. 4 The substrate I-V relation and the backgating characteristics of samples with and without proton bombardment. The test structure used is the same as in Fig. 2. The measurements were done with FET biased at $V_{DS} = 2$ V

backgating and the substrate conduction exhibit a threshold at $V_{BG} = -2.8$ V. With proton bombardment, no backgating and no threshold for substrate conduction were observed with V_{BG} up to -7 V. It is also interesting to see that with proton isolation, the leakage current between devices at low voltages is higher than without proton isolation, meaning the resistivity is actually lower in the proton bombarded sample. Because the substrate leakage current in proton bombarded samples never reaches a high enough value to cause any significant voltage drop across the FET's channel-substrate interface junction, no backgating is observed.

The improvement in device isolation by proton bombardment indicates that the current leakage path in the substrate is probably very close to the surface. One possible reason for such surface leakage is the outdiffusion of Cr or EL2 during the post-implantation annealing.[10,11] Because of fewer traps at the surface, the trap-fill-limited voltage or the threshold voltage for current conduction near the surface is smaller than in the bulk, therefore, the surface leakage is higher. With proton bombardment, the number of traps is increased so the threshold for current leakage is also increased. Besides the trap-filling model, there is another possible explanation for the improvement made by proton isolation. Because of the proton damaged layer, the potential distribution in the substrate may be changed and some local high field regions may be eliminated. So the instability phenomena associated with high fields and the resulting nonlinear substrate current conduction no longer exist.

LIGHT SENSITIVITY

The effects of light illumination on substrate conduction, backgating and circuit performance have also been studied. Because of the increased number of free carriers generated by light, a low resistance path is created near the substrate surface and surface leakage is greatly enhanced. The substrate I-V relation is usually linear with high current leakage and no threshold behavior is observed. The backgating characteristics, shown in Fig. 5, measured in both dark and light conditions with the test structure of Fig. 2, correlate well with the observed substrate conduction behaviors. In the dark, the backgating does not occur until a threshold voltage of about -3 V is reached, while under light illumination no threshold exists and the FET's current, I_{DSS}, starts to decrease even at very low backgating voltages. Based on the model presented earlier, such light sensitivity in backgating is easily explained by the light induced change in substrate conduction. The high leakage with no threshold substrate conduction behavior under light illumination results in a voltage drop appearing across the FET's channel-substrate interface junction and, therefore, backgating at even very low backgating voltages.

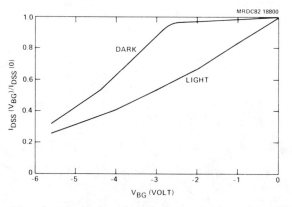

Fig. 5 The backgating characteristics of a FET measured in dark and light measurements. The test structure used is shown in Fig. 2.

Samples with proton isolation were also found to be very light sensitive. Under light illumination, carriers are excited from the proton induced traps and contribute to the current conduction. The backgating behavior for proton isolated samples in this condition is very similar to that of those without proton isolation.

CIRCUIT PERFORMANCE

The influence of substrate conduction and backgating on the eprformance of GaAs integrated circuits has been evaluated by using SDFL 9-stage ring oscillators. The speeds of the circuits were measured as functions of the negative voltage supply, V_{ss}. Ideally, if no backgating is present, the propagation delay (τ) should not depend on V_{ss} as long as V_{ss} is higher than the threshold of the switching FET. However, because backgating increases with V_{ss}, the circuit performance degrades when V_{ss} is increased.[12] We have compared such speed performances of circuits with several substrate conditions. Figure 6 shows the plot of τ vs. V_{ss} for ring oscillators with and without proton isolation. It is clear that for the circuit which is only substrate isolated the propagation delay increases at a much faster rate than it does for the proton isolated circuit. At high V_{ss} values speed improvement up to 50% has been observed for proton isolated circuits.

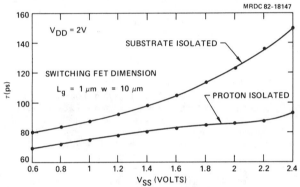

Fig. 6 Propagation delay (τ) versus negative voltage supply (V_{ss}) for ring oscillators with and without proton isolation.

When circuits are operated under light illumination, because backgating is stronger, the degradation of τ with increasing V_{ss} is enhanced. Figure 7 shows the τ vs V_{ss} plot of the same two

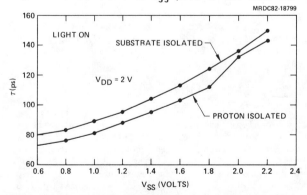

Fig. 7 Propagation delay (τ) versus negative voltage-supply (V_{ss}) for ring oscillators with and without proton isolation under light illumination.

circuits used for Fig. 6 operated under light illumination. Both circuits show enhanced degradation of τ vs V_{ss}; no significant difference was observed between proton isolation and substrate isolation. This result again demonstrates that the advantage of proton isolation shows up only when circuits are operated in dark environments.

CONCLUSION

Current leakage through semi-insulating substrate has been found to be directly related to the backgating effect. A model that uses a nonlinear resistor for the substrate and a reverse biased diode for the FET's channel-substrate interface has been used to explain the observed phenomena. Proton isolation has been found to icnrease the threshold for both substrate conduction and backgating. As a result, the ring oscillator performance is greatly improved and is less sensitive to the negative supplying voltage. Under light illumination, substrate conduction and backgating are enhanced, and the ring oscillator speed degrades with increasing negative voltage supply.

REFERENCES

1. C. P. Lee, S. J. Lee and B. M. Welch, IEEE Elect. Dev. Lett., EDL-3, p. 97, 1982.
2. H. Trandue, P. Rossel, J. Graffeuil, C. Azizi, G. Nuzillat and G. Bert, Rev. Phys. Appl., Vol. 13, p. 655, 1978.
3. I. Itoh and H. Yanai, IEEE Trans. Elect. Dev., ED-27, p. 1037, 1980.
4. N. Yokoyama, A. Shibatomi, S. Ohkawa, M. Fukuta and H. Ishikawa, Gallium Arsenide and Related Compounds, Conf. Ser. No. 366, p. 201, 1976.
5. C. Kocot and C. Stolte, IEEE Trans. Elec. Dev., ED-29, p. 1059, 1982.
6. C. P. Lee, Proc. of 2nd Conf. on Semi-Insulating III-V Matlerials, Evian, France, 1982.
7. S. J. Lee and C. P. Lee, Elec. Lett., Vol. 17, p. 760, 1981.
8. M. A. Lampert and P. Mark, Current INjection in Solids, Academic, N.Y. 1970.
9. D. C. D'Avanzo, IEEE Trans. Elec. Dev., ED-29, p. 1051, 1982.
10. A. M. Huber, G. Morillot, N. T. Linh, P. N. Favemec, B. Deveaud and B. Toulouse, Appl. Phys. Lett., Vol. 34, p. 858, 1979.
11. s. Makrom-Ebeid, D. Gautard, P. Devillard and G. M. Martin, Appl. Phys. Lett., vol. 40, p. 161, 1982.
12. C. P. Lee, B. M. Welch and R. Zucca, IEEE Trans. Electron Device, ED-29, p. 1103, 1982.

Invited Paper

RADIATION EFFECTS IN GaAs INTEGRATED CIRCUITS:
A COMPARISON WITH SILICON

Mayrant Simons

Research Triangle Institute

Research Triangle Park, NC 27709

ABSTRACT

This paper reviews the hardness capability of contemporary GaAs devices and logic circuits in terms of the four major nuclear and space radiation threat categories--neutron effects, total dose effects, dose rate effects, and single particle phenomena. The basic interaction mechanisms for each threat area are briefly described, with emphasis given to potential problem areas and to fundamental differences between gallium arsenide and silicon. Existing and projected hardness levels characteristic of GaAs devices are compared with corresponding levels for silicon LSI technologies.

INTRODUCTION

There is considerable interest in the radiation hardness capability of contemporary GaAs integrated circuits since their performance properties make them attractive candidates for future military and space systems applications. Contemporary GaAs IC technology is based principally on the MESFET or JFET structure which is fabricated in epitaxially grown layers or implanted directly into the semi-insulating GaAs substrate material. Among the more popular logic families are SDFL and BFL, which employ depletion mode MESFETs, and DCFL, which is based on the EFET. GaAs linear ICs are limited mainly to monolithic microwave amplifiers. Although the library of available radiation response data for GaAs ICs is relatively small in comparison with the large silicon data base, enough has been published to enable a general characterization of the present technology. Existing GaAs IC data are based principally on tests of discrete FETs and logic gates of SSI complexity, some monolithic microwave amplifiers, and, more recently, several MSI level digital circuits.

It should be noted that, in general, the task of maintaining a particular hardness level becomes more difficult and requires tighter parameter control as circuit complexity increases toward the LSI and VLSI levels and the number of devices that must remain functional on a single chip increases.

In considering the effects of nuclear and space radiation on semiconductor circuits, it is convenient to separate the effects into four categories that can be treated independently (neglecting synergistic effects that may occur when simultaneously approaching failure levels from two or more threats). The four categories are:

1. Neutron Effects - permanent damage resulting from cumulative neutron exposure; units are in n/cm^2.

2. Total Dose Effects - permanent damage arising from cumulative ionizing radiation exposure to X-rays, gammas, electrons, or protons; units are in rads (material).

3. Dose Rate or Photocurrent Effects - transient upset, soft error, or latchup phenomena resulting from the ionizing dose rate usually associated with a gamma pulse; permanent damage (burnout) may occur at very high intensities; units are in rads (mat)/s.

4. Single Event Upset (SEU) Phenomena - soft errors (and maybe latchup) arising from α-particles, protons, or heavy ions incident on an individual cell or logic element; device susceptibility expressed as an upset cross section or particle fluence per upset/bit.

In the following sections, the basic interaction mechanisms associated with each of these threats are summarized and approximate failure thresholds for contemporary GaAs ICs are established from a review of available test data. These values are then compared with corresponding levels characteristic of the major silicon LSI technologies.

NEUTRON EFFECTS

High energy (E > 10 keV) neutrons originating from a nuclear event or reactor interact with a semiconductor through elastic collisions with atoms in the crystal lattice. The resulting atomic displacements from the equilibrium lattice sites produce defect centers in the semiconductor bandgap that can affect majority carrier concentration and mobility and minority carrier generation and recombination lifetimes. Gallium arsenide and silicon are affected in much the same manner by displacement damage. Since traps are introduced near midgap, p- and n-type material in both semiconductors are compensated by neutron exposure and, at very high fluences, tend to become intrinsic. Carrier removal rates depend on a number of factors (such as carrier type and concentration, temperature, etc.) but are on the order of 10 cm^{-1} for GaAs and Si (1). Majority carrier mobility and minority carrier lifetimes also decrease in GaAs and Si as additional ionized scattering centers and generation/recombination centers are introduced by atomic displacements.

Minority carrier devices suffer mainly from gain degradation and increased leakage currents resulting from lifetime reduction while majority carrier device parameters are affected primarily by reductions in carrier concentration and mobility. Since the onset of significant minority carrier lifetime changes typically occur at lower fluences than do changes in majority carrier concentration and mobility, FET technologies (whether Si or GaAs) are inherently more resistant to neutron irradiation than bipolar technologies. The principal FET parameters that are affected by displacement damage are the maximum channel current (I_{DSS}), the maximum transconductance (g_{max}), the pinchoff voltage (V_p), and the cutoff frequency (f_c). All decrease monotonically with increasing neutron exposure (2).

Neutron test data reported for discrete GaAs JFETs (3) and MESFETs (4) show that the onset of parameter degradation occurs in the 10^{14} to 10^{15} n/cm^2 range; however, at an optimum channel doping level of 10^{17} cm^{-3}, changes were small after 10^{15} n/cm^2 (2,4). GaAs MSI logic devices based on either JFET or MESFET technologies should thus be capable of operating up to about 10^{15} n/cm^2 without suffering failure or significant performance degradation. However, as these technologies mature, failure thresholds for MSI devices are projected to extend into the 10^{15} to 10^{16} n/cm^2 regime. Linear devices, on the other hand, can be expected to be somewhat less tolerant as indicated by data showing increased noise figure in GaAs MESFET microwave amplifiers beginning at 10^{14} n/cm^2 (5). Significantly lower failure thresholds have been reported for GaAs CCDs which, like Si CCDs, suffer unacceptably large increases in transfer inefficiency between 10^{12} and 10^{13} n/cm^2 as a result of the introduction of shallow trapping levels (6).

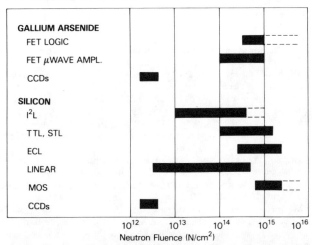

Figure 1. Reported (solid) and Projected (dotted) Displacement Damage Thresholds.

For the reason previously mentioned, both GaAs FET and Si MOS ICs are characterized by a higher neutron tolerance than Si bipolar circuits. This is reflected in the comparison of neutron damage thresholds for GaAs and Si (7,8) IC technologies depicted in Figure 1.

TOTAL DOSE EFFECTS

The most dramatic difference in the radiation tolerances of GaAs and Si devices occur in their sensitivities to cumulative ionizing radiation exposure. In silicon devices, ionizing radiation produces a positive space charge in SiO_2 insulating layers and interface states at the SiO_2-Si interfaces. MOS technologies are particularly sensitive to this "surface damage" and ultimately fail because of gate threshold voltage shifts or increased leakage currents. However, since thermally grown SiO_2 is an integral part of the silicon planar process, bipolar ICs are also sensitive to surface damage phenomena (leakage channel formation, gain degradation, etc.), although they are usually characterized by higher failure thresholds than MOS devices. (Various hardening techniques have greatly improved the tolerances of bipolar and MOS technologies to total dose effects.)

Since unpassivated GaAs surfaces exhibit good stability and are generally insensitive to slowly accumulated total dose effects, JFET and MESFET technologies that employ no dielectric isolation or surface passivation should be almost immune to surface damage problems. However, since dielectric passivation is used in some planar GaAs FET fabrication processes (9), the potential for surface problems in such devices cannot be completely dismissed; obviously, any sensitivity to total dose effects would be determined by the properties of the passivation material and its proximity to active device regions. (The total dose susceptibility of GaAs MISFET devices would also be governed by the properties of the gate dielectric.)

While surface damage does not now appear to pose a major threat to contemporary GaAs ICs, GaAs devices do not have unlimited tolerance to cumulative ionizing radiation exposure. At high doses displacement damage becomes the limiting mechanism (as in the case of neutrons).

Co 60 radiation test data taken on discrete EJFET devices have shown that relatively small changes in threshold voltage and mobility occur at 10^8 rad (GaAs), provided the channel doping level is about 10^{17} cm^{-3}; however, at a doping level of 10^{16} cm^{-3} significant changes were seen at 10^7 rads (GaAs) (2). A 256-bit EJFET DCFL RAM of MSI complexity has also been reported to remain functional after 5×10^7 rads (GaAs) with only a 10% increase in power dissipation (10). Small changes in MESFET parameters have been measured up to a level of 8×10^7 rads (Si) (11), and a Schottky diode FET logic (SDFL) circuit (150 gate complexity) was tested up to 10^8 rads while continuing to function and requiring only a small adjustment in the pulldown supply voltage (12). GaAs microwave MESFETs (13) and an MMIC amplifier (14) have also retained functionality up to 10^8 rads or more, although RF gain (13) and noise figure (5) begin to degrade between 10^7 and 10^8 rads.

Based on the above data, it is concluded that both digital and linear GaAs ICs fabricated with contemporary FET technologies are capable of operating well into the 10^7 to 10^8 rad regime without suffering significant performance degradation. A comparison of total dose failure thresholds for contemporary GaAs and silicon technologies (7,8) is shown in Figure 2. Although total dose data have not been published on GaAs CCDs, they are expected to tolerate levels well above a megarad, making them appreciably harder than Si CCDs (4×10^4 to 10^6 rads).

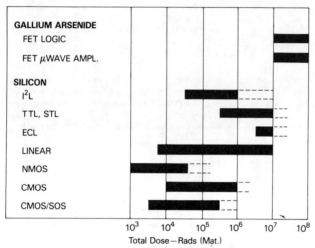

Figure 2. Reported (solid) and Projected (dotted) Total Dose Failure Thresholds.

DOSE RATE EFFECTS

Dose rate or photocurrent effects are produced in ICs as the excess carriers generated by the incident ionizing radiation are collected by p-n junctions. These effects are manifested in a circuit by large current or voltage perturbations whose magnitude and duration depend on the peak dose rate, pulse duration (t), junction area (A), collection volumes, carrier lifetimes (τ), and circuit/device time constants (especially when > t). For the simplest case of a p-n junction exposed to a square pulse of dose rate $\dot{\gamma}$ and duration $t \gg \tau_n, \tau_p$, the primary photocurrent is given by

$$I_{pp} = qA(W + L_n + L_p)g_o\dot{\gamma}$$

Here W is the depletion width and g_o is the pair generation constant which is about 7×10^{13} pairs/cm^3-rad for GaAs and 4×10^{13} pairs/cm^3-rad for silicon. Although g_o is slightly higher for GaAs than for Si, GaAs is characterized by much shorter diffusion lengths (L_n, L_p) and carrier lifetimes.

Obviously, photocurrent phenomena become quite complex in ICs where, in addition to primary photocurrents, transistors may contribute larger secondary photocurrents due to current or voltage amplification. The net effect of photocurrent transients in an IC is to disrupt linear circuit operation with large signal swings and produce logic errors in digital devices. Logic errors will appear as transient excursions or upsets in combinational logic elements and soft errors or bit-flips in bistable elements such as latches, registers, and memory cells.

Widespread differences exist in the upset sensitivities of the various silicon technologies and even among different types of devices belonging to the same technology. Variations can also be anticipated among GaAs circuits and logic families. Certain hardening techniques, such as photocurrent compensation, are available to the circuit designer to increase transient upset hardness; however, some of the most effective hardening methods are intimately related to the technology itself and include small device geometries and thin active layers to minimize photocurrent collection volumes and junction areas, dielectric isolation (D.I.) rather than junction isolation, and the use of low lifetime material. In view of these factors, GaAs IC technology would therefore appear relatively attractive in terms of dose rate hardness.

Unfortunately, semi-insulating GaAs substrate material can reduce the dose rate hardness of GaAs ICs by two different mechanisms. The first is via substrate photocurrents that can flow between contact pads or metallization placed directly on chip surfaces. Measurements have shown that such currents can be larger by several orders of magnitude than device photocurrents and, in fact, can dominate circuit response (15); however, it was also found that circuit response could be reduced by an order of magnitude by placing the bonding pads and metal interconnects on an insulating layer. The second substrate response mechanism is charge trapping in the deep levels that are characteristic of the semi-insulating substrate material (16). Conduction in FET structures can be severely affected (even cut off) by the trapped charge which decays with time constants that range from milliseconds to seconds. Recent work has shown though that this backgating-like effect can be appreciably reduced (and perhaps even eliminated) by the use of high quality LEC substrates and/or by various FET structural modifications, such a p-layer implanted beneath the n-channel (17). The effect can also be minimized by operating at high current levels.

Transient response data reported for GaAs digital ICs of MSI complexity have shown a broad range of upset thresholds. SDFL circuits have demonstrated thresholds ranging from 1×10^8 to 2×10^{10} rads/s (18), while the EJFET 256-bit RAM functioned without soft errors up to dose rates of 6×10^9 to 1×10^{10} rads (GaAs)/s (10). While these circuit upsets apparently resulted from photocurrent phenomena, disruptions in BFL logic gate and ring oscillator performance for tens of milliseconds have been observed by the author following 1 μsec LINAC exposures at total doses between about 10^2 and 10^3 rads (10^8 to 10^9 rads/s) and after 3 ns FXR pulses at the 100 rad level (3×10^{10} rads/s), all as a result of the backgating problem. Radiation-induced backgating has also been reported in power MESFETs and in a monolithic amplifier operating at X-band (19), although transient-free operation of a microwave MESFET was observed up to a dose rate of 3×10^{10} rad (Si)/s (20).

As indicated by the data above and illustrated in Figure 3, short pulse (< 100 ns) upset thresholds reported for GaAs ICs compare quite favorably with those characteristic of silicon LSI technologies (7,8). Long pulse (> 1 µs) thresholds decrease somewhat for silicon devices (because of the long lifetimes) but should not change appreciably for GaAs devices in the absence of severe "backgating" problems. Moreover, GaAs FET circuits are not susceptible to latchup associated with extraneous 4-layer paths as are many bipolar and CMOS/bulk silicon ICs. With respect to burnout or survivability thresholds (not discussed in this paper), most GaAs and Si ICs should exceed 10^{11} rads/s. Extremely low upset thresholds can be expected for both GaAs and Si CCDs (10^6 to 10^8 rad/s and lower).

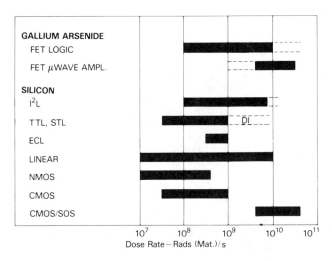

Figure 3. Reported (solid) and Projected (dotted) Short Pulse Upset Thresholds.

SINGLE EVENT UPSETS

Single event or single particle upset represents a relatively new radiation threat for digital microcircuits and has arisen because of the trend to smaller device dimensions and faster switching speeds (21). The SEU problem involves the introduction of soft errors in the bistable elements that comprise circuit memories, registers, and latches. Single particle upset can occur when the charge collected, Q_{COL}, by a sensitive cell node exceeds the critical charge, Q_{CRIT}, required at the node for cell upset. Obviously, as devices become smaller and faster, the energy and charge associated with a bit of stored data (and hence Q_{CRIT}) decreases. The charge deposited by an incident particle depends on the product of the characteristic energy loss (dE/dx) of the particle and its path length, dl; Q_{COL} will be collected in a fraction of a nanosecond along that part of the path length that extends through the depletion region to the depth of the funnel beneath the depletion region. Charge funneling can increase Q_{COL} in silicon devices by several times that expected from the ionization path in the depletion region alone, and can thereby significantly increase device sensitivity to SEU. Although dielectric isolation and SOS eliminate funneling from the substrate, recent experiments with 5.1 MeV alpha particles indicate that charge funneling can extend into the semi-insulating GaAs substrate underlying an epilayer device (22).

The radiation environment producing the SEU threat consists of heavy ions from cosmic rays and solar flares and also trapped protons in the radiation belts. Heavy ions produce ionization paths directly in the target material. Protons, on the other hand, do not produce upsets directly but undergo elastic scattering or nuclear reactions with the semiconductor atoms; the heavy recoil neuclei or byproduct alpha particles then produce ionization tracks (23).

There are insufficient experimental data presently available to enable a meaningful assessment of the SEU susceptibility of GaAs technology in general. In fact, the only information published to date describes measurements made on the 256-bit EJFET RAM when exposed to 40 MeV protons (24). The proton upset data point for this device, expressed in terms of an inverse cross section, is shown in Figure 4 along with data taken on several silicon RAMs in the same environment. Not shown in the figure are the results of proton experiments conducted on some other CMOS bulk silicon RAMs and two CMOS/SOS RAMs in which no upsets were observed (25).

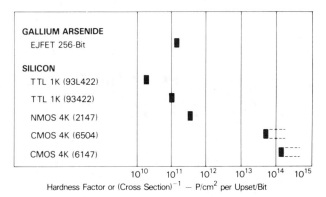

Figure 4. 40 meV Proton Upset Data for Static RAMs (24).

Since no heavy ion upset data are available for GaAs, one can only speculate as to how GaAs RAMs will compare with silicon RAMs in the cosmic ray environment. It is anticipated that in the absence of specific SEU hardening techniques, GaAs devices will be characterized by a range of sensitivities corresponding approximately to those for bulk silicon devices but below the range for CMOS/SOS. Whether or not circuit hardening techniques that can successfully eliminate SEU susceptibility in CMOS RAM cells (26) can be applied to GaAs cells remains to be seen.

SUMMARY

Contemporary GaAs ICs are characterized by a neutron hardness capability exceeding that of most silicon bipolar circuits and comparable to that of MOS circuits; the neutron hardness of GaAs FET logic devices is projected to improve. Gallium arsenide circuits exhibit a clear superiority over silicon

ICs in terms of their total dose susceptibility due to the absence of radiation sensitive dielectric layers such as SiO_2. Transient upset thresholds of GaAs ICs vary widely, corresponding approximately to those of the harder silicon bipolar devices; increased dose rate hardness toward the CMOS/SOS level is expected in some GaAs logic families as the technology improves. The SEU sensitivity of GaAs technology cannot be accurately established from presently available data; the SEU threat is clearly a potential problem area that needs further study.

REFERENCES

(1) Microwave Semiconductor Devices: Fundamentals and Radiation Effects by R.J. Chaffin, John Wiley and Sons, New York, 1973.

(2) R. Zuleeg and K. Lehovec, IEEE Trans. on Nucl. Sci., NS-27, 1343, Oct. 1980.

(3) R. Zuleeg, J.K. Notthoff, and K. Lehovec, IEEE Trans. on Nucl. Sci., NS-24, 2305, Dec. 1977.

(4) R.J. Gutmann and J.M. Borrego, IEEE Trans. on Rel., R-29, 232, Aug. 1980.

(5) J.M. Borrego, R.J. Gutmann, and S.B. Moghe, IEEE Trans. on Nucl. Sci., NS-26, 5092, Dec. 1979.

(6) W.C. Jenkins and J.M. Killiany, Naval Research Lab, Washington, DC, July 1982 (Private Communication).

(7) M. Simons, R.P. Donovan, and J.R. Hauser, AFAL-TR-76-194, Jan. 1977, RTI, Res. Tri. Pk., NC.

(8) D.M. Long, IEEE Trans. on Nucl. Sci., NS-27, 1674, Dec. 1980.

(9) B.M. Welch, Y. Shen, R. Zucca, R.C. Eden, S.I. Long, IEEE Trans. on Elec. Dev., ED-27, 1116, June 1980.

(10) J.K. Notoff, R. Zuleeg, and G.L. Troeger, 1983 IEEE NSRE Conf., Gatlinburg, TN, July 1983.

(11) J.M. Borego, R.J. Gutmann, S.B. Moghe, and M.J. Chudzicbi, IEEE Trans. on Nucl. Sci., NS-25, 1436, Dec. 1978.

(12) E.R. Walton, Rockwell International, Thousand Oaks, Cal., Aug. 1983 (Private Communication).

(13) D.M. Newell, P.T. Ho, R.L. Mencik, and J.R. Pelose, IEEE Trans. on Nucl. Sci., NS-28, 4403, Dec. 1981.

(14) Y. Kadowaki, Y. Mitsui, T. Takebe, O. Ishihara, and M. Nakatani, Technical Digest 1982 GaAs IC Symposium, 82CH17640, 83, Nov. 1982.

(15) R. Zuleeg, J.K. Notthoff, and G.L. Troeger, 1983 IEEE NSRE Conf., Gatlinburg, TN, July 1983.

(16) M. Simons, E.E. King, W.T. Anderson, and H.M. Day, J. Appl. Phys., 52, 6630, Nov. 1981.

(17) W.T. Anderson, M. Simons, E.E. King, H.B. Dietrich, and R.J. Lambert, IEEE Trans. on Nucl. Sci., NS-29, 1533, Dec. 82.

(18) E.R. Walton, W.T. Anderson, R. Zucca, and J.K. Notthoff, 1983 IEEE NSRE Conf., Gatlinburg, TN, July 1983.

(19) W.T. Anderson and S.C. Binary, 1983 IEEE NSRE Conf., Gatlinburg, TN, July 1983.

(20) J G. Castle, Sandia Laboratories, Albuquerque, NM, July 1983 (Private Communication).

(21) For recent papers on SEU see IEEE Trans. on Nucl. Sci., NS-29, 2018-2100, Dec. 1982.

(22) M.A. Hopkins and J.R. Srour, 1983 IEEE NSRE Conf., Gatlinburg, TN, July 1983.

(23) E.L. Petersen, IEEE Trans. on Nucl. Sci., NS-27, 1494, Dec. 1980.

(24) P. Shapiro, A.B. Cambell, J.C. Ritter, R. Zuleeg, and J.K. Notthoff, 1983 IEEE NSRE Conf., Gatlinburg, TN, July 1983.

(25) W.E. Price, D.K. Nichols, and C.J. Malone, 1983 IEEE NSRE Conf., Gatlinburg, TN, July 1983.

(26) S.E. Diehl, A. Ochoa, P.V. Dressendorfer, R. Koga, W.A. Kolasinski, IEEE Trans. on Nucl. Sci., NS-29, 2040, Dec. 1982.

GAMMA RAY RADIATION EFFECTS ON MMIC'S ELEMENTS

K.Aono, O.Ishihara, K.Nishitani, M.Nakatani, K.Fujikawa, M.Ohtani* and T.Odaka*

LSI Research and Development Laboratory, Mitsubishi Electric Corporation,
4-1, Mizuhara, Itami, Hyogo 664, Japan. Tel;(0727) 82-5131
*Materials Engineering Laboratory, Mitsubishi Electric Corporation,
1-1, Tsukaguchi-honmachi, 8-chome, Amagasaki, Hyogo 661, Japan. Tel;(06) 491-8021

ABSTRACT

Three experimental studies concerning radiation effects of gamma ray on GaAs MMIC's elements are reported.

Firstly, the effects of gamma ray on passive elements such as MIM capacitors and SBDs are discussed. No significant change has been observed for these elements up to 10^8 rads.

Secondly, performance of GaAs FETs heated at 200 °C is monitored during gamma ray irradiation up to 10^9 rads. The smaller change in DC parameters has been observed comparing to samples without heating. No annealing effects have been observed for the samples irradiated at high temperature.

Finally, degradation of Au/Ni/AuGe ohmic electrodes caused by gamma ray irradiation up to 10^9 rads in various ambient gases is discussed. It has been clarified that H_2O plays an important role in the enhancement of the inter-diffusion among the ohmic metal layers.

INTRODUCTION

GaAs monolithic microwave ICs (MMICs) will become one of the most widely utilized solid state microwave devices in the next generation of radar and microwave communication equipments.

Informations on the behavior of MMIC's elements during the exposure to severe radiation environments are of considerable significance from practical view points.

Degradation of electrical performance of MMICs caused by gamma ray irradiation was previously discussed[1] by noting on the GaAs FET that is the main component of the MMIC.

In this work, firstly, the radiation effects have been investigated for the MMIC's passive elements such as MIM (Metal-Insulator-Metal) capacitors and SBDs (Schottky Barrier Diodes). Secondly, DC characteristic change of GaAs FETs exposed to radiation at high temperature has been discussed along with annealing effects. Finally, inter-diffusion of Au/Ni/AuGe ohmic metallized layers induced by gamma ray has been analyzed by means of Auger electron spectroscopy.

PASSIVE ELEMENTS

MIM Capacitors

MIM capacitors were formed on a semi-insulating GaAs substrate. Silicon nitride (Si_3N_4) films having 4000 Å thickness are used as the insulator. Figure 1 shows relative change in capacitance versus total dose of gamma ray. The capacitance before radiation was 2.2 pF. The capacitance decreased about 3% by the irradiation of 10^8 rads. Hoewver, this change is small enough and practically tolerable in circuits.

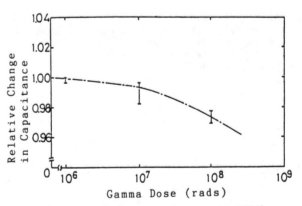

Fig.1 Relative change in MIM's capacitance

SBDs

The SBDs used for this work have planer type structure. Anode and cathode electrodes are arranged in parallel on an epitaxial layer having carrier concentration of 3×10^{17} cm^{-3}.

Figure 2 shows C-V characteristics of a SBD before and after gamma ray irradiation of 10^9 rads. Capacitance of the depleted (deeply reverse biased) SBD decreased about 50% by the irradiation. While, change of the capacitance in low bias region is small. As for forward-biased SBDs, DC characteristics as well suffered negligible change. Since SBDs are usually operated at no bias or small bias conditions, the change shown in Fig.2 is not significant.

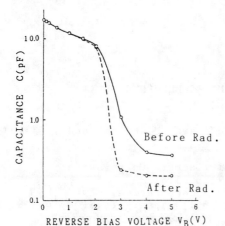

Fig.2 C-V characteristics of a SBD before and after gamma ray irradiation of 10^9 rads.

HEAT EFFECTS

Radiation effects at high temperature

GaAs FETs having gate length of 0.7 μm and gate width of 300 μm are used in this experiment. During the irradiation up to 10^9 rads, the devices were being heated at 200°C. The DC parameters of the FETs were measured at room temperature after the irradiation of 10^6, 10^7, 10^8 and 10^9 rads.

Figure 3 and 4 show changes of drain saturation current (I_{dss}) and transconductance (g_m) along with the results for room temperature (R.T) irradiation. It has been clarified that the devices kept at high temperature show smaller characteristic changes compared to those at room temperature.

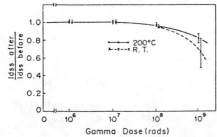

Fig.3 Gamma ray irradiation effects on I_{dss}.

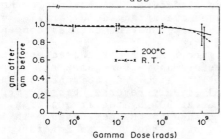

Fig.4 Gamma ray irradiation effects on g_m

Annealing effects

After the gamma ray irradiation of 10^9 rads, the GaAs FETs were annealed at various temperatures and the I_{dss} was monitored.

Figure 5 shows recovery rate [A] versus annealing time [t] with annealing temperature [T] as a parameter for the samples irradiated at room temperature. The recovery rate is defined as the following equation.

$$A=(I-I_0')/(I_0-I_0') \qquad (1)$$

where, I_0; I_{dss} before irradiation
I_0'; I_{dss} after irradiation of 10^9 rads (before annealing)
I; I_{dss} during annealing.

Solid lines in this figure are fitting curves expressed by a following equation.

$$A=16.24\exp(-0.149/kT)(1-\exp(t/\tau)) \qquad (2)$$

where, k ; Boltzman's constant (eV/K)
T ; annealing temperature (k)
t ; annealing time (sec)
τ ; time constant (sec).

Time constant and final recovery rates (A at t=∞) for various annealing temperatures are summarized in table 1.

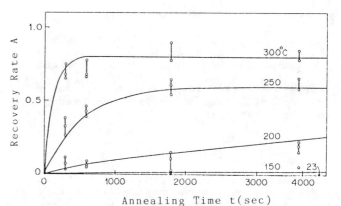

Fig.5 Annealing effects for irradiation at room temperature.

Table 1 Time constants and final recovery rates

Annealing Temperature T (°C)	Time Constant τ (sec)	Recovery Rate A at t=∞
27	7.66×10⁸	0.052
100	1.16×10⁶	0.159
150	5.11×10⁴	0.275
200	4349	0.423
250	593	0.600
300	114	0.800

In contrast to the samples irradiated at room temperature, no annealing effects were observed for the samples irradiated at high temperature (200 °C), as shown in Fig.6. The heat treatment during the irradiation has been found to be effective to reduce radiation damages.

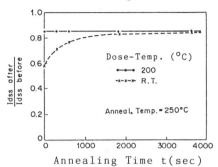

Fig.6 Annealing effect of GaAs FETs irradiated at high temperature.

DEGRADATION OF OHMIC ELECTRODES

Samples

Ohmic electrodes studied were constructed with Au(500 Å)/Ni(100 Å)/AuGe(500 Å) multilayers which were made by successive evaporation on GaAs(100) substrates.

Irradiation of gamma ray was carried out in well controlled ambient gases which are summarized in table 2.

The samples exposed to radiation are observed by optical microscope and analyzed by Auger electron spectroscopy.

Table 2 Ambient gases

Ambient gas	Composition
wet-air	R.H=59% (Temp.=23°C)
O_2-H_2O	P_{O_2}=500 Torr, P_{H_2O}=1 Torr
N_2/O_2-H_2O	P_{N_2}=591 Torr, P_{O_2}=133 Torr P_{H_2O}=1 Torr
Dry O_2	P_{O_2}=500 Torr
Dry N_2/O_2	P_{N_2} 592 Torr, P_{O_2}=158 Torr
H_2O	P_{H_2O}=1 Torr

Surface degradation

Figure 7 shows surfaces of the samples exposed to gamma dose up to 10^9 rads in wet air. As the gamma dose becomes greater, the surface degradation grows up.

Auger analysis

Figure 8 shows Auger depth profile of the sample before irradiation and Fig.9 shows those after gamma ray irradiation of 10^9 rads in the different ambient gases shown in table 2.

Figure 9 indicates that Ni, AuGe and the surface of GaAs layers were oxidized, when gamma ray irradiation was carried out in dry N_2/O_2, H_2O-O_2 and H_2O-N_2/O_2. However, such oxidation was not observed in

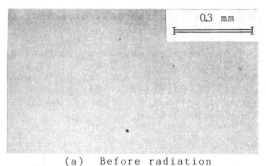
(a) Before radiation

(b) 10^6 rads

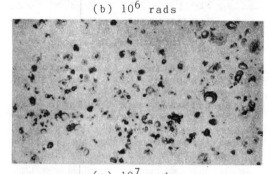
(c) 10^7 rads

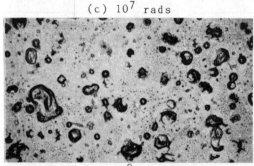
(d) 10^8 rads

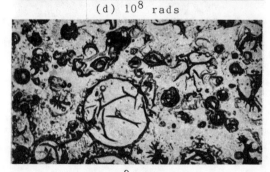
(e) 10^9 rads

Fig.7 Surface degradation

dry O_2 and pure H_2O. The degree of oxidation depends upon the ambient gases; the strongest for H_2O-N_2/O_2 followed by H_2O-O_2 and dry N_2/O_2.

These results show that oxidizing agents such as HNO_3 or H_2O_2 may be produced in H_2O-N_2/O_2, H_2O-O_2 and dry N_2/O_2 under gamma ray irradiation. Especially, H_2O plays an important role in the oxidation and the inter-diffusion.

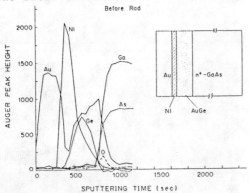

Fig.8 Auger depth profile before irradiation.

CONCLUSION

Radiation effects of gamma ray on the MMIC's elements have been investigated from various view points.

The followings are concluded.
(1) No significant degradation has been observed for MIM capacitors and SBDs.
(2) Heat treatment during the irradiation is effective to reduce radiation damages.
(3) The inter-diffusion among the ohmic metal layers has been observed, depending strongly on species in the ambient gas. It is clarified that H_2O plays an important role in the enhancement of the diffusion.

<u>Reference</u> (1) Y.Kadowaki et al.;"Effects of Gamma Ray Irradiation on GaAs MMICs" Tech. Dig., 1982 GaAs Symp., p.83.

ACKNOWLEDGEMENT

The experiments of gamma ray irradiation have been carried out using Co^{60} source facility of Radiation Center of Osaka Prefecture. Authors are grateful to Dr. M. Kitagawa et al. for their arrangements.

This work was performed under the management of the R&D Association for Future Electron Devices as a part of the R&D Project of Basic Technology for Future Industries, sponsored by Agency of Industrial Science and Technology, MITI.

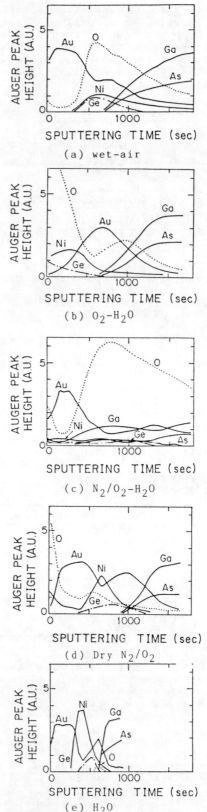

(a) wet-air

(b) O_2-H_2O

(c) N_2/O_2-H_2O

(d) Dry N_2/O_2

(e) H_2O

Fig.9 Auger depth profile after irradiation of 10^9 rads in various ambient gases.

Part III
Monolithic Circuit Applications

ONE of the most impressive features of the MMIC approach is the large variety of circuits to which this technology has been applied. Included in this part are papers describing low-noise and power amplifiers, multi-octave bandwidth amplifiers, voltage-controlled oscillators, transmit/receive modules, as well as examples of special-purpose components. The frequency range represented by these circuits extends from UHF well into the millimeter bands, a span of three decades!

Section III-A
Low-Noise Amplifiers and Other Receiver Circuits

THIS section describes various low-noise and low-power circuits for the DBS home receiver market, for military applications in active phased-array radar antennas, as well as in the instrumentation field.

The DBS market represents potentially the largest market for GaAs MMIC's, and the major consumer application. Should this market develop, as many anticipate, it will represent the best test for low-cost MMIC's. The strong interest in this application of MMIC's is reflected in the number of reprint papers included in this volume dealing with this topic.

GaAs Monolithic MIC's for Direct Broadcast Satellite Receivers

SHIGEKAZU HORI, MEMBER, IEEE, KIYOHO KAMEI, KIYOYASU SHIBATA, MIKIO TATEMATSU, KATSUHIKO MISHIMA, AND SUSUMU OKANO

Abstract —A 12-GHz low-noise amplifier (LNA), a 1-GHz IF amplifier (IFA), and an 11-GHz dielectric resonator oscillator (DRO) have been developed for DBS home receiver applications by using GaAs monolithic microwave integrated circuit (MMIC) technology. Each MMIC chip contains FET's as active elements and self-biasing source resistors and bypass capacitors for a single power supply operation. It also contains dc-block and RF-bypass capacitors.

The three-stage LNA exhibits a 3.4-dB noise figure and a 19.5-dB gain over 11.7–12.2 GHz. The negative-feedback-type three-stage IFA shows a 3.9-dB noise figure and a 23-dB gain over 0.5–1.5 GHz. The DRO gives 10-mW output power at 10.67 GHz, with a frequency stability of 1.5 MHz over a temperature range from -40–$80\,°$C. A direct broadcast satellite (DBS) receiver incorporating these MMIC's exhibits an overall noise figure of $\leqslant 4.0$ dB for frequencies from 11.7–12.2 GHz.

I. INTRODUCTION

TELEVISION BROADCAST SYSTEMS via 12-GHz direct broadcast satellites (DBS) are scheduled to enter into operation in various countries in the mid 1980's. Success of such systems, however, depends heavily on the availability of low-noise 12-GHz receivers in large quantities at an acceptable price. In view of the potential of large-scale production and low cost, the GaAs monolithic microwave integrated circuit (MMIC) seems to be the most viable candidate for the receiver application.

Extensive efforts are being directed in various laboratories toward the development of GaAs MMIC's for application to outdoor units of DBS receivers [1]–[3]. As shown in Fig. 1, a typical outdoor unit is composed of a GaAs FET low-noise amplifier (LNA), a bandpass filter, a mixer, an IF amplifier (IFA), and a dielectric resonator oscillator (DRO). This paper describes three kinds of FET-based GaAs MMIC's (LNA, IFA, and DRO) that have been developed for actual operation in an experimental outdoor unit. Each MMIC chip has been designed to operate under a single power supply by incorporating self-biasing resistors and capacitors and to have dc-block and RF-bypass capacitors.

The LNA and IFA chips are mounted in ceramic packages for easy handling, and the DRO chip is mounted in a hermetically sealed housing, together with a dielectric resonator, to avoid a moisture effect. The FET's and resistors in the IFA and DRO chips are fabricated by

Manuscript received April 28, 1983; revised July 10, 1983.
The authors are with the Microwave Solid-State Department, Electronics Equipment Division, Toshiba Corporation, 1 Komukai-Toshiba-cho, Saiwai-ku, Kawasaki 210, Japan.

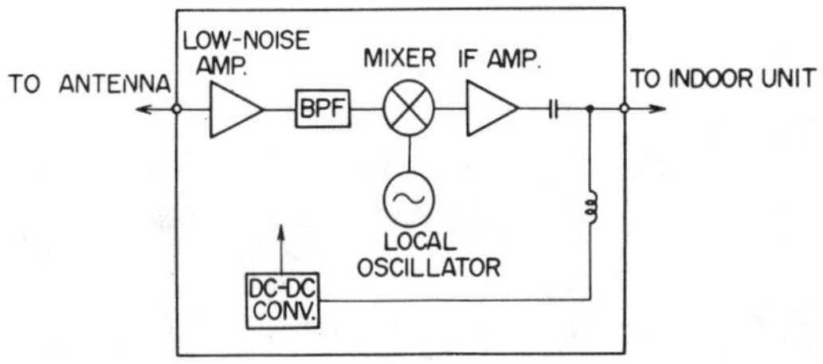

Fig. 1. Typical block diagram of an outdoor unit of a DBS receiver.

selective direct ion implantation into a semi-insulating GaAs substrate. These MMIC's have been successfully employed in the outdoor unit of a DBS home receiver. The following sections describe circuit design, fabrication, packaging, and RF performance of the MMIC's.

II. CIRCUIT DESIGN

A. Low-Noise Amplifier

The design goal was to build a low-noise amplifier with a noise figure of $\leqslant 3.5$ dB and a gain of $\geqslant 20$ dB at 12 GHz. In order to fulfill this goal, a three-stage FET amplifier was chosen. Prior to the three-stage design, however, a single-stage amplifier was built and thoroughly tested.

Figs. 2 and 3 show the circuit diagram and the top view of the single-stage LNA chip, respectively. The chip measures 1.5×1.5 mm. The noise figure of an FET usually improves with decreasing gate length, as previously demonstrated by some authors with a quarter-micrometer gate GaAs FET fabricated using electron-beam lithography [4]. In the present work, however, the FET gate length was chosen to be 0.4 μm by making tradeoffs between noise figure and device yield in the future production phase. The gate is also designed to have a width of 200 μm and a single pad to minimize the gate–source overlay parasitic capacitance. The source electrodes outside the FET are tapered to reduce the source inductance.

In order to obtain the optimum source and load impedances for gain and noise figures of the FET, gain and noise parameters of a discrete FET equivalent to the FET to be employed in the MMIC were measured. Fig. 4 shows the constant noise figure (solid lines) and gain (dotted lines) circles for the source impedance at 12 GHz, where the inductance of 0.2 nH for the tapered grounding pattern is taken into account. It is found that the minimum noise figure of 1.9 dB and the associated gain of 7.5 dB are

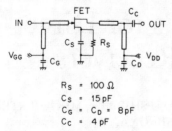

Fig. 2. Circuit diagram of a single-stage low-noise amplifier.

R_S = 100 Ω
C_S = 15 pF
C_G = C_D = 8 pF
C_C = 4 pF

Fig. 3. Single-stage low-noise amplifier chip (1.5 × 1.5 mm).

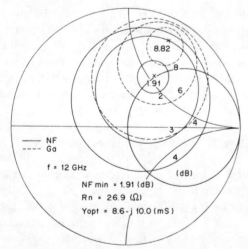

Fig. 4. Constant noise figure and gain circles of a discrete FET.

f = 12 GHz
NF min = 1.91 (dB)
Rn = 26.9 (Ω)
Yopt = 8.6 − j10.0 (mS)

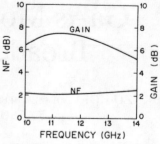

Fig. 5. Calculated noise figure and gain of a single-stage LNA as a function of frequency.

substrate has been chosen to be 300 μm by considering matching circuit loss, ease of chip handling, and less performance changes due to thickness variation.

The source resistor R_s and RF-bypass capacitor C_s are incorporated for single power supply operation. R_s is chosen to be 100 Ω to the drain current to 7 mA, at which the noise figure of the FET reaches a minimum. An external gate-bias terminal is also included in order to evaluate RF performance as a function of drain current. The capacitances of RF-bypass capacitors C_G and C_D, source capacitor C_s, and dc-block capacitor C_c are 8, 15, and 4 pF, respectively.

Fig. 5 shows the calculated RF performances of the single-stage LNA by taking into account the losses of microstrip lines and RF-bypass capacitors in the matching circuits. The minimum noise figure of 2.2 dB and the gain of 7.0 dB are predicted at 12 GHz.

After evaluation of the single-stage LNA, the three-stage LNA was designed. The circuit diagram and the top view of the three-stage chip with a size of 1.5 × 3.0 mm are shown in Figs. 6 and 7, respectively. The input and output matching networks of each stage have been optimized for a source and load impedance of 50 Ω. The input-matching circuits have been designed to optimize noise figure in the first and second stages and to optimize gain in the third stage. The source capacitor C_s and the resistor R_s are employed in every stage so that the LNA can be operated using a single power supply. The capacitors C_s, C_D, and C_c and the resistors R_s have the same values as the corresponding ones in the single-stage LNA. From the single-stage LNA performance, the RF performance of the three-stage LNA was predicted to have a minimum noise figure of 2.5 dB and a gain of 22 dB at 12 GHz.

B. IF Amplifier

The IFA has been designed to give a noise figure of ⩽ 3.5 dB, a gain of ⩾ 20 dB, and an input and output VSWR of ⩽ 2.0 for frequencies from 0.5–1.5 GHz. Several approaches to amplifier designs in this frequency range have been proposed to date [6], [7]. However, most of them utilize a gate–drain resistive feedback around an individual FET with a relatively wide gate width. In order to reduce the gate width for lower dc power consumption, we have designed a three-stage monolithic IF amplifier in which the feedback resistor is connected between the input and output ports of the amplifier. The circuit diagram and the top

obtained for the optimum source admittance of Y_{opt} = 8.6 − j10.0 mS.

After the evaluation and analysis of the discrete FET, the input and output matching circuits of the single-stage LNA were designed to optimize a noise figure and realized by shunt- and series-connected microstrip lines with a characteristic impedance of 70 Ω. The shunt microstrip lines are terminated by RF-bypass capacitors. The length of each microstrip line is compensated by taking into account the effective line length reduction due to the coupling between lines [5]. The thickness of the GaAs

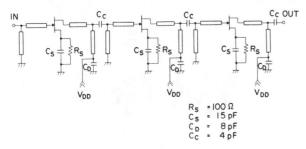

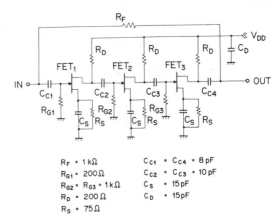

Fig. 6. Circuit diagram of a three-stage low-noise amplifier.

$R_S = 100 \, \Omega$
$C_S = 15 \, \text{pF}$
$C_D = 8 \, \text{pF}$
$C_C = 4 \, \text{pF}$

$R_F = 1 \, \text{k}\Omega$ $C_{C1} = C_{C4} = 8 \, \text{pF}$
$R_{G1} = 200 \, \Omega$ $C_{C2} = C_{C3} = 10 \, \text{pF}$
$R_{G2} = R_{G3} = 1 \, \text{k}\Omega$ $C_S = 15 \, \text{pF}$
$R_D = 200 \, \Omega$ $C_D = 15 \, \text{pF}$
$R_S = 75 \, \Omega$

Fig. 8. Circuit diagram of a 1-GHz IF amplifier.

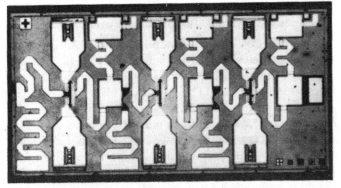

Fig. 7. Three-stage low-noise amplifier chip (1.5 × 3.0 mm).

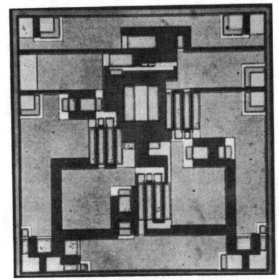

Fig. 9. IF amplifier chip (1.5 × 1.5 mm).

view of the IFA chip with a size of 1.5 × 1.5 mm are shown in Figs. 8 and 9, respectively.

In the amplifier design, consideration was also given to ease of assembly and operation in DBS receivers. The source resistors R_s and bypass capacitors C_s are incorporated as in the LNA for single power supply operation. Furthermore, the drain resistors R_D and bypass capacitor C_D are connected to each drain and at the bias feed terminal, respectively. The drain resistors R_D together with the bypass capacitor C_D make it possible to eliminate the RF choke otherwise necessary outside the chip.

The gate of each FET has a length of 1 μm. Its width as well as the gate resistors R_{G1}, R_{G2}, R_{G3}, the drain resistors R_D, and feedback resistor R_F are optimized by a computer simulation in terms of gain flatness and input and output VSWR's. Fig. 10 shows the calculated noise figure as a function of the FET gate width [6]. The noise figure becomes lower as the gate width becomes wider. Its width is determined to be 600 μm by tradeoffs between noise figure and dc power consumption. The capacitor values are determined such that the performance does not degrade at the lowest frequencies of interest.

The predicted performance of the IFA shows a noise figure of ≤ 3.2 dB, a gain of 22 ± 0.1 dB, and input and output VSWR's of ≤ 2.0 at frequencies from 0.5–1.5 GHz.

C. Dielectric Resonator Oscillator

The local oscillator for DBS receiver applications is required to have a frequency stability of ±1 MHz at a center frequency of 10.7 GHz over a temperature range of −40–80 °C. Among several types of MMIC oscillators proposed to date [8], [9], the dielectric resonator oscillator configuration is the most promising candidate for the local

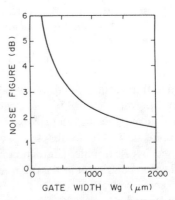

Fig. 10. Gate width dependence of an IFA noise figure.

oscillator application in view of size and cost. We have designed a dielectric resonator oscillator which consists of an MMIC oscillator chip and a dielectric resonator circuit. The oscillator circuit diagram and the top view of the oscillator chip with a size of 1.5 × 1.5 mm are shown in Figs. 11 and 12, respectively.

Hybrid MIC technology is used to form the resonant circuit for frequency stabilization. The dielectric resonator is mounted on alumina substrate and coupled to a micro-

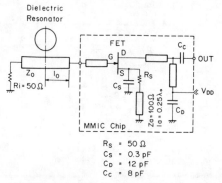

Fig. 11. Circuit diagram of a dielectric resonator oscillator.

$R_s = 50\ \Omega$
$C_s = 0.3\ \text{pF}$
$C_D = 12\ \text{pF}$
$C_C = 8\ \text{pF}$

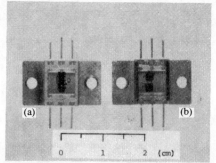

Fig. 13. Inside view of packaged (a) LNA and (b) IFA.

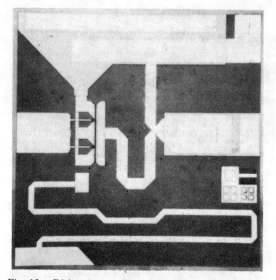

Fig. 12. Dielectric resonator oscillator chip (1.5 × 1.5 mm).

strip line terminated by a 50-Ω load. The dielectric resonator used has an unloaded Q of 7400, a relative dielectric constant of 36.3, and a resonant-frequency temperature coefficient of +6 ppm/°C.

The MMIC oscillator chip has a common-source configuration with a feedback capacitor C_s. An FET with a gate length of 1 μm and a width of 300 μm is used to obtain an oscillator output power of 10 dBm. The circuit design has been performed using the measured S-parameters of the equivalent discrete FET. The feedback capacitor C_s of 0.3 pF is determined by a computer simulation so as to make the output reflection coefficient at the drain terminal maximum under a given reflection coefficient of the resonant circuit with a loaded Q of 1000. The output matching circuit is optimized by a nonlinear analysis based upon the measured large signal impedance of the discrete FET [10]. It is composed of a series microstrip line and a shunt microstrip line terminated by the RF-bypass capacitor C_D.

The resistor R_s with a quarter-wavelength shunt stub was used for single power supply operation. A resistance of 50 Ω was chosen to make the drain current 20 mA. The drain-bias circuit is included in the output matching circuit. The bypass capacitor C_D and dc-block capacitor C_c have a capacitance of 12 pF and 8 pF, respectively.

III. Device Packaging

For evaluation of the LNA and IFA MMIC's in a DBS receiver, a specially designed universal hermetic package has been provided. It is a rectangular stripline package with a copper flange. The packaged LNA and IFA are shown in Fig. 13 (a) and (b), respectively, with top covers removed. The LNA contains a three-stage LNA chip. The IFA contains two cascaded three-stage IFA chips. The two center leads are for RF input and output. While other leads can be used for dc biasing, only one of them is employed in the present application since the MMIC chips are designed for single power supply operation. The dc-bias connection to the chips is done via a dielectric standoff. The DRO chip has been mounted together with a dielectric resonator in a hermetically sealed MIC housing with outer dimensions of 25 × 25 × 12 mm.

IV. Device Fabrication

The LNA chip was fabricated using an epitaxial wafer with active and buffer layers successively grown on Cr-doped semi-insulating substrate by a metal-organic chemical vapor deposition method. The carrier concentration and thickness of the active layer are 2.0×10^{17} cm^{-3} and 0.5 μm, respectively. Mesas are formed to define FET active areas and resistors.

The IFA and the DRO MMIC's were fabricated by selective Si ion implantation into undoped semi-insulating GaAs substrates. A resist/SiO$_2$ film is used as a mask for the ion implantation. The acceleration energy and dose are 70 keV and 3.5×10^{12} cm^{-2} for FET active layers. For FET contact layers and resistor layers, Si ions are dually implanted at a dose of 2×10^{13} cm^{-2} and energies of 250 and 120 keV. After ion implantation and removal of the resist/SiO$_2$ film, the wafers are annealed at 850°C for 15 min in AsH$_3$/Ar atmosphere without encapsulants.

The gates of the FET's in the LNA are defined to a length 0.4 μm by using electron-beam lithography. The gate is recessed in order to attain a low noise figure. The gate length of FET's in the IFA and DRO MMIC's is 1 μm, and the widths are 600 and 300 μm, respectively. They were delineated by conventional photolithography.

The Schottky-barrier gate electrodes are formed of Al with a thickness of 6000 Å. The ohmic electrodes are formed by alloying Pt/AuGe at 450°C. The first-level metallization of the MIM capacitor is a 0.8-μm-thick Al.

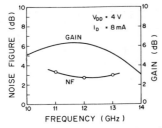

Fig. 14. Measured noise figure and gain of a single-stage low-noise amplifier as a function of frequency.

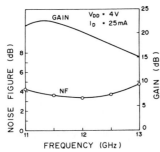

Fig. 15. Measured noise figure and gain of a three-stage low-noise amplifier as a function of frequency.

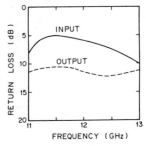

Fig. 16. Measured input and output return loss of a three-stage low-noise amplifier as a function of frequency.

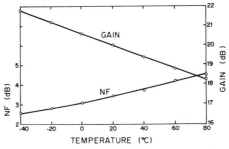

Fig. 17. Temperature dependence of noise figure and gain of a three-stage LNA.

CVD SiO_2 film with a thickness of 3,500 Å is used as the capacitor dielectric. The Au/Pt/Ti metal systems are used for the top plates of the capacitors, the bonding pad, and the interconnection metals. All of the metal patterns are formed by a liftoff process.

The MMIC chips are all glassivated with a CVD SiO_2 film, except for the bonding area. After lapping the substrate to a thickness of 300 μm, the backside of the wafer is metallized.

V. RF Performance

A. Low-Noise Amplifier

Fig. 14 shows the measured frequency response of the noise figure and gain of the single-stage LNA at a drain voltage $V_{DD} = 4$ V and a drain current $I_D = 8$ mA. The MMIC chip is mounted on a coplanar test fixture. A minimum noise figure of 2.8 dB and a gain of 6.5 dB are obtained at 12 GHz, including the test fixture loss of 0.3 dB. The discrepancy between predicted and measured noise figures is mainly due to the performance variation of the FET employed in the MMIC.

The three-stage LNA chip was evaluated after mounting into the hermetic ceramic package as shown in Fig. 13 (a). Fig. 15 shows the measured frequency response of the noise figure and gain of the LNA operated at $V_{DD} = 4$ V and a total current $I_D = 25$ mA. A minimum noise figure of 3.4 dB and a gain of 20 dB were obtained at 12 GHz, and a noise figure $\leqslant 3.4$ dB and a gain $\geqslant 19.5$ dB were obtained at frequencies from 11.7–12.2 GHz. These measured results also include the test fixture loss of 0.4 dB.

The minimum noise figure and the gain are in good agreement with the values predicted from the single-stage LNA performance. The frequency of maximum gain, however, is shifted towards a lower frequency. This might be caused by an interstage matching problem.

Fig. 16 shows the measured frequency response of input and output return losses of the three-stage LNA. An input return loss $\geqslant 6$ dB (VSWR $\leqslant 3.0$) and an output return loss $\geqslant 12$ dB (VSWR $\leqslant 1.7$) were obtained at frequencies from 11.7–12.2 GHz. The input VSWR characteristic is similar to that of the single-stage LNA, measured using the coplanar test fixture. However, the output VSWR is worse than that of the single-stage LNA. The disagreement is due to the dimensional limitation of the ceramic package. The inner dimension of the package is about one and half times longer than the MMIC chip length. Since the MMIC chip has been mounted at the input terminal side of the package, long bonding wires are used to connect the output pad of the chip and output terminal of the package. The bonding wire has an inductance of ~1.0 nH, causing the output VSWR discrepancy between the measurement and design.

Fig. 17 shows the temperature dependence of the noise figure and the gain for the packaged three-stage LNA at 12 GHz. The noise figure of $\leqslant 4.4$ dB and the gain of $\geqslant 18.4$ dB were obtained over the temperature range of -40–$80\,°$C.

B. IF Amplifier

The three-stage IFA chip was also evaluated after mounting in the hermetic ceramic package. Fig. 18 shows the measured frequency response of the noise figure and gain of the IFA operated at a drain voltage $V_{DD} = 7$ V and a total current $I_D = 39$ mA. A noise figure $\leqslant 3.9$ dB and a gain 23.5 ± 0.1 dB were obtained from 0.5–1.5 GHz. The measured gain performance shows good agreement with the simulated result. However, the noise figure is somewhat

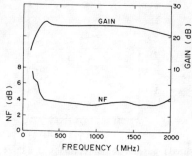

Fig. 18. Measured noise figure and gain of a 1-GHz IF amplifier as a function of frequency.

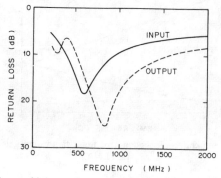

Fig. 19. Measured input and output return loss of a 1-GHz IF amplifier as a function of frequency.

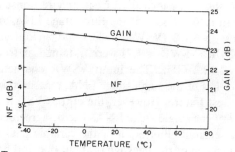

Fig. 20. Temperature dependence of the noise figure and gain of IFA

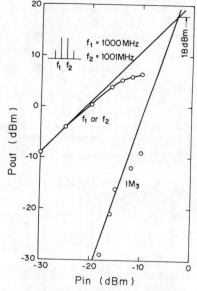

Fig. 21. Third-order intermodulation products of IFA.

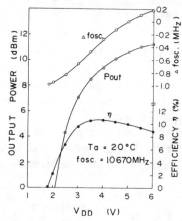

Fig. 22. Measured output power (P_{out}), oscillation frequency deviation (Δf_{osc}), and efficiency (η) of a dielectric resonator oscillator as a function of drain voltage (V_{DD}).

poorer than the predicted performance. This may be due to insufficient noise modeling of the FET's, whose noise parameters were obtained by simple scaling from the measured ones of the 300-μm gate width FET.

Fig. 19 shows the measured frequency response of input and output return loss. An input return loss $\geqslant 7$ dB (VSWR $\leqslant 2.6$) and an output return loss $\geqslant 11$ dB (VSWR $\leqslant 1.8$) were attained in the same frequency band. Although the IFA has been designed to minimize the input and output VSWR's at a center frequency of 1 GHz, the measured minimum input and output VSWR's are shifted towards lower frequencies. A computer simulation can explain this discrepancy if we assume that the gate–source capacitance of the FET is larger, by a factor 1.3, than that used in the design.

Fig. 20 shows the temperature dependence of the noise figure and the gain for the IFA at 1 GHz. A noise figure of $\leqslant 4.5$ dB and a gain of $\geqslant 23$ dB were obtained over the temperature range of -40–$80\,^\circ$C. No major changes are observed for gain flatness or input and output VSWR's. Fig. 21 shows the third-order intermodulation products of the IFA. It is found that the output power at the 1-dB gain compression point and the third-order intercept point are measured to be 8 and 18 dBm, respectively.

Since an IF amplifier gain of $\geqslant 40$ dB was required to operate the outdoor unit, two IFA chips were cascaded in the package as shown in Fig. 13(b). The measured gain of the cascaded IF amplifier was 45 ± 0.3 dB over 0.5–1.5 GHz, and the input and output VSWR's are similar to those of the single-chip IFA.

C. Dielectric Resonator Oscillator

Fig. 22 shows the measured drain voltage V_{DD} dependence of output power P_{out}, oscillation frequency deviation Δf, and efficiency (η) of the DRO with a center oscillation frequency of 10.67 GHz. An output power of 10.5 dBm with 10-percent efficiency was obtained at a drain voltage $V_{DD} = 5$ V. The frequency pushing is ~ 0.2 MHz/V. Since

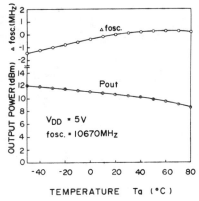

Fig. 23. Measured temperature dependence of the oscillation frequency deviation (Δf_{osc}) and output power (P_{out}) of a dielectric resonator oscillator.

the measured frequency pulling for VSWR = 1.5 is ~ 400 kHz, the external Q of the DRO is calculated to be ~ 10 000.

Fig. 23 shows the temperature dependence of frequency deviation (Δf) and the output power (P_{out}). It is found that the frequency variation is 1.5 MHz (1.2 ppm/°C) and the output power variation is 4 dB over a temperature range from −40–80°C.

VI. DBS Receiver Application

An outdoor unit of a DBS receiver has been modified to accommodate the developed LNA, IFA, and DRO MMIC's. The packaged LNA containing a three-stage chip and the packaged IFA containing two cascaded three-stage chips have been employed. The outdoor unit has exhibited an overall noise figure of ⩽ 4 dB for frequencies from 11.7–12.2 GHz.

VII. Conclusion

Three kinds of GaAs MMIC's have been developed for the DBS receiver. The low-noise amplifier has a noise figure of 3.4 dB and a gain of 19.5 dB for frequencies from 11.7–12.2 GHz. The IF amplifier has a noise figure of 3.9 dB and a gain of 22 dB for frequencies from 0.5–1.5 GHz. The DRO incorporating an MMIC chip and a dielectric resonator gives an output power of 10 mW at 10.67 GHz and a frequency stability of 1.5 MHz over the temperature range of −40–80°C. All MMIC chips incorporate self-biasing source resistors and bypass capacitors for single power supply operation, and are contained in hermetic ceramic packages or housings. The DBS receiver using three MMIC chips exhibits an overall noise figure of ⩽ 4.0 dB for frequencies from 11.7–12.2 GHz.

Acknowledgment

The authors wish to thank Dr. M. Ohtomo and S. Makino for continuous encouragement and helpful discussions, and N. Tomita, S. Watanabe, N. Kurita, T. Soejima, and H. Kawasaki for valuable contributions through this work.

References

[1] C. Kermarrec, P. Harrop, C. Tsironis, and J. Faguet, "Monolithic circuits for 12 GHz direct broadcasting satellite reception," in *1982 IEEE MTT Monolithic Circuit Symp. Dig.*, pp. 5–10.
[2] R. A. Pucel, "Design considerations for monolithic microwave circuits," *IEEE Trans. Microwave Theory Tech.*, vol. MTT-29, pp. 513–534, June 1981.
[3] L. C. Liu, D. W. Maki, M. Feng, and M. Siracusa, "Single and dual stage monolithic low noise amplifiers," in *1982 GaAs IC Symp. Dig.*, pp. 94–97.
[4] K. Kamei, S. Hori, H. Kawasaki, and T. Chigira, "Quarter micron gate low noise GaAs FETs operable up to 30 GHz," in *IEDM Tech. Dig.*, 1980, pp. 102–105.
[5] T. Bryant and J. Weiss, "Parameters of microstrip transmission lines and of coupled pairs of microstrip lines," *IEEE Trans. Microwave Theory Tech.*, vol. MTT-16, pp. 1021–1027, Dec. 1963.
[6] K. Honjo, T. Sugiura, T. Tsuji, and T. Ozawa, "Low-noise, low-power-dissipation GaAs monolithic broadband amplifiers," in *1982 GaAs IC Symp. Dig.*, pp. 87–90.
[7] S. Hori, K. Kamei, M. Tatematsu, T. Chigira, H. Ishimura, and S. Okano, "Direct-coupled GaAs monolithic IC amplifiers," in *1982 IEEE MTT Monolithic Circuit Symp. Dig.*, pp. 16–19.
[8] J. S. Joshi, J. R. Cockrill, and J. A. Turner, "Monolithic microwave gallium arsenide FET oscillators," *IEEE Trans. Electron Devices*, vol. ED-28, pp. 158–165, Feb. 1981.
[9] B. N. Scott and G. E. Brehm, "Monolithic voltage controlled oscillator for *X* and *Ku*-band," in *1982 IEEE MTT-S Int. Microwave Symp. Dig.*, pp. 482–485.
[10] Y. Tajima, B. Wrona, and K. Mishima, "GaAs FET large-signal model and its application to circuit designs," *IEEE Trans. Electron Devices*, vol. ED-28, pp. 171–175, Feb. 1981.

12-GHz-Band Low-Noise GaAs Monolithic Amplifiers

TADAHIKO SUGIURA, HITOSHI ITOH, TSUTOMU TSUJI, AND KAZUHIKO HONJO, MEMBER IEEE

Abstract —One- and two-stage 12-GHz-band low-noise GaAs monolithic amplifiers have been developed for use in direct broadcasting satellite (DBS) receivers. The one-stage amplifier provides a less than 2.5-dB noise figure with more than 9.5-dB associated gain in the 11.7–12.7-GHz band. In the same frequency band, the two-stage amplifier has a less than 2.8-dB noise figure with more than 16-dB associated gain. A 0.5-μm gate closely spaced electrode FET with an ion-implanted active layer is employed in the amplifier in order to achieve a low-noise figure without reducing reproducibility. The chip size is 1 mm $\times$ 0.9 mm for the one-stage amplifier, and 1.5 mm $\times$ 0.9 mm for the two-stage amplifier.

I. INTRODUCTION

RECENT ADVANCES in GaAs technology have made monolithic microwave integrated circuits (MMIC's) more practical. Promising applications for this technology include inexpensive receiver front ends for direct broadcasting satellite (DBS) systems [1], [2]. This paper describes design considerations, the fabrication process, and performances for newly developed one- and two-stage 12-GHz-band low-noise GaAs monolithic amplifiers for use in DBS receivers. For MMIC's used in DBS receivers, reproducibility improvement and chip size reduction are essential in order to achieve low cost. A low noise figure is also required for the amplifiers, because it determines the overall receiver noise figure. In this work, most efforts were focused on achieving these requirements.

II. FET DESIGN

The main reason for poor MMIC reproducibility is the variation in FET characteristics caused by nonuniformity of active layers. To improve uniformity, an ion-implantation technique was employed to form the active layers, although epitaxially grown active layers are believed to be better for low-noise FET's. In conventional MMIC's, a recessed gate structure has been widely used for reducing unfavorable source resistance [5]. The gate-recessing process, however, degrades uniformity of active layers. To overcome this difficulty, a closely spaced electrode (CSE) FET structure [3], [4] was introduced. In the CSE FET, source–gate and drain–gate spacings are shortened to 0.5

Manuscript received May 2, 1983; revised July 20, 1983.
The authors are with the Microelectrics Research Laboratories, NEC Corporation, Miyazaki, Miyamae-ku, Kawasaki 213, Japan.

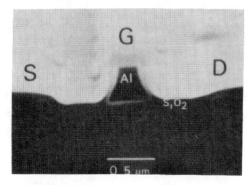

Fig. 1. CSE FET cross-sectional SEM photograph.

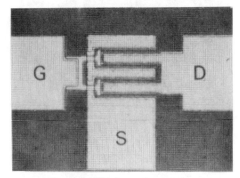

Fig. 2. FET electrode pattern.

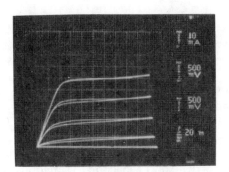

Fig. 3. FET static characteristics.

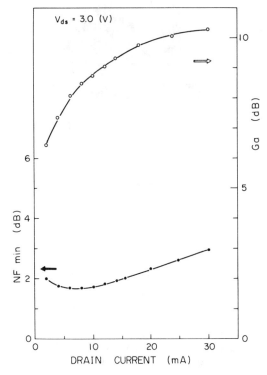

Fig. 4. FET minimum noise-figure and associated gain characteristics.

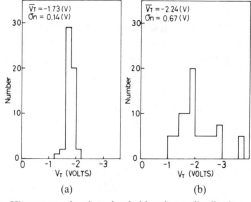

Fig. 5. Histograms showing threshold voltage distributions for CSE FET and deeply recessed epi-FET. (a) Ion-implanted wafer. (b) Epitaxial wafer.

μm, so that source resistance can be reduced sufficiently without recessing the gate. Fig. 1 is a cross-sectional SEM photograph of the FET. The gate was formed 0.5 μm long by side-etching from a 1.5-μm-long photoresist mask. The mask was also utilized to form ohmic electrodes. Because the ohmic electrodes are formed by lifting-off technique, source–drain spacing becomes 1.5 μm long and the gate is formed at the center of the spacing. Therefore, the gate and the ohmic electrodes were self-aligned.

Although a bar-shaped gate pattern is usually used in discrete low-noise FET's [5], an interdigital electrode pattern has been employed because this pattern uses less space. As shown in Fig. 2, the FET has four gate fingers. Each finger is 70 μm long. The total gate width is 280 μm. An FET threshold voltage V_t has been chosen as -1.7 V. Fig. 3 shows static characteristics for the FET. The saturated drain current I_{dss} is 60 mA. Observed transconductance g_m at a 10-mA drain current is 30 mS, which corresponds to 105 mS/mm. Gate breakdown voltage V_{bg} is -8.5 V. Source resistance R_s and source-to-gate capacitance C_{sg} at 0-V bias voltage are 4 Ω and 0.23 pF, respectively. Microwave characteristics were measured at 12 GHz. Results are shown in Fig. 4. The minimum noise figure $NF_{\min}$ and associated gain Ga at a 10-mA drain current are 1.7 and 8.8 dB, respectively. Maximum available gain (MAG) at a 30-mA drain current is 11.5 dB. These characteristics are almost the same as for the deeply recessed epi-FET [5].

In order to study orientation effects [6], FET's oriented in both [011] and [01$\bar{1}$], crystal directions were fabricated and measured. However, both for static and microwave characteristics, no significant difference was observed.

In Fig. 5, histograms showing V_t distributions on a wafer for the CSE FET and the deeply recessed epi-FET are comparatively presented. Averaged threshold voltage $\bar{V}_t$

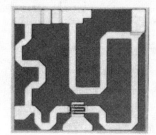

Fig. 6. One-stage amplifier chip photograph.

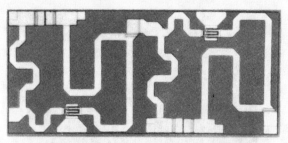

Fig. 8. Cascaded version of the two-stage amplifier chip photograph.

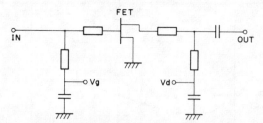

Fig. 7. One-stage amplifier equivalent circuit.

Fig. 9. Modified version of the two-stage amplifier chip photograph.

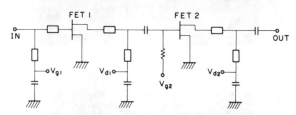

Fig. 10. Modified version of the two-stage amplifier equivalent circuit.

and standard deviation σ_n for the CSE FET are -1.73 and 0.14 V, respectively. For the epi-FET, they are -2.24 and 0.67 V, respectively. A sample variation coefficient defined by $\sigma_n/\overline{V}_t$, which is a measure for achieving uniformity, is 0.08 for the CSE FET and 0.30 for the epi-FET. Therefore, the uniformity in the CSE FET is improved by a factor of approximately four compared with the epi-FET.

When ion-implanting conditions are fixed, reproducibility in FET threshold voltages for two wafers sliced from different ingots is rather poor, because physical parameters, especially Cr concentration, for each ingot are greatly different. Also, in one ingot, the Cr concentration at the top and bottom positions are considerably different, because it varies along the crystal growth direction. Therefore, the implanting conditions have been experimentally determined by test implantation into several wafers sampled from an ingot. By this procedure, the desired FET threshold voltage can be realized with good reproducibility for all the remaining wafers in the ingot.

III. Amplifier Design

Fig. 6 shows a chip photograph for the one-stage amplifier. Fig. 7 shows its equivalent circuit. The chip size is 1 mm $\times$ 0.9 mm, and the wafer thickness is 150 μm. One-section parallel and series microstrip lines are used for matching circuits. These lines can also be utilized as dc-bias feed lines. This arrangement allows a great savings in the chip area. To retain a high Q value, a 2.5-μm metallization thickness was chosen. The measured Q value for a 50-μm-wide line, which was mostly used in the amplifier, is $30 \sim 40$ at 12 GHz. This value is somewhat lower than the value estimated from data described in [7]. As shown, the microstrip lines were folded in order to reduce the chip size. To avoid parasitic couplings, the spacings between adjacent lines were designed to be as large as possible.

Capacitors are of the MIM type, where dielectric material is SiO_2, with a relative dielectric coefficient ϵ_r of 4.8. Measured Q values for a 1.42-pF capacitor are 31.3 at 2.5 GHz, 28.8 at 5.4 GHz, and 27.1 at 8.6 GHz. For a 4.2-pF capacitor, they are 22.6 at 2.1 GHz, 17.2 at 5.2 GHz, and 17.0 at 8.4 GHz. The measurement was carried out by employing a resonant method [8]. All capacitors are used as dc-block or RF-short capacitors. Therefore, capacitance has been chosen larger than 2 pF and thickness control for SiO_2 film is not critical.

By using measured S-parameters for a discrete FET, element values in a FET equivalent circuit were derived. Based on these values, the amplifier circuit parameters were optimized by a CAD program. In the desired frequency band, which is from 11.7 to 12.7 GHz, more than a 9-dB gain was predicted.

There are two different versions of the two-stage amplifier. One is constructed through a cascaded connection between two identical one-stage amplifiers, as shown in Fig. 8. Therefore, the chip is double sized at 2 mm $\times$ 0.9 mm. In this version, the impedance locus from the first- to second-stage FET passes through a 50-Ω point. This route is obviously redundant. Therefore, chip size reduction can be expected by modifying the interstage matching circuit, as the route becomes shorter. In the other version, the modification was carried out by using the CAD program. A chip photograph and its equivalent circuit for the modified version are shown in Fig. 9 and Fig. 10, respectively. As shown, the chip size is reduced to 1.5 mm $\times$ 0.9 mm,

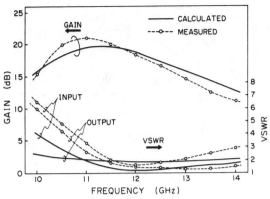

Fig. 11. Comparison between calculated and measured characteristics in modified version of the two-stage amplifier.

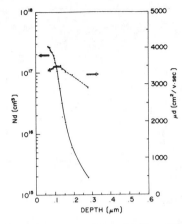

Fig. 13. Carrier concentration and drift mobility profiles.

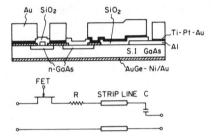

Fig. 12. Circuit elements cross-sectional view and equivalent circuit.

although the matching element sensitivity becomes slightly higher. Gate-bias voltage for the second-stage FET is supplied through a resistor. The resistance was chosen larger than 5 kΩ to prevent causing dissipation loss.

Fig. 11 shows a comparison between calculated and measured characteristics for gain and VSWR's in the modified version two-stage amplifier. As shown, quantitative designability is poor, although qualitative tendencies agree well. The poor designability is caused by inaccuracy in FET S-parameter measurement and by parasitic couplings between matching elements. Therefore, in actual design, the matching circuit pattern layout, including noise-matched operation, has been determined by experimental modification from original gain-matched CAD data. This layout modification can be accomplished by changing only one photomask level at the final fabrication process. Usually, desired characteristics are obtained after one to two modifications.

IV. Fabrication Process

Fig. 12 shows a cross-sectional view for various circuit elements together with their equivalent circuit. A Cr-doped semi-insulating HB-grown GaAs wafer is selectively implanted with $^{30}Si^+$ to form FET active layers. Resistive layers are formed at the same time. Implanting conditions for realizing $V_T = -1.7$ V are a 70-keV acceleration energy and a 3.2×10^{12} cm^{-2} dose. The wafer is then annealed with a 0.2-μm-thick CVD-SiO$_2$ cap at 800° C for 20 min in a H$_2$ ambient. Donor carrier concentration N_d and drift mobility μ_d profiles in the ion-implanted layer are shown in Fig. 13. They were measured by using a 250-μm-long gate FET [9]. The peak carrier concentration is 2.7×10^{17} cm^{-3} and the mobility is about 3500 cm^2/V·s.

Al, which is used as FET gates and capacitor lower electrodes, is deposited by vacuum evaporation to 0.4-μm thickness and etched to form the gates. Ohmic electrodes for the FET's and the resistors are then formed by lifting off a AuGe-Ni film and alloying it at 400° C. In GaAs IC fabrication, the gate-forming process is usually most difficult. In this process, however, it is very simple because the gates and the ohmic electrodes are self-aligned, as previously mentioned. After ohmic electrodes are formed, the Al is gain etched to form the capacitors lower electrodes. Then, SiO$_2$ for FET passivation and capacitor dielectric material is chemically vapor-deposited to a 0.2-μm thickness and etched to form contact vias. Next, Ti for the electroplating feeder is evaporated onto the whole wafer. Microstrip lines and capacitor upper electrodes are then formed by Ti-Pt-Au liftoff and thickened to 2.5 μm by selective Au plating. Topside processing is completed by etching off the feed metal Ti. The wafer is thinned and the rear is metallized by AuGe-Ni-Au evaporation. Amplifier chips can be obtained by cleaving the wafer.

V. Microwave Performance

Amplifier chips were chosen for microwave evaluation on the basis of visual inspection and dc testing. The selected chips were mounted on Au-plated copper carriers using AuSn solder. The carriers were then mounted on test fixtures and tested in a 50-Ω system. Needless to say, no external bias tee is necessary, because bias circuits are included on the chips.

Fig. 14 shows gain and noise figure characteristics for the one-stage amplifier. In the 11.7–12.7-GHz band, which is the desired frequency band, the amplifier provides a less than 2.5-dB noise figure with more than 9.5-dB associated gain. The maximum gain and the minimum noise figure in the band are 12 and 2.2 dB, respectively. Bias conditions are 2.5-V drain voltage and 10-mA drain current. Although the gain can be increased by increasing the drain voltage, as well as the drain current, when this increase is made, the noise figure is degraded. The noise figure was measured at 70-MHz IF frequency using a mixer. Input and output VSWR's are less than 3, and less than 2.5 in the desired frequency band, respectively.

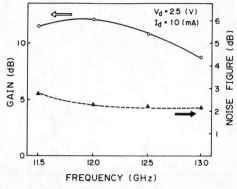

Fig. 14. One-stage amplifier gain and noise-figure characteristics.

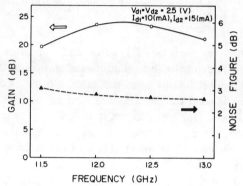

Fig. 15. Cascaded version of the two-stage amplifier gain and noise-figure characteristics.

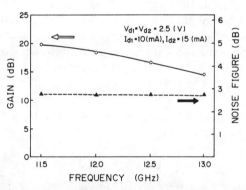

Fig. 16. Modified version of the two-stage amplifier gain and noise-figure characteristics.

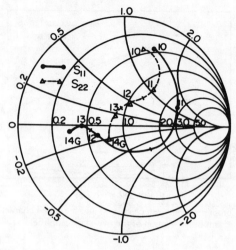

Fig. 17. Modified version of the two-stage amplifier input and output impedances.

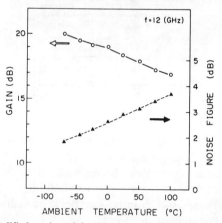

Fig. 18. Modified version of the two-stage amplifier temperature characteristics.

Gain and noise figure characteristics for the cascaded version two-stage amplifier are shown in Fig. 15. More than a 20-dB gain with a 24-dB maximum, and less than a 3-dB noise figure with 2.6-dB minimum are obtained in the desired band. The input VSWR is less than 4, and the output VSWR is less than 2.5. These characteristics mostly agree with the value predicted from the one-stage amplifier, because this version is constructed simply by the cascaded connection. For two-stage amplifiers, the drain current for the second-stage FET is set at 15 mA.

Fig. 16 shows gain and noise figure characteristics for the modified version two-stage amplifier. In the desired frequency band, the amplifier has less than a 2.8-dB noise figure with more than a 16-dB associated gain. Although the gain is degraded, compared with the cascaded version, the noise figure is improved. Input and output impedances for this amplifier are shown in Fig. 17. In the figure, reference planes are chosen at cleaved edges of the amplifier chip. Less than 2.5 input VSWR and less than 2 output VSWR are obtained in the 11.7–12.7-GHz band.

Fig. 18 shows gain and noise figure versus ambient temperature characteristics at a 12-GHz frequency for the modified version two-stage amplifier. In the measurement, drain currents, which varied with the ambient temperature, were set to standard values at room temperature. At $-70°$ C, the noise figure is 1.8 dB, with a 20-dB associated gain. While at 100° C, the noise figure is 3.7 dB with a 16.9-dB associated gain. Between these temperatures, the gain and noise figure are changed linearly according to changes in ambient temperature. These tendencies are almost the same for frequencies other than 12 GHz. Therefore, frequency characteristics both for the gain and the noise figure are shifted in parallel when varying the ambient temperature.

VI. Conclusion

The design considerations, fabrication process, and performances for newly developed one- and two-stage 12GHz-band low-noise GaAs monolithic amplifiers for use in DBS receivers have been described. By introducing the ion-implanted CSE FET, both a low noise figure and improved uniformity, which implies high reproducibility, can be achieved. Chip size reduction is also accomplished by employing compact matching circuits. The measured microwave performances are well within the acceptable range for DBS receivers. Although the investigation on yields has not been sufficiently carried out, it is believed that this approach has the potential to obtain good results. This work has made a cost-effective one-chip front end for DBS systems more realistic.

ACKNOWLEDGMENT

The authors would like to thank K. Suzuki for fabricating photomasks and T. Ozawa for ion implanting. They also would like to thank H. Muta and Y. Takayama for their constant encouragement throughout this work.

REFERENCES

[1] K. Kermarrec, P. Harrop, C. Tsironis, and J. Faguat, "Monolithic circuits for 12 GHz direct broadcasting satellite reception," in *Microwave and Millimeter-Wave Monolithic Circuits Symp., Dig. Papers*, June 1982, pp. 5–10.

[2] L. C. Liu, D. W. Maki, M. Feng, and M. Siracusa, "Single and dual stage monolithic low noise amplifiers," in *GaAs IC Symp. Tech. Dig.*, Nov. 1982, pp. 94–97.

[3] T. Furutsuka, T. Tsuji, F. Katano, A. Higashisaka, and K. Kurumada, "Ion-implanted E/D type GaAs IC technology," *Electron. Lett.*, vol. 17, no. 25/26, pp. 944–945, Dec. 1981.

[4] K. Honjo, T. Sugiura, T. Tsuji, and T. Ozawa, "Low-noise low-power-dissipation GaAs monolithic broadband amplifiers," in *GaAs IC Symp. Tech. Dig.*, Nov. 1982, pp. 87–900.

[5] K. Ohata, H. Itoh, F. Hasegawa, and Y. Fujiki, "Super low-noise GaAs MESFET's with a deep-recess structure," *IEEE Trans. Electron Devices*, vol. ED-27, pp. 1029–1034, June 1980.

[6] C. P. Lee, R. Zucca, and B. M. Welch, "Orientation effect on planar GaAs Schottky barrier field effect transistors," *Appl. Phys. Lett.*, vol. 37, no. 3, pp. 311–313, Aug. 1980.

[7] A. Higashisaka and T. Mizuta, "20-GHz band monolithic GaAs FET low-noise amplifier," *IEEE Trans. Microwave Theory Tech.*, vol. MTT-29, pp. 1–6, Jan. 1981.

[8] R. E. DeBrecht, "Impedance measurements of microwave lumped elements from 1 to 12 GHz," *IEEE Trans. Microwave Theory Tech.*, vol. MTT-20, pp. 41–48, Jan. 1972.

[9] R. A. Pucel and C. F. Krumm, "Simple method of measuring drift-mobility profiles in thin semiconductor films," *Electron. Lett.*, vol. 12, no. 10, pp. 240–242, May 1976.

MONOLITHIC CIRCUITS FOR 12 GHz DIRECT BROADCASTING SATELLITE RECEPTION

C. Kermarrec, P. Harrop, C. Tsironis and J. Faguet

LABORATOIRES D'ELECTRONIQUE ET DE PHYSIQUE APPLIQUEE (LEP)
3, avenue Descartes, 94450 LIMEIL-BREVANNES, FRANCE

ABSTRACT

This paper describes the design, fabrication and performances of gallium arsenide monolithic circuits of each of the principal microwave functions of a 12 GHz DBS receiver. The technology includes the use of Czochralski grown semi-insulating substrates, ion implanted active layers and localised growth of lines and interdigital capacitances. The low noise amplifier presents a 3,6 dB noise figure with 7,3 dB gain in the r.f. band. A dual gate mixer is presented with 6,5 dB noise figure and 2 dB conversion gain. The stable local oscillator has 32 mW output power and a stability of ± 0,3 ppm/K.

INTRODUCTION

A number of countries have recently indicated their intention to pursue their projects to launch 12 GHz direct broadcasting satellites in the mid 80's. Although exact system specifications may vary from country to country, the receiver requirements may generally be considered unchanged thus giving rise to the largest single market for microwave equipment to date. Although satisfactory hybrid front-ends already exist and are in advanced states of development[1], a number of research centres[2-4] are investigating the possibility of producing MMIC's for this application in view of the required large quantity production.

The advantages offered by the GaAs MESFET in fulfilling the principal roles in a downconverter have been described elsewhere[1,5]. In order to render these functions monolithic, the necessary passive circuitry must be fully compatible with sub-micron gate FET processing and must be capable of reproducible results with a high yield.

This paper will outline the technology employed, the modelling of the lumped elements and give examples of monolithic MESFET subassemblies of each r.f. function ; 12 GHz low noise amplifiers, dual gate mixers and stable local oscillators designed to reply to the specifications of a DBS receiver.

LUMPED ELEMENTS

Detailed characterisation of the passive elements used to present the required impedances to the active components is fundamental to the design of MMICs. These elements include inductances (straight line, loop and spiral) and interdigital capacitances and their characterisation must include the element itself as well as its associated parasitic elements.

Inductances

An inductive line can be considered as a lumped element under certain conditions. Consider the normalised impedance at the input of a line loaded by z'

$$z = \frac{z' + j \tan \beta l}{1 + j z' \tan \beta l} \quad (1)$$

when $z' \tan \beta l \ll 1$,
$$z = z' + j \beta l, \quad (2)$$
that is the load impedance and a series inductance βl. Thus two conditions are necessary to consider the element as a lumped element : l must be small and z' must not be too high.

Another consideration is the equivalent inductance of the inductive term $j \beta l Z_o$,

$$L = Z_o \sqrt{\varepsilon_{eff}} \cdot \frac{1}{c} \quad (3)$$

This term represents the inductance of a metal line in the presence of an earth plane at a distance h. This analysis is only valid for heights h less than h_{max} beyond which the equivalent inductance is greater than the lumped inductance of a line which is not perturbed by the presence of an earth plane. For typical line dimensions this maximum height would be about 300 microns thus severely restricting the application of a line model for substrates greater than this thickness.

Linear and loop inductances

A substrate thickness of 300 microns has been selected so as not to degrade the Q value of the lumped elements. Under these conditions the ground plane modifies only slightly the inductances and the following expressions remain valid[6].

$$L = 2 l (2.3 \log_{10} (2\pi l/W) - 1 + W/l\pi)$$
for a straight inductance and, $\quad (4)$

$$L = 12,57 a (2.3 \log_{10} (8\pi a/W) - 2)$$
for loop inductances, $\quad (5)$

where a = mean radius in cms
W = line width in cms
l = line length in cms

Both straight and loop inductances have been characterized in reflection from 2 to 14 GHz. The results confirm the values of inductance in equations (4) and (5). Measurements carried out in transmission on identical elements enable the construction of an equivalent circuit for the inductance :

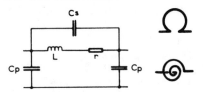

Fig. 1 : equivalent circuit of loop and spiral inductances

where C_s, C_p and r are the parasitic elements.

The Q value of a typical 1 nH inductance with a 20 micron track width and 3 micron metallisation thickness is approximately 25 at 10 GHz.

Spiral inductances

This element fulfils two rôles in our requirements ; that of matching element at frequencies around 1 GHz and that of r.f. choke for biasing.

Using a similar technique to that detailed above, equivalent circuits have been deduced for inductances whose values vary from 2 to 20 nH. Typically, the r.f. choke used in the oscillator circuit has 6 turns 10 micron track width and gap and a parallel resonant frequency at 12 GHz.

Interdigital capacitances

The interdigital capacitance for this type of periodic structure can be deduced from a closed form expression[7]. However, precise characterisation is required over a wide bandwidth in order to generate accurate estimations of the parasitic elements of the structure.

Measurements in reflection and in transmission have been effectuated from 2 to 14 GHz and the following equivalent circuits have been deduced, where L_s, L_p, C_p and r represent the parasitic elements.

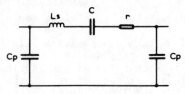

Fig. 2a : Equivalent circuit of interdigital series capacitances

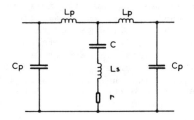

Fig. 2b : Equivalent circuit of interdigital parallel capacitances

Throughout circuit optimisation of the functions presented in this article, the complete equivalent circuit was used for each passive element in the matching circuits.

As a means of verifying the validity of the modelling of the lumped elements, a pass band filter was processed and designed to have more than 20 dB rejection in the image band of a 12 GHz receiver. The frequency response of this filter was sufficiently close to the predicted curve to conclude that the equivalent circuits were adequate for use in matching circuits.

CIRCUIT TECHNOLOGY

Two complementary technologies have been developed. The first, "semi-monolithic", involves the deposition of all necessary passive circuitry onto the gallium arsenide and subsequent bonding of the device into the circuit. This technology enables simple and rapid characterisation of the lumped elements and circuit optimisation. The second technology is purely monolithic and will be described in more detail.

The material used throughout this work is in-house Czochralski GaAs (resistivity $> 10^7$ ohm-cm) generating wafers of about 2" diameter. After crystallographic alignment the wafers are implanted, capped and annealed to produce uniform active layers. The wafers are then polished to a 300 μm thickness. This latter is chosen in order to maintain useful Q values of the passive elements.

The transistor technology is based on a self-aligned process[8] which produces gates of 0,7 micron lengths in a 2 micron drain-source spacing. Devices have gate widths of either 150 microns or 300 microns. The active areas are defined by a mesa etching isolation step down to the semi-insulating substrates onto which the circuit is grown by selective gold plating.

In order to minimise skin effect contributions and to generally increase the Q of the passive matching elements, a metallisation thickness of 3 microns was chosen. The processing that has been developed is based on selective growth through polyimide film and is capable of producing interdigital capacitances with 5 micron gaps and 10 micron finger widths. The first step involves an evaporation of a continuous layer of Ti/Pt/Au on the GaAs, on the pads of the active components and the MIM capacitance contacts. Three microns of polyimide are then spun onto the wafer and polymerised after which the polyimide is capped, masked and plasma etched. Three microns of gold are then grown on the circuit pattern.

Finally, the remaining polyimide guide is removed by plasma etching and the adhesive layer of Ti/Pt/Au is removed by ion milling down to the semi-insulating GaAs substrate. Since the growth surface of the gold pattern remains constant throughout the process, the technique is controllable and has given reproducible interdigital capacitances with 5 micron interfinger gaps. (figure 3).

Fig. 3 : Interdigital capacitance : 10 micron finger width, 5 micron gaps.

Fig. 4 : Spiral inductor and underpass interconnection : 10 micron track width, 10 micron gap.

Second level interconnections are necessary at a number of points in the circuits : between the source pads of the oscillator transistor, over the MIM capacitances and from the centre of the spiral inductances to the outside circuit. An example of this latter is shown in figure 4. This interconnection is an underpass beneath the turns of the inductor and is effectuated by using a negative photoresist that may subsequently removed to leave small air bridges.

MONOLITHIC SUB-ASSEMBLIES

Low noise amplifier

The noise figure of the downconverter is mainly fixed by that of the preamplifier and for that reason our target specification is a noise figure less than 4 dB with a gain of 7 dB per stage in the 11,7 - 12,5 GHz bandwidth.

In order to estimate an eventual dispersion in device parameters a number of ion implanted devices of the same geometry was characterised both in "S" parameters and in noise parameters. This characterisation was carried out using a "peeling" routine[9]. The measurements show typical dispersions of less than 5 % in amplitude and in phase for all four "S" parameters whereas the noise parameters present dispersions between 10 and 15 %.

Resulting from this characterisation a number of amplifiers were designed using a maximum of two elements at the input and output of the device ; a configuration which is sufficient to cover the required bandwidth.

The conception of the amplifier was based on the analysis and characterisation of the lumped elements presented above. Care was taken in the use of inductive elements so as to use the equivalent electrical length of the element as a function of the impedance to be matched, particularly at the output of the device. Circuit optimisation was carried out both in noise and gain using COMPACT.

Amplifier performance

Both semi-monolithic and monolithic versions of these amplifiers were processed and measured between two coplanar lines in order to reduce access capacitances to the circuit. The ground contact was assured by a mesh down to the ground plane.

Under these conditions the best results for the single stage amplifier are a noise figure less than 3,6 dB and a gain more than 7,3 dB between 11,5 and 12,5 GHz (figure 5). VSWR is less than 1,9 at the input and less than 1,7 at the output. Figure 6 shows the circuit configuration of one of the amplifiers.

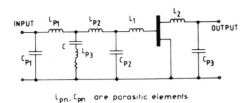

L_{pn}, C_{pn} are parasitic elements

Fig. 6 : Complete amplifier circuit topology.

Figure 7 shows the monolithic amplifier, the chip size is 1,6 mm^2.

Monolithic dual gate MESFET mixer

Single gate MESFET mixers have been successfully operated with as much as 6 dB conversion gain over octave I.F. bandwidths, however, their use in MMIC's is limited by their need for space consuming active or passive couplers for optimum operation. The dual gate MESFET provides an attractive alternative since the r.f. signal and the local oscillator frequencies can be applied to two separate gates and the intermediate frequency extracted from the drain circuit.

Optimum performance in terms of noise figure and stability occurs for bias points near pinch-off and by injecting the local oscillator modulation into the second gate with the signal injected into the first gate. Under these bias conditions non-linearities are provoked in the transconductances associated with both gates of the device, the effect being more pronounced in g_{m1} of the first gate. Furthermore there is a varistor-type modulation of the output conductance g_{d1} associated with the first gate.

Although optimisation of the performance of the device as a mixer by applying a non-linear, frequency domain analysis on each port is not yet completed, results of such an analysis on a single gate mixer[10] have been applied to the design of a dual gate mixer. These results indicated that optimum conversion gain was obtained with r.f. and local oscillator short circuits presented to the output of the mixer and an intermediate frequency short circuit presented at the input to the device. These conditions should be satisfied whilst maintaining power matching on the input at r.f. and local oscillator frequencies and i.f. matching at the output.

The totality of these conditions are difficult to fulfil in monolithic technology, however, an adequate r.f./local oscillator short at the mixer output can be obtained by using a parallel capacitance at its series resonant frequency directly at the drain terminal of the device. (C_R in figure 8).

This capacitance subsequently forms an integral part of the i.f. matching network at frequencies around 1 GHz.

After characterisation of the dual gate devices, matching circuits are calculated based on typical device parameters in the bands 11,7 - 12,5 GHz on the first gate, 10,8 GHz in the second gate and the corresponding i.f. band on the drain. This operation and limiting the number of passive matching elements gives rise to the circuit of figure 8.

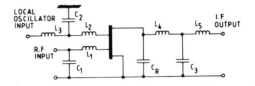

C_R presents a series resonance at 12GHz

Fig. 8 : Dual gate mixer circuit

An example of this type of mixer realised in monolithic form is shown in figure 9. Chip size is 2,4 x 1,4 mm^2.

Mixer performance

The mixer was characterised in a 50 ohms coplanar circuit. The local oscillator was fixed at 10,8 GHz and the drive to the mixer's second gate was about 11 dBm. The r.f. input match presented a VSWR better than 2 across a 2 GHz bandwidth centred at 11,8 GHz. The corresponding intermediate frequency VSWR is less than 3,5 across the band.

Noise figure and gain performances are summarized on figure 10. Noise figures of 6,5 dB have been measured associated with maximum conversion gains of 2 dB.

Monolithic stable local oscillator

The monolithic GaAs FET oscillators reported up to now[11,12] were based on a common gate configuration with inductive series feedback. The circuit presented here is a common source FET oscillator with series capacitive feedback stabilised using a dielectric resonator (DRO). This type of circuit demonstrated superior performance for temperature stable oscillator application. Stabilisation is obtained connecting the dielectric resonator circuit to the gate terminal. The circuit is shown in fig. 11 and is designed for frequencies around 11 GHz.

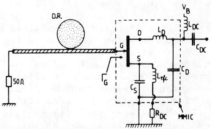

Fig. 11 : Complete stabilised monolithic oscillator topology

The FET has a 0,7 x 300 µm² gate. The source series feedback and the drain tuning circuit were determined on microstrip for an equivalent chip FET using an experimental oscillator design method : it consists in measuring the device line into the gate and optimizing the oscillator power ; calculated from the injected and reflected power at the gate and the power leaving the drain-port, following equations (6) and (7) :

$$P_{osc.} = P_D + P_i (|\Gamma|_G^2 - 1) \quad (6)$$

$$\Gamma_G = \Gamma_G (P_i) = \Gamma_R^{-1} \quad (7)$$

where P_D and P_i are the drain power and injected gate power, Γ_G the reflexion coefficient at the gate ($|\Gamma|_G > 1$) and Γ_R the reflexion coefficient of the dielectric resonator stabilization circuit at the resonance frequency (fig. 11). The obtained impedances on the source and drain terminals were then synthesized using lumped elements and the design rules described in previous sections. The spiral bias inductance L_{rfc} is designed to present a maximum impedance (parallel resonance) at 11 GHz.

Oscillator performance

The monolithic oscillator chip shown in fig. 12 measures 1,2 x 1,4 mm². It has been tested in a 50 Ω microstrip test fixture. The performances of the monolithic oscillator, stabilized using a dielectric resonator, are given in table I. The output power can be linearly controlled by the bias supply from 0 to 26 mW.

Output power (V_{DS} = 4,9V)	32 mW (15 dBm)
Chip efficiency (V_{DS} = 4 V)	20 %
Frequency pushing (1,5 V ≤ V_{DS} ≤ 5 V)	≤ ± 500 kHz
Frequency pulling (VSWR out = 3) (depends on coupling -factor β - of diel. resonators)	(a) < 20 MHz ($\beta \approx 10$) (b) < 1,5 MHz ($\beta \approx 4$)
Frequency stability (- 20°C ÷ 80°C)	< 1 ppm/K
Output power variation with temp.	< ± 0,75 dB

Table I : Performance of monolithic FET DRO

Temperature stabilization was obtained using a dielectric resonator which has a linear resonance frequency variation with temperature with a slope of around + 6 ppm/K and an unloaded Q factor of 1500 to 2000 ; it allows a better compensation of temperature drift of the active device parameters compared with nonlinear $f_r(T)$ characteristics (figure 13). This is a consequence of equation (8) that describes the oscillator's frequency drift with temperature[13].

$$S_o = S_R + \frac{\beta + 2}{4 Q_r} \cdot S_D \quad (8)$$

Here, $S_o = \frac{df}{fdt}$, $S_R = \frac{df_r}{f_r dT}$ and $S_D = \frac{\partial \phi_D}{\partial T}$ ($\approx$ 2800 ppm/K) are the temperature drifts of oscillation frequency, resonance frequency of dielectric resonator and phase of the active circuit at the stabilization port, β and Q_r the coupling factor between the microstrip line and the resonator and the resonator's unloaded Q respectively.

CONCLUSIONS

MMICs have been developed which fulfil the requirements of the discrete sub-assemblies of a 12 GHz downconverter for domestic satellite reception. A summary of the analysis and characterisation of the lumped elements used in the circuit design is presented along with the technology used in their processing. Low noise amplifiers with 3,6 dB noise figure and 7,3 dB gain across the r.f. band, dual gate mixers with 6,5 dB noise figure and 2 dB conversion gains as well as stable local oscillators with 32 mW output power and stabilities of 1 ppm/K have been presented.

It is believed that the MMIC's presented in this article may form the basis for the complete monolithic integration of 12 GHz receivers.

Acknowledgements

The authors would like to express their gratitude for valuable contributions towards the success of this work to : A. Collet, C. Mayousse, P. Lesartre, P. Kaikati, A. Villegas Danies, J.L. Gras and V. Pauker.

References

1. R. Dessert, P. Harrop, B. Kramer and T. Vlek, "All FET Front-End for 12 GHz Satellite Broadcasting Reception", Proc. Eu.M.C., p. 638, 1978, Paris
2. P. Harrop, A. Collet and P. Lesartre, "GaAs Integrated All FET Front-End at 12 GHz", GaAs I.C. Symposium, Las Vegas, 1980
3. J.A. Turner, R.S. Pengelly, R.S. Butlin and S. Greenhalgh, GaAs I.C. Symposium, San Diego, 1981
4. D. Maki, R. Esfandiari and M. Sirakusa, "Monolithic low noise amplifiers", Microwave J. p. 103-106, 1981, Oct.
5. C. Tsironis, "12 GHz Receiver with Self-Oscillating Dual Gate MESFET Mixer", Electronics Letters, 17, 617, 1981
6. F.W. Grover, Inductance Calculations van Nostrand, N.Y. 1946
7. Yu Chin Lim and R.A. Moore, "Properties of Alternately charges coplanar strips by conformal Mapping", IEEE Electron Devices, ED-15, March 1968.
8. P. Baudet, M. Binet and D. Boccon-Gibod "Sub-micron self-aligned GaAs MESFET", IEEE Trans. MTT-24, 372, 1976
9. M. Parisot, M. Binet and A. Rabier, "Caracterisation automatique en hyperfrequence du TEC", Acta Electronica 23, 2, 1 1980

10 P. Harrop and T.A.C.M. Claasen, "Modelling of a FET Mixer", Electronics Letters 14, 369, 1978
11 J.S. Joshi, J.R. Cockrill, J.A. Turner "Monolithic Microwave Gallium Arsenide FET Oscillators", IEEE Trans. on Electron Devices, ED-28, pp. 158-162, (1981).
12 H.Q. Tserng and H.M. Macksey "Performance of monolithic GaAs FET oscillators at J-band", IEEE Trans. on Electron Devices, ED-28, pp. 163-165, (1981)
13 C. Tsironis, and P. Lesartre, "Temperature stabilisation of GaAs FET Oscillators using Dielectric Resonators", to be published.

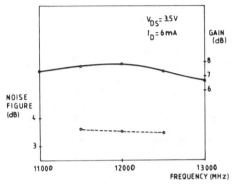

FIG. 5 : Gain and noise figure of single stage amplifier

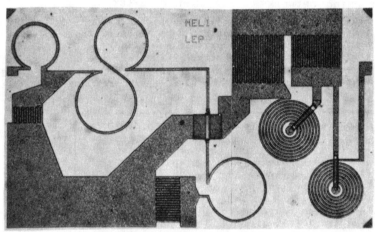

FIG. 7 : Monolithic one stage amplifier

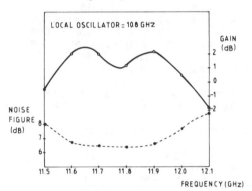

FIG. 10 : Mixer noise figure and conversion gain

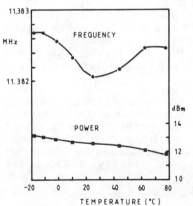

FIG. 13 : Temperature performance of stabilised monolithic oscillator

FIG. 12 : Monolithic oscillator chip

FIG. 9 : Monolithic dual gate mixer chip

A LOW COST GaAs MONOLITHIC LNA FOR TVRO APPLICATIONS

S Moghe, T. Andrade, H. Sun
Avantek Incorporated
3175 Bowers Avenue
Santa Clara, Calif. 95051

ABSTRACT

A low noise amplifier has been developed for the 3.7 to 4.2 GHz frequency band. This two stage amplifier chip provides 21.0 ± 1 dB gain and a 1.3 dB maximum noise figure across the design frequency band. The chip contains all the DC bias and most of the RF matching circuitry and is fabricated within a 22 x 35 mil^2 area. The amplifier input matching network is realized in external circuitry.

INTRODUCTION

The primary use of a 3.7 - 4.2 GHz low noise amplifier (LNA) is for applications in the television receive only (TVRO) market. This commercial market is characterized by cost pressures which have reduced complete 50dB gain amplifier prices to the $100 range. These LNAs are manufactured using hybrid microwave techniques and the production process includes tuning of the amplifier. A monolithic approach will minimize the labor cost in assembly and tuning, and makes integration attractive for LNA design in the TVRO market (1). Currently most of the MMIC LNA efforts (2, 3, 4) have been focused on the 12GHz DBS band. This paper will present a cost-effective MMIC LNA for the 4GHz TVRO market.

CIRCUIT DESIGN

To achieve the high gain (20dB minimum) and low noise required, a two-stage amplifier design was employed. The first stage provides the low noise match, while the second stage provides the high gain, at the desired frequency range (3.7 - 4.2GHz). The circuit uses a common source cascaded configuration with the first MESFET biased for lowest noise figure (Ids=10% to 20% Idss) and the second MESFET biased for highest linear gain (Ids=50% Idss).

The input matching network is pivotal to the successful implementation of an LNA chip. For the TVRO application, a reactively matched amplifier input is the only viable circuit topology. Yet two problems have to be solved before a monolithic approach can be utilized. First, to realize the best noise figure performance, the monolithic matching networks used to provide a noise match to the first stage FET input should have very low loss. Second, the process induced variation of the device parameters and hence the variation of the LNA performance, has to be minimized in the design to make this approach feasible. The design approach taken here is to realize the input matching external to the chip. This approach reduces the chip area and maximizes the chip RF yield. To further ease the external input matching requirements and minimize tuning, a 500μm input MESFET gate periphery was chosen. This MESFET gate width was determined to be optimum for best noise figure at 4 GHz. A proprietary computer design program which automatically scales all the MESFET parameters with gate periphery was used for this LNA design.

The interstage matching network is a low pass L-C circuit which provides the highest gain over a bandwidth several times the input stage design bandwidth. To accomodate the process variations, some amount of negative feedback was applied on the second stage. These factors have given uniform circuit performance over a wide process variation range. The second stage feedback amplifier has also resulted in a low output VSWR, without any external matching. The gate periphery of the second stage was selected to be 500μm to realize low output VSWR.

External bypass chip capacitors were used for the FET source RF grounding. Because of the relatively low operating frequency of the LNA, the source inductance incurred by the bond wires is tolerable while the use of external capacitors greatly reduces the chip size.

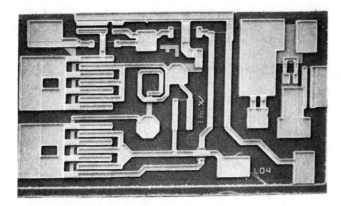

Figure 1. Scanning Electron Micrograph of the LNA Chip

FABRICATION

A scanning electron micrograph of the unmatched LNA chip is shown in figure 1. The manufacturing process for this MMIC is identical to that for a previously described amplifier(5). The 0.5μm gate length MESFETs are fabricated on caplessly annealed, direct Si ion implanted wafers. GaAs mesa resistors use the same 500Ω/□ active layer as the mesa MESFETs. MIM capacitors use the same plasma assisted CVD silicon nitride as is used to passivate the MESFETs. By making maximum use of the MESFET process steps, the MMIC process is only slightly more complicated and hence only slightly more expensive than the discrete MESFET process.

It should also be noted that the wraparound ground technique is used only for DC bias return and grounding of the interstage shunt capacitor. This grounding process could also be eliminated in favor of a bonding pad. On redesign, an additional 30% chip area reduction is possible by elimination of the networks designed to allow the chip to function in other applications. The presently utilized area is approximately 600 mil^2 of the 22 x 35 mil^2 chip. By comparison, two discrete 500μm gate periphery MESFETs would require approximately 500 mil^2.

MEASURED PERFORMANCE

After wafer fabrication, each LNA chip is DC tested. A probe pad has been added at the second MESFET gate node. This pad allows all circuit nodes to be automatically probed resulting in a rapid and thorough DC screening procedure. The DC test yield is comparable to the yield of an ion implanted 1mm gate periphery discrete MESFET.

The DC qualified chips were mounted on 15 mil thick alumina test carriers with 50 Ω input and output transmission lines ending 10 mils from the edge of the grounded chip attach pad. The unmatched LNA chip and source bypass capacitors were attached to the chip attach pad using Au/Sn solder. Bond wires of 0.7 mil gold wire were then attached to provide input noise matching as well as to connect the LNA output and source capacitors. The chip power was provided through a bias tee integral to the test apparatus.

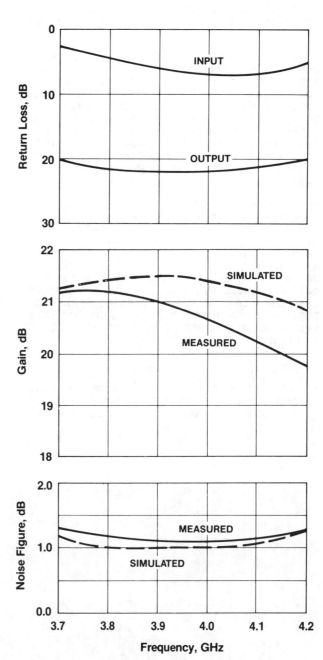

Figure 2. LNA Performance in a 50Ω System

Figure 2 shows sequentially the amplifier input and output return loss, the gain, and the noise figure. For this data, the DC power supplied to the chip was 8V and 60 mA. The simulated performance of the gain and noise figure are also shown. The agreement between the simulations and the measurements is gratifying but expected as the circuit element models are quite accurate near 4GHz. Additionally distributed circuit effects are less important at 4GHz than at higher frequencies.

A minimum 3dB input return loss is measured across the frequency band. This low valve for the input return loss is expected for this noise matched amplifier. The output return loss is greater than 20dB across the entire frequency band. The second stage MESFET has an output resistance of 120 Ω at the operating bias condition. This resistance plus the effects of resistive feedback provide the high output return loss.

The LNA gain and noise figure depend on the loop height of the input bond wire. Modern assembly equipment can control this height well enough that this inductance is reproducable to better than ±10%. This bond wire height may be tuned to vary the LNA performance, but this was not done for the data presented in figure 2. When the chip is assembled in a discrete transistor package together with the two bypass capacitors, the performance is comparable to that reported in figure 2.

SUMMARY

An LNA chip for TVRO applications has demonstrated 1.3dB maximum noise figure and 21±1dB gain. The input bond wire is used as the input series inductance to minimize matching element losses. This technique plus the use of feedback second stage provides high RF yield and could yield over 5,000 die per 2 inch diameter wafer.

ACKNOWLEDGMENT

The authors wish to thank V. Hersey for assistance with assembly.

REFERENCES

1. Daniel Chen, "DBS High Volume Market for GaAs MMIC", Microwave Journal, Feb 1983, pp. 116-122.

2. C. Kermarrec, P. Harrop, C. Tsironis, and J. Faguet, "Monolithic Circuits for 12 GHz Direct Broadcasting Satellite Reception," IEEE 1982 Microwave and Millimeter-Wave Monolithic Circuits Symposium Digest, pp. 5-10.

3. L.C. Liu, D.W. Maki, M. Feng, M. Siracusa, "Single and Dual Stage Monolithic Low Noise Amplifiers", IEEE 1982 GaAs IC Symposium Digest, pp. 94-97.

4. T. Sugiura, H. Itoh, T. Tsuji and K. Honjo, "12 GHz Band Low Noise GaAs Monolithic Amplifiers", IEEE Trans. Electron Devices, Vol E.D.-30 Dec. 1983, pp.1861-1866.

5. S. Moghe, T. Andrade, H. Sun, C. Huang, "A Manufacturable GaAs MMIC Amplifier with 10 GHz Bandwidth, "IEEE 1984 Microwave and Millimeter-Wave Monolithic Circuits Symposium Digest, pp. 37-40.

12 GHz-BAND GaAs DUAL-GATE MESFET MONOLITHIC MIXERS

Tadahiko SUGIURA, Kazuhiko HONJO and Tsutomu TSUJI

Microelectronics Research Laboratories
NEC Corporation
Miyazaki, Miyamae-ku, Kawasaki 213, Japan

ABSTRACT

12 GHz-band GaAs dual-gate MESFET monolithic mixers have been developed for use in direct broadcasting satellite receivers. In order to reduce chip size, a buffer amplifier has been connected directly after a mixer IF port, instead of employing an IF matching circuit. The mixer and the buffer were fabricated on separate chips, so that individual measurements could be achieved. Chip size is 0.96 x 1.26 mm for the mixer and 0.96 x 0.60 mm for the buffer. A dual-gate FET for the mixer, as well as a single-gate FET for the buffer, has a closely-spaced electrode structure. Gate length and width are 1 μm and 320 μm, respectively. The mixer with the buffer provides 2.9 ± 0.4 dB conversion gain with 12.3 ± 0.3 dB SSB noise figure in the 11.7 to 12.2 GHz RF band. LO frequency is 10.8 GHz. A low-noise converter was constructed by connecting a monolithic pre-amplifier, an image rejection filter and a monolithic IF amplifier to the mixer. The converter provides 46.8 ± 1.5 dB conversion gain with 2.8 ± 0.2 dB SSB noise figure in the same frequency band.

INTRODUCTION

Recent advances in GaAs technology have made monolithic microwave integrated circuits (MMICs) more practical. Promising applications for this technology include inexpensive receiver front-ends for direct broadcasting satellite (DBS) systems[1]-[4]. 12 GHz-band low-noise GaAs monolithic amplifiers for use in DBS receivers have already been developed and reported[4]. This paper describes design considerations and performances for newly developed 12 GHz-band GaAs dual-gate MESFET monolithic mixers.

For MMICs used in DBS receivers, simple and high reproducible fabrication process and small chip size are required for meeting low cost objectives. For the first requirement, an IC process using ion implanted closely-spaced electrode FETs, which was successfully employed in fabricating monolithic amplifiers[4][5], can be applied to mixers. For the second requirement, it is necessary to realize a suitable circuit for monolithic mixers.

For reducing chip size, a dual-gate MESFET is attractive as a mixing device, because filtering circuits can be greatly simplified due to its built-in isolation effect among electrodes. In a mixer for the DBS receivers, a multi-section matching circuit is necessary for an IF port, because a large bandwidth ratio is usually required for the IF band. Element size itself for the IF matching ciruict is also larger than that for the RF matching ciruict, because a lower frequency is used. Therefore, the IF matching circuit requires the most area on a chip, other than the filtering circuit. In this work, chip size reduction is accomplished by introducing a buffer amplifier following directly after the mixer IF port, instead of employing the IF matching circuit. In addition to the chip size reduction, this approach is suitable for MMICs, because characteristics on the IF port become less sensitive to variation in element parameters, as compared with using the matching circuit.

CIRCUIT DESIGN

Fig. 1 shows an equivalent circuit for the mixer with the buffer. RF and LO powers are injected into the first and the second gates, respectively, and IF power is extracted from the drain. The RF port is matched by one-section parallel and series microstrip lines. For the LO port, the lines are shortened by loading a parallel capacitor. A quarter wavelength microstrip stub is employed at the drain to realize a LO short circuit. Gate length and width for the dual-gate FET are 1 μm and 320 μm, respectively.

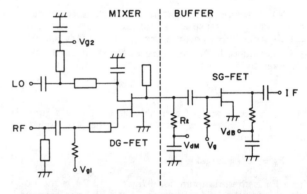

Fig. 1 Equivalent circuit for mixer with buffer

The buffer amplifier is a one-stage resistor-capacitor coupled amplifier. Input resistance for the buffer, or load resistnace for the mixer, is determined by a resistor shown as Rl in Fig. 1. An equivalent circuit for mixer-buffer interstage is shown in Fig. 2, where I represents an equivalent current source for mixer IF output signal, Gd is drain conductance for the mixer FET, Gl is conductance for the resitor Rl, and Cg is gate capacitance for the buffer FET.

Voltage V across conductance Gl becomes

$$V = I/(Gd + Gl + j\omega Cg). \qquad (1)$$

The design objective is to increase V without large

frequency dependence. For this purpose, Cg must be small, or a narrow gate FET is preferable for the buffer. However, gain and noise figure values for the buffer become better when a wider gate FET is used. To compromise between these requirements, the gate width was chosen at 320 μm, which is the same as for the mixer dual-gate FET. The load resistance was designed at 250Ω (Gl=4 mS), taking Gd=2 mS and Cg=0.5 pF into consideration.

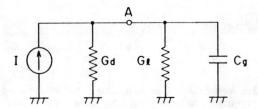

Fig. 2 Equivalent circuit for mixer-buffer interstage

When matching is performed, voltage Vm across load conductance G becomes

$$Vm = I\sqrt{G/Gd}/(2G + jwCg). \qquad (2)$$

Therefore, power loss L due to imperfect matching is given by

$$L = |Vm/V|^2 = G[(Gd+Gl)^2 + (wCg)^2]/Gd[(2G)^2 + (wCg)^2]. \qquad (3)$$

When G=Gl=4 mS, L at the center IF of 1.15 GHz is 1.3 (1.1 dB). In a usual case, however, G is chosen as 20 mS (50Ω). For this case, L becomes 0.30 (-5.2 dB), which means matching to 50Ω is worse than direct connecting to the buffer with higher input resistance.

Noise figure for mixer is independent from the IF port matching, because an output circuit does not affect noise figure. Thus, noise figure for the buffer amplifier will be considered. Noise figure Fl for input conductance Gl is given by

$$Fl = (Gd + Gl)/Gd, \qquad (4)$$

which is a reciprocal of available gain. Noise figure F_f for the FET is given as follows, under the low frequency approximation[6].

$$F_f = Fo + Rn(Gd + Gl), \qquad (5)$$

where Fo is the minimum noise figure and Rn is the noise resistance for the FET. Because Gl and FET are cascaded, composite noise figure F can be derived by using the Friis formula, as follows.

$$F = Fl F_f = (Gd + Gl)[Fo + Rn(Gd + Gl)]/Gd. \qquad (6)$$

Substituting measured values Fo=1.1 (0.4 dB), Rn=60Ω, Gd=2 mS and Gl=4 mS into (6), F is calculated as 4.4 (6.4 dB). When the IF output port is matched to a conductance G, Gd becomes G in (6). Thus, when matching to Gl(4 mS) is performed, F becomes 3.2 (5.0 dB). When matching to 50Ω, substituting Gd=Gl=20 mS into (6), F becomes 7.0 (8.5 dB). Therefore, the noise figure is also degraded by matching to 50Ω.

From these results, it is concluded that matching to 50Ω has no merit as long as the resistor-capacitor coupled type IF amplifier is employed, and that the input resistance for the amplifier should be chosen as large as possible within the limitation for keeping gain flatness. When choosing the large resistance, the mixer performances are somewhat improved by IF port matching. The improvement, however, is a minor problem as compared with chip size expansion in a practical point of view, because a low-noise pre-amplifier is employed in front of the mixer.

DEVICE DESIGN

The mixer and the buffer were fabricated on separate chips, so that individual measurement could be achieved. Fig. 3 and Fig. 4 show chip photographs for the mixer and the buffer, respectively. Chip size is 0.96 x 1.26 mm for the mixer and 0.96 x 0.60 mm for the buffer. Both chips are 150 μm thick. A Cr-doped semi-insulating HB-grown GaAs wafer is used as the substrate. The doped Cr concentration is 0.5 wt-ppm, which corresponds to 3×10^{16} cm^{-3}.

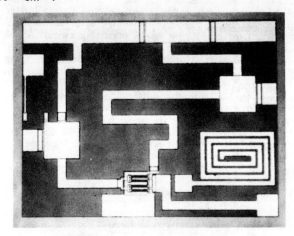

Fig. 3 Mixer chip photograph

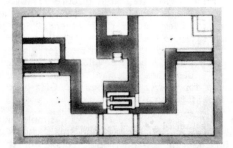

Fig. 4 Buffer chip photograph

The dual-gate FET for the mixer, as well as the single-gate FET for the buffer, has a closely-spaced electrode structure[4][5]. The electrodes have an interdigital pattern with 80 μm long Al gate fingers. Gate length and width are 1 μm and 320 μm, respectively. Ohmic electrodes are made of AuGe-Ni. The FET is passivated by SiO_2 film.

The FET active layers were formed by selective $^{30}Si^+$ ion implantation. Implanting conditions are 70 keV acceleration energy and 3.3×10^{12} cm^{-2} dose. Then, annealing for activation was accomplished with a SiO_2

cap at 800 °C for 20 min in a H$_2$ ambient. Fig. 5 shows static characteristics for the single-gate FET. Threshold voltage V$_t$ is -1.0 V and transconductance gm at the saturated drain current Idss is 37 mS (116 mS per mm gate width).

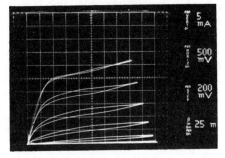

Fig. 5 Single-gate FET static characteristics

The resistive layers were also formed by ion implantation. Implanting conditions are 130 keV energy and 3.8 x 10^{12} cm^{-2} dose, which provide 1 kΩ sheet resistance for a square resistor. The capacitors are of the MIM type, where dielectric material is 300 nm thick CVD-SiO$_2$, whose relative dielectric coefficient εr is 4.8. The microstrip lines were formed by Ti-Pt-Au lifting-off and thickened to 2.5 μm by Au plating.

MICROWAVE PERFORMANCES

The IC chips were mounted on chip carriers and tested in a 50Ω system. Designed RF band is 11.7 to 12.2 GHz, which is the DBS system frequency band. The LO frequency is 10.8 GHz. Therefore, IF band becomes 0.9 to 1.4 GHz.

Fig. 6 shows gain and noise figure versus frequency characteristics for the mixer chip alone. It provides 6.1 ± 1.5 dB conversion loss with 11.1 ± 0.6 dB SSB noise figure in the desired frequency band. Bias conditions are shown in the figure. The optimal bias point for the first gate was near pinch-off[1]. For the second gate, performances were scarcely varied in the vicinity of 0V bias voltage. Therefore, measurements were carried out while grounding the second gate.

LO power was set at 13 dBm, where the optimal performance was obtained. When 13 dBm LO power was supplied, leaked LO power was -5.3 dBm at the RF port and -11.7 dBm at the IF port. Input VSWR for the RF port is within 1.5 in the 11.7 to 12.2 GHz band. For the LO port, however, it is about 6 at the LO frequency.

Fig. 7 shows frequency characteristics when connecting the buffer chip. It provides 2.9 ± 0.4 dB conversion gain with 12.3 ± 0.3 dB SSB noise figure in the same frequency band.

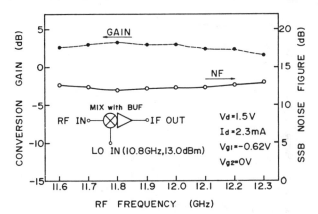

Fig. 7 Frequency characteristics for mixer with buffer

Fig. 8 shows gain and noise figure versus LO power characteristics for the mixer with the buffer. Bias conditions were set at the same values as for the Fig. 7 case. RF is 12.0 GHz. The optimal performance is obtained when the LO power is 13 to 15 dBm.

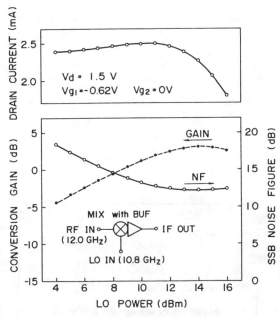

Fig. 8 LO power characteritics for mixer with buffer

Fig. 9 shows RF input power versus IF output power characteristics. The IF output power at 1 dB gain compression is -1.2 dBm. This value is determined by the mixer chip, and is about 5 dB lower than that for the same gate width FET amplifier case. One reason for the lower saturation power may be due to the near pinch-off biasing

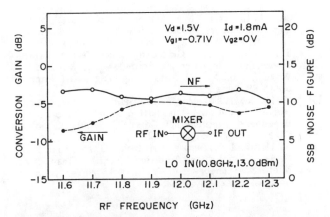

Fig. 6 Frequency characteristics for mixer without buffer

for the first gate.

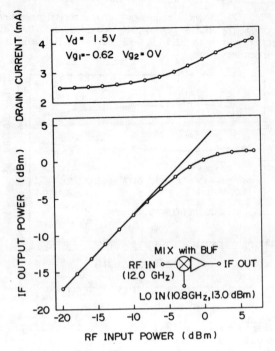

Fig. 9 RF power characteristics for mixer with buffer

To examine the performances as a low-noise converter, a monolithic pre-amplifier, an image rejection filter and a monolithic IF amplifier were connected to the mixer. A block diagram for the converter is shown in Fig. 10. The pre-amplifier is realized by cascade connection between one- and two-stage amplifiers. The one-stage amplifier has more than 9.5 dB gain with less than 2.5 dB noise figure, and the two-stage amplifier has more than 16 dB gain with less than 2.8 dB noise figure in the 11.7 to 12.7 GHz band[4]. As the image rejection filter, a commercially available band pass filter was used. The filter provides more than 69 dB image rejection. The IF amplifier has 16 dB gain with 3 dB noise figure in the 9 MHz to 3.9 GHz band[5].

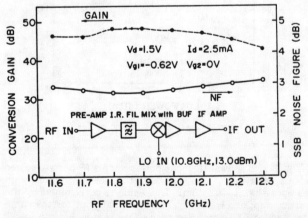

Fig.10 Frequency characteristics for low-noise converter

Fig. 10 shows gain and noise figure versus frequency characteristics for the converter. It provides 46.8±1.5 dB conversion gain with 2.8±0.2 dB SSB noise figure in the 11.7 to 12.2 GHz desired frequency band. These performances are well within the acceptable range for DBS receivers.

CONCLUSION

The design considerations and microwave performances for newly developed 12 GHz-band GaAs dual-gate MESFET monolithic mixers for use in DBS receivers have been described. Small chip size was realized by removing the IF matching circuit. IF circuit effects on mixer performances were investigated. The results show that matching to 50Ω does not improve performances.

The mixer was fabricated by well established IC process, using an ion implanted closely-spaced electrodes FET. The mixer provides 2.9±0.4 dB conversion gain with 12.3±0.3 dB SSB noise figure. A low-noise converter was constructed by connecting a monolithic pre-amplifier, an image rejection filter and a monolithic IF amplifier to the mixer. The conveter provides 46.8±1.5 dB conversion gain with 2.8±0.2 dB SSB noise figure. These performances are well within the acceptable range for DBS receivers. This work has made a cost-effective one-chip front-end for DBS systems more realistic.

ACKNOWLEDGEMENT

The authors would like to thank T. Ozawa for ion implanting and H. Itoh for taking SEM photographs. They would also like to thank Y. Takayama and H. Muta for their constant encouragement throughout this work.

REFERENCES

(1) C. Kermarrec, P. Harrop, C. Tsironis and J. Faguet, "Monolithic Circuits for 12 GHz Direct Broadcasting Satellite Reception," Microwave and Millimeter-Wave Monolithic Cirucits Symp. Digest of Papers, pp.5-10, June 1982.

(2) L.C. Liu, D.W. Maki, M.Feng and M. Siracusa, "Single and Dual Stage Monolithic Low Noise Amplifiers," GaAs IC Symp. Technical Digest, pp.94-97, Nov. 1982.

(3) S. Hori, K. Kamei, K. Shibata, M. Tatematsu, K. Mishima and S. Okano, "GaAs Monolithic MICs for Direct Broadcast Satellite Receivers," Microwave and Millimeter-Wave Monolithic Circuits Symp. Digest of Papers, pp.90-95, June 1983.

(4) H. Itoh, T. Sugiura, T. Tsuji, K. Honjo and Y. Takayama, "12 GHz-Band Low-Noise GaAs Monolithic Amplifiers," Microwave and Millimeter-Wave Monolithic Circuits Symp. Digest of Papers, pp.85-89, June 1983.

(5) K. Honjo, T. Sugiura, T. Tsuji and T. Ozawa, "Low-Noise Low-Power-Dissipation GaAs Monolithic Broadband Amplifiers," GaAs IC Symp. Technical Digest, pp.87-90, Nov. 1982.

(6) K. Honjo, T. Sugiura and H. Itoh, "Ultra-Broad-Band GaAs Monolithic Amplifier," IEEE Trans. on Microwave Theory and Tech., vol. MTT-30, No. 7, pp.1027-1033, July 1982.

AN 8-18 GHz MONOLITHIC TWO-STAGE LOW NOISE AMPLIFIER*

L. C. T. Liu, D. W. Maki, C. Storment, M. Sokolich and W. Klatskin

Torrance Research Center
Hughes Aircraft Company
3100 W. Lomita Boulevard
P.O. Box 2999
Torrance, CA 90509

ABSTRACT

A wideband monolithic low noise amplifier which covers the frequency band from 8 to 18 GHz has been designed and fabricated. The amplifier has a noise figure less than 4.3 dB and an associated gain of 8.5 dB across the entire band. A revised version of the amplifier which has a design goal of sub-four dB noise figure and an associated gain of 12 dB has been processed. Measured data will be presented at the conference.

INTRODUCTION

GaAs monolithic microwave integrated circuits (MMICs) have become practical components for various system applications due to recent advances in material processing technology. The monolithic low noise amplifier is an essential functional chip for a number of applications as it determines the overall system noise figure. Recent papers have demonstrated the potential of the monolithic low noise amplifier and shown good results at X-band.[1-2]

There is growing interest within the government for multifunction radar systems which cover much wider bandwidths than the conventional 10 percent. A typical application is multiband radar covering both X- and Ku-band. Monolithic ICs offer significant advantages in wideband circuits since the matching can be done very closely to the device.

In this paper, details of an 8-18 GHz two-stage monolithic low noise amplifier will be presented. The circuit design and fabrication process will be discussed and the RF performance will be shown. Finally, an analysis of the measured results and a revised design of the amplifier will be discussed.

CIRCUIT DESIGN

The broadband monolithic low noise amplifier design is based on Hughes low noise GaAs MESFETs operating over X- and Ku-band.[3] These FETs have a gate dimension of 0.5 by 300 μm, and a minimum noise figure of 1.7 dB with an average associated gain of 9.5 dB at 12 GHz. An equivalent circuit of the device is derived from the measured S-parameters. Based on this equivalent circuit and the noise model given by Podell, Ku and Liu,[4] the minimum noise figure and optimum source admittance are obtained at the frequency band of interest.

An input matching network is synthesized[5] to provide the optimum source admittance for noise figure to the input port of the device over the full band. An interstage matching network is then synthesized to obtain the gain match between the output port of the first device to the input port of the second device. Finally, an output matching network is used to match to the 50 ohm load.

Figure 1 illustrates the schematic of this low noise amplifier. The characteristic impedances of all the transmission lines are between 47 and 56 ohms to facilitate the realization of the amplifier. The line lengths are references at 1 GHz, and are all appropriate for monolithic layout. In the amplifier design, a very simple configuration has been chosen to assure that high yield can be achieved in the wafer processing. The calculated performance of this amplifier is illustrated in Figure 2. The noise figure is below 4 dB across the entire frequency band with an average associated gain of 12 dB.

A completed amplifier chip is shown in Figure 3. As can be seen from the figure, the amplifier uses two 0.5x300 μm FETs as active devices. MOM capacitors are employed for both RF bypass and DC blocking applications. High airbridges fabricated on 6 μm thick photoresists are used to connect the source pads and to interconnect the microstrip lines to the top plates of the MOM capacitors. Microstrip transmission lines are used due to their low loss, low dispersion and useful impedance range. The amplifier was designed to use via hole grounding but was laid out to use over the edge (OTE) grounds if desired. The chip size is 1.52 x 2.0 x 0.1 mm.

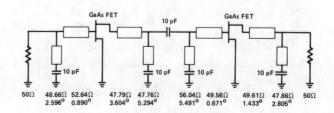

Fig. 1   Schematic of 8-18 GHz amplifier.

---

*This work was supported in part by Naval Research Laboratory on contract N00014-82-R-MT12, monitored by Dr. B. Spielman.

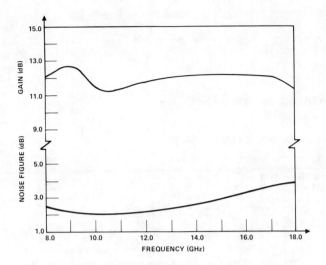

Fig. 2 Calculated performance of 8-18 GHz amplifier.

WAFER FABRICATION

Horizontal Bridgeman grown Cr-doped semi-insulating substrates are used for the amplifier fabrication. Also used are Cr-doped substrates with a 2 μm-thick unintentionally doped VPE buffer layer. The latter material is used to improve the uniformity of the active layer. The active channel layer is formed using silicon ion implantation with a dose of $6 \times 10^{12} cm^{-2}$ at 100 KeV. Isolation between FETs is achieved by either proton bombardment or mesa etch. Source-drain ohmic contacts are deposited and alloyed. Slightly recessed gates are defined either using a conventional photolithography or E-beam direct write system to give a gate length of 0.5 to 0.7 μm.

The source-drain overlay metal is deposited to a thickness of 1.5 μm to form the RF circuitry and the bottom plates of the MOM capacitors. A silicon dioxide layer with a thickness of 2000 Å is sputtered for the overlay capacitor. Cr-Au top metal layers and airbridges for interconnection are then formed. The wafers are thinned to 0.1 mm and via holes are etched from the back of the wafer. Finally, via holes and the back of the wafer are metallized. Amplifier chips can be obtained by sawing the wafers.

RF PERFORMANCE

A number of wafers have been processed for this 8-18 GHz monolithic low noise amplifier and amplifier chips have been extensively tested. In the characterization, each FET on the amplifier chip can be biased individually. When both FETs are biased for low noise operation, i.e., $V_{D1} = V_{D2} = 3V$, $I_{D1} = I_{D2} = 12$ mA, the amplifier has an average gain of 8.5 dB over the frequency band from 8 to 18 GHz. The noise figures of the amplifier are 4.3 and 4.2 dB at 18 and 12 GHz respectively. The associated gain is lower than the 12 dB predicted by computer simulation. The measured noise figure and gain response are shown in Figure 4.

As described before, both via hole and over the edge grounds can be used for this amplifier. One wafer of amplifiers, LNA 266, was split into two pieces, one receiving via hole, the other using wire bond grounds. The gain of the via holed device was consistently 1 to 2 dB greater than that of the circuits using wire bonds.

AMPLIFIER DIAGNOSIS

FETs from the amplifier chips were scribed free and tested as discrete devices from 2 to 18 GHz using an automatic network analyzer. Comparison of the measured S-parameters of the FET from wafer LNA 290 to those of a discrete FET which was used in the amplifier design shows that the magnitude of S_{21} of the FET from the amplifier chip is lower than that of the discrete FET. In turn, this causes the gain in the amplifier to drop. Also, this excised FET has a larger output reflection coefficient (S_{22}) than that of the discrete FET. Figure 5 shows the measured S-parameters of both FETs. We believe that the variation in the FET is caused first by the redistribution of the Cr in the Cr-doped GaAs substrate, which changes the doping profile of the active layer and secondly, by the possible damage to the active channel during the proton bombardment process because a photoresist layer was too thin to protect the active region. The

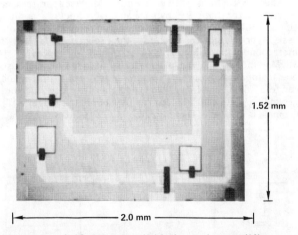

Fig. 3 8-18 GHz monolithic low noise amplifier.

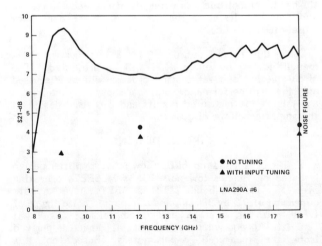

Fig. 4 Gain and noise figure of the 8-18 GHz LNA.

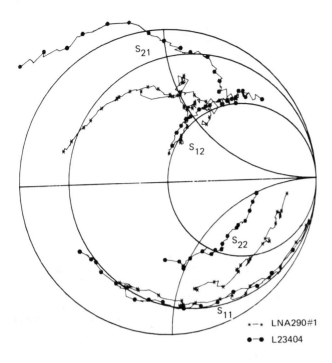

Fig. 5 S-parameters of the 0.5 μm GaAs FET.

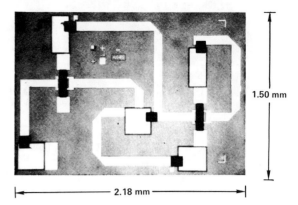

Fig. 6 Revised version of 8-18 GHz monolithic low noise amplifier.

first problem (Cr redistribution) can be solved by fabrication active layers using ion implantation into a VPE buffer. Tighter control of the photoresist thickness will eliminate the second problem (channel damage).

A second version of the amplifier has been designed and is being fabricated. In the new design, via holes are the only means of obtaining grounding. This eliminates the need to bring all ground pads to the edge of the chip, and it makes layout of the circuitry more flexible. Two via holes are employed for each FET to reduce the source inductance. Figure 6 shows a photograph of an amplifier chip from this wafer. Data from the amplifiers will be presented at the conference.

CONCLUSIONS

An 8 to 18 GHz monolith low noise amplifier has been developed successfully. The RF performance of this amplifier satisfies a broad range of applications and can be used as an inexpensive gain block.

ACKNOWLEDGMENTS

The authors would like to thank Dr. H. J. Kuno for his continuing support and encouragement. The contributions of L. E. Perry, L. Marich, D. Hynds and J. Burpo are also appreciated.

REFERENCES

1. L.C.T. Liu, D.W. Maki, M. Feng and M. Siracusa, "Single and Dual Stage Monolithic Low Noise Amplifiers," Technical Digest, 1982 GaAs IC Symposium, pp. 94-97, November 1982.

2. D.W. Maki, R. Esfandiari, H. Yamasaki, M. Siracusa and W.F. Marx, "A Monolithic Low Noise Amplifier," Proceedings of 8th Biennial Cornell Conference, Vol. 8, pp. 27-35, August 1981.

3. M. Feng, V.K. Eu, H. Kanber, E. Watkins, J.M. Schellenberg and H. Yamasaki, "Low Noise GaAs MESFETs Made by Ion Implantation," Applied Physics Letters, 409., May 1, 1982.

4. A. Podell, W. Ku and L. Liu, "Simplified Noise Model and Design of Broadband Low Noise MESFET Amplifiers," Proceedings of 7th Biennial Cornell Conference, Vol. 7, pp. 429-443, August 1979.

5. L.C.T. Liu and W. Ku, "Computer-Aided Synthesis of Monolithic Microwave Integrated Circuits (MMICs)," Proceedings of 8th Biennial Cornell Conference, Vol. 8, pp. 283-295, August 1981.

SINGLE AND DUAL STAGE MONOLITHIC LOW NOISE AMPLIFIERS

L.C. Liu, D.W. Maki, M. Feng, and M. Siracusa

Torrance Research Center
Hughes Aircraft Company
3100 W. Lomita Boulevard
Torrance, CA 90509

ABSTRACT

Both single and dual stage GaAs monolithic low noise amplifiers have been developed which have demonstrated state-of-the-art performance at X-band. The measured noise figure of the single stage amplifier is 2.2 dB with an associated gain of 10.9 dB at 12 GHz. At the same frequency, the two stage amplifiers have achieved 2.5 dB noise figure with an associated gain of 22.0 dB. We believe these to be the lowest noise figures for monolithic amplifiers reported to date. FETs with 300x0.5 μm gate are used as active devices. Active layers fabricated both by direct ion implantation into GaAs semi-insulating substrates and ion implantation into VPE buffer layers are employed. The amplifier chips measure 1.27x1.27x0.2 and 1.27x2.54x0.2 mm for the one and two stage circuits, respectively.

I. Introduction

Advances in material technology and wafer processing in the last several years have made GaAs monolithic microwave integrated circuits (MMICs) more practical (1)(2). Obvious applications for this technology include X-band trasmit-receive modules for airborne phased-array systems and inexpensive receivers for satellite direct broadcast systems. The monolithic low noise amplifier is an essential component in both the above applications as it determines the overall system noise figure. The development of monolithic low noise amplifier is somewhat easier than a higher power circuit due to decreased periphery, higher device impedances and reduced thermal problems. It is also the most attractive X-band monolithic component from a cost standpoint due to small die size and potentially high yield. Recent papers have demonstrated the potential of the monolithic low noise amplifier and shown good results (3).

In this paper details of single and dual stage monolithic low noise amplifiers will be presented. The circuit design, and fabrication process will be discussed and the RF performance and yield data will be shown.

II. Circuit Design

The monolithic low noise amplifier design is based on the X-band low noise discrete GaAs MESFETs developed at Hughes (4). These FETs have a gate dimension of 0.5 by 300 μm, and an average minimum noise figure of 1.7 dB with an average associated gain of 9.5 dB at 12 GHz. A simplified equivalent circuit of the FET is derived from the measured S-parameters. Based on this equivalent circuit and the noise model given by Podell et al. (5), the minimum noise figure and optimum source admittance are obtained at the frequency band of interest. A distributed input matching network is then synthesized to provide the optimum source admittance for the FET. For a single stage amplifier, an output matching network is synthesized to match the output port of the cascade of the input matching network and the FET. An interstage matching network is derived to provide the needed matching between the output of the first stage and the input of the second stage for the two stage amplifier. An output matching network is then synthesized for the output of the second FET.

In the realization of the amplifier, microstrip transmission lines are used for RF matching due to their low loss, low dispersion and useful impedance range. Metal-oxide-metal overlay capacitors are used for RF bypass and DC blocking applications. To interconnect the FET source pads and to connect the transmission lines to the top plates of the capacitors, airbridges are employed. To obtain higher microstrip Q and improved performance, a chip thickness of 0.2 mm was chosen, requiring the use of wirebonding over the edge of the chips to obtain grounds and giving rise to scribing and dicing problems.

The redesign of amplifiers with a 0.1 mm substrate and via hole grounding is underway. The amplifier chips measure 1.27x1.27x0.2 and 1.27x2.54x0.2 mm for the single and dual stage circuits as shown in Figures 1 and 2, respectively.

III. Wafer Fabrication

Two types of GaAs material are used in this work: (1) horizontal Bridgeman grown Cr-doped semi-insulating substrate, (2) 2 μm-thick unintentionally doped VPE buffer material grown on Cr-doped substrate. The latter material is used to improve the uniformity of the active layer. The selection of good material is a critical step in assuring the success of the final monolithic amplifier. This is accomplished by fabrication and RF testing of low noise discrete FETs from every boule under consideration. The active channel layer is then formed using silicon ion implantation with a dose of 6×10^{12} cm^{-2} at 100 keV. Mesas are etched and source-drain ohmic contacts are deposited and alloyed. Recessed aluminum gates are defined either optically or using E-beam direct writing to give a gate length of 0.5 to 0.7 μm.

The source-drain overlay metal is deposited through a photoresist-aluminum mask (6) to a thickness of 1.5 μm to form the RF circuitry and the bottom plates of the overlay capacitors. The dielectric layer for the overlay capacitor is then formed by sputtered silicon dioxide with a thickness of 2000 Å. Cr-Au top metal layers and

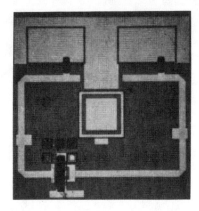

Figure 1   Monolithic single stage low noise amplifier.

airbridges for interconnection are then formed. The wafers are thinned to 0.2 mm and back metallized. Amplifier chips can be obtained by scribing or sawing the wafers. The above processing sequences are summarized in Figure 3, and are routinely carried out at our processing laboratory.

IV. RF Performance

A number of wafers have been processed for both one and two stage monolithic low noise amplifiers and amplifier chips have been extensively tested. Figure 4 shows the test fixture used in noise figure and gain response test. For single stage amplifiers, DC bias is applied using bias T's. For the two stage amplifier, two additional bias terminals are connected through the bottom of the test fixture. The gain response of a typical run of the single stage monolithic amplifier is shown in Figure 5. The variation among several chips is also shown in this figure. The average gain is 11.0 dB at 12 GHz with a bias of V_D = 3 V and I_D = 12 mA. The measured noise figure is shown in an expanded curve covering the 11.7 to 12.2 GHz frequency band in Figure 6. The noise figure is less than 2.5 dB across the frequency band with a minimum noise figure of 2.2 dB at 11.9 GHz. The associated gain is 10.8 ± 0.2 dB. The noise figure and associated gain as functions of drain current are shown in Figure 7. At I_D = 8 mA, a minimum noise figure of 2.3 dB is achieved with an associated gain of 10.1 dB. Higher gain can be obtained at the cost of higher noise figure, e.g., at I_D = 25 mA, an

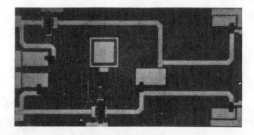

Figure 2   Monolithic two-stage low noise amplifier.

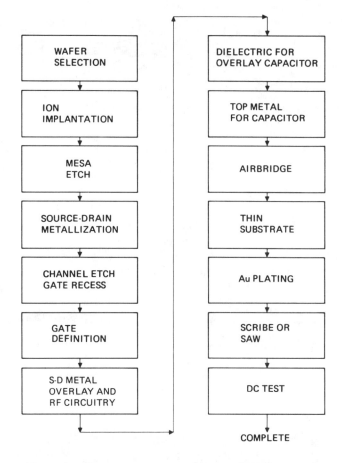

Figure 3   Processing sequence for monolithic low noise amplifier.

associated gain of 12.4 dB is achieved with a noise figure of 3.1 dB.

The gain response of a two stage monolithic amplifier is shown in Figure 8. The gain is 16.6 dB at a center

Figure 4   Test fixture for monolithic amplifiers.

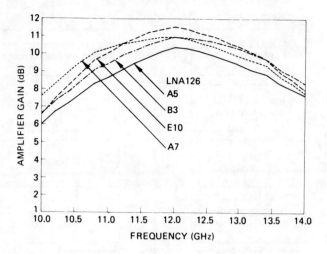

Figure 5 Gain Response of the single stage monolithic amplifier.

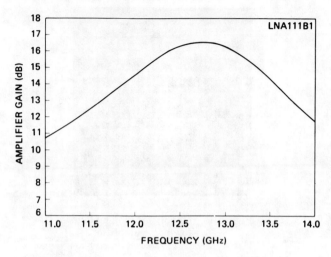

Figure 8 Gain response of two stage monolithic amplifier.

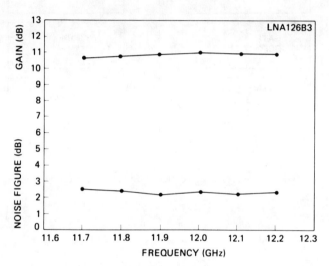

Figure 6 Noise figure and associated gain of single stage monolithic amplifier.

frequency of 12.8 GHz with a 1-dB bandwidth of 1 GHz. The shift of the center frequency away from the desired 11.9 GHz is caused by the variation of the FETs and some small variation in passive components. Some external tuning is used to compensate these variations and also to explore the potential of the amplifier. A noise figure of 2.9 dB with an associated gain of 22.5 dB is obtained with this off chip tuning. Figure 9 shows an expanded curve of both measured noise figure and associated gain from 11.7 to 12.2 GHz, without tuning. Among the amplifier chips measured, the best noise figure is 2.5 dB with off chip turning.

The measured data of both single and dual stage monolithic low noise amplifier from different wafers are summarized in Tables 1 and 2.

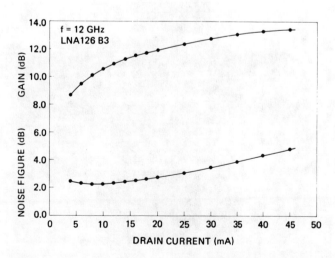

Figure 7 Noise figure and associated gain as functions of drain current.

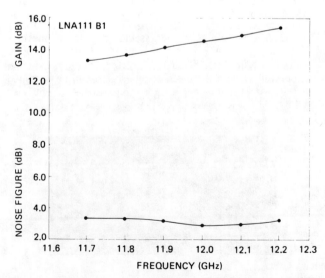

Figure 9 Noise figure and associated gain of the two stage monolithic amplifier.

TABLE 1. SUMMARY OF SINGLE STAGE MONOLITHIC LOW NOISE AMPLIFIER

LOT NO.	AVERAGE N.F. (dB)	$\bar{\sigma}$ (dB)	BEST N.F. (dB)	AVERAGE Ga (dB)	$\bar{\sigma}$ (dB)	BEST Ga (dB)
LNA126	2.36	0.05	2.30	10.96	0.47	11.54
LNA96	2.43	0.23	2.19	10.41	0.64	11.11
LNA78	3.29	—	3.29	7.13	2.34	8.78

TABLE 2. SUMMARY OF THE TWO STAGE MONOLITHIC LOW NOISE AMPLIFIER

LOT NO.	AVERAGE N.F. (dB)	$\bar{\sigma}$ (dB)	BEST N.F. (dB)	AVERAGE Ga (dB)	$\bar{\sigma}$ (dB)	BEST Ga (dB)
LNA127	3.19	0.38	2.92	15.59	0.64	16.04
LNA111(A)	2.97	0.07	2.91	15.66	0.53	16.13
LNA95	4.34	0.48	3.80	13.16	1.03	14.24
WITH OFF-CHIP TUNING						
LNA111(A)	2.65	0.19	2.48	22.67	0.51	23.13
LNA95	3.45	0.36	3.12	18.19	0.89	19.04

V. Yield

Yield may be one of the most important parameters in monolithic circuit development. In this experimental stage, the current yield cannot reflect the final result in a production environment since so many different approaches have been used in the processing for different wafers. However, it can be used to project the yield in a latter production stage. Table 3 shows some of the DC yield information on the wafer fabrication in our laboratory. The average DC yields for single and dual stage amplifiers are 35 and 23%, respectively. These yields will improve when all the processing approaches are finalized. A preliminary study shows that the primary mechanism limiting the yield of monolithic circuit are gate defects.

TABLE 3. YIELD OF MONOLITHIC AMPLIFIER

CIRCUIT TYPE	LOT NO.	YIELD (%)
SINGLE STAGE AMPLIFIER	LNA126	42.5
	LNA96	27.6
TWO STAGE AMPLIFIER	LNA127	16.7
	LNA111(B)	19.0
	LNA111(A)	25.0
	LNA103	20.4
	LNA95	32.0
	LNA89	18.8
	LNA86	29.0

VI. Conclusion

Both single and dual stage monolithic low noise amplifiers have been developed successfully with useful yield. The RF performance of the above amplifiers satisfies a broad range of applications, including direct broadcast satellite receivers, active array radar module receivers and, in a balanced configuration, inexpensive gain blocks.

VII. Acknowledgements

The authors would like to thank Dr. T. A. Midford for his continuing support and encouragement. The contributions of L. Cochran, L. Marich and H. Baker are also appreciated.

REFERENCES

1. W.R. Wisseman, "GaAs Technology in the 80's," Microwave Journal, Vol. 24, No. 3, pp 16-18, March 1981.

2. R.A. Pucel, "Design Considerations for Monolihic Microwave Circuits," IEEE Transactions on MTT, Vol. MTT-29, No. 6, pp. 513-534, June 1981.

3. D.W. Maki, R. Esfandiari, H. Yamasaki, M. Siracusa and W.F. Marx, "A Monolithic Low Noise Amplifier," Proceedings of 8th Biennial Cornell Conference, Vol. 8, pp. 27-35, August 1981.

4. M. Feng, V.K. Eu, H. Kanber, E. Watkins, J.M. Schellenberg and H. Yamasaki, "Low Noise GaAs MESFET Made by Ion Implantation," Applied Physics Letters, 40 (9), May 1, 1982.

5. A. Podell, W. Ku and L. Liu, "Simplified Noise Model and Design of Broadband Low Noise MESFET Amplifiers," Proceedings of 7th Biennial Cornell Conference, Vol. 7, pp. 429-443, August 1979.

6. M. Siracusa, Z.J. Lemnios and D.W. Maki, "Advanced Processing Techniques for GaAs Monolithic Integrated Cicuits," IEDM Technical Digest, Washington, D.C., December 1980.

10 GHz MONOLITHIC GaAs LOW NOISE AMPLIFIER WITH COMMON-GATE INPUT

R.E. Lehmann, G.E. Brehm, D.J. Seymour and G.H. Westphal

Texas Instruments Incorporated
Dallas Texas

ABSTRACT

An X-band monolithic three-stage LNA employing a common-gate FET at the input has been developed. The common-gate FET, used for active matching, permits single-ended operation by providing low input VSWR and low noise figure simultaneously. A 3.8 dB noise figure, 16 dB gain and an input VSWR of 1.8:1 have been achieved at 10 GHz. All matching and bias circuitry is implemented monolithically on the chip. Each stage employs a 300 μm gate-width FET with gold-plated air bridges to minimize and control grounding parasitics. Dimensions of the chip are 1.3 mm x 2.5 mm and 0.15 mm.

DESIGN APPROACH

This is the first monolithic demonstration of common-gate active matching with a high device transconductance to achieve both low input VSWR and low noise figure. For bandwidths on the order of 10-30% this design technique allows single-ended performance without degrading noise figure. For most radar and satellite receiver applications only 10-20% bandwidths are required, thus giving an excellent opportunity to use single-ended LNAs with good input VSWRs and low noise figures. The FET used for active matching occupies less GaAs real estate than a passive matching circuit (distributed transmission lines and capacitors) such as might be used in a conventional common-source input LNA.

In conventional FET LNAs, the input stage is a noise-matched FET in the common-source configuration(1). This approach results in good noise figure, but high (3:1 or 4:1) input VSWR. In many applications this high input VSWR, can be reduced by using a balanced amplifier incorporating a 3 dB hybrid (coupler) at the input and output. The balanced configuration requires a matched pair of amplifiers as well as 3 dB hybrids at the input and output. This approach, when implemented monolithically, consumes a great deal of GaAs real estate.

The "active-matching" approach(2-4) uses an input FET in a common-gate configuration. This design achieves good input VSWR, but at the expense of degrading the noise figure. The common-gate FET is designed and biased to provide a device transconductance, g_m, of approximately 20 mS. Because the input impedance of a common-gate FET is $\approx 1/g_m$, the input impedance is nominally 50 Ω. In this case, input VSWR is optimized over a broad band of frequency. Noise figure can be considerably improved over a narrower bandwidth if the FET is designed and biased for higher values of transconductance, and, in fact, noise figures as good as those for optimum common-source operation are predicted.

Using the latter approach, we have designed and demonstrated a monolithic low noise amplifier that uses a circuit design concept to achieve good input VSWR and excellent noise figure simultaneously. The LNA employs a common-gate FET for active matching at the input, operated at an optimum transconductance to achieve low noise figure. The device is integrated monolithically on a GaAs substrate to achieve tighter control of circuit and device parasitics than can be achieved with a hybrid (discrete FET) implementation. Figure 1 is a photograph of the monolithic LNA.

A 50 Ω input impedance is achieved by proper loading at the drain of the common-gate FET. By taking advantage of the drain-source capacitance, C_{ds}, of the monolithic FET, a wide range of input impedances can be realized. C_{ds}, a repeatable, device geometry-related element, provides the necessary feedback mechanism. For this application at 10 GHz a first-stage load impedance, Z_L, was chosen which yields a large voltage gain. Because an FET operated in a common-gate configuration is a unity current gain device, the voltage gain, A_v, of the first stage is simply the ratio of the load impedance to the input impedance:

$$A_v = \frac{Z_L}{Z_{in}}$$

Maximizing the voltage gain of the first stage is imperative in achieving low amplifier noise figure. A load impedance is thus chosen that yields an input impedance of 50 Ω while simultaneously providing a high voltage gain through the first stage. The common-gate FET is operated at an optimum transconductance to achieve minimum device and amplifier noise figure. This technique can easily be employed at other frequencies and bandwidths or to achieve input impedances other than 50 Ω. The major performance tradeoff involved in this procedure is bandwidth versus noise figure.

PROCESSING

The three-stage LNA employs a common-gate FET input stage, followed by two common-source FETs. Each stage utilizes a 300 μm gate-width, 0.5 μm gate-length FET with gold-plated air bridges to minimize and control grounding parasitics. Starting material for the monolithic LNA consists of a Cr-doped GaAs substrate with a high resistivity buffer layer 1-2 μm thick and an anodically thinned n-layer doped near 3×10^{17} cm^{-3}. Epitaxial layers were deposited by conventional VPE.

The monolithic LNA was fabricated using both mesa etching and unannealed boron implantation to define the FET active areas. This procedure ensures that no parasitic buffer layer capacitance will degrade circuit performance. Gold-germanium nickel ohmic contacts were formed by conventional evaporation and lift-off. Titanium/platinum/gold FET gates were defined by e-beam lithography. First-level metallization (inductors and capacitor bottom plates) is 1.6 μm thick Ti/Au. Plasma deposited Si_3N_4 serves as the capacitor dielectric and passivates and stabilizes the FETs. Capacitor top plates are evaporated Ti/Au, followed by plated gold. The inductors and capacitors are defined by two rather simple photomasks that can be rapidly changed by e-beam mask-making techniques. After the desired substrate thickness has been achieved by lapping (0.15 mm for the current design), the backside is metallized and the wafer sawed into individual circuits.

The monolithic circuitry consists of high impedance microstrip trans-mission lines, metal-insulator-metal (MIM) capacitors, and AuGeNi resistors. The transmission lines are used for rf matching and dc biasing of the FETs. Bypass capacitors are included on the chip to provide rf grounding for the shunt inductive elements and bias lines. Monolithic resistors (≈ 10 kΩ) are used to apply bias to the gates of the second and third stage FETs. Bias points for each stage are brought to the edge of the chip to facilitate dc probing and bonding. The chip size, including all bias circuitry, is 1.3 mm x 2.5 mm x 0.15 mm.

CAD ANALYSIS

Computer-aided-design (CAD) tools were used extensively in the characterization of FETs and the design of the amplifier. Discrete and monolithic FETs were characterized by means of S-parameters and noise figure and gain performance. The FET S-parameters and simulated monolithic circuit components were used as the basis for a multistage amplifier design. Figure 2 illustrates the projected gain and VSWR performance of a three-stage LNA with a common-gate input. The circuit schematic is shown in Figure 3. An integrated circuit program, SPICE2, was utilized to predict an amplifier noise figure of 2.5 dB at 10 GHz. A complete stability analysis was also performed.

RF PERFORMANCE

The three-stage LNA has achieved a 3.8 dB noise figure, 16 dB gain, and input VSWR of 1.8:1 at 10 GHz. Gain and noise figure performance is illustrated in Figure 4. The input and output VSWR response at minimum noise figure bias is shown in Figure 5. To illustrate the potential of the active matching, the bias current to the common-gate FET is reduced to 10 mA. The dramatic improvement in input VSWR (less than 1.3:1 from 8 to 11 GHz) is illustrated in Figure 6.

Load pull testing was done to examine the stability of the amplifier under various input and output loading conditions. A double-stub coaxial tuner was tuned for many randomly selected magnitudes and phases of impedance at both the input and output of the LNA. No oscillations were observed, with or without rf input drive.

Differences in predicted and measured rf performances are due primarily to a variation in FET device characteristics. Figure 7 compares the measured performance with computer-calculated performance of the circuit with monolithic FET S-parameters similar to those in the actual LNA. As shown, the computer-calculated performance is in good agreement with the measured data, indicating that the actual monolithic circuit components and the modeling techniques employed are accurate.

SUMMARY

An X-band monolithic three-stage LNA has been demonstrated which provides 16 dB gain, 3.8 dB noise figure, and a 1.8:1 input VSWR at 10 GHz. The common-gate FET input stage permits single-ended operation by providing low input VSWR and low noise figure simultaneously.

ACKNOWLEDGMENT

The authors wish to thank D.S. Bolding and A.J. Stinedurf for technical assistance.

REFERENCES

1. R.E., Lehmann, G.E. Brehm, and G.H. Westphal, "10 GHz 3-Stage Monolithic 4 dB Noise Figure Amplifier," <u>1982 IEEE International Solid-State Circuits Conference Digest of Papers</u>, Feb. 11, 1982, pp. 140-141.

2. W.C. Petersen, A.K. Gupta, D.R. Decker, and D.R. Ch'en, "Monolithic GaAs Microwave Analog Integrated Circuits," <u>Interim Report for ERADCOM</u>, Sept. 1980.

3. D.B. Estreich, "A Wideband Monolithic GaAs IC Amplifier," *1982 IEEE International Solid-State Circuits Conference Digest of Papers*, Feb. 11, 1982, pp. 194-195.

4. R.S. Pengelly, J.R. Suffolk, J.R. Cockrill, and J.A. Turner, "A Comparison Between Actively and Passively Matched S-Band GaAs Monolithic FET Amplifiers," *1981 MTT Symposium Digest*, pp. 367-369.

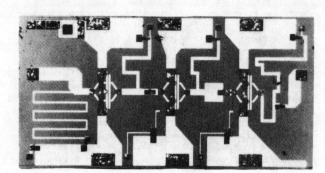

Figure 1 Monolithic Three-Stage Common-Gate Input LNA

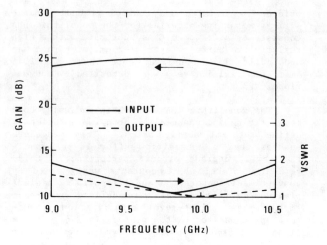

Figure 2 Computer-Predicted Performance of a Monolithic Three-Stage Common-Gate Input LNA

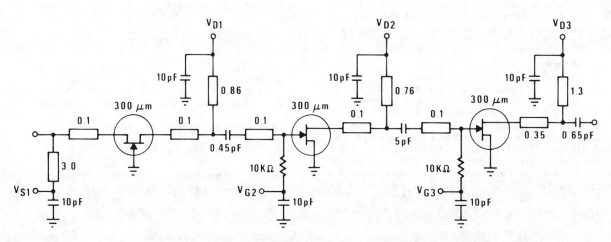

TRANSMISSION LINE LENGTHS ARE IN mm

Figure 3 Circuit Schematic of a Common-Gate Input LNA

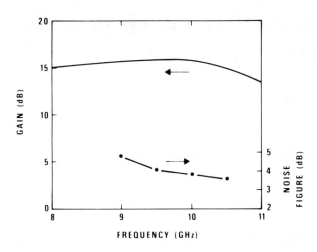

Figure 4 Gain and Noise Figure Response of a Monolithic Three-Stage Common-Gate Input LNA

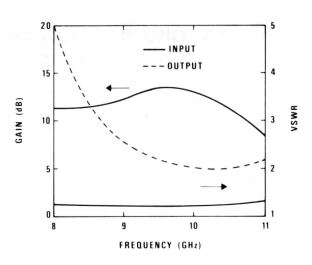

Figure 6 VSWR Response at Reduced Current Level

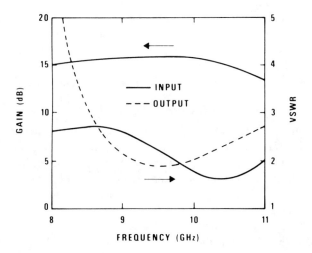

Figure 5 VSWR Response at Minimum Noise Figure Bias

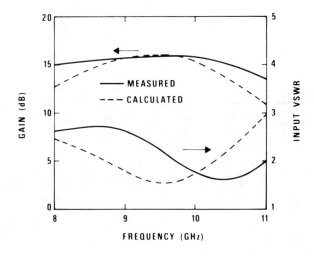

Figure 7 Measured and Calculated Performance of Monolithic LNA

Monolithic Microwave Gallium Arsenide FET Oscillators

J. S. JOSHI, J. R. COCKRILL, AND JAMES A. TURNER

Abstract—Results of the first monolithic microwave GaAs FET oscillator [1] are presented. The oscillator design philosophy is outlined. Design procedures for the oscillator and techniques for realization of various circuit components on the semi-insulating GaAs substrate are indicated. Performance of the oscillator is described and commented upon.

INTRODUCTION

THE 1960's saw the development of the first practical active microwave solid-state sources such as transferred-electron device and IMPATT diode oscillators and the extension of bipolar transistor oscillators to microwave frequency. In the 1970's, with the advancements in gallium arsenide material and device technology, it was possible to realize GaAs Schottky-gate field-effect transistors (GaAs MESFET's) with noise figures of the order of 2 dB at 10 GHz. Intensive effort has been devoted towards the development of low-noise and broad-band amplifiers using the GaAs FET. Some effort has also been devoted in the recent years for other applications like oscillators and mixers. The GaAs MESFET's oscillators generally provide a higher dc to RF conversion efficiency (typically 10-20 percent) and do not have any threshold current requirement. Being a three-terminal structure, the GaAs FET is an extremely versatile active oscillator circuit element, and by making use of this feature it should be easily possible to control the behavior of the FET to provide frequency modulation, temperature compensation, stabilization, etc.

Due to these inherent advantages, the GaAs FET oscillator activity has received much attention recently. High-efficiency free-running GaAs FET oscillators have been reported in the literature up to 25 GHz [2]-[6]. Electronic tuning of FET oscillators by YIG resonator and varactor diode has been researched and greater than octave tuning range has been achieved. Almost all the oscillators have been realized on low-Q microstrip circuit using hybrid film techniques.

Semi-insulating gallium arsenide ($\rho \approx 10^7$-$10^9 \Omega \cdot cm$) has a slightly higher dielectric constant ($\epsilon_r = 13.0$) than alumina and has a low-loss tangent ($\sim 10^{-3}$). It is, therefore, comparable to alumina as far as microwave-circuit properties are concerned. Additionally, the gallium arsenide material is an excellent base for deposition of low-resistivity layers by epitaxy or ion implantation. This has been successfully used for fabrication of devices like the FET. Being a planar device, the FET is ideally suited for monolithic circuit work. In monolithic microwave circuits, the properties of the semi-insulating gallium arsenide material are used to realize the associated circuit components on the chip itself.

In this paper, techniques related to a monolithic microwave GaAs FET oscillator are described. First, the oscillator design philosophy is outlined, followed by the design procedure for the oscillator and realization of circuit components by monolithic techniques. Performance characteristics of the oscillator are detailed and commented upon.

Such a microwave GaAs FET oscillator on a chip would help in subsystem integration when used in conjunction with other similar components. It would also result in a source of microwave power in an extremely small outline and would be easily reproducible and cost effective. One important area for application would be in monolithic all-FET front ends.

DESIGN PHILOSOPHY

In order to produce sustained steady-state oscillations at microwave frequencies, the active solid-state device must possess negative resistance. For two-terminal devices, like the transferred electron or Gunn diode, IMPATT, and tunnel diode, etc., the negative-resistance condition can be obtained by simply applying dc bias to the device. Such devices are inherent negative-resistance devices. The rest of the circuit can then be properly designed to make use of the negative-resistance condition for obtaining oscillations at the desired frequency. The bipolar devices and the FET's, on the other hand, do not possess this property and the negative-resistance condition has to be simulated by suitably coupling the input and output ports of these devices.

Basically, the coupling can be of two types, as shown in Fig. 1. In the series-feedback arrangement, the feedback element is the common current-carrying element between the input and output ports while in the parallel arrangement, it is the common voltage element between the two ports. More complicated arrangements containing both types of feedback and higher order interconnections can also be envisaged. When the feedback element(s) is appropriately selected it is possible to simulate the negative-resistance condition at one or both the device ports. Furthermore, if one device port is appropriately terminated, the whole combination can be represented as a negative-resistance device, and its associated reactance. Such a condition is shown in Fig. 2 in which a three-terminal device with feedback and matching elements produces a negative resistance $-R_D(w)$ and the associated reactance $jX_D(w)$. This,

---

Manuscript received March 25, 1980; revised September 12, 1980. Some of the work described in this paper has been carried out with the support of the Procurement Executive, Ministry of Defence, sponsored by DCVD.

The authors are with the Allen Clark Research Centre, Plessey Research (Caswell) Limited, Caswell, Towcester, Northants., England.

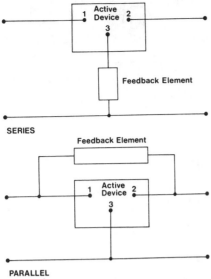

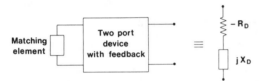

Fig. 1. Basic feedback arrangements.

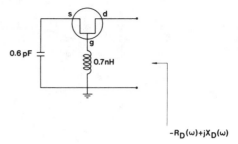

Fig. 3. Monolithic oscillator schematic diagram.

Fig. 2. Simulation of negative resistance with a two-port device.

in turn, is connected to a load of resistance $R_L(w)$ and reactance $jX_L(w)$. In general, the device elements are not only functions of frequency but of amplitude also, but for the sake of simplicity the amplitude variation is usually ignored. For sustained steady-state oscillation, the net resistance in the circuit should be zero and the device reactance should be resonant with the load reactance at the oscillation frequency. These are the so-called oscillation conditions given by (1)

$$R(w) = -R_D(w) + R_L(w) = 0$$
$$X(w) = X_D(w) + jX_L(w) = 0. \tag{1}$$

Although the oscillation conditions given by (1) may be satisfied, the oscillations obtained at that frequency may not be stable. In order to obtain stable oscillations at that frequency, Edson's [7] stability criteria must also be satisfied which state that

$$\frac{\partial R}{\partial w} > 0 \quad \frac{\partial X}{\partial w} > 0. \tag{2}$$

Circuit Design and Realization

The characteristics of a GaAs FET at microwave frequencies are normally specified by S parameters. In order to study the influence of simple series- or parallel-feedback arrangement it is most effective to use computer-aided-design techniques. One of the functions most useful in this aspect is the so-called mapping function routine in COMPACT [8]. In this, a general impedance element is embedded in a two-port network and the resultant modified S parameters of the original two-port network are "mapped" as a function of the embedding impedance.

This technique has been utilized here to study the influence of series feedback on the S parameters of a Plessey GAT5 device at 12 GHz. It was seen that when the feedback element is inductive, both the input and output reflection coefficients of the circuit are greater than zero. In other words, negative-resistance conditions exist at both the device ports. It was decided to use a feedback inductance of 0.7 nH as this value can be easily realized by lumped-element techniques on the semi-insulating GaAs substrate. The influence of source termination on the output reflection coefficient was further investigated by computer-aided-design techniques. It was seen that a source termination of 0.6 pF is suitable for the desired oscillation frequency of 12 GHz. This capacitance can also be easily realized on a semi-insulating GaAs substrate. The oscillator circuit is shown in Fig. 3. The computed output reflection coefficient versus frequency for the monolithic oscillator circuit (Fig. 3) is shown in Fig. 4 on a compressed Smith chart. It can be seen that at 8 GHz the output reflection coefficient is just greater than unity. As the frequency is increased, the magnitude of the output reflection coefficient increases further and reaches the maximum amplitude of 9.3 at 12 GHz. The impedance variation is also plotted as a function of frequency in Fig. 5. It can be seen that if the circuit of Fig. 3 is directly terminated with a 50-Ω load, both the oscillation condition and stability criteria of (1) and (2) will be satisfied near 12.0 GHz.

Circuit Realization

It is quite important at the circuit design stage that only those circuit elements which can be easily obtained by monolithic processing techniques be selected. This constraint limits the choice and range of available circuit elements one can employ in the design. These can be realized either by distributed- or by lumped-element techniques. Because of its compactness and, therefore, optimum use of gallium arsenide "real estate," a lumped-element technique has been adopted here.

The gate feedback inductance is realized in the form of a loop inductor with straight sections on either end for connection to gate metallization pad and the external dc bias to the gate terminal. The inductance of the loop inductor is given by [9]

$$L = 1.257 a \left[\ln\left(\frac{8\pi a}{w}\right) - 2 \right] \text{nH} \tag{3}$$

where a is the mean loop radius and w is the conductor width (Fig. 6), both in millimeters. This formula is in close agreement with others reported in the literature [10] and is gener-

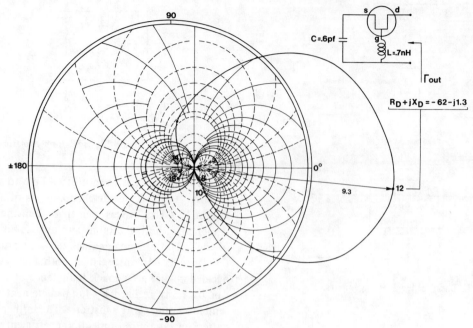

Fig. 4. Output reflection coefficient versus frequency on a compressed Smith chart.

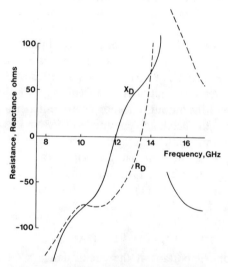

Fig. 5. Output impedance versus frequency.

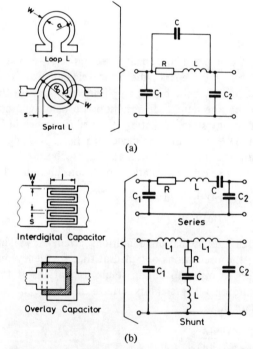

Fig. 6. Lumped-circuit-element geometries. (a) C, C_1, C_2, R are parasitic elements of lumped inductors. (b) C_1, R, L, L_1, C_2 are parasitic elements of lumped capacitors.

ally adequate for $2a/w \geqslant 5$. In practice, part of the loop inductor must be removed for connection to the straight sections of the same width w. The inductance of a straight section of width w millimeters and infinitesimally small thickness t and length l millimeters is given by [9]

$$L = 0.2\, l \left[\ln \frac{2l}{w} + 0.5\right] \text{ nH}. \quad (4)$$

Using these two equations, a feedback inductance of 0.7 nH was obtained.

The capacitive source termination can be realized by either interdigital or overlay capacitors. The choice between the two depends upon the capacitance value to be realized and the processing technology available. Usually for values of less than 1 pF, interdigital capacitors can be used while for values greater than 1 pF, overlay techniques are generally used to minimize the overall size.

The interdigital capacitor is a periodic structure (Fig. 6) and the total capacitance between N fingers each of length l (mm) can be given by

$$C_T = C(N-1)\, l \text{ pF} \quad (5)$$

where C, the capacitance between two successive strips of

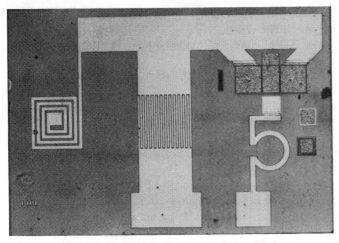

Fig. 7. Monolithic microwave oscillator chip.

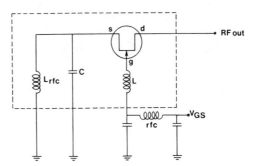

Fig. 8. Complete oscillator circuit including gate bias and grounding arrangements.

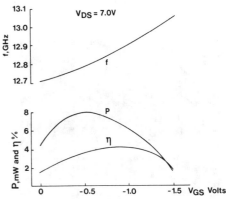

Fig. 9. Oscillator performance versus gate bias.

width w and spacing s, is given by [11]

$$C = \epsilon_0 (1 + \epsilon_r) \frac{K(k)}{K'(k')} \quad (6)$$

where $K(k)$ and $K'(k')$ are elliptic integrals of kernel k and k', respectively, with

$$k = \tan^2 \frac{a\pi}{4b}$$

$$k' = \sqrt{1 - k^2} \quad (7)$$

where $a = w/2$ and $b = (w + s)/2$.

Using (5) and (6), the interdigital capacitance geometry for a given value can be obtained.

A square spiral inductor was also designed according to the design equations provided in [9]. This spiral inductor is connected to the source lead and acts both as an RF choke and a dc return for the source. In order to maintain the lumped nature of these circuit components, the maximum linear dimension should be less than 5 percent of the operating wavelength.

It is worth noting that the preceding equations only define the primary circuit component. Parasitic components are associated with each of these. Resistive parasitic components arise due to finite metallization thickness used in the circuit. Capacitive parasitic components arise due to finite parallel-plate capacitance between the metallization and the ground plane. In loop inductors there is parasitic edge capacitance between the ends of the loop and in interdigital capacitors there is parasitic inductance due to finite length of the individual digits (Fig. 6). Detailed discussion on parasitics is treated elsewhere [12].

A photograph of the monolithic microwave FET oscillator chip is shown in Fig. 7. It measures 1.8×1.2 mm. The FET uses a buffer-layer arrangement and has n^+ ohmic contacts. The gate geometry of 0.8×300 μm is obtained using an etched channel technology. The source and drain ohmic contacts are gold-germanium while the gate metallization is aluminum. The circuit elements are realized on the semi-insulating gallium arsenide material. For experimental purposes, the chip is mounted on a 50-Ω trough line jig. As dictated by the design procedure, no output-matching elements are used and the output from the drain is directly connected to a 50-Ω load with a bond wire. Drain bias is externally provided by a bias tee. A π-type low-pass filter circuit consisting of external 20-pF chip capacitors and an inductor is used for gate bias. Fig. 8 shows the complete circuit. The components outside the dashed area are external to the chip.

Circuit Performance and Discussion

The output power, frequency, and dc to RF conversion efficiency as functions of gate bias at a fixed drain bias of 7.0 V are shown in Fig. 9. No external "tweaking" of the circuit was required. As the gate bias is varied from 0 to -1.5 V, the oscillator frequency varies from nearly 12.7 to 13.0 GHz. This compares favorably with the predicted performance. No spurious oscillations or frequency switching was observed at any setting. The maximum output power was 8 mW and the maximum efficiency at this drain bias level was 4 percent. Fig. 10 shows the variation of these parameters with drain bias at -0.5-V gate bias. The frequency variation with gate voltage and drain voltage is of the order of 200 and 50 MHz/V, respectively. Fig. 11 shows the output spectrum of the oscillator. It is quite typical of low-Q microwave oscillators.

It has been seen that the design procedure and realization of individual circuit elements by lumped techniques have resulted in a monolithic oscillator operating frequency close to its predicted value. However, when compared to a similar hybrid

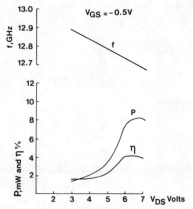

Fig. 10. Oscillator performance versus drain bias.

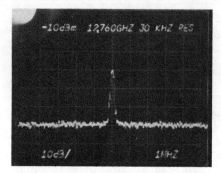

Fig. 11. Oscillator spectrum.

MIC oscillator, the output power and efficiency of this oscillator are inferior.

It should be recalled that in this monolithic GaAs FET oscillator the design procedure dictated that the drain terminal be directly connected to the 50-Ω load through a bond wire. Thus the drain port did not have any tuning elements to maximize the power transfer to the load. It is generally found with hybrid MIC oscillators that the output port must be fine tuned to maximize the power and thus improve the efficiency of the oscillator.

It is also worth noting here that the quality factor of lumped circuit elements on semi-insulating GaAs is dependent upon the metallization thickness used. These Q factors are lower than those obtainable with distributed MIC components on alumina substrates by about an order of magnitude. This would imply a slight reduction in oscillator power in monolithic circuits compared to similar hybrid MIC ones.

Summary

Results of the first GaAs FET "oscillator-on-a-chip" using grounded gate arrangement with series feedback have been presented here. It has been shown that it is possible to obtain sources of microwave power in an extremely small outline. It should be possible to "marry" a varactor diode to the circuit to obtain a monolithic VCO.

It is believed that this work will influence the microwave oscillator scene from simple intruder alarm applications to highly sophisticated microwave systems.

Acknowledgment

The authors are grateful to the directors of Plessey Research (Caswell) Ltd. for permission to publish this paper.

References

[1] J. S. Joshi, J. Cockrill, and J. A. Turner, "Monolithic microwave GaAs FET oscillator," in *Dig. GaAs IC Symp.* (Lake Tahoe, NV), Sept. 1979.

[2] M. Maeda, K. Kimura, and H. Kodera, "Design and performance of X-band oscillators with GaAs Schottky gate field-effect transistors," *IEEE Trans. Microwave Theory Tech.*, vol. MTT-23, no. 8, pp. 661–667, Aug. 1975.

[3] M. Omori and C. Nishimoto, "Common gate GaAs FET oscillator," *Electron. Lett.*, vol. 11, no. 16, pp. 369–371, Aug. 1975.

[4] J. S. Joshi and J. A. Turner, "High peripheral power density GaAs FET oscillator," *Electron. Lett.*, vol. 15, no. 5, pp. 163–167, Mar. 1979.

[5] H. Tserng, H. Macksey, and V. Sokolov, "Performance of GaAs MESFET oscillators in the frequency range 8-25 GHz," *Electron. Lett.*, vol. 13, no. 3, pp. 85–86, Feb. 1977.

[6] K. M. Johnson, "Large signal GaAs MESFET oscillator design," *IEEE Trans. Microwave Theory Tech.*, vol. MTT-27, no. 3, pp. 217–227, Mar. 1979.

[7] W. Edson, *Vacuum Tube Oscillators.* New York: Wiley, 1953, pp. 430–450.

[8] *COMPACT User Manual*, Compact Engineering Inc., U.S.A.

[9] F. W. Grover, *Inductance Calculations: Working Formulas and Tables.* New York: Dover, 1946.

[10] C. S. Aitcheson *et al.*, "Lumped circuit elements at microwave frequencies," *IEEE Trans. Microwave Theory Tech.*, vol. MTT-19, Dec. 1971.

[11] Y. C. Lim and R. A. Moore, "Properties of alternately charged coplanar parallel strips by conformal mapping," *IEEE Trans. Electron Devices*, vol. ED-15, no. 3, pp. 173–180, Mar. 1968.

[12] R. S. Pengelly and M. G. Stubbs, to be published.

Monolithic Voltage Controlled Oscillator for X- and Ku-Bands

BENTLEY N. SCOTT, MEMBER, IEEE, AND GAILON E. BREHM, MEMBER, IEEE

Abstract — A GaAs voltage controlled oscillator circuit that tunes from 11.15 to 14.39 GHz and 16 to 18.74 GHz has been designed and fabricated. The 1.1 mm × 1.2-mm chip includes two varactors, a 300-μm FET, bypass capacitors, tuning inductors, and isolation resistors. Wide-band circuit design techniques will be described. Varactor and circuit effects causing the noncontinuous bandwidth will be discussed showing the capability of continuous 11 to 18 GHz tuning using a single GaAs chip.

I. INTRODUCTION

MAJOR EMPHASIS in the past 5 years in monolithic GaAs circuits has been in amplifier, mixer, T/R switch, and phase shifter networks while oscillator development has continued to be hybrid oriented. The first microwave monolithic oscillator above *L*-band including both the power generating and tuning elements on a single chip was previously reported by the authors [1]. This circuit covered 8.8 to 10 GHz. The present work allows two varactors to be implemented, one in the source and one in the gate circuits has given a tuning bandwidth covering 11.15 to 14.39 GHz and 16 to 18.74 GHz from a single chip. This circuit demonstrates the capability to cover *X*- and *Ku*-bands continuously with one monolithic circuit. The design technique used and a description of the unique varactor are summarized with an explanation relating to the varactor Q which created the frequency gap between 14.39 and 16 GHz.

II. CIRCUIT DESIGN

The negative impedances required to cause oscillations in a MESFET are a result of positive feedback. If the correct feedback reactance is added to a properly selected device configuration, oscillations can occur from very low frequencies to approximately f_{max} of the active device. After comparing the possible combinations both analytically and experimentally, the common gate configuration utilizing an inductive reactance between gate and ground as the regenerative feedback element was selected due to its inherent broad-band negative resistances and ease of analysis. A varactor was placed in series with the inductor to tune the negative impedances across a wide bandwidth.

In order to make design tradeoffs between the tuning circuit topologies and arrive at element values which would produce a wide bandwidth oscillator, several simplifying

Manuscript received March 23, 1982; revised June 4, 1982. This work was supported in part by the Naval Air Systems Command, under NRL Contract N00173-79-C-0048.

The authors are with Texas Instruments, Inc., Central Research Laboratories, Dallas, TX 75265.

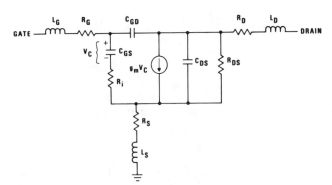

Fig. 1. FET model schematic.

TABLE I
FET MODEL ELEMENT VALUES

Element	Value
L_G	0.15 nH
R_G	7.43 Ω
C_{GS}	0.338 pF
R_i	0.100 Ω
C_{GD}	0.028 pF
C_{DS}	0.097 pF
R_{DS}	215 Ω
R_D	1.04 Ω
L_D	0.244 nH
R_S	4.9 Ω
L_S	0.077 nH
g_m	37 m℧
τ	2.65 ps

assumptions were required. Since oscillations build up from small signal conditions, small signal *S*-parameters were used to characterize the FET, and this data was then used to predict tuning bandwidths. Experimental verification showed reasonable correlation to the *S*-parameter analysis. Nonlinear device behavior due to large signal operation of the device was assumed to be primarily a change in source–drain resistance which does not appreciably change the phase angle or magnitude of the negative impedances appearing at the source with the drain appropriately loaded as in the case of this common gate design. Use of a large signal model confirmed this assumption while showing that there is a shift downward in frequency due to large signal

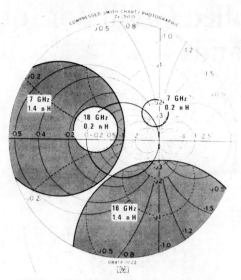

Fig. 2. Stability circles for 0.2-nH and 1.4-nH gate inductance.

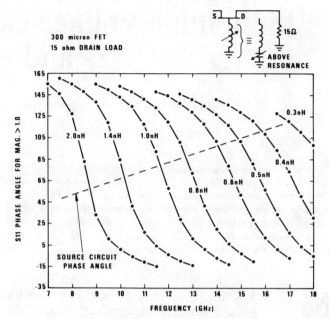

Fig. 3. FET source reflection coefficient.

TABLE II
SOURCE REFLECTION COEFFICIENT FOR SPECIFIED GATE INDUCTANCES

Frequency	FET Source Reflection Coefficient for Specified Gate Inductances		
	.2 nH	.6 nH	1.0 nH
8.0 GHz			0.7 ∠163
9.0			1.0 ∠155
10.0			1.6 ∠143
11.0		0.9 ∠153	2.8 ∠119
12.0		1.2 ∠144	3.9 ∠58
13.0		1.7 ∠130	2.3 ∠12
14.0		2.5 ∠102	1.4 ∠-7
15.0		2.8 ∠52	1.1 ∠-13
16.0	0.9 ∠134	1.8 ∠13	0.9 ∠-17
17.0	1.2 ∠121	1.1 ∠-7	
18.0	1.5 ∠99	0.7 ∠-19	

operation; however, this downward shift in frequency does not appreciably affect bandwidth analysis. A design technique that allowed representative bandwidth and element value predictions was developed that entails characterization of the device by S-parameter measurements, stability circle analysis, drain circuit definition, computer generation of composite S-parameters utilizing the selected drain circuit, and source circuit phase angle graphs which allow bandwidth and element value selection.

To perform this bandwidth analysis, first S-parameters are taken on a common source MESFET and a model is fit to the data. Fig. 1 shows the common source FET model used, and Table I gives the associated element values for a 300-μm device after a fit has been performed on measured S-parameter data. Using computer aided design, the model is transformed to a common gate device with a series gate feedback inductor and stability circles are generated for varying values of this gate inductance as shown in Fig. 2. This family of circles defines the region of impedances that the drain must see to maintain negative source resistances as the gate-feedback inductance is varied. A 15-Ω drain load will allow negative resistances to appear at 11 GHz with 1.2-nH net inductance in the gate and will also allow negative source resistances at 18 GHz if that net gate inductance is changed to 0.2 nH.

Maintaining negative impedances at the source makes it possible to terminate that port using an appropriate passive phase cancellation network that will allow oscillation to occur. Tuning is accomplished by varying the impedance (net inductive reactance) from gate to ground. The frequency at which negative impedance is seen at the source for a given gate inductance and drain matching network is the frequency at which free-running oscillations will occur (assuming an appropriate phase cancellation network is used to terminate the source). The relative magnitude of that negative impedance is an indication of the amount of tuning element (varactor) loss that can be tolerated and the amount of output power that can be obtained. Table II gives the source reflection coefficients of the common gate FET model as described earlier with the gate feedback inductance values of 0.2 nH, 0.6 nH, and 1.0 nH. The drain was loaded into 15-Ω real. The change in gate inductance clearly gives a change in the frequency at which negative impedances occur and in the phase angle.

The family of curves generated using the above analysis is given in Fig. 3. The abscissa is the phase angle of the source reflection coefficient (when it is greater than unity) with the drain loaded into a 15- to 50-Ω two-section transformer. From these data it was determined that a capacitive termination is required in the source network to cancel the imaginary part of the FET source impedance. The circuit then oscillates at that frequency where the source phase angle is cancelled and negative impedance is seen at the source. The bandwidth may be determined by calculation of the terminating network's phase angle slope;

the intersection of this slope and the varying gate inductance family (given by the net gate inductance value as the series varactor capacitance changes) gives the tuning bandwidth.

Several hybrid oscillator circuits were built using this analytical approach. Results showed that selection of varactor values to achieve a particular bandwidth was possible. The frequency of oscillation was also predictable from use of this method for other circuit element value selection.

Oscillator circuits utilizing one varactor in the gate and optimally terminated at the other two ports will give maximum tuning bandwidths of 4 GHz at X-band [2], [3]. The tuning curves were used to study improvements in the source network which would increase tuning bandwidth. A varactor was added to the source network so that the phase angle could be tuned radically versus frequency, thus improving the obtainable VCO bandwidth by more than 2.5 GHz.

Fig. 4 shows phase versus frequency data for the source network shown in the figure. As the varactor is tuned, the phase changes rapidly. If Fig. 5 is considered to exemplify the FET phase angle requirements for oscillation then at 10.5 GHz with 1.0 nH in the gate, a phase angle of approximately $-130°$ is required at the source circuit. As the source varactor (gate varactor value held constant) is tuned, the source would follow the dotted line on Fig. 4 since the oscillator would ride the 1.0-nH line given in Fig. 5. At a source varactor value equal to 0.3 pF the frequency of the oscillation would be 12.5 GHz. If the gate varactor is now tuned (source varactor held constant while the gate varactor is tuned) the frequency of oscillation will follow the source phase curve given by the 0.3-pF line in Fig. 4. This source and gate tuning path is given by the dotted line in Fig. 5 which shows the source tuning the oscillator from 10.5 GHz to 12.5 GHz (with a source varactor swing of 0.7 pF and 0.28 pF) and the gate varactor tuning the oscillator from 12.5 GHz to 18 GHz (with a gate varactor swing of 1.0 pF to 0.10 pF). Table III gives the values of gate inductance and source circuit varactor capacitance for selected frequencies. This example simply illustrates the basic tuning mechanisms and does not include all circuit parameter considerations required for proper analysis.

This varactor analysis which was developed at Texas Instruments has allowed TI to establish state-of-the-art tuning bandwidths for hybrid varactor tuned VCO's [4]. This technique allows a practical estimation of varactor swing and the circuit element values required for a given bandwidth.

Recently, efforts have been made to incorporate the FET, varactors, and capacitor on one piece of GaAs, thus reducing the present 6-mm × 6-mm hybrid oscillator circuit at X-band to 1.1 mm × 1.2 mm. This reduced the number of bond wires and assembly steps required by a factor of five to ten times.

A circuit to obtain oscillations in X- and Ku-band was generated for monolithic implementation with the schematic given in Fig. 6. the 15- to 50-Ω transformer was not placed on the mask for the GaAs chip in order to maximize the number of oscillator circuits obtained from each slice. Isolation resistors were included on the chip to reduce the

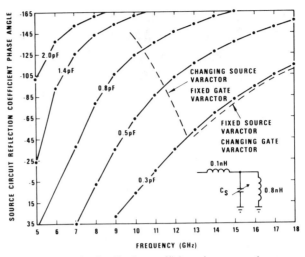

Fig. 4. Source circuit reflection coefficient phase versus frequency.

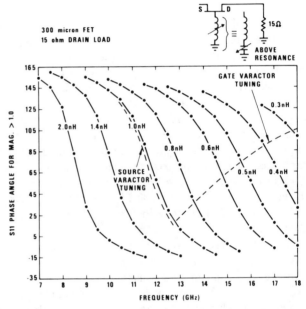

Fig. 5. Two varactor oscillator source reflection coefficient.

TABLE III
GATE INDUCTANCES AND SOURCE CIRCUIT VARACTOR CAPACITANCES FOR SELECTED FREQUENCIES

Frequency (GHz)	L_G^* (nH)	C_S (pF)
10.5	1.0	0.9
12.0	1.0	0.4
12.5	1.0	0.3
13.0	0.9	0.28
15.0	0.58	0.28
17.0	0.36	0.28
18.0	0.29	0.28

* L_G is actually a series L-C network where the capacitor is a varactor and the inductor is fixed.

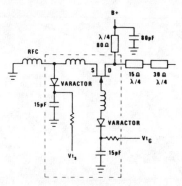

Fig. 6. VCO schematic.

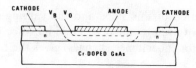

Fig. 7. Monolithic varactor.

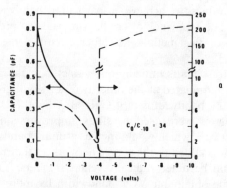

Fig. 8. Theoretical $C-V$ and Q for a monolithic varactor.

chance of low frequency oscillations and to serve as current limiting resistors if the varactor diodes should be accidently forward biased.

III. Monolithic Varactor Diode

Conventional varactor diodes, particularly those with large tuning ratios (hyperabrupt diodes) require highly conductive substrate material and relatively thick epitaxial layers (>1 μm). These materials requirements are not compatible with those for GaAs FET-based monolithic microwave integrated circuits (MMIC's) which require a thin (<0.05 μm), uniformly doped active layer on a semi-insulating substrate. To integrate the conventional hyperabrupt diode on a semi-insulating substrate requires a very complicated selective epitaxial deposition. The materials required to fabricate the diode discussed here are the same as or very similar to those for the FET, so this device type will be extremely important in monolithic voltage controlled oscillators.

Van Tuyl [5] has reported the use of a GaAs varactor diode in an MMIC which is similar in design and materials requirements to that discussed here. This device does not however, provide the same wide-band capacitance tuning characteristics. This wide tuning range ($C/C_0 > 10$) is essential for many microwave applications.

The device discussed here is an interdigitated Schottky-barrier diode and was reported on by the authors previously [6]. The key feature of the diode is that an arbitrarily high capacitance ratio is achieved by the change in the effective junction area as the depletion layer punches through on the semi-insulating substrate. A simple, uniformly doped n-layer on a Cr-doped or other semi-insulating GaAs substrate can be used. The layer thickness and/or the amount of anode recess is chosen to allow punch-through before breakdown. The diode consists of one or more Schottky-barrier anode fingers spaced betweeen ohmic cathode regions. An n-type GaAs layer is defined by mesa etching and other means so that it is only under the active area. The bond pads (or interconnects to other parts of a monolithic circuit) are located on the semi-insulating substrate for minimum parasitic capacitance and conductance.

The key feature of the device is illustrated in Fig. 7. The doping-thickness product of the n-layer under the anode is selected so that punch-through to the substrate occurs prior to breakdown. A rapid drop in capacitance occurs at punch-through because the effective area of the diode is reduced to that of the depletion layer sidewalls alone. The fractional drop in capacitance is related to the ratio of the anode length (direction parallel to current flow) and the layer thickness. The series resistance is related to the anode length and anode-cathode spacing.

The region beneath the anode prior to punch-through can be regarded as a distributed RC network. On that basis the equivalent terminal impedance can be calculated. The equivalent resistance R_e and capacitance C_e are given by

$$R_e = 1/2 \sqrt{\frac{R_0}{2\omega C_0}} \left(\frac{\sinh\theta - \sin\theta}{\cosh\theta - \cos\theta} \right) + 1/2 R_s \quad (1)$$

$$\frac{1}{\omega C_e} = 1/2 \sqrt{\frac{R_0}{2\omega C_0}} \left(\frac{\sinh\theta + \sin\theta}{\cosh\theta - \cos\theta} \right) \quad (2)$$

where $\theta = L\sqrt{1/2\omega R_0 C_0}$, R_s is the parasitic series resistance to the cathode on each side, and R_0 and C_0 are the resistance and capacitance per unit length under the anode. These equations do not include the sidewall capacitance, which were added in separately for biases near punch-through.

Using (1) and (2), the $C-V$ characteristic, the series resistance and $Q(=1/\omega R_e C_e)$ at 10 GHz have been calculated for several appropriate diode designs. Fig. 8 shows the $C-V$ characteristics and Q for a 6-μm $\times$ 100-μm anode device. In this case the Q is 6 at zero bias and drops below 2 before punch through at approximately 3.8 V. It is this sharp reduction in Q and a high surface leakage path on this slice of monolithic oscillators that caused the drop in power output in the center of the frequency band.

IV. Experimental Results

A. Tuning Bandwidth

The varactor diodes, a 300-μm FET, and the associated circuitry discussed earlier were implemented on a single GaAs 1.1-mm $\times$ 1.2-mm chip as shown in Fig. 9. The

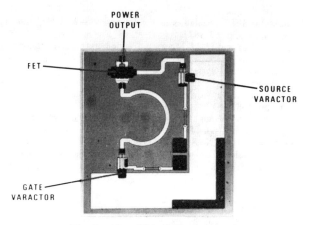

Fig. 9. Monolithic VCO photograph.

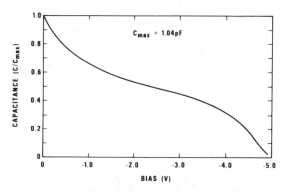

Fig. 10. Experimental $C-V$ characteristics for a monolithic varactor.

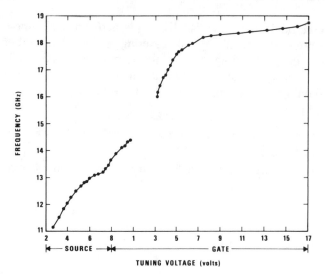

Fig. 11. Monolithic VCO tuning curve.

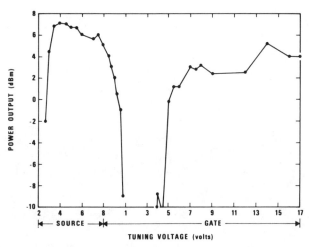

Fig. 12. Monolithic VCO power curve.

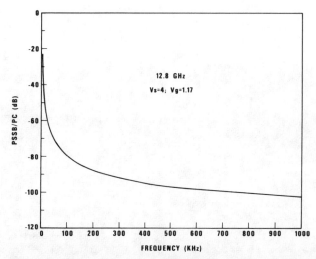

Fig. 13. SSB phase noise of monolithic VCO (1-Hz bandwidth).

varactor diodes were designed for a 1.0-pF starting value and a capacitance ratio greater than 15:1. Fig. 10 gives the $C-V$ characteristics for the actual diode. Figs. 11 and 12 show the frequency and power respectively versus tuning voltage of an oscillator as first the varactor in the source was tuned from 0 to 9 V, and then the varactor in the gate was tuned 0 to 17 V. The actual voltage across the varactor was less in each case due to the voltage drop across the 2-kΩ resistor as the varactor became leaky. The oscillator tuned smoothly from 11.15 to 14.39 GHz where the oscillations ceased due to the low Q and excessive surface leakage across the diode. As the tuning voltage was increased on the gate to 3.2 V the oscillation began once again at 16.0 GHz and continued tuning to 18.74 GHz.

B. Noise Performance

Phase noise in oscillators is an important system consideration. Although these monolithic oscillators have low Q varactors which will be improved by process refinements, data is included for these circuits as a reference point.

A dual source measurement method was used with a very clean reference oscillator as the second source. The mixed output was then fed to a frequency discriminator which then drives a wave analyzer. Fig. 13 is the single-sideband phase noise in a 1-Hz bandwidth.

V. Conclusion

A design technique has been developed that allows oscillator bandwidth to be predicted and tradeoffs to be performed between circuit configurations and varactor ratio. Monolithic voltage tuned oscillators have been built on a 1.1-mm × 1.2-mm chip which included the 300-μm FET, two varactors, and all of the associated circuit elements except the 15- to 50-Ω output matching transformer. These circuits exhibited a tuning performance of 11.15 to 14.39

GHz and 16.0 to 18.74 GHz. This performance demonstrates the feasibility of covering octave bandwidths continuously using monolithic technologies. Assembly problems, size, and repeatability of present hybrid VCO's should be vastly improved once this technology matures.

REFERENCES

[1] B. N. Scott, G. E. Brehm, and F. H. Doerbeck, "X-Band GaAs monolithic voltage controlled oscillators," in *ISSCC-82 Dig.*, pp. 138–139.

[2] H. Q. Tserng, H. M. Macksey, and V. Sokolov, "Performance of GaAs MESFET oscillators in the frequency 8–25 GHz," *Electron. Lett.*, vol. 13, p. 85, 1977.

[3] H. Q. Tserng and H. M. Macksey, "Wide-band varactor-tuned GaAs MESFET oscillators at X- and Ku-Bands," in *1977 Int. Microwave Symp. Dig. Tech. Papers*, pp. 267–269.

[4] B. N. Scott, G. E. Brehm, D. J. Seymour, and F. H. Doerbeck, "Octave-band varactor-tuned GaAs FET oscillator," in *ISSCC-81 Dig.*, pp. 138.

[5] R. Van Tuyl, "A monolithic GaAs FET of signal generation chip," in *ISSCC-80 Dig.*, pp. 118.

[6] G. E. Brehm, B. N. Scott, D. J. Seymour, W. R. Frensley, W. N. Duncan, and F. H. Doerbeck, "High capacitance ratio monolithic varactor diode," in *1981 Cornell Microwave Conf. Dig.*, pp. 53–63.

A Monolithic GaAs IC for Heterodyne Generation of RF Signals

RORY L. VAN TUYL, SENIOR MEMBER, IEEE

Abstract – An integrated heterodyne signal-generating GaAs chip is reported. This circuit contains: an on-chip local oscillator with external inductors tunable by means of on-chip variable capacitors from 2.1 to 2.5 GHz; a doubly balanced mixer with associated drive circuitry; and an IF preamplifier. The circuit delivers +6 dBm (equivalent 50 Ω) into the designed load impedance of 200 Ω with −30-dBc harmonic distortion over a 1.4-GHz 3-dB bandwidth. Circuit elements presented include: a unique variable-threshold limiter, a self-biasing push–pull oscillator, doubly balanced mixer, and a self-biasing unity–gain phase splitting amplifier.

INTRODUCTION

FULL UTILIZATION of GaAs integrated circuits in the RF and low-microwave-frequency ranges will depend upon the integration of as many functions as possible on a single chip, and the minimization of chip interconnections. In addition to testing ease, and the obvious economy of material usage associated with this high-level integration approach, there is a substantial saving in power consumption which results from direct interconnection of circuit functions without resorting to low-impedance 50-Ω lines. Bandwidths of several gigahertz are achievable in the high-impedance environment (hundreds to thousands of ohms) typical of small GaAs FET's, due to the very low parasitic capacitance associated with the semi-insulating GaAs substrate. Impedance-matching techniques are not practical in integrated form at these relatively low frequencies and high-impedance levels, so low-frequency direct-coupled circuit-design approaches are used to realize circuit functions such as oscillators, mixers, modulators, phase splitters [1], amplifiers [2], and frequency dividers [3]. When inductive elements are required for purposes of peaking or resonating, external elements must be used.

The integrated devices included in the GaAs IC process are GaAs FET's which can function as amplifiers, loads, current sources, series or shunt switches, and controlled limiters; Schottky diodes, which can be used as rectifiers, low dynamic impedance voltage sources, switches, and voltage-variable capacitors; resistors, which are useful as loads, attenuators, and bias connectors. The subject of this paper is the use of this GaAs IC process to fabricate a circuit which incorporates a large part of a heterodyne signal-generation system on a single chip.

THE SIGNAL-GENERATION SYSTEM

A local oscillator (LO), switching modulator, doubly balanced mixer, RF phase splitter, and IF amplifier on the IC chip can be combined with external inductors, a low-pass filter, post

Manuscript received August 5, 1980.
The author is with Hewlett-Packard Company, Santa Rosa, CA, 95404.

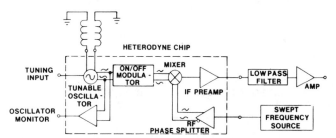

Fig. 1. Block diagram of an RF signal generation system consisting of a heterodyne GaAs IC chip and external hybrid circuit elements. The low-level variable-frequency source is external.

amplifier, and swept-frequency source to form a complete system, as shown in Fig. 1. The purpose of the external inductors is to provide the inductive portion of a parallel resonant circuit for the on-chip tunable oscillator. The low-pass filter between on-chip IF preamplifier and separate post amplifier is a 200-Ω characteristic-impedance, singly terminated, low-pass filter. The purpose of this filter is to reject unwanted feedthrough such as local oscillator, RF signal, and undesired mixing products.

The functions of the on-chip elements are as follows. The *local oscillator*, which is tunable for purposes of frequency adjustment, phase locking, or FM generation, drives the *doubly balanced mixer* through an ON/OFF *modulator*. The purpose of this modulator is to switch off the LO drive to the mixer without disturbing the amplitude or frequency of the oscillator itself, thereby enabling pulsed modulation of the IF output. The RF *phase splitter* converts an approximately −10-dBm external RF signal into 0° and 180° phase components for driving the mixer, which requires these phases for balancing of the RF feedthrough. The IF *preamplifier* is used to boost the IF signal to approximately +6-dBm equivalent (the voltage equivalent of +6 dBm into 50 Ω) into the 200-Ω impedance level of the IF filter. The *oscillator monitor* is used to observe an attenuated replica of the *local oscillator's* output. We will discuss the operation of each of these on-chip subcircuits separately.

THE LOCAL OSCILLATOR

The LO design used here is a positive-feedback push–pull configuration which is well suited to this application because of its broad-band negative resistance and because it has 0° and 180° phase outputs which are required for the mixer. Fig. 2(a) illustrates the basic circuit diagram and simplified equivalent circuit of the oscillator. The balanced input admittance generated by the cross-coupled positive-feedback circuit consists of a real

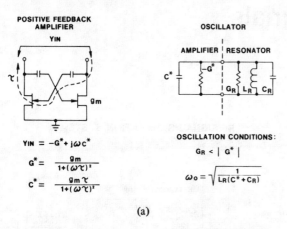

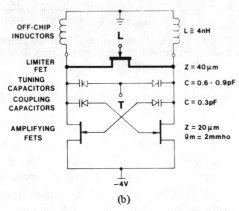

Fig. 2. (a) Basic push–pull FET oscillator and equivalent circuit. (b) Circuit schematic of actual oscillator used in the heterodyne chip.

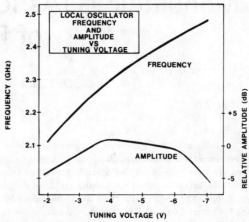

Fig. 3. Tuning characteristics of on-chip oscillator.

part approximately equal to the transconductance of each FET, and an imaginary part due to loop time delay caused by internal FET phase shift, and RC product time delays within the circuit. For oscillation to occur, the real part of a parallel resonator's shunt conductance must be less than the magnitude of the real part of the active circuit's negative conductance. The oscillation frequency is somewhat decreased below the external circuit's resonance by the capacitive portion of the active circuit's admittance.

The actual-circuit diagram of the integrated oscillator is shown in Fig. 2(b). The amplifying FET's gate width of 20 μm results in a transconductance of 20 mS, which is capable of producing oscillations with a resonator "Q" of greater than 10 in the frequency range studied here. The cross-coupling capacitors in this integrated realization are reverse-biased Schottky diodes. Since the oscillator is in continuous operation, the self-biasing effect produced by the balance between FET gate rectification current and diode-capacitor leakage current is sufficient to bias the amplifying FET's, making bias resistors unnecessary. A major portion of the parallel resonator's capacitive component is provided by on-chip voltage-variable capacitors. Applying a negative potential to the constant voltage terminal "T" reverse biases the tuning diode capacitors, reduces their capacitance, and causes the oscillator's frequency to increase. Due to the balanced nature of the circuit, little RF signal appears at node "T," so it can be accessed for tuning without RF decoupling. The oscillator also includes a voltage-variable limiter FET which can be used to regulate the oscillator's output amplitude. When the control terminal "L" is biased below -4.0 V, the limiter is nonoperational and the tuning characteristic of Fig. 3 is obtained. This curve shows that the oscillator tunes from nearly 2.1 to 2.5 GHz as the tuning control voltage on node "T" is varied from -2 to -7 V. The amplitude falloff at the low-frequency end is due to decrease of the external inductor "Q" as frequency decreases. The external inductors used in this test were a combination of high-frequency probe inductance and the inductance of a loop of bonding wire in series with the probe. Total inductance was about 4 nH.

Although oscillation amplitude could only be measured indirectly through the attenuating oscillator monitor, the maximum amplitude was calculated to be about 2 V peak-to-peak, the voltage equivalent of +10 dBm (50 Ω). In a case where less amplitude is desired, or where complete shutoff of the oscillator is required, the limiter FET can be activated by applying a more positive bias to terminal "L." The principle of operation for this limiter FET is illustrated in Fig. 4(a). This figure shows what could be termed the balanced common-gate I–V characteristics of an FET, that is, the I–V curves which result when source voltage V_S and drain voltage V_D are applied 180° out-of-phase and gate voltage V_G is held constant. For example, if V_G is held at the pinchoff voltage of the FET, -2.5 V in this case, the device is pinched off and the drain-to-source resistance is infinite. As V_D goes positive and V_S goes negative, we eventually reach the point where $V_S = V_G = -2.5$ V and $V_D = +2.5$ V, which means $V_{GS} = 0$ V and $V_{DS} = +5$ V. The current which flows under these conditions is, by definition, I_{DSS}. If V_G is then reset to a more negative potential, I_{DS} decreases at all values of V_{DS}. A family of symmetrical I_{DS} versus V_D, V_S curves is generated for various values of V_G, as shown in Fig. 4(a). The principal feature of interest in these curves is the voltage-variable onset of conduction which can be used as a variable clipping device in any balanced-signal circuit. The effect of this limiter on the relative output amplitude of the push–pull oscillator is shown in Fig. 4(b). Here, the oscillator runs at an equilibrium amplitude governed by its amplifying FET nonlinearities when the control voltage on node "L" is more negative than -4.0 V. As this voltage is moved positive, the relative amplitude decreases over a 10:1 range (20 dB) before oscillation ceases at -2.1 V on node "L." This limiting process introduces a frequency reduction of

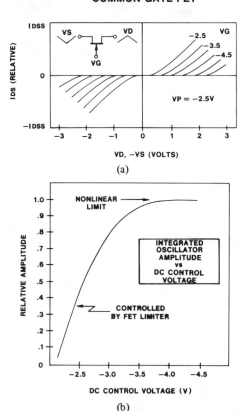

Fig. 4. (a) Balanced common-gate I-V characteristics of a GaAs FET. When V_D and V_S are applied in opposite phases, the device behaves as a clipping nonlinearity which is controllable by gate voltage. (b) The effect of variable clipping nonlinearity on oscillator output amplitude. When the limiter control terminal is more negative than −4 V, the oscillator operates at an amplitude determined by its device non-linearities. The oscillation amplitude can be reduced by a factor of ten by application of appropriate control voltages.

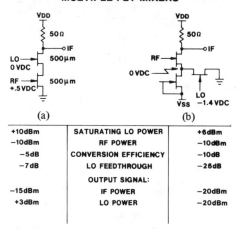

Fig. 5. Multiple FET mixers (gate width = 500 μm). The two-FET, or dual-gate mixer (a) has best conversion efficiency, but large LO feed-through. The three-FET version (b) suppresses LO feedthrough at the expense of conversion efficiency.

about 20 MHz, an amount which can easily be corrected by an offsetting decrease of tuning voltage at node "T."

As would be expected, the low-"Q" oscillator has relatively poor phase noise performance. Observed on a spectrum analyzer at 50-kHz offset from resonance in a 1-kHz bandwidth, the noise is 53 dB below carrier. For comparison, a frequency synthesizer[1] operating at the same frequency and measured the same way, shows a noise sideband 77 dB below carrier. The observed oscillator noise-frequency spectrum shows an amplitude proportional to $1/(\text{frequency offset})^3$, indicating that the spectrum is dominated by $1/f$ noise. This is not too surprising in light of the fact that our GaAs FET's have measured $1/f$ noise intercepts with white-noise floor at frequencies of up to 100 MHz. Three possibilities exist for decreasing the oscillator's phase noise: increasing the resonator "Q," decreasing the $1/f$ device noise, or locking the oscillator to an external standard (using phase lock with digital frequency division or injection locking techniques). A power of −25-dBm incident on the tuning terminal "T" enabled the oscillator to lock to a frequency synthesizer and replicate the synthesizer's output noise performance. Phase locking with GaAs frequency dividers should also be possible.

[1] HP model 8672A.

The Mixer

There are various ways to combine the amplifying and switching properties of the GaAs FET to achieve mixing. One way is to use two FET's in series, or a dual-gate FET, as shown in Fig. 5(a). Here, the lower FET is biased as an amplifier, but the upper FET is used to switch the lower FET's drain voltage from an unsaturated to a saturated condition, thereby effecting a switched gain change. The measured performance of such a two-FET mixer (at F = 1.0 GHz) is shown in Fig. 5(a) to have a 5-dB conversion loss into 50-Ω load with an associated LO feedthrough to IF port of −7 dB. For typical RF input of −10 dBm, this results in LO feedthrough of +3 dBm for IF power of −15 dBm. In a narrow-band IF situation, this presents no problem, since the LO feedthrough is rejected by the IF filter. For a broad-band IF situation, however, a large LO feedthrough can overload the IF amplifier if there is no pre-amplifier filtering to reject the feedthrough. Since it was desired to incorporate the IF preamplifier on the same chip as the mixer (a situation which precluded IF filtering before amplification), LO rejection in the mixer was required. To accomplish this, a three-FET mixer, shown in Fig. 5(b), was developed. This mixer uses an FET as a shunt switch to alternately open and close a path to ground from an amplifier FET's source. A constant current source to a negative supply voltage biases the source of this FET to a quiescent voltage of zero, so that application of an LO signal to the shunt switch results in no current flow in the absence of an RF signal. Consequently, the LO feedthrough for this three-FET mixer is reduced to 26 dB below LO input. The three-FET mixer has a conversion loss of −10 dB into 50 Ω at 1 GHz, resulting in −20-dBm IF output at typical RF input of −10 dBm, which is identical to the saturated level LO feedthrough. Since the IF preamplifier is always operated below saturation, this reduction of LO feedthrough is sufficient to prevent the LO feed-through overload problem.

Greater conversion efficiency as well as further suppression of RF and LO feedthrough can be obtained by operating two of the three-FET mixers in the balanced configuration shown in Fig. 6. In this mixer, the actual circuit used on the hetero-

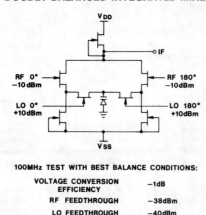

Fig. 6. An active doubly balanced mixer can be realized by combining two three-FET mixers in parallel with the two sides driven 180° out of phase.

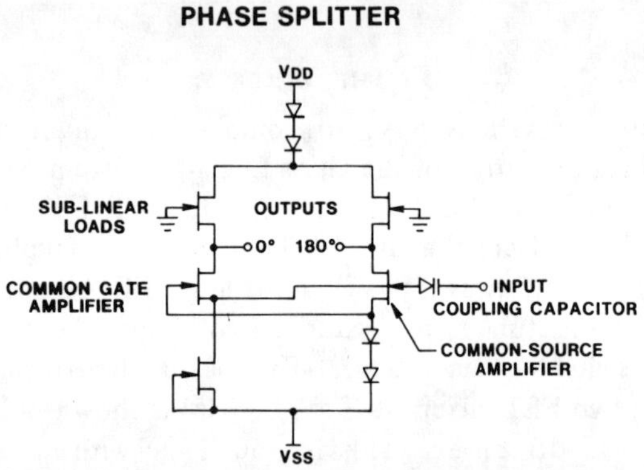

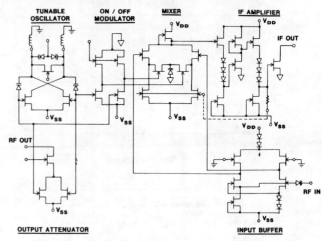

Fig. 8. Complete circuit schematic for heterodyne IC chip.

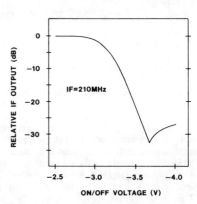

Fig. 7. Active phase splitter. Two unity gain amplifiers, one common-source, the other common-gate, are driven in parallel to convert a single-ended input to balanced output.

Fig. 9. Effect of ON/OFF modulation voltage on IF output. Greater than 30-dB dynamic range can be obtained without disturbing the local oscillator.

dyne chip, the outputs of two mixers (one of which is driven at 0°, the other at 180° RF and LO) are summed into an active load device. The central ground node is an ac ground through a diode capacitor. An integrated mixer cell tested at 100 MHz with optimized dc bias conditions showed -1-dB voltage conversion efficiency, a 6-dB improvement due to balancing, and an additional 3 dB due to higher relative load impedance, as compared to the discrete FET breadboard three-FET mixer. Both RF and LO feedthrough were greatly suppressed by the balancing. At best bias, RF feedthrough was 28 dB below RF input and LO feedthrough was nulled to about 50 dB below input. The depth of this LO feedthrough null was such that a 25-mV offset from best balance resulted in a 25-dB increase in feedthrough, so best balanced operation of this active doubly balanced mixer, which also depends on good amplitude match between 0° and 180° components, was difficult to maintain on the complete chip.

Phase Splitter

The 0° and 180° LO and RF signals required to drive the mixer are derived from unity-gain amplifiers. In the case of the low-level (typically -10-dBm) RF signal, a phase-splitting buffer is used, so that a single-phase input can be delivered to the chip. This circuit, shown in Fig. 7, inverts the input in a common-source amplifier while amplifying the same signal in noninverted fashion through a common-gate FET. The input impedance of the circuit, determined by the 20-μm gatewidth common-gate FET's input impedance, is nominally 500 Ω. The input is capacitively coupled, and the circuit requires no external dc bias. The load elements are common gate "sublinear" load FET's so called because their input impedance to the source node is nonlinear, with current at each operating point less than that of a linear resistor of equivalent average value. This load nonlinearity complements the transconductance nonlinearity of the amplifying FET's, leading to improved harmonic distortion.

ON/OFF Modulator

The 0° and 180° LO signals are derived from the push–pull oscillator, inverted by a dual, non-phase-splitting unity gain amplifier, and fed to the mixer's LO ports. This circuit, shown in the overall chip schematic (Fig. 8) is also used to inject an externally controlled dc bias to the LO port. When this bias is offset negatively with respect to optimum bias, up to 30-dB reduction in mixer efficiency is obtained (Fig. 9). This offset is used to produce ON/OFF modulation of the IF signal without disturbing the local oscillator.

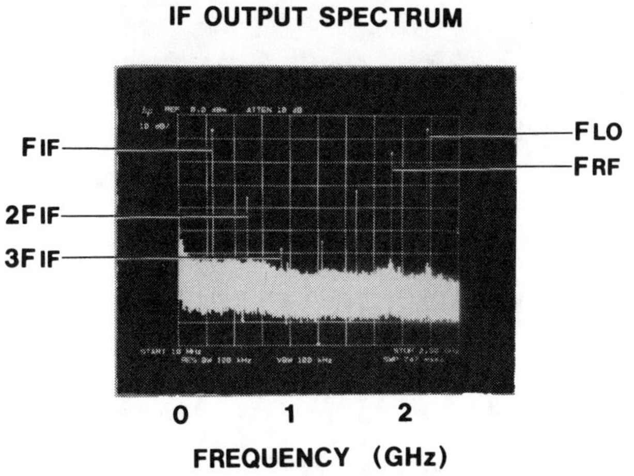

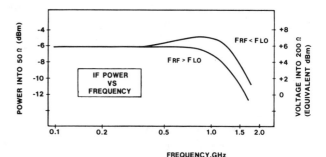

Fig. 10. Frequency spectrum of IF output.

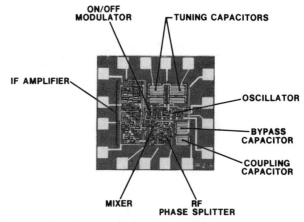

Fig. 11. Amplitude response of IF output as external RF signal is swept above and below LO frequency.

The Complete Chip

A schematic diagram of the complete IC is shown in Fig. 8. In addition to the subcircuits previously discussed, the chip contains an RF attenuator for monitoring the oscillator's output, and an IF preamplifier of the type discussed in [2]. Note that the doubly balanced mixer is also, in effect, the first stage of the two-stage preamplifier. DC bias stabilization of the preamplifier is accomplished by a low-pass filtered external feedback loop indicated by the dotted line in Fig. 8. The dc component of the second-stage output is sensed through a high-value on-chip resistor, compared to an externally set reference voltage at the input to an operational amplifier, and the error signal applied to the constant-current sources in the mixer.

The chip was characterized in a high-frequency probe system with 50-Ω impedance lines at all ports. The IF response spectrum is shown in Fig. 10. In this case, LO frequency is 2.2 GHz, RF frequency is 1.9 GHz, and IF signal is at 300 MHz. The preamplifier output is observed directly into a 50-Ω-input spectrum analyzer, without the intended 200-Ω low-pass filter of Fig. 1. Consequently, the LO and RF feedthrough, as well as spurious mixing products are observed without filtering. The RF input amplitude was adjusted to a point where second-harmonic distortion is 30 dB below the IF fundamental. At this output power level (which is arbitrarily considered the maximum level, as determined by harmonic-distortion requirements) the IF output power is plotted against frequency, Fig. 11. As the RF is swept at frequencies below the 2.2-GHz LO, a slight peaking is observed in the IF response at 1 GHz. In the more typical case, where the swept RF frequency is greater than the LO frequency, the overall IF response is flatter, with a 3-dB bandwidth of 1.4 GHz.

Fig. 12. Chip photo. Active area is 400 μm × 350 μm; chip size is 650 μm × 600 μm.

The circuit, shown in Fig. 12, was fabricated on a 600-μm × 650-μm test chip with a standardized 16 bonding-pad layout. The active circuit area is 350 μm × 400 μm, and contains 29 FET's, 8 diodes, 6 capacitor diodes, and 1 resistor. A large portion of the chip is occupied by capacitors, since the FET sizes are small (10 to 100 μm), and the layout of the active area quite compact. The nominal on-chip power dissipation is 330 mW.

Acknowledgment

The authors wish to thank D. Harkins, D. Hornbuckle, and Carol Coxen for their contributions to this work.

References

[1] R. L. Van Tuyl, "A monolithic GaAs FET RF signal generation chip," in *ISSCC Dig. Tech. Papers*, pp. 118–119, Feb. 1980.
[2] D. B. Hornbuckle and R. L. Van Tuyl, "Monolithic GaAs direct-coupled amplifiers," this issue, pp. 175–182.
[3] R. L. Van Tuyl, C. A. Liechti, R. E. Lee, and E. Gowen, "GaAs MESFET logic with 4-GHz clock rate," *IEEE J. Solid-State Circuits*, vol SC-12, no. 5, Oct. 1977.

Section III-B
Power Amplifiers

MONOLITHIC microwave integrated circuit development of power amplifier circuits has been spurred, primarily, by the potential military applications of MMIC's in active phased-array modules for air- and space-borne radar systems as well as ECM systems. The following papers represent a sampling of such applications of MMIC power amplifiers.

Design, Fabrication, and Characterization of Monolithic Microwave GaAs Power FET Amplifiers

HUA QUEN TSERNG, MEMBER, IEEE, H. MICHAEL MACKSEY, MEMBER, IEEE, AND STEPHEN R. NELSON, MEMBER, IEEE

Abstract—The design, fabrication, and characterization of three- and four-stage monolithic GaAs power FET amplifiers are described. Each of the amplifier chips measures 1 mm × 4 mm. Procedures for characterizing these monolithic amplifiers are outlined. Output powers of up to 1 W with 27-dB gain were achieved with a four-stage design near 9 GHz. The circuit topologies used were flexible enough to allow external bondwires to be used as shunt inductors for amplifier operation at C- or S-bands. An output power of 2 W with 28-dB gain and 36.6-percent power-added efficiency was achieved at 3.5 GHz, using a modified four-stage amplifier.

I. INTRODUCTION

THE DESIGN and performance of monolithic microwave GaAs FET amplifiers have been covered in a number of scientific publications over the past few years. The emergence of monolithic circuit technology is a consequence of the combined results of improved GaAs material quality, proven discrete device performance, and increased understanding of device–circuit interactions. Microwave system requirements are also responsible for the various developmental efforts in this exciting area. Power combiners and low-noise, push–pull, and IF amplifiers fabricated on monolithic chips have been reported [1]–[8] as has the amplifier performance of up to two monolithically cascaded stages [7]. In this paper, the design, fabrication, and characterization of three- and four-stage single-ended monolithic GaAs power FET amplifiers are described. The three-stage amplifier performance goal is 0.5-W output power with 24-dB gain from 9 to 10 GHz. These amplifiers are designed primarily for phased-array radar applications requiring small-size and lightweight modules.

II. AMPLIFIER DESIGN

For the amplifier design, pertinent device equivalent circuit parameters derived from experimentally measured S-parameters were used. Fig. 1 shows a simple FET equivalent circuit model. Circuit element values for 300- and 1200-μm gate-width devices are also shown. The device parameters shown were extensively verified using hybrid MIC amplifier designs up through Ku-band. Large signal R_d values were also obtained (knowing the loads for optimum match) from these amplifier measurements.

Manuscript received September 5, 1980; revised October 14, 1980. This work was supported in part by the U.S. Air Force Avionics Laboratory, Wright Patterson Air Force Base, Ohio, under Contract F33615-78-C-1510.
The authors are with Texas Instruments, Inc., Central Research Laboratories, Dallas, TX 75265.

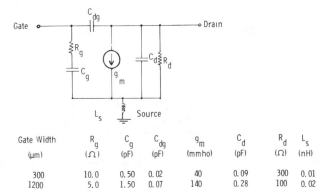

Gate Width (μm)	R_g (Ω)	C_g (pF)	C_{dg} (pF)	g_m (mmho)	C_d (pF)	R_d (Ω)	L_s (nH)
300	10.0	0.50	0.02	40	0.09	300	0.01
1200	5.0	1.50	0.07	140	0.28	100	0.02

Fig. 1. Simplified device model for amplifier design.

Using the equivalent circuit shown in Fig. 1, several cascaded three-stage amplifiers were designed with a CAD technique (TI-CAIN). Since the effects of L_s are small at X-band, it was neglected in the amplifier design. Fig. 2 shows the optimized results of the computer calculation and the circuit topology using distributed high-impedance transmission lines. A characteristic impedance of ~80 Ω (corresponding to a w/h ratio of 0.25 for GaAs) was used for all but the two input lines. Capacitors (8 pF) were used for dc blocking and RF bypass. In the gain–frequency optimization routine, a constant gain with minimum input/output VSWR's was imposed. Although not shown, a similar three-stage design using lumped inductors was also optimized over the same frequency band. Since it is difficult to realize a lumped inductor with transmission line longer than 0.1λ (this is especially true when the high-impedance transmission line is terminated by a relatively high-impedance load such as the output of an FET), the distributed design approach was adopted for experimental realization of the amplifier. The fact that the high-impedance line when loaded by the output of the FET behaves in a distributed fashion will be shown in Section IV, covering characterization of individual FET stages.

III. AMPLIFIER FABRICATION

A. Chip Layout

A photograph of two completed amplifier chips is shown in Fig. 3. These amplifiers have three or four stages with the first two being 300-μm gate-width devices and the third a 1200-μm gate-width device. The last stage of the four-stage amplifier has 2400-μm gate width. The 300-μm devices have 75-μm-wide gate fingers for high gain, but the use of 75-μm fingers on the

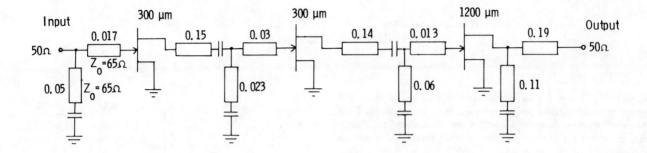

Fig. 2. Circuit topology and calculated gain-frequency response of a three-stage monolithic amplifier using a distributed matching approach.

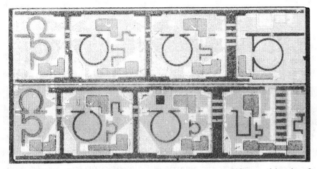

Fig. 3. Circuit layout of three- and four-stage amplifiers; chip size for each amplifier: 1 mm × 4 mm.

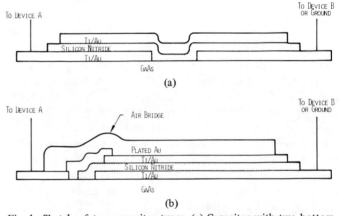

Fig. 4. Sketch of two capacitor types. (a) Capacitor with two bottom plates. (b) Air bridge capacitor.

1200- and 2400-μm devices would be unacceptable as too many fingers would be required. This would make the devices too large for the chip size so these devices have 150-μm fingers. Performance is not seriously degraded since the latter stages operate with considerable gain compression. Both amplifier chips are 1 mm × 4 mm. The amplifier inputs are to transmission lines on the left side of the photograph. The series input transmission line on the mask was inadvertently made too large, and has been corrected on the amplifiers by shorting out the loop gap with a bond wire, providing a good input match (see Fig. 2). The various loops in the photograph are all transmission lines. The FET sources are all interconnected by plated Au air bridges which are grounded by the bars along both sides of the chip. Low source lead inductances are achieved by expanding the sources as they leave the device active regions and by employing a low inductance mesh grounding all along both sides of the chip. Measurements indicate that gain degradation due to source lead inductance is less than 1 dB compared with similar discrete devices where source wires are bonded to ground right at the FET.

Both blocking and bypass capacitors have two bottom plates separated from a single top plate by the dielectric so they are really two capacitors in series as seen in Fig. 4(a). The two portions of the circuit being capacitively coupled are connected to the two bottom plates which are separated by 10 to 20 μm. The bypass capacitors are all near one side of the chips in Fig. 3 and one of the bottom plates is just the large grounded source pad. With this design, it is never necessary to make a connection to the top plate. A disadvantage is the reduced capacitance of this configuration. The capacitors contain bonding pads on both bottom plates so that the transmission lines can be replaced by bond wires to quickly investigate the effects of different inductance values. The gates are biased through the pads connected to the ungrounded bottom plates of the bypass capacitors. An improved capacitor to be discussed later is shown in Fig. 4(b).

B. Fabrication Process

The amplifier fabrication process is very similar to that employed in our laboratory for discrete devices [9], [10]. The GaAs active layers are produced by either vapor phase epitaxy [11] or ion implantation [12]. When vapor phase epitaxy is employed, two layers are grown sequentially on a Cr-doped substrate: An undoped buffer layer 1 to 3 μm thick is followed by the sulfur-doped n-type active layer. The active layer is grown thicker than necessary and then thinned anodically. The ion-implanted active layers are produced by Si^{28} implantation directly into Cr-doped substrates. These layers are proximity annealed [12] at 850°C for 30 min. The peak doping level of both types of active layers is in the range 1 to 2×10^{17} cm^{-3}.

Following device isolation, the AuGe/Ni source and drain ohmic contacts are defined by liftoff. The source-drain separation is 5 μm for all FET's in the amplifier. All of the FET gates are defined by an electron beam machine [13] which automatically aligns the gates to AuGe/Ni alignment marks defined along with the source and drain. Gate lengths of 0.5 to 1 μm have been employed for these amplifiers (all devices on a slice have the same gate length). The gate is 0.05 μm Ti/ 0.05 μm Pt/0.4 μm Au and is recessed 0.05 to 0.1 μm below the epitaxial surface in order to increase device output power [9].

Following gate definition, a 0.05-μm Ti/0.5-μm Au layer is evaporated onto the sources and drains to improve their conductivity. At the same time the gate bonding pads, the inductors, and the capacitor lower plates are defined. All of the inductors are on a separate photomask and although this increases the complication slightly by requiring two exposures of the photoresist, it permits changing the inductor lengths by changing only one coarse-geometry photomask. A 0.4-μm silicon nitride layer is then plasma deposited. This layer serves as the capacitor dielectric and also provides device protection. Capacitor top plates of 0.05 μm Ti/1.0 μm Au are then lifted off. The sources are then interconnected by 2- to 4-μm-thick plated Au air bridges. The first amplifiers were found to have excessive gain degradation due to high resistance of the inductors so the layout was changed slightly to permit plating up of the inductors. This is done simultaneously with the air bridge plating and results in 2- to 4-μm-thick inductors having much lower resistance. If desired, the capacitor top plates can also be plated up by this process (as they were for the devices of Fig. 3).

Some slices have had open capacitors with the present design due to poor step coverage by the top plate at the junction of the two bottom plates. In addition, the capacitance values are only about 2.5 pF, and larger values would improve impedance matching. Consequently, the amplifiers are now being redesigned so that the capacitors have a single bottom plate with inductors connected to the top plate by a plated Au air bridge as shown in Fig. 4(b). The new capacitors should be about 8 pF and be more producible and reliable and have lower losses. All amplifier results reported here involve the old style capacitor.

Following Au plating, the amplifier slices are lapped to 150 μm and the backs are metallized with Cr/Au. The circuits are scribed apart and soldered to Au-plated Cu carriers with 4 percent Ag/96 percent Sn solder (MP 221°C). AuSn solder could not be used due to chip cracking during cooling caused by the thermal expansion of these large chips. To check for other possible amplifier thermal problems, nematic liquid crystals provided a quick, high-resolution method to study the three-stage monolithic amplifier's temperature distribution under dc bias [14]. Nematic liquid crystals having an isotropic temperature of 70°C were placed on top of a three-stage monolithic amplifier; the amplifier was biased at $V_D = 9$ V, $I_D = 300$ mA, $V_G = -1$ V for 30 min and observed. Ambient temperature was 22°C. The 70°C or higher temperature distribution was restricted solely to the three FET's (300, 300, and 1200 μm) and did not extend underneath any of the passive components. No severe amplifier thermal problems have been found to date. Further studies must be made concerning amplifier pulse and CW performance, and mechanical stress, over temperature.

IV. Characterization and Performance

During the early design stage of a monolithic amplifier, several schemes can be used to characterize the amplifier prior to the final amplifier realization. This is especially true when unexpected parasitics and deviation of the active (FET) or passive component values from the design values cause the amplifier performance to differ from the design goals in either bandwidth or gain-output power. In this section, several characterization methods for monolithic circuits in general and a three-stage amplifier in particular are described. These characterization methods have led to performance improvements of three- and four-stage amplifiers fabricated since the first iteration of the mask set design. It is anticipated that with the necessary design changes, final amplifiers can be fabricated to meet the design goals.

A. Discrete FET's

FET's in a monolithic amplifier can be characterized in terms of their power-gain performance, and their impedance levels determined from S-parameter measurements. Discrete FET's can either be incorporated in the amplifier mask as a test pattern or they can be cut from the actual amplifier to perform the above measurements. In the work described in this paper, discrete FET's cut from the amplifier were used for characterization.

As described in Section III, a 300-μm gate-width FET was used in both the first and second stages. A 1200-μm gate-width FET was used in the third stage. For the fourth stage, shown in Fig. 3, a 2400-μm gate-width FET was used. If the discrete FET power-gain performance was found to be satisfactory, FET S-parameters were then measured to determine input/output impedance levels and deviations from the design window.

The input/output S-parameters for discrete FET's taken from a three-stage monolithic amplifier are shown in Fig. 5. These S-parameters are well within the range of experimentally observed variations of different slices and compare favorably with the simplified device model shown in Fig. 1.

B. Passive Components

Transmission lines and inductors and capacitors of various dimensions, including metal and dielectric thicknesses, were constructed on different thickness GaAs substrates and characterized to aid in monolithic amplifier design. Spiral inductors for possible future use as RF bias chokes were also constructed and characterized on alumina and GaAs. Passive component transmission and reflection measurements were made on four

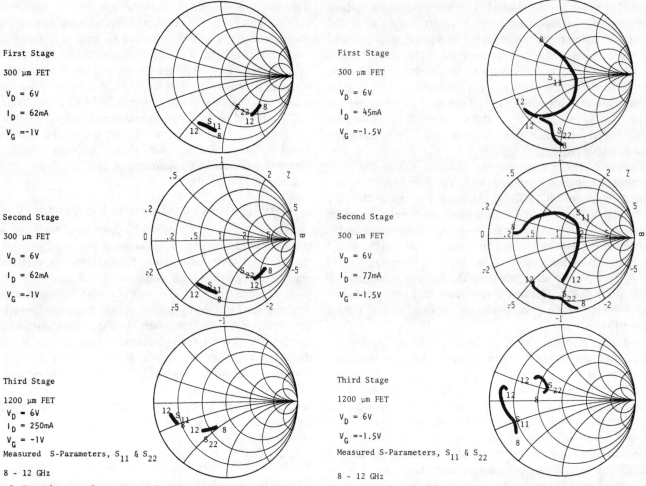

Fig. 5. Input/output S-parameters for discrete FET's taken from a three-stage monolithic amplifier.

Fig. 6. Input/output S-parameters for discrete stages (matching networks and FET's) taken from a three-stage monolithic amplifier.

equipment setups: 1) an automatic network analyzer (L, C measurements), 2) reduced-height waveguide–diode package and microstrip series-resonant DeLoach setups (R, L, C, and Q measurements) [15], 3) a swept amplitude analyzer (insertion and return loss measurements at various power levels), and 4) a time domain reflectometer (L, C, and Z_0 measurements). DC current leakage and breakdown voltage measurements were made on a curve tracer.

Numerous articles can be found in the literature dealing with microstrip lines and various loss mechanisms [16], with single-loop and spiral inductors [7], [17], and MIM capacitors.

Major inductor, transmission line, and capacitor losses have centered around conductor thickness. For conductor thicknesses of 0.5 to 1.0 μm and inductance values between 0.2 to 1.5 nH, inductor Q's were measured at 10 or below at X-band. Increasing conductor thickness to 2 to 3 μm or higher, these same inductors exhibited Q's in the 20 to 30 range. Gold inductors and transmission lines are normally evaporated onto semi-insulating GaAs substrates to a thickness of 0.5 μm and then gold plated to 2-μm thickness or more to reduce RF skin losses. Plating, however, increases conductor surface roughness. Ground-plane roughness and substrate conductance also enter into the loss picture. Experimental optimization regarding these factors is not complete.

Linewidths between 25 and 40 μm were used in the inductor/transmission line experiments. Substrate thicknesses of 0.11, 0.15, 0.20, and 0.33 mm were also used in the tests. Keeping inductor linewidths and lengths constant, inductance and Q increased with substrate thickness less than theoretically expected. Theory and experiment have yet to agree within 10 percent. Substrate thickness was chosen at 0.15 mm as a compromise between chip mechanical strength, device thermal resistance, and inductance and Q.

Q's greater than 50 at X-band have been achieved with small ($\sim$1-pF) MIM plated capacitors and it is expected that the new larger capacitors will repeat this performance. Unplated $\sim$2.5-pF capacitors cut from actual amplifiers exhibited Q's of 30 or less.

Silicon nitride (4000 Å thick) breakdown voltages have generally exceeded 150 V. Good GaAs substrate breakdown voltages have been well above 50 V.

C. Discrete FET Stages

Initially, the monolithic three-stage amplifiers as fabricated, generally had gains of only between 17 and 19 dB with +3-dBm input, even though individual devices (without matching networks) cut from the slices showed adequate gains and output powers. To determine the cause(s) for the lower than expected gains, the S-parameters (specifically, S_{11} and S_{22}) of the individual stages, including input/output matching networks, were measured and are shown in Fig. 6. Output resistances of $\sim$10 Ω measured for the first and second stages were mainly due to the distributed nature of the output matching transmission lines. To determine the quality of the interstage matching under actual operating conditions, a light-emission study, to be discussed later, was conducted on the three-stage amplifier.

Passive component and FET sensitivity analyses were performed on the distributed matching model three-stage monolithic GaAs FET amplifier during its initial design and at later intervals. These analyses were done using the TI-CAIN computer program, but could have been performed equally well on COMPACT.

Amplifier gain and input/output VSWR sensitivity to ±10- and ±25-percent changes in single or grouped FET internal components and transmission line parameters were studied over 8 to 11 GHz. Nominal calculated amplifier peak gain was approximately 30 dB. The most gain-sensitive passive components were transmission-line impedances and lengths surrounding the second-stage FET. Amplifier gain sensitivity to FET internal components began with g_m, followed in decreasing order of importance by C_g, R_d, and R_g. Changes of ±10 percent in g_m of any one FET resulted in approximately a ±1-dB amplifier gain change. Component changes in the first FET had the strongest effect on amplifier gain. As expected, first- and third-stage components or transmission lines had the strongest effects on input and output VSWR's, respectively.

Sensitivity analyses are important during the design stage of any amplifier. Choosing low sensitivity (gain, VSWR, etc.) matching networks for monolithic amplifiers is challenging because of amplifier size restrictions.

D. Light Emission

Light emission from microwave power GaAs FET's under RF operating conditions has been reported by Tserng et al. [18]. It was shown that emitted light intensity could be correlated with FET RF input drive levels and output power saturation characteristics. This observation has been used to investigate the interstage matching of a three-stage monolithic amplifier.

Multistage amplifier FET light-emission study holds promise as a method of determining the quality of interstage matching, as well as providing information about FET compression characteristics and gate finger uniformity and defects. Cases exist, however, where light emission might be misleading. FET's having high gate-drain breakdown voltage do not emit light as readily as lower gate-drain breakdown FET's under similar dc bias and RF drive levels. This must be considered in analyzing the light-emission results.

Initial unmodified three-stage monolithic amplifiers showed only 22-dB gain with -3 dBm in, 19-dB gain with +3 dBm in, and 16-dB gain with +7 dBm in at 10 GHz and $V_D = 6$ V, $V_G = -1.5$ V dc bias. Light emission studies on these amplifiers showed the second-stage 300-μm FET's to emit light with +3-dBm amplifier input power; the light emission became more intense with +7 dBm in to the amplifier. Third-stage 1200-μm FET's emitted little or no light under either amplifier input power condition.

Discrete 300-μm FET's were found to normally begin emitting visible light under similar bias conditions at 12-dBm input (8- to 9-dB gain), while discrete 1200-μm FET's normally began emitting light at +20-dBm input (7-dB gain).

To be driven into light emission, the second-stage FET's, if good, were receiving between 12 and 14 dBm in from the first-stage FET's (+3 dBm in, 9- to 11-dB gain). Apparently, the first- and second-stage FET's were well matched. However, since the third-stage FET's did not emit light, further improvement was needed on the second-third interstage matching network. These conclusions were verified experimentally from single-stage S-parameter measurements, two-stage monolithic amplifier performance results, and third-stage matching network modifications which increased gain and improved third-stage FET light emission.

To isolate the matching problem between the second and the third stage, the gain-frequency response of a two-stage monolithic amplifier was obtained by cleaving off the third-stage FET and the gate shunt inductors (at plane AA', Fig. 7(a)). With no modification (except the output tuning required to match the 50-Ω test system), and with +3-dBm input power, a gain of 17 to 18 dB (output power of 100 to 125 mW) was achieved. This result substantiated the light-emission study described above.

The design gain and output power of the two-stage amplifier were usually obtained with drain voltages of between 5 and 6 V. The output resistance of the second-stage (including the loop transmission line and the series blocking capacitor) was ~10 Ω at mid X-band. This resistance level was inferred from the load impedance required for maximum gain and output power. The S-parameter data shown in Fig. 6 also confirm these results. The circuit topology for a two-stage amplifier including the capacitors and estimated parasitic capacitances are shown in Fig. 7(a). The calculated gain-frequency performance (including a computed output resistance of ~10 Ω at plane AA'), across the 9- to 10-GHz band is shown in Fig. 7(b) and (c). With a 50-Ω load at the second-stage output, a two-stage amplifier gain of ~19 dB was calculated. Reducing the output VSWR by placing an external 20-Ω transformer at the output, increased gain to ~21 dB. The experimental result of 19.5-dB gain is shown in Fig. 8(a). Discrepancy between theory and experiment is due to the low value bypass capacitors and FET deviations from the FET models used.

E. RF Performance

The measurements described in the previous paragraphs demonstrated that the reduced gain of the three-stage amplifier was due to nonoptimum transmission-line lengths at the input of the third stage. These limits were replaced with band wires. Fig. 9 shows the gain-frequency response of the modified three-stage amplifier with -3-dBm input. A gain of 27.6 dB was obtained with 290-mW output at 9.2 GHz. With increased input drive, 400-mW output with 23-dB gain was achieved at 9.2 GHz. The power-added efficiency was 15 percent. This result compares favorably with a hybrid amplifier using similar FET's [19].

Few noise measurements have been made on an entire three-stage monolithic amplifier to date. Although low-noise operation was not a design goal, initial noise measurements on the monolithic amplifier compared favorably with those of similar hybrid amplifiers. At 10 GHz and a dc bias of $V_D = 3.85$ V, $I_D = 247$ mA, $V_G = -1.26$ V, one amplifier has been measured to have a noise figure of 6.4 dB with 24.8-dB (small-signal)

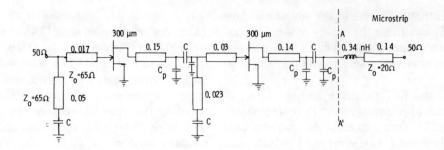

FREQUENCY	GAIN(DB)	PHASE	IN VSWR	RIN	XIN	OUT VSWR	ROUT	XOUT
9.000	19.43	-117.9	1.008	50.65	32.01	4.893	10.82	-11.88
9.200	19.40	-138.0	1.529	50.98	21.57	4.698	11.17	-10.87
9.400	19.17	-156.0	1.369	49.25	15.63	4.627	11.22	-9.49
9.600	19.02	-172.0	1.250	48.00	10.93	4.592	11.16	-7.0
9.800	19.01	171.5	1.138	47.02	5.994	4.484	11.28	-5.238
10.00	18.85	156.2	1.067	46.97	0.9052	4.266	11.76	-2.707

(b)

FREQUENCY	GAIN(DB)	PHASE	IN VSWR	RIN	XIN	OUT VSWR	ROUT	XOUT
9.000	21.55	176.4	1.933	43.66	30.61	1.755	33.31	16.19
9.200	21.65	152.5	1.756	45.28	26.73	1.498	36.92	14.12
9.400	21.44	130.0	1.599	48.95	23.43	1.262	46.29	10.57
9.600	21.32	109.2	1.397	54.22	16.97	1.111	55.03	2.259
9.800	21.10	86.07	1.177	56.22	6.001	1.357	59.07	-13.97
10.00	20.41	64.65	1.102	52.40	-4.329	1.765	51.47	-29.30

(c)

Fig. 7. Circuit topology and calculated performance of a two-stage amplifier. (a) Circuit topology. (b) Calculated gain and output impedance (at plane AA'). (c) Calculated gain with 20-Ω microstrip impedance transformer at the output.

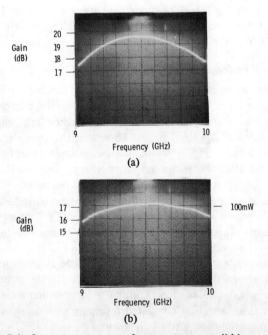

Fig. 8. Gain–frequency response of a two-stage monolithic amplifier. (a) RF input = –3 dBm, V_D = 4.6 V, I_D = 100 mA, V_G = –1 V. (b) RF input = +3 dBm, V_D = 5.5 V, I_D = 110 mA, V_G = –1 V.

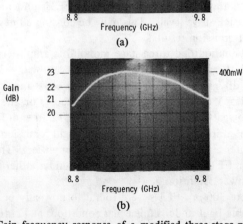

Fig. 9. Gain–frequency response of a modified three-stage monolithic amplifier. (a) RF input = –3 dBm, V_D = 7.0 V, I_D = 340 mA, V_G = –1 V. (b) RF input = +3 dBm, V_D = 7.7 V, I_D = 346 mA, V_G = –1 V.

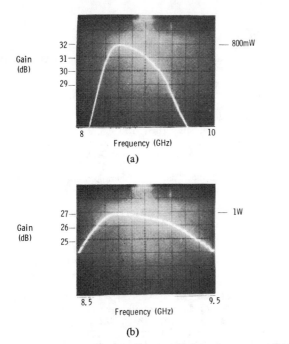

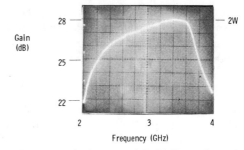

Fig. 11. Performance of a four-stage hybrid discrete/monolithic amplifier. RF input = +5 dBm, V_D = 9.3 V, I_D = 588 mA, power-added efficiency = 36.6 percent.

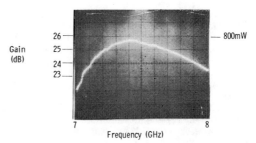

Fig. 10. Gain–frequency response of a modified four-stage monolithic amplifier. (a) RF input = −3 dBm, V_D = 7.8 V, I_D = 780 mA, V_G = −1.7 V. (b) RF input = +3 dBm, V_D = 9.0 V, I_D = 650 mA, V_G = −2.4 V.

Fig. 12. Performance of a four-stage hybrid discrete/monolithic amplifier. RF input = +3 dBm, V_D = 6.0 V, I_D = 570 mA, V_G = −2 V, power-added efficiency = 22 percent.

associated gain. With circuit (low-noise match to first stage) and FET modifications, amplifier noise figures less than 5 dB should be possible.

The circuit layout for the four-stage amplifier (Fig. 3) is similar to that of the three-stage amplifier. A mistake in the mask design was made in the bypass capacitors in the fourth stage, resulting in a strong oscillation near 6 GHz. This was due to the inadvertent use of a common top plate for the bypass capacitors of the fourth-stage gate and drain. Because of the feedback effect, an oscillation resulted. This was corrected by scratching off the drain shunt inductor of the fourth stage, which necessitated the use of an external microstrip transformer for matching to the 50-Ω load. Fig. 10 shows the performance of the modified four-stage amplifier. An output power of 800 mW was obtained with 32-dB gain at 8.7 GHz. With increased drive at +3-dBm input, 1-W output was obtained with 27-dB gain at 8.9 GHz. The 1-dB bandwidth was 600 MHz (8.6 to 9.2 GHz). Further increase of the input power to +5 dBm resulted in an output power slightly over 1 W with a 1-dB bandwidth of 8.6 to 9.5 GHz.

Using similar monolithic amplifier chips but with additional modifications, the amplifiers were made to operate at other frequency bands. For example, a four-stage monolithic chip was modified by scrubbing off all the shunt elements involving shunt capacitors. External shunt inductors using gold bond wires were used in each of the drain circuits along with external RF bypass capacitors. Fig. 11 shows the gain-frequency response of such an amplifier. An output power of 2 W with 28-dB gain and 36.6-percent power-added efficiency was achieved at 3.5 GHz. The 1-dB bandwidth was approximately 1 GHz. When the external inductors were readjusted for 7- to 8-GHz operation using another four-stage chip, the gain–frequency response of Fig. 12 resulted. Since the interstage matchings probably were still not optimum, the output power was lower than expected. An output power of 750 mW with 26-dB gain was obtained at 7.5 GHz. The power-added efficiency was 22 percent. This amplifier might have application in the satellite communication band.

Performances of amplifiers shown in Figs. 11 and 12 at S- and C-bands illustrate the flexibility of the monolithic amplifier design to be shifted to different operating frequency bands. Design information obtained as previously discussed is useful prior to finalizing circuit-element values to cover different design frequency bands. At lower frequencies, required matching inductors might be prohibitively large to be monolithically integrated with the FET's. The above results have shown that some sort of hybrid discrete/monolithic approach might be attractive from the standpoints of design flexibility and minimization of amplifier size. This is especially true at frequencies below C-band.

V. Conclusions

In this paper, the design, fabrication, and characterization of multistage single-ended monolithic GaAs FET amplifiers were described. Procedures for characterizing monolithic amplifiers were outlined, using these multistage amplifiers as examples. These procedures are particularly useful during the early design stage, when unaccounted for parasitics and device and/or circuit parameter variations might cause amplifier performance to deviate from design goals. As a result of this characterization process, improved performances of three- and four-stage amplifiers were obtained at the design frequency band. Output powers of up to 1 W with ∼27-dB gain were achieved with a four-stage amplifier design near 9 GHz. As shown, the circuit topologies used were flexible enough to allow external bond wires to be used as shunt inductors for amplifier operation at C- or S-bands. Truly monolithic amplifier versions could then be designed using circuit topology information thus obtained.

Second-generation photomasks incorporating the necessary design changes for the *X*-band monolithic amplifiers described here have been generated and should result in improved performance without modifications.

ACKNOWLEDGMENT

The authors wish to thank W. R. Wisseman, V. Sokolov, R. E. Williams, G. E. Brehm, and M. R. Namordi for helpful discussions and process support. They also wish to thank F. H. Doerbeck and W. M. Duncan for supplying the GaAs material, and S. F. Goodman and L. P. Graff for technical assistance.

REFERENCES

[1] R. S. Pengelly and D. C. Rickard, "Design, measurement, and application of lumped elements up to J-band," in *7th Europ. Microwave Conf. Proc.* (Copenhagen, Denmark), pp. 460-464, 1977.
[2] R. A. Pucel, P. Ng, and J. Vorhaus, "An X-band GaAs FET monolithic power amplifier," in *1979 MTT-S Internat. Microwave Symp. Dig.*, pp. 387-389, Apr./May 1979.
[3] D. A. Abbott, J. Cockrill, R. S. Pengelly, M. G. Stubbs, and J. A. Turner, "Monolithic gallium arsenide circuits show great promise," *Microwave Syst. News*, pp. 73-96, Aug. 1979.
[4] R. S. Pengelly, "Monolithic GaAs ICs tackle analog tasks," *Microwaves*, pp. 56-65, July 1979.
[5] R. A. Pucel, "Performance of GaAs MESFET power combiner," *Microwave J.*, pp. 51-56, Mar. 1980.
[6] J. E. Degenford, M. Cohn, R. R. Freitag, and D. C. Boire, "Processing tolerance and trim considerations in monolithic FET amplifiers," in *Dig. Techn. Papers, 1980 IEEE Int. Solid-State Circuits Conf.*, pp. 120-121.
[7] V. Sokolov and R. E. Williams, "Development of GaAs monolithic power amplifiers in X-band," *IEEE Trans. Electron Devices*, vol. ED-27, pp. 1164-1171, June 1980.
[8] D. R. Chen and D. R. Decker, "MMIC's—The next generation of microwave components," *Microwave J.*, pp. 67-78, May 1980.
[9] H. M. Macksey, F. H. Doerbeck, and R. C. Vail, "Optimization of GaAs power MESFET device and material parameters for 15 GHz operation," *IEEE Trans. Electron Devices*, vol. ED-27, pp. 467-471, Feb. 1980.
[10] G. E. Brehm, F. H. Doerbeck, W. R. Frensley, H. M. Macksey, and R. E. Williams, "High yield reproducible process techniques for microwave GaAs FETs," in *Proc. 1979 Cornell Conf. on Active Microwave Semiconductor Devices and Circuits*, pp. 157-163, 1980.
[11] F. H. Doerbeck, "Materials technology for X-band power GaAs FETs with uniform current characteristics," *Inst. Phys. Conf. Ser.*, no. 45, pp. 335-341, 1979.
[12] F. H. Doerbeck, H. M. Macksey, G. E. Brehm, and W. R. Frensley, "Ion-implanted GaAs X-band power FETs," *Electron. Lett.*, vol. 15, pp. 576-578, 30 Aug. 1979.
[13] T. G. Blocker, H. M. Macksey, and F. H. Doerbeck, "Electron beam fabrication of submicron gates for GaAs FETs," *J. Vac. Sci. Technol.*, vol. 15, pp. 965-968, May/June 1978.
[14] H. M. Macksey, unpublished results.
[15] B. C. DeLoach, "A new microwave measurement technique to characterize diodes and an 800 Gc Cutoff frequency varactor at zero volts bias," *IEEE Trans. Microwave Theory Tech.*, vol. MTT-12, pp. 15-20, Jan. 1964.
[16] E. J. Denlinger, "Losses of microstrip lines," *IEEE Trans. Microwave Theory Tech.*, vol. MTT-28, pp. 513-522, June 1980.
[17] M. Caulton, "Lumped elements in microwave integrated circuits," in *Advances in Microwaves*, vol. 8, L. Young, Ed. New York, Academic Press, 1974, ch. 4.
[18] H. Q. Tserng, W. R. Frensley, and P. Saunier, "Light emission of GaAs power MESFETs under RF drive," *IEEE Electron Device Lett.*, vol. EDL-1, pp. 20-21, Feb. 1980.
[19] H. Q. Tserng, "Design and performance of microwave power GaAs FET amplifiers," *Microwave J.*, pp. 94-100, June 1979.

MONOLITHIC BROADBAND POWER AMPLIFIER AT X-BAND

A. Platzker, Raytheon Missile Systems Division, Bedford, MA 01730

M.S. Durschlag, J. Vorhaus, Raytheon Research Division, Lexington, MA 02173

ABSTRACT

Single-ended broadband power amplifier for X-band operation was designed and fabricated on GaAs with chip dimensions of 4.3 x 2.1 x 0.1 mm. The amplifier exhibited over 2 GHz of 1 db small signal bandwidth with more than 1.6 W at 9 db gain at mid-band CW operation with 20% power added effieicnty. The recently developed, potentially high yielding, Ta_2O_5 capacitor technology enabled the small chip size.

CHIP DESCRIPTION

We have fabricated a single-ended cascaded two-stage power amplifier for X-band broadband operation on GaAs. The chip dimensions are 4.3 x 2.1 x .1 mm. The FET devices have total peripheries of 2 mm and 4 mm with unit gate width of 200μm. The nominal gate to gate and source to drain separations are respectively 30μm and 6.8μm. A photo of a chip is shown in Fig. 1.

The 1μm gate is recessed with a contact N^+ layer .3μm thick doped to 2×10^{18} ions/cm³ reaching to the edge of the narrow recess. The active channels are processed in an epi layer .4 μm thick doped to 9×10^{16} ions/cm³ grown on a buffer layer 1.6μm thick on semi-insulating Cr-doped GaAs substrate. The individual source fingers are connected by air bridges which are raised 2.5μm above a .5μm thick polyimide passivating film. Via holes are used to ground the two ends of the sources as well as the bypass capacitors. The gate metallization is Ti/Pt/Au while the transmission lines are Au plated to a nominal thickness of 3μm.

The amplifier is designed to allow power-combining of up to four units. To this end it is symmetric along its input/output axis; DC biases can be applied from either side. Separate gate and drain bias pads are provided to each stage. Three Si_3N_4 decoupling/matching capacitors of values 1.55, .63 and 1.36 pF are provided in the input, interstage and output matching networks. Eight bypassing Ta_2O_5 capacitors, 50 pF each, are also provided.

POWER AMPLIFIER DESIGN

The unique feature of a power amplifier, in contrast to a small signal amplifier, is the non-linear behavior of the active FET devices. Both the input and output loads required for optimum power match are nonlinear functions of drive level. Two basic approaches to the design problem are possible.

One straightforward approach is to measure the output load presented to the FET as a function of the power delivered to the load. This is the load-pull method. Its main disadvantages are two: first, the measurements are tedious with inherently low accuracy, and second and most important, the method is very difficult to implement for the design of a broadband multistage power amplifier. Since no analytical model is associated with the load-pull measurements, analytical performance analysis, needed for design with actual non-unilateral FETs with fabrication tolerances and variations, is extremely difficult.

The second approach which we have adopted, is to develop a large signal, physically rooted model of the FET devices and use analytical methods to design the two-stage amplifier. The model which we used is the one presented by A. Platzker and Y. Tajima[1] which in turn is an extension of an earlier model by Y. Tajima et al[2]. The model which is described by a set number of parameters, is based on small signal S-parameter measurements, DC I-V curves and the physics of the FET devices. An in-house CAD program, LSFET, was developed to analytically calculate the power performance of discrete FET devices as well as multi-stage (up to two stages) power amplifiers as a function of power. This program as well as the commercially available COMPACT program, were used extensively in our design.

The steps which we follow in designing two-stage power amplifiers begin with the output and progress toward the input. Upon selecting the periphery of the output FET, in accordance with its capability to deliver the required power, its optimum output impedance is determined by LSFET.

In the next step, the output matching stage is designed. The function of any of the three matching networks, input, interstage and output is to match the output impedance which terminates it, to the required input impedance while at the same time allowing for the introduction of the proper DC biases. This dual function should be performed with minimum interference, i.e. with maximum isolation between the RF and DC circuitry. The cut off frequency should be low enough to suppress the natural tendency of circuits with even medium periphery devices to oscillate at far out of band low frequencies. In the case of the output matching

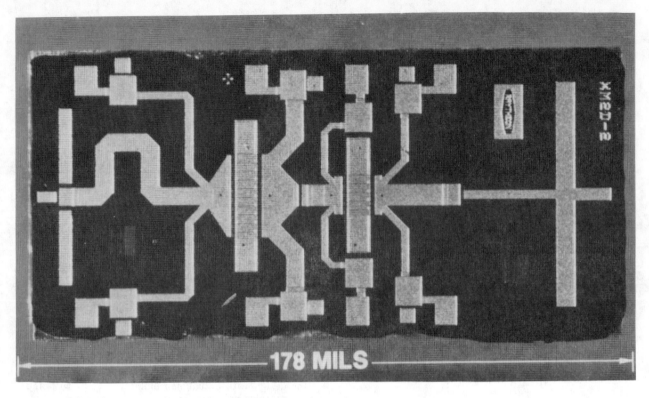

Fig. 1. Photograph of X-band power amplifier chip.

network, it should be designed to present to the output FET the optimum impedance it requires while the network is terminated with 50Ω. This step is best performed by a program such as COMPACT. Once a circuit with realizable element values is obtained, we proceed to the next design step which is the determination of the required input impedance to be presented to the output FET. This is done again with LSFET.

The next few steps are a repitition of the previously described ones but with respect to the driver FET in the first amplifier stage.

The last step in the amplifier design is the determination of the input matching network. This is done by optimizing the overall small signal performance and input match of the amplifier. During this last step, the interstage and output elements are naturally kept constant. A final verification of the power performance of the circuit is done with LSFET.

In selecting the topology of any of the matching networks, proper attention should be given to the special constraints presented by the MMIC technology. In particular, line impedances are limited to the range of 20-80Ω and capacitor values cannot be produced with tolerances of less than ± 15% over a large wafer. Two particular requirements arise in the case of large periphery devices. DC current carrying lines should be capable of handling the maximum allowable mil spec current density of .5 Mega Amp/cm^2 and large bypass capacitors on the gates and drains should be used. In our case we used Ta_2O_5 dielectric capacitors whose specific capacity is 10 times larger than the commonly used Si_3N_4 capacitors. Their availability enabled us the usage of 50 pF bypass capacitors which insured negligible bias interference. The minimum acceptable values for adequately bypassing the 4 mm periphery devices were found to be 30 pF.

For decoupling purposes, however, Ta_2O_5 capacitors cannot be used since by the very nature of their high capacity, their minimum practical value is 10 pF. Where smaller values were needed, Si_3N_4 capacitors were used. In our circuit, we used three Nitride decoupling capacitors and eight Tantalum bypassing ones. The details of the Tantalum process were given by M. Durschlag and J. Vorhaus[3].

AMPLIFIER PERFORMANCE

The amplifier was designed to operate over the 8.5 - 10.5 GHz range. The small signal performance was measured over the range of 6-12 GHz while the power performance was measured over the 8.5 - 11 GHz range. The measured power performance at small signal and at the 1 db compression points are presented in Fig. 2. The maximum power of 32.1 dbm at 9 db gain was obtained at 9 GHz tapering to 31.5 db at 8.5 GHz and 9.5 GHz with the same gain. The 1 db compression at 10 GHz was at 30.4 dbm with a gain of 7 db. At 6 db gain, 31.3 dbm were achieved at 10 GHz. All the numbers quoted above pertain to an assembled amplifier unit with jig losses estimated conservatively at 1 db not subtracted. The actual chip performance should be increased by 1 db in gain and 1/2 db in power. The

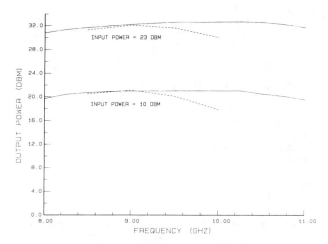

Fig. 2. Small signal and power performance. Solid line - original design, dashed-measurement.

power added efficiency of the CW performance was 20% with drain biases of ~8.5 V and a total DC current of 800 mA.

The small signal performance is shown in Fig. 3. As can be seen in the figure, the small signal 1 db

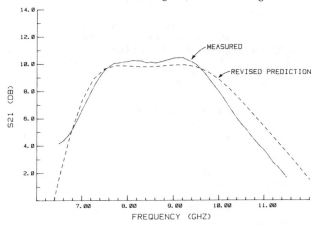

Fig. 3. Small signal performance.

bandwidth is over 2 GHz with a gain of 10 db. It can also be seen from Figs. 2 and 3 that the performance is shifted downwards in frequency with the operating band somewhat smaller than originally designed.

Several causes contributed to the observed deviation in the amplifier performance. The combined effect of these causes led to the revised performance perdiction shown in Fig. 3. The excellent agreement between it and the observed performance is an object lesson in the power and importance of computer analysis in MMIC design.

The amplifier was originally designed with non-passivated devices which were characterized and modelled. When test devices, processed on the same wafers as the amplifiers, were analyzed, it was found that their element values changed somewhat with the main difference, as anticipated, in the values of S12, specifically, the magnitude increased by roughly a factor of two. No agreement, however, was obtained when the new FET models were used in the circuit. Increasing the values of the Si_3N_4 capacitors by roughly 40% contributed towards closing the gap between prediction and observation. In subsequent measurements we were able to determine that a subtle unforeseen new processing step was indeed responsible for this increase in the coupling capacitors -- an increase well outside the tolerances allowed for the variation in these components. A third contributor was the fact that the wafers were erroneously thinned to 75 μm instead of 100 μm. A final contributor was the test jig. With all four contributions, the revised performance prediction is seen to be, as already stated, in excellent agreement with the actual observation.

CONCLUSIONS

We have demonstrated the feasibility of processing broadband single-ended high power monolithic amplifiers for X-band operation with high power added efficiency. The sources of deviations of performance between predictions and acutal observation were identified and corrective steps were being taken to eliminate them in future circuits.

REFERENCES

1. A. Platzker, Y. Tajima, "A Large Signal GaAs Amplifier CAD Program," 1982 IEEE MTT-S International Microwave Symposium Digest, pp. 450-452.

2. Y. Tajima, et al, "GaAs FET Large-Signal Model and Its Application to Circuit Designs," IEEE Trans. ED-28, No. 2, pp. 171-175 (1981).

3. M. Durschlag, J.L. Vorhaus, "A Tantalum-Based Process for MMIC On-Chip, Thin-Film Components," GaAs IC Symposium, New Orleans, pp. 146, (1982).

WIDEBAND 3W AMPLIFIER EMPLOYING CLUSTER MATCHING

R. G. Freitag, J. E. Degenford, D. C. Boire, M. C. Driver*,
R. A. Wickstrom* and C. D. Chang*

Westinghouse Defense and Electronics Center
Baltimore, Maryland

*Westinghouse R&D Center
Pittsburgh, Pennsylvania

ABSTRACT

High power, broadband monolithic amplifiers have inherent performance limitations and require specialized fabrication and design techniques to ensure optimal performance and high yield. The "cell cluster matching" design approach is presented as a solution to this power-bandwidth problem along with a detailed discussion of the high yield fabrication techniques used to implement it. In particular, a 3 W 8-12 GHz amplifier, employing the cluster matching technique, is discussed and some initial test results are presented.

INTRODUCTION

A fundamental concern to monolithic IC amplifier designers is determining the maximum power output-bandwidth product achievable consistent with acceptable yield and good performance. For the 3 W 8-12 GHz amplifier discussed herein (see Figure 1) the output stage requires a total gate periphery of $6400 \mu m$. With such a large FET, one in faced with several inherent performance limiting problems. First, if the $6400 \mu m$ FET were composed of 32 directly combined $200 \mu m$ cells, the low input impedance (1 to 2 ohms) and optimum output load impedance (5 to 6 ohms) would result in a difficult impedance transforming problem over a broad bandwidth (8-12 GHz) and substantial circuit losses. More importantly, in a FET this large, phase differences between the cells close to the center feed point and those farthest away from the feed point cause a significant reduction in FET gain and power output. This is particularly true at X-band frequencies and above. In addition, odd mode signals can be generated due to process variations inevitable across such a large FET structure. These odd modes contribute to further gain reduction and to the possibility of intercell oscillations. Careful circuit design considerations must be made to minimize these deleterious effects and maximize the power bandwidth product.

CELL CLUSTER MATCHING

An alternative to direct cell combining, as mentioned above, is to use a "cell cluster matching" design approach in conjunction with planar power combining techniques (see Figures 3 & 5). In this approach, the $200 \mu m$ cells are grouped together into eight separate $800 \mu m$ cell clusters. Note that matching is done at the $800 \mu m$ cell cluster level where impedance levels are high. Each cell cluster is partially matched before combining, thereby resulting in an overall higher impedance level at the combining points and a corresponding improvement in performance over a broad bandwidth. Each cell cluster is first "double tuned" to a "convenient" resistive impedance level and combined with another partially matched cell cluster. A $\lambda/4$ transformer then transforms the combined resistive impedance level of each of the four cluster pairs to 200Ω. A "convenient" impedance level is defined as one which allows a practical line impedance to be used as the $\lambda/4$ transformer while not severely complicating the initial double tuning circuitry. The four cell cluster pairs, now transformed to 200Ω, are combined to produce the desired 50Ω match.

Note that since the cell cluster matching circuits are identical, the phase angles of the wavefronts reaching each cluster are identical thereby eliminating intercell phasing problems. Figure 2 shows an example of the impedance progression just described on the output or drain side of a single stage, 3 W amplifier using four $1600 \mu m$ cell clusters. The intermediate impedance level in this design is 17Ω corresponding to a 59Ω $\lambda/4$ transformer.

By placing resistors, whose values are equal to the resistive impedance level looking into each partially matched cell cluster pair, at the ends of each $\lambda/4$ line and tying the other end of each resistor to a common node, a four-way Wilkinson power combiner/splitter is formed. This type of combiner has good isolation characteristics over reasonably broad bandwidths. Implanted resistors are used and are fabricated at the same time as the FET active channel. This isolation feature is used to eliminate odd mode oscillation tendencies and/or make the circuit tolerant to processing variations. Unfortunately, on a planar structure such as a monolithic amplifier, long inductive bondwires and airbridges must be used to connect the isolation resistors together. This inductance causes the isolation of the combiner to decrease with increasing frequency. To eliminate this effect, series capacitors have been added at the ends of the resistors (Figure 3) to resonate out the inductive effects at midband (10 GHz) and better approximate a short circuit. The isolation characteristics of this "resonant node" structure were calculated for the various port to port combinations. The isolation calculated is that seen by the FETs (i.e., FET capacitance and the initial matching circuitry are included). For the worst case (two outer ports) of the input and output combiners, isolations between 25 and 50 dB were obtained across the band.

AMPLIFIER FABRICATION TECHNIQUES

A direct selective ion implantation process is used at Westinghouse for fabricating power FETs and IC's and is routinely yielding power FETs with power outputs of 0.5 - 0.7 W/mm of gate periphery at 12 GHz. An outline of this fabrication procedure is given in the following paragraphs.

Material Growth

The gallium arsenide semi-insulating substrates used for the fabrication of monolithic integrated circuits are produced in-house from <100> oriented, single crystals grown by the liqud encapsulated Czochralski (LEC) technique using a Melbourn high-pressure puller (Metal Res. Ltd.). The crystals are pulled from the melt contained in a pyrolytic boron nitride (PBN) crucible following compounding in-situ; a liquid B_2O_3 encapsulant and inert gas over-pressure are employed to prevent As sublimation which would result in low resistance, nonstoichio metric crystals.

Careful elimination of electrically active impurities, paricularly Si which can be introduced by SiO_2 crucibles, eliminates the need to counter-dope with Cr to compensate residual shallow donors. These crystals exhibit state-of-the-art chemical purity and minimize residual impurity activation during processing. As grown, the material exhibits sheet resistivities (R_s) of $3 \times 10^8 \Omega$/square and mobilities of 5000 cm^2/volt-sec.

Ion Implantation

Implantation processing originates with plasma deposition of a 900 Å Si_3N_4 primary encapsulation layer on the front surface. This layer remains on the surface through processing to prevent mechanical damage and/or chemical contamination as well as to prevent dissociation during the implant anneal. 2500 Å of phosphosilicate glass (PSG) deposited after the Si_3N_4 serves as both an implantation mask and a registration layer; the PSF layer remains on the wafer through the implant anneal and is plastic at the annealing temperature, thereby reducing stress.

Contact photolithography and ion milling are used to open windows to the PSG/Si_3N_4 interface for ion implantation. Implants are made at two different Si^{29} ion energies (260 KeV and 125 KeV) through the 900 Å thick Si_3N_4 at ambient temperature to provide a flat profile whose depth is $\approx$ 2750 Å.

After the stripping of the photoresist and the deposition of a second PSG layer the wafers are annealed through controlled cycle to 860°C for 15 minutes to activate the implant.

Metalization Technology

Following the implantation and activation of the Si^{29} implant, ohmic contacts for the sources and drains of the FETs are deposited by a liftoff process on to the gallium arsenide using the alignment marks ion milled into the surface during the ion implantation processing. The metal used for the ohmic contacts consists of 1100 Å gold 12% germanium alloy, 500 Å nickel and 400 Å of platinum. This metal system is alloyed at 490°C for 10 secs in an argon-10% hydrogen atmosphere and the contacts thus formed are very reproducible from run to run and across the 2" wafers with values of contact resistance less than $3 \times 10^{-6} \Omega-cm^2$, monitored routinely using the TLM method.

The gates of the FETs are formed using contact lithography and masks made by E-beam on 4-inch, 90 mil thick quartz plates. Two photoresist systems have been employed. The first of these is AZ1350J photoresist in combination with near-UV (405 nM) radiation and has resulted in gates down to 0.7 μm long. A chlorobenzene soak of the AZ1350J photoresist provides an overhang which greatly assists the liftoff of the gate metallization which consists of 500 Å titanium, 400 Å platinum and 6500 Å of gold. Dimensions down to 0.5 μm have been achieved using a double layer structure of a co-polymer of polymethyl methacrylate (PMMA) and 25% methacrylic acid (MAA) on top of a lyer of PMMA and exposed by deep UV (210 nM) radiation.

Prior to deposition of the gate metals the gate region is wet chemically etched to a depth of 1000 Å to provide a recessed gate structure.

Following testing of the now gated transistors, the circuit metallization is defined using a thick layer of AZ1350J (3.7 μm). The pattern places inductors on the surface of the gallium arsenide in the form of microstrip lines which are plated up to a thickness of 5 μm later in the fabrication process when the air bridges are formed. In addition, the bottom plates of metal-insulator-metal (MIM) capacitors are defined together with the drain interconnections on the FETs.

The circuit metalization is 11500 Å of metal which is composed of chromium (500 Å), palladium (1000 Å) and gold (10000 Å). For the M-I-M capacitor circuit designs an additional layer of chromium (500 Å) is added to provide good adhesion of the insulating layer (sputtered silicon dioxide). The dielectric for the M-I-M capacitors is bias-sputtered SiO_2, 3000 Å thick, that is patterned by lift-off using 2.5μm of AZ1350J photoresist.

Interconnections of the sources of the FETs and the top plates of the MIM capacitors are formed by "air bridges" as shown in the scanning electron micrograph of Figure 6. The process consists of depositing a lower layer of AZ1375 photoresist (4μm thick), which is then opened to expose the areas to be connected by the bridge. 500 Å of titanium and 500 Å of gold are then evaporate over the whole wafer and a second layer of AZ1375 (4μm) is then used to define the bridge itself. The opened areas are plated up with gold to a thickness of 5μm, the top photoresist removed and the thin gold and titanium layers chemically etched away. Removal of the lower resist layer completes the process and results in high yield (> 99%) rugged interconnections.

The front of the wafer is now complete and the remaining steps are thinning of the slice, formation of vias and metalizations of the back of the wafer. The slice is thinned from 500μm down to 100μm to provide the correct spacing between the microstrip lines on the front of the wafer and the ground plane on the back. Thinning is accomplished by a combination of lapping in a semi-automatic jig and a final chem-mechanical polishing. A rough surface finish of the back of the wafer has been shown to have a deleterious effect on circuit losses.

Air bridges and via holes are used, respectively, to interconnect the FET source pads and to provide low inductance source grounds. The via holes also improve circuit layout flexibility by permitting grounding of circuit elements interior to the chip.

Vias are formed by wet chemical etching using a slightly preferential etch to produce coneshaped holes through the 100μm substrate. The vias and the back of the wafer are then coated with Cr-Au-Ni-Au metalization, the nickel acting as a barrier to the gold-tin alloy used to bond the finished chips to the microwave test carrier.

The slice is then sawed into chips and those chips which have acceptable DC FET and passive characteristics are mounted and bonded onto RF carriers for testing.

3 W 8-12 GHz SINGLE STAGE AND FOUR STAGE AMPLIFIERS

The cell cluster matching technique has been used to design a 3 W, 8-12 GHz single stage amplifier in addition to the aforementioned four stage amplifier. The first iteration of this 6400μm FET single stage amplifier (Figure 3) produced 2.58 Watts (cw) out at 8 GHz with ≈ 5 dB associated gain and a 1 dB bandwidth of 1.56 GHz (Figure 4). As far as is known, this is the highest reported output power for an X-band monolithic amplifier.

The complete four stage amplifier has also been fabricated (Figure 5) and is now in the process of evaluation. The complexity of this chip is best illustrated by its parts count; i.e., 60 cells (total gate width 12 mm), 30 vias, 39 overlay capacitors, and 140 air bridges, as well as its overall size which is 6.9 mm x 5.3 mm. In spite of this complexity, DC yield on the first run is 12 chips out of 100. RF results will be presented at the conference.

CONCLUSIONS

The cell cluster matching technique was described as a solution to the three main problems associated with large periphery power FET amplifiers. Namely, the problems of: 1) matching to low impedance levels over broad bandwidths, 2) ensuring equal power and phase to each FET cell, and 3) minimizing inter-cell interactions through the use of the "resonant node" isolation structure.

High yield fabrication techniques were also discussed. In particular, direct selective ion implantation into high purity LEC grown semi-insulating substrates was discussed as a key fabrication technology.

Finally, a 3 W, 8-12 GHz four stage and single stage amplifier were discussed as an implementation of the cell cluster matching and fabrication techniques. An initial result of 2.58 W at 8 GHz with 5 dB gain was obtained from the single stage amplifier.

ACKNOWLEDGEMENTS

The authors would like to acknowledge the support of DARPA and NAVAIR (contract nos. N00014-78-C-0268 and N00014-81-C-0247) in this continuing program, and the encouragement and technical direction of M. N. Yoder, K. Davis, S. A. Roosild, G. Cudd, and L. R. Whicker. We would also like to acknowledge the technical contributions of M. Cohn and G. Eldridge and the skillful assembly and testing contributions of D. Nye and W. Stortz.

REFERENCES

1. Degenford, J. E., Freitag, R. G., Boire, D. C. and Cohn, M, "Broadband Monolithic MIC Power Amplifier Development", Microwave Journal, pp. 89-96, March, 1982.

2. Wilkinson, E. J., "An N-way Hybrid Power Divider", IRE Trans. MTT-8, January, 1960, pp. 116-118.

3. Driver, M. C., Eldridge, G. W. and Degenford, J. E.," Broadband Monolithic Integrated Power Amplifiers in Gallium Arsenide", Microwave Journal, pp. 87-94, November, 1982.

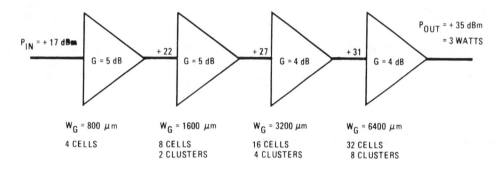

Figure 1. Block diagram of a four stage, 3 W, 8-12 GHz amplifier.

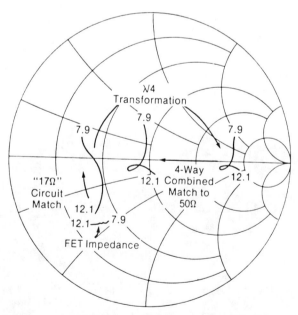

Figure 2. Impedance progression for a cell cluster matched, 3 W single stage amplifier design using four 1600 μm cell clusters.

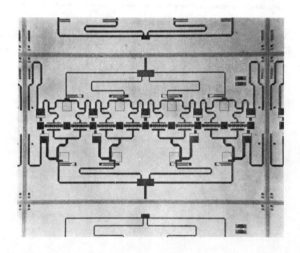

Figure 3. 8-12 GHz, 3 W single stage amplifier. Eight 800 μm cell clusters are used.

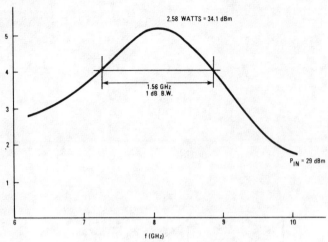

Figure 4. Measured gain vs. frequency plot for the amplifier in Figure 3.

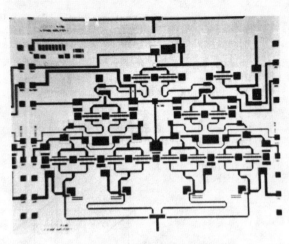

Figure 5. 8-12 GHz, 3 W four stage amplifier.

Figure 6. Air bridge interconnection on a GaAs monolithic circuit (5 μm high, 6 μm thick and 100 μm long).

A 2-8 GHz 2 WATT MONOLITHIC AMPLIFIER

by

John Dormail, Yusuke Tajima, Robert Mozzi,
Mark Durschlag, Anna Marie Morris

Raytheon Company
Research Division
Lexington, Massachusetts

S.A. McOwen

Raytheon Company
Electromagnetic Systems Division
Goleta, California

ABSTRACT

A 2-8 GHz, 2 watt 10 dB gain power amplifier has recently been designed and tested which utilizes a new and unique GaAs power FET device design, a slow-wave transmission line matching structure and large capacitance (1340 pF/mm^2) tantalum pentoxide capacitor processing technology. The two-stage power amplifier with input, interstage, and output matching as well as bias circuitry is included on a 7.7 mm^2 × 0.1 mm chip.

Initial results from amplifier chips designed with approximated FET S-parameters show 10 to 13 dB gain from 1.7 to 8 GHz. Maximum output power of 2 watt was achieved at 2 dB compression with 20% power-added efficiency.

INTRODUCTION

There are various limitations on maximizing the power output from a monolithic broadband amplifier. They include (1) high-yield fabrication of large-periphery FETs, (2) a compact layout of high-power FETs to make an easy transition from matching circuits, (3) broadband impedance matching to a diminishing impedance, and (4) accurate characterization of large-periphery devices for large-signal conditions. For the frequency region of interest (2-8 GHz) there is an additional technical problem because the large wavelength makes it difficult to keep the circuit size small.

In this paper, we will discuss some approaches that are designed to overcome these problems. We will show a new FET design which has an advantage for large gate periphery over the conventional structure. The new FETs have shown a higher fabrication yield and have a compact pad size which made an easy transition to matching circuits. We will also discuss the compact circuit design technique using capacitively loaded lines.

Two kinds of amplifiers were designed and fabricated using new and conventional FETs. They are both two-stage amplifiers designed for power operation. Both of them performed successfully showing 2 watts and 10 dB gain in the 3.5 to 8 GHz band.

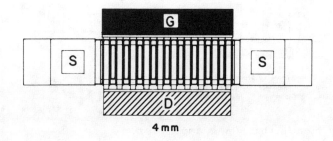

Figure 1a. Conventional FET layout.

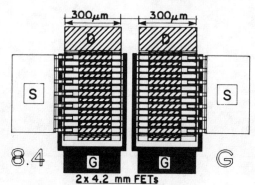

Figure 1b. New FET layout.

Reprinted from *IEEE Gallium Arsenide Integrated Circuit Symp. Tech. Dig.*, 1983, pp. 115-118.

FET DESIGN

A new approach was taken in the layout of high-power FETs (Fig. 1b). The device is rotated 90° relative to conventional power FET structures (Fig. 1a). The gate fingers are fed from both ends via a "U" shaped bus bar. The drain pads are air-bridged away from the input and the source pads are air-bridged to the side and grounded by via holes.

Some of the features of this new structure are: (1) the double feed structure minimizes the probability of a flawed transistor due to broken or pinched gates. (2) Input/output discontinuities are relatively small and independent of periphery scale-up because the small pad size does not change with the number of gate fingers. (3) Source inductance is minimized by providing ground contact immediately adjacent to each channel of the FET.

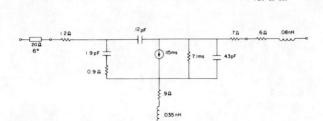

Figure 2. 2-mm FET equivalent circuit.

The equivalent circuit of the device was derived by fitting the S-parameter data to a device model as shown in Fig. 2. It was found that a short section of transmission line was necessary at the input terminal to represent the gate bus bar.

TRANSMISSION LINES

Generally, most of the monolithic circuit chip area is consumed by passive matching circuit elements which have become a major cost-driving factor. The circuit size can become unbearably large for low-frequency circuits where the wavelength is large and lumped elements cannot provide high Q components.

Here, a different approach--the capacitively loaded transmission line--was tested to provide compact broadband matching

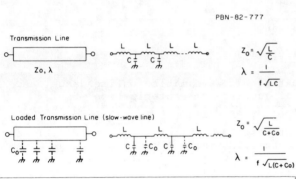

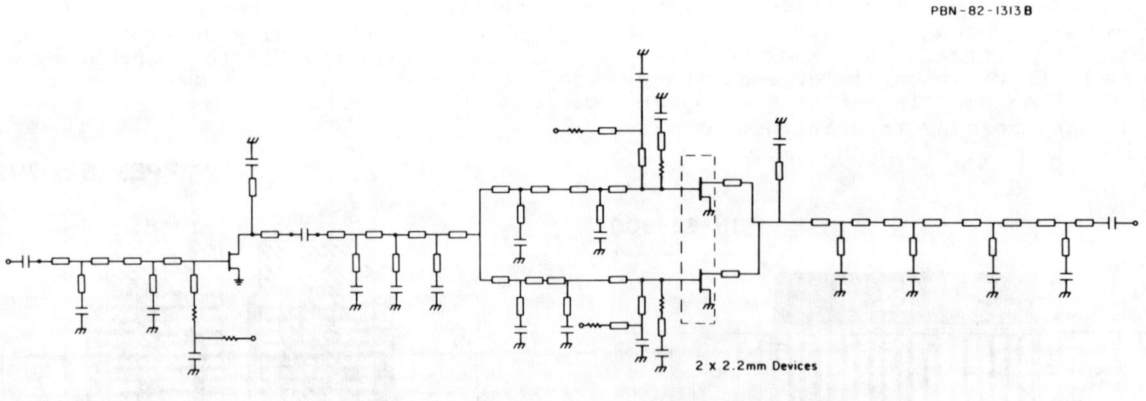

Microstrip Line	Width (μm)	Distributed Elements (nH/cm)	(pF/cm)	Loading Capacitors (pF/cm)	Wavelength (cm)
Z_0 = 70	30	6.57	1.31	0	1.079
Z_0 = 50	75	4.77	1.93	0	1.043
Z_0 = 15	500	1.65	6.76	0	0.945
Slow-Wave Line					
Z_0 = 50	30	6.57	1.31	1.32	0.76
Z_0 = 15	30	6.57	1.31	27.89	0.228

Figure 3. Comparison of distributed microstrip and capacitively loaded microstrip transmission lines.

Chip Size: 3.24mm × 2.57mm (101 mils × 126 mils)
Total Chip Capacitance (Si_3N_4) = 17.7 pf
Total Chip Capacitance (Ta_2O_5) = 196 pf

Figure 4. Schematic diagram of the two-stage monolithic GaAs amplifier.

elements. A schematic comparison between a microstrip line and capacitively loaded microstrip line is shown in Fig. 3. Note that the capacitively loaded 15 Ω line has a wavelength approximately 1/4 that of the microstrip. The capacitively loaded line segment is particularly useful for providing broadband matching circuits for large-periphery FETs where low-impedance lines are needed.

AMPLIFIER DESIGN

Two amplifier circuits are designed for 2-8 GHz bandwidth, 10 dB gain, and 2 W output power. The first circuit (A) includes the conventional structure FET and the second (B) utilizes the new FET. The first circuit uses open stub loading while circuit B uses lumped capacitor loading. Because of the difference in the approach, the chip size of the first circuit is larger by 44%.

In the design of the amplifier the large-signal program, LSFET,[1,2] was fully utilized. The FET was characterized and modeled for both small and large signal operation. As a result of design iterations, a 2 mm gate periphery FET was chosen to drive a 4.4 mm FET in order to achieve the 2 W output power with maximum efficiency. The circuit schematic of amplifier B is shown in Fig. 4. The predicted power performance of the amplifier is an output power of 1.8-2.2 W with 10 dB gain and 16-20% power-added efficiency in the 2-8 GHz band (Fig. 5).

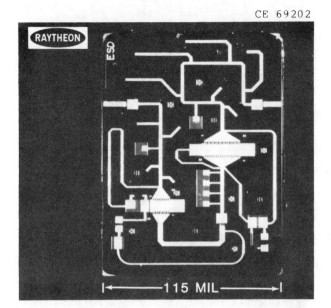

Figure 6a. Circuit A.

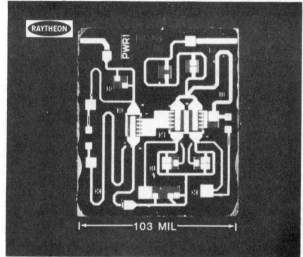

Figure 6b. Circuit B.

CIRCUIT FABRICATION

The circuits are fabricated on epitaxial material with an active layer doping of 10^{17} cm^{-3}. The gates are 1 μm in length and pattern generated by E-beam lithography.

Two kinds of capacitors are used in the circuits. Since capacitors for transmission line loading and for matching are relatively small in value, Si_3N_4 film is used. The bypass capacitors ranging from 30-45 pF were made from Ta_2O_5 film. The fabrication of the Ta_2O_5 capacitors was reported in Ref (3).

Figures 6a and 6b show completed chips of circuit A and B, respectively.

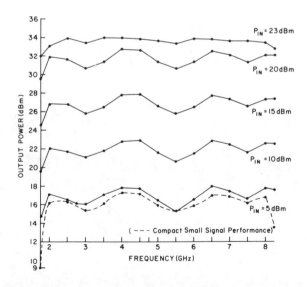

Figure 5. Predicted power performance of the monolithic GaAs FET amplifier.

225

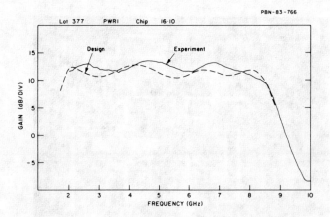

Figure 7. Small signal RF performance. Lot 377, PWR1 chip 16.10.

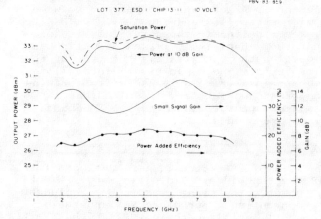

Figure 8. Monolithic power amplifier, measured RF performance. Lot 377, ESD1 chip 13.11.

TEST RESULTS

The small-signal gain, measured from several chips, of circuit B shows up to 2 dB more gain than predicted as shown in Fig. 7. This was due to higher gain devices from the wafer than those used in the design model. Overall resemblance of measured data to the designed data was excellent. The power data of circuit A is shown in Fig. 8. A 2 watt power level was measured between 3.5 and 8 GHz with 10 dB associated gain. Power-added efficiency exceeded 20% in the same band.

CONCLUSIONS

We have demonstrated solutions to the limitations of broadband power amplifier design discussed earlier. We have shown a compact layout of high power FET's with suitable transition to matching circuits. We have accurately characterized 2 mm and 4 mm gate width devices both large and small signal. By use of a double feed structure and E-beam lithography, we have demonstrated high-yield fabrication of large gate periphery FETs. Also employing suitable broadband matching structure, we have fabricated and tested a 2 watt, 10 dB gain amplifier.

REFERENCES

[1] Y. Tajima et al., "GaAs FET large-signal model and its application to circuit design," IEEE Trans. ED-28 pp. 171-175 (1981).

[2] A. Platzker et al., "Large-signal GaAs FET amplifier CAD program," 1982 IEEE MTT-S, Int. Microwave Symp. pp 450-452 (June 1982).

[3] M. Durschlag et al., "A tantalum-based process for MMIC on-chip thin film components," GaAs IC Symposium, New Orleans, p 146 (1982).

ACKNOWLEDGEMENT

This work was supported in part by the Naval Air Systems Command (NRL Contract No. N00014-82-C-2458).

A GaAs MONOLITHIC 6—18 GHz MEDIUM POWER AMPLIFIER

C.D. Palmer, P. Saunier, and R.E. Williams

Texas Instruments Incorporated
P.O. Box 226015
Dallas, Texas 75266

ABSTRACT

A monolithic two-stage medium power amplifier, fabricated on GaAs and designed to cover the 6-18 GHz frequency band, is described. Amplifier circuit topology, process sequence, and measured performance results are presented. Chips from several epitaxial and ion implanted slices have demonstrated good output power and gain performance across the 7.0-17.0 GHz band, achieving an average of 27.3 dBm (540 mW or 0.45 watts per millimeter of gate width) at 18.8 percent power-added efficiency, with an average gain of 10.6 dB, at 1 dB gain compression.

INTRODUCTION

The success of a new generation of electronic warfare systems requires the development of high performance solid-state microwave power amplifiers suited for volume production at low cost. GaAs monolithic microwave integrated circuit (MMIC) technology offers the potential of meeting this challenge. This paper addresses the design and initial performance results of a broad bandwidth, medium power MMIC amplifier intended for EW application. Performance goals for the amplifier design are shown in Table I.

TABLE I. MONOLITHIC MEDIUM POWER AMPLIFIER PERFORMANCE GOALS

Frequency Range	6.0 - 18.0 GHz
Small-signal Gain	7.0 dB, minimum
Gain Ripple	2.0 dB, maximum
Output Power at 1 dB Gain Compression (CW)	400 mW, minimum
Power-added Efficiency at 1 dB Gain Compression	20% minimum

DESIGN APPROACH

A single-ended, bandpass, two-FET cascade was chosen as the basic circuit topology for the monolithic medium power amplifier. A symmetrical circuit topology was selected to give the option of applying bias supply voltages from either side of the chip. Incorporation of integral bias networks necessitates the bandpass matching network approach. The two-stage single-ended cascade represents a compromise between maximizing the gain per monolithic chip and obtaining a reasonable quantity of functional chips per slice, based on current GaAs processing yields.

A 900-micron gate width FET was used for the input device and two 600-micron FETs in parallel, for a total of 1200-micron, were used for the output device. The "split" 1200-micron device was chosen to decrease source inductance and to diffuse heat concentration in the device. A 0.10-millimeter GaAs substrate thickness was selected for these same reasons.

Computer-aided-design (CAD) programs were used extensively throughout the design of the 900-1200 micron monolithic amplifier. SUPER-COMPACT was used for the analysis and optimization of the various circuit elements. An in-house program was utilized to predict power performance and, in conjunction with SUPER-COMPACT, to design matching circuits for maximum amplifier output power.

Models for the 900-micron and 600-micron FETs were derived from data obtained from discrete GaAs FETs and previously built GaAs FET hybrid power amplifiers. The various FET model parameters were optimized using CAD techniques to match the measured data.

The design of the matching circuits progressed from amplifier output to input. After the initial optimization of the matching network was complete, microstrip discontinuities were added, and a final optimization was performed. Due to the constraints placed on the amplifier in the initial planning stages, each network layout was completed following its design. This ensured that the network was realizable, while meeting the performance goals of the amplifier; if not, the network design was re-iterated. Finally, an analysis was performed to determine the sensitivity of the amplifier performance to fabrication tolerances. If an element was found to be too sensitive, adjustments were made, and the network was re-optimized.

Achieving a producible monolithic circuit design requires an understanding of the performance sensitivities to process variations. Extensive sensitivity analyses, involving Monte Carlo statistical techniques, provide insight into the expected performance distribution in volume production and the critical RF yield drivers. Sensitivity analyses were a major component of the 900-1200 micron amplifier design process.

Figure 1 shows the circuit schematic of the 900-1200 micron amplifier. A photograph of the chip is shown in Figure 2, and the predicted gain and power responses are presented in Figure 3.

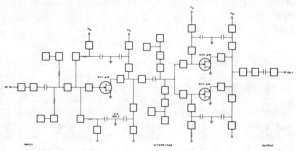

Figure 1. 900-1200 Micron Amplifier Circuit Schematic

MMIC PROCESSING

The 900-1200 micron amplifier was fabricated using both epitaxial and ion implanted material, with comparable performance results. In both cases, two-inch LEC substrates were used as the starting material. For epitaxial work, the substrates are etched and the $AsCl_3$ process is used to grow the buffer and active layers. The active layer is doped to approximately 1.5-1.8E17 in both types of material.

The first step in the process sequence is patterning and etching the active layer for device isolation. Next, the source-drain metallization and resistors are formed using alloyed AuGeNi, with a resistivity of approximately 1.9 ohms per square. All levels except gate definition are formed using optical contact photolithography, usually with the well known chlorobenzene technique to enhance liftoff. The recessed TiPtAu gates are defined using E-beam lithography, yielding a 0.5-micron gate length.

First level metallization forms the transmission lines and overlays the device "bond pads," increasing the conductivity over the AuGeNi. A 0.4-micron plasma-deposited layer of silicon nitride forms the capacitor dielectric and provides passivation. Fabrication of the top capacitor plates follows the nitride deposition.

The air-bridge/plating sequence consists of the following steps. Resist is applied and patterned, and the exposed nitride etched away. Then, a thin layer of gold is sputtered onto the slice, and another layer of resist is applied and patterned with the air-bridge mask. The transmission lines, bonding pads, capacitor top plates, as well as the air-bridges, are plated.

The slice is lapped to 0.1-millimeter and via holes are formed using reactive-ion-etching. The backside is metallized and plated, and the slice is sawed into separate bars.

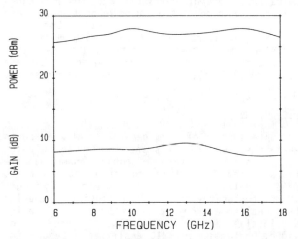

Figure 3. Predicted Amplifier Performance at 1 dB Gain Compression

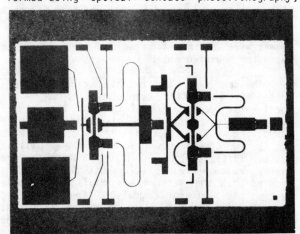

Figure 2. Monolithic 900-1200 Micron Amplifier

PERFORMANCE RESULTS

The first design iteration achieved the performance results described in this section. No tuning was performed either on the MMIC or the test fixture. Figure 4 shows measured gain and output power at 1 dB gain compression obtained from typical 900-1200 micron medium power amplifiers fabricated on epitaxial and ion implanted substrates. Measured amplifier bandwidth is less than predicted, covering the 7.0-17.0 GHz frequency band, with gain rolloffs at both band edges. This is due to differences between the actual FETs and circuit elements and the models used in the design.

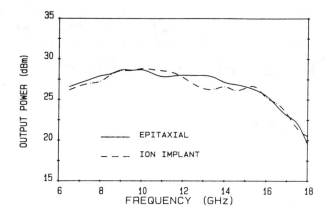

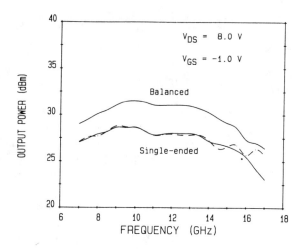

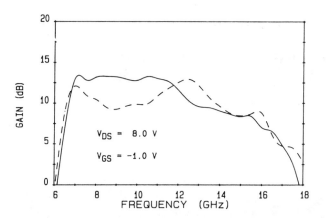

Figure 4. Measured amplifier performance (1 dB Gain Compression)

Figure 5. Balanced Amplifier Performance at 1 dB Gain Compression. The Chips Are From Two Different Slices

Measured power performance, at 1dB gain compression, of the epitaxial and ion implanted 900-1200 micron amplifiers averaged 27.4 dBm (550 mW) and 27.2 dBm (530 mW) across the 7.0 to 17.0 GHz band, which translates to 0.46 and 0.44 watts per millimeter of FET gate width. Under the same measurement conditions, average power-added efficiencies were 18.7 and 18.9 percent, with average gains of 11.0 and 10.2 dB, respectively.

The 900-1200 micron amplifier was designed to be used in a balanced configuration. Figure 5 shows the combining efficiency achieved when balancing two chips from different slices. The chips were matched strictly from device I-V data, obtained prior to circuit assembly. The output power of each separate chip is shown, as well as the power obtained from the balanced pair. Across the 7.0-17.0 GHz frequency band, the balanced amplifier delivered an average of 30.1 dBm (1020 mW) at 1 dB gain compression. This represents an average combining efficiency of 2.5 dB, which compares well with results achieved using chips selected from the same slice.

CONCLUSION

First design iteration monolithic 900-1200 micron medium power amplifiers, from a number of epitaxial and ion implanted slices, have demonstrated excellent gain and output power performance across the 7.0-17.0 GHz frequency band. An average of 27.3 dBm (540 mW or 0.45 watts per millimeter of gate width), at 18.8 percent power-added efficiency, with an average gain of 10.6 dB, was achieved at 1 dB gain compression.

Section III-C
Broad-Band Amplifiers

NO area of MMIC endeavor has shown such remarkable progress as that of broad-band amplifier performance. In the relatively short period of three years, bandwidths spanning over four octaves have been achieved!

Broad-band designs in MMIC format generally speaking have followed four unrelated approaches. These are 1) traveling-wave or distributed amplifiers, 2) resistive-loaded amplifiers, 3) resistive-feedback amplifiers, and 4) direct-coupled amplifiers.

The last three approaches utilize the cascade principle of interconnecting the active devices. The traveling-wave approach, on the other hand, is based on a parallel distribution of active devices, in this case FET's, along two transmission lines, one across the input of the devices, the other across the output.

The performance characteristics of these four approaches are compared in the papers that follow. Examples of monolithic implementation of these approaches are described, which cover the frequency spectrum from below 100 MHz to above 30 GHz. There is no reason to doubt that in the near future the upper frequency limit will be extended to beyond 60 GHz.

Multi-Octave Performance of Single-Ended Microwave Solid-State Amplifiers

KARL B. NICLAS, SENIOR MEMBER, IEEE

(*Invited Paper*)

Abstract —The computed performances of multi-stage single-ended GaAs MESFET amplifiers are compared when employing one and the same transistor type. The circuit principles studied are of the reflective match, the lossy match, the feedback, the distributed, and the active-match amplifier variety. It was found that the gain characteristics of the single-stage modules using either passive or active matching do not conclusively identify the optimum circuit type in the band of interest (2–18 GHz). For the case of multistage devices, however, the gain and the VSWR performance clearly favor the distributed amplifier principle.

In addition to the data reported in the literature, the paper discusses recent experimental results obtained from a 3–17.5-GHz reflective match module, a two-stage 2–18-GHz and a four-stage 0.5–18.5-GHz feedback amplifier, as well as a two-stage 2–20-GHz and a four-stage 2–18-GHz distributed amplifier.

I. INTRODUCTION

THE CONCEPT of the balanced reflective match amplifier has dominated the design of microwave solid-state amplifiers for nearly two decades [1]. Up to this day, quadrature hybrids of the type invented by J. Lange [2] are almost exclusively occupying the position of the signal combiner and divider yielding excellent performance, regardless of the mismatch presented by the two identical single-ended modules. However, the bandwidth of the balanced reflective match amplifier is limited to two octaves, at best. Extending the frequency band beyond this 4:1 ratio requires a minimum of three 90° couplers in tandem configuration [3]–[5]. Unfortunately, these tandem couplers are not only complicated, but space-consuming and costly to manufacture. Due to these reasons and their cost effectiveness, multi-octave single-ended amplifiers are gaining more and more in importance. Four circuit design principles exhibiting excellent ultra-wideband characteristics are now challenging the concept of the balanced reflective match amplifier [6]–[30]. Hence, the five competitors are:

1) the reflective match amplifier,

2) the lossy match amplifier,
3) the feedback amplifier,
4) the distributed amplifier, and
5) the active match amplifier.

While the first four circuit types employ passive elements to improve the input and output matches, the fifth principle makes use of active elements to achieve the same goal [9]–[13]. Characterized by their simplicity, compact size, and low cost, single-ended amplifiers represent attractive options whenever an economical solution to wide-band amplification is of primary concern.

Finding the optimum solution from these five circuit types poses, however, a difficult problem and has no simple answer. In an attempt to compare the performance characteristics of these alternatives, one needs to establish certain conditions to arrive at a meaningful solution. To keep matters simple, we chose only the following two:
1) all amplifiers use identical active devices independent of the circuit type employed, and
2) the frequency band of interest is 2–18 GHz.

The limiting conditions are necessary since, among other factors, the electrical performance of each circuit principle depends greatly on the transistor type and the desired bandwidth. Either factor may have a decisive impact on the choice of the optimum design concept. Confining the comparison to a specific device type and a particular frequency band may seem to lack the depth expected from such an analysis. However, the removal of these restrictions would render our study rather unmanageable and, therefore, beyond the scope of this paper.

II. Computed Results

A. Matching with Passive Elements

In the following, we compare the performance characteristics of the reflective match (RM), the lossy match (LM), the feedback (FB), and the distributed amplifier (DA) based on computed results. The individual circuits are optimized for gain, gain flatness, and reflection coefficients. All passive elements are realizable, although some high-impedance lines may be difficult to manufacture. As already pointed out, the frequency band extends from 2 to 18 GHz and identical GaAs MESFET's are being used in all circuit types. The transistor's model and its element values are presented in Fig. 1. The latter have been obtained from the measured S-parameters of a GaAs MESFET with a 0.5×300-μm gate and a $2 \cdot 10^{17}$ cm^{-3} carrier concentration.

The topologies of the amplifier modules and the values of their components are presented in Fig. 2. Biasing can be accomplished easily through any of the short-circuited shunt elements. However, this causes a loss in efficiency in case a resistor is part of the biasing network. The values of all passive circuit components have been optimized for best gain performance and do not represent the optimum conditions for noise figure. The positions of the active device in both the lossy match and the feedback amplifier are occupied by two GaAs MESFET's in parallel. This is due to

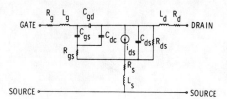

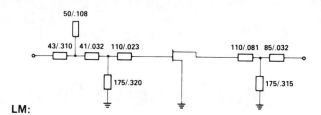

Fig. 1. Half-micron gate FET model and its element values.

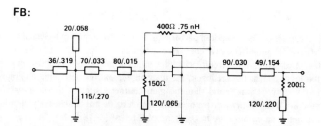

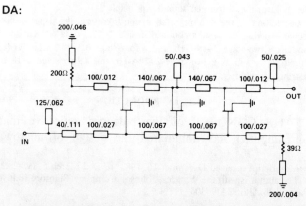

Fig. 2. Circuit topologies of the single-ended amplifier modules (lengths of transmission lines in inches for air dielectric).

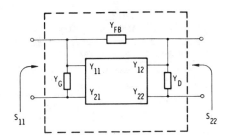

$$S_{21} = -\frac{Y_0}{Y_{12} - Y_{FB}} \left[\sqrt{1 + \frac{(Y_{21} - Y_{FB})(Y_{12} - Y_{FB})}{Y_0^2}(1+S_{11})(1+S_{22})} - 1 \right]$$

Fig. 3. Two-port with lossy match and feedback components.

the insufficient gain produced by the single device of Fig. 1 for these types of circuits, as will be further explained below.

The insertion gain of a single-ended module making simultaneous use of lossy matching and feedback (Fig. 3) may be expressed by [6]

$$\text{GAIN} = \left| \frac{Y_0}{(nY_{12} - Y_{FB})} \left[\sqrt{1 + \frac{(nY_{21} - Y_{FB})(nY_{12} - Y_{FB})}{Y_0^2}(1+S_{11})(1+S_{22})} - 1 \right] \right|^2 \quad (1)$$

when n parallel transistors are employed. The module's reflection coefficients are

$$S_{11} = \frac{(Y_0 - Y'_{11})(Y_0 + Y'_{22}) + (nY_{21} - Y_{FB})(nY_{12} - Y_{FB})}{(Y_0 + Y'_{11})(Y_0 + Y'_{22}) - (nY_{21} - Y_{FB})(nY_{12} - Y_{FB})} \quad (1a)$$

$$S_{22} = \frac{(Y_0 + Y'_{11})(Y_0 - Y'_{22}) + (nY_{21} - Y_{FB})(nY_{12} - Y_{FB})}{(Y_0 + Y'_{11})(Y_0 + Y'_{22}) - (nY_{21} - Y_{FB})(nY_{12} - Y_{FB})} \quad (1b)$$

$$Y'_{11} = nY_{11} + Y_G + Y_{FB} \quad (1c)$$

$$Y'_{22} = nY_{22} + Y_D + Y_{FB} \quad (1d)$$

$$Y_0 = Z_0^{-1}. \quad (1e)$$

At low frequencies ($B_{FB} \ll G_{FB}$, $Y_{21} \simeq g_m$) and $nY_{12} \ll G_{FB}$, the set of equations (1) reduces to

$$\text{GAIN} \simeq \left[\frac{Y_0}{G_{FB}} \left[\sqrt{1 - (ng_m - G_{FB})G_{FB}Z_0^2(1+S_{11})(1+S_{22})} - 1 \right] \right]^2 \quad (2)$$

$$S_{11} \simeq \frac{(Y_0 - Y'_{11})(Y_0 + Y'_{22}) - (ng_m - G_{FB})G_{FB}}{(Y_0 + Y'_{11})(Y_0 + Y'_{22}) + (ng_m - G_{FB})G_{FB}} \quad (2a)$$

$$S_{22} \simeq \frac{(Y_0 + Y'_{11})(Y_0 - Y'_{22}) - (ng_m - G_{FB})G_{FB}}{(Y_0 + Y'_{11})(Y_0 + Y'_{22}) + (ng_m - G_{FB})G_{FB}} \quad (2b)$$

$$Y'_{11} \cong G_G + G_{FB} \quad (2c)$$

$$Y'_{22} \cong nG_{ds} + G_D + G_{FB}$$

(G_{ds}—drain-to-source conductance.) (2d)

For $n = 2$, $g_m = 28$ mS, $G_{ds} = 3.7$ mS, $G_{FB} = 2.5$ mS, $G_G = 6.7$ mS, and $G_D = 5$ mS [Fig. 2 (FB)] we calculate with (2) a gain of $G = 5.3$ dB and the reflection coefficients $|S_{11}| = 0.21$ and $|S_{22}| = 0.02$. The corresponding parameters for $n = 1$ are $G = 0.3$ dB, $|S_{11}| = 0.28$, and $|S_{22}| = 0.20$, representing an unacceptably low gain response. Similar results exist for the LM circuit of Fig. 2, for which (1) assumes the following low-frequency expressions [6]

$$\text{GAIN} \cong \frac{1}{4}[ng_m Z_0 (1+S_{11})(1+S_{22})]^2 \quad (3)$$

$$S_{11} \cong \frac{Y_0 - G_G}{Y_0 + G_G} \quad (3a)$$

$$S_{22} \cong \frac{Y_0 - (nG_{ds} + G_D)}{Y_0 + (nG_{ds} + G_D)}. \quad (3b)$$

For $n = 2$, $g_m = 28$ mS, $G_G = 13.5$ mS, and $G_D = 5$ mS, we calculate with (3) $G = 6.2$ dB, $|S_{11}| = 0.19$, and $|S_{22}| = 0.24$. In the case of $n = 1$, the gain is reduced to $G = 1.2$ dB with $|S_{11}| = 0.19$ and $|S_{22}| = 0.40$. As discussed elsewhere, (3) also represents the gain and the reflection coefficients of a distributed amplifier at low frequencies when n links are employed [7]. In order to achieve an equivalent gain with the distributed amplifier, three links are required.

In the following, we will show that similar gain performance may be obtained with the four circuits illustrated in Fig. 2. Their small signal gains, noise figures, and reflection coefficients are plotted in Fig. 4 across the band of interest. While the average gains of all four amplifier types remain within 1.6 dB of each other, the reflection coefficients exhibit vast differences. The latter, more than any other parameters, dictate the feasibility of the design principle in the case of multistage operation. The average noise figures of the four types stay within 1.1 dB of each other. Table I summarizes the performance characteristics of the 2–18-GHz modules. Comparing the data, the distributed amplifier demonstrates the best gain flatness, the lowest reflection coefficients, and the highest stability factors. Its maximum noise figure, however, exceeds those of the other modules. The lossy match amplifier shows the best overall noise figure in addition to excellent gain performance. The feedback amplifier trails both the lossy match and the distributed amplifier in gain, but has the advantage of lower reflection coefficients over the LM unit. In contrast, the RM module is unstable at frequencies below 9 GHz and, therefore, not very well suited for cascading.

As already pointed out, the choice of the circuit type is mostly dictated by the reflection coefficients, for they represent the most critical parameters. The importance of the modules' input and output VSWR becomes very much apparent when cascading several units. The impact on gain, gain flatness, maximum VSWR, and stability factor is summarized in Table II. While the gain characteristics of both the feedback and the distributed amplifier may be

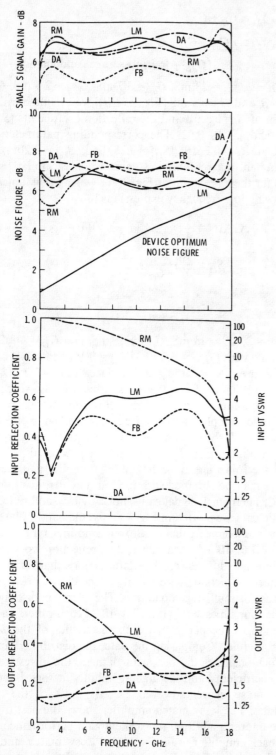

Fig. 4. Computed performance characteristics of the 2–18-GHz amplifier types employing the active device of Fig. 1.

TABLE I
SINGLE-STAGE AMPLIFIER PERFORMANCE

TYPE	SS GAIN dB	NOISE FIGURE dB	MAX. VSWR		MIN. REVERSE ISOLATION dB	MIN. STABILITY FACTOR
			INPUT	OUTPUT		
RM	6.6 ± 1.0	6.7 ± 1.6	∞	7.3	−20.6	0.21
LM	6.8 ± 0.5	6.6 ± 0.5	4.7	2.6	−21.6	1.74
FB	5.4 ± 0.5	6.9 ± 0.7	3.4	2.0	−21.3	2.90
DA	7.0 ± 0.5	7.7 ± 1.5	1.3	1.4	−21.6	2.75

TABLE II
MULTISTAGE AMPLIFIER PERFORMANCE

STAGES	TYPE	SS GAIN dB	MAX. VSWR		MIN. K-FACT.
			INPUT	OUTPUT	
1	RM	6.6 ± 1.0	∞	7.3	0.21
	LM	6.8 ± 0.5	4.7	2.6	1.74
	FB	5.4 ± 0.5	3.4	2.0	2.80
	DA	7.0 ± 0.5	1.3	1.4	2.75
	AM	5.8 ± 0.7	2.6	2.4	6.0
2	RM	15.5 ± 5.6	—	—	−23.8
	LM	14.3 ± 1.7	6.9	2.7	5.50
	FB	11.2 ± 1.0	3.4	2.1	18.1
	DA	14.6 ± 1.0	1.5	1.6	13.5
	AM	13.8 ± 2.2	2.5	2.6	40.2
3	RM	—	—	—	—
	LM	20.9 ± 3.0	7.5	2.7	18.8
	FB	17.2 ± 1.8	3.4	2.1	108.8
	DA	21.8 ± 1.5	1.5	1.6	75.7
	AM	20.0 ± 1.5	2.6	2.8	260.0
4	RM	—	—	—	—
	LM	27.9 ± 4.6	7.6	2.7	64.6
	FB	22.7 ± 3.0	3.4	2.1	672.7
	DA	29.0 ± 2.0	1.5	1.6	408.2
	AM	27.1 ± 2.8	2.6	2.8	>1000
6	RM	—	—	—	—
	LM	43.8 ± 9.8	7.6	2.7	>1000
	FB	34.1 ± 5.0	3.4	2.1	>1000
	DA	43.5 ± 3.0	1.5	1.6	>1000
	AM	40.2 ± 6.4	2.6	2.8	>1000

acceptable up to three stages, as long as gain flatness is of concern only the distributed amplifier principle appears to be usable above three stages. The data in Table II clearly demonstrates the distributed amplifier's superior input and output match performance, making it the logical choice for most high-gain 2–18-GHz applications. However, any appreciable reduction of this frequency band may render the feedback or the lossy match amplifier the best suited candidate, while, for frequency bands below two octaves, the reflective match principle may survive its opponents in the competition for the best overall performance.

So far, we have determined that the best overall performance of our multistage 2–18 GHz amplifier with passive matching elements will be achieved when using the *DA* principle. However, comparing the noise figures of the modules of Fig. 2, we find from Fig. 4 that the distributed amplifier module exhibits the highest maximum noise figure. Hence, the question arises whether and how the noise characteristics can be improved without significantly impairing the other performance parameters.

A closer look at Fig. 4 reveals that the lossy match amplifier (*LM*) achieves the best overall noise performance. From this, we should expect that a modification of the *DA* module towards the *LM* module would bring about an improvement of the *DA* noise figure. Fortunately, the principle of the lossy match and the distributed amplifier are somewhat related, for, when removing the elements linking together the transistors of a distributed amplifier, it turns into a lossy match amplifier. Therefore, reducing the lengths of the linking transmission lines from those in Fig. 2 (*DA*) should improve the noise figure at the high end of the frequency band, possibly compromising the reflection coefficients and gain flatness.

It is known from the literature that the resistive components of the idle ports' terminations have their greatest

k-th LINK	SIGNAL MATRIX (A_{Fk})	NOISE MATRIX (B_{Fk})
(a)	$\begin{bmatrix} 1 & 0 & 0 & 0 \\ Y_{22k} & 1 & Y_{21k} & 0 \\ 0 & 0 & 1 & 0 \\ Y_{12k} & 0 & Y_{11k} & 1 \end{bmatrix}$	$\begin{bmatrix} 0 & 0 & 0 & 0 \\ Y_{21k} & 0 & 0 & 0 \\ 0 & 0 & 0 & 0 \\ Y_{11k} & 1 & 0 & 0 \end{bmatrix}$
(b)	$\begin{bmatrix} 1 & 0 & 0 & 0 \\ Y_{FBk} & 1 & -Y_{FBk} & 0 \\ 0 & 0 & 1 & 0 \\ -Y_{FBk} & 0 & Y_{FBk} & 1 \end{bmatrix}$	$\begin{bmatrix} 0 & 0 & 0 & 0 \\ 0 & 0 & Y_{FBk} & 0 \\ 0 & 0 & 0 & 0 \\ 0 & 0 & -Y_{FBk} & 0 \end{bmatrix}$
(c)	$\begin{bmatrix} 1 & 0 & 0 & 0 \\ (Y_{22k}+Y_{FBk}) & 1 & (Y_{21k}-Y_{FBk}) & 0 \\ 0 & 0 & 1 & 0 \\ (Y_{12k}-Y_{FBk}) & 0 & (Y_{11k}+Y_{FBk}) & 1 \end{bmatrix}$	$\begin{bmatrix} 0 & 0 & 0 & 0 \\ Y_{21k} & 0 & Y_{FBk} & 0 \\ 0 & 0 & 0 & 0 \\ Y_{11k} & 1 & -Y_{FBk} & 0 \end{bmatrix}$

Fig. 5. Signal matrix $[A_{FK}]$ and noise matrix $[B_{FK}]$ of the distributed amplifier's basic link when employing feedback.

impact on the noise figure at the low end of the frequency band [8]. In this range, their share in the amplifier's noise figure may be reduced by partial or total elimination of these resistive components whose gain equalizing and matching functions may be performed by using negative feedback. In this case, the computation of the noise figure, the gain, and the reflection coefficients can be accomplished by using formulas published in the literature substituting, however, the signal matrix $[A_{FK}]$ and the noise matrix $[B_{FK}]$ of the active device (Fig. 5(a)) with that of the active device employing feedback (Fig. 5(c)) [8].

The circuits in Fig. 6 represent three of many combinations that are worth exploring: the distributed amplifier with complex terminations (LMDA); with negative feedback (FBDA); and with both complex terminations and negative feedback (LMFBDA). A comparison of these circuits with that of the distributed amplifier in Fig. 2 (DA) shows a significant reduction in the lengths of the linking transmission lines, especially in the gate line. The performance results of the modules illustrated in Fig. 6 are plotted in Fig. 7 and summarized in Table III. The maximum noise figure of the LMDA is reduced by $\Delta_{NF} = 1.4$ dB, an improvement that has been paid for by a partial deterioration of the input match and, to a lesser degree, by an increase in the gain variation. In general, the FBDA and the LMFBDA circuit do not measure up to the LMDA's overall performance and, in addition, are more complicated and, therefore, of less practical interest. Since the LMDA of Fig. 6 is identical to the DA of Fig. 2, except for the dimensions of the individual circuit elements, it emerges as the optimum choice when the noise figure is of more concern than the input match and the gain variation (Table

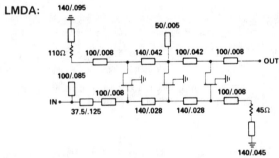

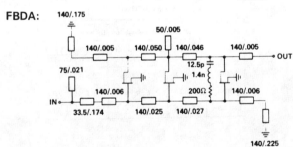

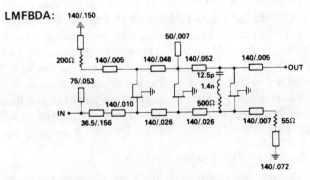

Fig. 6. Proposed distributed amplifier topologies to reduce noise figure (lengths of transmission lines in inches for air dielectric).

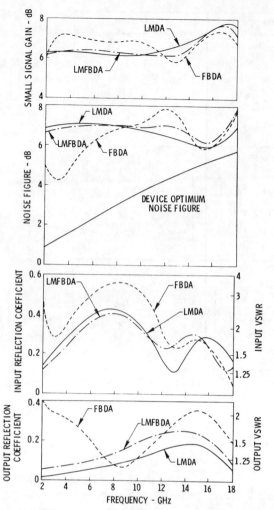

Fig. 7. Computed performance characteristics of the distributed amplifier circuits proposed in Fig. 6.

TABLE III
SUMMARY OF THE SINGLE-STAGE DISTRIBUTED AMPLIFIER CHARACTERISTICS

DISTRIBUTED AMPLIFIER PERFORMANCE				
TYPE	SS GAIN dB	NOISE FIGURE dB	MAXIMUM VSWR	
			INPUT	OUTPUT
LMDA	7.0 ± 0.8	6.5 ± 0.6	2.5	1.5
FBDA	6.6 ± 0.8	6.1 ± 1.9	3.5	2.3
LMFBDA	6.9 ± 0.8	7.0 ± 0.9	2.3	1.7

IV). A comparison of the LM circuit's performance in Fig. 4 with that of the LMDA circuit in Fig. 7 reveals identical maximum noise figures and a somewhat better gain flatness of the *LM* module. However, the LMDA circuit clearly exhibits, by far, the better input and output VSWR's, making it more suitable for multistage operation. In the final analysis, the specifications of the amplifier will dictate which approach to take unless added complexity or other reasons rule out the expenditure for a lower noise figure.

B. Matching with Active Elements

In the preceding part, we have compared a class of amplifiers whose ports are matched to a 50-Ω system by

TABLE IV
SUMMARY OF THE MULTISTAGE DISTRIBUTED AMPLIFIER CHARACTERISTICS

LMDA – MULTISTAGE AMPLIFIER PERFORMANCE			
STAGES	SS GAIN dB	MAXIMUM VSWR	
		INPUT	OUTPUT
1	7.0 ± 0.8	2.5	1.5
2	14.0 ± 1.4	2.6	1.4
3	21.0 ± 2.1	2.7	1.4
4	28.0 ± 2.8	2.6	1.4
6	41.9 ± 4.2	2.6	1.4

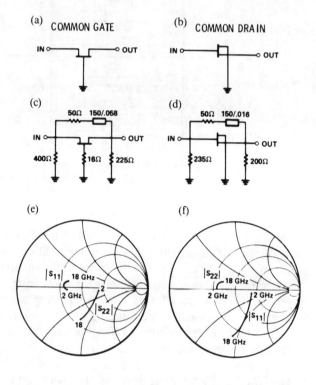

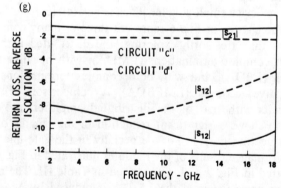

Fig. 8. Common-gate input stage and common-drain output stage (lengths of transmission lines in inches for air dielectric).

means of networks that exclusively consist of passive elements, i.e., components that are not capable of amplifying a signal. In the following, we will briefly discuss a group of amplifiers employing active elements, namely MESFET's, which are able to accomplish the same task, i.e., provide acceptable input and output impedances. This class of amplifiers, which has recently found its way into the design of microwave solid-state amplifiers, may be collectively described as active-match amplifiers [9]–[13].

It is well known that a GaAs MESFET operated in "common-gate" configuration as shown in Fig. 8(a) is capable of providing a good input match. The significance of the common-gate input circuit can be easily understood by examining its scattering parameters. To make matters simple, we determine the S-parameters for those frequencies for which the MESFET's parasitics are negligible and, in addition, neglect all resistive elements except for the drain source conductance G_{ds}. For a source impedance of Z_1 and a load impedance of Z_2, we find

$$S_{11} = \frac{1 - g_m Z_1 + G_{ds}(Z_2 - Z_1)}{1 + g_m Z_1 + G_{ds}(Z_2 + Z_1)} \quad (4a)$$

$$S_{12} = \frac{2 G_{ds}\sqrt{Z_1 Z_2}}{1 + g_m Z_1 + G_{ds}(Z_2 + Z_1)} \quad (4b)$$

$$S_{21} = \frac{2(g_m + G_{ds})\sqrt{Z_1 Z_2}}{1 + g_m Z_1 + G_{ds}(Z_2 + Z_1)} \quad (4c)$$

$$S_{22} = \frac{1 + g_m Z_1 - G_{ds}(Z_2 - Z_1)}{1 + g_m Z_1 + G_{ds}(Z_2 + Z_1)}. \quad (4d)$$

It can be easily seen from (4) that, for $g_m Z_1 = 1 + G_{ds}(Z_2 - Z_1)$, the input reflection coefficient becomes $|S_{11}| = 0$ and the associated gain is $G = |S_{21}|^2 = Z_2/Z_1$, i.e., 0 dB for $Z_1 = Z_2 = Z_0$. Under the condition that $R_{ds} \gg Z_2 + Z_1$, the output reflection coefficient approaches $|S_{22}| = 1$. Thus, a tradeoff exists between the input match on one hand and the gain and the output match on the other. For the practical device characterized in Fig. 1 and $Z_1 = Z_2 = Z_0$, the maximum magnitudes of the reflection coefficients are $|S_{11}| = 0.138$ and $|S_{22}| = 1.36$ between 2 and 18 GHz which compares to $|S_{11}| = 0.145$ and $|S_{22}| = 0.87$, respectively, when calculated with (4a) and (4d). Furthermore, the approximations of (4) which yield acceptable magnitudes of the scattering parameters up to 4 GHz are independent of frequency and therefore do not reflect the frequency dependence of the scattering parameters' phase angles. Over the 2–18-GHz band, these angles vary between $-109° \geqslant \angle S_{11} \geqslant -179°$, $-4° \geqslant \angle S_{22} \geqslant -69°$ and $-7° \geqslant \angle S_{21} \geqslant -89°$, a fact that has to be taken into account in the design of the active match input stage.

A similar situation exists for the device in "common-drain" configuration, only the roles of the ports are reversed, as can be easily seen from the low-frequency S-parameters

$$S_{11} = 1 \quad (5a)$$

$$S_{12} = 0 \quad (5b)$$

$$S_{21} = \frac{2 g_m \sqrt{Z_1 Z_2}}{1 + (g_m + G_{ds}) Z_2} \quad (5c)$$

$$S_{22} = \frac{1 - (g_m + G_{ds}) Z_2}{1 + (g_m + G_{ds}) Z_2}. \quad (5d)$$

In the absence of parasitics, the input terminal presents an open circuit and the input and the output are totally isolated from each other. For $(g_m + G_{ds}) Z_2 = 1$, the output reflection coefficient becomes $|S_{22}| = 0$ while the gain is $G = |S_{21}|^2 = g_m^2 Z_1 Z_2$, i.e., $G \geqslant 0$ dB when $g_m \geqslant 20$ mS and $Z_1 = Z_2 = Z_0$. Comparing the magnitudes of the reflection coefficients computed for the device of Fig. 1 with those calculated with (5), we find $1 \leqslant |S_{11}| \leqslant 1.27$ versus $|S_{11}| = 1$ and $0.218 \leqslant |S_{22}| \leqslant 0.233$ versus $|S_{22}| = 0.226$ when $Z_1 = Z_2 = Z_0$. As was the case for the common-gate configuration, the phase angles of the practical device vary significantly from the constant angles determined with (5) over the 2–18-GHz frequency range.

At this point, a brief discussion on the influence of the source and load impedance of the common-gate FET (CGF) and the common-drain FET (CDF) on their individual S-parameters is in order. As can be seen from (4) and (5), the choices of Z_1 and Z_2 have an appreciable impact on the gain and the reflection coefficients of both FET configurations, and this is true for the common-source FET (CSF) as well. For a demonstration, let us assume we cascade three idealized devices characterized by $g_m = 20$ mS and $R_{ds} = 272$ Ω ($G_{ds} = 3.68$ mS) in accordance with the simple active match amplifier circuit of Fig. 9. Choosing $R_1 = 160$ and $R_2 = 750$ Ω, the three-stage amplifier's computed gain is $G = 14.0$ dB and the computed maximum VSWR's are 1.34:1 for the input and 1.18:1 for the output port. As in all of our studies, the source and load impedance of the amplifier are $Z_0 = 50$ Ω. For comparison, the idealized versions of the individual devices when operating in a 50-Ω system yield gains of $G = 0$ dB for the CGF, $G = 4.55$ dB for the CSF, and $G = -0.76$ dB for the CDF module. The computed parameters above are independent of frequency due to the choice of the device model. As one might expect, the technique of providing a set of source and load impedances over multi-octave bands loses its strength at frequencies where the parasitics of the actual device exert a strong influence on the S-parameters. When replacing the idealized model with the transistor of Fig. 1, and choosing $R_1 = 125$ and $R_2 = 400$ Ω, our three-stage amplifier's gain deteriorates from $G = 14.1$ dB at 2 GHz to $G = 5.8$ dB at 7 GHz. However, across the same frequency band, the maximum input and output VSWR's do not exceed 1.4:1 and 1.5:1, respectively. Of course, there are means of extending the band coverage by introducing additional circuit elements. However, it appears rather difficult to extract appreciable gain and simultaneously achieve superior matching from either the CGF or the CDF module when the frequency band is 2–18 GHz, unless we succeed in significantly reducing the parasitics of our devices.

While as we have demonstrated, the common-gate FET (CGF) serves as a good input match and the common-drain FET (CDF) as a good output match, the CGF output port and the CDF input port present unacceptable and, over the upper part of the frequency band, often negative impedances to the amplifying section to be placed between them. Shunt resistors, as well as parallel and series feedback, may

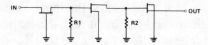

Fig. 9. Simple active match amplifier circuit.

be used to improve the reflection coefficients of the ports facing the actual amplifier section and simultaneously provide unconditional stability for both active matching circuits. The schematics of the circuits which result from such stabilizing techniques when employing the device of Fig. 1 are shown in Fig. 8(c) and 8(d), while the curves of the circuits' reflection coefficients are plotted in Fig. 8(e) and 8(f). Finally, the loss and reverse isolation of both networks are shown in Fig. 8(g). The results presented have been computed for a source and load impedance of $Z_1 = Z_2 = 50\ \Omega$. The common-gate circuit of Fig. 8(c) produces a flat gain response and an acceptable input VSWR brought about by the use of the parallel and the series ohmic feedback, as well as the lossy shunts. However, the noise figure is impaired, primarily due to the noise injected by the series feedback resistor. The elimination of the latter is therefore beneficial for low-noise operation.

Since in many practical cases the output match of the last amplifier stage needs little improvement, an active match output may not be needed or not be able to provide any appreciable improvement, especially over multi-octave bandwidths. The schematic of an active match amplifier module that employs the device of Fig. 1 in both the active-match input stage and the subsequent amplifier stage is shown in Fig. 10(a), while Fig. 10(b) represents the gain, the noise figure, and the reflection coefficients of this module between 2 and 18 GHz. For reasons of lower noise figures, we have not made use of the series feedback, even though this measure accounted for a deteriorated input match. The amplifier stage is of the lossy match type and, because of it, has an acceptable output VSWR. The input VSWR shows significant improvement (2.6:1) over the lossy match module of Fig. 2 (4.7:1). However, a comparison of Fig. 10(b) and Fig. 4 reveals a ±0.7-dB gain variation and a 9.6-dB maximum noise figure of the active match amplifier versus a ±0.5-dB gain variation and a 7.1-dB maximum noise figure of the lossy match module. The output VSWR's are 2.4:1 for the AM unit and 2.6:1 for the LM unit. Replacing the common-gate input stage of Fig. 10 with that of Fig. 8(c) and, in addition, optimizing the second stage for gain results in an improved input VSWR (2.0:1) and similar gain performance (6.0±0.75 dB). However, the maximum noise figure increases from 9.6 to 13.0 dB, primarily due to the noise contributed by the series feedback resistor.

The element values of the module shown in Fig. 10(a) need to be reoptimized in order to achieve an acceptable gain performance when more than one LM gain stage follows the active match input stage. The characteristics of multistage amplifiers consisting of n LM gain stages preceded by an AM input stage are also summarized in Table II. The results presented are based on element values

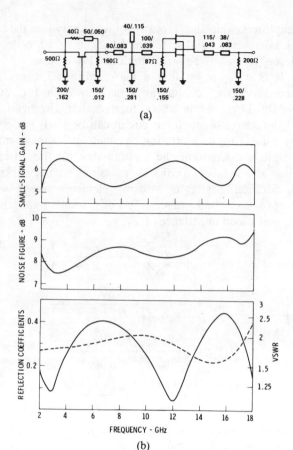

Fig. 10. (a) Active match amplifier schematic and (b) performance (—— input, ---- output refl. coeff.)

obtained from the optimization of the three-stage unit ($n = 3$). The data reveals that our active match amplifier incorporating n stages has a gain performance similar to that of the LM amplifier with significantly improved input VSWR and comparable output VSWR. Furthermore, a better gain flatness is obtained for $n > 2$. However, as were the LM and FB amplifiers, it is not a serious competitor to the distributed amplifier in the 2–18-GHz frequency band when using the device of Fig. 1. In addition, biasing the common-gate input stage via the lossy match networks on either side of the transistor results in significant voltage drops and efficiency losses.

III. Experimental Results

In the preceding chapter, we have concentrated our efforts on discussing the designs of multistage 2–18-GHz amplifiers and have compared their computed performances. Now we will attempt to assemble some of the supporting experimental results.

A. Previously Reported Data

A number of multi-octave single-ended solid-state amplifiers employing lossy match, feedback, distributed, and active match circuits have been described in the literature [6]–[30]. While the frequency coverage of the first three amplifier types has been extended into Ku-band and, in case of the distributed amplifier, beyond 18 GHz, the design efforts of the active match amplifiers have almost exclusively been confined to frequencies below 5 GHz, with one known exception [13].

Until now, no data has come to our attention on the successful design of a practical 2–18-GHz reflective match amplifier. On the other hand, H. Q. Tserng and co-workers have reported small-signal gains of $G = 6.4 \pm 1.0$ dB from 3.2 to 18 GHz in a single-ended module designed for 6–18-GHz operation [14]. From 3 to 2 GHz, this module's gain increased rather rapidly to $G > 10$ dB; yet, when driven with $P_{in} = 20$ dBm, a less varying gain of $G = 5 \pm 2.3$ dB was observed between 2 and 18 GHz.

A. M. Pavio demonstrated for the first time the feasibility of a 2–18-GHz feedback amplifier and was able to achieve $G = 11.0 \pm 2.1$ dB across this band, as well as $G = 14.4 \pm 2.6$ dB from 2.8 to 18 GHz after self-biasing was added [15]. Since feedback amplifiers, even at low frequencies, can be realized on very small substrates, they offer an economical design option to monolithic technology. P. A. Terzian et al. measured a small-signal gain of $G = 6.0 \pm 0.2$ dB between 1 and 7 GHz in a monolithic feedback amplifier using lumped elements [16]. The input and output VSWR were 2.3:1 and 1.7:1, respectively. In a similar approach, W. O. Camp et al. realized $G = 7.0 \pm 0.7$ dB on a 0.76-mm² GaAs chip with a maximum input VSWR of 3.6:1 and a maximum output VSWR of 2.4:1 [17]. R. N. Rigby and co-workers achieved $G = 5.8 \pm 0.6$ dB between 0.6 and 6.1 GHz, realizing maximum VSWR's of 3.2:1 for the input and 2:1 for the output port [18]. K. Honjo et al. reported on a two-stage monolithic amplifier using negative feedback and self-biasing for the second stage [19]. The authors achieved a 3-dB bandwidth gain of $G = 13.5$ dB from 500 kHz to 2.8 GHz. The unit exhibited an input VSWR of $\leq 2.5:1$ from 500 kHz to 2.1 GHz and an output VSWR of $\leq 1.6:1$ between 500 kHz and 4.5 GHz.

The concept of the lossy match solid-state amplifier at microwave frequencies was studied by two teams. In France, T. Obregon et al. demonstrated $G = 12.0 \pm 2.1$ dB in a self-biased three-stage unit operating between 150 MHz and 16 GHz, and measured a maximum noise figure of $NF = 8.5$ dB from 1–12.4 GHz [20]. At almost the same time, K. Honjo and Y. Takayama in Japan obtained an 8.6-dB gain over the 3-dB bandwidth from 800 kHz to 9.5 GHz [21]. The input VSWR was better than 4:1 over the 1 MHz to 10-GHz frequency range, while a 2:1 VSWR was obtained between 2 MHz and 9 GHz. The amplifier had an output power at 1-dB gain compression of 12 dBm over the 2 MHz to 9-GHz range. A maximum noise figure of 8 dB was observed from 50 MHz to 6 GHz. The widest band coverage reported in a lossy match amplifier was achieved by M. Mamodaly and co-workers, who were able to obtain $G = 4.5 \pm 1.5$ dB in a 100 MHz to 17 GHz two-stage amplifier with gain control by way of dual-gate GaAs MESFET's [22]. Maximum input and output VSWR's were 3.5:1 and 2.5:1, respectively, and the unit's gain could be varied by 15 dB.

Concerning distributed amplifiers, Y. Ayasli et al. succeeded in realizing $G = 11.6 \pm 1.6$ dB in 2–20-GHz two-stage monolithic amplifiers on $2.2 \times 5.5 \times 0.1$ mm GaAs substrates [23]. The same team was successful in combining the powers of two distributed amplifiers by devising a unique circuit that consists of two distributed amplifiers sharing a common drain line [24]. The amplifier which uses a power divider to feed two input ports produced a nominal output power of 250 mW at an input power of 20 dBm from 2–18 GHz.

The activities in the field of active-match amplifiers have mostly been concentrated at frequencies below 5 GHz. The only exception known to this author is the work done by W. C. Peterson et al. who reported on a 0.1–10.0-GHz monolithic four-stage amplifier design consisting of a common-gate active match input stage, two common-source lossy match amplifier stages, and a common-drain output stage [13]. Manufactured on a 2.5-mm² chip, the amplifier yielded $G = 7.2 \pm 1.2$ dB of small-signal gain between 0.7 and 9.0 GHz. The input and output reflection coefficients over this band were better than 2:1. In contrast, at lower frequencies, where the stability of the circuits can more easily be obtained, a number of researchers have made use of active matching. In 1978, R. L. Van Tuyl first described his monolithic integrated 4-GHz amplifier [9]. This unit went far beyond the concept of active matching, for Van Tuyl replaced passive with active elements throughout the amplifier. While a common-drain circuit was employed as the output stage, the resistor, normally used in parallel feedback, was replaced by a MESFET. Furthermore, the resistive load had given way to an active load in order to improve the unit's large-signal performance. As the basic amplifying element, a common-source MESFET was employed and biasing of this direct coupled amplifier was accomplished by means of level shifting diodes. Thus, a single amplifier stage used five FET functions. Many refinements have altered this circuit type in subsequent years, among them a resistively loaded common-gate input stage [10], [11]. A typical voltage gain of 26 dB between 5 MHz and 3.3 GHz with less than 1.3:1 of input VSWR have been achieved in a 0.4×0.65 mm gain block incorporating 17 MESFET's. Using the circuit techniques described by R. L. Van Tuyl, D. P. Hornbuckle, and D. B. Estreich, a nominal gain of $G = 20$ dB, as well as an input and output VSWR of 1.5:1 and 2:1, respectively, were measured by V. Pauker and M. Binet in their 0.11–3.2-GHz amplifier. However, in their unit the feedback was performed by a resistor [12].

B. Recent Experimental Data

In this section, we shall add some of our own experimental data recently obtained from a 3–17.5-GHz reflective match module, a two-stage and a four-stage 2–18-GHz feedback amplifier, as well as a two-stage and a four-stage 2–18-GHz distributed amplifier. While the results support what has been discussed in Section II, they are not meant to represent the experimental proof to the computed results of the circuits in Fig. 2. In particular, the *RM* module was intended to operate in the 4–18-GHz band, while the *FB* amplifiers were designed for a single GaAs MESFET capa-

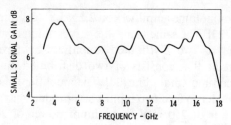

Fig. 11. Measured gain of a 3–17.5-GHz reflective match gain module.

TABLE V
ELEMENT VALUES OF THE SUB HALF-MICRON GATE GaAs MESFET

INTRINSIC ELEMENTS		EXTRINSIC ELEMENTS	
g_m	= 53 mS	R_g	= 2 ohm
τ_0	= 3.2 psec	L_g	= .097 nH
C_{gs}	= .345 pF	R_s	= .95 ohm
C_{gd}	= .035 pF	L_s	= .016 nH
C_{dc}	= .011 pF	C_{ds}	= .115 pF
R_{gs}	= 4.7 ohm	R_d	= 2.2 ohm
R_{ds}	= 213 ohm	L_d	= .177 nH

ble of replacing the two parallel devices shown in the schematic of Fig. 2 (FB) and described in Fig. 1. However, except for the distributed amplifier's biasing circuitry, the types of the passive elements and their locations are identical to those shown in the schematics of Fig. 2. Only their values have been altered in order to satisfy specific requirements.

1) Reflective Match Amplifier: While no attempt was made to design a 2–18-GHz reflective match amplifier, we have studied the feasibility of a 4–18-GHz module using the GaAs MESFET described in Fig. 1. The rather limited effort was confined to the design of a single-stage module and was terminated with the measurements of its gain and reflection coefficients. The gain performance is plotted in Fig. 11, demonstrating $G = 6.8 \pm 1.1$ between 3 and 17.5 GHz. The measured reflection coefficients range from a maximum of $|S_{11}| = 0.99$ and $|S_{22}| = 0.89$ at 3 GHz to a minimum of $|S_{11}| = 0.33$ and $|S_{22}| = 0.20$ at 18 GHz. Obviously, the extremely high reflection coefficients at the low end of the frequency band and the necessity of using tandem couplers due to the multi-octave bandwidth make it quite a challenge to realize a reliable 2–18-GHz *RM* amplifier performance and therefore invite the consideration of alternate approaches.

2) Feedback Amplifier: Encouraged by the computed results shown in Fig. 4 and the multi-stage characteristics of Table II, it was decided to study the feasibility of a two-stage and a four-stage feedback amplifier. Since, however, the use of two parallel transistors in a feedback amplifier is somewhat impractical, the decision was made to replace the two devices with a single sub half-micron gate GaAs MESFET of matching characteristics. The element values of its equivalent circuit are presented in Table V. Except for the short-circuited shunt element of the input matching network, which was omitted, all stages are characterized by the schematic in Fig. 2 (FB) and connected by a T-shaped interstage matching circuit consisting of an open-circuited shunt stub flanked by transmission lines on both sides. In contrast to the circuit diagram of Fig. 2 (FB), the actual amplifiers incorporated the following resistors: $R_G = 475$ Ω, $R_D = 220$ Ω, $R_{FB1} = 240$ Ω, $R_{FB2} = 500$ Ω in case of the two-stage unit and $R_G = 475$ Ω, $R_D = 235$ Ω, $R_{FB1} = 200$ Ω, $R_{FB2} = R_{FB3} = R_{FB4} = 500$ Ω for the four-stage unit. The overall dimensions of the two circuits are $0.308 \times 0.120 \times 0.015$ in and $0.530 \times 0.120 \times 0.015$ in, respectively. Alumina was used as substrate material.

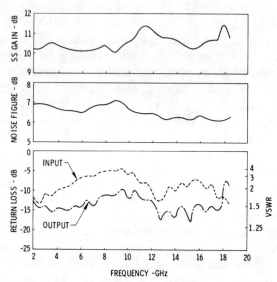

Fig. 12. Measured small-signal gain, noise figure, and return loss of the 2–18-GHz two-stage feedback amplifier.

Fig. 12 shows the curves of the small-signal gain, the noise figure, and the return loss of the two-stage amplifier between 2 and 18.5 GHz. A gain of $G = 10.8 \pm 0.7$ dB and maximum return loss of -4.4 dB (VSWR of 4:1) for the input port and -9.5 dB (VSWR of 2:1) for the output port were measured between 2 and 18 GHz. Across the same frequency band, a maximum noise figure of $NF = 7.1$ dB was recorded. The performance characteristics of the four-stage amplifier are plotted in Fig. 13 between 0.5 and 18.5 GHz. In this case, a gain of $G = 23.1 \pm 1.1$ dB and a maximum return loss of -4.0 dB (VSWR of 4.4:1) for the input port and -7.5 dB (VSWR of 2.5:1) for the output port were measured across the 18.0-GHz bandwidth. The associated maximum noise figures were $NF = 7.9$ dB from 0.5–18.5 GHz and $NF = 7.0$ dB from 2.0–18.5 GHz. As in all of our studies, the emphasis was put on gain flatness and no effort was made to improve either the noise figure or the return loss of the input or the output port. Nevertheless, the above measurements mark, to the best of our knowledge, the lowest instantaneous noise figure reported to date across the respective frequency bands. The minimum output power at the 1-dB compression points was $P_{\text{out}} = 13.5$ dBm.

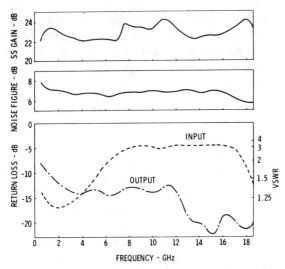

Fig. 13. Measured small-signal gain, noise figure, and return loss of the 0.5–18.5-GHz four-stage feedback amplifier.

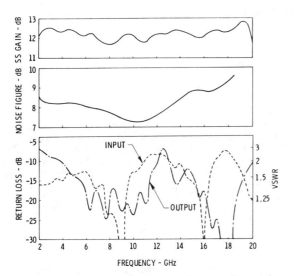

Fig. 14. Measured small-signal gain, noise figure, and return loss of the 2–18-GHz two-stage distributed amplifier.

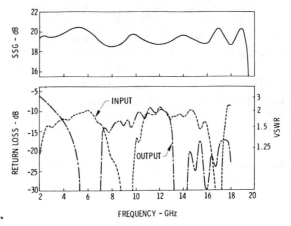

Fig. 15. Measured small-signal gain and return loss of the 2–18-GHz four-stage distributed amplifier.

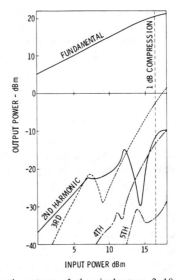

Fig. 16. Harmonic output of the single-stage 2–18-GHz distributed amplifier ($f_0 = 2$ GHz).

3) Distributed Amplifier: The data reported here was taken on a two-stage and a four-stage amplifier whose individual stages are essentially built to the schematic of Fig. 2 (*DA*), with the exception of the drain bias circuitry and the resistance of the drain termination. Since each stage operated at a drain current of approximately 120 mA, a voltage drop of 24 V would have occurred across the 200-Ω termination resulting in 2.9 W of power dissipation in the 3×12 mils tantalum nitride resistor. To avoid a decrease in reliability due to overheating and a loss in efficiency, we chose to bias the drains directly through a high-impedance short-circuited shunt stub located parallel to the termination resistor. The latter was changed from 200 to 125 Ω for best gain flatness. The fabrication and dimensions of the modules have been described elsewhere [7].

The gain, the noise figure, and the return loss of the two-stage unit are plotted in Fig. 14. A gain of $G = 12.3 \pm 0.55$ dB and a maximum return loss of -8 dB (VSWR of 2.3:1) for the input and -7 dB (VSWR of 2.6:1) for the output terminal were measured from 2.0–20.0 GHz, while the maximum noise figure was $NF = 9.6$ dB between 2 and 18 GHz. The curves for gain and return loss of the four-stage amplifier are shown in Fig. 15. This unit exhibits a gain of $G = 19.4 \pm .9$ dB, while a maximum input return loss of -7.5 dB (VSWR of 2.5:1) and output return loss of -6 dB (VSWR of 3.0:1) were achieved between 2.0 and 18.0 GHz. Across the same band, a maximum noise figure of $NF = 11$ dB was measured. It should be reemphasized that both amplifiers were tuned for best gain flatness, compromising noise figure as well as optimum gain performance. Thus far, no attempt has been made to improve the noise figure by implementing the theoretical findings discussed earlier. Finally, Fig. 16 represents the harmonic output power curves of a single-stage module when driven by an input signal of $f = 2$ GHz at various power levels. They show a 23-dB separation between fundamental and the dominant harmonic output power at the 1-dB compression point.

IV. Conclusion

The computed performance characteristics of the reflective match, the lossy match, the feedback, the distributed, and the active match amplifier have been compared across the 2–18-GHz frequency band. In addition, a set of formulas has been developed that demonstrates the significant interdependence of an amplifier's gain and reflection coefficients when feedback, lossy matches, or both are being employed. When utilizing one and the same type of active device in all five circuit types, the computed results reveal gain characteristics that make it difficult to favor one concept over the others. However, when the gain specifications require the cascading of two or more gain modules, as is the case in most practical applications, the reflection coefficients of the input and output ports become of major significance and the choices narrow down with the number of cascaded stages. As demonstrated in Table II, for more than three stages, the distributed amplifier principle is clearly the favorite option. However, one should be quick to point out that matters are not as clear-cut when reducing the bandwidth requirement at the high end of the frequency range.

In search for a solution to improve the noise figure of the distributed amplifier for a given transistor, it was found that any improvement in noise figure impairs the amplifier's gain flatness. Of the three proposed circuit configurations discussed, that of the lossy match distributed amplifier (LMDA) appears to be the most practical solution with the best prospects in noise reduction.

In order to provide an overview of the accomplishments in the field of single-ended amplifiers, some of the previously reported data has been briefly reviewed. In addition, new test results have been presented in support of the computed data. Of the amplifiers tested recently, the reflective match gain module exhibited 6.8 ± 1.1 dB of small-signal gain from 3–17.5 GHz. However, cascading of single-ended modules was not attempted for reasons of instability. A gain of $G = 10.8 \pm 0.7$ dB and a maximum noise figure of $NF = 7.1$ dB were demonstrated in the two-stage feedback amplifier employing lossy match biasing networks. A four-stage feedback amplifier operated between 0.5 and 18.5-GHz exhibited $G = 23.1 \pm 1.1$ dB of small-signal gain and 7.9 dB of maximum noise figure. Above noise figures, though not optimized, are believed to represent state-of-the-art performance in the 2–18.5-GHz and 0.5–18.5-GHz frequency bands. The gain and the maximum noise figure measured in the two-stage distributed amplifier were $G = 12.3 \pm 0.55$ dB and $NF = 9.6$ dB, respectively. A maximum return loss of -7 dB was recorded in this direct-biased unit. With the four-stage distributed amplifier, we were able to demonstrate 19.4 ± 0.9 dB of gain and 11.0 dB of maximum noise figure. Higher order harmonic output power of a single gain module did not exceed -23 dBc up to the 1-dB compression point for a $f = 2$ GHz fundamental input signal.

In conclusion, we have found that the optimum circuit type of a GaAs MESFET amplifier depends to a great degree on the frequency band of interest, the characteristics of the active devices, and the required gain level. For the 2–18-GHz frequency band and transistors with characteristics similar to those of Fig. 1, however, the optimum multistage gain performance is offered by the distributed amplifier.

Acknowledgment

The author wishes to thank R. R. Pereira, who performed all measurements and whose skills in tuning the amplifiers greatly contributed to the success of our studies. In addition, thanks go to J. Martin and M. Lozada, who assembled the circuits. The author is indebted to W. T. Wilser, who kindly edited the manuscript, and to R. Perry, who typed the formulas. Special thanks are due to B. A. Tucker, who modified the existing computer program to include the noise contributed by feedback as expressed by the matrices of Fig. 5. Finally, the author would like to express his appreciation for the constant support and encouragement afforded by W. K. Kennedy during the course of this work.

References

[1] R. S. Engelbrecht and K. A. Kurakawa, "A wideband low noise L-band balanced transistor amplifier," *Proc. IEEE*, vol. 53, pp. 237–247, Mar. 1965.
[2] J. Lange, "Interdigitade stripline quadrature hybrid," *IEEE Trans. Microwave Tech.*, vol. MTT-17, pp. 1150–1151, Dec. 1969.
[3] J. P. Shelton and J. A. Mosko, "Synthesis and design of wideband equal ripple TEM directional couplers and fixed phase-shifters," *IEEE Trans. Microwave Theory Tech.*, vol. MTT-14, pp. 462–473, Oct. 1966.
[4] Y. Tajima and S. Kamihashi, "Multi-conductor couplers," *IEEE Trans. Microwave Theory Tech.*, vol. MTT-26, pp. 795–801, Oct. 1978.
[5] K. B. Niclas and R. R. Pereira, "Feedback applied to balanced FET amps," *Microwave Syst. News*, pp. 66–69, Nov. 1980.
[6] K. B. Niclas, "On design and performance of lossy match GaAs MESFET amplifiers," *IEEE Trans. Microwave Theory Tech.*, vol. MTT-30, pp. 1900–1907, Nov. 1982.
[7] K. B. Niclas, W. T. Wilser, R. T. Kritzer, and R. R. Pereira, "On theory and performance of solid-state microwave distributed amplifiers," *IEEE Trans. Microwave Theory Tech.*, vol. MTT-31, pp. 447–456, June 1983.
[8] K. B. Niclas and B. A. Tucker, "On noise in distributed amplifiers at microwave frequencies," *IEEE Trans. Microwave Theory Tech.*, vol. MTT-31, pp. 661–668, Aug. 1983.
[9] R. L. Van Tuyl, "A monolithic integrated 4-GHz amplifier," *1978 ISSCC, Dig. Tech. Pap.*, pp. 72–73, Feb. 1978.
[10] D. Hornbuckle and R. L. Van Tuyl, "Monolithic GaAs direct coupled amplifiers," *IEEE Trans. Electron Devices*, vol. ED-28, pp. 175–182, Feb. 1981.
[11] D. B. Estreich, "A wideband monolithic GaAs IC amplifier," *1982 ISSCC, Dig. Tech. Papers*, Feb. 1982, pp. 194–195.
[12] V. Pauker and M. Binet, "Wideband high gain small size monolithic GaAs FET amplifiers," *1983 Microwave Symp. Dig.*, June 1983, pp. 81–84.
[13] W. C. Petersen, D. R. Decker, A. K. Gupta, J. Dully, and D. R. Chen, "A monolithic GaAs 0.1 to 10 GHz amplifier," *1981 Microwave Symp. Dig.*, June 1981, pp. 354–355
[14] H. Q. Tserng, S. R. Nelson, and H. M. Macksey, "2–18 GHz, high efficiency, medium-power GaAs FET amplifiers," *1981 Microwave Symp. Dig.*, June 1981, pp. 31–33.
[15] A. M. Pavio, "A network modeling and design method for a 2–18 GHz feedback amplifier," *1982 Int. Microwave Symp. Dig.*, June 1982, pp. 162–165.
[16] P. A. Terzian, D. B. Clark, and R. W. Waugh, "Broadband GaAs monolithic amplifier using negative feedback," *IEEE Trans. Microwave Theory Tech.*, vol. MTT-30, pp. 2017–2020, Nov. 1982.
[17] W. O. Camp, Jr., S. Tiwari, and D. Parsons, "2–6 GHz monolithic microwave amplifier," *1983 Microwave Symp. Dig.*, June 1983, pp. 76–80.

[18] P. N. Rigby, J. R. Suffolk, and R. S. Pengelly, "Broadband monolithic low-noise feedback amplifiers," *1983 Microwave Symp. Dig.*, June 1983, pp. 41–45.

[19] K. Honjo, T. Sugiura, and H. Itoh, "Ultra-broadband GaAs monolithic amplifier," *IEEE Trans. Microwave Theory Tech.*, vol. MTT-30, pp. 1027–1033, July 1982.

[20] T. Obregon and R. Funk, "A 150 MHz–16 GHz FET amplifier," *1981 ISSCC, Dig. Tech. Papers*, Feb. 1981.

[21] K. Honjo and Y. Takayama, "GaAs FET ultrabroad-band amplifiers for Gbit/s data rate systems," *IEEE Trans. Microwave Theory Tech.*, vol. MTT-29, pp. 629–636, July 1981.

[22] M. Mamodaly, P. Quentin, P. Dueme, and J. Obregon, "100 MHz to 17 GHz dual gate variable gain amplifier," *IEEE Trans. Microwave Theory and Tech.*, vol. MTT-30, pp. 918–919, June 1982.

[23] Y. A. Ayasli, L. D. Reynolds, J. L. Vorhaus, and L. Hanes, "Monolithic 2–20 GHz GaAs traveling wave amplifier," *Electron. Lett.*, vol. 18, pp. 596–598, July 1982.

[24] Y. Ayasli, L. D. Reynolds, R. L. Mozzi, J. L. Vorhaus, and L. K. Hanes, "2–20 GHz GaAs traveling-wave power amplifier," *1983 MTT Monolithic Circuits Symp. Dig.*, June 1983, pp. 67–70.

[25] E. Ulrich, "Use of negative feedback to slash wideband VSWR," *Microwaves*, pp. 66–70, Oct. 1978.

[26] K. B. Niclas, W. T. Wilser, R. B. Gold, and W. R. Hitchens, "The matched feedback amplifier: Ultrawide-band microwave amplification with GaAs MESFETs," *IEEE Trans. Microwave Theory Tech.*, vol. MTT-28, pp. 285–294, Apr. 1980.

[27] E. L. Ginzton, W. R. Hewlett, J. H. Jasberg, and J. D. Noe, "Distributed amplifier," in *Proc. IRE*, vol. 36, Aug. 1948, pp. 956–969.

[28] Y. Ayasli, R. L. Mozzi, J. L. Vorhaus, L. D. Reynolds, and R. A. Pucel, "A monolithic GaAs 1-13 GHz traveling-wave amplifier," *IEEE Trans. Microwave Theory Tech.*, vol. MTT-30, pp. 976–981, July 1982.

[29] Y. Ayasli, J. L. Vorhaus, R. L. Mozzi, and L. D. Reynolds, "Monolithic GaAs traveling wave amplifier," *Electron. Lett.*, vol. 15, pp. 413–414, June 1981.

[30] E. W. Strid, K. R. Gleason, and J. Addis, "A dc-12 GHz GaAs FET distributed amplifier," *Res. Abstracts 1981 Gallium Arsenide Integrated Circuit Symp.*, Oct. 1983, p. 47.

A Monolithic GaAs 1–13-GHz Traveling-Wave Amplifier

YALCIN AYASLI, MEMBER, IEEE, ROBERT L. MOZZI, JAMES L. VORHAUS, LEONARD D. REYNOLDS, AND ROBERT A. PUCEL, FELLOW, IEEE

Abstract —This paper describes a monolithic GaAs traveling-wave amplifier with 9-dB gain and ±1-dB gain flatness in the 1–13-GHz frequency range. The circuit is realized in monolithic form on a 0.1-mm GaAs substrate with 50-Ω input and output lines. In this approach, GaAs FET's periodically load input and output microstrip lines and provide the coupling between them with proper phase through their transconductance. Experimental results and the circuit details of such a structure are discussed. Initial results of a noise analysis and predictions on the noise performance are also given.

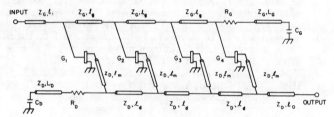

Fig. 1. Schematic representation of four-stage FET traveling-wave amplifier.

I. INTRODUCTION

THE POTENTIAL of traveling-wave or distributed amplification for obtaining gains over wide frequency bands has long been recognized. There is a vast amount of literature on the subject and therefore only two representative references are given [1], [2]. In this approach, the input and output capacitances of electron tubes or transistors are combined with inductors to form two lumped-element artificial transmission lines. These artificial transmission lines are coupled by the transconductance of the active devices.

In actual circuits, however, the extreme bandwidths predicted by a first-order theory are modified by several factors, such as capacitive and inductive couplings, loading of the lumped-element transmission lines due to grid and coil losses, lead inductance, and parasitic capacitances associated with the coil windings. In circuits which employ FET's as the active elements, the gate and drain loading plays a very significant role in the operation and the high-frequency performance of the amplifier.

A new approach to traveling-wave amplification, which is more suitable for obtaining wide-band gain at microwave frequencies, was reported earlier [3]–[5]. In this approach, GaAs FET's are used as the active elements, and the input and output lines are periodically loaded microstrip transmission lines. With such an arrangement, the factors mentioned above as degrading the expected performance are either completely eliminated or their effect is included in the design.

II. AMPLIFIER DESIGN CONSIDERATIONS

A simplified equivalent-circuit diagram of the amplifier is shown in Fig. 1. In this circuit, microstrip lines are periodically loaded with the complex gate and drain impedances of the FET's, forming lossy transmission line structures of different characteristic impedance and propa-

Manuscript received December 30, 1981; revised February 3, 1982.
The authors are with Raytheon Research Division, 28 Seyon Street, Waltham, MA 02254.

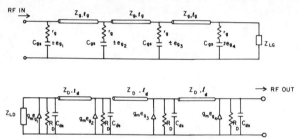

Fig. 2. Simplified equivalent-circuit diagram of FET traveling-wave amplifier.

gation constant. The resultant effective input and output propagation structures are referred to as the gate and drain lines.

An RF signal applied at the input end of the gate line travels down the line to the other end, where it is absorbed by the terminating impedance. However, a significant portion of the signal is dissipated by the gate circuits of the individual FET's along the way. The input signal sampled by the gate circuits at different phases (and generally at different amplitudes) is transferred to the drain line through the transconductance of the FET's. If the phase velocity of the signal at the drain line is identical to the phase velocity of the gate line, then the signals on the drain line add. The addition will be in phase only for the forward-traveling signal. This can readily be verified by examining the various possible signal paths between the input and output terminals. Any signal which travels backward, and is not quite cancelled by the out-of-phase additions, will be absorbed by the complex drain impedance.

A simplified equivalent-circuit diagram for the amplifier is shown in Fig. 2. In conventional amplifiers, one cannot increase the gain-bandwidth product by paralleling FET's, because the resulting increase in transductance g_m is compensated for by the corresponding increase in the input and output capacitances. The distributed amplifier overcomes this difficulty by adding the individual g_m's of the FET's without adding their input and output capacitances. If the spacing between the FET's is small compared with the wavelength, the characteristic impedances of the gate and drain lines shown in Fig. 2 can be approximated as

$$Z_g \simeq [L_g/(C_g + C_{gs}/l_g)]^{1/2} \quad (1)$$

and

$$Z_d \simeq [L_d/(C_d + C_{ds}/l_d)]^{1/2} \quad (2)$$

where L_g, C_g and L_d, C_d are the per-unit-length inductance and capacitance of the gate and drain lines, respectively. C_{gs} is the input gate-to-source capacitance and C_{ds} is the output drain-to-source capacitance of the unit cell FET; l_g and l_d are the lengths of the unit gate- and drain-line sections, respectively (see Fig. 2). The effect of resistive components r_g and r_d are neglected. The impedance expressions in (1) and (2) are clearly independent of the number of FET's used in the circuit.

Using the simplified equivalent-circuit of Fig. 2 and approximating the gate and drain lines as continuous structures, the gain expression for an n-cell circuit can be derived as

$$G = g_m^2 Z_d Z_g \left| \frac{\gamma_g l_g [\exp(-\gamma_g l_g n) - \exp(-\gamma_d l_d n)]}{\gamma_g^2 l_g^2 - \gamma_d^2 l_d^2} \right|^2 \quad (3)$$

where

$$Z_g \simeq \left[\frac{L_g}{C_g + \frac{C_{gs}}{l_g}} \right]^{1/2}$$

$$Z_d \simeq \left[\frac{L_d}{C_d + \frac{C_{ds}}{l_d}} \right]^{1/2}$$

and

$$\gamma_g \simeq j\omega \sqrt{L_g \left(C_g + \frac{C_{sg}}{l_g} \right)}$$

$$+ \frac{1}{2} \frac{r_g \omega^2 C_{sg}^2}{l_g} \sqrt{\frac{L_g}{\left(C_g + \frac{C_{sg}}{l_g} \right)}} \equiv j\beta_g + \alpha_g$$

$$\gamma_d \simeq j\omega \sqrt{L_d \left(C_d + \frac{C_{ds}}{l_d} \right)}$$

$$+ \frac{1}{2} \frac{1}{R_D l_d} \sqrt{\frac{L_d}{\left(C_d + \frac{C_{ds}}{l_d} \right)}} \equiv j\beta_d + \alpha_d.$$

Under normal operating conditions, the signals in the gate and drain lines are near synchronism ($\beta_g l_g \simeq \beta_d l_d$). If it is further assumed that $|\gamma_g| \simeq |\beta_g|$, and $|\gamma_d| \simeq |\beta_d|$, and $Z_g \simeq Z_d \equiv Z_0$, (3) can be simplified to

$$G = \frac{g_m^2 Z_0^2}{4} \frac{[\exp(-\alpha_g l_g n) - \exp(-\alpha_d l_d n)]^2}{(\alpha_g l_g - \alpha_d l_d)^2}. \quad (4)$$

This expression clearly shows that, as the number of cells n is increased, gain does not increase monotonically. This conclusion is in contrast to early tube distributed amplifier theories. In fact, as n gets large, gain approaches zero in the limit.

For values of $\alpha_g l_g n \leqslant 1$ and when drain-line losses are neglected compared with gate-line losses, (4) can be rewritten as

$$G \simeq \frac{g_m^2 n^2 Z_0^2}{4} \left(1 - \frac{\alpha_g l_g n}{2} + \frac{\alpha_g^2 l_g^2 n^2}{6} \right)^2. \quad (5)$$

We note that, in this operating regime, gain can be made proportional to n^2.

The gain expression derived above represents the actual circuit response reasonably well and can be used as a useful design tool. For instance, if a practical upper limit for the gate-line attenuation is assumed as $\alpha_g l_g n \leqslant 1$, then using the expression for the gate-line attenuation constant α_g, we find

$$r_g \omega^2 C_{sg}^2 Z_0 n \leqslant 2. \quad (6)$$

Thus we observe that, for a given FET, the maximum number of cells n that can be employed in the traveling amplifier can be determined from (6). Since the gate periphery of a unit cell is known, (6) also brings an upper limit to the total gate periphery that can be used in a practical amplifier. Clearly, this upper limit should be determined at the high end of the frequency band where satisfying the inequality of (6) is most difficult.

The performance of the traveling-wave amplifier was examined in a 2–12-GHz design band as a function of the more significant FET parameters such as r_g and R_D and the number of cells n. For this study, the simplified equivalent circuit of a FET traveling-wave amplifier shown in Fig. 2 was used and typical 1-μm gate length GaAs FET parameters assumed.

In Fig. 3, gain is plotted as a function of frequency as the number of cells or sections is varied. Clearly there is an optimum for n; going above that optimum value actually degrades the gain, starting from the high frequency end of the design band where gate and drain loading is the most severe.

In Fig. 4, the effect of the gate loading is examined as the gate resistance r_g used in the design (22 Ω) is set to zero or doubled. One can redesign the circuit elements to regain the flat gain performance, as shown in Fig. 5. This figure clearly indicates that the resistive part of the gate loading typically results in 3-dB gain reduction and doubling the effect of this loading gives an additional 3-dB reduction in gain. However, it is satisfying to see that the gain response can still be flattened despite assumed large attenuation on the gate line.

RF voltage variation along the gate line at individual

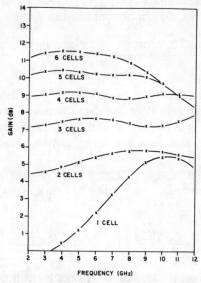

Fig. 3. Gain versus frequency, as the number of cells is varied.

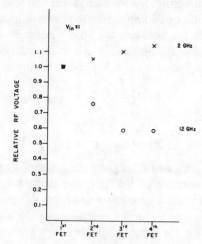

Fig. 6. Variation of the microwave signal along the gate line, at actual gate positions.

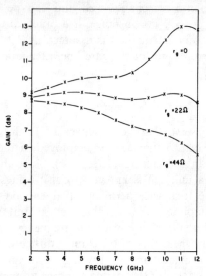

Fig. 4. Effect of the gate loading on gain versus frequency response.

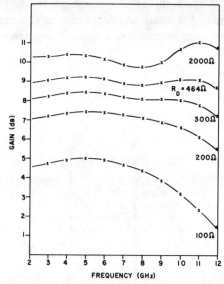

Fig. 7. Effect of drain loadings on gain versus frequency response.

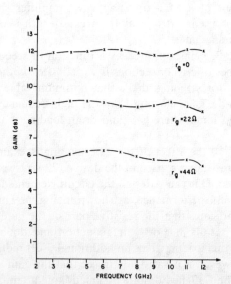

Fig. 5. Flat gain design at three different gate loadings.

gate terminals is shown in Fig. 6 for the four-cell 9-dB gain design. The decrease of the signal level at the actual gates is less than one would calculate from the $\exp(-\alpha_g l_g n)$ expression obtained on the basis of a single propagating wave in a continuous-line model. In such an approximation, reflections from the gate and drain-line impedances are not taken into account. For this reason, gain formulas (4) and (5) tend to underestimate the actual achievable gains by 1–2 dB.

In Fig. 7, the effect of the resistive drain loading on the performance of the four-cell design is examined. As the drain resistance is increased to values much larger than the 464 Ω used in the design, gain increases by about 1 dB. Note that, compared with the effect of gate loading, the effect of the drain loading is smaller. As the loading is increased further by decreasing R_D, gain at first comes down uniformly, then faster at the high band end. Since the attenuation constant α_d is independent of frequency to first order, this behavior indicates that FET's first in line contribute to the high-frequency performance more than the later stages. This is not surprising when we consider that the later FET stages are not excited as well as the first few stages at the high end of the frequency band.

III. Circuit Fabrication

Circuits are processed on vapor-phase epitaxy layers grown by the $AsCl_3$ system on semi-insulating GaAs substrates. The three layer structures consist of a high-doped contact layer ($n > 2 \times 10^{18}$ cm^{-3}, $t = 0.2$ μm), an active layer of moderate doping ($n = 9 \times 10^{16}$ cm^{-3}, $t = 0.3$, μm), and an undoped buffer region ($n < 5 \times 10^{13}$ cm^{-3}, $t = 2.0$ μm). Device isolation is achieved with a combination of a shallow mesa etch and a damaging $^{16}O^+$ implant.

Ohmic contacts are formed by alloying the standard Ni/AuGe metalization into the surface. The ohmic metal also forms the bottom plates of the thin-film capacitors. The gates, which are recessed, consist of a Ti/Pt/Au (1000/1000/3000 Å) metalization and are nominally 1 μm long.

The capacitor dielectric is a plasma-assisted CVD silicon nitride layer with a nominal thickness of 5000 Å and relative dielectric constant of 6.8. The thin-film resistor material is titanium. During deposition, the film thickness is monitored using a four-point resistance setup to assure a final sheet resistance of about 6.7 $\Omega/\square$.

The final frontside processing steps define the transmission-line structures, capacitor top plates, and air-bridge interconnects. All of these are fabricated out of plated gold about 3–4 μm thick. The air-bridges are used to connect from the GaAs surface to the top plates of the MIM capacitors without having to cross the dielectric step and risk shorting of the structure.

After plating, the wafer is lapped to its final thickness of 100 μm by first mounting it upside down on an alumina substrate. Via-holes are etched through the wafer to ground points on the frontside. The via-holes are aligned by looking through the slice with infrared optics to see the frontside pattern. Finally, a chip-dicing grid is defined in the back by alignment to the via-hole pattern and the region between the grid lines (the chip back) is plated to a thickness of 12–15 μm with gold. The grid lines are etched through the frontside, the wafer dismounted, and the chips allowed to simply fall apart.

IV. Circuit Description and Experimental Results

The circuit shown in Fig. 1 is realized in monolithic form on 0.1-mm GaAs with 50-Ω input and output lines, as shown in Fig. 8. The chip dimensions are 2.5 mm $\times$ 1.65 mm. The total gate periphery is 4×300 μm with nominal 1-μm gate length. Devices typically have -2-V pinchoff voltages. The design calls for 9-dB gain in the 2–12-GHz frequency band. The experimental performance in the 0.5–14-GHz frequency band is shown in Figs. 9 and 10. In Fig. 10, the predicted gain points are also included, indicating good agreement with the 9-dB$\pm$1-dB experimental gain performance obtained in the 2–13-GHz band. The input–

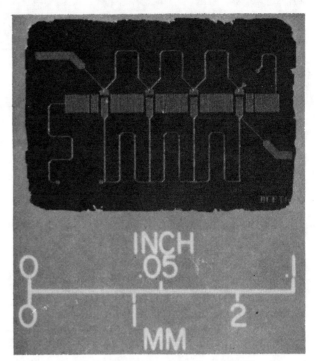

Fig. 8. GaAs monolithic traveling-wave amplifier chip.

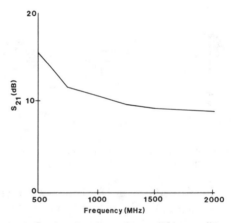

Fig. 9. Experimental performance of monolithic traveling-wave amplifier at 0.5–2-GHz frequency range.

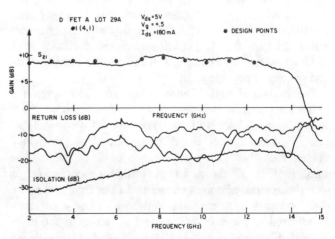

Fig. 10. Experimental performance of monolithic traveling-wave amplifier at 2–14-GHz frequency range.

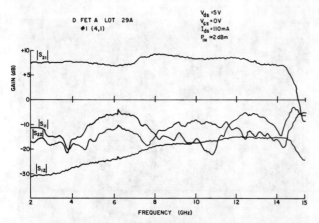

Fig. 11. Experimental performance of the traveling-wave amplifier at zero gate bias. Note the increase in bandwidth up to 14.5 GHz.

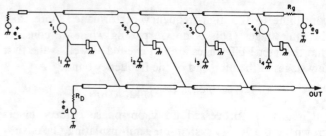

Fig. 13. Schematic representation of the relevant noise sources in a four-cell traveling-wave amplifier.

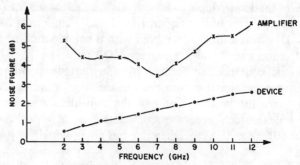

Fig. 14. Calculated noise performance of the amplifier in comparison to the noise performance of the devices used.

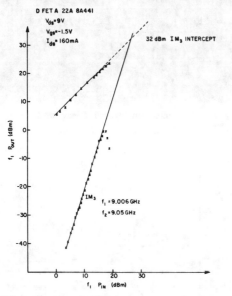

Fig. 12. Third-order intermodulation measurement results for an amplifier with -4-V pinchoff devices.

output isolation is better than 18 dB across the band. In Fig. 11, performance at 0-V gate bias is shown. The gain is 1-dB lower at 8 dB $\pm$1 dB, but the high-frequency end of the gain curve now extends to 14.5 GHz, in agreement with (6).

To observe the power performance, circuits with 4-V pinchoff voltages were processed. These circuits gave power outputs of 300 mW at the 1-dB gain compression point with 6-dB gain and 17 percent power added efficiency at 10 GHz. In Fig. 12, the results of third-order intermodulation measurements are presented.

The measured noise figure of the amplifier when the FET's are biased at the minimum noise condition is 3.9 dB at 4 GHz and 5.4 dB at 10 GHz with 5.4-dB associated gain. We have also developed a computer program which can predict the noise performance of a traveling-wave amplifier. Fig. 13 shows a typical amplifier configuration with the associated noise generators. In this figure, e_j and i_j with $j = 1$ to 4 represent the intrinsic FET noise generators, all referred to the gate side; e_g and e_d represent the noise generated by the gate- and drain-line load resistances, respectively. Given the noise parameters R_n, Y_{opt}, and F_{min} of the FET's used in the circuit, the program calculates the contribution of all the individual noise generators to the overall noise figure of the amplifier, including the correlations between e_j's and i_j's. We have applied the program to a 2–12-GHz amplifier design using the low-noise S-parameters of 300-μm periphery devices. The noise performance of such an amplifier with an associated gain of 5.5 dB is shown in Fig. 14.

V. Conclusion

We have described a traveling-wave amplifier with 9-dB $\pm$1-dB gain over a 1–13-GHz bandwidth, demonstrating that traveling-wave amplification in microwave frequencies is realizable with GaAs FET's and distributed input and output lines. The experimental results are in excellent agreement with the theoretical predictions. The complete amplifier is realized with monolithic circuit technology on a 2.5-mm $\times$ 1.65-mm $\times$ 0.1-mm chip.

We have examined the microwave performance of such a structure as a function of the important FET and amplifier parameters, derived gain expressions including the effect of gate- and drain-line loading, and given initial results on the noise performance of such a structure.

References

[1] E. L. Ginzton, W. R. Hewlett, J. H. Jasburg, and J. D. Noe, "Distributed amplification," in *Proc. IRE*, vol. 36, pp. 956–969, 1948.
[2] W. Jutzi, "A MESFET distributed amplifier with 2-GHz bandwidth," in *Proc. IEEE*, vol. 57, pp. 1195–1196, 1969.
[3] Y. Ayasli, J. L. Vorhaus, R. Mozzi, and L. Reynolds, "Monolithic GaAs traveling-wave amplifier," *Electron. Lett.* vol. 17, p. 12, 1981.
[4] Y. Ayasli, R. Mozzi, J. L. Vorhaus, L. Reynolds, R. A. Pucel, "A monolithic GaAs 1 to 13 GHz travelling-wave amplifier," presented at GaAs IC Symp. (San Diego, CA), Oct. 27–29, 1981.
[5] Y. Ayasli, "Monolithic GaAs travelling-wave amplification at microwave frequencies," presented at 8th Biennial Conf. Active Microwave Semiconductor Devices and Circuits, (Cornell University), Aug. 11–13, 1981.

2 TO 30 GHz MONOLITHIC DISTRIBUTED AMPLIFIER

James M. Schellenberg, Hiro Yamasaki, and Peter G. Asher

Hughes Aircraft Company
Torrance Research Center
Torrance, CA 90509

ABSTRACT

A low noise distributed amplifier is demonstrated over the 2 to 26.5 GHz band. The device exhibits a gain of 6 ± 0.3 dB over this band with a maximum noise figure of 5.4 dB and a maximum input/output VSWR of 1.7. The output power at the 1 dB gain compression point is typically 13 dBm. The 7-section distributed amplifier is fabricated on a chip with dimensions of 1.1x3.2 mm using optical lithography.

INTRODUCTION

Distributed amplifiers (DAs), while inherently broadband, cannot match the narrow band performance in noise figure or power of reactively matched amplifiers. However, for sheer bandwidth, the DA approach appears to be unparalleled. To date, DAs have demonstrated broadband low noise performance[1,2] and power levels of up to 250 mW over a decade (2 to 20 GHz) bandwidth.[3]

In addition, DAs are relatively insensitive to FET device and circuit parameter variation, exhibit good input/output match, and are stable devices with little tendency to oscillate. The DA may yet prove to be the most useful as a "universal" amplifier gain block which can be used in the intermediate stages of either a low noise or power amplifier. This paper presents the first DA spanning the 2 to 30 GHz band.

CIRCUIT DESCRIPTION

The DA circuit configuration is shown in Figure 1, and the parameters of the circuit are summarized in Table 1. Here, an array of active devices (GaAs FETs) couple an input transmission line to an output tranmission line. In effect, a single large FET with its associated matching networks, is replaced by an array of smaller devices which are distributed along the transmission lines. The gate and drain capacitive elements of the FET devices are incorporated into the parameters of the input/output lines, thereby forming matched, lumped-distributed transmission lines. As a result, this approach exhibits extremely broad bandwidth capability. However, due to the lumped nature of the FETs attached to the transmission lines, this structure exhibits a low-pass type frequency response which is characterized by an upper cutoff frequency. This upper cutoff frequency is predominantly determined by the size (gate periphery) of the individual constituent FET devices. In order to realize a cutoff frequency of approximately 35 GHz (required for 30 GHz operation), a gate capacitance of approximately 0.09 pf is required. With a 0.5 μm gate structure, this corresponds to a gate periphery of 68 μm and a total gate periphery of 476 μm for 7 sections which are required to realize optimum gain performance over the 2 to 30 GHz band.

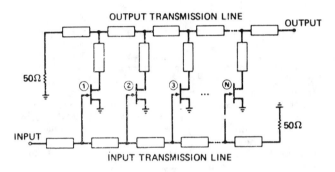

Figure 1   Distributed amplifier configuration.

The transmission line segments in the drain circuits serve two basic purposes. First of all, they provide capacitive compensation in the drain circuit equalizing the FET gate and drain capacitances. Secondly, this transmission line element is used to compensate for inherent gain rolloff due primarily to the input transmission line losses which increase as ω^2. The terminations and bias networks are included off chip in order to conserve chip area and provide maximum flexibility for testing and diagnostics.

TABLE 1

DISTRIBUTED AMPLIFIER DESIGN PARAMETERS

Device Gate Length	0.5 μm
Device Gate Width	68 μm
Gate Capacitance	0.09 pf
TL Z_0	85 OHMS
TL Length (26.5 GHz)	73°
Number of Sections	7

CIRCUIT FABRICATION

A photograph of the 7-section DA chip is shown in Figure 2.[1] The circuit is fabricated on a 100 μm thick GaAs substrate with dimensions of 1.1x3.2 mm. Via holes are etched from the back of the chip to form the grounded source contacts. Proton bombardment is employed to isolate the circuit elements from the active area.

The FET device detail is shown in Figure 3. The device geometry consists of two Ti/Pt/Au gate fingers with dimensions of 0.4x34 μm which are formed by optical lithography. The gate is recessed by approximately 0.2 μm. The GaAs active layer is formed by VPE with a thickness of 0.4 μm and a doping level of $2.4 \times 10^{17} cm^{-3}$. The device g_m is typically 160 mS/mm.

PERFORMANCE

The gain performance of a typical amplifier chip is shown in Figure 4, and the corresponding noise figure is shown in Figure 5. The bias conditions are summarized as $V_G = 0$, $V_D = 2.5$ V and $I_D = 45$ mA. Over the 2 to 26.5 GHz band, this chip has demonstrated a gain level of 6.0 ± 0.3 dB. This was measured in two separate test fixtures - one covering the 2 to 18 GHz band and the other the 18 to 26.5 GHz band. A similar chip from the same wafer exhibited a gain level of 4 dB at 32 GHz with the gain falling rapidly above 32 GHz. As shown in Figure 5, the noise figure increases gradually from 4.2 dB at 8 GHz to 5.4 dB at 26.5 GHz. The noise figure was not measured below 8 GHz, but is assumed to be nearly constant. The input/output VSWR was typically better than 1.5 with a worst case of 1.7 at 22 GHz. The output power was typically 13 dBm at the 1 dB gain compression point.

CONCLUSION

The first 2 to 30 GHz monolithic DA has been demonstrated with excellent noise and gain performance. This distributed approach has yielded excellent repeatability and a tolerance for device and circuit parameter variations. By cascading several of these chips, it should be possible to achieve a multiple stage amplifier spanning the 2 to 30 GHz band with a maximum noise figure of 7 dB.

ACKNOWLEDGEMENTS

The authors would like to thank A. Gomez and L. Marich for their contributions to this work. This work was supported in part by contracts N00014-82-C-0833

Figure 2   7-section distributed amplifer chip.

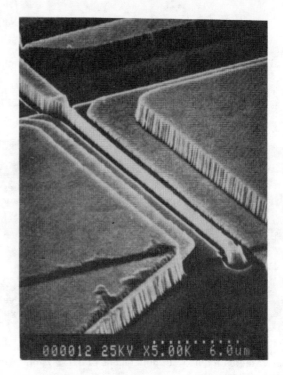

Figure 3   FET device structure.

from the Office of Naval Research and N00014-82-C-2493 from the Naval Research Laboratory.

REFERENCES

(1) D.E. Dawson, M.J. Salib and L.E. Dickens; "Distributed Cascade Amplifier and Noise Figure Modeling of an Arbitrary Amplifier Configuration;" ISSCC Sym. Dig., pp. 78-79; February 1984.

(2) W. Kennan, T. Andrade and C. Huang; "A Miniature 2-18 GHz Monolithic GaAs

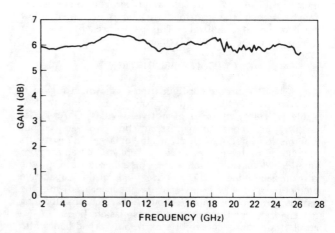

Figure 4   Gain performance of distributed amplifier.

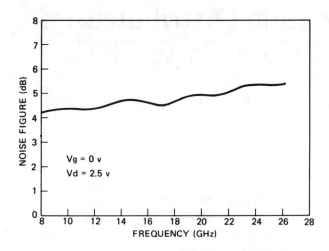

Figure 5 Noise figure performance of distributed amplifier.

Distributed Amplifier;" IEEE Monolithic Circuits Symposium; May 1984.

(3) V. Ayasli, L.D. Reynolds, R.L. Mozzi and L.K. Hanes; "2-20 GHz GaAs Traveling-Wave Power Amplifier;" IEEE Trans. Microwave Theory Tech. MTT-32, pp. 290-295; March 1984.

A 2–18-GHz Monolithic Distributed Amplifier Using Dual-Gate GaAs FET's

WAYNE KENNAN, THOMAS ANDRADE, MEMBER, IEEE, AND CHARLES C. HUANG, MEMBER, IEEE

Abstract—This paper describes a 2–18-GHz monolithic distributed amplifier with over 6-dB gain, ±0.5-dB gain flatness, and less than 2.0:1 VSWR. Measured noise figure is below 7.5 dB, and power output capability is greater than 17 dBm. The amplifier is designed with dual-gate GaAs FET's instead of single-gate FET's for maximum gain over the design bandwidth. Cascaded amplifier performance will also be presented.

I. INTRODUCTION

IN THE PAST several years, distributed amplification [1] has enjoyed a renaissance due to the GaAs FET [2]–[4]. Applied originally to electron tubes, this amplification technique has the unique capability of adding device transconductance without adding device parasitic capacitance. This is accomplished by linking the parasitic shunt capacitance of the devices with series inductors to form an artificial low-pass transmission line. By terminating these links with resistive loads, the unwanted signals are dissipated while the desired signals are added in-phase at the output of the amplifier. The result is unprecedented gain-bandwidth product with flat gain and low VSWR. The structure is shown in Fig. 1.

II. CIRCUIT DESIGN

The topology of the 2–18-GHz distributed amplifier is shown in Fig. 2, and a SEM photograph appears in Fig. 3. In this design, there are essentially three features which distinguish it from previous distributed amplifiers. First and most important, the design uses dual-gate GaAs FET's in place of the more traditional single-gate devices. The contribution of dual-gate FET's to distributed amplification is equivalent to that of cascode-connected single-gate devices [5]. The dual-gate FET, which is in fact modeled as a cascode connection of single-gate FET's, has an input impedance comparable to single-gate devices but much higher isolation and output impedance. This is evident from the equivalent circuit models shown in Fig. 4, which were derived from measured S-parameters. (Note that the models include a common inductance of 0.04 nH to account for ground path inductance in the monolithic chip.) High reverse isolation in the device is necessary for high amplifier isolation and often extends the amplifier's bandwidth. High device output impedance, on the other hand, improves gain flatness and output VSWR, and increases gain. This is because the single-gate FET's output resistance is relatively low (250 Ω for a 250-μm device) and a significant load on the drain transmission line.

At low frequencies where the output resistances of the devices are virtually in parallel, a four-section design would result in a 62.5-Ω resistive load on the 50-Ω impedance transmission line. This problem is completely eliminated with the dual-gate (or cascode-connected) device.

The second feature of this design is the distribution of total gate width among the individual devices. In order to achieve a minimum gain of 7 dB with devices scaled from the model of Fig. 4 and 8-μm cm-wide transmission lines,

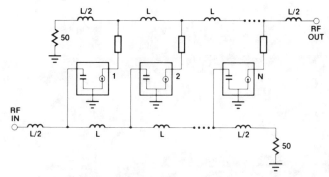

Fig. 1. Simplified distributed amplifier structure.

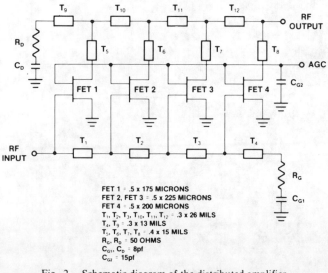

Fig. 2. Schematic diagram of the distributed amplifier.

Manuscript received May 3, 1984. This work was supported in part by the Office of Naval Research under Contract N00014-81-C-0101.
The authors are with Avantek, Inc., Santa Clara, CA 95951.

Fig. 3. SEM micrograph of the distributed amplifier.

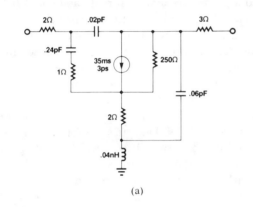

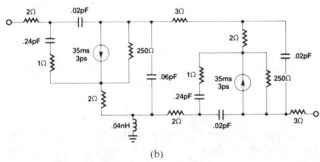

Fig. 4. (a) Single-gate and (b) dual-gate FET equivalent circuit models (250 μm).

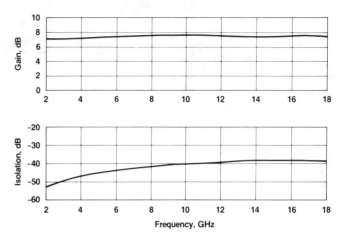

Fig. 5. Predicted gain and isolation of the dual-gate FET distributed amplifier.

which are fairly lossy, the amplifier requires at least 800 μm of gate width. This gate width could be theoretically partitioned into nearly any number of sections, but four sections prove optimal in many respects. First, it was desired to absorb the input and output bond wires into the distributed amplifiers input and output $L/2$ sections. For an input and output impedance of 50 Ω with 0.3-nH bond wires, this means

$$\sqrt{L/C} = 50 \text{ or } C = L/2500 = 0.3 \text{ nH}/1250 = 0.24 \text{ pF}.$$

This capacitance corresponds to a device gate width of 250 μm, but a choice of 200 μm leaves margin for error and the shunt capacitance of the high impedance transmission lines. Secondly, the four-section design offers a good compromise in gain flatness, VSWR, and noise figure over other 800-μm designs. This was determined by simulating the alternatives with a microwave analysis program. The four-section design is also more area efficient than its alternatives and results in a nearly square chip for ease of handling. As a final touch, the four individual device gate widths were optimized for VSWR and gain flatness. As shown in Fig. 2, the first and last sections are smaller than the two internal devices. This is primarily to absorb the parasitic capacitance of the input and output bonding pads.

The third feature of the design is its small area. At 0.75 mm × 0.85 mm the chip area is 0.64 mm², which yields a potential of over 2500 amplifiers per 2-in-diam wafer. This is chiefly a result of the single-turn inductors and wraparound ground. The single-turn inductor is modeled with lengths of coupled transmission line to account for coupling between the two major lengths and coupling to the FET sources (ground). If these inductors were laid out in a straight line rather than coiled one turn, the chip height would be unchanged at 0.75 mm but its length would be increased by 1.5 mm to 2.35 mm. The resulting layout would occupy nearly three times the area of the present one. The wraparound ground is also helpful in reducing chip area since the perimeter of the chip is normally not used. Via-hole grounds, on the other hand, require prime chip area and may be significant in size.

The amplifier contains three capacitors which are used for RF bypass to ground. CG1 and CD are used to bypass the input and output terminating resistors, respectively, so that dc power is not dissipated in these elements. The third capacitor, CG2, is used to bypass the second gates of the dual-gate FET's. This provides isolation from external dc circuitry and insures that the dual-gate FET operates as a cascode circuit. The FET sources are all grounded, thus requiring two bias voltages—one positive for the drain and one negative for the gate.

The simulated gain and isolation of the dual-gate distributed amplifier are shown in Fig. 5. From 2–18-GHz, the predicted gain is 7.25 ± 0.22 dB with greater than 35-dB isolation. Predicted return loss is shown in Fig. 6 and is

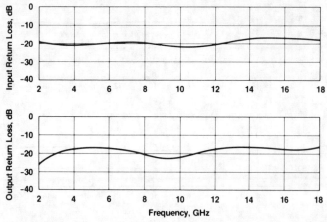

Fig. 6. Predicted input and output return loss of the dual-gate FET distributed amplifier.

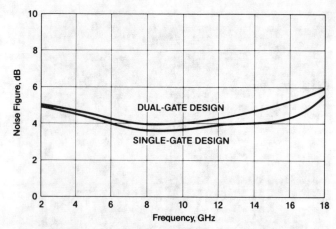

Fig. 9. Predicted noise figure of single-gate FET and dual-gate FET distributed amplifiers.

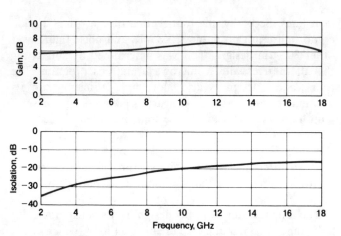

Fig. 7. Predicted gain and isolation of the single-gate FET distributed amplifier.

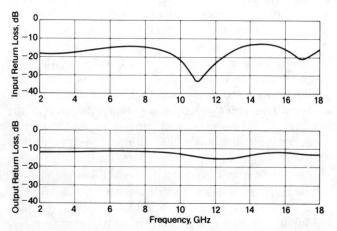

Fig. 8. Predicted input and output return loss of the single-gate FET distributed amplifier.

greater than 14 dB (1.5:1 VSWR) at both the input and output. Figs. 7 and 8 show the predicted performance of the same amplifier when the dual-gate FET models are replaced with the single-gate FET model from Fig. 4. It is clear that gain, gain flatness, isolation, output VSWR, and bandwidth are all degraded as expected. Noise figure, however, is lower in the single-gate design as shown in Fig. 9.

III. Circuit Fabrication

Ion-implanted GaAs is used as the starting material for the IC's due to its excellent uniformity and controllability. After implantation, the wafers are annealed at 800 °C until the active layer sheet resistance drops to approximately 500 Ω/square. This layer is then selectively etched to form mesas for the FET's and resistors. Later, the resistor mesas are trimmed to 800 Ω/square with a process that is controllable to a standard deviation of 15 percent.

The FET's in the IC are fabricated with the same process used for discrete FET's. The gates are formed on a nominally 0.5-μm-long base of TiW/Au, which is gold-plated to 0.7 μm. The resulting structure achieves very short gate length with large gate cross-sectional area for high device transconductance with low parasitic capacitance and resistance [6]. Source and drain ohmic contacts are formed with a AuGe/Ni/Au alloy.

Parallel-plate dielectric capacitors and surface passivation are provided by a thin layer of plasma-enhanced CVD silicon nitride. This process achieves a capacitance density of 390 pF/mm^2 with a standard deviation of less than 40 pF/mm^2.

Metallic interconnections are achieved with a two-level wiring process which provides surface connections, crossovers, and air bridges. The top level is situated 3 μm above the GaAs surface and is gold-plated to 1.5 μm. The bottom level rests directly on the semi-insulating substrate and is 0.8 μm thick. For additional thickness, the top level is deposited directly on the bottom level to achieve a thickness of 2.3 μm. These lines are designed for a dc current density not to exceed 5×10^5 amps/cm^2.

Wraparound ground technology is chosen over via-hole ground technology for low parasitic inductance and improved area efficiency. To form the wraparound ground, metal is first deposited and gold-plated to 2 μm on the frontside of the chips. The wafer is then lapped to 115 μm and backside metallized to complete the wraparound ground connection.

IV. Circuit Performance

Before backside-lapping and die-separation, the GaAs wafer is stepped and dc-probed for saturated current,

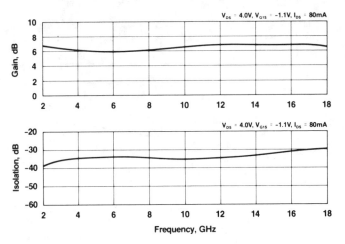

Fig. 10. Measured gain and isolation of the distributed amplifier.

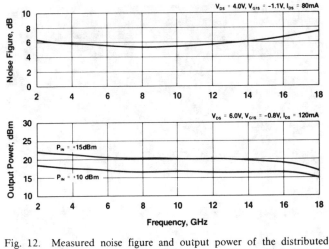

Fig. 12. Measured noise figure and output power of the distributed amplifier.

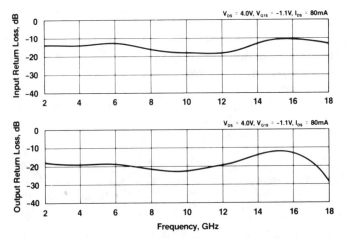

Fig. 11. Measured input and output return loss of the distributed amplifier.

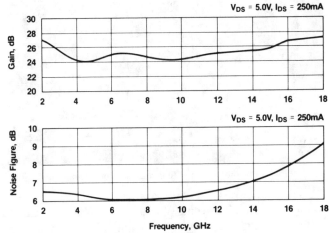

Fig. 13. Measured gain and noise figure of a four-stage distributed amplifier.

pinchoff voltage, and transconductance. Devices which are open-circuited, short-circuited, or otherwise fail the dc test are identified with an ink spot and later discarded. Data on passed devices is presented in summary form and may also be formatted into histograms for statistical analysis. The chips are then separated, visually inspected, and prepared for assembly into packages and thin-film hybrid circuits.

For RF evaluation, the amplifiers are mounted on 15-mil-thick alumina substrates. The substrate includes 50-Ω transmission lines, bias resistors, and plated through slots for ground. The data reported in this paper was measured on IC's mounted on the substrate with input and output bonding wires and no tuning. Bias was injected through external bias tee's.

Fig. 10 shows gain and isolation measured on a typical amplifier fabricated within process specifications. The gain is 6.3 dB $\pm$ 0.5 dB with greater than 25-dB isolation. Fig. 11 shows input and output return loss for the same chip. The worst case VSWR is 2.0:1 although it is less than 1.5 over most of the band. The device is biased at 4.0-V VDS, 80-mA IDS, which is half the saturated current level. Higher gain may be achieved with increased drain current, but gain flatness degrades slightly.

Noise and power performance are illustrated in Fig. 12. Noise figure is typically less than 6 dB and rises to 7.5 dB at 18 GHz. This can be reduced with an adjustment in bias but with a corresponding loss in associated gain. Output power is plotted from 2-18 GHz with constant input power levels of 10 dBm and 15 dBm. The device is capable of 20-dBm power over most of the band, but it degrades to 17 dBm at 18 GHz. Gain compression is more severe at the higher frequencies, as can be determined from the two plots.

Cascaded performance is demonstrated in Fig. 13. This data was measured on a four-stage amplifier consisting of two alumina substrates, four IC's, and ten bypass capacitors. The amplifier includes bias circuitry and measures only 5.1 mm $\times$ 12.7 mm $\times$ 0.38 mm (Fig. 14). This assembly shows that, even without integrating the blocking capacitors and bias circuitry on chip, a linear gain density of 50 dB per inch is easily achieved over the full 2-18-GHz band. With more compact hybrid layouts, this number could easily double.

The last figure (Fig. 15) illustrates a number of possible applications for the IC. As an AGC amplifier, gain variation is very flat over the full 2-18-GHz range when gate 2

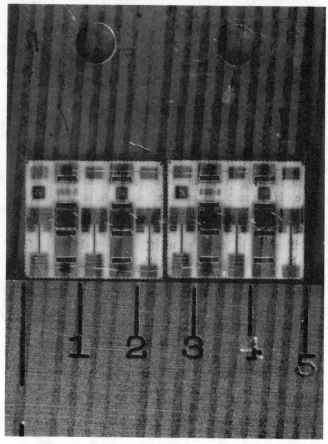

Fig. 14. Photograph of four-stage distributed amplifier.

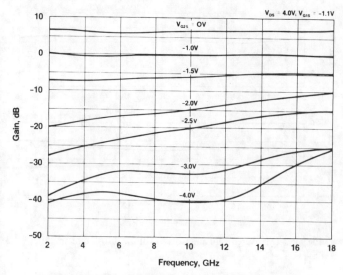

Fig. 15. Measured gain/loss as a function of gate 2 voltage.

voltage is varied between 0 V and −1 V. When the gate 2 voltage is made more negative, the amplifier becomes lossy and may be used as a limiter. Finally, when the voltage is increased to −4 V, the drain current drops to nearly zero and the amplifier provides over 25-dB isolation. This property could be used in switching applications since VSWR and reverse isolation remain less than 2:1 and greater than 25 dB, respectively, under all bias conditions.

V. Conclusion

A 2–18-GHz monolithic GaAs distributed amplifier with over 6-dB gain has been described. Dual-gate FET distributed amplifiers were compared to single-gate FET distributed amplifiers and were shown to provide more gain with better flatness, VSWR, and bandwidth. This is demonstrated in a four-stage amplifier which achieved 25.5-dB ± 1.5 dB gain from 2–18 GHz. The dual-gate FET distributed amplifier may also be used for many control functions by adjusting the gate 2 voltage.

References

[1] E. L. Ginzton, W. R. Hewlett, J. H. Jasburg, and J. D. Noe, "Distributed amplification," *Proc. IRE*, vol. 36, pp. 956–969, 1948.

[2] E. W. Strid and K. R. Gleason, "A dc-12 GHz monolithic GaAs FET distributed amplifier," *IEEE Trans. Microwave Theory Tech.*, vol. MTT-30, pp. 969–975, July 1982.

[3] Y. A. Ayasli, L. D. Reynolds, J. L. Vorhaus, and L. Hanes, "Monolithic 2–20 GHz GaAs traveling-wave amplifier," *Electron. Lett*, vol. 18, pp. 596–598, July 1982.

[4] K. B. Niclas, W. T. Wilser, T. R. Kritzer, and R. R. Pereira, "On theory and performance of solid-state microwave distributed amplifiers," *IEEE Trans. Microwave Theory Tech.*, vol. MTT-31, pp. 447–456, June 1983.

[5] D. E. Dawson, M. J. Salib, and L. E. Dickens, "Distributed cascode amplifier and noise figure modeling of an arbitrary amplifier configuration," in *1984 IEEE Int. Solid-State Circuits Conf. Dig.*, pp. 78–79.

[6] C. Huang, A. Herbig, and R. Anderson, "Sub-half micron GaAs FETs for applications through K-band," *IEEE 1981 Microwave Symp. Dig.*, pp. 25–27.

2–20-GHz GaAs Traveling-Wave Power Amplifier

YALCIN AYASLI, MEMBER, IEEE, LEONARD D. REYNOLDS, ROBERT L. MOZZI, AND LARRY K. HANES

Abstract —Power amplification in FET traveling-wave amplifiers is examined, and the mechanisms which limit power capability of the amplifier are identified. Design considerations for power amplification are discussed. A novel single-stage and two-stage monolithic GaAs traveling-wave power amplifier with over 250-mW power output in the 2–20-GHz frequency range is described.

I. INTRODUCTION

THE WIDE BANDWIDTH capability of distributed or traveling-wave amplifiers is well known [1]–[3]. Traveling-wave amplification by adding the transconductance of several FET's without paralleling their input or output capacitances looks very promising for achieving wide-band microwave amplification. Already 2–20-GHz decade band amplification with 30-dB gain has been reported with GaAs FET's in monolithic form [4]. The relative insensitivity of the amplifier performance with respect to transistor and circuit parameter variations, good input and output match, and stable operation of these devices makes them very attractive for future commercial and military applications. Because of these potential applications, the power performance of the device is also of great interest; however, to our knowledge this problem has not yet been addressed in the literature.

This work discusses the power-limiting mechanisms in a GaAs FET traveling-wave amplifier and describes a new circuit approach which decreases the effect of some of these limiting mechanisms. In particular, design and performance of a 2–20-GHz power amplifier are presented.

Manuscript received June 14, 1983; revised December 15, 1983. This work was supported in part by the Air Force Wright Aeronautical Laboratories, Avionics Laboratory, Air Force Systems Command, U.S. Air Force, Wright-Patterson Air Force Base, OH.

The authors are with Raytheon Research Division, 131 Spring St., Lexington, MA 02173.

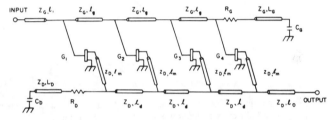

Fig. 1. Schematic representation of a four-cell FET traveling-wave preamplifier.

II. TRAVELING-WAVE POWER AMPLIFICATION CONSIDERATIONS

Schematic representation of a four-cell FET traveling-wave amplifier is shown in Fig. 1. The design considerations and microwave performance of such an amplifier with GaAs MESFET's as active devices have been described in our earlier paper [5], where it was shown that when drain losses are small compared with gate-line losses, the small-signal gain expression for the amplifier can be written approximately as

$$G = \frac{g_m^2 n^2 Z_0^2}{4}\left(1 - \frac{\alpha_g l_g n}{2}\right)^2 \quad (1)$$

where

- g_m transconductance per FET,
- n number of FET's,
- Z_0 input and output characteristic impedance,
- α_g effective gate-line attenuation per unit length,
- l_g length of gate transmission line per unit cell.

This expression is derived using simplified circuit and device models. As such, it is not intended to be used as a design equation. However, despite its simplicity, (1) is very

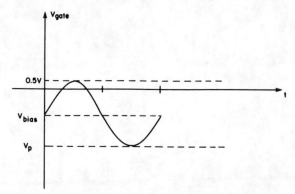

Fig. 2. Maximum RF voltage swing allowed on the gate line.

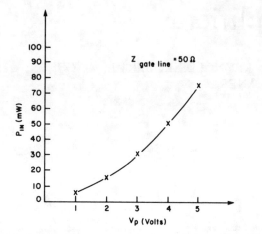

Fig. 3. Maximum allowable input power to a FET traveling-wave amplifier as a function of device pinchoff voltage.

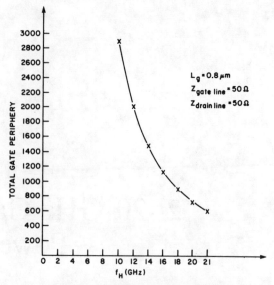

Fig. 4. Maximum total gate periphery per gain stage as a function of the highest frequency of operation.

useful in bringing out the effect of almost all the salient design parameters on the amplifier gain, allowing us to investigate various parameter tradeoffs. For a full circuit design, the analysis gets so complicated that computer-aided design techniques must be employed.

For power amplification, there are additional constraints. In fact, one can identify four separate power-limiting mechanisms in microwave traveling-wave power amplifiers.

In discussing these constraints, we will consider large-signal operation of FET's only in basic terms. Our intention here is to define the issues involved in distributed power amplification and identify the limiting mechanisms.

The first power-limiting mechanism is the finite RF voltage swing that can be allowed on the input gate line. This swing is limited on the positive RF cycle by the forward conduction of the gate and on the negative cycle by the pinchoff voltage of the device, as shown in Fig. 2. Hence, for a 50-Ω input impedance amplifier with −4-V pinchoff voltage FET's and, assuming the devices are biased at a drain current $I_{dss}/2$, the maximum input RF power to the amplifier is limited by

$$P_{in,max} \cong \frac{(4+0.5)^2}{8 \times 50} = 0.051 \text{ W}.$$

Thus, maximum output power from the amplifier cannot be larger than Gain × $P_{in,max}$ under any circumstances. The quantity $P_{in,max}$ is plotted as a function of device pinchoff in Fig. 3.

The second power-limiting mechanism is the maximum total gate periphery that can be included in a single-stage design. Referring to (1), we note that the total attenuation on the gate line has to be kept below a certain value to maximize gain-per-stage per total FET periphery. In fact, from the simplified gain expression of (1), one can show that $\partial G/\partial n = 0$ at $\alpha_g l_g n = 1$. Other factors which also reduce gain but are not included in (1) frequently force the term $\alpha_g l_g n$ to be chosen less than 1. Hence, the following inequality has to be satisfied for a given design if one intends to employ the FET's in a single-stage design most efficiently:

$$\alpha_g l_g n \leqslant 1. \qquad (2)$$

Relating the effective gate-line attenuation constant α_g to the FET input parameters r_g and C_{gs}, we find

$$r_g \omega^2 C_{gs}^2 Z_0 n \leqslant 2 \qquad (3)$$

where

r_g gate resistance,
C_{gs} gate-source capacitance.

In (3), r_g varies inversely and C_{gs} varies directly with periphery for a given FET geometry. Hence, in terms of the periphery w per FET, (3) becomes

$$nw\omega^2 \leqslant \text{const.} \qquad (4)$$

Thus, for a specified maximum frequency of operation and for a given FET, there is an upper limit to the maximum total periphery nw that can be employed in a single-stage design. This maximum periphery determines the gain and consequently the output power of a single-stage traveling-wave amplifier.

Fig. 4 shows the variation of maximum periphery per stage as a function of the highest frequency of operation f_H for a given 0.8-μm gate-length GaAs FET. Note that for

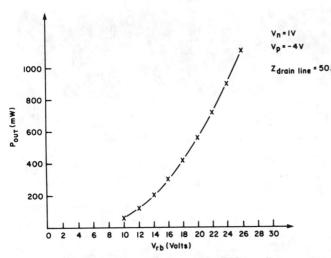

Fig. 5. Maximum output power for a GaAs FET traveling-wave amplifier as a function of gate drain reverse breakdown voltage.

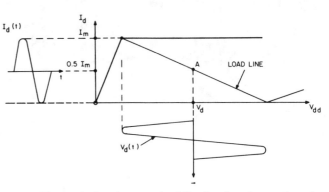

Fig. 6. The required optimum ac load line for class A operation of a discrete FET.

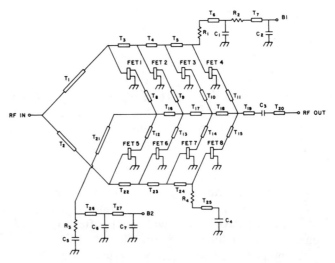

Fig. 7. Power amplifier with eight 150-μm unit cells.

this particular FET, maximum total stage periphery per stage is around 600 μm for $f_H = 20$ GHz.

The third power-limiting mechanism is the gate–drain breakdown voltage of the FET's. The drain terminals must be able to sustain the amplified RF voltage swings on the output transmission line. This voltage is given by

$$V_{\max}, \text{peak-to-peak} \leq V_{\text{breakdown}} + V_{\text{pinch-off}} - V_{\text{knee}}. \quad (5)$$

Using this equation, the maximum saturated power output per stage of the amplifier can be calculated for a given output impedance. The results of such a calculation are plotted in Fig. 5.

Note from this figure that for a typical K-band FET with 15-V breakdown voltage, approximately 250 mW of output power can be expected. Our experience with actual amplifiers indicates that the values obtained from Fig. 5 are conservative.

Note that reduction of output impedance is not a clear-cut solution for increased output power because of the corresponding gain reduction associated with it.

The fourth power-limiting mechanism is related to the optimum ac load-line requirements. Fig. 6 illustrated the required optimum ac load line for class A operation of a discrete FET. For a traveling-wave amplifier, the load line that each individual FET sees is predetermined by the drain-line characteristic impedance, the only flexibility left in the design is the periphery of the unit FET. However, the total periphery is also predetermined from gate-loading considerations. For example, we have established that the total periphery allowed is around 600 μm for the 2–20-GHz amplifier. Hence, for the four-cell design, each FET has a 150-μm periphery. A typical optimum load line R_L for such a device is 280 Ω, representing a significant mismatch to a 50-Ω range output impedance.

Such an impedance mismatch has a significant effect on the output power and consequently on the efficiency of the amplifier stage. Assuming that maximum gain per stage $G_{\max}$ is always less than $g_m^2 n^2 Z_0^2 / 4$ (see (1)), the maximum theoretical limit for the power-added efficiency of the amplifier can be calculated. It can be shown that this theoretical maximum value is

$$\eta_{\text{power added}_{\max}} < \left(1 - \frac{1}{G}\right) \frac{1}{8} n \frac{Z_0}{R_L}. \quad (6)$$

Equation (6) indicates that increased periphery per stage helps to increase efficiency if this increased periphery does not adversely affect the gain at f_H.

On the basis of this analysis of power amplification in traveling-wave amplifiers, we will next describe a novel approach for increased power.

III. TRAVELING-WAVE POWER AMPLIFIER DESIGN

Some of the problems outlined above have been addressed in the development of 2–20-GHz power amplifiers. Consider the power amplifier circuit design shown in Fig. 7. In this circuit, the adverse effects of three of the mechanisms identified as limiting the maximum output power are reduced.

First, the input power is equally divided into the gate lines, each employing 4×150-μm FET's, using a Wilkinson power divider without the isolation resistor; we are able to obtain decade bandwidth performance from a single-section Wilkinson divider because of the good input match characteristics of the amplifiers.

Second, the FET's excited from two separate gate lines are combined on a single drain line, effectively giving a

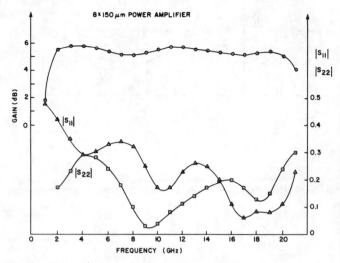

Fig. 8. The predicted performance of the traveling-wave power amplifier in the 2–20-GHz frequency range.

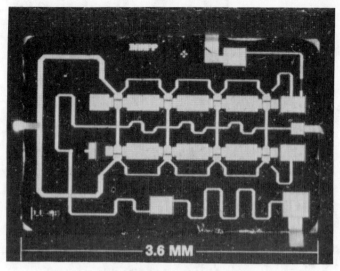

Fig. 9. Single-stage 2–20-GHz power amplifier chip.

4×300-μm drain periphery. Thus, the total gate periphery is doubled without affecting the loading on the gate lines. In this way, the gate loading is limited to a 600-μm gate periphery, whereas the output power will be determined by a 1200-μm drain periphery.

Third, the required load-line impedance is halved, since we have twice the drain periphery on the output line. This brings its value closer to the optimum load-line impedance. Since gain per stage is not significantly affected, the efficiency of the amplifier is effectively doubled.

This amplifier configuration does nothing for the drain-line voltage breakdown problem. However, in this particular design, the improvements outlined above are able to bring the output power only to the point where drain-line breakdown will start to be a limiting factor. Hence, the circuit is optimized with respect to all the constraints described above. In the design, -4-V pinchoff voltages are assumed for the FET's. This allows 50-mW input power per gate line and twice that for the amplifier. With 5-dB small-signal gains, approximately 300 mW can be expected at the output. Drain-line breakdown effects should start showing up at around 250-mW output power levels, with gate–drain breakdown voltages in the 15-V range.

The predicted performance of the amplifier designed for 50-Ω input/output impedance is shown in Fig. 8. Gain is 5 ± 0.5 dB. Input and output return loss is in the neighborhood of 10 dB at all frequencies except at 2 GHz, where the input return loss is only 7 dB. The increased input mismatch at the low end is due to the fact that the input impedance transformer is becoming less and less effective as its electrical length gets smaller and smaller at the low end of the band.

A dc blocking capacitor is included in the design of the output drain line. This amplifier is intended as one of the stages in a chain of cascaded amplifiers. Hence, dc blocking on one side of the amplifier is sufficient.

The two-stage version of this amplifier is also designed with a five-FET stage driving the power stage, with a total of 10-dB small-signal gain.

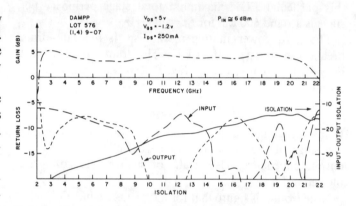

Fig. 10. Measured small-signal performance of the 2–20-GHz power amplifier.

IV. Experimental Performance

A photograph of the finished single-stage power amplifier chip is shown in Fig. 9. The chip size is 2.31×3.64 mm (91×143 mils) on 0.1-mm (4-mil) GaAs substrate.

Thin-film capacitors on the chip add up to a total of 34 pF. The dielectric material is plasma-enhanced CVD silicon nitride. The thin-film resistor material is titanium; it is evaporated by electron beam and patterned by photoresist liftoff.

The total gate periphery on the chip is 1200 μm. The amplifier design was completed using 0.8-μm gate-length, -4-V pinchoff voltage FET models. However, the actual gates on the wafer turned out to be 1 μm long.

The measured small-signal performance of the amplifier is shown in Fig. 10. Its gain is 4 ± 1 dB in the 2–21-GHz frequency band, which is about 1-dB lower than predicted. The input and output return loss curves are about 2 dB higher than predicted.

Despite the lower gain, the amplifier achieved over 250-mW power output in the 2–20-GHz frequency band. Power performance of the amplifier is shown in Fig. 11. Power-added efficiencies in the 7–14-percent range have been recorded.

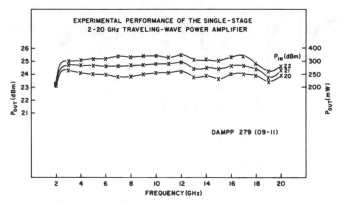

Fig. 11. Experimental power performance of the 2–20-GHz amplifier

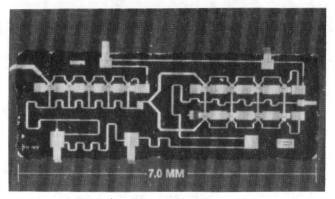

Fig. 13. Two-stage amplifier chip.

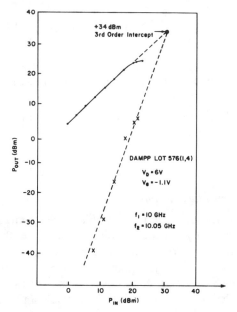

Fig. 12. Third-order intermodulation measurement results at midband.

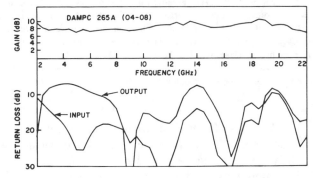

Fig. 14. Two-stage power amplifier gain and return loss performance in the 2–22-GHz frequency band.

Third-order intermodulation products for the amplifier have also been measured. Fig. 12 shows the measured data at midband. The intercept point is at +34 dBm.

Fig. 13 is a photograph of the two-stage chip. The chip size is 2.31×6.95 mm (91×274 mils). Total gate periphery is 1950 μm with thirteen 150-μm FET's. Both stages are biased from a single-gate and a single-drain bias terminal. A typical bias point would be -1 V on the gates with $+6$ V on the drains.

Experimental performance is 9 dB ± 2 dB in the 2–21-GHz frequency range, as shown in Fig. 14. The gain is 1-dB lower than the predicted gain of 10.25 ± 1.25 dB in the same frequency band.

V. Conclusion

A GaAs traveling-wave microwave amplifier is examined in terms of its large-signal power amplification capabilities; several mechanisms which may limit the output power are identified. A new circuit configuration which doubles the output power of the amplifier is described in relation to a 2–20-GHz power amplifier design. The experimental performance of the amplifier with 4-dB and 9-dB gain and over 250-mW power output is presented.

References

[1] W. S. Percival, British Patent Specification No. 460 562, July 24, 1936.
[2] E. L. Ginzton, W. R. Hewlett, J. H. Jasburg, and J. D. Noe, "Distributed amplification," *Proc. IRE*, vol. 36, pp. 956–969, 1948.
[3] W. Jutzi, "A MESFET distributed amplifier with 2-GHz bandwidth," *Proc. IEEE*, vol. 57, pp. 1195–1196, 1969.
[4] Y. Ayasli, L. D. Reynolds, J. L. Vorhaus, and L. Hanes, "2–20 GHz GaAs traveling-wave amplifier," in *IEEE Gallium Arsenide Integrated Circuit Symp.*, (New Orleans, LA), Nov. 1982.
[5] Y. Ayasli, R. L. Mozzi, J. L. Vorhaus, L. D. Reynolds, and R. A. Pucel, "A monolithic GaAs 1–13-GHz traveling-wave amplifier," *IEEE Trans. Microwave Theory Tech.*, vol. MTT-30, pp. 976–981, July 1982.

On Noise in Distributed Amplifiers at Microwave Frequencies

KARL B. NICLAS, SENIOR MEMBER, IEEE, AND BRETT A. TUCKER

Abstract —Formulas for the noise figure, and the minimum noise figure of a multi-link distributed amplifier have been developed. In addition, a relatively simple approximation formula has been devised that predicts the minimum noise figure of a practical amplifier design with good accuracy up to frequencies of 9 GHz. Finally, after the dependence of the noise characteristics on the circuit parameters is discussed, the noise figures of a 2–18-GHz three-link module are computed and compared with those measured on an actual amplifier. The measured data across the 2–18-GHz band compare favorably with the computed results. Measurements and theory agree that only small improvements in noise figure may be achieved, when noise matching the module's input impedance.

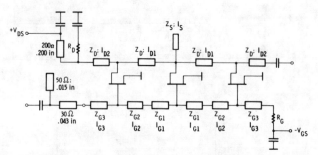

Fig. 1. Schematic of a three-link distributed amplifier.

I. INTRODUCTION

THE PRINCIPLE of distributed amplification, when applied at microwave frequencies, has yielded extremely encouraging results. Monolithic [1] as well as hybrid amplifiers [2] have exhibited respectable gain performances over the frequency band 2–20 GHz and, due to the concept of additive amplification [3], have demonstrated good power-handling capabilities. Another parameter that is of great significance to designers and users alike is the noise figure. Hence, the dependence of the noise figure on the circuit elements has become of major interest in the design of distributed amplifiers.

Little has been reported on the noise characteristics of distributed amplifiers. In their pioneering paper on distributed amplification, E. L. Ginzton *et al.* addressed the noise behavior for the case of gridded electron tubes [3]. By adding the noise powers generated in the active devices and the grid-line terminations, the authors developed an expression for the noise figure that is based primarily on a qualitative evaluation of the amplifier's noise sources. Recently, Y. Ayasli *et al.* published the computed noise performance of a 2–12-GHz amplifier. However, except for the schematic containing the relevant noise sources, the authors did not give any details of the theory their computer program is based on [4].

It is the purpose of this paper to provide some understanding of the noise phenomenon in distributed amplifiers and to quantitatively examine the factors that determine its magnitude. Based on a rigorous analysis, it is demonstrated that the difference between the noise figure and the optimum noise figure is rather small when designing for a flat gain response across a multi-octave band. This fact was first discovered in experiments which proved it to be virtually impossible to improve the noise figure of a three-link amplifier module by means of outside tuning.

II. NOISE PARAMETERS AND NOISE FIGURE

A. Elementary Amplifier

The formulas derived in this paper contain the noise generated by the active devices and the thermal noise agitation injected by the terminations of both idle ports. The part that may be contributed by lossy transmission line elements or lossy inductors, as well as capacitors, is considered to be comparatively small and has therefore been neglected.

The circuit diagram of a practical three-link distributed amplifier is shown in Fig. 1 [2]. The linking components between the transistors of this particular network are composed of transmission line elements of equal electrical lengths in both the gate and the drain line ($\theta_{G1} + \theta_{G2} = \theta_{D1}$). For the analysis of an amplifier such as that of Fig. 1, it is convenient to divide the module into functional blocks such as the input matching circuit, the elementary amplifiers and, if necessary, the output matching network. Hence, Fig. 2 shows the k th elementary amplifier as it will be used in the analysis of the unit's noise parameters. It incorporates the active device which is located between the gate and the drain line and described by a Π-shaped equivalent circuit. The MESFET's noise sources are characterized by the voltage v_1 and the current i_1 at the input terminal of the transistor. The input and output links of the drain and gate line may either consist of transmission line or lumped circuit elements. For reasons of simplifying the mathematical problem, it is convenient to combine both input and output links to a single input and a single output four-port and apply the concept of cascaded four-ports to the MESFET's and shunt elements as well. All circuit elements

Manuscript received January 25, 1983; revised March 17, 1983.
The authors are with the Watkins-Johnson Company, Palo Alto, CA.

Reprinted from *IEEE Trans. Microwave Theory Tech.*, vol. MTT-31, pp. 661–668, Aug. 1983.

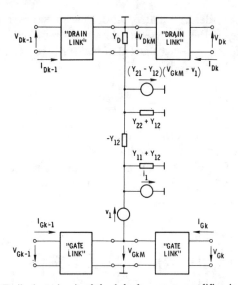

Fig. 2. Equivalent circuit of the kth elementary amplifier, including the noise sources v_1 and i_1 of the active device.

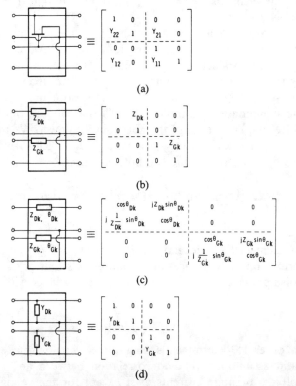

Fig. 3. Transformation matrices of the four-ports typically employed in distributed amplifier design.

as they are typically employed in a distributed amplifier can be represented by one of the four-ports shown in Fig. 3. It should be noted that the matrix for the building block of Fig. 3(d) may be used for either lumped or transmission line elements.

Now that the analytical tools have been assembled, we are able to find the voltages and currents at the drain and the gate input of the elementary amplifier of Fig. 2. They depend on the passive circuit elements, the active device and its internal noise sources (v_1 and i_1) as well as the output voltages and currents. Voltages and currents may then be expressed in form of the chain matrix equation

$$\begin{bmatrix} V_{Dk-1} \\ I_{Dk-1} \\ V_{Gk-1} \\ I_{Gk-1} \end{bmatrix} = A_k \begin{bmatrix} V_{Dk} \\ -I_{Dk} \\ V_{Gk} \\ -I_{Gk} \end{bmatrix} + B_k \begin{bmatrix} -v_{1k} \\ i_{1k} \\ 0 \\ 0 \end{bmatrix} \quad (1a)$$

where

$$A_k = A_{1k} A_{Fk} A_{2k} \quad (1b)$$

$$B_k = A_{1k} B_{Fk} \quad (1c)$$

$$B_{Fk} = \begin{bmatrix} 0 & 0 & 0 & 0 \\ Y_{21} & 0 & 0 & 0 \\ \hline 0 & 0 & 0 & 0 \\ Y_{11} & 1 & 0 & 0 \end{bmatrix}. \quad (1d)$$

$[A_{1k}]$ is the matrix of the input link and $[A_{2k}]$ is that of the output link of Fig. 2, while $[A_{Fk}]$ constitutes the MESFET's chain matrix (Fig. 3(a)). All voltages and currents in (1) contain both the signal and the noise components. The matrix equation (1a) transforms the transistor's noise voltage v_1 and noise current i_1 to the input port of the elementary amplifier in Fig. 2. The transformation of $[B_{Fk}]$ depends solely on Y_{21} and Y_{11} of the active element.

Finding the relationship that exists between the input and output quantities and the internal noise sources of a multi-link module becomes now a mere exercise in matrix algebra. However, before we engage in this task, it seems appropriate to separate the signal from the noise quantities. Since we always assume to operate under linear conditions, total voltages and currents are simply the sum of their signal and noise components. The latter are symbolized by lower case letters. We therefore define for the gate side $(\ )_{Gk}$ and the drain side $(\ )_{Dk}$

$$V_{Dk,Gk} = V'_{Dk,Gk} + v_{Dk,Gk} \quad (2a)$$

$$V_{Dk-1,Gk-1} = V'_{Dk-1,Gk-1} + v_{Dk-1,Gk-1} \quad (2b)$$

$$I_{Dk,Gk} = I'_{Dk,Gk} + i_{Dk,Gk} \quad (2c)$$

$$I_{Dk-1,Gk-1} = I'_{Dk-1,Gk-1} + i_{Dk-1,Gk-1}. \quad (2d)$$

Substituting (2) into (1a) results in two matrix equations, one for the signal and one for the noise parameters. The former has been treated elsewhere [2] and is of no further concern in this paper. The noise behavior of a single link is then accurately described by

$$\begin{bmatrix} v_{Dk-1} \\ i_{Dk-1} \\ v_{Gk-1} \\ i_{Gk-1} \end{bmatrix} = A_k \begin{bmatrix} v_{Dk} \\ -i_{Dk} \\ v_{Gk} \\ -i_{Gk} \end{bmatrix} + B_k \begin{bmatrix} -v_{1k} \\ i_{1k} \\ 0 \\ 0 \end{bmatrix}. \quad (3)$$

B. Multi-Link Amplifier

1) Exact Solution: Let us now consider the amplifier's boundary conditions. The unit's idle ports are terminated with the admittances Y_{DT} on the drain and Y_{GT} on the gate side. In addition, we connect a signal source with an

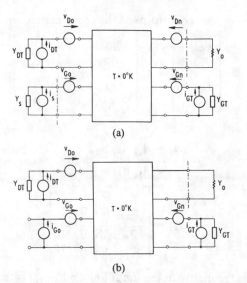

Fig. 4. Termination conditions and equivalent noise sources of the distributed amplifier for (a) the noise output power method and (b) the noise input power method.

internal admittance Y_S to the amplifier's input terminal and a load of $Y_0 = Z_0^{-1}$ to its output terminal. Since all four terminations contain a finite conductance, they inject thermal noise, which in the case of Y_S, Y_{GT}, and Y_{DT} contributes to the noise power at the amplifer's output terminal. Furthermore, those components generated by the MESFET's have a strong influence on the amplifier's noise behavior and may be represented by voltage sources at each of the four terminals. This is accomplished by transforming the transistors' individual noise sources to the four-port's terminals as illustrated in Fig. 4(a) and thereby making the four-port itself free of noise.

Applying the boundary conditions to (3) in accordance with the directions of the voltages and currents indicated in Fig. 4(a), and assuming a number of n elementary amplifiers, we derive the following matrix equation:

$$\begin{bmatrix} v_{D0} \\ -(Y_{DT}v_{D0} - i_{DT}) \\ v_{G0} \\ -(Y_s v_{G0} - i_s) \end{bmatrix} = D \begin{bmatrix} v_{Dn} \\ Y_0 v_{Dn} \\ v_{Gn} \\ (Y_{GT}v_{Gn} - i_{GT}) \end{bmatrix} + I \begin{bmatrix} v_{DF} \\ i_{DF} \\ v_{GF} \\ i_{GF} \end{bmatrix}$$

(4a)

where

$$D = \prod_{k=0}^{n} A_k \qquad (4b)$$

$$\begin{bmatrix} v_{DF} \\ i_{DF} \\ v_{GF} \\ i_{GF} \end{bmatrix} = \sum_{k=1}^{n} \left\{ E_k \begin{bmatrix} -v_{1k} \\ i_{1k} \\ 0 \\ 0 \end{bmatrix} \right\} \qquad (4c)$$

$$E_k = \left\{ \prod_{m=0}^{k-1} A_m \right\} B_k \qquad (4d)$$

$$A_0 \equiv I. \qquad (4e)$$

($[I]$ is the identity matrix and the links are numbered beginning at the input end.) For the simple case that all links and active devices are identical ($[A_k] = [A]$ and $[B_{Fk}] = [B]$), (4b) and (4d) take the form

$$D = A^n \qquad (5a)$$

$$E_k = A^{k-1}B. \qquad (5b)$$

While (4) expresses the relationship that exists between the noise parameters at the amplifier's input and output ports, it does not directly formulate the unknown voltages ($v_{D0}, v_{G0}, v_{Dn}, v_{Gn}$) as functions of the known quantities ($v_1, i_1, i_{GT}, i_{DT}, i_S$). This is accomplished in (6):

$$\begin{bmatrix} v_{D0} \\ v_{G0} \\ v_{Dn} \\ v_{Gn} \end{bmatrix} = (K)_0 \begin{bmatrix} 0 \\ 0 \\ i_{GT} \\ i_{DT} \\ i_s \end{bmatrix} + \sum_{k=1}^{n} (K)_k \begin{bmatrix} v_{1k} \\ i_{1k} \\ 0 \\ 0 \\ 0 \end{bmatrix}. \qquad (6)$$

(The elements of $[K]_0$ and $[K]_k$ are contained in the Appendix.)

Up to this point it was convenient to treat the problem with the technique of cascading four-ports. However, since the noise parameters responsible for the amplifier's noise figure can now be determined with (6) we abandon the four-port representation. Aside from the noise input power, the computation of the amplifier's noise figure only requires the knowledge of the noise output power. We therefore need to extract the noise voltage v_{Dn} from (6) which can be expressed by the two-port equation

$$v_{Dn} = \sum_{k=1}^{n} \left[(K_{31})_k v_{1k} + (K_{32})_k i_{1k} \right] + (K_{33})_0 i_{GT}$$
$$+ (K_{34})_0 i_{DT} + (K_{35})_0 i_s. \qquad (7)$$

Its elements K_{ij} depend on D_{ij} and E_{ij} of (4) and the termination admittances at the amplifier's four ports. The noise power at the unit's output is in accordance with Fig. 4(a)

$$N_{Dn} = Y_0 \overline{|v_{Dn}|^2}. \qquad (8)$$

In order to determine its magnitude, we now make the valid assumption that no correlation between any noise voltages or noise currents exist except for the voltage v_1 and the current i_1 of the same active device. It is usually expressed in terms of the correlation admittance Y_{cor} [5]

$$i_1 = i_n + Y_{\text{cor}} v_1 \qquad (9)$$

where i_n represents that part of the current i_1 that is not correlated with v_1. By substituting (9) into (7) we are in a position to express the amplifier's noise output voltage in terms of parameters that are not correlated with each other.

$$v_{Dn} = \sum_{k=1}^{n} \left[(K'_{31})_k v_{1k} + (K_{32})_k i_{nk} \right] + (K_{33})_0 i_{GT}$$
$$+ (K_{34})_0 i_{DT} + (K_{35})_0 i_s \qquad (10a)$$

with
$$(K'_{31})_k = (K_{31})_k + (Y_{\text{cor}})_k (K_{32})_k. \tag{10b}$$

Taking into consideration that noise correlation exists only between sources of the same active device, we obtain with (8) and (10)

$$N_{Dn} = Y_0 \Bigg[\sum_{k=1}^{n} \left\{ |K'_{31}|_k^2 \overline{|v_1|_k^2} + |K_{32}|_k^2 \overline{|i_n|_k^2} \right\}$$
$$+ |K_{33}|_0^2 \overline{|i_{GT}|^2} + |K_{34}|_0^2 \overline{|i_{DT}|^2} + |K_{35}|_0^2 \overline{|i_s|^2} \Bigg]. \tag{11}$$

Replacing the voltages and currents of (11) with the MESFETs' equivalent noise parameters as well as the noisy components of the termination admittances leads to

$$N_{Dn} = 4kT_0 \Delta f Y_0 \Bigg[\sum_{k=1}^{n} \left\{ |K'_{31}|_k^2 R_{nk} + |K_{32}|_k^2 G_{nk} \right\}$$
$$+ |K_{33}|_0^2 G_{GT} + |K_{34}|_0^2 G_{DT} + |K_{35}|_0^2 G_s \Bigg]. \tag{12}$$

With the available input noise power
$$N_S = kT_0 \Delta f \tag{13}$$
and the unit's overall gain
$$\text{Gain} = 4|K_{35}|_0^2 G_S Y_0 \tag{14}$$
we are now ready to formulate the noise figure of the multi-link distributed amplifier

$$F = 1 + \frac{1}{|K_{35}|_0^2 G_s} \times \Bigg[\sum_{k=1}^{n} \left\{ |K'_{31}|_k^2 R_{nk} + |K_{32}|_k^2 G_{nk} \right\}$$
$$+ |K_{33}|_0^2 G_{GT} + |K_{34}|_0^2 G_{DT} \Bigg]. \tag{15}$$

The computation of the minimum noise figure $F_{\min}$, the equivalent noise resistor R_n^I, the equivalent noise conductance G_n^I, and the correlation admittance $Y_{\text{cor}}^I = G_{\text{cor}}^I + jB_{\text{cor}}^I$ of the amplifier may then be accomplished by computing the noise figures (15) for four arbitrary but different input admittances Y_s and use the resulting noise figures to determine these quantities. Since the method is straightforward, it is not reported here.

While the formula (15) of the noise figure is based on the noise output power (12), it may as well be derived by transforming all noise sources in accordance with the equivalent circuit of Fig. 4(b) to the amplifier's input. Such a transformation leads to the direct formulation of the noise parameters R_n^I, G_n^I, and Y_{cor}^I which in turn can be used to calculate the noise figure F and the minimum noise figure $F_{\min}$. In pursuing this method we found the results to be in total agreement with those of the "output power method" proving its validity when applied to four-ports. Both approaches lead to identical results and neither seems to have advantages over the other regarding the computer programming.

2) Low Frequency Model: Due to the complexity of the formulas presented so far, it might be beneficial to analyze a simplified model of the multi-link amplifier. For this purpose we choose the low frequency model for which the transforming characteristics of the linking elements may be neglected. In other words, we treat the amplifier as a lossy match amplifier with n numbers of parallel transistors. The noise characteristics of this amplifier model are described by the parameters [6]

$$R_n^I = \frac{1}{n} \left[R_n + \frac{G_{DT}}{n|Y_{21}|^2} \right] \tag{16a}$$

$$G_n^I = G_{GT} + n \left[G_n + |Y_{11} - Y_{\text{cor}}|^2 \frac{G_{DT} R_n}{n|Y_{21}|^2 R_n + G_{DT}} \right] \tag{16b}$$

$$Y_{\text{cor}}^I = Y_{GT} + n \left[Y_{\text{cor}} + (Y_{11} - Y_{\text{cor}}) \frac{G_{DT}}{n|Y_{21}|^2 R_n + G_{DT}} \right]. \tag{16c}$$

If we further assume that
$$G_{DT} \ll n|Y_{21}|^2 R_n \tag{17a}$$
$$G_{DT} \ll \frac{|Y_{21}|^2}{|Y_{11} - Y_{\text{cor}}|^2} (G_{GT} + nG_n) \tag{17b}$$
$$G_{DT} \ll |Y_{21}|^2 R_n \frac{G_{GT} + nG_{\text{cor}}}{G_{11} - G_{\text{cor}}} \tag{17c}$$

which are satisfied in most practical cases the formula for the approximate minimum noise figure of a multi-link distributed amplifier takes the simple form

$$F_{\min} \cong 1 + 2 \Bigg[R_n \left(\frac{G_{GT}}{n} + G_{\text{cor}} \right)$$
$$+ \sqrt{R_n \left(\frac{G_{GT}}{n} + G_n \right) + R_n^2 \left(\frac{G_{GT}}{n} + G_{\text{cor}} \right)^2} \Bigg]. \tag{18}$$

It clearly indicates that the minimum noise figure of a distributed amplifier at low frequencies may be reduced by increasing the number of links. We will briefly investigate in Section III to what extent the approximation formula (18) may be used to determine the minimum noise figure of a multi-link distributed amplifier.

III. COMPUTED NOISE FIGURES AND GAINS OF PRACTICAL AMPLIFIER DESIGNS

In this chapter we discuss the computed noise figure, minimum noise figure, and gain for a number of practical amplifier designs as they depend on various circuit parameters. The noise figure computations are based on (15), while the small signal gain has been computed by means of (14).

A. Cascading of Identical Links

Let us first study the case of an amplifier that consists of identical elementary blocks (Fig. 5) cascaded to a multi-link chain and terminated with R_G and R_D at the idle ports.

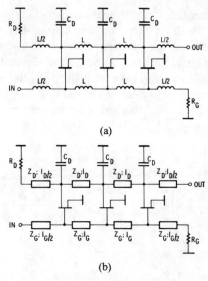

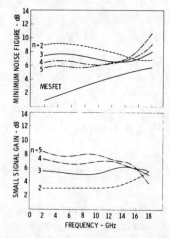

Fig. 5. Schematic of a three-link distributed amplifier with (a) lumped elements and (b) distributed line elements.

Fig. 7. Dependence of the minimum noise figure and the small-signal gain of a transmission line element amplifier with identical links on the number of links ($Z_G = Z_D = 125$ Ω, $l_G = 0.037$ in and $l_D = 0.057$ in, $C_D = 0$, $R_G = 38$ Ω, $R_D = 125$ Ω).

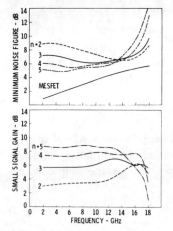

Fig. 6. Dependence of the minimum noise figure and the small signal gain of a lumped element amplifier with identical links on the number of links ($L/2 = 0.3125$ nH, $C_D = 0.159$ pF, $R_G = 38$ Ω, $R_D = 125$ Ω).

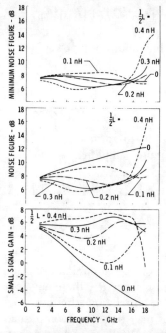

Fig. 8. Influence of the link inductance on minimum noise figure, noise figure, and small signal gain of a three-link lumped element amplifier ($C_D = 0.159$ pF, $R_G = 38$ Ω, $R_D = 125$ Ω).

The characteristics of the MESFET used in all of the theoretical and practical studies reported here have been described in the literature [2]. The minimum noise figure and the small signal gain of a distributed amplifier employing identical links composed of lumped circuit elements in accordance with the schematic of Fig. 5(a) are shown in Fig. 6. The circuit's components $L/2$, C_D, R_G, and R_D have been optimized in order to obtain a flat gain response between 2 and 18 GHz for the case of a three-link unit. The curves of the minimum noise figure clearly demonstrate the influence of the number of elementary amplifiers at lower frequencies. However, at frequencies above 13 GHz, the minimum noise figure does not experience any improvement by cascading additional blocks. To the contrary, the parasitics of the MESFET's cause a deterioration of the overall noise performance.

Similar characteristics are exhibited in Fig. 7 when replacing the lumped elements by transmission line elements for $C_D = 0$ in accordance with Fig. 5(b). While for this case the minimum noise figure shows slightly higher values at low frequencies, it is appreciably lower at the high frequency end of the band. The dependency of the noise characteristics and the small-signal gain on the inductance $L/2$ are displayed in Fig. 8 for a three-link module. The curves of the noise figure and gain include the case of our low frequency model for which we chose $L = 0$. They demonstrate that in this particular example (18) offers an acceptable approximation for the amplifier's minimum noise figure up to frequencies of 16 GHz. In addition, the plotted curves express the significance of the linking element's inductance $L/2$ on the amplifier's noise figure and its gain. Similarly, when employing transmission line elements, we find that for $n = 3$, $R_G = 38$ Ω, $R_D = 125$ Ω, $Z_G = Z_D = 125$

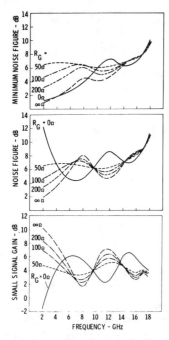

Fig. 9. Influence of the gate termination resistance on the minimum noise figure, noise figure, and small-signal gain of a three-link lumped element amplifier ($L/2 = 0.3125$ nH, $C_D = 0.159$ pF, $R_D = 125$ Ω).

Ω, $l_G = 0.037$ in, and $l_D = 0.057$ in, the minimum noise figure computed with (15) deviates from its approximation (18) by only ± 0.85 dB between 2 and 18 GHz.

Let us now investigate the dependence of the amplifier's performance parameters on the magnitude of the resistors terminating the idle ports. The influence of the gate resistance on minimum noise figure, noise figure, and small-signal gain is plotted in Fig. 9 for a three-link module ($\frac{1}{2}L = 0.3125$ nH, $C_D = 0.159$ pF, and $R_D = 125$ Ω). The curves demonstrate that the magnitude of R_G has its greatest impact on all three parameters at the low frequency end of the band. Since R_G determines to a great extent the magnitude of the gain variation across the band, the requirement of a flat gain response over a very wide band leaves little freedom to improve the noise figure by means of R_G. As expected, the amplifier's noise figures experience little change with the terminating resistance R_D. Both, minimum noise figure and noise figure do not differ by more than 1.7 dB for $0 \leqslant R_D \leqslant \infty$ over the 2–18-GHz band. However, the right choice of R_D is important for a flat gain performance at lower frequencies. The performance characteristics of the distributed amplifier employing high impedance transmission line elements and their dependence on the individual circuit parameters are essentially very similar to those incorporating lumped elements.

B. The Equal Line Lengths Amplifier

In this section we examine the noise characteristics of an amplifier that employs transmission lines of equal lengths between the active elements [2]. This approach has the advantage of placing MESFET's between two parallel straight lines resulting in a very simple structure. The schematic of a three-link module designed for the 2–18-GHz frequency band is shown in Fig. 1. The magnitudes of its elements are listed in Table I while a detailed description of the unit's fabrication can be found elsewhere [2]. The amplifier's computed, noise parameters R_n, G_n, and Y_{cor} are plotted in Fig. 10 which also contains the noise parameters of the MESFET. As expected, at low frequencies, we are essentially paralleling the noise resistances R_n of the three transistors, while the noise conductance G_n is mainly dependent on the resistance R_G of the gate line's termination. Similarly, G_{cor} depends almost entirely on R_G at low frequencies. It is also apparent from these curves that the noise parameters of the amplifier are confined to a much narrower range than those of the transistor, suggesting much less variation of the amplifier's minimum noise figure versus frequency. This is easily discernable when comparing the optimum noise figures of the three-link

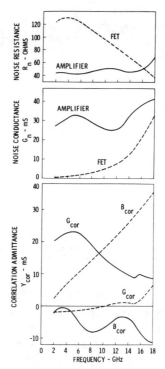

Fig. 10. The equivalent noise parameters of the three-link equal line lengths amplifier of Fig. 1 and Table I (solid curves) and the MESFET (dashed curves).

TABLE I
ELEMENT VALUES OF THE THREE-LINK MODULE WITH EQUAL LINE LENGTHS

Z_{G1}	= 65 Ω	Θ_{G1}	= 12.2°	l_{G1}	= .020 in
Z_{G2}	= 87 Ω	Θ_{G2}	= 20.7°	l_{G2}	= .034 in
Z_{G3}	= 87 Ω	Θ_{G3}	= 15.2°	l_{G3}	= .025 in
Z_{D1}	= 140 Ω	Θ_{D1}	= 32.9°	l_{D1}	= .054 in
Z_{D2}	= 140 Ω	Θ_{D2}	= 6.7°	l_{D2}	= .011 in
		R_G	= 38 Ω		
		R_D	= 125 Ω		

(DEGREES AT 20 GHz)

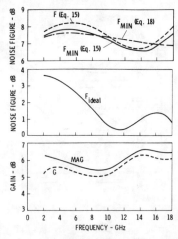

Fig. 11. Noise figures and small signal gains of the three-link equal line lengths amplifier (Fig. 1 and Table I).

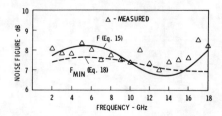

Fig. 12. Comparison of the exact (15), the approximated (18), and the measured noise figures of the three-link equal line lengths amplifier.

device (Fig. 11) with those of the MESFET (Fig. 6). As evidenced by the curves in Fig. 11, there is little difference between the minimum noise figure and the noise figure as computed for a 50-Ω system. The same is true for the maximum available gain and the insertion gain measured when operating between a 50-Ω source and a 50-Ω load. The former can be easily verified in an experiment designed to improve the unit's noise figure by means of noise matching. As a matter of fact, it was the result of such an experiment that aroused our curiosity and prompted the investigation reported in this paper. For the designer, it is of interest to know the extent to which the terminations of the idle ports are contributing to the overall noise figure. For this reason we have computed the noise figure of the amplifer incorporating ideal MESFET's, i.e., for devices that are totally noiseless. As demonstrated by F_{ideal} in Fig. 11, the influence of mainly R_G on the amplifier's noise figure at low frequencies is significant. The curve explains why the unit's noise figure at low frequencies is so much higher than that of the individual transistor. Finally, a comparison of the optimum noise figures as computed with the exact formula (15) and the approximation formula (18) demonstrates the validity of (18) at frequencies up to 9 GHz for this design (Fig. 11).

IV. Experimental Verification

In order to support the theoretical results, a three-link amplifier module was constructed and its noise figure measured. The topology of this unit is identical to that of Fig. 1 with the element values listed in Table I. Its fabrication is described in a previous paper [2] which also discusses the module's small-signal gain, reflection loss, and reverse isolation characteristics. The noise parameters of the MESFET are plotted in Fig. 10 as dashed curves. They are derived from measurements performed on a single transistor biased at a drain voltage and a drain current ($V_{ds}=4$ V, $I_{ds}=42$ mA) representative for the MESFET's operation in the amplifier. While painstaking efforts were taken to measure the data leading to the computation of the MESFET's noise parameters, the accuracy of the re-

sults was only fair and was especially compromised in the area of correcting the noise figure for the tuner losses. Considering these errors and taking into account the performance tolerances of the devices, the agreement between theoretically predicted (15) and measured noise figures is surprisingly close as evidenced by the comparison of Fig. 12. It is interesting to note how well the approximation formula of the minimum noise figure (18) predicts the actual computed and measured noise figures. The reason is based on two observed facts. First, there is little difference between the minimum and the actual noise figure in a distributed amplifier. Second, the transformation properties of its linking elements seem to have nearly the same effect on the amplifier's noise figure as noise matching has on the noise figure of the equivalent lossy-match amplifier.

V. Conclusion

A set of formulas has been developed that makes it possible to compute the noise parameters and noise figures of distributed amplifier modules consisting of n arbitrary links. Based on these theoretical expressions, the noise figures' dependence on the circuit parameters was studied. As a result, it was found that improving the noise figure by altering individual circuit parameters compromises the gain performance of the amplifier whenever ultra-broad-band performance is desired. Due to the complexity of the accurate formulas, an attempt was made to generate approximation formulas designed to replace the accurate noise figure expressions under low frequency conditions. These formulas predict the noise figure of a practical amplifier design up to 9 GHz with good accuracy and, most important, contribute to the understanding of the noise phenomenon at lower frequencies.

Even though the accuracy in our measurements of the device's optimum noise figure leading to the equivalent noise parameters is fair, the agreement between measured and computed data was very encouraging. Finally, our initial measurements indicating little difference between the noise figure and the minimum noise figure of a practical three-link amplifier design were theoretically supported by the computed results.

Appendix
The Elements of [K]

The noise voltages at the four terminals of the multi-link amplifier are expressed by (6) as functions of the active devices' noise parameters v_1 and i_1, the noise currents generated by the idle ports' terminations i_{GT} and i_{DT}, as well as the noise current of the source impedance i_s. The output noise voltage v_{Dn} is formulated in (7) and its elements are

$$(K_{31})_k = \frac{1}{C}\{C_1[(E_{21})_k + Y_{DT}(E_{11})_k] - C_2[(E_{41})_k + Y_s(E_{31})_k]\} \quad \text{(A1a)}$$

$$(K_{32})_k = -\frac{1}{C}\{C_1[(E_{22})_k + Y_{DT}(E_{12})_k] - C_2[(E_{42})_k + Y_s(E_{32})_k]\} \quad \text{(A1b)}$$

$$(K_{33})_0 = \frac{1}{C}\{C_1[D_{24} + Y_{DT}D_{14}] - C_2[D_{44} + Y_s D_{34}]\} \quad \text{(A1c)}$$

$$(K_{34})_0 = \frac{C_1}{C} \quad \text{(A1d)}$$

$$(K_{35})_0 = -\frac{C_2}{C} \quad \text{(A1e)}$$

$$C_1 = D_{43} + Y_{GT}D_{44} + Y_s(D_{33} + Y_{GT}D_{34}) \quad \text{(A1f)}$$

$$C_2 = D_{23} + Y_{GT}D_{24} + Y_{DT}(D_{13} + Y_{GT}D_{14}) \quad \text{(A1g)}$$

$$C = C_1[D_{21} + Y_0 D_{22} + Y_{DT}(D_{11} + Y_0 D_{12})] - C_2[D_{41} + Y_0 D_{42} + Y_s(D_{31} + Y_0 D_{32})]. \quad \text{(A1h)}$$

ACKNOWLEDGMENT

The authors wish to thank R. R. Pereira, who performed the measurements and the tuning of the amplifiers. Thanks go also to W. T. Wilser in whose Department the GaAs MESFET's were fabricated. The authors are indebted to R. Mendiola who typed the complicated formulas.

REFERENCES

[1] Y. A. Ayasli, L. D. Reynolds, J. L. Vorhaus, and L. Hanes, "Monolithic 2–20 GHz GaAs traveling-wave amplifier," *Electron. Lett.*, vol. 18, pp. 596–598, July 1982.
[2] K. B. Niclas, W. T. Wilser, T. R. Kritzer, and R. R. Pereira, "On theory and performance of solid-state microwave distributed amplifiers," *IEEE Trans. Microwave Theory Tech.*, vol. MTT-31, June 1983.
[3] E. L. Ginzton, W. R. Hewlett, J. H. Jasberg, and J. D. Noe, "Distributed amplification," *Proc. IRE* vol. 36, pp. 956–969, Aug. 1948.
[4] Y. A. Ayasli, R. L. Mozzi, J. L. Vorhaus, L. D. Reynolds, and R. A. Pucel, "A monolithic GaAs 1–13 GHz traveling-wave amplifier," *Trans. Microwave Theory Tech.*, vol. MTT-30, pp. 976–981, July 1982.
[5] H. Rothe and W. Dahlke, "Theory of noisy four-poles," *Proc. IRE*, vol. 44, pp. 811–818, June 1956.
[6] Karl B. Niclas, "The exact noise figure of amplifiers with parallel feedback and lossy matching circuits," *Trans. Microwave Theory Tech.*, vol. MTT-30, pp. 832–835, May 1982.

X, Ku-BAND GaAs MONOLITHIC AMPLIFIER*

Y. Tajima, T. Tsukii[†], E. Tong, R. Mozzi, L. Hanes, B. Wrona
Research Division, Raytheon Company
131 Spring Street
Lexington, Massachusetts 02173

[†]Electromagnetic Systems Division
Goleta, California 93017

Abstract

A two-stage X-, Ku-band monolithic FET amplifier has been developed. Initial results indicate a gain of 7-10 dB across the 8-20 GHz band with a typical rf power output of 100 mW. A balanced amplifier consisting of two two-stage amplifiers and a pair of Lange couplers yielded 10.5 ± 1 dB gain from 7.5 to 18 GHz and an output power of 150-250 mW in Ku-band.

Introduction

Broadband GaAs monolithic amplifiers operating in X-band or below have been reported[1-5] and a few papers discuss monolithic amplifiers which operate over a narrow range in Ku-band. In this paper, we will describe the design, fabrication, and performance of broadband amplifiers which operate in both X- and Ku-band. These results are the first demonstration of broadband monolithic amplifiers in this frequency range.

Amplifier Design

The circuit topology of the two-stage amplifier is shown in Fig. 1. The amplifier consists of input matching section, first FET, interstage matching circuit, followed by second FET and output matching section.

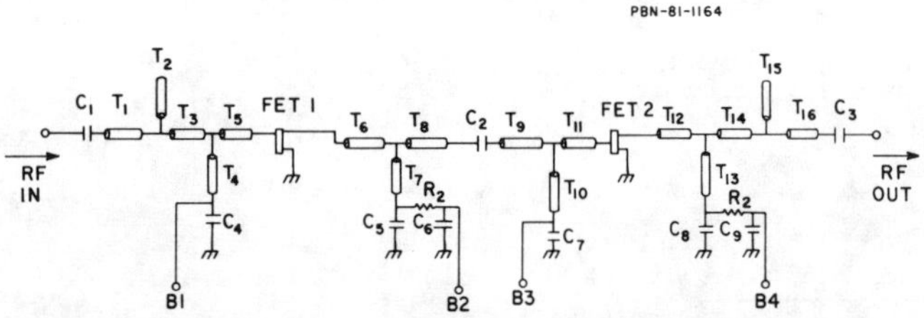

Fig. 1 Circuit topology of the two-stage amplifier.

First and second FETs have been selected as 200 μm and 400 μm respectively to achieve 100 mW of rf level across 7-18 GHz frequency band.

Equivalent circuit model of both FETs were derived from the extrapolation of other submicron gate discrete devices previously developed.

*This work was partially supported by the U.S. Air Force Systems Command under Contract No. F33615-81-C-1413.

Restrictions imposed upon the monolithic integrated circuit for selecting values of each passive element were carefully considered and incorporated into the amplifier design. For example, each capacitor varies from sub-picofarad to several picofarad based upon dc blocking or rf bypass requirements and upon their compatability with other adjacent passive element interfaces. By the same token, individual microstrip transmission impedance varied from 44 to 80 ohms.

Computer aided design program, COMPACT, has been used in optimizing matching elements coupled with passive elements.

Thin-film Ti resistors were used to improve the gain flatness in the band; i.e., large amounts of available gain were absorbed in the resistors in the low-frequency region and these resistors had hardly any effect at high frequencies. Thin film Si_3N_4 capacitors were used for matching elements as well as dc blocking and bypass capacitors. Rf shorted stubs terminated by large bypass capacitors were used for dc bias lines.

Figure 2 is the photograph of a two-stage amplifier chip. It should be noted that the signal path has been laid out in a meander pattern to make a dense circuitry. Such a layout also offers flexibility in determining chip size with a goal of achieving low aspect ratio in mind. The chip size is 2.6 × 2.3 mm.

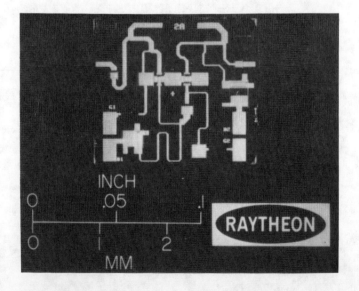

Fig. 2 Photograph of a two-stage amplifier chip.

Circuit Fabrication

The circuits were made on vapor-phase-deposited n-type epilayers grown on semi-insulating substrates. The epilayers consisted of a 0.2 μm thick n⁺ contact layer (10^{18}-10^{19} carriers/cm^3) over a 0.4 μm thick n active layer (1.7×10^{17} carriers/cm^3) on a 2 μm thick buffer layer. Source and drain contacts for the FETs, and the bottom metal of dielectric capacitors consisted of an alloyed NiAuGe composite. Transmission line capacitor tops were TiAu films gold-plated to a thickness of about three microns. Contact pads and air bridges were plated at the same time. After thinning the backside of the wafer, via-holes were etched to the FET source contacts and a ground plane evaporated and gold-plated. Completed wafers were diced into chips, and each chip with acceptable dc parameters was die-bonded on a carrier and placed on a test fixture to make rf measurements.

Two kinds of FETs with total gate periphery of 200 μm and 400 μm respectively were used in these amplifiers. All the gate fingers (TiPtAu) were 0.7-0.8 μm long and 100 μm wide and were recessed through the contact layer and partially into the active layer. An air bridge was used to connect source contacts on the 400 μm periphery FET. An SEM photograph of this type FET is shown in Fig. 3.

Fig. 3 SEM photograph of a 400 μm FET used in the amplifier.

Rf Performance

Figure 4 illustrates the small-signal gain performance of a two-stage amplifier, along with input/output return loss over the 6-21 GHz range. There is 7-10 dB gain across 8-20 GHz with a good match to the output and an input match monotonically improving with increasing frequency. The overall gain, however, is 1 dB lower than the predicted design data, and the amplifier gain has a tendency to dip around 10 GHz. Both discrepancies are due to the difference between GaAs FET models used for the design and actual device parameters which were later derived from the discrete FET S-parameters obtained by de-embedding techniques.

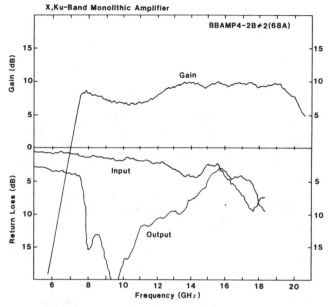

Fig. 4 Small signal performance of a two-stage monolithic amplifier.

A balanched amplifier was constructed by combining two monolithic chips and 3 dB hybrid couplers fabricated on 15 mil thick alumina substrates. A photograph of the balanced amplifier module is shown in Fig. 5.

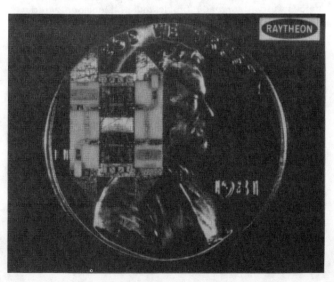

Fig. 5 Balanced amplifier module.

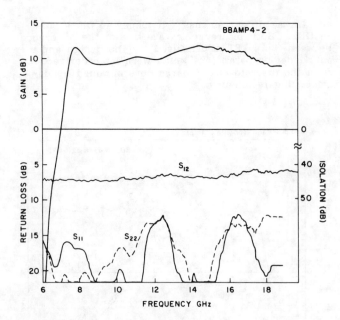

Fig. 6 Small signal performance of a balanced amplifier module.

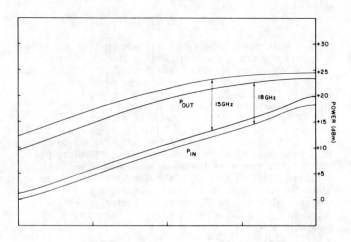

Fig. 7 Power performance of a balanced amplifier module.

The module showed excellent performance with 10.5 ± 1.0 dB gain across 7.5 to 18.6 GHz (Fig. 6). Note also the outstanding input/output VSWRs as well as the high reverse isolation. The module was exposed under swept power measurements which resulted in +21 to +23 dBm at 1 dB compression level for X- through Ku-band. Figure 7 shows the power transfer characteristics taken at 15 and 18 GHz. No tweaking was done on any of these chips and the chips were only selected based on dc data.

Conclusions

Monolithic, medium-power, two-stage amplifiers for X-, Ku-band frequency coverage have been successfully designed, fabricated and demonstrated. We have also proven that these monolithic chips can be integrated with hybrid couplers to form a balanced amplifier.

References

1. J. Degenford, "Design considerations for wideband monolithic power amplifiers," 1980 GaAs IC Symposium Paper 22.

2. W. C. Petersen, et al., "A monolithic GaAs 0.1 to 10 GHz amplifier," 1981 IEEE MTT-S International Microwave Symposium Digest, p. 354.

3. K. B. Niclas, et al., "A 2-12 GHz feedback amplifier on GaAs," 1981 IEEE MTTS International Microwave Symposium Digest, p. 356.

4. Y. Ayasli, et al., "A monolithic GaAs 7 to 13 GHz travelling-wave amplifier," The 1981 GaAs IC Symposium, Paper 46.

5. Y. Tajima, et al., "FET and circuit models for monolithic power amplifiers," The 1981 GaAs IC Symposium, Paper 48.

6. A. Higashisaka, "20 GHz band monolithic GaAs FET low noise amplifier," IEEE Trans. on MTT, Jan. 1981, MTT 29, No. 1, pp. 1-6.

The Matched Feedback Amplifier: Ultrawide-Band Microwave Amplification with GaAs MESFET's

KARL B. NICLAS, MEMBER, IEEE, WALTER T. WILSER, MEMBER, IEEE, RICHARD B. GOLD, MEMBER, IEEE, AND WILLIAM R. HITCHENS

Abstract—An ultrawide-band amplifier module has been developed that covers the frequency range from 350 MHz to 14 GHz. A minimum gain of 4 dB was obtained across this 40:1 bandwidth at an output power of 13 dBm. The amplifier makes use of negative and positive feedback and incorporates a GaAs MESFET that was developed with special emphasis on low parasitics. The transistor has the gate dimensions 800 by 1 μm. The technology and RF performance of the GaAs MESFET are discussed, as are the design considerations and performance of the single-ended feedback amplifier module.

I. INTRODUCTION

TWO-THIRDS of a century have gone by since, in 1913, Alexander Meissner was granted a patent on a feedback oscillator circuit by the German patent office. In the same year Edwin H. Armstrong presented a paper on regenerative circuits and in 1914 Lee DeForest, without whose earlier invention of the triode the feedback circuit would have been meaningless, filed a patent application on the regenerative circuit.

Ever since the invention of the original regenerative circuit, a great number of new feedback circuits with a multitude of applications have emerged. One particularly significant application is the use of negative feedback to control the gain and the input and output impedance of an amplifier [1]–[3]. It is, therefore, not surprising that the principle of negative feedback with its wide bandwidth potential, low input and output reflection coefficients, and good gain flatness has made its entry into the field of microwave amplifiers and, quite recently, into that of GaAs MESFET amplifiers [4].

This paper describes the use of both negative and positive feedback to extend the bandwidth of microwave amplifiers far beyond that reported to date. In order to accomplish this goal, a GaAs MESFET was developed with special emphasis on reduction of parasitics. Experimental amplifier modules exhibit a bandwidth of more than five octaves covering a frequency band from 350 MHz to 14 GHz. Minimum gain over this band is 4 dB at 13 dBm of output power. The amplifier makes use of "frequency controlled" feedback and simple matching techniques. The theory behind the basic feedback ampli-

Manuscript received September 4, 1979; revised November 19, 1979.
The authors are with Watkins-Johnson Company, Palo Alto, CA 94304.

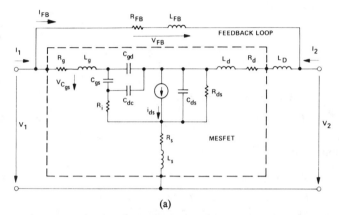

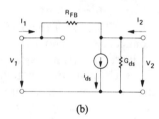

Fig. 1. Circuit diagram of the basic feedback amplifier. (a) High-frequency model. (b) Low-frequency model.

fier circuit is discussed in detail, as are the fabrication and performance of the amplifier modules. The technology, performance, and model of the GaAs MESFET are also described. Finally, gain, reflection coefficients, and reverse isolation of the computer model are compared to the measured data of two amplifier modules.

II. BASIC FEEDBACK AMPLIFIER CIRCUIT

The parasitic elements of a GaAs MESFET, as shown in the schematic of Fig. 1 (a), restrict the amplifier bandwidth capability. Minimization of these parasitics was a major goal of the device development, but there are obvious practical limitations. For this reason, we looked for supporting techniques to extend the bandwidth of the "conventional" negative feedback amplifier to higher frequencies. A practical answer was found in the introduction of the drain inductance L_D and the feedback induc-

TABLE I
ELEMENTS OF THE TRANSISTOR MODEL

INTRINSIC ELEMENTS		EXTRINSIC ELEMENTS	
g_m = 54 m mhos		R_g = 1.5 ohm	
τ_0 = 3.5 psec		L_g = .19 nH	
C_{gs} = .67 pF		R_s = .9 ohm	
C_{gd} = .017 pF		L_s = .151 nH	
C_{dc} = .032 pF		C_{ds} = .082 pF	
R_i = 4.4 ohm		R_d = 2 ohm	
R_{ds} = 200 ohm		L_d = .143 nH	

DC BIAS CONDITIONS

V_{DS} = 4V
V_{GS} = -1V
I_{DS} = 100 mA

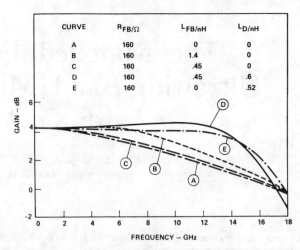

Fig. 2. Computed small-signal gain for various combinations of R_{FB}, L_{FB}, and L_D.

tance L_{FB}. These inductances and the parasitic elements of the GaAs MESFET are being employed to frequency control the feedback.

A. High-Frequency Model

The circuit diagram of the basic feedback amplifier making use of frequency controlled feedback is shown in Fig. 1(a). The amplifier's active device, the GaAs MESFET, is presented in form of its equivalent circuit, whose element values are listed in Table I [5]. The values of the reactive elements are small enough that at frequencies below 1.5 GHz all reactive elements can be neglected for determination of such quantities as gain, input and output VSWR and reverse isolation, i.e., up to 1.5 GHz the transistor can be represented by its dc model. However, the element values in Table I are such that aside from neglecting R_g, R_d, R_s, and maybe R_i, any further simplification of the high-frequency model of Fig. 1(a) leads to erroneous results at frequencies above 1.5 GHz. But even with $R_g = R_d = R_s = R_i = 0$, the admittance matrix of the basic feedback amplifier is so complicated that the computer provides the only efficient tool for obtaining solutions.

The basic feedback amplifier (Fig. 1 (a)) contains two series inductors, L_D in the drain line and L_{FB} in the feedback loop, that have been inserted to extend the amplifier's bandwidth. L_D was chosen to compensate for the capacitive portion of the output impedance at the upper band edge. It improves the output VSWR and eliminates the reactive component of the output impedance at this frequency. L_{FB} reduces the effectiveness of the negative feedback with increasing frequency.

The purpose of these two series inductors are best demonstrated by comparing their influence on insertion gain and bandwidth of the amplifier of Fig. 1. This comparison is presented in Fig. 2 for five selected combinations of R_{FB}, L_{FB}, and L_D. The curves clearly demonstrate the influence of L_D and L_{FB} on the bandwidth and gain response of the basic feedback amplifier. The inductor L_D is mainly responsible for the extended band coverage while the feedback loop provides the flat gain response.

For better understanding of the feedback amplifier's behavior, the vector diagrams of several important voltages as they appear across certain terminals of the amplifier have been drawn at selected frequencies. They are shown in Fig. 3(a) for the case R_{FB} = 160 Ω, L_{FB} = 0.45 nH, L_D = 0.6 nH, and in Fig. 3(b) for the case of R_{FB} = 160 Ω, $L_{FB} = L_D = 0$. All voltages are normalized to $V_S/2$, which is that portion of the source voltage V_S that appears across a 50-Ω load when the signal source is terminated with such a load. Comparing the vector diagrams of Fig. 3 (a) one notices that at 13.75 GHz V_1 and V_2 are in phase, while at very low frequencies they are 180° out of phase. The feedback current I_{FB} has advanced 180° with respect to the signal source current I_S, as is shown in Fig. 4. The signal source current is the current that flows from a 50-Ω source into a 50-Ω load. At 13.75 GHz the ratio V_2/V_1 reaches its maximum and this frequency marks the point of optimum positive feedback.

In order to obtain a more detailed comparison between the behavior of the "conventional negative feedback amplifier" and the amplifier that makes use of controlled feedback, we have plotted the input voltage V_1 and the output voltage V_2 in Fig. 5 as a function of frequency. It can be seen that the influence of L_D and L_{FB} on the magnitude of the input voltage is not very pronounced. Noticeable phase differences exist above 9 GHz, however, with a crossover point at 13.75 GHz. The output voltage of both the conventional negative feedback amplifier and that of the frequency controlled feedback amplifier are almost identical up to 4 GHz. The reactive elements L_D and L_{FB} maintain a nearly constant gain response to almost 14 GHz (curve C) however, which is the reason for the use of reactive control. Also shown in Fig. 5 (curve A) are the input (V_1) and output (V_2) voltages of the amplifier without feedback ($R_{FB} = \infty$). This curve in particular demonstrates the bandwidth potential of the transistor

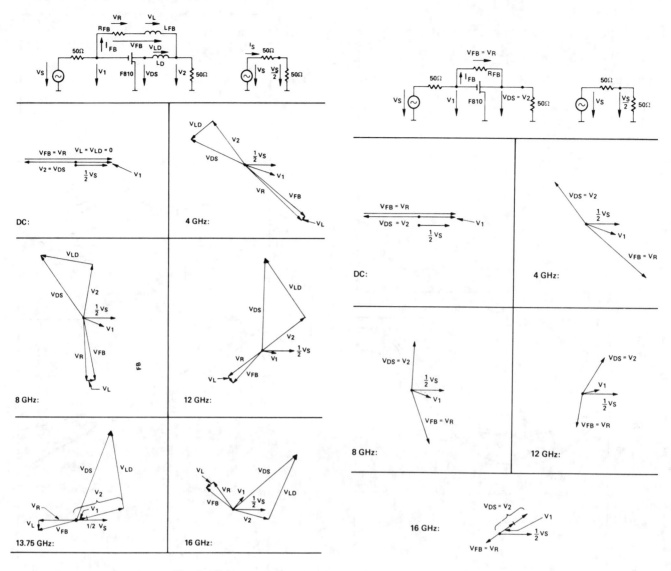

Fig. 3. Voltage vector diagrams of (a) the amplifier with frequency-controlled feedback, and (b) the conventional negative feedback amplifier.

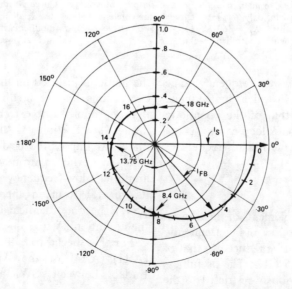

Fig. 4. Magnitude and phase of the normalized feedback current.

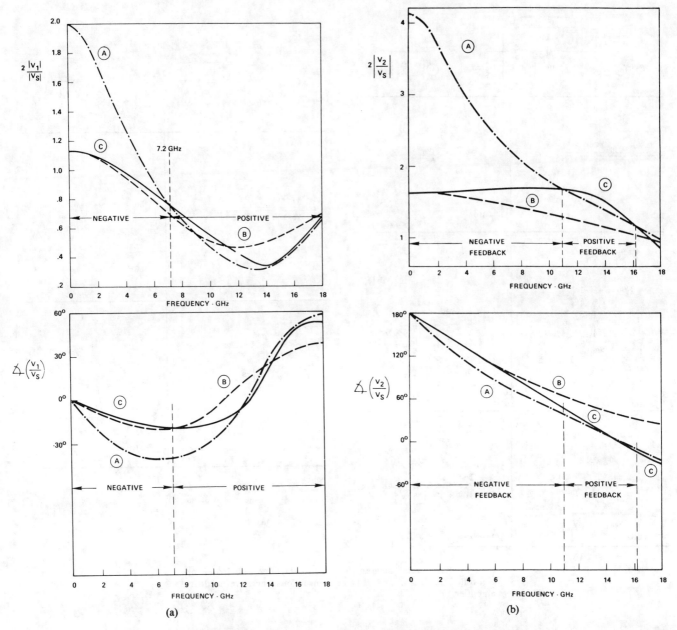

Fig. 5. Normalized input (a) and output (b) voltages. Curve A: the transistor only, $R_{FB} = \infty$; curve B: the conventional negative feedback amplifier, $R_{FB} = 160$ Ω, $L_{FB} = L_D = 0$; curve C: the amplifier with frequency controlled feedback, $R_{FB} = 160$ Ω, $L_{FB} = 0.45$ nH, $L_D = 0.6$ nH.

used in our experiments. It also shows the frequency range of positive feedback, i.e., the band in which the feedback elevates the insertion gain above that of the open-loop amplifier (11–16.2 GHz). The feedback loop causes such a "positive range" for the input voltage for all frequencies above 7.2 GHz.

A. Low-Frequency Model

As discussed earlier, the reactive elements of the transistor model under study are relatively low and consequently the circuit diagram of Fig. 1 (a) can be reduced to that of Fig. 1 (b) for frequencies below 1.5 GHz. The usefulness of the low-frequency model is obviously restricted to the very low end of the frequency band. However, the use of the model is justified on the basis that it yields two simple expressions for the feedback resistor R_{FB} ((A.10), (3a)) that we found to be extremely valuable for our amplifier design (see Section IV-A). The model also provides an understanding of the tradeoffs between match and gain. The scattering parameter matrix of the feedback amplifier's low-frequency model (A.4)–(A.9) is derived in its general form in the Appendix. Input and output VSWR become identical if the feedback resistor satisfies the condition (A.10)

For this special case the S parameters are presented in (A.11)–(A.14) of the Appendix. To demonstrate the tradeoff between VSWR and gain, we assume

$$G_{ds} Z_0 \ll 1; \qquad G_{ds} \ll g_m. \qquad (1)$$

Using the general set of equations (A.5)–(A.9) of the Appendix, we find the S parameters

$$S_{11} = S_{22} = \frac{1}{\Sigma}\left[\frac{R_{FB}}{Z_0} - g_m Z_0\right] \quad (2a)$$

$$S_{12} = \frac{2}{\Sigma} \quad (2b)$$

$$S_{21} = \frac{-2}{\Sigma}\left[g_m R_{FB} - 1\right] \quad (2c)$$

with

$$\Sigma = 2 + g_m Z_0 + \frac{R_{FB}}{Z_0}. \quad (2d)$$

They are plotted in Fig. 6, which shows the relationship between gain, VSWR, transconductance g_m, and feedback resistance R_{FB}. Only the lower section of the right half of the diagram is of practical interest. In this region the amplifier yields the highest-gain–lowest-VSWR combinations. The curves demonstrate that gain at low frequencies can be significantly increased due to an increase of the reflection coefficients. Ideal matching yields the lowest gain. The curves of Fig. 6 further reveal that by varying the feedback resistor the gain can be changed between total attenuation and maximum gain.

The condition for ideal match ($S_{11} = S_{22} = 0$) requires

$$R_{FB} = g_m Z_0^2. \quad (3a)$$

The associated gain is

$$G = 20\log(g_m Z_0 - 1). \quad (3b)$$

C. Circuit Objectives and Requirements

The circuit objective was to extend the bandwidth of the negative feedback amplifier from its upper frequency limit of about 6–14 GHz. The choice of the two series inductances L_D and L_{FB} made this possible. These values were determined as follows.

1) L_D was chosen to compensate for the capacitive component of the GaAs MESFET's output impedance so that resonance occurs at the upper band edge. This measure simultaneously results in a marked improvement in the output match.

2) L_{FB} was chosen, in cooperation with L_D, to adjust the S parameters of the feedback amplifier so that optimum positive feedback exists at the upper band edge. This condition coincides with the input (V_1) and output (V_2) voltage being in phase and the feedback current I_{FB} advanced by 180° with respect to its phase at very low frequencies. The degree of feedback is mainly controlled by the feedback resistor R_{FB}.

Once steps 1) and 2) are accomplished and the broad-band potential of the feedback amplifier shown in Fig. 1(a) is nearly exhausted, a third important step is added; i.e., a simple input and output matching network to further improve the input and output VSWR. In this case one follows published design techniques. Since we chose to use distributed rather than lumped elements, we were confined to short series transmission lines and short open-ended shunt stubs due to the enormous bandwidth we set out to cover. More details on the matching networks will be found in Section IV.

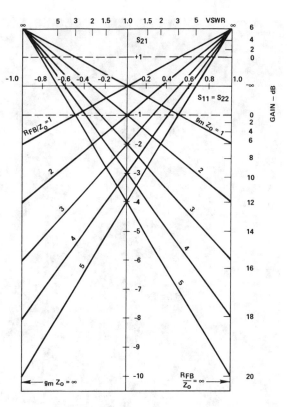

Fig. 6. Gain and VSWR of the low-frequency model for $G_{ds} = 0$ and $R_{FB} = g_m Z_0^2$.

III. Device Technology and Performance

A. Technology

The GaAs MESFET used in this study, the WJ-F810, is the 800-μm gate width device shown in Fig. 7. The chip size is 320 by 370 μm. The 1-μm long gate is centered in a 4-μm source–drain channel.

The GaAs MESFET's were fabricated using a self-aligned etched aluminum gate process described earlier [6]. The n-type active layer was grown by liquid phase epitaxy on Cr-doped substrates The epitaxial layer was doped with Sn to a concentration of 1.0×10^{17} cm^{-3}.

Special emphasis was paid to the reduction of parasitic elements, particularly capacitances which would limit broad-band performance of the transistor. The gate and drain pads were made as small as practical to minimize the input, output, and feedback capacitances. By using 0.5-mil wire for all bond connections we were able to easily bond to these small pads. The minimal size of the source pad is important for other applications which use the common-gate configuration. In this common source application, however, the source pad size is of little concern because both the source pad and the back of the chip are at RF ground. To further minimize the capacitances, a gate–drain spacing of 1.5 μm was chosen, in contrast to 0.9 μm for an earlier device [6]. This spacing is a com-

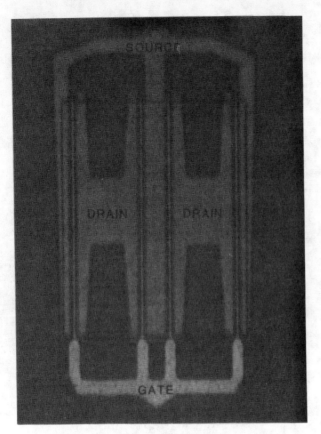

Fig. 7. GaAs MESFET chip (0.32×0.36 mm).

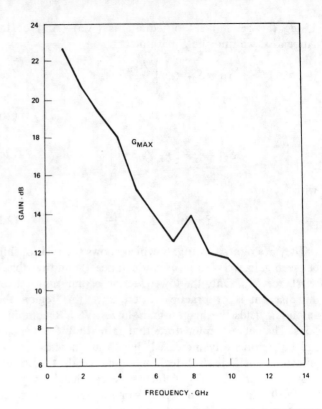

Fig. 8. Measured maximum available gain of the GaAs MESFET.

promise between increased resistance and decreased capacitance.

Most low-noise GaAs FET's have a gate width of about 300 μm. Larger gate widths lead to a linear increase in the transconductance g_m as well as intrinsic capacitances; the cutoff frequency f_t is only slightly affected. The 800-μm gate width was chosen to give substantially higher g_m than available in standard low-noise devices. This high g_m has proven to be an essential element in circuit designs for feedback amplifiers.

B. Transistor Performance and Device Model

The GaAs MESFET has a saturated drain-source current I_{DSS} of 170–240 mA, gate-source pinchoff voltage V_p of 5–7 V, and dc transconductance g_m at $1/2\ I_{DSS}$ of 50–60 mmhos. The range of the drain-source bias voltage was between 4 and 6 V, depending on output power requirements. Fig. 8 presents the maximum available gain of the device between 1 and 14 GHz.

The device model is shown in Fig. 1 (a) and the quantities of the model elements are given in Table I. The agreement between the measured S parameters and those computed using the model was excellent. Table II compares the parasitic elements of the intrinsic transistor model of the device discussed in this paper with two of our GaAs MESFET's which have been described elsewhere ([6], [7]). The element values are normalized to the gate width for reasons of comparison. It can be seen that the topology of the WJ-F810 has lead to a significant

TABLE II
ELEMENTS OF TRANSISTOR MODELS NORMALIZED TO GATEWIDTH

PARAMETER	UNIT	WJ-F110	WJ-F1010	WJ-F810
W_{GATE}	μm	300	1000	800
L_{GATE}	μm	1	1	1
g_m/W	mmhos/mm	73	68	68
C_{gs}/W	pF/mm	1.37	.94	.84
C_{gd}/W	pF/mm	.033	.055	.021
C_{dc}/W	pF/mm	.097	.047	.040
C_{ds}/W	pF/mm	.41	.24	.10

reduction in the magnitudes of the normalized capacitive elements making this device highly suitable for broadband applications.

IV. AMPLIFIER DESIGN AND PERFORMANCE

A. Small-Signal Design

The first step in designing an ultrabroad-band feedback amplifier is the selection of the feedback resistor R_{FB}. According to Table I the transconductance g_m and the drain-source conductance G_{ds} of our device are 0.054 mhos and 0.005 mho, respectively. If it is desired that input and output VSWR be identical, we find from (A10)–(A14) that, at very low frequencies, a gain of 4.2 dB and an input and output VSWR of 1.2:1 will be obtained for $R_{FB} = (g_m + G_{ds})Z_0^2 = 147.5\ \Omega$. Raising the

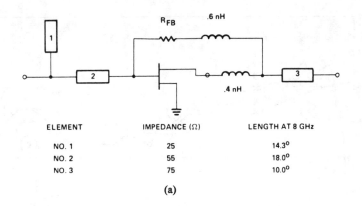

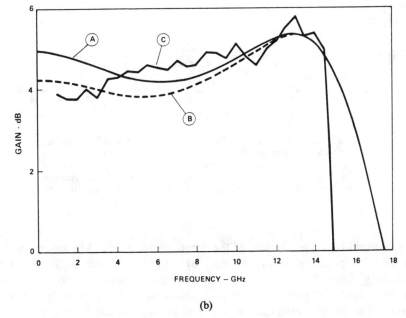

Fig. 9. Schematic (a) and gain curves (b) of the matched feedback amplifier. Curve A: computed for $R_{FB} = 180\ \Omega$; curve B: computed for $R_{FB} = 160\ \Omega$; curve C: measured for $R_{FB} = 160\ \Omega$ ($V_{DS} = 4$ V, $I_{DS} = 53$ mA).

feedback resistor to $R_{FB} = 160\ \Omega$ increases the gain to 4.6 dB as calculated with (A.7). Using (A.5) and (A.8) the input VSWR for this case computes to 1.27:1 and the output VSWR to 1.13:1.

The selection of L_D and L_{FB} as well as the influence of these two inductors on gain has been described in detail in Section II. The gain response of the basic feedback amplifier shown in Fig. 1 (a) with $R_{FB} = 160\ \Omega$ is nearly flat to almost 14 GHz (Fig. 2, curve D). However, the corresponding input reflection coefficient is rather poor above 8 GHz. In order to improve the input match over the upper portion of the band, we employed an open-circuit shunt stub and a series transmission line. Due to the feedback, these two components had a negative influence on the output match, so we inserted a series transmission line connected to the output terminal of the basic amplifier to counteract the degradation of the output reflection coefficient. The introduction of the input and output matching networks led to the matched feedback amplifier.

The schematic of the matched feedback amplifier is shown in Fig. 9 (a). The selection of $L_{FB} = 0.6$ nH and $L_D = 0.4$ nH deviates from the optimum values of $L_{FB} = 0.45$ nH and $L_D = 0.6$ nH discussed in Section II. This change was made to attain a more practical circuit layout. Since L_{FB} bridges most of the physical distance between the transistor's gate terminal and the node between L_D and the series transmission line, it becomes somewhat impractical to make L_{FB} smaller than L_D. In addition, a higher feedback resistor of $R_{FB} = 180\ \Omega$ was inserted to obtain optimum gain flatness for the new set of inductors. These measures constitute a necessary tradeoff between practicality and optimum performance. The computed small signal gain of the amplifier between 50-Ω impedances is plotted as curve A of Fig. 9 (b), while curve B represents the computed small signal gain for $R_{FB} = 160\ \Omega$. The computed reflection coefficients of the practical version of the matched amplifier (Fig. 9 (a)), the negative feedback amplifier (Fig. 3 (b)), and the basic feedback amplifier (Fig. 3 (a)) are plotted in Fig. 10 as curves A, B,

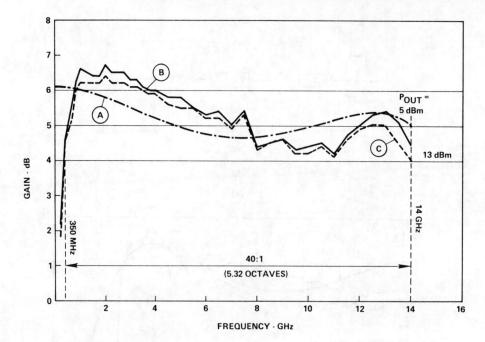

Fig. 10. Input and output reflection coefficients. Curve A: matched feedback amplifier of Fig. 9 (a) with $R_{FB}=180$ Ω, $L_{FB}=0.6$ nH, $L_D=0.4$ nH; curve B: conventional negative feedback amplifier of Fig. 3 (b) with $R_{FB}=160$ Ω, $L_{FB}=L_D=0$; curve C: basic feedback amplifier of Fig. 3 (a) with $R_{FB}=160$ Ω, $L_{FB}=0.45$ nH, $L_D=0.6$ nH.

and C, respectively. The input reflection coefficient of the matched amplifier shows a significant improvement brought about by the matching circuits.

However, the output reflection coefficient has experienced a slight degradation due to the influence of the input matching circuit on the output impedance brought about by the feedback loop. This is particularly pronounced in the area of relatively strong feedback, i.e., above 11 GHz. The amplifier is unconditionally stable up to 13.8 GHz. Reverse isolation computed between dc and 18 GHz has a minimum value of 9.7 dB at 17 GHz for $R_{FB}=160$ Ω and 10 dB at 16 GHz for $R_{FB}=180$ Ω.

B. Amplifier Fabrication and Performance

Fused silica, 0.015 in in thickness, was used as substrate material for the input and output circuits. The circuit pattern was etched into a thin gold film, while the feedback resistor was subsequently etched into a tantalum nitride film which was deposited below the gold.

The measured small-signal gain of the amplifier is plotted in Fig. 9 (b) (curve C). The feedback resistor of this unit measured $R_{FB}=160$ Ω instead of the desired 180 Ω. A comparison of the measured small-signal gain and the gain computed for $R_{FB}=160$ Ω (curve B) shows excellent agreement at frequencies up to 14.5 GHz. Beyond this frequency the actual measured gain dropped rather abruptly. Below 1 GHz the drop in gain was due to the influence of the internal dc biasing network (not shown in the schematic of Fig. 9 (a)). The measured reverse isolation had its minimum value of $|S_{12}|^2_{\min}=14$ dB at 1 GHz. This isolation is somewhat better than the 11.2 dB computed with (A.6). Maximum measured reverse isolation was $|S_{12}|^2_{\max}=21$ dB at 10.5 GHz which compares to the maximum computed value of 22.4 dB at 9 GHz. The reflection coefficients of the actual amplifier did not exceed $|S_{11}|_{\max}=0.38$ for the input and $|S_{22}|_{\max}=0.37$ for the output terminal between 2 GHz and 14.4 GHz. The computed values were $|S_{11}|_{\max}=0.42$ and $|S_{22}|_{\max}=0.37$, respectively.

The measured small-signal gain of an identical amplifier module except for $R_{FB}=225$ Ω is plotted as curve B in Fig. 11. Biasing of this amplifier was accomplished by means of external bias networks. The drop in the measured gain below 1 GHz was caused by the 50-pF dc blocking capacitor in the feedback loop and by the bias networks. The capacitor serves the purpose of separating the drain bias from the gate bias potential. The computed gain is plotted as curve A, while curve C shows the gain at 13 dBm of output power. This module covers a 40:1 bandwidth ranging from 350 MHz to 14 GHz, or almost 5 1/3 octaves. Maximum input and output reflection coefficients between 350 MHz and 14 GHz were $|S_{11}|_{\max}=0.45$ and $|S_{22}|_{\max}=0.54$, respectively. A minimum reverse isolation of $|S_{12}|^2_{\min}=11.5$ dB was measured at 14 GHz. Computed and measured reflection coefficients and minimum reverse isolations are in good agreement.

V. Conclusion

The design of an ultrawide-band amplifier has been described which makes use of both negative and positive feedback. The frequency dependence of the feedback is controlled by two inductors, one in series with the feed-

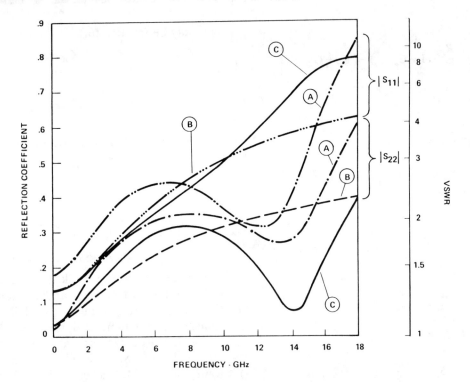

Fig. 11. Gain of the matched feedback amplifier for $R_{FB}=225\ \Omega$. Curve A: computed small signal gain; curve B: measured gain at 5 dBm of output power ($V_{DS}=5$ V, $I_{DS}=72$ mA); curve C: measured gain at 13 dBm of output power ($V_{DS}=5$ V, $I_{DS}=72$ mA).

back resistor and the other in series with the output of the GaAs MESFET. A detailed study of the effect of the inductors on the characteristics of the feedback amplifier has been made and guidelines have been presented on how to determine their magnitudes. In addition to the feedback circuitry, simple matching networks have been employed to improve the input and output reflection coefficients. The GaAs MESFET used in the experiments was designed with major emphasis on reducing its parasitics. These efforts resulted in a device with ultrawide bandwidth capability.

Two experimental amplifier modules have been described that cover the frequency band 1.0–14.5 GHz and 350 MHz–14.0 GHz with 3.8 dB and 4.2 dB of minimum gain, respectively. Maximum reflection coefficients of these single-ended units were $|S_{11}|_{max}=0.38$ and $|S_{11}|_{max}=0.45$ for the respective input terminals and $|S_{22}|_{max}=0.37$ and $|S_{22}|_{max}=0.54$ for the respective output terminals. Minimum reverse isolations across these bands were $|S_{12}|^2_{min}=14$ dB and $|S_{12}|^2_{min}=11.5$ dB, respectively. The agreement between the computed and the measured data of small-signal gain, reflection coefficients, and reverse isolation was very good.

The use of the matched feedback amplifier with GaAs MESFET's provides several advantages over usual techniques for broad-band microwave amplification. It has an exceptionally wide bandwidth and much lower reflection coefficients than regular single-ended amplifiers of comparable bandwidths. It can be constructed with a very simple and small circuit and is relatively easy to cascade.

These advantages make feedback amplifiers prime candidates for monolithic applications.

Appendix

The equivalent circuit of a GaAs MESFET feedback amplifier as shown in Fig. 1 (a) can be reduced to the model of Fig. 1 (b) for frequencies at which the reactive elements of the transistor and the matching networks may be neglected. A discussion of the limitations of the low-frequency model is found in Section II.2. Under these conditions and the assumption that R_d and R_s of Fig. 1 (a) are very small compared to the feedback resistor R_{FB} and the load resistor $R_L=Z_0$ voltages and currents are described by the simple conductance matrix

$$\begin{bmatrix} I_1 \\ I_2 \end{bmatrix} = \begin{bmatrix} G_{FB} & -G_{FB} \\ (g_m - G_{FB}) & (G_{FB} + G_{ds}) \end{bmatrix} \begin{bmatrix} V_1 \\ V_2 \end{bmatrix} \quad (A.1)$$

where

$$G_{FB} = R_{FB}^{-1} \quad (A.2a)$$

$$G_{ds} = R_{ds}^{-1} \quad (A.2b)$$

and

$$i_{gs} = g_m V_{gs}. \quad (A.3)$$

Using elementary algebra, the matrix (A.1) converts into the scattering parameter matrix

$$S_{ij} = \begin{bmatrix} S_{11} & S_{12} \\ S_{21} & S_{22} \end{bmatrix}. \quad (A.4)$$

Its elements are

$$S_{11} = \frac{1}{\Sigma}\left[\frac{R_{FB}}{Z_0}(1+G_{ds}Z_0) - (g_m + G_{ds})Z_0\right] \quad (A.5)$$

$$S_{12} = \frac{2}{\Sigma} \quad (A.6)$$

$$S_{21} = \frac{-2}{\Sigma}\left[g_m R_{FB} - 1\right] \quad (A.7)$$

$$S_{22} = \frac{1}{\Sigma}\left[\frac{R_{FB}}{Z_0}(1-G_{ds}Z_0) - (g_m + G_{ds})Z_0\right] \quad (A.8)$$

with

$$\Sigma = 2 + (g_m + G_{ds})Z_0 + \frac{R_{FB}}{Z_0}(1+G_{ds}Z_0). \quad (A.9)$$

For the condition

$$\frac{R_{FB}}{Z_0} = (g_m + G_{ds})Z_0 \quad (A.10)$$

we find

$$S_{11} = -S_{22} = \frac{G_{ds}Z_0^2}{\Sigma}(g_m + G_{ds}) \quad (A.11)$$

$$S_{12} = \frac{2}{\Sigma} \quad (A.12)$$

$$S_{21} = -\frac{2}{\Sigma}\left[g_m(g_m + G_{ds})Z_0^2 - 1\right] \quad (A.13)$$

$$\Sigma = 2 + (g_m + G_{ds})(2 + G_{ds}Z_0)Z_0. \quad (A.14)$$

The reflection coefficients S_{11} and S_{22} improve with decreasing drain-source conductance G_{ds}. The general (A.5)–(A.9) and the special (A.10)–(A.14) equations for the S parameters demonstrate that gain, input and output VSWR, and reverse isolation are all fixed quantities once the value for the feedback resistor R_{FB} is chosen.

The ideal matching condition

$$S_{11} = S_{22} = 0 \quad (A.15)$$

can only be satisfied for

$$G_{ds} = 0$$

and

$$\frac{R_{FB}}{Z_0} = g_m Z_0. \quad (A.16)$$

In this case we find the S parameters

$$S_{11} = S_{22} = 0 \quad (A.17)$$

$$S_{12} = \frac{1}{g_m Z_0 + 1} \quad (A.18)$$

$$S_{21} = -(g_m Z_0 - 1). \quad (A.19)$$

Acknowledgment

The authors wish to thank R. Pereira, who was responsible for the measurement and the tuning of the amplifiers. Thanks are also due to A. Hallin, K. Lutz, and K. Lindstedt, who fabricated the GaAs MESFET devices and to J. Martin, who assembled the circuits. The authors are indebted to S. Rose, who was responsible for device testing, and to M. Walker for many helpful discussions.

References

[1] H. S. Black, "Stabilized feedback amplifiers," *Elec. Eng.*, vol. 53, pp. 114–120, Jan. 1934.
[2] F. E. Terman, *Radio Engineer's Handbook*. New York: McGraw-Hill, 1943, pp. 395–406.
[3] M. S. Ghausi, *Principles and Designs of Linear Active Circuits*. New York: McGraw-Hill, 1965, pp. 363–370.
[4] E. Ulrich, "Use of negative feedback to slash wideband VSWR," *Microwaves*, pp. 66–70, Oct. 1978.
[5] R. Dawson, "Equivalent Circuit of the Schottky-barrier field-effect transistor at microwave frequencies," *IEEE Trans. Microwave Theory Tech.*, vol. MTT-23, pp. 499–501, June 1975.
[6] K. B. Niclas, R. B. Gold, W. T. Wilser, and W. R. Hitchens, "A 12–18 GHz medium power GaAs MESFET amplifier," *IEEE J. Solid State Circuits*, vol. SC-13, pp. 520–527, Aug. 1978.
[7] K. B. Niclas, W. T. Wilser, R. B. Gold, and W. R. Hitchens, "Application of the two-way balanced amplifier concept to wideband power amplification using GaAs MESFET's," *IEEE Trans. Microwave Theory Tech.*, to be published.

A Monolithic GaAs DC to 2-GHz Feedback Amplifier

WENDALL C. PETERSEN, MEMBER, IEEE, ADITYA GUPTA,
AND
D. R. DECKER, SENIOR MEMBER, IEEE

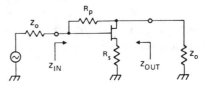

Fig. 1. Series and shunt resistance feedback applied to an ideal FET.

Abstract —Resistive feedback in low-frequency FET amplifiers is an attractive method of simultaneously attaining gain flatness and excellent input–output VSWR over wide bandwidths. Combined with simple matching circuitry, the feedback approach allows the design of general-purpose utility amplifiers requiring much less chip area than when conventional matching techniques are used. The 1.5- by 1.5-mm chip described in this paper provides 10-dB ± 1-dB gain, excellent input and output VSWR, and saturated output power in excess of +20 dBm, from below 5 MHz to 2 GHz. The noise figure is approximately 2 dB when biased for minimum noise, with an associated gain of 9 dB.

I. INTRODUCTION

Negative feedback amplifiers have found wide acceptance in the marketplace for low-frequency bipolar transistor designs and have recently been introduced as the first commercially available GaAs FET monolithic microwave integrated circuit [1]. Continuing work on this fruitful design technique will lead to improved amplifier performance on smaller chips while incorporating more bias and signal-processing circuitry on the chip. The low-frequency FET feedback amplifier described in this paper provides resistive bias isolation for ease of use, but also allows the direct application of drain bias for efficiency sensitive applications. A resistively isolated gate bias line allows operation of the FET under either low-noise or high-power bias conditions for further versatility.

II. THEORY

The design of resistive-feedback FET amplifiers is based on the near-ideal voltage-controlled current-source characteristics of a microwave GaAs FET operated at low frequencies. Application of series and shunt resistive feedback as shown in Fig. 1 provides simultaneous input and output match, while maintaining flat gain from dc to a frequency determined by the parasitic elements of the FET and circuit elements. For an ideal FET with transconductance g_m, gain of the circuit shown in Fig. 1 is given by

$$G_T = \frac{2Z_0(1+g_m R_s) - 2g_m Z_0 R_p}{g_m Z_0^2 + 2Z_0(1+g_m R_s) + R_p(1+g_m R_s)}$$

while input and output impedance are given by

$$Z_{in} = Z_{out} = \frac{(R_p + Z_0)(1 + g_m R_s)}{1 + g_m(R_s + Z_0)}$$

where Z_0 is the system impedance. Under perfect match conditions

$$Z_{in} = Z_{out} = Z_0$$

which implies that

$$R_p = \frac{g_m Z_0^2}{1 + g_m R_s} \quad \text{and} \quad G_T = \frac{Z_0 - R_p}{Z_0}.$$

In order to have amplification, it is necessary that $|G_T| > 1$ which under matched conditions reduces to $g_m > 2/(Z_0 - 2R_s)$ for positive g_m, Z_0, and R_s. Since g_m is proportional to FET width, which must be kept at a minimum, $R_s = 0$ should be selected. With $R_s = 0$ and perfect input and output match, the well-known formula

$$R_p = g_m Z_0^2 \quad \text{and} \quad G_T = 1 - g_m Z_0$$

is obtained. However, it is often advantageous to allow a slight VSWR degradation in order to increase amplifier gain. Assuming $R_s = 0$ and assuming an input and output VSWR of $K:1$ is acceptable, selecting

$$R_p = KZ_0(1 + g_m Z_0) - Z_0$$

will yield the allowed VSWR and a gain of

$$G_T = \frac{2(1 - Kg_m Z_0)}{1 + K}.$$

As an example, if $g_m = 80$ mS, under matched conditions R_p would be 200 Ω resulting in a gain of 9.54 dB, but if a 1.5 to 1 VSWR is acceptable, R_p becomes 325 Ω and gain is increased to 12.04 dB. However, additional tradeoffs are also inherent in this gain-enhancement technique. Reducing the amount of negative feedback applied allows the effects of parasitic elements to become apparent at lower frequencies thus reducing amplifier bandwidth. The dominant parasitic elements are the FET gate-to-source capacitance C_{gs} and the gate-to-drain capacitance C_{gd}. Continuing the above example, with $C_{gs} = 1.5$ pF and $C_{gd} = 0$, the 3-dB corner frequency is 4.3 under matched conditions and 3.5 GHz with a VSWR of 1.5:1. Also, with $C_{gs} = 0$ and $C_{gd} = 0.15$ pF, the corner frequencies become 9.0 and 6.9 GHz, respectively. When both capacitances are considered at the same time, their interaction results in corner frequencies of 3.1 and 2.5 GHz, respectively. The output capacitance C_{ds} has a less significant effect at these frequencies, but the drain resistance R_{ds} can reduce the effective load impedance and, therefore, can reduce amplifier gain. Input and output match are also degraded by R_{ds} under matched conditions ($K = 1$), but the output match can actually be improved when $K > 1$. As much as 3 dB of gain can be lost due to R_{ds} and once again the sensitivity depends on the amount of negative feedback applied.

The gain roll-off due to the parasitic capacitance of the FET is accompanied by a degradation of input VSWR. The output VSWR remains acceptable due to the effect of R_{ds}. Therefore, amplifier performance can be significantly enhanced by adding an input-matching network to the amplifier as shown in Fig. 2. It is essential to use a low-pass matching structure to maintain acceptable low-frequency performance. Standard tables can be

Manuscript received June 2, 1982.
The authors are with Rockwell International, Microelectronics Research and Development Center, Thousand Oaks, CA 91360.

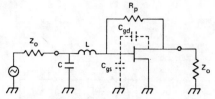

Fig. 2. Feedback amplifier with low-pass input-matching network.

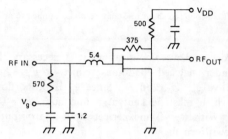

Fig. 3. Feedback amplifier schematic.

used to determine the values of L and C once the effective input impedance is known, and computer optimization is used as the final design step to account for the remaining parasitic elements.

III. AMPLIFIER DESIGN

The design procedure described above was used to design a negative-feedback amplifier to cover the 5-MHz to 2-GHz band. A 1200-μm-wide FET, consisting of four 300-μm-wide fingers was selected to obtain a transconductance of 80 mS at low-noise bias. The effective value of the feedback resistor was increased to 375 Ω to maintain a gain of at least 10 dB when the effects of bias networks and R_{ds} were considered. The feedback resistor R_p was split into four parallel resistors of 1500 Ω each and distributed between the four gate fingers, to minimize parasitic inductance and capacitance. An alternate way of looking at the four gate fingers and their associated feedback resistors is to view them as four separate 200-Ω feedback stages wired in parallel. The lumped-element input-matching network is placed at the 50-Ω side of the connection to conserve chip area and further reduce parasitic effects. With the FET parameters used above, $C = 2$ pF and $L = 5$ nH are needed to bring the input impedance back to 75 Ω (1.5:1 VSWR) at 2 GHz. After optimization across the dc to 2-GHz band with the simplified FET model, the element values become $C = 1.4$ pF and $L = 5.3$ nH. Further optimization including a full FET device model, parasitic interconnection and bias elements, and the higher value of R_p results in final matching element values of $C = 1.2$ pF and $L = 5.4$ nH. The capacitor is implemented as a metal-insulator-metal parallel-plate capacitor and the inductor is implemented as a spiral inductor to conserve chip area. With proper modeling of the parasitics, both function to at least 2 GHz. A full schematic of the amplifier and bias circuitry is shown in Fig. 3.

IV. FABRICATION

The design is implemented on a semi-insulating (SI) GaAs substrate containing lumped elements, ion-implanted resistors, spiral inductors, MIM capacitors, and FET's. Both Cr-doped and undoped SI substrates grown by the horizontal Bridgeman and the liquid-encapsulated Czocharalski (LEC) techniques have been used for device fabrication. A preselection test for bulk SI GaAs substrates, involving qualification of the entire GaAs ingot by sampling the front and the tail of each boule, is first employed in order to select the ingot to be used. The qualification procedure

Fig. 4. 1.5-$\times$1.5-mm amplifier chip.

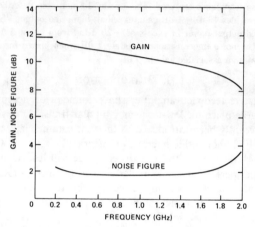

Fig. 5. Measured gain and noise figure.

assesses the ability of the SI substrate to withstand high-temperature (850 °C) processing and to yield device quality active layers by ion implantation. Direct implantation of Si$^+$ in selected areas, defined photolithographically, is used for forming the FET and resistor active areas. The wafer is then coated with reactively sputtered Si$_3$N$_4$ and annealed at 850°C in an H$_2$ ambient, resulting in active layers of $\sim$1000-$\Omega/\square$ sheet resistivity and 4000-5000 cm^2 V·s Hall mobility at 1×10^{17} cm^{-3} doping concentration. AuGe/Ni is used to form the ohmic contacts. The 1-μm-long gates and the first-level metallization are defined by conventional photolithography and lift-off process. Gate metal is Ti-Pt-Au for good reliability. A dielectric layer of Si$_3$N$_4$ is used for the insulation between the first level and the second level interconnections and dielectric for the circuit MIM capacitors. Typically, 130-pF/mm^2 capacitance is obtained. Capacitance uniformity and reproducibility can generally be maintained to within $\pm$5 percent. Reactive ion etching is used to open via holes in the dielectric wherever the first level metallization needs to be accessed. Second-level metallization is gold plated to a thickness of 2-3 μm, to reduce RF losses in the passive circuitry. The GaAs wafer is thinned to 125 μm and metallized on the back to complete the ground plane.

V. RESULTS

Fig. 4 is a photograph of the 1.5- by 1.5-mm chip which is 125 μm thick. (The test structures at the bottom of the chip are not part of the amplifier circuitry.) Measured gain and noise figure are shown in Fig. 5 when the amplifier is biased for low-noise operation ($V_{ds} = 2.6$ V, $I_{ds} = 80$ mA). A separate low-frequency measurement yielded a measured gain of almost 12 dB as indicated in Fig. 5. Input match is excellent at 2 GHz but at low frequencies return loss is 7 dB due to the 375-Ω feedback resistor and a lower than expected transconductance at low-noise bias.

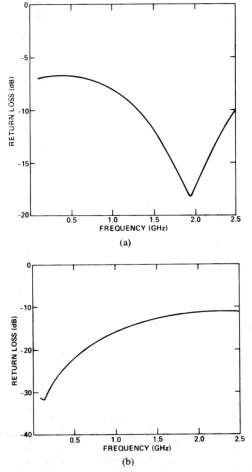

Fig. 6. (a) Measured input match. (b) Measured output match.

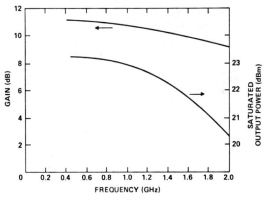

Fig. 7. Measured small-signal gain and saturated power at power bias.

Output return loss is better than 12 dB across the band and is better than 16 dB below 1 GHz. Measured input and output match are shown in Fig. 6. When the amplifier is biased for maximum gain, the output match remains excellent and the input return loss is better than 10 dB across the band due to higher transconductance. Gain shape remains the same, but the gain is increased slightly. At this bias point, saturated power in excess of +20 dBm is obtained across the band while +23 dBm is obtained below 1 GHz, as shown in Fig. 7.

VI. Conclusion

In conclusion, the application of negative resistive feedback around a 1200-μm-wide GaAs FET has led to the fabrication of a low-noise wide-band amplifier suitable for use as a utility amplifier or an IF amplifier. The high dynamic range amplifier is useful as both a discrete component and part of a larger monolithically integrated circuit. Potential future enhancements of the circuit include higher frequency performance, active loads for higher large signal efficiency, and a level shifting circuit to enable dc cascading of the amplifiers.

References

[1] H. P. Weidlich, J. A. Archer, E. Pettenpaul, F. A. Pety, and J. Huber, "A GaAs monolithic broadband amplifier," in *ISSCC Dig. Tech. Papers*, Feb. 1981, 192–193.

A 2.2dB NF 30-1700MHz Feedback Amplifier

Masahiro Nishiuma, Shin-ichi Katsu, Shutaro Nambu, Masahiro Hagio, Gota Kano

Matsushita Semiconductor Laboratory

Osaka, Japan

TO OBTAIN LOW-NOISE CHARACTERISTICS in a monolithic GaAs FET broadband amplifier covering the VHF and UHF bands, the negative feedback approach has been shown to offer the most effective circuit[1,2,3]. However, this circuit dissipates a large current due to the wide gate width required for achieving a large open loop gain ($\propto g_m$) through which a low NF and a high gain are obtainable.

This paper will offer a GaAs monolithic broadband amplifier design (Figure 1) in which the following approach has been taken to resolve the problem. The use of a cascade connection of three GaAs FET's with high input/output impedances affords a decrease of the dissipation current under a fixed open loop gain. The open loop gain G of a single FET amplifier can be expressed approximately as

$$G \approx g_m \frac{R_d R_L}{R_d + R_L},$$

where R_L is the load resistance and R_d is the output resistance of the FET (= $1/g_d$). Therefore, the high open loop gain can be obtained by increasing RL without increasing g_m. The input/output VSWR is reduced by connecting the drain of the third-stage FET and the gate of the first-stage FET through a shunt negative feedback resistor R_f. The increase of the input/output impedances of each FET leads to a decrease of the bandwidth, though the open loop gain is increased. The role of the feedback resistors R_2 and R_3 of the second- and third-stage FETs, respectively, is to suppress the input/output impedances of the respective FETs, increasing the bandwidth of the amplifier. Automatic gain control (AGC) is provided by using dual-gate FETs in the first- and second-stages.

The performance of the amplifier was simulated based on the SPICE 2 program. It was found that the gain and the input/output VSWR are principally functions of a set of R_f, R_2 and R_3 and a set of R_f, g_{m1} and g_{m3}, respectively.

The NF value of the amplifier was calculated using a simple equivalent circuit[1]. It was found that the NF is determined predominantly by g_{m1} and R_f. Figure 2 shows the result of the NF value calculated as a function of g_{m1} for various R_f values. The NF decreases with increasing g_{m1} (determined by the gate width W_1 of FET 1) and R_f. On the other hand, the dissipation current and the input/output VSWR increase with g_{m1} and R_f, respectively.

The foregoing simulation results helped to design the broadband amplifier. Device parameters chosen are listed in Table 1. Essential parameters g_{m1} and R_f were determined so as to satisfy the target specification that NF $\leq$ 2.5dB and input/output VSWR $\leq$ 2.5. The feedback resistances R_2 and R_3 of FET 2 and FET 3 were determined so as to obtain a gain higher than 25dB over a 30 to 1500 MHz frequency range. The gate widths W_2 and W_3 of FET 2 and FET 3, respectively, were designed to be as small as possible to reduce the total dissipation current. This reduction of W_2 and W_3 has little effect on the values of the NF and the gain. The simulated characteristics of the amplifier are shown in Table 2.

The amplifier was fabricated using a VPE grown active layer. Ohmic contacts and Schottky-barrier gates were formed with AuGe/Au and Cr/Pt/Au, respectively. The feedback resistors were made of epitaxial islands and the capacitors of Schottky-junctions. A chip pattern of the amplifier is shown in Figure 3.

Figure 4 shows the NF and the gain of the fabricated amplifier. The calculated result is also shown by the dashed line. The NF of 1.7 to 2.2dB and the gain of 25 to 28dB with a power dissipation of 200mW (V_{DS} = 5V, I_{DD} = 40mA) were obtained over a 30 to 1700 MHz frequency range. It can be seen that a low NF value can be obtained with a very low power dissipation. The input/output VSWR is below 3 in the same frequency range. The characteristics of the amplifier are summarized in Table 3.

Acknowledgments

The authors wish to thank H. Mizuno and I. Teramoto for continuous encouragement.

[1] Archer, J.A., Weidlich, H.P., Petternpaul, E., Petz, F.A. and Huber, J., "A GaAs Monolithic Low-Noise Broad-Band Amplifier", *IEEE J. Solid-State Circuits*, p. 648-652; Dec., 1981.

[2] Estreich, D.B., "A Wideband Monolithic GaAs IC Amplifier", *ISSCC DIGEST OF TECHNICAL PAPERS*, p. 194-195; Feb., 1982.

[3] Nishiuma, M., Nambu, S., Hagio, M. and Kano, G., "A GaAs Monolithic Low-Noise Wideband Amplifier", *Int. Symp. GaAs and Related Compounds*, Japan, p. 425-430; 1981.

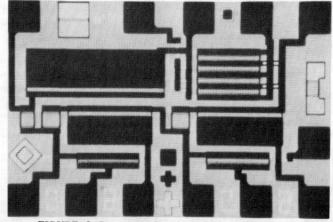

FIGURE 3–Photomicrograph of the amplifier chip: chip size = 1.1 x 0.75mm.

Transconductance (Gate width)	FET1	g_{m1}	80 mS
		W_1	800 μm
	FET2	g_{m2}	20 mS
		W_2	200 μm
	FET3	g_{m3}	20 mS
		W_3	200 μm
Feedback Resistance		R_f	1000 Ω
		R_2	750 Ω
		R_3	750 Ω
Feedback Capacitance		C_f	30 pF
Interstage Capacitance		C_1	15 pF
		C_2	15 pF

TABLE 1—Designed device parameters of the amplifier.

Gain	29 dB
Bandwidth	10 – 2200 MHz
NF	2.0 dB
VSWR	2.5

TABLE 2—Calculated performance of the amplifier: $V_{DS} = 5V$.

Gain	28 dB
Bandwidth	30 – 1700 MHz
NF	< 2.2 dB
VSWR ($Z_0 = 50 Ω$)	< 3.0
Dissipation Current	40 mA ($V_{ds} = 5V$)

TABLE 3—Performance of the GaAs monolithic amplifier.

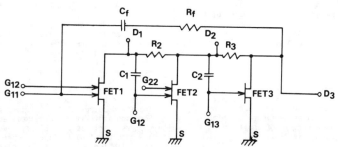

FIGURE 1—Circuit diagram of the GaAs monolithic amplifier.

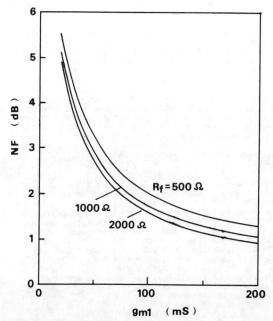

FIGURE 2—Calculated NF of the amplifier versus g_{m1} for various values of R_f. (Other parameters are equal to the values shown in Table 1).

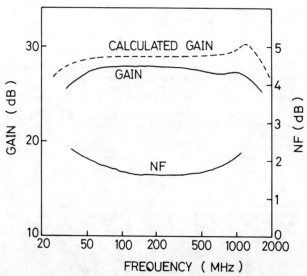

FIGURE 4—Measured NF and small signal gain: $V_{DS} = 5V$, $I_{DD} = 40$mA.

BROADBAND MONOLITHIC LOW-NOISE FEEDBACK AMPLIFIERS

P.N. Rigby, J.R. Suffolk and R.S. Pengelly

Plessey Research (Caswell) Limited,
Allen Clark Research Centre
Caswell, Towcester, Northants, England.

ABSTRACT

A 0.6 to 6 GHz monolithic GaAs FET low-noise feedback amplifier has been developed. This amplifier chip has a gain of 6 dB and a noise figure of around 4 dB over the bandwidth. Gains of 8 dB have been achieved at $\frac{1}{2}I_{dss}$ with 1 dB gain compression points of 21 dBm over the band. This paper discusses the design of such amplifiers using as an example a 1 to 10 GHz two stage monolithic amplifier chip presently under development which is capable of being cascaded up to total gains of 50 dB or so with ± 1.5 dB ripple.

INTRODUCTION

The amplification of microwave signals in a single ultra-wideband amplifier is attractive to various systems designers involved in electronic countermeasures and surveillance. In addition such amplifiers, featuring low noise figure and high 1 dB output compression points find a variety of applications as general purpose components in microwave systems.

The purpose of this paper is to describe the design of decade bandwidth monolithic amplifiers developed for the .6 to 6 GHz or 1 to 10 GHz frequency bands. The paper summarises the requirements for the MESFETs used in these amplifiers as well as giving details of the design techniques used. Monolithic chip realisations of the 0.6 to 6 GHz amplifier are described in detail together with the design and fabrication details of a 1 to 10 GHz two-stage chip.

MESFET DEVICE REQUIREMENTS

The parasitic elements of a GaAs MESFET, restrict the performance of amplifiers fabricated using the transistor. The gate-to-source and drain-to-source capacitances restrict the bandwidth of the amplifier whilst the gate-to-drain capacitance restricts the useful upper cut-off frequency of the amplifier. By applying controlled amounts of external drain-to-gate feedback to the MESFET with frequency, the gain of the MESFET can be made almost constant with frequency. In addition by correct choice of feedback resistance, feedback inductance and drain inductance the terminal VSWRs and noise figure of the amplifier can be made acceptably low for a given FET structure (1,2).

To a first approximation the forward gain of a feedback amplifier is given by

$$S_{21} = \frac{2(1 - g_m R_{FB})}{2 + (g_m + 1/R_o) Z_o + \frac{R_{FB}}{Z_o}(1 + \frac{Z_o}{R_o})}$$

For a 900 micron gate width MESFET, $g_m \approx 80$ mS at low noise bias and R_o (drain to source resistance) is 120 ohms. It is easier to match the output of the MESFET to 50 ohms than the input for a particular high g_m transistor. For $S_{22} = 0$, $S_{21} = 1-(g_m + 1/R_o)Z_o$ and

$$S_{11} = \frac{(g_m + 1/R_o) Z_o^2}{(1 + g_m Z_o) R_o} \quad \text{where} \quad R_{FB} = \frac{g_m + 1/R_o}{1 - Z_o/R_o} Z_o^2,$$

being the feedback resistor.

Thus, $R_{FB} = 378$ ohms
$S_{11} = 0.37$

and $S_{21} = 5.3$ dB.

For $S_{11} = S_{22}$, at low frequencies, $R_{FB} = 221$ ohms resulting in an S_{21} of 4.2 dB and $S_{11} = S_{22} = 0.145$.

GAIN BANDWIDTH OF FEEDBACK AMPLIFIERS

In order for the bandwidth of the feedback amplifier to be maximised for a given transistor an inductance or high impedance transmission line L_D in the drain line and a feedback inductance or high impedance transmission line L_{FB} are required (Fig. 1). L_D compensates for the drain-to-source capacitance of the MESFET at the upper band edge whilst L_{FB} adjusts the S-parameters of the feedback amplifier so that optimum positive feedback occurs at the upper band edge. Further amplifier performance improvements can be produced by using simple matching networks at the input and output of the FET to supplement the feedback network. For monolithic amplifiers these additional networks should be kept as simple as possible to reduce GaAs usage as well as being low-pass enabling the low frequency performance of the amplifier to be limited only by d.c. blocking and feedback capacitances.

The best performance from such amplifiers has been found to occur for MESFETs whose S_{11} and S_{22} phase angles never exceed -180^0 at the highest frequency of operation (3).

For this reason several low parasitic, high g_m FETs were designed and their S-parameters, noise figure parameters and power handling capabilities measured up to 18 GHz.

Four MESFETs shown in Table 1 have been compared for their parasitic component characteristics. The Z584 type MESFET is a physically much smaller device than the GAT6/3, the latter being the Plessey commercially available device paralleled up three times in a monolithic format. For a given carrier concentration and total gate width, the gate-to-source C_{gs}, and gate-to-drain C_{dg}, capacitances are proportional to the gate length. Comparing the Z584, 1 micron and 0.5 micron gate length versions it can be seen that C_{gs} is proportional to gate length. C_{dg} is however dominated by the geometry of the device and the contribution due to covering the FET channels with polyimide dielectric prior to the source interconnect stage. A truly airbridged source FET, the SOFET, has a C_{dg} per mm of approximately 60% that of the Z584 design indicating the dependence on MESFET layout and construction.

The importance of the value of C_{dg} can be clearly seen by inspection of Fig. 2. This figure shows the maximum available gain of the Z584/0.5 FET having a gate length of 0.5 micron as a function of feedback capacitance. The MAG at a feedback capacitance of 0.095 pF is only 3.5 dB at 18 GHz. Increasing the physical spacing between the gate and drain electrodes and removing the polyimide from the channel area results in a decreased feedback capacitance of 0.055 pF giving an MAG of 6 dB at 18 GHz. Fig. 3 shows the measured MAG of the Z584/0.5 device up to 18 GHz as well as that for the Z584/1, the 1 micron gate length version of the same basic design. The effect of C_{gs}, C_{ds} and C_{dg} can be seen clearly. Fig. 4 shows the equivalent circuits of these two MESFETs.

NOISE FIGURE OF FEEDBACK AMPLIFIERS

A feedback amplifier having no other passive matching depends for its noise figure on the MESFET and the value of the feedback resistor in the feedback path. In the limit where the feedback resistor is not present the noise figure of the amplifier becomes that of the MESFET working into a characteristic impedance, Z_o (50 ohms). A considerable number of papers have been written on calculating the noise figure of feedback amplifiers having both series and shunt feedback paths. This paper attempts to provide a simple means of estimating the noise figure of amplifiers using various geometry FETs. The feedback amplifier can be considered as two noise blocks. The noise figure of block 1 (the feedback resistor) is given by (4)

$$F_1 = 1 + \frac{|\frac{1}{R_o} + g_m|^2 R_o}{(g_m - \frac{1}{R_{FB}})^2 R_{FB}}$$

where R_o is the drain to source resistance
g_m is the transconductance, and
R_{FB} is the feedback resistor.

Considering block 2 alone (the device) having a noise figure F_2 where F_{50} is defined as the 50 ohm noise figure of the MESFET, it can be shown that:

$$\frac{F_2 - 1}{F_{50} - 1} \simeq 1.5$$

The 50 ohm noise figure is directly related to the noise resistance of the MESFET. From Fukui's theory (5)

$$R_n \propto \frac{1}{g_m^2} \propto \frac{1}{W_g^2} (\frac{aL_g}{N})^{2/3}$$

where W_g is the total gate width, L_g is the gate length, a is the effective channel thickness and N is the carrier concentration in the channel.

The minimum noise figure is given by

$$F_{min} - 1 \propto f\, C_{gs} \sqrt{\frac{R_g + R_s}{g_m}}$$

$$\propto f\, L_g \sqrt{g_m(R_g + R_s)}$$

The total noise figure of the transistor with resistive shunt feedback is given by

$$F_T = 1 + \sum_{i=1}^{2} (F_i - 1)$$

Table 2 compares the estimated feedback amplifier noise figures at 3 GHz for three MESFETs of gate lengths 0.5, 0.7 and 1 micron. In order to allow maximum bandwidth with acceptable gain ripple and input and output VSWRs the 1 micron gate length MESFET can only have a total gate width of 450 micron resulting in a low-noise biased g_m of 40 mS. The effect of such a gate width device on feedback amplifier noise figure is demonstrated clearly. Table 2 also indicates the penalty in noise figure brought about by choosing the feedback resistor value to give the lowest simultaneous S_{11} and S_{22}. This simple method of calculating amplifier noise figure has been used to check two previously published results (2,6) at 3 GHz. These are also given in Table 2. Agreement is acceptable. The noise figure can also be estimated as a function of frequency where it is assumed that the noise resistance R_n is frequency independent. The noise figure so calculated will be a maximum figure particularly at the highest frequency since the reactive matching, particularly the drain transmission line, present in an actual amplifier will lower the noise figure (2).

Fig. 5 shows a theoretical plot of noise figure versus frequency for a 1200 micron gate width device for 0.5 and 1.25 micron gate length versions. Measured noise figures are also included showing the good agreement with theory. For the monolithic 0.6 to 6 GHz feedback amplifier described later in this paper a noise figure of 3.5 dB is estimated at 1 GHz whilst at 6 GHz this has risen to 4.2 dB. Fig. 6 shows a comparison of the gains and noise figures of two feedback amplifiers employing MESFETs of different total gate widths and lengths.

AMPLIFIER DESIGNS

In order to obtain acceptably low noise figures, 900 micron gate width FETs employing gate lengths of 0.7 and 0.5 micron have been used in monolithic feedback amplifiers enabling bandwidths of greater than 6 and 10 GHz to be realised respectively.

0.6 to 6 GHz Monolithic Amplifier

The device selected for this amplifier was the GAT6/3, which is three Plessey GAT6s connected in parallel, modified for use in monolithic circuits by removing the extra source pad area, used in the discrete device for bonding. In the monolithic version the sources are all interconnected by an air bridge. The circuit diagram for this amplifier is shown in Fig. 7. Lengths of high impedance transmission line have been used to realise 'inductors'. The feedback loop comprises such a length of line in series with a cermet resistor and an overlay silicon nitride capacitor. The smaller value shunt capacitors in the input and output matching circuits have been realised using an interdigital structure. The gate bias is fed in through a high value cermet resistor. A similar method was considered for the drain bias, but a resistor capable of dissipating sufficient heat (up to 350 mW) would have occupied a significant proportion of the total chip area. After considering alternative methods for supplying the drain bias it was found that any satisfactory method would require an unreasonably large area of GaAs, and so it was decided to have part of the bias circuit off-chip. A 5 nH spiral inductor is included on the chip and a further 10 nH are off-chip. The overall chip size is 2.8 mm x 1.8 mm and a photograph of the chip is shown in Fig. 8.

The circuits were fabricated on 200 micron thick GaAs with epitaxially grown buffer and active layers the latter being doped to 1.5×10^{17} cm^{-3}. The active device uses 0.7 micron long recessed gates and comprises six identical cells each one having two 75 micron wide gate stripes making a total gate width of 900 micron. The source and drain contacts are defined using a float off process, as is the lower metallisation, used for the interdigital capacitors in the input and output matching circuits and the bottom plates of the overlay capacitors. A thin layer of silicon nitride, used as the dielectric for the overlay capacitors, is deposited using plasma enhanced CVD and defined by plasma etching.

A layer of polyimide is spun over the wafer through which via holes are plasma ashed to form contacts between the cermet resistors and the lower metallisation. The cermet is deposited by r.f. sputtering and defined by ion beam milling using the polyimide layer as a barrier. A second polyimide layer is spun on top of the first thus sandwiching the cermet and forming a protective layer. Coincident via holes are ashed through the separate polyimide layers to enable interconnections to be made between the lower and upper metallisations, the latter being deposited by r.f. sputtering to a thickness of 3 micron and defined by ion beam milling. This upper metallisation is used as the top plate for the silicon nitride capacitors and also the lengths of transmission line and the spiral inductor in the drain bias circuit. An air bridge technology is used to connect to the centre of this spiral and also to connect together the individual sources of the FET. Finally, the polyimide is removed over the interdigital capacitors to improve their Q-factors.

The measured performance of this circuit is shown in Fig. 9 for two different bias levels. At I_{dss} the gain is 7.9 ± 0.6 dB from 1 GHz to 5.7 GHz. The output VSWR is better than 2:1 over this frequency range and the input VSWR rises to 3.2:1 at the top end of the bandwidth. It is a characteristic of these broadband feedback amplifiers that the input is more difficult to match resulting in higher input than output VSWRs. At low noise bias the measured gain is 5.8 ± 0.6 dB from 0.6 GHz to 6.1 GHz and the noise figure rises from 4.0 to 4.4 dB over this frequency range. (These results are obtained from circuits whose devices had lower than expected values of g_m and it is reasonable to assume that an improvement in gain flatness and high frequency response will be obtained from further batches currently being processed). The power gain transfer curve for this amplifier at 5 GHz is shown in Fig. 10(a) indicating a 1 dB power compression point of 21 dBm. The use of feedback also improves the intermodulation distortion and Fig. 10(b) shows the level of the third order intermodulation distortion products at 2 GHz, where the feedback is fully effective. For input signals at 2.0 GHz and 2.1 GHz and an output power of 10 mW the level of these IMD products is 50 dB below the carriers. The corresponding third order intercept point was 27 dBm.

1 to 10 GHz Monolithic Amplifier

Fig. 11 shows the predicted gain and VSWR performance of a single stage feedback amplifier employing a 450 micron gate width, 1 micron gate length MESFET. Although amplifiers based on such devices would be preferred for yield and d.c. power consumption reasons their terminal characteristics with the simple matching networks shown in Fig. 12 do not allow cascades of chips to be used without incurring unacceptable gain ripples for total gains of more than 20 dB. The only concession to off-chip tuning allowed in the amplifiers is a variation in the length of 50 ohm microstrip lines between the chips. The input VSWR in particular is clearly unacceptable (Fig. 11). In contrast Fig. 12 shows the predicted response of a similar cascade of chips employing the 900 micron, 0.5 micron gate length transistor discussed earlier in this paper. To aid gain flatness small amounts of series feedback are included in the FET sources as well as gate inductances. The circuit diagram of the two stage chip is also shown in Fig. 12. Each chip contains five silicon nitride MIM capacitors, two 5 pF values in the feedback paths and 5 and 10 pF values for d.c. blocking. Polyimide dielectric MIM capacitors are used for the tuning capacitors where tolerances + 10% can be achieved from batch to batch. Such a tolerance is acceptable for the amplifier. Because of the small overall size of the transistor, shown in close-up in Fig. 13, only a single ground is used enabling a simple feedback topology with no need for through-GaAs vias. Lumped inductances are replaced with meandered high impedance (85 ohm) transmission lines. The feedback and gate resistors are cermet types, the gate resistors being 2 Kohm in value. The ohms per square of the resistors is 50 ohm/sq. The drain bias to each stage is provided by two on-chip 5 nH spiral chokes having resonant frequencies beyond 10 GHz. These employ underpassed centre arm connections with polyimide separation between metal layers. The remainder of the bias network consisting of a further 10 nH inductor and capacitor are off-chip.

A composite drawing of the various mask layers is shown in Fig. 14. The two stage chip size is 1.5 x 4 mm. This chip is undergoing processing at the time of writing. It is believed that this is the first ultrabroadband low noise two stage monolithic amplifier which has been designed specifically for cascadability up to 50 dB gain.

SUMMARY

This paper has described the design, fabrication and results of monolithic feedback amplifiers covering 0.6 to 6 GHz and 1 to 10 GHz. It has been shown that with optimized MESFET geometries bandwidths up to 18 GHz are feasible. Feedback amplifiers, although not offering the lowest noise figures are capable of providing general purpose and ultrawideband units which can be readily cascaded.

ACKNOWLEDGEMENTS

The authors would like to thank their colleagues at Plessey Research (Caswell) Ltd. particularly R. Butlin, D. Parker and Z. Jackson for processing the wafers. Part of this work has been carried out with the support of Procurement Executive, Ministry of Defence, sponsored by DCVD.

REFERENCES

1. Niclas, K.B., Wilser, W.T., Gold, R.B. and Hitchens, W.R., 'The matched feedback amplifier: ultrawide-band microwave amplification with GaAs MESFETs' IEEE Trans. on Microwave Theory and Techniques, Vol. MTT-28, No. 4, April 1980, pp.285-294.

2. Niclas, K.B., 'Noise in broad-band GaAs MESFET amplifiers with parallel feedback' IEEE Trans. on Microwave Theory and Techniques Vol. MTT-30, No. 1, January 1982, pp.63-70.

3. Pengelly, R.S., 'Application of feedback techniques to the realisation of hybrid and monolithic broadband low-noise and power GaAs FET amplifiers' Electronics Letters, 15 Oct. 1981, Vol. 17, No. 21, pp.798-799.

4. Honjo, K., Sugiura, T. and Itoh, H., 'Ultra broadband GaAs monolithic amplifier' IEEE Trans. on Microwave Theory and Techniques, Vol. MTT-30, No. 7, July 1982, pp.1027-1033.

5. Fukui, H., 'Design of microwave GaAs MESFETs for broadband low noise amplifiers' IEEE Trans. on Microwave Theory and Techniques, Vol. MTT-27, No. 7, July 1979, pp.643-650.

6. Peterson, W.C., Gupta, A.K. and Decker, D.R., 'A monolithic GaAs DC to 2 GHz feedback amplifier' IEEE 1982 Microwave and Millimetre-wave Monolithic Circuits Symposium Digest of Papers, pp.20-22.

TABLE 1: Normalised Equivalent Circuit Parameters for 4 Different MESFETs

Parameter		MESFET Type			
		GAT6/3	Z584/1	Z584/0.5	SOFET
W_g	(µm)	900	900	900	900
L_g	(µm)	0.7	1	0.5	1.2
G_m/W	mS/mm	80	89	80	60
C_{gs}/W	pF/mm	1	1.17	0.58	0.95
C_{gd}/W	pF/mm	.05	0.13	0.1	.08
C_{ds}/W	pF/mm	0.4	0.2	0.2	0.275

TABLE 2: Calculated Noise Figures at 3 GHz for Feedback Amplifiers as a Function of MESFET Type

MESFET	Feedback Resistor (ohms)	Noise Resistance, R_n	Noise Figure dB	Gate Width µm	Gate Length µm
GAT6/3	250	8	3.9	900	0.7
Z584/0.5	197*	6.4	4.5	900	0.5
Z594/0.5	300	6.4	3.9	900	0.5
Z584/1	110*	40.6	8.8	450	1.0
Z584/1	328	40.6	5.8	450	1.0

MESFET	Freq. GHz	Feedback Resistor (ohm)	G_m mS	Calculated Noise Figure (dB)	R_n ohms	Published Noise Figure (dB)	Gate Width µm	Gate Length µm
WJ-F810	3	200	57	4.5	22	4.0	800	1
Rockwell	2	375	75	3.4	7	3.6	1200	1

*Feedback resistor values for lowest simultaneous $S_{11} = S_{22}$

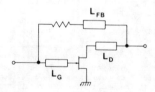

FIG. 1. Basic parallel feedback amplifier

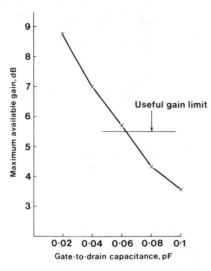

FIG. 2. Maximum available gain of MESFET as a function of gate-to-drain capacitance (W_g = 900 micron, L_g = 0.5 micron)

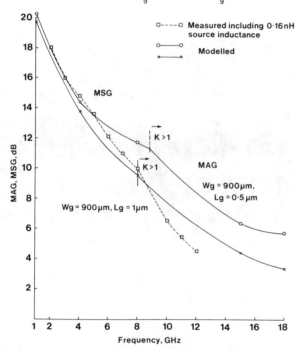

FIG. 3. Measured and modelled MAG for 900 micron gate width MESFETs as a function of gate length

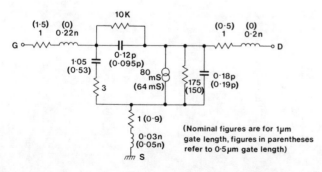

FIG. 4. Equivalent circuits of 900 micron gate width MESFETs

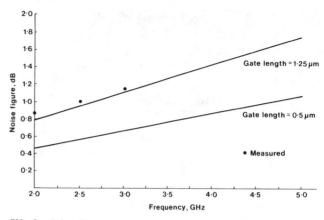

FIG. 5. Noise figure variation vs gate length for 1200 micron MESFET

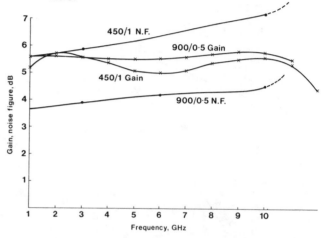

FIG. 6. Gain and maximum noise figure of feedback amplifiers

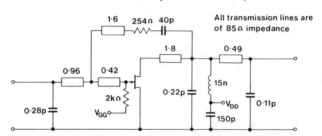

FIG. 7. Circuit diagram for monolithic 0.6 to 6 GHz amplifier

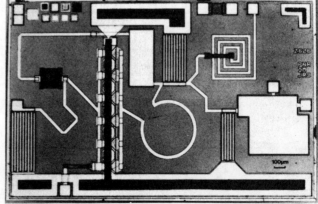

FIG. 8. Photomicrograph of 0.6 to 6 GHz chip amplifier

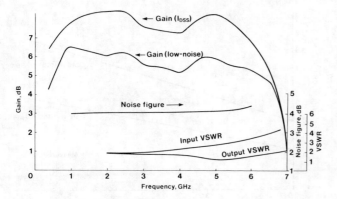

FIG. 9. Measured performance of monolithic 0.6 to 6 GHz amplifier

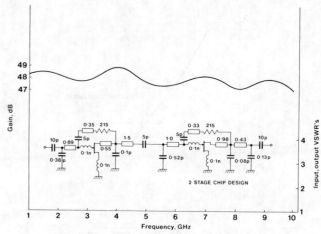

FIG. 12. Response of 4, 2 stage chips directly cascaded (2 stage chip design inset)

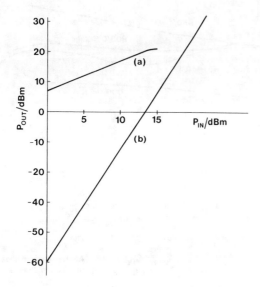

FIG. 10. (a) Power transfer characteristic and (b) 3rd order IMDs of monolithic 0.6 to 6 GHz amplifier

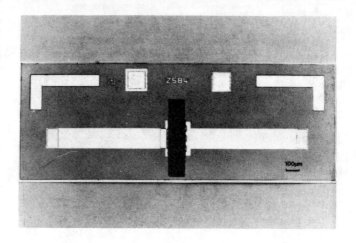

FIG. 13. High g_m, low parasitic monolithic FET

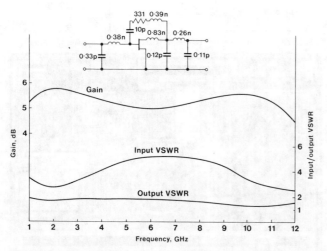

FIG. 11. Circuit diagram for 1-11 GHz feedback amplifier (W_g = 450 micron, L_g = 1 micron)

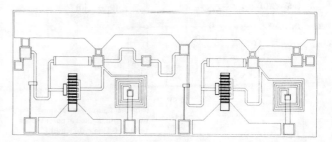

FIG. 14. Chip layout of 1 to 10 GHz low-noise monolithic amplifier.

A MONOLITHIC MULTI-STAGE 6-18 GHz FEEDBACK AMPLIFIER

A.M. Pavio, S.D. McCarter and P. Saunier

Texas Instruments Incorporated
P.O. Box 226015
Dallas, Texas 75266

ABSTRACT

A design approach and circuit modeling method, which includes parasitic effects, will be presented for a two-stage monolithic feedback amplifier with enhanced high frequency performance. Using the proposed approach, a 6-18 GHz amplifier has been demonstrated with low input/output VSWR, respectable noise figure (6 dB) and a minimum gain of 8 dB.

SUMMARY

The GaAs FET amplifier has become the basic building block of modern microwave systems and as such, usually carries the burden of establishing system bandwidths, sensitivities and dynamic range. These parameters are especially important in ECM and surveillance systems. However, unlike discrete microwave amplifiers, which tend to be bandwidth limited by circuit and assembly techniques, monolithic feedback amplifiers exhibit excellent bandwidth and dynamic range characteristics, and can be directly cascaded. Thus, the size, density and performance of present day EW systems can be enhanced by employing this technology.

Recently, several monolithic feedback amplifiers have been reported with frequency limits of approximately 8 GHz [1,2], but what follows is a method to design monolithic multi-stage feedback amplifiers with high frequency performance extending above 18 GHz. The design and performance of a two-stage 6-18 GHz amplifier aids in the illustration of the proposed technique.

AMPLIFIER DESIGN TECHNIQUES

One of the prime areas of difficulty in designing cascaded MESFET amplifiers is controlling the impedance match between devices [3]. Shunt feedback can be used to reduce the magnitude of S_{11} and S_{22} at the terminals of the active elements, thus enabling the circuit designer to synthesize wideband matching networks. Flat gain versus frequency and greatly improved amplifier stability, especially at lower microwave frequencies, are also desirable byproducts of feedback[4,5]. This improved circuit performance is not without cost; it is obtained at the expense of reduced transducer gain.

In practice, when the feedback element values are lowered to reduce FET maximum available gain, the gain-versus-frequency response begins to exhibit an upward slope. This effect is due to the fact that uniform amounts of negative feedback cannot be applied due to the finite phase shift of the feedback loop and the increasing phase shift of the device as a function of frequency. Hence, there will exist an optimum gain level obtainable with a particular FET and feedback loop circuit.

At frequencies above 14 GHz, the amplifier performance degradation, due to the effects mentioned above, is pronounced. However, a monolithic circuit realization can aid in alleviating some of the drawbacks encountered in hybrid design with extended high-frequency performance. To begin the amplifier design, a monolithic FET model must be developed. This is not an easy task in that S-parameters for monolithic devices are not usually available since direct device measurements must be made. The formulation of the model is further complicated by the fact that an accurate value of the angle of S_{21}, as well as the magnitude, must be known. The FET size is also an important consideration in that the associated feedback loop and input/output matching network element's absolute values are scaled directly with the input and output impedances of the FET.

With the above constraints being considered, a 300-micron, high gain FET with 75-micron-long fingers and 0.5 micron gate length was chosen as the active device. A circuit model for the device includes the source via inductance (Figure 1). The element values for the model were extrapolated from measured S-parameter data of discrete FET's. However, the values of input and output capacitances

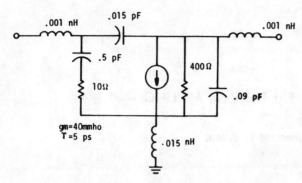

Figure 1. Lumped element 300 μm FET model.

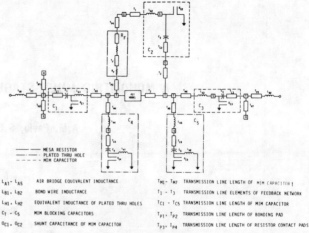

Figure 2. Single-stage amplifier circuit model.

were modified based on geometry differences between the discrete and monolithic devices, and the source lead inductance was changed to account for mounting differences.

Once an appropriate FET model has been chosen, the desired performance of the composite FET and feedback loop can be determined by employing CAD techniques and careful modeling [6]. The input and output networks are now added to complete a single stage design. A typical circuit model is shown in Figure 2.

With a single stage amplifier designed, multistage amplifiers can be constructed by cascading several gain stages. The reduced bandwidth and accentuated gain ripple resulting from directly cascading gain stages can be reduced or eliminated by adjusting the interstage networks (input/output network) to obtain an amplifier with an equal ripple performance. It is not uncommon that the gain and bandwidth of a properly designed two stage amplifier will usually compare favorably with the performance obtained with a single stage design. With larger cascade designs, it may be necessary to design an interstage network that has a completely different ripple characteristic in order to obtain the optimum gain performance.

MONOLITHIC CIRCUIT DESCRIPTION

The above design philosophy was employed in the fabrication of a two stage, monolithic amplifier. The amplifier devices are interdigitated 300 micron FET's which use plated-through source vias to establish RF and DC ground. Vias, with a diameter of 0.084 mm, centered on a 0.2-mm square pad, were chosen in order to minimize the length of the feedback path. All capacitors are of the metal-insulator-metal fabrication type and are used to perform DC blocking as well as RF bypassing functions. Series and shunt resistors in the gate bias circuitry are used to prevent device damage from static and to provide a measure of over-voltage protection. Matching networks of the highpass form were also employed to minimize amplifier length and to provide decoupling for DC bias (gate and drain).

AMPLIFIER CIRCUIT FABRICATION

The two stage feedback amplifier was constructed on a 0.1-mm-thick, semi-insulating, GaAs substrate. The active layers of the circuit have a doping level of approximately 2×10^{17} cm^{-3}. The FET source via holes were fabricated using reactive ion etching methods.

The bypass and decoupling capacitors used in the amplifier are of the metal/silicon nitride/metal type, with a 400 nm layer of silicon nitride as the dielectric. With this thickness, the MIM capacitors have about 150 pF/mm^2; thus, 1 pF capacitors are about 80 microns X 80 microns, and 10 pF capacitors are about 250 microns X 250 microns.

The high impedance transmission lines used as inductors in the monolithic amplifier are several skin-depths thick to reduce the resistance as much as possible. In order to produce transmission lines three to four microns thick, extra gold was plated during the air bridge process.

Mesa resistors were used throughout the amplifier and were fabricated during the first photo-etching of the active layer. The width of the resistors is determined by the etching, and the length is determined by the placement of the gold contact pads. Surface resistivities in the order of 400 ohms/square are obtainable, which allows resistor values between 10 ohms and 1,000 ohms to be realized. The final two-stage amplifier is shown in Figure 3.

MEASURED PERFORMANCE

The initial selections of two stage amplifiers were based on DC probing and visual inspection. The amplifiers were further screened by measuring individual FET characteristics such as I_{dss}, V_p, and gm. The chips that were determined "DC good" were then mounted on carrier plates with associated input/output 50 ohm transmission lines fabricated on alumina.

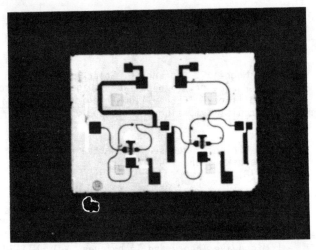

Figure 3. Monolithic two stage amplifier chip.

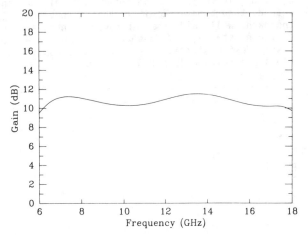

Figure 4. Computed gain performance of modeled two-stage amplifier.

The computed gain performance, which was calculated using circuit and FET models, is shown in Figure 4. The measured gain performance and a new computed response in which the actual values of the feedback resistances and measured FET S-parameters were used in the circuit description, are shown in Figure 5. The difference between the measured and computed gain is due in part to circuit modeling errors and process variations in FET characteristics.

The measured performance indicates that respectable amplifier gain can be obtained at frequencies extending through 18 GHz. It was also found that several two stage amplifiers could be cascaded without any degradation in performance. The noise figure was also measured and is shown in Figure 6.

Although the output network on the amplifier was designed for low VSWR, the FET is terminated so that a modest amount of output power can be obtained. The compression characteristics that are shown in Figure 7, indicate that a minimum output power of 18 dBm is obtainable over the entire frequency range of 6-18 GHz.

CONCLUSION

The illustrated design method allows the microwave engineer to synthesize and analyze broadband monolithic feedback amplifiers. The parasitic elements that degrade ultimate high-frequency performance of feedback amplifiers constructed with hybrid fabrication methods are reduced substantially with monolithic circuit implementations. The resistor lengths and shunt capacitance associated with mounting pads and DC blocking capacitors can be made quite small. Thus, phase shift of the overall loop can also be minimized because it can be placed closer to the active area of the FET than with conventional circuit designs. By properly selecting the device size, the monolithic circuit designer can optimize the input VSWR and the power output capabilities of the resulting amplifier. The excellent stability and impedance control char-

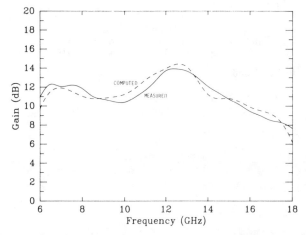

Figure 5. Measured versus computed gain performance using actual circuit and device parameters.

acteristics allow amplifier gain chains to be constructed with cascaded multistage feedback gain blocks, certainly an important factor in reducing the cost, complexity, and size over conventional designs employing cascaded, hybrid-

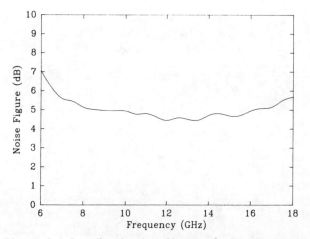

Figure 6. Broadband noise figure of two-stage monolithic amplifier

coupled gain stages. It is hoped that this design method, though more complex than a conventional approach, will aid the microwave engineer in designing stable, single-ended amplifiers, where costly design iterations and balanced configurations are neither desirable nor possible.

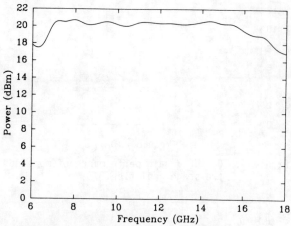

Figure 7. Amplifier power output at 1 dB gain compression point.

REFERENCES

1) P. N. Rigby, J. R. Suffolk, and R. S. Pengelly, "Broadband Monolithic Low Noise Feedback Amplifiers," IEEE Microwave and Millimeter-wave Monolithic Circuits Symposium Digest, pp. 71-75, June 1983.

2) S. Moghe, T. Andrade, G. Policky, and C. Huang, "A Wideband Two Stage Miniature Amplifier," IEEE Gallium Arsenide Integrated Circuits Symposium Digest, pp. 7-10, October 1983.

3) E. Ulrich, "The Use of Negative Feedback to Slash Wideband VSWR," Microwaves, October 1978.

4) K. B. Niclas, W. T. Wilser, R. B. Gold and W. R. Hitchens, "The Matched Feedback Amplifier Ultra-wideband Microwave Amplification With GaAs MESFET's," IEEE Transaction on Microwave Theory and Techniques, Vol. MTT-28, No. 4, April 1980.

5) A. M. Pavio, "A Network Modeling and Design Method for a 2-18 GHz Feedback Amplifier," IEEE International Microwave Symposium Digest, MTT-S 1982, pp. 162-165.

6) A. M. Pavio and S. D. McCarter, "Network Theory and Modeling Method Aids Design on a 6-18 GHz Monolithic Multi-Stage Feedback Amplifier," Microwave Systems News, Vol. 12, No. 12, December 1982.

GaAs FET Ultrabroad-Band Amplifiers for Gbit/s Data Rate Systems

KAZUHIKO HONJO AND YOICHIRO TAKAYAMA

Abstract—A novel ultrabroad-band amplifier configuration suitable for GaAs FET's has been developed. The developed amplifier circuit operates as a capacitor–resistor (C–R) coupled amplifier circuit in the low-frequency range in which $|S_{21}|$ for the GaAs FET's is constant. It also operates as a lossless impedance matching circuit in the microwave frequency range in which $|S_{21}|$ for the GaAs FET has a slope of approximately -6 dB/octave. Using this configuration technique, 800-kHz to 9.5-GHz band (13.5 octaves), 8.6-dB gain GaAs FET amplifier modules have been realized. The amplifier module has 40-ps step response rise time. It also has low input and output VSWR. By cascading two-amplifier modules, 19-dB gain over the 800-kHz to 8.5-GHz range and 50-ps step response rise time were obtained. NF is lower than 8 dB over the 50-MHz to 6-GHz range.

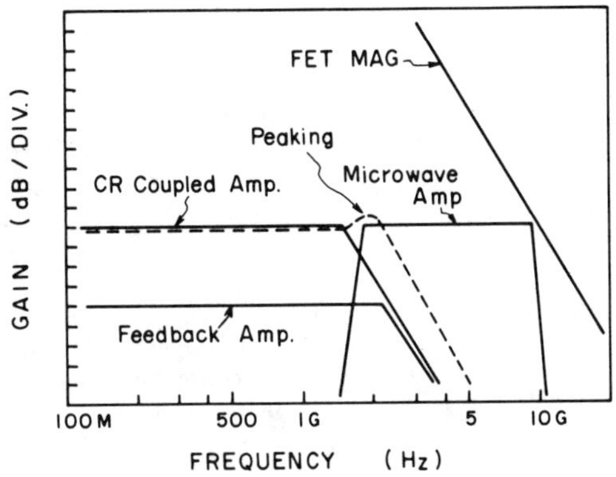

I. INTRODUCTION

SIGNIFICANT advances in GaAs FET's have made it possible to realize gigabit systems. The gigabit-per-second data rate systems need amplifiers which exhibit flat gain of 20 dB or more over the frequency range from below several hundred kilohertz to above several gigahertz. In addition to these performances, low input and output voltage standing wave ratio (VSWR) is also required, especially for communication applications such as ultrahigh-speed pulse-code modulation (PCM) and optical communication systems.

A conventional approach to achieving these requirements is to use capacitor–resistor (C–R) coupled amplifier configuration, with feedback and/or peaking circuits, as the case may be.

Qualitative frequency-gain behaviors for the C–R coupled amplifier, the C–R coupled amplifier with the negative feedback circuit and the C–R coupled amplifier with the peaking circuit are shown comparatively in Fig. 1. $|S_{21}|$ of a conventional GaAs field-effect transistor (FET) having from 0.5- to 1.5-μm gate length is constant below the frequency range from 0.5 to 2 GHz and exhibits a 6-dB/octave rolloff above that frequency range. Accordingly, the bandwidth of the simple C–R coupled GaAs FET amplifiers, including direct coupled amplifiers, cannot be extended above about 2 GHz.

To achieve wider bandwidth, additional techniques, such as negative feedback and peaking techniques, have been used [1], [2]. However, there are two major disadvantages

Manuscript received November 4, 1980; revised January 26, 1981.
The authors are with the Basic Technology Research Laboratory, Nippon Electric Company, Ltd., 1-1, Miyasaki, Yonchome, Takatsu-ku, Kawasaki, Japan.

Fig. 1. Qualitative frequency-gain behaviors for a capacitor–resistor coupled amplifier, a peaking circuit, a negative feedback amplifier, and a microwave amplifier.

in negative feedback amplifiers. These are 1) degradation of amplifier gain, since the negative feedback amplifier has a constant value for gain–bandwidth product, and 2) added design difficulty in impedance matching. For the peaking technique, the bandwidth cannot be extended significantly since it is used in the C–R coupled amplifier.

Meanwhile, as shown in Fig. 1, the gain of the C–R coupled amplifier is low, compared with the maximum available gain (MAG) of the GaAs FET.

In conventional microwave amplifiers, lossless circuit elements such as lumped-element capacitors, inductors, and distributed transmission lines are usually used for impedances matching [3] or positive feedback [4] to achieve MAG at the upper band edge. However, not only do impedances for these lossless circuit elements depend upon frequencies, but also the number of sections of the matching network is limited from a practical point of view. Consequently, ultrabroad-band impedance matching using these lossless circuits is very difficult. Usually bandwidths for the broad-band multistage microwave amplifier are from 1 to 3 octaves. As a matter of fact, the bandwidths of these microwave amplifiers are too narrow to use for baseband pulse amplification in the gigabit data rate systems.

If the C–R coupled amplifier low-frequency characteristics and the lossless matched microwave amplifier characteristics are combined, the GaAs FET high-frequency capability can be utilized to obtain a ultrabroad-band amplifier.

The purpose of this paper is to present a novel ultra-

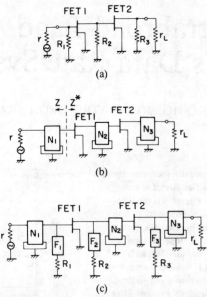

Fig. 2. Equivalent circuits for (a) conventional C–R coupled amplifier; (b) conventional microwave amplifier; and (c) newly developed amplifier.

broad-band amplifier configuration suitable for GaAs FET's, and to demonstrate the performance of developed ultrabroad-band GaAs FET amplifiers. The novel amplifier circuit operates as the C–R coupled amplifier circuit in the low-frequency range in which $|S_{21}|$ for the GaAs FET's is constant. It also operates as the lossless impedance matching circuit in the microwave frequency range in which $|S_{21}|$ for the GaAs FET's has a slope of approximately −6 dB/octave. Using this circuit configuration, an 800-kHz to 9.5-GHz band, 8.6-dB gain amplifier module, in which 13.5-octave bandwidth has been achieved, has been developed. The amplifier module has 40-ps step response rise time. It also has low input and output VSWR. By cascading two amplifier modules, 19-dB gain over the 800-kHz to 8.5-GHz range and 50-ps step response rise time have been obtained. A 14-dB gain, 700-kHz to 6-GHz band amplifier module has also been developed. The noise characteristics are discussed.

II. Circuit Design

A. Configuration

An impedance matching technique for multistage amplifiers which have interstage matching networks is much more difficult than that for single-stage amplifiers. In order to obtain high gain, however, a multistage amplifier circuit configuration is necessary. Accordingly, design considerations have been made on two-stage amplifiers.

Schematic diagrams for a conventional C–R coupled amplifier, a conventional microwave amplifier and a newly developed amplifier are shown in Fig. 2. In the figure, source grounded GaAs FET's are used. All coupling (dc blocking) capacitors and RF bypass capacitors are omitted for convenience, because these capacitors only affect a low-cutoff frequency. Fig. 2(a) shows the C–R coupled amplifier which is generally used for baseband pulse amplification. In the low-frequency range, the input impedances for source grounded GaAs FET's are very high, compared with the signal source impedance, which is usually 50 Ω. Accordingly, by selecting R_1 to be r, low VSWR at the input port is achieved. Load resistance for the first stage FET (FET 1) is decided mainly by $R2$. Considering the output impedance for the second-stage FET (FET 2), R_3 is chosen to achieve low VSWR at the output port. Load resistance for the amplifier r_L is usually 50 Ω.

Fig. 2(b) shows a typical two-stage microwave amplifier using lossless low-pass impedance matching networks N_1, N_2, and N_3. The impedance is matched to obtain MAG at the upper band edge.

Fig. 2(c) shows a newly developed amplifier schematic diagram. In the figure, N_1, N_2, and N_3 are low-pass lossless impedance matching networks, and F_1, F_2, and F_3 are low-pass lossless impedance transformers. Resistors R_1, R_2, and R_3 have the same values, respectively, as in Fig. 2(a) and N_1, N_2, and N_3 have the same values, respectively, as in Fig. 2(b). By means of transformers F_1, F_2, and F_3, resistors R_1, R_2, and R_3 are transformed into high impedances in the microwave frequency range so that these resistors do not affect microwave impedance matching. The microwave impedance matching is achieved by N_1, N_2, and N_3, just like the circuit in Fig. 2(b). Meanwhile, N_1, N_2, N_3, F_1, F_2, and F_3 in Fig. 2(c), which are all low-pass form elements, operate as circuits having short electrical length in the low-frequency range. The low-frequency range gain is determined by R_1, R_2, and R_3.

Amplifier gains both in the low-frequency range and in the microwave frequency range can be established individually. To apply the circuit to flat-gain ultrabroad-band amplifiers, the gains in both frequency ranges are designed to be the same. A gain ripple which may occur in a crossover frequency range between the low and the microwave frequency ranges can be flattened by proper design

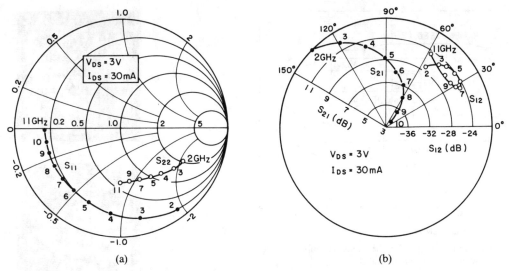

Fig. 3. S parameters for V-218 FET. (a) S_{11} and S_{22}. (b) S_{12} and S_{21}.

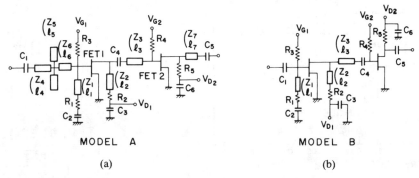

Fig. 4. Equivalent circuits for amplifier modules. (a) Model A. (b) Model B.

TABLE I
DESIGNED CIRCUIT PARAMETERS FOR MODEL A AND MODEL B

	(Ω)				(Ω)(mm)							(pF)	
	R_1	R_2	R_3,R_4	R_5	Z_1/ℓ_1	Z_2/ℓ_2	Z_3/ℓ_3	Z_4/ℓ_4	Z_5/ℓ_5	Z_6/ℓ_6	Z_7/ℓ_7	C_1,C_4,C_5	C_2,C_3,C_6
Model A	50	37	3000	75	100/1.7	100/1.7	48/1.6	26/5.2	40/1.4	100/1.7	100/0.5	3300	21300
Model B	50	50	3000	300	100/5	100/4	33/5.6	—	—	—	—	3300	21300

using a computer simulation. The simulation results are demonstrated in the next section.

B. Ultrabroad-Band Amplifier Module Design

The GaAs FET's used in the amplifier modules are V-218 FET's (NEC). The FET gate length and total gate width are 1.0 and 400 μm, respectively. The FET has two-cell and recessed gate structure. Saturated drain current I_{DSS} is 120-mA and pinchoff voltage V_p is -2.2 V.

S parameters for a V-218 FET are shown in Fig. 3. The S parameters in the figure are extracted using the FET equivalent circuit element values which are determined by computer data fitting techniques for the measured S parameters.

Two categories (Model A and Model B) of amplifier modules were designed. Equivalent circuits for the Model A and the Model B are shown in Figs. 4(a) and (b), respectively.

Model A was designed to have a wider bandwidth but a lower gain compared to Model B. Using microstrip line type single section impedance transformers, the low-pass transformers (F_1, F_2, and F_3) in Fig. 2(c) are realized. Microstrip lines and stubs are also adopted for input, interstage and output low-pass matching networks (N_1, N_2, and N_3).

F_3 in Fig. 2(c) is omitted in Model A and N_1, N_3, and F_3 in Fig. 2(c) are omitted in Model B. Resistances R_3 and R_4 in gate bias voltage supply circuits are high enough not to affect the low frequency and the microwave circuits. C_1, C_4, and C_5 are coupling (dc blocking) capacitors. C_2, C_3, and C_6 are RF bypass capacitors. Designed circuit parameters are shown in Table I.

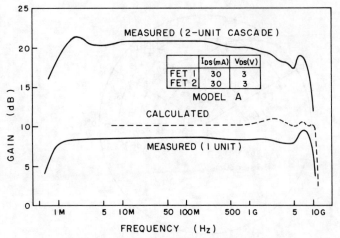

Fig. 5. Gain-frequency characteristics for the Model A amplifier module and the cascade amplifier.

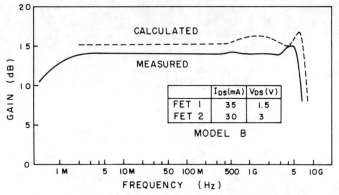

Fig. 6. Gain-frequency characteristics for the Model B amplifier module.

Fig. 7. Amplifier module photographs. (a) Model A. (b) Model B.

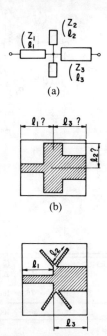

Fig. 8. Equivalent circuit and configurations for an input microwave matching network. (a) Equivalent circuit. (b) Conventional configuration. (c) Configuration used in this paper.

A simulated frequency-gain characteristic for the Model A amplifier module is shown in Fig. 5. That for the Model B amplifier module is shown in Fig. 6. As shown in the figures, the Model A amplifier module band reaches 10.5 GHz with 10-dB gain and the Model B amplifier module band reaches 7 GHz with 15-dB gain. Low-cutoff frequencies of the amplifier modules are due to the coupling capacitors. For the matching network design in the microwave frequency range, computer-aided design optimization was used.

C. Circuit Description

Photographs of Model A and Model B amplifier modules are shown in Fig. 7. Coupling and RF bypass capacitors are multilayer high dielectric constant ceramic capacitors. These capacitors were tested in a 50-Ω system, in advance, over the 2-GHz to 8-GHz range. The VSWR for these capacitors is less than 1.5 over the frequency range. Resistors and microstrip lines are fabricated on 0.635-mm thick alumina ceramic plates having a Au–Cr–Ta_2N metal system. The rated sheet resistance for the Ta_2N film is 50 Ω.

Fig. 8 shows an equivalent circuit for an input microwave matching circuit in Model A, and two microstrip configurations for the equivalent circuit. In the circuit configuration in Fig. 8(b), when microstrip-conductor widths of the stubs are not negligibly narrow compared with the microstrip line lengths (characteristic impedances for the stubs are low), it is difficult to realize equivalent electrical angles γl_1, γl_2, and γl_3 (where γ is propagation constant) which are shown in Fig. 8(a).

Y parameters for a transmission line having characteristic impedance Z_0, propagation constant γ, and length l are given by

$$(Y) = \begin{pmatrix} \dfrac{\cosh \gamma l}{Z_0 \sinh \gamma l} & \dfrac{-1}{Z_0 \sinh \gamma l} \\ \dfrac{-1}{Z_0 \sinh \gamma l} & \dfrac{\cosh \gamma l}{Z_0 \sinh \gamma l} \end{pmatrix}. \quad (1)$$

Accordingly, Y parameters (Yn) for n parallel transmission lines are calculated as

$$(Yn) = \begin{pmatrix} \dfrac{\cosh \gamma l}{(Z_0/n) \sinh \gamma l} & \dfrac{-1}{(Z_0/n) \sinh \gamma l} \\ \dfrac{-1}{(Z_0/n) \sinh \gamma l} & \dfrac{\cosh \gamma l}{(Z_0/n) \sinh \gamma l} \end{pmatrix} \quad (2)$$

where n is a positive integer.

As shown in (2), the n parallel transmission lines for characteristic impedance Z_0 are equivalent to a transmission line with characteristic impedance Z_0/n, having the same line length. The relation between microstrip-conductor width W for characteristic impedance Z_0 and strip conductor width W_n for characteristic impedance Z_0/n is as shown in the following:

$$\frac{W}{W_n} \ll \frac{1}{n}.$$

By substituting a stub with characteristic impedance Z_2 and length l_2 by two parallel stubs with characteristic impedance $Z_2/2$ and length l_2, the sum of microstrip-conductor widths for the stubs can be reduced. Therefore, the circuit configuration in Fig. 8(c) is more suitable than that in Fig. 8(b), for the stubs having low characteristic impedances. All circuit components were mounted on a metal carrier measuring $3\text{ cm} \times 3\text{ cm}$.

IV. Performance

A. Frequency-Domain Characteristics

The gain-frequency characteristic measured in a 50-Ω system for the Model A amplifier module is shown in Fig. 5. An 8.6-dB gain is obtained over the 3-dB bandwidth from 800 kHz to 9.5 GHz. A 13.5-octave bandwidth is achieved. Fig. 9 shows input and output VSWR for the 800-kHz to 9.5-GHz amplifier module. The input VSWR is lower than 2 over the 2-MHz to 1-GHz range and is lower than 4 over the 1-MHz to 10-GHz range. For the output VSWR, less than 2 can be obtained over the 2-MHz to 9-GHz range. Input-output power responses for the amplifier module, measured at 0.5 GHz, 1 GHz, 4 GHz, and 8 GHz, are shown in Fig. 10. The amplifier module has a 12-dBm power output at 1-dB gain compression over the frequency range, from 0.5 GHz to 8 GHz.

Gain-frequency characteristic for the Model B amplifier module is also shown in Fig. 6. A 14-dB gain is obtained over the 3-dB bandwidth from 700 kHz to 6 GHz. For pulse amplifications, a linear phase (nondispersive) char-

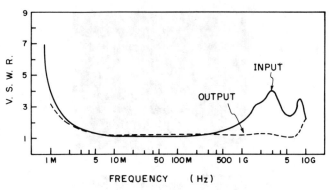

Fig. 9. Input and output VSWR for the 800-kHz to 9.5-GHz amplifier module.

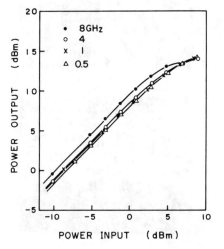

Fig. 10. Input–output power response for 800-kHz to 9.5-GHz amplifier module.

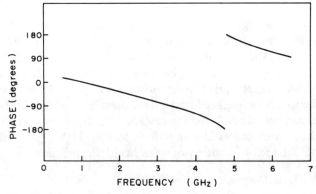

Fig. 11. Phase-frequency characteristic for Model B amplifier module.

acteristic as well as a flat-gain characteristic are required. Fig. 11 shows a measured phase-frequency characteristic for the Model B amplifier module. As seen in the figure, the amplifier modules has approximately linear phase across the frequency band.

By cascading two 800-kHz to 9.5-GHz amplifier modules without any external matching, 19-dB gain is obtained over the 800-kHz to 8.5-GHz band. The result is also

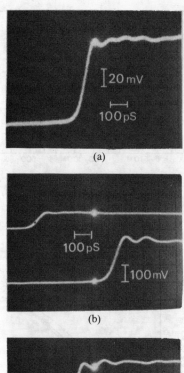

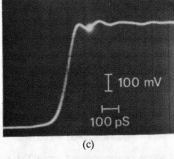

Fig. 12. Step responses for the 800-kHz to 9.5-GHz amplifier module and the cascade amplifier. (a) Input waveform. (b) Input and output waveforms for the module. (c) Output waveform for the cascade amplifier.

shown in Fig. 5. The gain-frequency characteristic has a slight gain slope caused by R_2 of the second amplifier module. The R_2 value measured for the fabricated second amplifier module was higher than the designed value. Adjusting the R_2 value lower, so that the gain of the low-frequency range is reduced a flat gain will be achieved. The experimental result for the cascade amplifier demonstrates that, by cascading the amplifier modules, higher gain can be obtained without serious bandwidth degradation.

B. Time-Domain Characteristics

Step responses for the 800-kHz to 9.5-GHz amplifier module and the cascade amplifier are shown in Fig. 12. Fig. 12(a) shows the input waveform having a 10-percent to 90-percent rise time of 75 ps. Both input and output waveforms for the 800-kHz to 9.5-GHz amplifier module are shown in Fig. 12(b). The input waveform is the same as in Fig. 12(a). Rise time for the output waveform is 85 ps. A well-known approximate relation involving real step response rise time for the amplifier output t_a measured rise time for the input waveform t_i and measured rise time for the output waveform t_o is given by

$$t_a^2 \simeq t_o^2 - t_i^2. \quad (3)$$

Using above relation, t_a is calculated as 40 ps. There is another approximate relation between amplifier high-cutoff frequency f_c and t_a, as

$$f_c t_a \simeq \tfrac{1}{3}. \quad (4)$$

From this relation, $t_a = 35$ ps is estimated. Results from (3) and (4) are in good agreement.

Fig. 12(c) shows a measured step response for the cascade amplifier, where the input waveform is the same as in Fig. 12(a). t_a is estimated as 50 ps. No serious rise time degradation in the cascade amplifier is observed. This can also be predicted from the gain-frequency characteristic for the cascade amplifier.

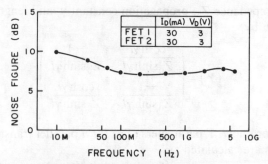

Fig. 13. Noise figure for the 800-kHz to 9.5-GHz amplifier module.

C. Noise

Since the developed amplifier module has 12-dBm power output at 1-dB gain compression, it can be used as a main amplifier in some systems. Noise characteristics matter little for main amplifiers. However, when using the amplifier module as a preamplifier, its noise figure (NF) is very important. Also, it has been reported that GaAs FET's have $1/f$ noise below several hundred megahertz [5]. Accordingly, noise characteristics for the 800-kHz to 9.5-GHz amplifier module were measured.

Fig. 13 shows the NF over the 10-MHz to 6-GHz range. Better than 8-dB NF was observed across the 50-MHz to 6-GHz range. As shown in the figure, NF has a -0.8 dB/octave slope across the 10-MHz to 50-MHz range.

To observe the noise characteristics below 10 MHz, output noise spectra for the amplifier module were measured using the spectrum analyzer. Fig. 14(a) shows the output noise spectra under bias supplied condition. Fig. 14(b) shows the output noise spectra without bias supply. As seen, amplifier output noise spectra have a -3 dB/octave slope below 4 MHz. This shows the existence of amplifier $1/f$ noise below 4 MHz. The NF variation caused by ± 33-percent drain current variation was within ± 0.1 dB.

In the amplifier module, parallel 50-Ω resistor R_1 in Fig. 2 is used to reduce input VSWR in the low-frequency range. However, this resistor degrades the amplifier NF in

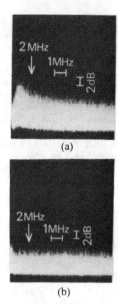

Fig. 14. Noise spectra for the 800-kHz to 9.5-GHz amplifier module. (a) With bias supply. (b) Without bias supply.

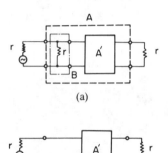

Fig. 15. Block diagram for NF explanation. (a) With parallel resistor. (b) Without parallel resistor.

the low-frequency range. To explain the magnitude of the NF degradation by R_1, Fig. 15 is presented. As shown in Fig. 15(a), amplifier module A is divided into two sections A' and B. NF (F) of section A' is given by [6]

$$F = F_0 + \frac{R_n}{(2/r)}\left[\left(\frac{2}{r} - G_0\right)^2 + B_0^2\right] \simeq 2 \cdot \frac{R_n}{r} \quad (5)$$

where minimum NF for section A' is F_0, and source admittance, which gives F_0 is $Y_0 = G_0 + jB_0$. R_n is the equivalent noise resistance for section A'. NF (F') for amplifier module A is given by

$$F' = F_B + \frac{F-1}{G_{av}} = 2 \cdot F \simeq 4 \cdot \frac{R_n}{r} \quad (6)$$

where F_B is the NF for section B ($F_B = 2$) and G_{av} is the available gain in section B ($G_{av} = \frac{1}{2}$). Meanwhile, the NF for amplifier A' in Fig. 15(b) is given by

$$F'' = F_0 + \frac{R_n}{(1/r)}\left[\left(\frac{1}{r} - G_0\right)^2 + B_0^2\right] \simeq \frac{R_n}{r}. \quad (7)$$

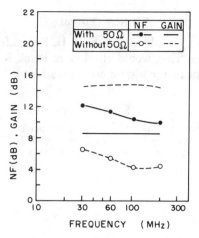

Fig. 16. NF degradation caused by parallel resistor.

Assuming $F_0 \ll (R_n/r)$, $(1/r) \gg G_0$, and $1/r \gg B_0$, which is reasonable for source grounded GaAs FET's in the low-frequency range, NF ratio F'/F'' is approximated as

$$\frac{F'}{F''} \simeq \frac{4R_n/r}{1R_n/r} = 4. \quad (8)$$

Consequently, the addition of section B causes nearly 6-dB NF degradation. To validate the result, NF values for the second 800-kHz to 9.5-GHz amplifier module, with and without R_1, were measured. As shown in Fig. 16, amplifier NF and gain without R_1 are about 6 dB better than amplifier NF and gain with R_1. However, input VSWR without R_1 becomes large.

Consequently, circuit configuration, which requires no parallel resistor, should be considered if NF is a matter of concern.

V. Conclusions

A novel ultrabroad-band amplifier design technique has been developed. The developed amplifier circuit operates as a $C-R$ coupled amplifier circuit in the low-frequency range. It also utilized a lossless matching circuit in the microwave frequency range. Using this design method, an 800-kHz to 9.5-GHz bandwidth (13.5 octaves) amplifier module with 8.6-dB gain has been realized. The amplifier module has 40-ps step response rise time. It also has low input and output VSWR. By cascading two amplifier modules, 19-dB gain over the 800-kHz to 8.5-GHz range and 50-ps step response rise time were obtained. It was demonstrated that, by cascading the modules, higher gain can be obtained without degrading the bandwidth and the rise time. NF for the amplifier module is better than 8 dB over the 50-MHz to 6-GHz range.

The design examples presented in this paper have a simple circuit configuration. By increasing number of matching circuit sections and transformer sections, and by using shorter gate length FET's, wider bandwidth can be realized.

Acknowledgment

The authors would like to thank H. Kohzu for supplying GaAs FET's. They would also like to thank K. Ayaki and H. Katoh for their constant encouragement throughout this work.

References

[1] D. Hornbuckle, "GaAs IC direct-coupled amplifiers," in *'80 MTT-S Int. Microwave Symp. Dig. Tech. Papers*, pp. 387–389, May 1980.
[2] R. V. Tuyle, "A monolithic integrated 4-GHz amplifier," in *'78 Int. Solid-State Circuits Conf., Dig. Tech. Papers*, pp. 72–73, Feb. 1978
[3] H. Q. Tserng and H. M. Macksey, "Ultra-wideband medium-power GaAs MESFET amplifiers," in *'80 Int. Solid-State Circuits Conf., Dig. Tech. Papers*, pp. 166–167, Feb. 1980
[4] K. B. Niclas, W. T. Wilser, R. B. Gold, and W. R. Hitchens, "A 350 MHz–14 GHz MESFET amplifier using feedback," in *'80 Int. Solid-State Circuits Conf., Dig. Tech. Papers*, pp. 164–165, Feb. 1980
[5] C. P. Snapp, "Microwave bipolar transistor Technology-present and prospects," in *9th European Microwave Conf. Proc.*, pp. 3–12, Sep. 1979
[6] W. R. Atkinson *et al.* "Representation of noise in linear twoports," *Proc. IRE*, vol. 48, pp. 69–74, Jan. 1960

Monolithic GaAs Direct-Coupled Amplifiers

DERRY P. HORNBUCKLE, MEMBER, IEEE, AND RORY L. VAN TUYL, SENIOR MEMBER, IEEE

Abstract — Monolithic GaAs dc-coupled amplifiers with bandwidths up to 5 GHz are described. The multistage amplifiers include designs having 25-dB gain with 2-GHz bandwidth and 10-dB gain with 5-GHz bandwidth. Analysis of gain, bandwidth, and noise agrees with measurements. Distortion mechanisms are discussed, along with the performance of a low-distortion amplifier.

Introduction

MONOLITHIC GaAs integrated amplifiers have demonstrated impressive performance capabilities, achieving broad-band frequency coverage through 18 GHz, and narrow-band power output levels above 2 W [1], [2]. Most of the reported amplifier circuits have relied on extensive use of passive matching components, with active devices occupying only a few percent of the chip area [1]-[5].

This paper describes an alternative approach to GaAs monolithic amplifier design which makes use of active devices in place of passive elements, resulting in between one and two orders of magnitude reduction in chip size (to 300 μm × 650 μm). Frequency response is limited to about half the unity current-gain frequency f_t of the MESFET's in the circuit, as is typical for unmatched transistor amplifier stages.

The objectives of this work were to develop general-purpose medium-power amplifiers with several combinations of gain and bandwidth, including 2-GHz upper corner frequency with 25-dB gain and 4 GHz (5 GHz was achieved) with 10-dB gain into 50 Ω. An output power capability of 10 dBm into 50 Ω was desired, with harmonic distortion as low as possible. Noise

Manuscript received August 6, 1980; revised September 12, 1980.
The authors are with Hewlett-Packard Corporation, Santa Rosa, CA 95404.

performance was not a consideration for any of the anticipated applications.

In the following sections, circuit function will be described, followed by comparison of theoretical and measured gain, bandwidth, distortion, and noise.

CIRCUIT FUNCTION

The GaAs MESFET's which are the active elements in these amplifiers are similar to those described in [6], except that the MESFET's are fabricated on selectively implanted pockets of n-type dopant instead of chemically etched mesas. Silicon-29 ions are implanted through a photoresist mask at an energy of 230 keV and a dose of $5.5 \times 10^{12}/\text{cm}^2$; the wafers are capped with chemical-vapor-deposited silicon dioxide and annealed at 850°C for 30 min. The resulting doping profile has a peak concentration of $2 \times 10^{17}/\text{cm}^3$ and a sheet resistance of 360 Ω/ square. MESFET's are processed with etch-recessed gate channels which trim the pinchoff voltage to -2.5 ± 0.5 V. MESFET electrical characteristics are similar to those in [6].

A common-source FET was chosen as the basic amplifying element since it provides voltage and current gain along with reverse isolation, a combination not available from common-gate or common-drain (source-follower) configurations.

An active-load FET was used rather than a resistor load because it results in better large-signal performance for a given small-signal gain, as illustrated in Fig. 1. With the active load it is possible to choose a quiescent operating point near half the saturation current I_{dss} of the common-source FET. This circuit can, therefore, source or sink nearly equal currents into a load capacitance, an improvement over the poor sourcing capability of the resistor load. The common-source FET and active-load FET are shown in Fig. 2(a) and are referred to as the "inverter" or "inverter and active load" portion of the amplifier.

The inverter circuit of Fig. 2(a) could be cascaded, with ac coupling, to produce a multistage amplifier with higher gain. However, bandwidth would be reduced substantially due to the heavy capacitive loading of the succeeding stage's gate capacitance, which is many times larger than the drain-to-source capacitance for the inverter and active load.

The capacitive loading is greatly reduced by use of a buffer circuit, as shown in Fig. 2(b). In addition to the buffering provided by the source-follower FET, 4 V of level shifting is provided by the four diodes and their associated series resistance (no level shifting occurs in the source follower, since it is the same size as the current source and thus operates at $V_{gs} = 0$ V). A +3-V inverter drain voltage at the buffer input results in a -1-V level at its output, which is suitable for direct coupling to the input of the next inverter.

The inverter and buffer circuits are shown together in Fig. 3(a); they will be referred to as a single amplifier stage, although they could, as well, be treated as two stages, common source followed by common drain. This circuit has 12 to 14 dB of gain and 2.5-GHz bandwidth when lightly loaded [7], as discussed in the next section. It is referred to as an "open-loop" or "SA-1" (Single-ended Amplifier type 1) stage.

It is possible to add feedback to the SA-1 amplifier stage to

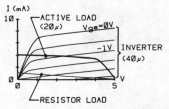

Fig. 1. Comparison of resistor load to FET active load for a 40-μm inverting amplifier circuit.

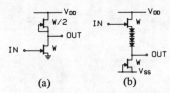

Fig. 2. (a) Inverter and active load. (b) Buffer/level-shift circuit.

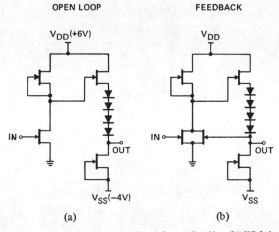

Fig. 3. GaAs amplifier stages: (a) Open loop (SA-1). (b) With internal feedback (SA-2).

improve its bandwidth while reducing its gain. Active feedback, using an FET with typically one-quarter the gate width of the inverter (Fig. 3(b)), is superior to resistive feedback for improving bandwidth. This "feedback" or "SA-2" stage has 6 to 7 dB of gain and 4-GHz bandwidth when lightly loaded.

These open-loop and feedback amplifier stages have been combined in a modular fashion to produce a number of multistage amplifiers. Three of these amplifiers will be discussed in this paper, along with a fourth design which uses inductor peaking to improve bandwidth. The four schematics are shown in Fig. 4. Their performance is best analyzed in terms of that of the component SA-1 and SA-2 stages. In the following sections, the performance of these stages and of multiple-stage amplifiers will be analyzed in detail.

GAIN

The gain of the open-loop and feedback amplifier stages of Fig. 3(a) and (b) can be calculated as follows. First, the voltage gain of the SA-1 inverter and active load, G_{inv1}, is

$$G_{inv1} = g_m * R_{load} = g_m/(g_d + g_d/2) = (\tfrac{2}{3}) * (g_m/g_d)$$
$$= 6.66$$

$$G_{inv1} \text{ (dB)} = 20 * \log\left[(\tfrac{2}{3}) * (g_m/g_d)\right] = 16.5 \text{ dB}.$$

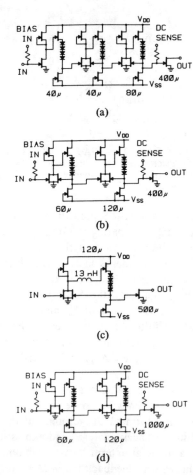

Fig. 4. Multistage GaAs IC amplifiers: (a) A20, (b) A40, (c) A60, (d) A21. Numbers indicate width of stage (gate width of inverter, or inverter and feedback FET's, and of source-follower and current-source FET's; active-load FET is half this width).

Nominal values of $g_m/W = 100$ mS/mm and $g_d/W = 10$ mS/mm have been used in this calculation, where g_m is the FET transconductance, g_d is its output (drain) conductance, and W is the gate width. The gain is independent of the inverter FET gate width, so long as the active-load FET remains half the width of the inverter FET, as assumed for this calculation.

The gain of the buffer/level-shift stage G_{buf} is roughly approximated by the usual source-follower formula

$$G_{buf} = (g_m * R_s)/(1 + g_m * R_s) = \tfrac{5}{6}$$

where

$$R_s = 1/(2 * g_d)$$

$$G_{buf} \text{ (dB)} = -1.5 \text{ dB}.$$

The series diode resistance R_{di} should be included for a more accurate gain estimate. For $R_{di} = 2$ ($\Omega * $ mm)/W per diode, the resulting gain is

$$G_{buf} = 0.776 \quad \text{or} \quad G_{buf} \text{ (dB)} = -2.2 \text{ dB}.$$

The calculated gain for the SA-1 amplifier stage of Fig. 3(a) is therefore

$$G(\text{SA-1}) = G_{inv1} \text{ (dB)} + G_{buf} \text{ (dB)} = 14.3 \text{ dB}.$$

This is similar to measured values of 12 to 14 dB; the measured variations in gain appear to be due primarily to drain conductance variations.

The feedback stage of Fig. 3(b) can be analyzed by determining the effect of the feedback FET on the load impedance at the inverter drain. The added conductance is $g_{df} + G_{buf} * g_{mf}$, where g_{df} and g_{mf} are the drain conductance and transconductance of the feedback FET. If the SA-2 inverter width is 80 percent of the width of the SA-1 inverter, and the feedback FET 20 percent, then the resulting gain can be written in terms of g_m and g_d of the SA-1 inverter FET as

$$G_{inv2} = 0.8 * g_m/(0.8 * g_d + g_d/2 + 0.2 * g_d$$
$$+ 0.776 * 0.2 * g_m)$$

$$G_{inv2} = 2.621$$

or

$$G_{inv2} \text{ (dB)} = 8.37 \text{ dB}.$$

When the 2.2-dB buffer loss is subtracted, the calculated gain for a feedback stage becomes 6.2 dB; measured values are 6 to 7 dB.

Multistage amplifiers described here utilize a single common-source FET in an open-drain configuration as an output device. The gain of this FET is $g_m/(1/R_L + g_d)$. R_L, the load impedance, is typically either 50 or 25 Ω. A 400-μm-gate FET produces -0.8-dB gain into 25 Ω for $g_m = 100$ mS/mm and $g_d = 10$ mS/mm.

Gain for a multistage amplifier can be calculated simply by taking the sum of the individual stage gains in decibels. Allowing 12 dB for each open-loop stage, and 6 dB for each feedback stage, and adding the gain of the output stage gives an estimate of the total amplifier gain. Thus the gain of the four-stage amplifier of Fig. 4(a), type A20, is estimated to be

$$G(\text{A20}) = 12 + 6 + 6 - 0.8 = 23.2 \text{ dB}$$

for operation into an external 25-Ω load. The measured value at 500 MHz is 22.6 dB, with a 2.4-dB standard deviation, for 102 chips from seven wafers. Similar agreement is obtained for other amplifiers.

BANDWIDTH

For bandwidth calculations, the SA-1 open-loop amplifier stage can be represented by the equivalent circuit shown in Fig. 5(a); diode series resistance has been ignored here for simplicity. It can be shown that the frequency response of this circuit is

$$G(\text{SA-1}) = v_{out}/v_1 = -g_{m1} * (s * C_{g2} + g_{m2})/(A * s^2 + B * s + C)$$

where s is the complex frequency variable; A, B, and C are defined below, and all other terms are shown in Fig. 5.

$$A = (C_{g2} * C_s + C_L * C_{g2} + C_L * C_s)$$

$$B = [C_{g2} * g_s + g_L * (C_{g2} + C_s) + C_L * (g_{m2} + g_s)]$$

$$C = g_L * (g_{m2} + g_s).$$

The frequency-response zero at $f = g_{m2}/(2 * \pi * C_{g2}) = f_t$ does not have much effect on performance compared to the

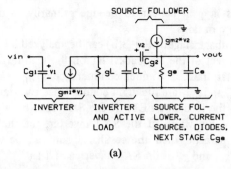

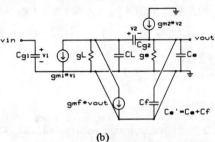

Fig. 5. Equivalent circuits for (a) SA-1 open-loop amplifier stage and (b) SA-2 internal-feedback amplifier stage.

poles, which are at frequencies much less than f_t. The pole frequencies are described by the quadratic equation solution

$$p = [-B \pm \sqrt{B*B - 4*A*C}]/(2*A).$$

To gain insight into the pole behavior, and hence the frequency response, it is helpful to write all conductances and capacitances in terms of a reference value. Typical values relative to C_{g2} and g_{m2} are: $g_{m1} = g_{m2}$; $g_L = 0.15 * g_{m2}$; $g_s = 0.2 * g_{m2}$; $C_L = 0.3 * C_{g2}$; and $C_s = n * C_{g2}$. Note that C_s has been expressed as a multiple, n, of the gate capacitance C_{g2} to allow for varying amounts of load capacitance. Typically, n will equal 2, rather than 1, for a succeeding stage equal in size to the stage under consideration. This is because of the drain-to-source capacitances of the source-follower and current-source FET's, and the capacitance to ground of the level-shift diodes. Using the above values, the discriminant of the pole expression is

$$B*B - 4*A*C = (0.225*n^2 - 0.723*n + 0.2881)$$
$$* (C_{g2} * g_{m2})^2.$$

This expression is negative for $0.40 < n < 31.7$, which includes all load capacitance ratios n of interest. So the SA-1 cell exhibits a complex-pole-pair frequency response.

What has happened is that the real-axis poles, which would occur at g_L/C_L and at $(g_{m2} + g_s)/C_s$ if C_{g2} were equal to zero, have been strongly coupled by C_{g2}. They interact and coalesce to form the complex pole pair.

The resulting pole locations, for several load-capacitance ratios n have been plotted in Fig. 6. Typical values of n are between 2 and 6, and result in real and imaginary pole components, p_r and p_i, of nearly equal magnitude. Such pole pairs have a frequency response which is 3 dB down at f_c, where

$$f_c = \sqrt{p_r*p_r + p_i*p_i}/(2\pi).$$

Unbuffered amplifiers, by comparison, have a response which is 3 dB down at the frequency of their single real pole. Thus

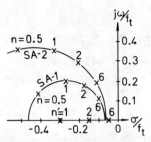

Fig. 6. Pole locations for SA-1 and SA-2 amplifier stages with various ratios n of load capacitance C_s to source-follower C_{gs} (normalized to the MESFET unity current-gain frequency f_t). Real-axis poles are for an unbuffered stage with $n' = C_L/C_{g1}$.

the improvement at the -3-dB point due to buffering can be determined by simply comparing the magnitude of the unbuffered real pole to the magnitude of the SA-1 complex pole in Fig. 6. Further improvements in frequency response can be achieved with feedback, as discussed next.

The equivalent circuit for the SA-2 feedback amplifier is shown in Fig. 5(b). The values of C_L and g_L are unchanged, since the inverter width plus the feedback FET width in the SA-2 are assumed equal to the width of the inverter in the SA-1 stage; g_{m1}' (for the SA-2) is less than g_{m1} (for the SA-1) by the amount of g_{mf}.

It can be shown, either by direct calculation or by feedback analysis, that the frequency response for the SA-2 feedback stage is of the same form as the response of the SA-1 open-loop stage. Using feedback analysis, and expressing the open-loop gain G as a numerator term N over a denominator D, the closed-loop gain G_{cl} with feedback factor H, becomes

$$G_{cl} = G/(1 + GH) = (N/D)/(1 + H*N/D) = N/(D + H*N).$$

For a frequency-invariant feedback factor H, additional poles are added only if the original numerator N is of higher order than the denominator D. In the SA-2 case, the numerator is of lower order, so poles are moved but no new poles are added.

The approximate effect of feedback is to increase the imaginary part of the pole without changing the real part, as can be shown by approximating the numerator of the SA-1 gain expression as a constant N, ignoring the zero at f_t. Then the term $H*N$ is a positive constant. The discriminant of the quadratic pole equation becomes $[B*B - 4*A*(C+H*N)]$, so the imaginary part of the pole increases. But the real part, $-B/(2*A)$, is unchanged under this approximation. Therefore, the magnitude of the pole and the amount of frequency-response peaking (determined by the ratio of imaginary to real part of the pole, p_i/p_r) both increase.

The exact SA-2 gain expression is

$$G(\text{SA-2}) = -g_{m1}' * (g_{m2} + s*C_2)/(A'*s^2 + B'*s + C')$$

$$A' = A \text{ (with } C_s \text{ replaced by } C_s')$$

$$B' = B + C_{g2} * g_{mf}$$

$$C' = C + g_2 * g_{mf}.$$

When the coefficients are written in terms of g_{m2} and C_{g2}, as before with C_s' being some multiple n of C_{g2}, the pole locations for the SA-2 stage are as shown in Fig. 6.

Calculated frequency response, using the above pole locations, is compared with measured response for the SA-1 and

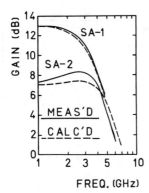

Fig. 7. Frequency response of lightly loaded ($n = 1.5$) SA-1 and SA-2 amplifier stages, as measured and as calculated from pole locations.

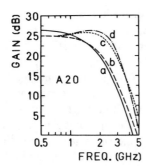

Fig. 8. A20 amplifier frequency response. (a) Measured. (b) Computer simulation with (b) and without (c) parasitic interstage capacitance. (d) Calculated from superposition of poles of each stage.

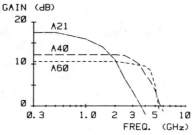

Fig. 9. Frequency response of A21, A40, and A60 amplifiers operating into 50-Ω load.

SA-2 stages in Fig. 7. Measurements were made as described in [7], and the load capacitance factor n was estimated to be 1.5, with $f_t = 12$ GHz. The measured SA-2 frequency response shows more peaking than calculated due to transit time delay in g_{mf}, which was not included above. The otherwise good agreement indicates that this frequency response expression is a valuable tool to understanding and improving the bandwidth of this type of amplifier.

In the ideal case, the bandwidth of multistage amplifiers could be analyzed by the superposition of the pole-zero patterns for each stage, since the high reverse isolation of the common-source FET makes the stages nearly independent.

In practice, however, there is a significant parasitic capacitance between stages which couples them together. This distributed parasitic is layout dependent, but generally can be represented by a single capacitor from the output of one stage to the output of the next. The value has been estimated, using coplanar transmission-line formulas, to be 20 fF for the four-stage A20 amplifier of Fig. 4(a). Fig. 8 shows the measured frequency response a for the amplifier; a computer simulation using 20 fF of parasitic interstage capacitance (curve b in Fig. 8) is very similar.

Fig. 8 also shows the frequency response calculated analytically, assuming pole superposition, and another computer simulation (curves c and d)—both without the parasitic. These two agree, again indicating the validity of the analytic expression. Unfortunately, neither agrees well with the measured response. Attempts to increase the agreement with a Miller-effect model of the parasitic capacitance provide only a modest improvement.

So the analytic expression for bandwidth presented here is useful for improving bandwidth of an individual amplifier stage; but computer simulations are needed for accurate modeling of multistage amplifiers when coupling between stages is significant.

From the preceding frequency-response analysis for SA-1 and SA-2 amplifier stages, two methods of improving bandwidth emerge. One is to decrease the load-capacitance factor n; the second is to increase the amount of feedback. Both of these improvements were made in the circuit of Fig. 4(b), type A40. The SA-1 input stage of the A20 design was eliminated, and each of the next two stages were increased in width by 50 percent, reducing load-capacitance factor from 6 to 4.3 for the stage driving the 400-μm output FET. In addition, for the output-driver stage the feedback FET was made 27 percent of the total inverter-plus-feedback gate width, instead of 20 percent. The resulting bandwidth was 4 GHz, as shown in Fig. 9, with 12 dB of gain into a 50-Ω load.

Bandwidth can also be improved by adding a peaking inductor, as shown in Fig. 4(c) (amplifier type A60). Analysis similar to that for the SA-1 amplifier stage shows that the inductor stage has a fourth-order response, resulting in 80 dB per decade of response rolloff above the upper band edge. However, the inductor can be chosen to provide peaking at what would be the upper corner frequency in a totally active amplifier; even a fairly lossy inductor will suffice in this application. The 13-nH inductor used in this design was made physically as small as possible, resulting in a Q of only about 3 at 4 GHz, but still it occupied more area than all of the active circuitry on the chip (see [8] for a chip photo). The inductor did provide the expected frequency response improvement, giving the 5-GHz bandwidth shown in Fig. 9.

Distortion

Distortion in a multistage amplifier is usually dominated by nonlinearities at the largest signal point in the circuit. Since this signal point is normally associated with the output stage, the design of this stage is critical if distortion is to be minimized. Other important properties of the output stage are output impedance, power consumption, and input capacitance (which can be a limiting factor for both bandwidth and distortion). Since the primary purpose of the output stage in these GaAs IC amplifiers is to faithfully transform the output signal of a high-impedance voltage amplifier to a low impedance (typically, but not necessarily, 50 Ω), the source-follower configuration would seem appropriate. It was decided, however, to use a common-source output stage for these amplifiers, since the output impedance of this configuration is typically higher than the desired output impedance, and various combinations of device width and external load resistor could be used to develop the output impedance required by a given application. Also, voltage gain is possible with the common-source configuration, whereas the source follower gives a factor-of-two attenuation for matched output. This can be important, because the distortion of the multiple-FET amplifier cells is usually worse than that of an individual FET for a given signal level, favoring the smaller voltage swing in the multiple-device amplifier cell made possible by a single-device output stage with gain.

The FET nonlinearities which affect distortion are primarily the nonlinear transconductance, output resistance, and input capacitance. Although the transconductance nonlinearity with gate-to-source voltage is the dominant distortion mechanism for low-gain amplifier stages, the output resistance nonlinearity can predominate in high-gain stages, because the drain-signal voltage amplitude is significantly larger than the gate signal amplitude. In certain cases of load resistance and device bias, the transconductance distortion and output resistance distortion are of comparable amplitude; they tend to flatten opposite sides of a sinusoidal output signal, and can actually cancel each other as far as second-harmonic generation. This second-harmonic nulling effect has been observed in discrete FET's which were tested at low frequency with various load impedances and operating biases.

Distortion is also introduced by the nonlinear input capacitance C_{gs} whose current is voltage dependent as well as frequency dependent. The nonlinear current develops a distorted voltage waveform across a nonzero linear input impedance; this effect has been verified for discrete FET's by observation of harmonics reflected from the gate which were not present in the incident signal. The effect increases for higher input impedances and higher frequencies; it is of opposite sign to the g_m nonlinearity, tending toward a frequency-dependent cancellation.

Further complication occurs in multiple-FET circuits. For example, the nonlinear I-V characteristic of an active-load FET increases third-harmonic distortion compared to a linear load.

Another serious source of distortion results from driving a capacitive load with a nonlinear current source, such as a source follower driving an output FET. A nonlinear voltage waveform is developed across the output FET's gate capacitance as greater current is demanded from the source follower at high frequencies. This gate voltage has second harmonic in phase with that of the output FET's g_m nonlinearity, so it increases the second-harmonic distortion. At very-high-signal levels, the source follower can alternate between a maximum available current at positive gate-to-source voltage and complete cutoff. The limited peak-to-peak current available from the source follower restricts the rate of voltage change at the output FET's gate to a value defined by $dV/dt = I/C_{gs}$. This results in a sawtooth-like waveform with third-order distortion which is less than that of the square-wave-like inverter output. This slew-rate effect does not necessarily result in degraded harmonic performance. However, it is the most important source of saturated output-power limitation at higher frequencies. Its clearest manifestation is a 6-dB per octave rolloff of saturated output power with frequency, beginning below the small-signal gain corner frequency.

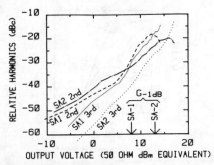

Fig. 10. Relative second- and third-harmonic distortion for SA-1 and SA-2 amplifier stages. Arrows indicate output voltage for 1-dB gain compression. V_{dd} = +6.0 V, V_{ss} = -4.0 V.

The sources of distortion discussed above have been verified both by computer simulation [9] and, in most cases, experimentally with discrete MESFET's. These results will be used below to make inferences about the distortion measured at the terminals of single-stage and multistage amplifiers. Computer simulations of the distortion in these circuits support the conclusions drawn here, but experimental testing inside the circuit is, of course, not generally practical.

Fig. 10 shows the measured second- and third-harmonic distortion for the single-stage SA-1 (open-loop) and SA-2 (internal feedback) amplifier cells, plotted as a function of output signal level. At low signal levels, the open-loop SA-1 cell has lower second-harmonic distortion. The reason for this is cancellation between g_m and r_{ds} nonlinearities, as inferred from the fact that the second-harmonic level is quite dependent on the positive supply voltage V_{dd}, passing through a null at V_{dd} = +6.25 V. The primary FET model element affected by V_{dd} is r_d, hence the conclusion that cancellation is occuring between the r_d nonlinearity and that of g_m, the other dominant FET nonlinearity. The internal feedback of the SA-2 cell reduces the impedance level at the drain node of the amplifying FET, thereby decreasing the interaction with the transconductance nonlinearity. So the SA-2 cell distortion is g_m dominated.

At high output levels, both cells show a decrease of second-harmonic distortion and a sharp increase of third harmonic. At low signal levels, a sinusoidal signal applied to the nonlinear voltage-transfer characteristic of the amplifier generates an unsymmetrically distorted waveform rich in second harmonic. At high signal levels, the strong saturating nonlinearities of the amplifier generate a more symmetrical waveform reminiscent of a square wave. Since such waveshapes are rich in odd-order distortion, the second harmonic decreases at the expense of the third harmonic.

The SA-2 feedback-type cell is lower in harmonics at the maximum output signal levels of interest due to a reduction in the saturating nonlinearities attributable to feedback. At 2-V peak-to-peak output (equivalent to +10 dBm if suitably buffered to drive a 50-Ω load), the SA-2 cell has second-harmonic distortion of -29 dB below the fundamental (dBc), 6 dB better than then SA-1. This lower distortion at high levels, coupled with the bandwidth-increasing lower output impedance of the SA-2 cell, have made it the choice for driving the common-source output FET.

Fig. 11 compares the distortion performance of two three-stage integrated amplifiers, types A40 and A21, shown in Fig. 4(b) and (d). The amplifiers are identical except for the output FET, which is 1000-μm wide for the A21 and 400 μm for the A40. The wider output device has higher gain, thereby allowing a lower signal level at the SA-2 cell for a given output power. The advantages of lower distortion, higher saturated output power, and higher gain for this design must be weighed against the disadvantages of greater power consumption, lower bandwidth, and greater frequency dependence of saturated output power.

The power saturation for these two amplifiers is shown in Fig. 11(a). The 1000-μm output device delivers higher saturated output power at low frequency; but the power rapidly decreases at higher frequency due to the slew-rate limiting effect associated with driving the 1000-μm output FET. Fig. 11(b) shows statistical data for the harmonic distortion at 0-dBm output power delivered to an external 50-Ω load with an internal 50-Ω reverse termination. In this case, typical of many applications, the larger output device gives a 5- to 10-dB improvement in harmonic distortion, providing better than -35-dBc average second harmonic at 1.5 GHz and better than -30 dBc at 2.5 GHz. Data are for a quantity of 18 A21 chips and 24 A40 chips, all from a single wafer to avoid wafer-to-wafer variations. Even so, the standard deviations of measured distortion are large, as indicated by the bars above and below the average points in Fig. 11(b). Both statistical and bias-dependent variations can be attributed to the complex interaction of separate nonlinearities within the amplifier, most particularly the FET output resistance.

Noise

Although noise was not a design consideration, noise figure was calculated and measured for verification of an FET noise model. Noise of the SA-1 amplifier stage can be adequately modeled as shown in Fig. 12. The equivalent input noise voltage v_{eq} is

$$(v_{eq})^2 = (e_{n1})^2 + (g_{m2} * e_{n2}/g_{m1})^2 + (e_{n3}/G_{inv1})^2$$
$$+ [e_{n4} * g_{m4}/(g_{m3} * G_{inv1})]^2$$
$$= (e_{n1})^2 * [1 + \tfrac{1}{2} + (1/6.67)^2 + (1/6.67)^2]$$
$$= (e_{n1})^2 * 1.545$$

where e_{n1} through e_{n4} are the equivalent input noise voltage generators for FET's 1 through 4, g_{m1} through g_{m4} are their transconductances, G_{inv1} is the gain of the inverter and active load (defined previously), and v_{eq} is the equivalent input noise voltage generator for the complete SA-1 circuit.

The equivalent input noise voltage for a 500-μm FET at 1 GHz has been measured as 1.5 nV per root-hertz [10]. Scaling by $1/\sqrt{W}$ gives 5.3 nV per root-hertz for a 40-μm FET, as is used in the A20 amplifier input. Therefore, the equivalent input noise voltage is 6.59 nV per root-hertz. The thermal noise voltage of a 50-Ω source is 0.911 nV per root-hertz, and the noise figure F is

$$F = 1 + (6.59/0.911)^2 = 53.3 \quad \text{or} \quad F(\text{dB}) = 17.3 \text{ dB}.$$

This compares well to the measured average of 19.3 dB at

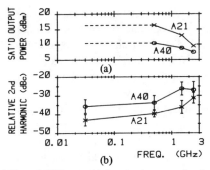

Fig. 11. (a) A40 and A21 saturated output power (typical single-chip data). (b) Relative second harmonic at 0 dBm out (averages for 24 A40 chips and 18 A21 chips; bars indicate standard deviations). Load impedance = 25 Ω for both (power into 50-Ω external load, disregarding power into 50-Ω reverse termination).

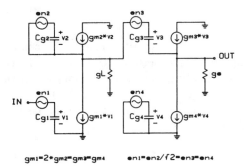

Fig. 12. Simplified noise equivalent circuit for SA-1 amplifier stage.

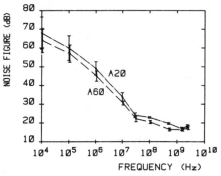

Fig. 13. A20 and A60 amplifier noise figures relative to 50-Ω input impedance. Bars indicate one standard deviation above and below average (28 A20 chips and 23 A60 chips).

1.5 GHz (1.3-dB standard deviation, for 151 chips from 8 wafers), even though such second-order effects as diode series resistance (which increases the contribution of e_{n4}), bias dependence of g_m and noise, and diode shot noise and thermal noise have not been included in the above computation.

The $1/f$ noise corner frequency for these amplifiers is in the range of 30 to 100 MHz, similar to that reported for discrete FET's [11]. Measured noise data are shown in Fig. 13.

Conclusion

Monolithic dc-coupled amplifier circuits with bandwidths up to 5 GHz have been developed. The chip size is as small as that for many discrete small-signal GaAs MESFET's, indicating a significant cost advantage over hybrid amplifiers in this frequency range. Analytic expressions for gain, bandwidth, and noise figure agree reasonably well with measured performance and can serve as valuable design tools; computer modeling is necessary for distortion analysis. Experimental results have demonstrated the usefulness of feedback and inductor peaking for bandwidth improvement, and the value of a large output FET for distortion improvement.

Important areas for further work include understanding and reducing the chip-to-chip variations in distortion and the magnitude of $1/f$ noise. High amplifier yield indicates that efforts should also be made to raise the level of integration by investigating and including on-chip other types of linear circuits to implement complete systems. Such circuits can be expected to have significant impact on the design of microwave instruments and systems of the future.

Acknowledgment

The authors wish to thank D. Estreich for contributions to noise and capacitance analysis, T. Taylor, D. D'Avanzo, and V. Kumar for help in process technology and device characterization, C. Coxen and A. Fowler for fabrication, and R. Fisher and C. Hart for testing, as well as P. Wang, D. Harkins, and J. Dupre for their support and advice.

References

[1] D. A. Abbott, J. Cockrill, R. S. Pengelly, M. G. Stubbs, and J. A. Turner, "Monolithic gallium arsenide circuits show great promise," *Microwave Syst. News*, pp. 73-96, Aug. 1979.

[2] R. A. Pucel, J. L. Vorhaus, P. Ng, and W. Fabian, "A monolithic GaAs X-band power amplifier," in *1979 IEDM Tech. Dig.*, paper 11.2.

[3] A. K. Gupta, J. A. Higgins, and D. R. Decker, "Progress in broad-band monolithic amplifiers," in *1979 IEDM Tech. Dig.*, paper 11.3.

[4] J. G. Oakes *et al.*, "Directly implanted GaAs monolithic X-band RF amplifier utilizing lumped element technology," in *1979 IEEE Gallium Arsenide Integrated Circuit Symp. Res. Abstr.*, paper 23.

[5] V. Sokolov, R. E. Williams, and D. W. Shaw, "X-band monolithic GaAs push-pull amplifiers," in *1979 ISSC Dig. Tech. Papers*, pp. 118-119.

[6] R. L. Van Tuyl, C. A. Liechti, R. E. Lee, and E. Gowen, "GaAs MESFET logic with 4-GHz clock rate," *IEEE J. Solid-State Circuits*, vol. SC-12, no. 5, pp. 485-495, Oct. 1977.

[7] R. L. Van Tuyl, "A monolithic integrated 4-GHz amplifier," in *1978 ISSC Dig. Tech. Papers*, pp. 72-73.

[8] R. L. Van Tuyl, D. Hornbuckle, and D. B. Estreich, "Computer modeling of monolithic GaAs IC's," in *1980 IEEE Int. Microwave Symp. Dig.*, pp. 393-394.

[9] D. P. Hornbuckle, "GaAs IC direct-coupled amplifiers," in *1980 IEEE Int. Microwave Symp. Dig.*, pp. 387-389.

[10] D. B. Estreich, Hewlett-Packard Co., private communication.

[11] E. Ulrich, "Use negative feedback to slash wide-band VSWR," *Microwaves*, vol. 17, no. 10, pp. 66-70, Oct. 1978.

A Monolithic Direct-Coupled GaAs IC Amplifier with 12-GHz Bandwidth

SANJAY B. MOGHE, MEMBER, IEEE, HORNG-JYE SUN, THOMAS ANDRADE,
CHARLES C. HUANG, MEMBER, IEEE, AND R. GOYAL

Abstract — A two-stage feedback amplifier with direct connection between two stages has been developed for the 0.1–12-GHz frequency band. The measured gain is 10 ± 1 dB with 2.5:1 input and 1.7:1 output VSWR. The measured minimum power output at 1-dB gain compression and maximum noise figure are 14 dBm and 9 dB, respectively, from 2 to 12 GHz. The chip size is less than 0.5×1.0 mm^2 and this area contains complete RF and bias circuitry. The amplifier also provides more than 25 dB of AGC capability.

I. INTRODUCTION

WIDE-BAND microwave amplifiers are important building blocks for modern electronic countermeasures and surveillance systems. Generally, two main circuit approaches, the traveling-wave amplifier [1], [2] and the feedback amplifier [3], [4], have been used to realize wide bandwidths from monolithic microwave integrated circuits (MMIC). For bandwidths greater than 10 GHz (two decades), both these techniques result in small gain (4–6 dB) per amplifier stage. To implement an amplifier with more than 10-dB gain over a 10-GHz bandwidth, two or more amplifier stages are required which results in a larger chip size. However, small chip size is important to ensure MMIC amplifier manufacturability with high fabrication yield and low cost.

This paper describes a two-stage feedback amplifier in which a high degree of miniaturization is achieved by using a novel direct-coupling scheme. The direct coupling of two stages eliminates the need for an interstage blocking capacitor and the associated extra bias circuitry. The complete matched amplifier is realized in a 0.5×1.0 mm^2 area resulting in an array of more than 3500 potential chips on a 2-in-diam wafer. The MMIC amplifier has 10 ± 1 dB measured gain, which is the highest gain per unit chip area reported to data for an amplifier of this bandwidth. The small amplifier size also allows packaging in conventional microwave transistor packages. The amplifier described here is a general purpose gain block which can also be used for AGC, limiting, and switching applications.

The following sections describe the circuit design, fabrication, packaging, and performance of this MMIC amplifier.

II. CIRCUIT DESIGN

The objective of this work was to develop a small size, low-cost, high-yield GaAs MMIC amplifier with at least 10-GHz bandwidth and 10-dB gain. Resistive feedback

Manuscript received May 7, 1984.
The authors are with Avantek, Inc., Santa Clara, CA 95051.

circuit topology was chosen to realize this objective because of its demonstrated tolerance to variations in device parameters [4]. Additionally, feedback amplifiers are well suited for monolithic realization since many device and circuit parasitics can be minimized in monolithic form. Another advantage of feedback amplifiers is their improved low-frequency stability. Feedback topology, along with direct-coupled stages, allows significant reduction in the chip area.

The basic amplifier circuit has two common-source FET stages with both series and shunt feedback. Two versions of the amplifier were realized. One version has single- and one has dual-power supply requirements. The single-power supply design, labeled $A1$, covers 2 to 12 GHz, whereas the dual-power supply design, labeled $A2$, covers 0.1 to 12 GHz and requires an external bypass capacitor.

A. DC Circuit

The circuit schematics for the two versions are shown in Fig. 1. Each amplifier stage has gate-to-drain resistive feedback which helps reduce input and output VSWR. The direct coupling of the two stages allows the drain current of the first FET to be supplied through the second-stage, feedback resistor. Thus, the drain current to both FET's is supplied through the output port. The difference between the one- and two-power supply designs is that $RS1 = 0$ Ω for the two-power supply design and the dc operating point of the first stage is set by adding negative gate bias. For the dual-power supply design, the bypass capacitor C_5 is not on-chip and is added externally. The $A2$ design provides flat amplifier gain down to 100 MHz since C_5 is the main component limiting the low-frequency gain response. For both amplifier versions, the V_{DD} and V_{GG} bias is applied through external bias tees or through off-chip chokes.

The selection of dc-bias resistors is done based on the following requirements. The first FET is generally biased at 0.15–0.20 I_{DSS} to achieve a lower amplifier noise figure. This is done by proper selection of $RS1$. The second FET is generally biased at 0.50 I_{DSS} to achieve maximum power output. This is realized by proper choice of $RF2$ and $RS2$. The increase in $RS2$ and $RF2$ decreases the second FET drain current I_{DS2}. As $RS2$ increases, V_{DS2} also increases until it reaches a maximum and decreases beyond that point. This is the optimum bias point when maximum V_{DS2} is desired. The selection of $RF2$ and $RS2$ also depends on the RF circuit requirements, i.e., gain, VSWR, etc. These will be discussed later. The detailed simulation of the bias

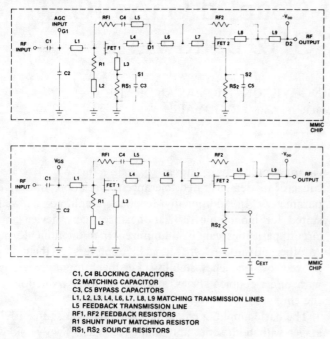

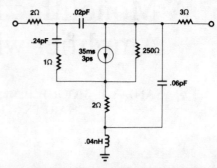

Fig. 2. A single-gate GaAs FET model for a 250-μm gatewidth device.

Fig. 1. Schematic of (a) single-power supply and (b) dual-power supply two-stage amplifier.

C1, C4 BLOCKING CAPACITORS
C2 MATCHING CAPACITOR
C3, C5 BYPASS CAPACITORS
L1, L2, L3, L4, L6, L7, L8, L9 MATCHING TRANSMISSION LINES
L5 FEEDBACK TRANSMISSION LINE
RF1, RF2 FEEDBACK RESISTORS
R1 SHUNT INPUT MATCHING RESISTOR
RS1, RS2 SOURCE RESISTORS

circuit was done using the SPICE JFET model. In these designs, the typical power supply requirements are +8 V and 55 mA for the one-supply design, and +8 V, −1.2 V, and 55 mA for the two-supply design.

Another advantage of this direct-coupling scheme is a decreased sensitivity of amplifier gain to variations in FET parameters. This invariance is due to the fact that changes in device I_{DSS} cause a change in I_{DS1}. However, since an increase in I_{DS1} causes a decrease in I_{DS2}, the net change in the amplifier gain due to a change in the FET bias point is small.

There are also a number of disadvantages of this biasing scheme. One limitation is that the first FET gatewidth cannot be too large, since that would result in higher I_{DS1} and consequently a large voltage drop across $RF2$. This requires V_{DD} to be very high. In this design with both FET's being 500 μm wide and with a V_{DD} of 8 V, the low I_{DS1} requirement is not a serious problem. Another limitation is the reduced amplifier RF power output capability. The amplifier RF power output cannot be increased without increasing the power dissipation in resistors $RS2$ and $RF2$. Therefore, this technique is not suited for high-efficiency applications. Additionally, the bias changes in the first FET due to gain compression adversely affect the gain compression characteristics of the second FET, resulting in an overall lower power output at 1-dB gain compression. In some cases, if the positive supply voltage is not large enough, the power output may be limited by the V_{DS} of the first FET. In this regard, the two-power supply design has slightly better power output capability since it allows higher V_{DS1}.

B. RF Circuit

The requirements for the RF circuit of this amplifier are similar to those in conventional feedback amplifiers with the exception that the dc-bias requirement, as discussed earlier, imposed a limitation on the range of FET gatewidth and feedback resistor $RF2$. For RF circuit analysis, an in-house computer analysis and optimization program (AMCAP) was used extensively in this design.

The equivalent circuit model of the FET was determined from S-parameter measurements taken on a number of 250-μm gatewidth devices from several wafers. Fig. 2 shows the average model parameter values for an FET biased at $I_{DS} = 0.2$ to 0.4 I_{DSS}. For a device with different gatewidths, these model values are scaled accordingly.

To optimize the performance of a high-frequency feedback amplifier, proper selection of the FET parameters is important. A number of papers have provided detailed analyses of GaAs FET feedback amplifiers [5]–[7]. To realize high forward gain, large FET transconductance is required. This implies large FET gatewidth for a given doping and gate length. Large FET gatewidth, however, is accompanied by larger Cgs and Cgd which tend to degrade the open loop gain at higher frequencies. Thus, there is an optimum gatewidth for a given gain, bandwidth, and VSWR.

The shunt feedback resistors $RF1$ and $RF2$ determine the maximum available gain and VSWR. If the feedback resistor is too large (small amount of feedback), the VSWR is generally poor. If the feedback resistor is too small, the amplifier gain is reduced [7]. In high-frequency amplifiers, the gain per stage is not very large. It is, therefore, important to realize the maximum possible gain with acceptable values of VSWR. Thus, there is an optimum value of $RF1$ and $RF2$ determined by the gain and VSWR requirements.

Based on the above discussion, the two FET's were chosen to have 500-μm gatewidth, which is close to optimum for 12-GHz operation [8]. This choice of FET's also agrees with the dc-bias requirement of low I_{DS1} for the direct-coupling scheme. In practice, this direct-coupling scheme is not ideal for feedback amplifiers below 2 GHz, due to the large FET gatewidth required [6]. $RF2$ was chosen to be 250 Ω to meet the low-output VSWR and high-gain requirements. A 2:1 output VSWR is realized easily without any additional matching elements. A low-input VSWR is difficult to achieve with feedback only. However, combined with a shunt $R-L$ network and a low-pass input matching network, the amplifier can achieve an input VSWR of better than 2:1. Without the input shunt $R-L$ network, the input VSWR at 6 GHz would be

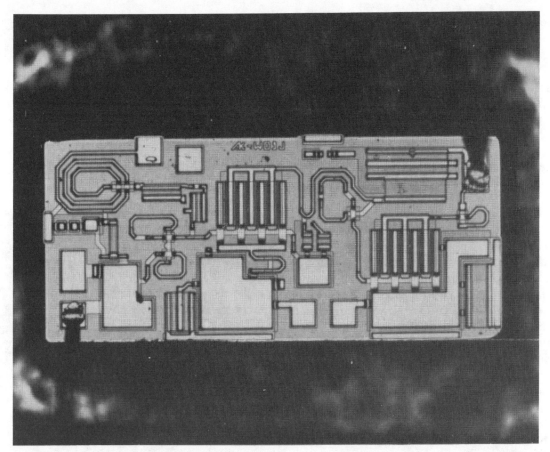

Fig. 3. A SEM photograph of the amplifier chip.

unacceptable (> 2:1). This frequency-dependent loss reduces the gain at 12-GHz only slightly (1 dB), but improves the VSWR at low frequencies significantly. The interstage matching network was optimized for maximum gain at 12 GHz. Overall gain of 10 ± 1 dB is obtained across 2 to 12 GHz.

An important feature of this circuit is its unconditional stability at all frequencies. This is the result of shunt and series feedback resistors which degrade the low-frequency gain significantly. No oscillations are observed, even if the input and output ports are not terminated with 50-Ω loads. This is important during wafer-level dc testing and screening of these IC's.

The RF circuit for the one- and the two-power supply designs are almost identical. The wider bandwidth of 0.1 to 12 GHz with 10 ± 1 dB gain is realized in the two-power supply design by bonding in an external capacitor C_5 of 50 pF. The bond wire inductance should be small to minimize gain degradation at 12 GHz. The power output performance of the two-power supply design is expected to be higher than the single-supply design due to higher V_{DS1}. The two-power supply design has the same chip area as the single-supply design, although it could be made smaller due to absence of on-chip bypass capacitors.

Fig. 3 shows the layout of the two-stage MMIC amplifier $A1$. The wraparound grounding technique allows easy access to RF ground on all four sides of the chip, minimizing the inductance to ground and the area used for grounding.

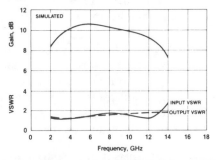

Fig. 4. Simulated gain and VSWR of the amplifier.

This chip has four different ground paths. In spite of the compact layout, care was taken to avoid coupling between certain critical circuit components and modes. Use of spiral inductors helped reduce the chip size. Inductors were modeled as high-impedance transmission lines that were fit to measurements taken on discrete inductors. [4].

Fig. 4 shows the computer-simulated results for the gain and VSWR of the MMIC amplifier. The gain at 2 GHz is limited mainly by the size of the source bypass capacitors ($C3$ and $C5$, being about 6 pF each). A larger source-capacitance value raises the gain at 2 GHz. Higher gain amplifier blocks are realized by directly cascading the chips. An input blocking capacitor of 4 pF is included for this purpose. This MMIC amplifier can also be used for AGC, limiting, and switching applications. By applying a negative voltage to the input of the amplifier, AGC operation with 25 dB of gain control is obtained.

III. Fabrication

Ion-implanted GaAs is used as the starting material for the IC's due to its excellent uniformity and controllability. After implantation, the wafers are annealed at 800 °C upon which the sheet resistance drops to approximately 500 Ω/square. This layer is then selectively etched to form mesas for the FET's and resistors. Later, the resistor mesas are etched to trim the sheet resistance to 800 Ω/square. In practice, this technique yields standard deviations in resistance of less than 15 percent.

The FET's in the IC are fabricated with the same process used for discrete FET's. The gates are formed on a nominally 0.5-μm-long base of TiW/Au which is gold plated to 0.7 μm. The resulting structure combines very short gate-length with large gate cross-sectional area for high device transconductance with low parasitics [9]. Source and drain ohmic contacts are formed with an AuGe/Ni alloy. Parallel-plate dielectric capacitors are fabricated with plasma-enhanced CVD silicon nitride and provide 390 pF/mm^2 of area. This procedure is also well controlled and achieves standard deviations of less than 40 pF/mm^2.

Metallic interconnections are achieved with a two-level wiring process which provides surface connections, crossovers, and air bridges. The top level is situated 3 μm above the GaAs surface and gold plated to 1.5 μm. The bottom level rests directly on a semi-insulating GaAs substrate and is 0.8 μm thick. These lines are normally designed for a current density not to exceed 5×10^5 amps/cm^2.

Wraparound ground technology is chosen over via-hole ground technology for low parasitic inductance and improved area efficiency. To form the wraparound, metal is first deposited and gold plated to 2 μm on the frontside of the chips. The wafer is then lapped to 115 μm and backside metallized to complete the wraparound ground connection.

IV. Measured Results

All the results presented in this section are for the single-power supply design, except when specified otherwise. The dual-power supply results are presented for wider bandwidth and higher power output.

A. DC Characteristics

Functional dice are selected by extensive automated dc wafer probing followed by visual inspection. Defective dice are indentified by measuring the drain current, transconductance, and pinchoff voltage of the individual transistors, as well as the values of the resistors (except the resistor $RF1$). Fig. 5 shows a histogram of the FET pinchoff voltage distribution. The average V_p and standard deviation for FET1 are -1.98 and 0.21 V, respectively. These values are similar to those reported previously [10]. This high degree of uniformity is due to the use of ion-implanted GaAs substrates. FET uniformity is very important in realizing high amplifier RF yield.

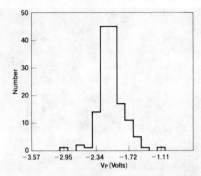

Fig. 5. Histogram showing V_p distribution of ion-implanted FET's.

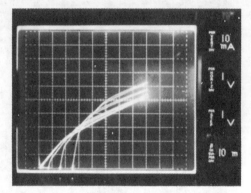

Fig. 6. Two-stage MMIC amplifier I/V characteristics.

The dc functional dice are then mounted on a metallized alumina test carriers which are via-hole grounded at the center and have 50-Ω input and output microstrip lines on each side. When the power supply is ramped from 0 to V_{DD}, a properly biased amplifier will exhibit the dc $I-V$ characteristic shown in Fig. 6.

For $V_{GG} = 0$ V, the dc supply current increases from zero at a rate of about 10 mA/V until the dc voltage supply approaches $+2.0$ V. Above 2.0 V, the dc current increases at about 5 mA/V due to the initiation of self-biasing of the second FET.

This two-section $I-V$ curve is the characteristic signature of a properly biased amplifier. The normal dc operating bias is $+8$V and 55 mA. Fig. 5 also shows the $I-V$ curves for negative V_{GG}.

B. Amplifier Performance

Fig. 7 shows the measured gain and VSWR of the MMIC amplifier. The measurements are in good agreement with the simulated results showing 10 ± 1-dB gain across the 2–12-GHz band. The gain flatness can be improved by using larger bypass capacitors $C3$ and $C5$. The maximum input VSWR is 2.5:1 across the band. This is slightly higher than the design goal of 2:1. This discrepancy is due to the FET Cgs being actually smaller than the modeled value. The output VSWR is better than 1.7:1 across the band.

The noise-figure performance of the amplifier is shown in Fig. 8. The noise-figure modeling was performed using an empirical FET noise model reported previously [11]. The measured noise figure is typically 6 dB with a maxi-

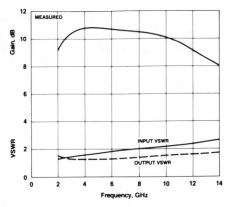

Fig. 7. Measured gain and VSWR of the amplifier.

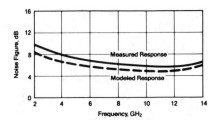

Fig. 8. Measured and modeled noise figure of the amplifier chip.

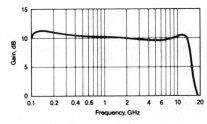

Fig. 9. Measured gain of 0.1 to 12-GHz amplifier.

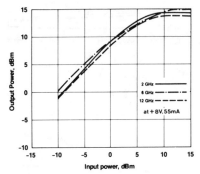

Fig. 10. P_{out} versus P_{in} of the amplifier.

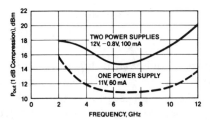

Fig. 11. P_{-1} versus frequency of the amplifier.

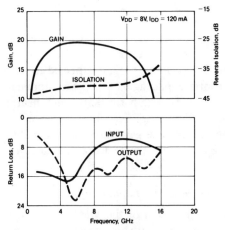

Fig. 12. Gain and return loss of two MMIC chips cascaded directly.

mum of 9 dB at 2 GHz. The measured noise figure is slightly higher than the modeled noise figure. This may be due to line loss in the input and interstage transmission lines being higher than originally modeled. The noise figure at 2 GHz is higher than that at 12 GHz due to extra loss in the input shunt $R-L$ network ($R1$ in Fig. 1) at lower frequencies.

Fig. 9 shows the measured gain of the $A2$ amplifier with a 50-pF external bypass capacitor. Gain of 10 ± 1 dB is obtained across the 0.1 to 12-GHz band. Four bond wires are used to minimize the source inductance from FET2 to the external capacitor. Still, this inductance degrades the gain by about 0.4 dB at 12 GHz. The VSWR for this amplifier is similar to that of Fig. 6 since the VSWR across 0.1 to 2-GHz band is always better than 1.7:1.

A plot of the output power versus input power of the $A1$ amplifier is shown in Fig. 10. At the nominal bias voltage of +8V, a saturated power output of +13 dBm is obtained across the band. At +12 V, a saturated power output of 15 dBm is obtained. Fig. 11 shows a plot of power output at 1 dB gain compression, P_{-1} versus frequency for $A1$ and $A2$ amplifiers. For the $A1$ amplifier, a minimum P_{-1} of 11 dBm is obtained. P_{-1} is less at 8 GHz than at 12 GHz, indicating that the output and the interstage match is not optimized for the best amplifier large-signal performance.

The bias for the FET's were optimized for maximum power output by adjusting $RS1$ and $RS2$ with external resistors. To prevent FET1 from going into the nonlinear region, the I_{DS1} and V_{DS1} should be large. This requires a larger voltage drop across $RF2$ and therefore larger V_{DD}. The $A2$ amplifiers have larger V_{DS1} for a fixed V_{DD} since $RS1 = 0 \ \Omega$. This is part of the reason why P_{-1} for the $A2$ is higher than that for the $A1$ amplifier. Another reason is that, for the same V_{DS1} and I_{DS1}, V_{D1} is lower on the $A2$ amplifier. This allows I_{DS2} to be adjusted to about 0.5 I_{DSS} rather than close to I_{DSS}. Thus, the dual-power supply design has the two main advantages of wider bandwidth and higher power output.

Higher gain amplifiers were realized by directly cascading two MMIC chips. Fig. 12 shows the performance of

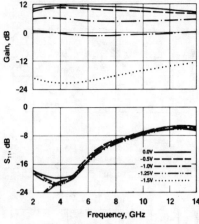

Fig. 13. AGC response of the amplifier.

this four-stage amplifier (two chips). A gain of 18 ± 1.5 dB is obtained across the 2–12-GHz band, indicating that the gain flatness does not degrade significantly. Input and output return loss are better than 6 and 8 dB, respectively.

Since the MMIC chip is less than 0.5×1.0 mm^2, it can fit easily into a variety of microwave transistor packages. These include the standard 70-mil flange, 70-mil stripline, and 100-mil flange package. The amplifier VSWR and gain flatness degrades slightly if the package leads do not have 50-Ω characteristic impedances. With improved gain flatness and VSWR, this is not expected to be a problem. The appropriate package for a given application is determined by the cost, hermeticity, and heat dissipation requirements. The amplifier performance was also tested over temperature. The gain decreases typically at the rate of 0.02 dB/°C over the temperature range of $-55°$ to $75°$C.

C. AGC Switching and Limiting Applications

The AGC operation of the MMIC amplifier is realized by applying either positive or negative voltage to the gate of the first FET. With positive V_{GG}, I_{DS1} increases causing a decrease in V_{DS1}. As V_{DS1} drops below 1.5 V, the amplifier gain starts to drop significantly. When negative V_{GG} is applied, the g_m of the first FET drops, causing a reduction in the amplifier gain. Of the two techniques, applying negative V_{GG} maintains better gain flatness across the 2–12-GHz band. With positive V_{GG}, the AGC response shows steeper slope with less gain decrease at higher frequencies. Fig. 13 shows the AGC performance of the amplifier. More than 25 dB of gain variation is realized without affecting the input VSWR significantly.

As a limiting amplifier, power output of up to 15 dBm can be realized. For lower power output, AGC response with positive or negative V_{GG} can be used, in which case the power output is limited by the first FET.

A broad-band single-pole single throw switch is realized by applying large (>2 V) negative V_{GG}. About 30 dB of on/off ratio can be obtained. For even greater (more than 40 dB) on/off ratios, V_{DD} can be turned off to 0 V. Even in this case, the input VSWR does not change significantly.

V. Conclusions

Design and performance of a two-stage, resistive feedback, direct-coupled amplifier is described. This amplifier provides 10 ± 1-dB gain across the 0.1 to 12-GHz band. The size of this MMIC amplifier chip is reduced to below 0.5×1.0 mm^2 by utilizing a novel direct-coupled interstage bias scheme. The resulting large number of die per wafer reduce the per-chip cost. The chip amplifier also provides AGC, switching, and limiting operation and allows multistage cascading and is compatible with conventional microwave transistor packaging.

Acknowledgment

The authors would like to acknowledge the help of V. Hersey for assembly and testing, N. Law and D. Lagman for processing assistance, and R. M. Malbon for helpful advice throughout this project.

References

[1] Y. Ayasli, R. L. Mozzi, J. L. Vorhaus, L. D. Reynolds, and R. A. Pucel, "A monolithic GaAs 1–13 GHz travelling-wave amplifier," *IEEE Trans. Microwave Theory Tech.*, vol. MTT-30, pp. 976–981, July 1982.

[2] E. W. Strid and K. R. Gleason, "A dc-12 GHz monolithic GaAs FET distributed amplifier," *IEEE Trans. Microwave Theory Tech.*, vol. MTT-30, pp. 969–975, July 1982.

[3] D. P. Hornbuckle and R. L. Van Tuyl, "Monolithic GaAs direct-coupled amplifiers," *IEEE Trans. Electron Devices*, vol. ED-28, pp. 175–182, Feb. 1981.

[4] S. Moghe, T. Andrade, G. Policky, and C. Huang, "A wideband two stage miniature amplifier," *IEEE 1983 GaAs IC Symp. Dig. Papers*, pp. 7–10.

[5] K. B. Niclas, W. T. Wilser, R. B. Gold, and W. R. Hitchens, "Matched feedback amplifier—Ultrawide-band microwave amplification with GaAs MESFET's," *IEEE Trans. Microwave Theory Tech.*, vol. MTT-28, pp. 285–294, Apr. 1980.

[6] W. C. Petersen, A. Gupta, and D. R. Decker, "A monolithic GaAs dc to 2 GHz feedback amplifier," *IEEE Trans. Electron Devices*, vol. ED-30, pp. 27–29, Jan. 1983.

[7] A. M. Pavio and S. D. McCarter, "Network theory and modeling method aids design of a 6–18 GHz monolithic multi-stage feedback amplifier," *Microwave Syst. News*, pp. 78–93, Dec. 1982.

[8] P. N. Rigby, J. R. Suffolk, and R. S. Pengelly, "Broadband monolithic low noise feedback amplifiers," in *IEEE 1983 Microwave and Millimeter-wave Monolithic Circuits Symp. Dig. Papers*, pp. 71–75.

[9] C. Huang, A. Herbig, and R. Anderson, "Sub-half micron GaAs FET's for applications through K-band," in *IEEE 1981 Microwave Symp. Dig. Papers*, pp. 25–27.

[10] T. Sugiura, H. Itah, T. Tsuji, and K. Honjo, "12 GHz band low noise GaAs monolithic amplifiers," *IEEE Trans. Electron Devices*, vol. ED-30, pp. 1861–1866, Dec. 1983.

[11] P. W. Chye and C. Huang, "Quarter-micron low noise FET's," *IEEE Electron Dev. Lett.*, vol. EDL-3, pp. 401–403, Dec. 1983.

A Monolithic Wide-Band GaAs IC Amplifier

DONALD B. ESTREICH, MEMBER, IEEE

Abstract—The design and performance of a general purpose, monolithic, wide-band GaAs IC amplifier is described. This amplifier features a high-voltage gain (26 dB), wide bandwidth (5 MHz to 3.3 GHz), and very low input VSWR (less than 1.3:1). No matching components are used on-chip, allowing for a small chip size of $\frac{1}{4}$ mm^2. The input stage consists of a 248 μm MESFET in a common-gate configuration with a noise figure under 10 dB (f = 1 GHz) with a 50 Ω source resistance. Noise figure limitations of both common-source and common-gate MESFET stages are discussed in detail. The amplifier uses 1 μm gate length MESFET's, GaAs Schottky diodes for level shifting, thin-film silicon nitride capacitors for ac coupling, and GaAs implanted resistors. The high gain and wide bandwidth make this amplifier useful for many signal processing and instrument/measurement applications.

I. INTRODUCTION

WIDE-BAND amplifiers are a commonly used component in most RF and microwave instrumentation and communication systems. Gallium arsenide is attractive for such amplifiers because of the high intrinsic performance available in Schottky-barrier gate field-effect transistors (FET's). The monolithic approach in amplifier design has the advantages of improved producibility and reliability, low cost, and high performance [1].

The design and performance of a general purpose, wideband, monolithic GaAs IC amplifier is described below [2]. There are numerous instrument applications for high-gain amplifiers in the 10 MHz to 3 GHz band. Furthermore, many of these applications work with sufficiently large signal levels so that very low noise figures are unnecessary (in contrast, communication system front ends usually put special emphasis on the best achievable noise figure). Designing a general purpose amplifier is a formidable task when considering the wide range of requirements which exist for different applications. Optimization for any particular application is difficult. The amplifier described below is an attempt to produce a gain block which would fit into a number of instrument assignments.

II. CIRCUIT DESIGN CONSIDERATIONS

The design of the amplifier is described in this section. Table I presents the principal design specifications and features which governed the design of the amplifier. The general purpose nature of the amplifier puts special importance on achieving a good input and output match to 50 Ω, the use of standard 5 V power supplies, low dc power dissipation, high RF output power, and a small chip size.

The configuration of the amplifier is as shown in Fig. 1, where three stages of gain are cascaded to an output driver

Manuscript received April 23, 1982; revised June 28, 1982.
The author is with the Hewlett-Packard Company, Santa Rosa, CA 95401.

TABLE I
AMPLIFIER SPECIFICATIONS AND FEATURES

PARAMETER	PERFORMANCE
Midband Voltage Gain (into 50 ohms)	$\geq$ 25 dB
Bandwidth	10 MHz to 3 GHz
Noise Figure	< 10 dB
Input VSWR	< 1.5 : 1
DC Power Dissipation	< 1 Watt

AMPLIFIER FEATURES:
(1) AC Coupled
(2) 650 μm by 440 μm Chip Size
(3) 5 Volt Power Supplies (2)
(4) Bias Compensated for Pinch-off Voltage Variation
(5) No Adjustments Needed for Operation

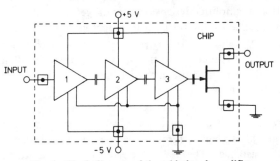

Fig. 1. Block diagram of the wide-band amplifier.

MESFET. The design of each block in Fig. 1 is discussed in the remainder of this section.

A. Input Stage Design

The design of the input stage is particularly important in the design of an amplifier because it controls the input matching characteristics and determines the noise figure. It is desirable to have high gain associated with the input stage to minimize the next stage's contribution to the noise figure. For this reason, a minimum voltage gain of 10 dB was required of the input stage.

Both the common-source and common-gate FET configurations are capable of greater than 10 dB voltage gain for the frequency range in Table I. Therefore, the choice of either common-source or common-gate will depend upon other factors such as noise figure, cell size, and impedance matching characteristics.

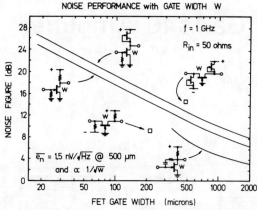

Fig. 2. Calculated noise figure versus FET gate width for five different input stages.

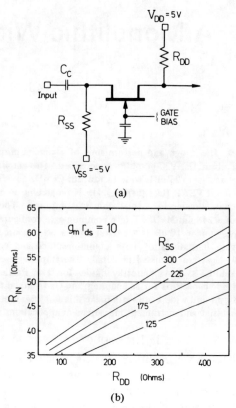

Fig. 3. (a) Common-gate, resistively loaded stage schematic, and (b) input resistance as a function of the load resistance R_{DD} and source resistance R_{SS}.

The input stage's noise performance merits special discussion. Fig. 2 compares the noise figure behavior of five different candidate input stages as a function of FET width (denoted here by W). This plot was constructed using the following rules. 1) A wideband 50 Ω input resistance was required, 2) the MESFET equivalent input noise voltage (denoted by $\overline{e_n}$) was assumed to be 1.5 nV/Hz$^{1/2}$ for $W = 500$ μm, 3) $\overline{e_n}$ was inversely proportional to the square root of W, 4) the source resistance R_S was 50 Ω, and 5) active loads had an assumed gate width of $W/2$ and passive loads had an assumed value of $r_{ds}/2$ where r_{ds} is the nominal output resistance of the amplifying FET. Experimental data [2], [3] from several different amplifier designs confirm the general validity of Fig. 2. The assumed value for $\overline{e_n}$ is the upper value for typically measured $\overline{e_n}$ values of 1-1.5 nV/Hz$^{1/2}$ for MESFET's fabricated with the IC process [4], and the scaling rule width W has been arrived at from noise measurements on FET's of different widths.

Consider first the common-source stages in Fig. 2. Either a feedback resistor from drain to gate or a 50 Ω shunt resistor from gate to source (ground) can be used to realize a 50 Ω input resistance. The use of the gate-to-source shunting resistor degrades the noise figure from 3 to 6 dB, being close to 6 dB for small gate widths ($W < 500$ μm) and close to 3 dB for large gate widths ($W > 5000$ μm). In general, it is obvious that the noise figure decreases as W increases because $\overline{e_n}$ decreases as $W^{-1/2}$. Active loads are noisier than resistive loads. Although not shown, this is also true for the resistive feedback configuration. The lowest noise figures are achieved with the drain-to-gate feedback. However, larger FET's are required and the blocking capacitor is large if low frequencies must be included, thereby rendering the resistive feedback configuration area-inefficient. Therefore, the resistive feedback stage is not compatible with the area requirement in Table I.

Consider next the common-gate stages shown in Fig. 2. Both the resistive-loaded and active-loaded cases are shown subject to the above stated assumptions. Only a single gate width is shown for each stage because of the 50 Ω input resistance requirement. This results because the input resistance is largely determined by the reciprocal of the transconductance g_m and g_m is proportional to W. The common-gate stage with active load is noisier than the resistive-loaded stage because of the large noise contribution of the smaller active load FET and the positive feedback present in the common-gate configuration. In fact, 60 percent of the noise power at the output of the stage with active load is generated by the load FET, whereas only about 5 percent of the output noise of the stage with resistive load is from the load resistor itself (assuming thermal noise only).

The data in Fig. 2 show that the lowest noise figures are achieved with the common-source configuration. However, given the noise figure design requirement (Table I) of < 10 dB, the common-gate stage with resistive load is a viable candidate for the input stage. Clearly, a smaller FET can be used with the common-gate configuration and this is a distinct advantage when chip area is important.

Voltage gains exceeding 10 dB are easily achieved with the common-gate stage. The midband voltage gain, denoted G_v, is given by

$$G_v = \frac{g_m r_{ds} + 1}{1 + (r_{ds}/R_{DD})} \qquad (1)$$

where R_{DD} is the load resistor [see Fig. 3(a)] and the other symbols have been defined above. As an example, taking a 250 μm MESFET ($g_m \cong 25$ mS, $r_{ds} \cong 400$ Ω), (1) gives $G_v = 4.23$ (12.5 dB) with $R_{DD} = 250$ Ω, and $G_v = 4.71$ (13.5 dB) with $R_{DD} = 300$ Ω.

It has long been recognized [5] that the common-gate stage provides a good input match at the frequencies spanned by this amplifier (see Table I). The low-frequency input resis-

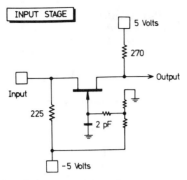

Fig. 4. Schematic of the common-gate input stage used in the wideband amplifier.

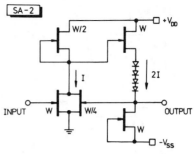

Fig. 5. Circuit schematic of a common-source gain stage which uses an active load and active feedback.

tance, denoted by R_{in}, is

$$R_{\text{in}} = \frac{(r_{ds} + R_{DD})R_{SS}}{(g_m r_{ds} + 1)R_{SS} + (r_{ds} + R_{DD})} \quad (2)$$

where R_{SS} and R_{DD} are defined in the circuit schematic of Fig. 3(a). The $1/g_m$ dependency dominates in determining R_{in}. Resistors R_{SS} and R_{DD} have a weaker influence on R_{in} as can be seen from Fig. 3(b). A 250 μm FET was assumed in the construction of Fig. 3(b). We see that the R_{DD} values used in the above voltage gain estimates are compatible with R_{in} being 50 Ω insofar that R_{SS} is about the same value as R_{DD} (this is convenient for setting the bias) and these are easily realized resistance values in implanted GaAs [4].

The midband noise figure F is estimated using

$$F = 1 + \frac{(\overline{e_n}^2)_0 + (\overline{e_{nss}}^2)_0 + (\overline{e_{ndd}}^2)_0}{(\overline{e_{ns}}^2)_0} \quad (3)$$

where

$$(\overline{e_n}^2)_0 = [g_m R_{DD}/D]^2 \, \overline{e_n}^2 \quad (4)$$

$$(\overline{e_{ns}}^2)_0 = [(g_m + g_{ds})(R_s \| R_{ss})(R_{DD}/R_s)/D]^2 \, \overline{e_{ns}}^2 \quad (5)$$

$$(\overline{e_{nss}}^2)_0 = [(g_m + g_{ds})(R_s \| R_{ss})(R_{DD}/R_{ss})/D]^2 \, \overline{e_{nss}}^2 \quad (6)$$

and

$$(\overline{e_{ndd}}^2)_0 = [((g_m + g_{ds})(R_s \| R_{ss}) + 1)/D]^2 \, \overline{e_{ndd}}^2 \quad (7)$$

are the mean-square noise voltages referred to the output of the first stage (Fig. 4). The source resistance is denoted by R_s, $g_{ds} = 1/r_{ds}$, and denominator D is

$$D = (g_m + g_{ds})(R_s \| R_{ss}) + (R_{DD} g_{ds}) + 1. \quad (8)$$

The equivalent input noise voltage of the FET is denoted by $\overline{e_n}$ and the thermal noise voltages (rms) associated with resistors R_S, R_{SS}, and R_{DD} are $\overline{e_{ns}}$, $\overline{e_{nss}}$, and $\overline{e_{ndd}}$, respectively.

The common-gate stage was chosen for the input stage primarily because of its smaller layout area in comparison to common-source stages with lower noise figures and its superior input matching capability. Fig. 4 shows the circuit schematic of the complete input stage. A 248 μm FET was used along with a nominal 270 Ω load resistor. The corresponding source bias resistor was 225 Ω. The resistors are implanted GaAs structures with a nominal 320 Ω/square sheet resistance [4]. At the quiescent point the electric field strength in the body of these resistors is approximately 450 V/cm and never exceeds 800 V/cm under maximum swing conditions. This is well below the velocity saturation field in GaAs. The approx-

imate quiescent point of the FET is I_{DS} = 12 mA and V_{DS} = 4 V. A 2 pF Si_3N_4 thin-film capacitor is used to RF-ground the gate. This capacitor is approximately ten times the FET's gate-source capacitance and hence, only a small fraction of the input signal is lost due to capacitive division. Finally, the common-gate stage has been found to have a high tolerance to burnout. Typically, burnout exceeds 32 dBm input drive.

B. Intermediate Stage Design

The second and third stages are common-source gain cells which use active loads and feedback for the required bandwidth. Fig. 5 shows a direct-coupled, common-source stage previously discussed by Van Tuyl and Hornbuckle [6], [7]. This stage, denoted as SA-2, has the required voltage gain and bandwidth needed for the intermediate gain cells. In the SA-2 the feedback is controlled by the width of the feedback FET, which is placed in parallel with the amplifying FET. A level-shift/buffer stage is included in the SA-2 to self-bias the feedback FET (and the input FET of a following stage in direct-coupled multistage amplifiers) and to increase the output current drive capability. One undesirable feature of the SA-2 is its high dc power dissipation in the level-shift/buffer network; it dissipates approximately 75–80 percent of the total dc power of the SA-2 because of the higher voltage and dc current (twice that of the amplifying section for the device widths indicated in Fig. 5).

By splitting the level-shift and buffer/driver functions the total power dissipation can be reduced. Consider the modified SA-2 stage shown in Fig. 6. Here the buffer/driver has only 45 percent of the dc power dissipation in the buffer/driver of the SA-2 (Fig. 5). Lower dc power dissipation is achieved in the output driver (note 1 of Fig. 6), principally because of the lower power supply voltage (typically $\frac{1}{2}$ that of the SA-2). For the level-shift stack (see note 2), power is conserved by operating at a much smaller current—approximately $\frac{1}{12}$ that of the level-shift/buffer in the SA-2. Four Schottky barrier diodes give a 3.4 V shift, thereby providing the bias to the feedback and amplifying FET's. Capacitor coupling allows this to be accomplished independently in the second and third stages.

Both dc and RF feedback are applied to the feedback FET as in the original SA-2. In addition, an 80 kΩ resistor provides dc feedback, but not RF feedback in the frequency band, to the amplifying FET (note 3 in Fig. 6). The inclusion of this resistor not only makes the entire modified SA-2 self-biasing, but also reduces the sensitivity of the bias to FET pinchoff voltage variations. It is important to hold the quiescent in-

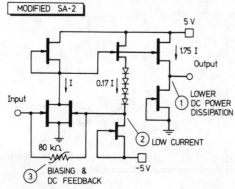

Fig. 6. Circuit schematic of the common-source amplifier in Fig. 5 modified to dissipate less dc power and to be less sensitive to pinchoff voltage changes. This circuit is used for the second and third stages.

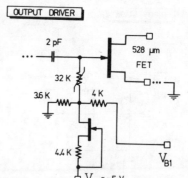

Fig. 7. Circuit schematic of the output driver of the wide-band amplifier.

ternal node voltage close to $\frac{1}{2}$ the positive supply voltage and the dual-path dc feedback gives an improvement here in desensitization to pinchoff voltage variations by a factor of six over the original SA-2. Computer simulations predict the quiescent internal node voltage (and quiescent output node voltage) to vary as −0.07 volt per volt increase in pinchoff voltage when the dual-path feedback scheme is used. Hence, the modified SA-2 offers improvements in providing for completely self-contained biasing, substantially better insensitivity to FET pinchoff voltage variations, and lower dc power dissipation, all without compromising the RF performance in terms of voltage gain and bandwidth. However, the reduced power supply voltage across the output driver does decrease the maximum output signal swing and requires tighter control on the quiescent point of the buffer/driver sections.

Holding the internal node voltage's quiescent level at midway between V_{DD} and ground also depends upon the level-shift diode stack producing a 3.4–3.5 voltage shift at its nominal operating current. Careful design of the Schottky barrier diode stack requires a reproducible barrier height and close attention to series resistance at the design stage.

The modified SA-2 cell in Fig. 6 has from 6 to 7 dB of voltage gain and a bandwidth exceeding 3 GHz. A comprehensive treatment of voltage gain, bandwidth, and distortion characteristics of the basic SA-2 gain cell has been given by Hornbuckle and Van Tuyl [7]. We refer the reader to these references for this information. Another reason for selecting the SA-2 type gain cell is that its gain response peaks slightly (approximately 1–1.5 dB) near 3 GHz [7] because of positive feedback from transit time delay and capacitive loading. This compensates for some falling off of the gain response of the first stage (which is perhaps 2 dB down at 3 GHz). Hence, this peaking provides for a flatter gain response across the bandwidth of the amplifier. It also helps to push out the frequency response of the amplifier.

C. Output Driver

A single common-source FET in an open-drain configuration is used at the output. This configuration was chosen because it can achieve a good wideband match [8] to a range of output impedances, including 50 Ω, and can provide voltage gain as well. It is desirable to choose an FET with as large a gate width as possible. A larger output FET is capable of driving smaller output resistances, allows greater current drive (which can tolerate higher output node parasitic capacitance), gives higher output saturated power, and can improve the harmonic distortion characteristics [7]. Of course, a practical limit on the size of the output driver is set by the drive capacity of the preceding stage and power dissipation limitations.

From the above considerations a 528 μm output driver FET was chosen. The voltage gain G is simply given by

$$G = \frac{g_m r_{ds} \cdot R_L}{r_{ds} + R_L} \tag{9}$$

where R_L is the load resistance connected to the drain node. Taking $g_m = 53$ mS and $r_{ds} = 185$ Ω as nominal values for the 528 μm FET, a load resistance of 50 Ω yields $G = 2.09$ (6.4 dB). For example, using an RF choke (inductor) to bias the drain of the FET and ac coupling to a 50 Ω transmission line allows this gain to be achieved. Of course, transmission lines of higher characteristic impedance could be used in this configuration to further increase the voltage gain. Another alternative is to bias the FET through a 69 Ω load resistor. This resistance in parallel with r_{ds} gives an output resistance of nominally 50 Ω. This connection allows for a good reverse termination to a 50 Ω transmission line. In this case the combination gives $G = 1.3$ (2.3 dB). Hence, by allowing the drain to be directly connected by the user provides many options with regard to the output driver's performance.

To realize lower dc power dissipation in the output driver, the gate-source voltage can be made more negative, resulting in a smaller drain current. Unfortunately, this also decreases g_m and hence, decreases the output driver's voltage gain. Fig. 7 shows the bias circuitry used to internally set the gate voltage of the driver FET. This network nominally sets the gate-source voltage at −1.4 V, which corresponds to a drain current of approximately 0.25 I_{DSS} (where I_{DSS} is the drain current for zero gate-source voltage in the saturated region of operation). The bias network partially compensates for pinchoff voltage variations and temperature changes. Note that an optional bias adjust (V_{B1}) has been provided for overriding the internally set gate-source voltage. This can be used to increase the overall amplifier voltage gain (by increasing g_m) or optimize the output harmonic performance.

D. Complete Amplifier

The complete ac-coupled amplifier schematic is shown in Fig. 8. All coupling capacitors between stages are 2 pF thin-film capacitors. Seventeen FET's are used in the design. Note that the output driver FET has a separate ground to reduce

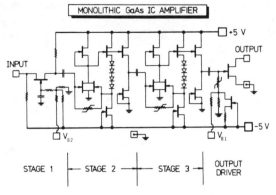

Fig. 8. Complete schematic of wide-band amplifier.

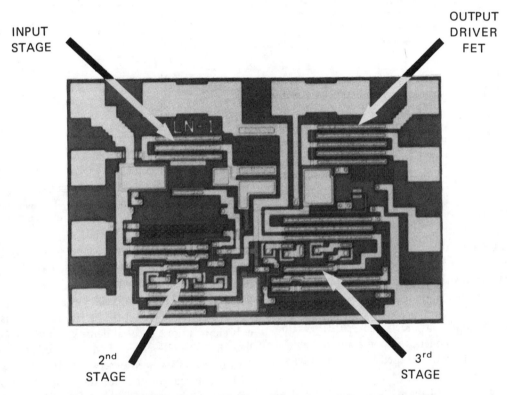

Fig. 9. Photomicrograph of the chip showing the layout of the wide-band amplifier.

common impedance problems. Two optional bias adjust pads are included for overriding the internal bias settings—V_{B1} for the output driver and V_{B2} for the input stage. Normally, these two bias adjust pads are left floating.

III. Amplifier Fabrication

The key features of the process [4] used for fabricating the amplifiers are

1) direct ion implantation into semi-insulating GaAs substrates,
2) planar proton isolation [9],
3) silicon nitride passivation and capacitor dielectric, and
4) a dual-level metallization system with polyimide interlayer dielectric.

The FET's are all 1 μm gate length devices with a recessed gate structure. Both resistors and FET channels are formed simultaneously—the nominal sheet resistance is 320 Ω/square. The proton isolation [9] provides excellent isolation between components and is necessary in the fabrication of pinch resistors for high resistance. The Schottky diodes used for level-shift applications are formed simultaneously with the recessed FET gates.

Fig. 9 shows a photomicrograph of the complete amplifier chip. It measures 440 $\times$ 650 μm. The location of each gain stage is noted in the photomicrograph. The bias resistors are not visible because of the planar nature of the process.

IV. Amplifier Performance

In this section the measured performance of the wideband amplifier is presented.

A. Voltage Gain and Bandwidth

Fig. 10 shows a typical measured voltage gain versus frequency plot. This data was taken without external bias adjustments (V_{B1} and V_{B2} floating). The midband voltage gain has a measured mean of 26 dB and a standard deviation of approxi-

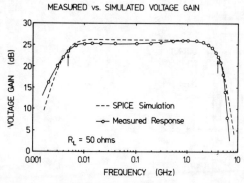

Fig. 10. Typical voltage gain versus frequency response of the wide-band amplifier. Arrows indicate 3 dB points on gain response.

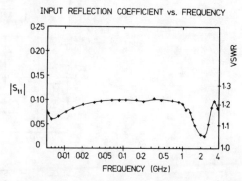

Fig. 12. Input reflection coefficient (and VSWR) versus frequency.

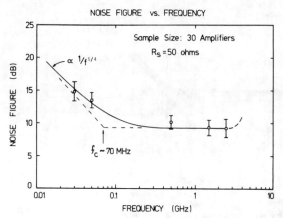

Fig. 11. Noise figure versus frequency for wide-band amplifier.

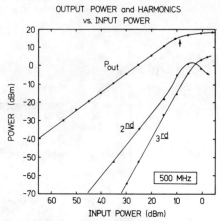

Fig. 13. Power output at fundamental and second and third harmonic versus input power. Arrow indicates 1 dB gain compression point.

mately 1.6 dB (based upon about 200 amplifier measurements). With V_{B1} optimized (positive voltage), an additional 3-4 dB increase in the midband voltage gain is achieved. This is primarily due to g_m improvement in the output driver.

The arrows in Fig. 10 indicate the −3 dB points on the response curve. These appear at 5 MHz and 3.30 GHz. Computer simulations of the frequency response are indicated by the dashed curve. The FET model used in the computer simulation is that reported by Van Tuyl et al. [10]. It was found that for accurate simulation of the frequency response, it was necessary to include both on-chip parasitic capacitances and inductances and probe (or wire bond) inductances. The simulation data shown here included 26 such parasitic components.

It might be mentioned that the gain versus frequency response is sensitive to common-terminal inductance in the V_{DD} and ground contacts. In fact, common-mode impedances at terminals are a well-known source of problems in high-gain amplifiers. Measurements on packaged amplifiers with excessive inductance from V_{DD} and ground pad wire bonds have been observed to push the −3 dB frequency out to approximately 5 GHz, with a peaking of as much as 4 or 5 dB at 4.6 GHz. Data reported here are for amplifiers with minimum practical packaging parasitics and with V_{B1} and V_{B2} floating.

B. Noise Figure

The expected noise figure performance of the common-gate FET was discussed in detail above. Fig. 11 shows measured noise figure data for several frequencies. The source resistance was 50 Ω for all noise measurements. At the higher frequencies the noise figure goal (cf. Table I) is achieved.

Especially noteworthy in Fig. 11 is the strong $1/f$ noise contribution to the noise figure. The inherently high $1/f$ noise levels in GaAs MESFET's have been previously noted by Ulrich [11], Pucel [12], and Archer et al. [13]. Measurements on discrete GaAs MESFET's fabricated using the IC process show a noise spectral density frequency dependence, below the $1/f$ noise corner frequency, of $1/f^\alpha$, where $\alpha \cong 5/4$. This behavior has been observed at frequencies as low as 5 Hz with no indication of a deviation from this slope. Assuming the same frequency dependence, a 70 MHz corner frequency is estimated as indicated in Fig. 11. However, even with this large amount of $1/f$ noise, it contributes less than four percent to the total integrated noise voltage fluctuation at the output of the amplifier.

C. Input Reflection Coefficient

As discussed above, the common-gate FET stage is expected to have an excellent input matching characteristic. This is verified by the data in Fig. 12 which plot the magnitude of the input reflection coefficient (and VSWR) versus frequency (shown from 5 MHz to 4 GHz). These data were taken using only the standard 5 V power supplies (V_{DD} and $-V_{SS}$) with V_{B2} floating (V_{B2} can often be adjusted for a better match). Although not shown, data have been taken to 10 GHz and the input VSWR remains below 1.5:1.

D. Power Output and Harmonics

Fig. 13 shows data of the fundamental output power, and second and third harmonic powers versus the input drive power. Typically, the 1 dB gain compression output power level is 15 dBm or greater. It has been measured to be as high as 18 dBm on some amplifiers.

Reference [7] gives an extensive discussion on GaAs FET distortion mechanisms and, in particular, describes the *SA-2* amplifier's distortion characteristics. Obviously, distortion in a multistage amplifier is dominated by component (usually FET) nonlinearities at those nodes where the largest signals appear. For the amplifier reported here the critical nodes are the connection between the buffer/driver of the third stage and the output driver FET and the drain node of the output driver FET. The principal FET nonlinearities are the transconductance, output resistance, and gate capacitance [7]. For example, distortion is generated when a nonlinear current source drives a nonlinear capacitance, e.g., a source follower driving a common-source output FET. The nonlinearity associated with the gate capacitance of the output driver FET results in an unsymmetrical voltage waveform (primarily second-harmonic distortion) which then feeds the nonlinear transconductance of the FET. The phase is such as to reinforce the second-harmonic distortion [7]. At the drain node of the output driver FET, the transconductance and output resistance nonlinearities dominate. The relative importance of the two nodes mentioned above in the distortion behavior depends upon the voltage gain of the output driver (and its loading). Low voltage gains tend to emphasize the distortion associated with the third stage's output node. As very high drive levels are approached, the second-harmonic distortion begins to decrease with an attendant increase in the third-harmonic component. This behavior is evident in Fig. 13. Saturation of the signal (symmetrical clipping) tends to produce a symmetrical waveform which is rich in third-harmonic distortion.

V. COMPARISON TO PREVIOUSLY REPORTED WIDE-BAND AMPLIFIERS

The amplifier described above is briefly compared to several recently reported GaAs monolithic wide-band amplifiers in Table II. The reader is of course aware that it is difficult to "directly" compare these amplifiers because of different design objectives and intended applications. However, this section is included as an aid to those who may not be familiar with the literature—the accompanying references can serve as a starting point.

Only amplifiers without on-chip matching components are included in Table II (actually, the Nippon Electric amplifier [8] is a minor exception insofar that a short 80 Ω transmission line interconnects the two stages). Table II lists the midband gain (dB), the bandwidth as defined by the 3 dB points, the maximum input VSWR over the 3 dB bandwidth, the minimum midband noise figure (dB), the gate width (μm) of the input FET as an aid to put the noise figure in perspective, and finally, the total chip area (mm^2).

With the exception of the amplifier reported here being common-gate, all amplifiers listed in Table II use common-source input stages. Furthermore, the Siemens [13], [14], Nippon Electric [8], Matsushita [15], [16], and Rockwell [18] amplifiers all use resistive shunt feedback [11] in their input

TABLE II
MONOLITHIC AMPLIFIER COMPARISON

REFERENCE	GAIN dB	BW MHz	INPUT VSWR	NF dB	GATE WIDTH μm	CHIP AREA mm^2
THIS WORK (HP) [2]	26	5-3300	<1.3	9	248	0.27
HEWLETT-PACKARD [3] - [7]	10.5 27	DC-5000 DC-1800	<1.4	16.5 19	96 40	0.2
SIEMENS [13] - [14]	24	930	<1.8	4.8	900	0.9
NIPPON ELECTRIC [8]	14	0.5-2800	<2.5	6	400	1.1
MATSUSHITA [a] [15] - [16]	10	50-2000	<2.5	1.6	1000	0.4
TOSHIBA [17]	27	DC-1400	high	1	1000	2.2
ROCKWELL [18]	11	DC-2000	<2.0	2	1200	2.2

[a]Also reported two-stage, 20 dB gain, 2 dB NF amplifier [16].

stages. This is clearly reflected in their lower noise figure performance; note also the greater gate widths reported for their input stage FET's. Furthermore, the low VSWR quoted for the two Hewlett-Packard common-source amplifiers [3], [7] use an external 50 Ω resistor to shunt the input (i.e., see Fig. 2 for simplified schematic of the common-source stage with active load).

Although not included in Table II, there are two other classes of amplifiers which should be mentioned. Distributed amplifiers have long been known [19] to be capable of producing high-gain wide-band performance. Ayasli *et al.* [20], [21] have reported a four FET (1 × 300 μm each) distributed amplifier with a nominal 9 dB gain up to approximately 14 GHz. Also, Strid *et al.* [22] recently reported on a distributed amplifier topology with a stage gain of 9 dB and a −3 dB bandwidth of 12 GHz (using 0.7 × 300 μm FET's). Of course, distributed amplifiers consume considerable chip area but can give extremely wide bandwidths.

Finally, a comment is in order on silicon, monolithic, wideband amplifiers [23], [24]. Recently, Kukielka and Snapp [24] reported on cascadable, feedback, silicon, bipolar gain blocks. They use a Darlington configuration and can be realized in a variety of gain/bandwidth combinations. For example, their *EO3* amplifier [24] gives 12.3 dB gain with a −3 dB bandwidth of approximately 3.2 GHz, 5.9 dB noise figure, and a 1 dB gain compression output power of +10 dBm. This is quite impressive performance compared to past results with silicon monolithic amplifiers.

VI. SUMMARY

The design and performance of a wide-band, monolithic GaAs IC amplifier has been described. The principal features of the amplifier are a 26 dB midband voltage gain, greater than 3 GHz bandwidth, a very low input VSWR of less than 1.3:1, and operation from standard 5 V power supplies. No on-chip matching components are used, allowing the amplifier to be placed on a 650 × 440 μm chip.

Acknowledgment

The author wishes to acknowledge the helpful suggestions and contributions given by R. Van Tuyl during the design phase of the amplifier. Also, V. Peterson and D. Hornbuckle aided in the amplifier characterization.

References

[1] R. A. Pucel, "Design considerations for monolithic microwave circuits," *IEEE Trans. Microwave Theory Tech.*, vol. MTT-29, pp. 513–534, June 1981.
[2] D. B. Estreich, "A wideband monolithic GaAs IC amplifier," in *ISSCC Dig. Tech. Papers*, 1982, pp. 194–195.
[3] D. P. Hornbuckle, "GaAs IC direct-coupled amplifiers," in *IEEE Int. Microwave Symp. Dig.*, May 1980, pp. 387–389.
[4] R. L. Van Tuyl et al., "A manufacturing process for analog and digital GaAs IC's," presented at the IEEE Gallium Arsenide Integrated Circuit Symposium, San Diego, CA, Oct. 1981; also in *IEEE Trans. Electron Devices*, vol. ED-29, pp. 1031–1038, July 1982.
[5] A. K. Gupta, J. A. Higgins, and D. R. Decker, "Progress in broad-band GaAs monolithic amplifiers," *IEDM Tech. Dig.*, pp. 269–272, Dec. 1979.
[6] R. L. Van Tuyl, "A monolithic integrated 4-GHz amplifier," *ISSCC Dig. Tech. Papers*, Feb. 1978, pp. 72–73.
[7] D. P. Hornbuckle and R. L. Van Tuyl, "Monolithic GaAs direct-coupled amplifiers," *IEEE Trans. Electron Devices*, vol. ED-28, pp. 175–182, Feb. 1981.
[8] K. Honjo, T. Sugiura, and H. Itoh, "Ultrabroadband GaAs monolithic amplifier," *Electron. Lett.*, vol. 17, pp. 927–928, Nov. 26, 1981.
[9] D. D'Avanzo, "Proton isolation for GaAs integrated circuits," presented at the IEEE Gallium Arsenide Integrated Circuit Symposium, San Diego, CA, Oct. 1981; also in *IEEE Trans. Electron Devices*, vol. ED-29, pp. 1051–1058, July 1982.
[10] R. L. Van Tuyl, D. Hornbuckle, and D. B. Estreich, "Computer modeling of monolithic GaAs IC's," in *IEEE Int. Microwave Symp. Dig.*, May 1980, pp. 393–394.
[11] E. Ulrich, "Use negative feedback to slash wideband VSWR," *Microwaves*, vol. 17, pp. 66–70, Oct. 1978.
[12] R. A. Pucel, "FET noise studies," U. S. Air Force, Office Sci. Res., Rep. S-2899, Mar. 1981.
[13] J. A. Archer, H. P. Weidlich, E. Pettenpaul, F. A. Petz, and J. Huber, "A GaAs monolithic low-noise broad-band amplifier," *IEEE J. Solid-State Circuits*, vol. SC-16, pp. 648–652, Dec. 1981.
[14] H. P. Weidlich et al., "A GaAs monolithic broadband amplifier," in *ISSCC Dig. Tech. Papers*, 1981, pp. 192–193.
[15] M. Nishiuma, S. Nambu, M. Hagio, and G. Kano, "A GaAs monolithic low-noise wideband amplifier," presented at the 1981 Int. Symp. GaAs Related Compounds, 1981.
[16] Anonymous, "GaAs UHF amp IC has low-noise, wide band, amplifier," *Electronics*, vol. 54, pp. 82–83, Feb. 24, 1981.
[17] S. Hori, K. Kamei, M. Tatematsu, T. Chigira, and H. Ishimura, "Direct-coupled GaAs monolithic IC amplifiers," in *IEEE 1982 Microwave Millimeter-Wave Monolithic Circuits Symp. Dig. Tech. Papers*, June 18, 1982, pp. 16–19.
[18] W. C. Peterson, A. K. Gupta, and D. R. Decker, "A monolithic GaAs dc to 2 GHz feedback amplifier," in *IEEE 1982 Microwave Millimeter-Wave Monolithic Circuits Symp. Dig. Tech. Papers*, June 18, 1982, pp. 20–22.
[19] E. L. Ginzton, W. R. Hewlett, J. H. Jasberg, and J. D. Noe, "Distributed amplification," *Proc. IRE*, vol. 36, pp. 956–969, Aug. 1948.
[20] Y. Ayasli, R. Mozzi, J. L. Vorhaus, L. Reynolds, and R. A. Pucel, "A monolithic GaAs 1 to 13 GHz traveling-wave amplifier," in *IEEE Gallium Arsenide Integrated Circuit Symp. Res. Abstracts*, Oct. 1981, paper 46
[21] Y. Ayasli, J. L. Vorhaus, R. Mozzi, and L. Reynolds, "Monolithic GaAs traveling-wave amplifier," *Electron. Lett.*, vol. 17, pp. 413–414, June 11, 1981.
[22] E. W. Strid, K. R. Gleason, and J. Addis, "A dc-12 GHz monolithic GaAs FET distributed amplifier," in *IEEE Gallium Arsenide Integrated Circuit Symp. Res. Abstracts*, Oct. 1981, paper 47.
[23] R. G. Meyer and R. A. Blauschild, "A 4-terminal wide-band monolithic amplifier," *IEEE J. Solid-State Circuits*, vol. SC-16, pp. 634–638, Dec. 1981.
[24] J. Kukielka and C. Snapp, "Wideband monolithic cascadable feedback amplifiers using silicon bipolar technology," late paper presented at the 1982 IEEE Microwave Millimeter-Wave Monolithic Circuits Symp., Dallas, TX, June 18, 1982.

INDUCTIVELY COUPLED PUSH-PULL AMPLIFIERS FOR LOW COST MONOLITHIC MICROWAVE ICs

S.A. Jamison, A. Podell[+], M. Helix, P. Ng[*] and C. Chao
Honeywell Corporate Technology Center, Bloomington, Minnesota 55420
and
G.E. Webber and R. Lokken
Honeywell Avionics Division, Minneapolis, Minnesota 55413

ABSTRACT

We report here the performance of inductively coupled gain stages near 4 GHz. Push-pull, interdigitated, common source FETs combined with transformers were used to design a two-stage balanced amplifier. The nearly 1:1 transformers used for coupling consist of two planar interwound spirals designed such that the secondary inductance matches the capacitive input of the FETs. The measured results were compared to those of a single push-pull FET to generate the gain of a transformer-FET combination. The cascaded gain is 9 dB at 4.4 GHz when the FETs are biased at 15% Idss for minimum noise. The chip area required for the two stage amplifier is only .5 X 1.0 mm, including current source FETs for biasing. This compact structure allows the realization of low cost, high performance MMICs for future generation systems.

INTRODUCTION

While it is generally accepted that MMICs offer the potential for cost reduction over their hybrid counterparts, this advantage cannot materialize without a large reduction in the amount of real estate devoted to each function. Monolithic circuits using microstrip techniques for coupling are inconsistent with this objective at frequencies below X-band. Lumped element designs previously reported are capacitively coupled requiring two spiral inductors for matching and biasing. Such capacitors are large in size and difficult to fabricate with high yield. In this paper we discuss an inductively-coupled two-stage amplifier (no on chip I/O match) employing an air-core spiral transformer. To our knowledge, this is the first time that a planar transformer has been extended to this frequency and realized monolithically on a GaAs substrate. The circuit design approach is balanced, with push-pull, common source FETs to minimize the RF source inductance and decrease packaging ground problems.

DESIGN

The circuit schematic of the inductively coupled amplifier is shown in Figure 1. The push-pull FETs (Fls) are interdigitated with 7 gate fingers each to reduce the effective gate metal resistance to $.86\Omega$. The common source stripes between gates implemented in layout decreases source inductance and source resistance. The current source F2 biases F1 to 15% Idss, hence only a single 5 volt supply is required. The gates of the first stage are DC grounded in the package. Power is supplied to the second stage through RF chokes connected to the drains leads.

The size of the spiral inductors used in the transformer design were based on $3\mu m$ line/space layout rules of the fabrication process. The secondary winding was increased to $6\mu m$ to reduce resistive losses and improve gain characteristics. Transformer designs were based on scale models tested at 43 MHz.

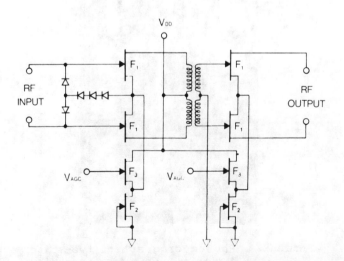

Figure 1. Circuit schematic for two-stage amplifier. FET gate widths are F1=490 micron, F2=170 micron and F3=220 micron.

Reprinted from *IEEE Gallium Arsenide Integrated Circuit Symp. Tech. Dig.*, 1982, pp. 91-93.

Coupling coefficient for 3μm spaces was expected to be K=.86, with this value falling off sharply for larger spaces, hence prohibiting large impedance transformation with continuous spirals. The nearly 1:1 transformers actually used do not allow a conjugate match between stages, however, sufficient gain is expected. The 1:1 design does greatly reduce potential resonance effects of inter-spiral coupling capacitance at or below the operating frequency. The 1.9nH inductance of the transformer secondary provided a match to the input capacitance of the second stage FET.

The current source FET (F3) provides AGC action by robbing the current from the F1s and forcing them into pinch-off.

FABRICATION

The circuits were fabricated using selective implantation into undoped LEC material. FET active channels were formed with a 360KeV Se implanted with the dose adjusted to give a V_T=-2.6 volts. An n^+ sulfer implant was used to improve the ohmic contacts and reduce FET source resistance. Ohmic contacts were formed using conventional AuGeNi metallization recessed into the wafer surface. The planar gate first level metallization was TiW/Au patterned using a dielectric assisted liftoff process. The interlevel dielectric was .5μm of silicon oxy-nitride deposited by plasma-enhanced CVD, and the second level metal was TiW/Au patterned using ion beam milling. The 1.5μm of Au used in this level resulted in a sheet resistance of .02ohms/□. A photograph of the two-stage amplifier (chip size: .5x1.0mm) is shown in Figure 2.

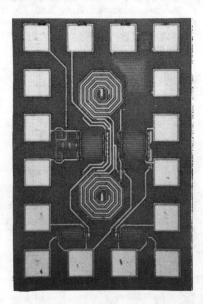

Figure 2. Photomicrograph of two-stage amplifier. Chip size is .5x1.0 mm.

RESULTS AND ANALYSIS

The FETs (F1) had a 1.2μm gate length, gm=30 mmhos, and a source resistance of 3.5 ohm. The measured transformer primary and secondary DC resistance was 22 ohms and 10 ohms, respectively. The expected AC resistance of the secondary is 16 ohms when skin and proximity effects are included.

The ICs were evaluated in the RF test fixture shown in Figure 3. The fixture incorporated wideband baluns and replaceable alumina chip carriers. No attempt was made to bypass the power supply lines at the IC.

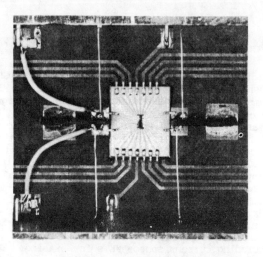

Figure 3. RF test fixture capable of testing balanced MMICs from 2 to 7 GHz.

The performance of the two-stage amplifier was compared to that of a single push-pull FET to generate the gain and isolation of a transformer-FET combination. The cascade S_{12} and S_{21} versus frequency is shown in Figure 4 with a peak gain of 9 dB at 4.4 GHz. The data was taken at V_{DD}=+5V, and the AGC input grounded. The cascade gain is 7.3 dB and -9 dB for V_{AGC} equal to 1.8 and 3.5 volts, respectively.

A signal analysis of the circuit was performed using SPICE with measured DC parameters as input. The simulation gave the observed frequency dependence but with 2.7 dB more gain. Since the RF resistance of the transformer windings was most in doubt, these values were modified to $R_{AC,p}$=55 ohms and $R_{AC,s}$=25 ohms to give the best fit. The potential discrepancy in transformer parameters is currently under investigation.

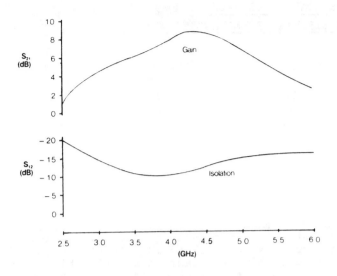

Figure 4. Cascade gain (S_{21}) and isolation (S_{12}) versus frequency for transformer-FET combination.

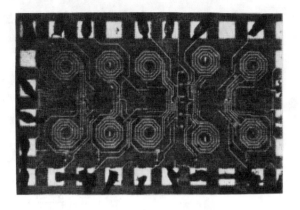

Figure 5. Photomicrograph of baseband receiver front-end utilizing the transformer-FET as a basic building block.

CONCLUSION

The reasonable add-on cascade gain and very compact size observed using balanced circuits with inductive coupling indicates that cost-effective sophisticated MMICs may be developed. Indeed, Figure 5 shows a complete baseband receiver front-end with three preamplifiers, double balanced mixer and two local oscillator buffer/amplifiers with a total IC size of 1.0x1.5 mm. This receiver IC is presently under evaluation.

ACKNOWLEDGEMENTS

We would like to acknowledge S. Hanka and M. Wilson for fabrication and packaging of the ICs.

+The author is with Podell Associates, Palo Alto, California.

*The author is no longer at Honeywell.

WIDEBAND MONOLITHIC CASCADABLE FEEDBACK AMPLIFIERS USING SILICON BIPOLAR TECHNOLOGY

By
Jose Kukielka
Craig Snapp

Avantek, Inc.
Microwave Semiconductor Division
3175 Bowers Avenue
Santa Clara, CA 95051

SUMMARY

Wideband amplifiers are needed in almost all modern communication and microwave instrumentation systems. They are used for general purpose, broadband and narrowband I.F., pulse and oscillator buffer amplification. These applications are currently met with a wide variety of thin film hybrid feedback amplifiers using high performance microwave bipolar transistors. In this paper, a new monolithic family of broadband cascadable Si MMIC amplifiers is described. The amplifier chips are characterized by very small die size, 50 ohms input and output impedances and usable gain up to 6 GHz.

The self-contained amplifiers are fabricated using a 9 GHz f_T silicon bipolar IC process with a chip size of only $0.4 \times 0.4 \text{ mm}^2$. Topologically, they are single-stage, Darlington-connected transistor pairs with a simple resistive bias network and shunt-series resistive feedback that sets the gain and terminal impedances (Figure 1). A Darlington connection was used because it has a higher S_{21} gain than a single device. This improved performance results from increased loop-gain in the Darlington's internal feedback loop. [1,2]

The amplifiers have a unique I-V characteristic between the power supply and ground terminals such that simple two terminal dc testing can detect faulty devices at both the wafer level and as packaged products (Figure 2). Moreover, correlation between dc tests and rf performance is very close. These characteristics, plus the small die size, make them extremely attractive for high volume microstrip and multistage hybrid products.

Table 1 summarizes the excellent performance of some of the family members.

Table 1

Amplifier	Gain	Bandwidth G_{-1dB}/2:1 VSWR	P_{-1dB}	Noise Figure	Bias	
E01	18.5dB	800MHz/4GHz	1.5dBm	5dB	5V-15V	18mA
E02	12dB	2.1GHz/3.2GHz	3dBm	6.1dB	5V-15V	30mA
E03	12.3dB	2.2GHz/3.5GHz	10dBm	5.9dB	5V-15V	40mA
E04 (GAIN CELL)	8.5dB	4.4GHz/6.0GHz	3dBm	-	Choke Bias	7V/20mA
E05 (GAIN CELL)	7.9dB	5.2GHz/4.3GHz	3dBm	-	Choke Bias	7V/20mA

The detailed performance of amplifier 3 is shown in Figure 3. The performance of the 7.9dB gain cell with off-chip biasing is shown in Figure 4.

The temperature stability of the amplifiers is also very good. For example, the gain variation of the 12dB/3dBm unit is ±45ppm/°C at midband and ±240ppm/°C at the high end over the full -55°C to 125°C temperature range. This translates to a gain variation of only ±0.1dB from 0.5 to 1GHz and ±.5dB at 2GHz. Measured noise figures were in the 5.0dB to 6.5dB range. No increase in noise figure was observed due to 1/f noise at 30MHz.

In summary, we have described a practical and cost effective Very Small Scale Integrated (VSSI) silicon bipolar amplifier family with performance that compares very favorably with larger GaAs MMIC feedback amplifiers. [3,4]

---

1. Meyer, R.S., et al, IEEE JSSC Vol. SC-9, pp. 167-175, Aug., 1974.

2. Meyer, R.S., and Blauschild, R.A., IEEE JSSC, Vol. SC-16, pp. 634-638, Dec., 1981.

3. Archer, J.A., et al, IEEE JSSC, Vol. SC-16, pp. 648-652, Dec., 1981.

4. Estreich, D.B., IEEE ISSCC Digest, pp. 194-195, Feb. 1982.

Presented at the *IEEE Microwave and Millimeter-Wave Circuits Symp.*, 1982.

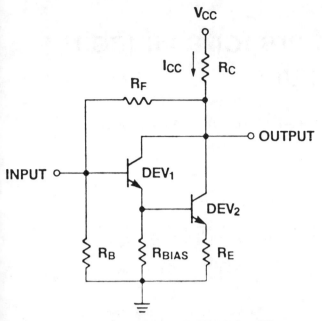

Figure 1A Schematic of Si MMIC Amplifier.

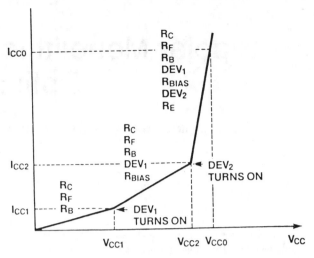

Figure 2 Amplifier I-V characteristic between power supply and ground.

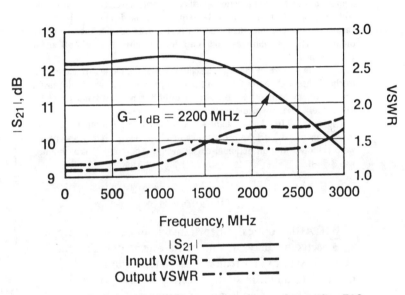

Figure 3 Gain and VSWR versus frequency of amplifier E03.

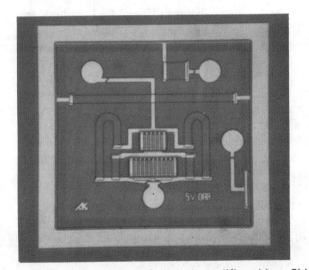

Figure 1B Photomicrograph of typical amplifier chip. Chip size is 0.4 × 0.4 mm².

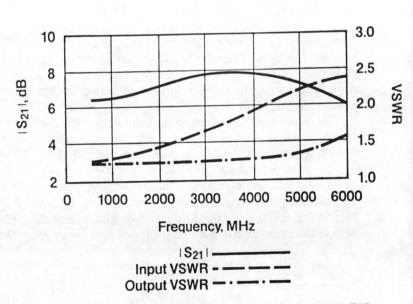

Figure 4 Gain and VSWR versus frequency of gain cell E05.

Bipolar Monolithic Amplifiers for a Gigabit Optical Repeater

MAMORU OHARA, YUKIO AKAZAWA, NOBORU ISHIHARA, AND SHINSUKE KONAKA

Abstract — A main amplifier IC, an AGC amplifier IC, and a preamplifier IC have been designed and fabricated using an advanced silicon bipolar process to provide the required characteristics of repeater circuits for a gigabit optical fiber transmission system.

The bipolar technology used was named SST-1A and had the special feature of a separation width of 0.3 μm between the emitter and the base electrode.

New circuit techniques were also employed. The differential type main amplifier has a peaking function which can be varied widely by means of dc voltage supplied at the outside IC terminal. A bandwidth which can be varied to about treble than that for a nonpeaking amplifier is easily obtained. The gain and maximum 3 dB down bandwidth were 4 dB and 4 GHz, respectively.

The main feature of the AGC amplifier is that the diodes are connected to the emitters of the differential transistor pair to improve the linearity. The maximum gain and 3 dB down bandwidth were 15 dB and 1.4 GHz, respectively, and a dynamic range of 25 dB was obtained.

The preamplifier has a shunt-series feedback configuration. Furthermore, a gain and 3 dB down bandwidth of 22 dB and 2 GHz, respectively, were achieved with an optimum circuit design. The noise figure obtained was 3.5 dB.

I. INTRODUCTION

GIGABIT optical fiber transmission systems are expected to play an important role in the information network system. To realize this system, stable GHz band monolithic amplifiers with small size, low power dissipation, and high reliability are essential.

The main circuits which constitute a repeater are the equalizing amplifier, timing, decision, and LD driver circuits. Especially among these, an equalizing amplifier requires a high gain, wide variable gain range, and very wide band performance. Therefore, we have first attempted the monolithic integration of an equalizing amplifier to discover the feasibility of its application to integrated repeater circuits. As shown in Fig. 1, the equalizing amplifier consists of a main amplifier, an AGC amplifier and a preamplifier. These amplifiers require wide-band characteristics of a few GHz, which is more than twice that for bipolar monolithic amplifiers [1]–[3] fabricated to date. For GaAs monolithic amplifiers, it has been reported [4] that 13.5 dB gain over a 3 dB down bandwidth, below 500 kHz to 2.8 GHz, is obtained using a two-stage construction. GaAs monolithic IC's are promising devices for use in

Manuscript received August 8 1983; revised December 6, 1983.
The authors are with the Atsugi Electrical Communication Laboratory, Nippon Telegraph and Telephone Public Corporation, Kanagawa 243-01, Japan.

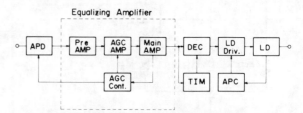

Fig. 1. Block diagram of an optical repeater.

ultra high-frequency amplifiers. However, as the fabricated dispersion of device parameters, such as threshold voltage, transconductance, etc., is comparatively large, realization of a direct-coupled circuit configuration in multiple stage amplifiers would require great effort. One chip monolithic integration in high gain baseband amplifiers is therefore very difficult for use in large interstage coupled capacitors. For Si monolithic IC's, high-frequency performance is inferior to GaAs IC's. However, the fabricated dispersion of the device parameters is extremely small. There are no problems when using the GaAs IC's mentioned above. Thus, we attempted to fabricate a monolithic amplifier of a few GHz bandwidth, using both new circuit techniques suitable for monolithic integration and an advanced silicon bipolar technology, named SST-1A, whose special feature is a separation width of 0.3 μm between the emitter and the base. This paper describes the design and performances of these very wide-band monolithic amplifiers, which are promising devices for application to integrated repeater circuits of a gigabit optical transmission system.

II. DEVICE TECHNOLOGY

The effects of device parameters on the frequency characteristics of an ordinary differential amplifier are shown in Fig. 2. It is clear that the 3 dB down bandwidth depends greatly on base transit time t_f and base resistance r_b. Therefore, obtaining a fast t_f and small r_b is the key to realizing this IC process. The SST-1A is a highly self-aligned process technique. Thus, an emitter width of 0.5 μm, a separation width of 0.3 μm between the emitter and the base electrode, and a boron doped polysilicon base electrode of 0.3 μm can be formed by conventional photolithography with a 2 μm mask pattern. This has resulted in the development of a process technique which forms a precise and stable fine base width of 1.7 μm. Therefore, base

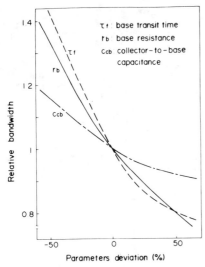

Fig. 2. Device parameter dependency of differential amplifier frequency characteristics.

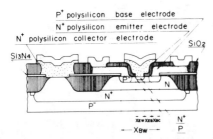

Fig. 3. Cross section of an integrated SST-1A transistor.

TABLE I
DEVICE PARAMETERS OF THE SST-1A TRANSISTOR
(TYPICAL CURRENT: 5 mA)

Emitter size	$(0.5 \times 25 \ \mu m^2) \times 5$ stripes
Emitter resistance (r_e)	1 Ω
Collector resistance (r_c)	28 Ω
Base resistance (r_b)	14 Ω
Collector–base capacitance (C_{cb})	0.065 pF
Base–emitter capacitance (C_{be})	0.13 pF
Collector–substrate capacitance (C_{cs})	0.06 pF
Cutoff frequency (f_T)	12 GHz

resistance and collector to base capacitance are greatly decreased, and high-speed operation can be expected. A cross-sectional view of an SST-1A transistor structure is shown in Fig. 3 and typical transistor parameters used in the simulation are listed in Table I.

III. MAIN AMPLIFIER

The main amplifier is a circuit which amplifies the AGC output signal and generates a 0.5–0.8 V output signal to drive the next decision circuit. Accordingly, wide-band and large output signal characteristics are required. A peaking technique is used to improve the frequency characteristics. In a differential amplifier, this is usually performed by applying series feedback resistance to each emitter and inserting capacitance between the emitters. However, the requirements of highly precise resistance and capacitance

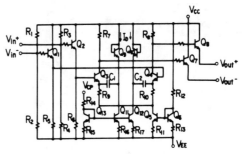

Fig. 4. Configuration of a variable differential resistance type amplifier in the main amplifier.

are not fitted for monolithic integration. Furthermore, it is quite difficult to estimate parasitic capacitance influences in monolithic circuits.

Thus, we have devised several new circuit configurations. One of these configurations is shown in Fig. 4, in which the peaking function is adjusted electrically and varied widely by means of outside IC terminal voltage. The features of the circuit are as follows:

1) Each emitter of transistors Q_3 and Q_4 which form a differential pair is grounded by ac through capacitance $C_1(C_2)$ and diode $Q_9(Q_{10})$. Accordingly, there are no dc bias design limitations.

2) The bias current I_D of diode $Q_9(Q_{10})$ can be adjusted through the current mirror circuit in which dc voltage is supplied at the outside control terminal V_{cp}. Namely, the variable peaking function is realized by changing the differential resistance r_d and diffusion capacitance C_d, and by adjusting the ac grounding condition by changing the current I_D.

Voltage gain A_v of this circuit is given by

$$A_v = Z_1(w)/Z_e(w)$$

where $Z_1(w)$ is load impedance, and $Z_e(w)$ is the parallel impedance between the series connection of $C_1(C_2)$ and $Q_9(Q_{10})$, and emitter resistance $R_9(R_{10})$. Here, $Z_e(w)$ is given by

$$Z_e(w) = R_9 // ((1 + jwr_d(C_1 + C_d))/jwC_1(1 + jwr_dC_d))$$

where r_d and C_d are functions of bias current I_D flowing through the diode, i.e., $r_d = KT/qI_D$ and $C_d = qI_Dt_f/KT$, and thus r_dC_d becomes constant ($= t_f$). When $1 \gg wr_dC_d$ is assumed, $Z_e(w)$ is given by

$$Z_e(w) = R_9(1 + jwr_dC_1)/(1 + jwC_1(R_9 + r_d)).$$

Thus, the pole of $Z_e(w)$ is determined by $C_1(R_9 + r_d)$, and the zero point can be adjusted by r_dC_1. Fig. 5 shows frequency characteristics of $Z_e(w)$ when I_D is changed. It is clear that $Z_e(w)$ varies very little at a low frequency, but varies greatly at a peaking frequency of 3–4 GHz with the change in I_D. Fig. 6 shows the performance of this main amplifier. The gain was 4 dB, and the 3 dB down bandwidth was 4 GHz, which is about treble that for a nonpeaking amplifier. Moreover, in this type of main amplifier, an output signal amplitude of 1 V, which is sufficient to drive

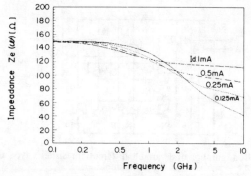

Fig. 5. Frequency characteristics of impedance $Z_e(w)$.

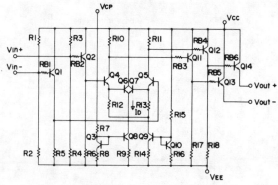

Fig. 8. Configuration of a variable diffusion capacitance type amplifier in the main amplifier.

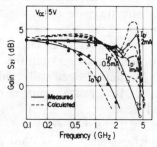

Fig. 6. Frequency response of a variable differential resistance type amplifier used in the main amplifier.

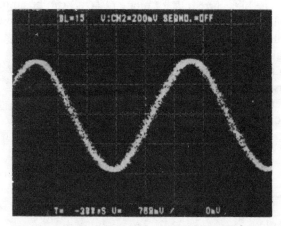

(100ps/div, 0.2V/div)

Fig. 7. Main amplifier output waveform at 1.8 GHz sine wave input.

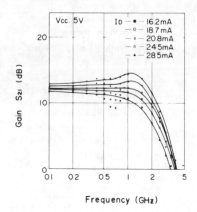

Fig. 9. Frequency response of a variable diffusion capacitance type amplifier used in the main amplifier.

the next decision circuit, was obtained without waveform distortion. The output waveform of the amplifier is shown in Fig. 7.

Another configuration in which peaking capacitance is adjusted electrically is shown in Fig. 8. In this circuit, the base–emitter capacitance of transistors Q_6 and Q_7 consists almost entirely of the diffusion capacitance and is used as the peaking capacitance. As the diffusion capacitance of a transistor is a function of the emitter current I_D, the variable peaking capacitance is achieved by controlling I_D through the current mirror circuit. As this circuit makes it possible to directly join the bases of transistors Q_6 and Q_7 to the emitters of the differential pair transistors Q_4 and Q_5, integrated thin film capacitors are not required. However, because the actual emitter feedback resistance varies with variations in the bias current I_D, there is some change in the dc gain. Fig. 9 shows the experimental results of this circuit. A gain of 12 dB and a 3 dB down bandwidth, which can be changed from 1.1 GHz to 2.2 GHz, were obtained.

In the main amplifier, moderate amplification and high amplitude characteristics are required. Therefore, a two-stage amplifier structure is used, in which the variable diffusion capacitance type amplifier, shown in Fig. 8, functions as the first stage and the variable differential resistance type amplifier, whose characteristics are shown in Fig. 5, acts as the second stage.

IV. AGC Amplifier

As for the AGC amplifier, wide-band, high dynamic range, and high linearity characteristics are required, and there have been several studies proposing monolithic integration [5], [6]. To improve the linearity, we devised a new circuit configuration. The main feature of the AGC amplifier is that the diodes are connected to the emitters of the differential transistor pair in such a way that the linearity of the conventional differential AGC amplifier [7] is improved by the dynamic range of the diodes. Fig. 10 shows a circuit configuration which gives wide-band, high dynamic range, and high linearity characteristics. This circuit consists of the differential pair of Q_4 and Q_8, and a dc bypass circuit containing Q_6 and Q_7. The gain control is effected by adjusting the dc current flowing through Q_6

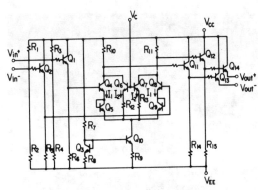

Fig. 10. AGC amplifier configuration.

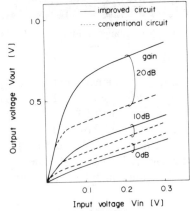

Fig. 11. Linearity characteristics of the AGC amplifier.

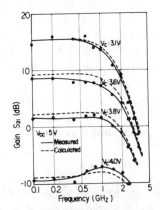

Fig. 12. AGC amplifier frequency response.

and Q_7 with the control voltage V_c and by changing the gain of the differential amplifier. The dc voltage gain A_v of this amplifier is given by

$$A_v = -R_1 I_1 / (V_T(r_b + R_e)/\beta I_1 + R_e I_1/V_T + 1),$$

where $R_e = V_t/I_1$, $V_T = q/KT$, and $R_L = R_{10}$ or R_{11}. Current gain and base resistance are represented by β and r_b, respectively.

From the above equation, it is clear that a high dynamic range is obtained for the AGC amplifier. Fig. 11 shows the linearity improvement effects obtained by applying diodes Q_5 and Q_9. Fig. 12 shows the performance of the monolithic integrated AGC amplifier. The maximum gain and 3 dB bandwidth are 15 dB and 1.4 GHz, respectively, and a 25 dB AGC dynamic range is obtained.

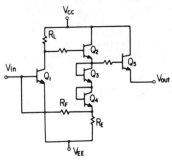

Fig. 13. Preamplifier configuration.

V. Preamplifier

As for the preamplifier, wide-band and lower noise characteristics than those for the main and the AGC amplifiers are required. The shunt-series feedback configuration [8], shown in Fig. 13, was used as the monolithic integrated wide-band and low noise amplifier. The preamplifier operates as a transimpedance amplifier which transforms the output current I_{APD} of the APD into output voltage V_0. Here, transimpedance gain A_z, 3 dB down bandwidth f_w and input noise current density I_n are given by

$$A_z = V_0/I_{APD} = R_F, \qquad f_w = 1/2R_{in}(C_{APD} + C_{in}), \quad \text{and}$$
$$I_n = (w(C_{APD} + C_{in}) + 1/R_F + 1/r_{in})E_{ni} + (4KT\Delta f/R_F)$$

where

I_{APD}: APD output current

V_0: output voltage of the preamplifier

R_F: feedback resistance of the preamplifier

R_{in}: input resistance of the preamplifier

C_{APD}: APD output capacitance

C_{in}: input capacitance of the preamplifier

r_{in}: input resistance of the open loop amplifier

E_{ni}: input noise voltage of the input transistor

K: Boltzmann constant

T: absolute temperature

Δf: noise bandwidth.

From the above equation, it is clear that the transimpedance gain is determined by the absolute value of feedback resistance R_F and that the bandwidth is inversely proportional to the input resistance R_{in} of the amplifier. The input current noise density decreases as feedback resistance R_F increases, but the bandwidth narrows. Fig. 14 shows bandwidth, transimpedance gain, and input current noise density as functions of feedback resistance R_f. The optimum R_F was determined from the values shown in Fig. 14. A feedback resistance R_F of 500 Ω was chosen. Frequency characteristics of the preamplifier when the output capacitance of APD varied is shown in Fig. 15. The output capacitance of the APD used is 2.1 pF. It is concluded that

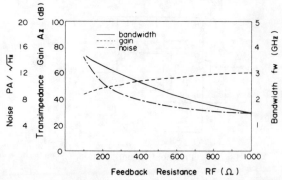

Fig. 14. Preamplifier characteristics as a function of the feedback resistance R_F.

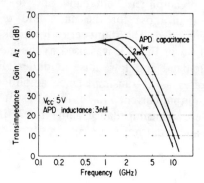

Fig. 15. APD capacitance dependency of preamplifier frequency characteristics.

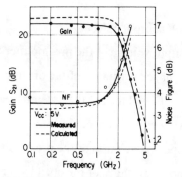

Fig. 16. Preamplifier frequency response.

a transimpedance of 55 dB and a 3 dB down bandwidth of 2.2 GHz are obtained. As it is difficult to directly measure transimpedance, frequency characteristics were estimated using $S21$ gain. The measured gain and noise characteristics are shown in Fig. 16, along with the theoretical performance values predicted by computer simulation. The gain and 3 dB bandwidth were 22 dB and 2 GHz, respectively, and a noise figure of 3.5 dB was obtained. The dynamic range of the preamplifier is designed with a wide range of APD output current, 24 to 315 μA. Simulation confirmed that good waveform performance is obtained over the entire range.

VI. Performance of the Equalizing Amplifier [9]

The equalizing amplifier shown in Fig. 1 was constructed with newly developed 5 IC's; preamplifier, buffer amplifier (using the circuit shown in Fig. 8), AGC amplifier and

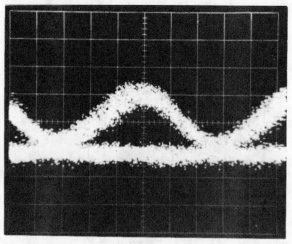

Fig. 17. Equalizing amplifier output waveform at 1.6 Gbit/s transmission experiment.

main amplifier, and performances were cleared. A $S21$ gain and 3 dB down bandwidth of 56 dB and 1 GHz, were obtained. Power dissipation was 630 mW at 5 V V_{cc}. Fig. 17 shows an output waveform [9] of the monolithic equalizing amplifier. For the detector connected to the preamplifier, a p^+–n type Ge-APD was used. The sensitivity at 10^{-11} bit error rate of the equalizing amplifier was -30.5 dBm.

VII. Measurement and Assembly

Measurements were performed using S parameters to avoid package parasitic effects. In order to achieve a good high-frequency performance and to estimate intrinsic IC characteristics as precise as possible, the main amplifier, AGC amplifier, and preamplifier IC chips are directly set on a strip line. To use these monolithic wide-band amplifier in an actual repeater circuit, IC packages must be well designed. The chip size is 1 mm square and photographs of these amplifier chips are shown in Fig. 18.

VIII. Conclusion

A monolithic integrated main amplifier, AGC amplifier, and preamplifier for repeater circuits of a gigabit optical fiber transmission system were fabricated using an advanced silicon bipolar process called SST-1A. In the main amplifier, new circuit configurations whose peaking functions are adjusted electrically by changing the diffusion capacitance and differential resistance of transistors has been devised, and a very wide-band amplifier has been realized. In addition to a high gain and wide dynamic range, high linearity was achieved in the AGC amplifier by adopting a new circuit configuration. In the preamplifier, a circuit design which optimizes gain, bandwidth, and noise characteristics was used.

Principal results of these IC's are summarized as follows.
1) The gain and 3 dB down bandwidth of the main amplifier are 12 dB, 2 GHz and 4 dB, 4 GHz, respectively.
2) The maximum gain, 3 dB down bandwidth and dynamic range of the AGC amplifier are 15 dB, 1.4 GHz, and 25 dB, respectively.

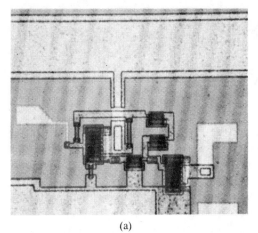
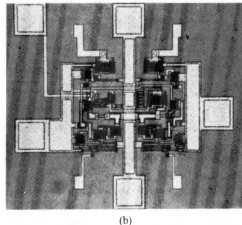
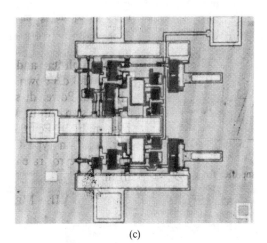

Fig. 18. Photographs of the wide-band amplifier chips. (a) Preamplifier. (b) AGC amplifier. (c) Main amplifier.

3) The gain and 3 dB down bandwidth of the preamplifier are 22 dB and 2 GHz, respectively. The noise figure obtained is 3.5 dB.

From the above results, the monolithic integration of an equalizing amplifier consisting of a main amplifier, AGC amplifier, and preamplifier appears quite promising. The remaining problem is the total assembly of the repeater circuits including a hybrid IC structure.

Acknowledgment

The authors are greatly indebted to Drs. T. Suzuki, H. Mukai, Y. Kato, H. Ariyoshi, N. Ohwada, T. Sakai, and H. Ikawa for their useful suggestions and comments. The authors also wish to thank those who helped fabricate these IC's. Acknowledgment is also expressed to K. Iwashita for his help in preparing this paper.

References

[1] R. G. Meyer et al., "A four-terminal wide-band monolithic amplifier," in *ISSCC Dig. Tech. Papers*, Feb. 1981, pp. 186–187.
[2] H. Hillbrand et al., "Computer aided design of a 1 GHz bandwidth monolithic integrated amplifier," in *ESSCIRC Dig. Tech. Papers*, Sept. 1977, pp. 122–124.
[3] M. Ohara et al., "Very wide-band silicon bipolar monolithic amplifiers," *Jap. J. Appl. Phys.*, suppl. 22-1, pp. 129–132, 1983.
[4] K. Honjo et al., "Ultra-broad-band GaAs monolithic amplifier," *IEEE Trans. Electron Devices*, vol. ED-29, pp. 1123–1129, July 1982.
[5] W. R. Davis et al., "A high-performance monolithic IF amplifier incorporating electronic gain control," *IEEE J. Solid-State Circuits*, vol. SC-16, pp. 408–416, Dec. 1968.
[6] W. M. C. Sansen et al., "An integrated wide-band variable-gain amplifier with maximum dynamic range," *IEEE J. Solid-State Circuits*, vol. SC-9, pp. 159–166, Aug. 1974.
[7] S. Tsutsumi et al., "A study on integrated 400 Mbit/s optical repeater," *Tech. Group. CS IECE Jap.*, vol. CS-62, 1982.
[8] A. B. Grebene, *Analog Integrated Circuit Design*. New York: Van Nostrand-Reinhold, 1972, chap. 8.
[9] K. Iwashita et al., in *Proc. IECE Japan Nat. Conv.*, Sept. 1983, paper 371.

Section III-D
Transmit/Receive Modules

TRANSMIT/RECEIVE or T/R modules, that is, modules consisting of a low-noise receiver, power amplifier, phase-shifter, and assorted microwave switches for phased-array antenna applications are "where it all began." It was this application which stimulated the rapid re-emergence of the MMIC approach in 1978.

The T/R module represents potentially the largest nonconsumer market for MMIC's, and reigns as the single most important military application of MMIC's. The following papers, which represent only a fraction of those published in this field, concern themselves with the design of the various functional circuit blocks comprising a complete T/R module. In addition, initial attempts to integrate all of these functional blocks on one chip are described.

At this time, the active T/R module, with its high level of circuit complexity, represents the most ambitious attempt at integration of a monolithic microwave circuit on a single chip.

GaAs monolithic microwave circuits for phased-array applications

R.S. Pengelly, M.Sc., C.Eng., M.I.E.E.

Indexing terms: Semiconductor materials, Microwave circuits, Semiconductor devices, Radar, Amplifiers, Switches, Transmitters, Schottky-barrier devices, Field-effect transistors

Abstract: The use of gallium arsenide as a material for monolithic microwave circuits where active devices, such as f.e.t.s, diodes etc., are integrated onto the same piece of material as passive components is now receiving considerable attention in Europe and the USA. The paper concentrates on the specific role of monolithic circuits in phased-array-radar applications with descriptions of the use of f.e.t.s in switches, phase shifters, attenuators, receivers and transmitters. A summary of GaAs f.e.t. device performance is included allowing some insight into the noise figures, output powers and efficiencies obtained from low-noise and power amplifiers, respectively, at frequencies in S- and X-band. Some examples of GaAs monolithic circuit designs are given and methods of using active as against passive matching to achieve higher packing densities are described. The yield and cost of monolithic techniques is reviewed in the light of present and predicted circuit design and technology improvements. The impact of such costs on the realisation of phased-array systems with large numbers of elements is reviewed.

1   Introduction

The use of gallium arsenide as a material for monolithic microwave circuits (m.m.c.s) where active devices, such as f.e.t.s, diodes etc, are integrated onto the same piece of material as passive components is now receiving considerable attention in Europe and the USA. This paper will concentrate on the specific role of monolithic circuits in phased-array-radar applications with descriptions of the use of f.e.t.s in switches, phase shifters, attenuators, receivers and transmitters.

A summary of GaAs f.e.t. device performance is included allowing some insight into the noise figures, output powers and efficiencies obtained from low-noise and power amplifiers, respectively, at frequencies in S- and X-band.

Many phased-array-radar systems for the future, presently being conceived, will require large numbers of receive and transmit elements. The ability to be able to include integrated circuits within the antenna elements themselves is also receiving considerable attention. Such systems will require low-cost components which also have the attributes of reproducible performance, a high mean time to failure, small size and the efficient use of d.c. power. For these reasons gallium-arsenide monolithic circuit technology may well have advantages over other, more conventional, methods. As with Si i.c.s, relatively complicated r.f. and i.f. functions can be included on one chip. Only where very high powers are envisaged, will the GaAs monolithic approach have fundamental limitations. However, for many applications output powers per element up to 30 W peak power is quite sufficient.

Monolithic circuits using GaAs are in their infancy. This paper is intended as a review of some of the current work aimed at introducing hardware to the phased-array-radar-systems engineer in the next few years.

2   GaAs f.e.t. device performance

The GaAs f.e.t. is now well established as a low-noise device

Paper 849F, first received 15th April and in revised form 5th June 1980
The author is with Plessey Research (Caswell) Limited, Allen Clark Research Centre, Caswell, Towcester, Northants, England

up to 20 GHz. Currently device noise figures of 1·7 dB at 12 GHz with 10 dB associated gain have been achieved whilst noise figures of 3 dB with 8 dB associated gain are possible at 18 GHz with 0·5 μm gate length f.e.t.s.[1] At frequencies below 4 GHz device noise figures of less than 1 dB can be achieved with 15 dB associated gain. Low noise 2·7 to 3·2 GHz amplifiers, for S-band radar applications, have been built with overall noise figures of 2·5 dB (including limiter loss). At X-band, radar front-end amplifiers having noise figures of 3·3 dB are currently being produced.

Thus it may be appreciated that the GaAs f.e.t. is capable of being included in phased-array type systems requiring low-noise figure receiver preamplifiers.

The GaAs Schottky barrier f.e.t. (also known as the m.e.s.f.e.t.) has been widely reported as a power amplifier. Currently, development work concentrates on producing higher powers with compact device and circuit structures. For example power outputs in S-band of 15 W c.w. have been reported with power-added efficiencies of 28%.[2] In order to produce such output power, f.e.t. chip and individual cell combining techniques are used and recent results[3] suggest that bandwidths of up to one octave are realisable with output powers of the order of 1 W in X-band. The GaAs f.e.t. offers the best overall performance when power output, bandwidth and efficiency are considered simultaneously. Recently it has become apparent that under pulsed conditions the f.e.t. device can achieve useful increases in output power.[4]

At S-band the advantages of the GaAs f.e.t. over the Si bipolar transistor are somewhat more unclear although it would appear that, under c.w. operation, the GaAs m.e.s.f.e.t. can be more successfully matched over broader bandwidths than the Si bipolar device. Performance improvements with Si bipolar-junction-transistor power devices are being attained with ion-implanted base and emitter profiles; the device geometries having a higher emitter periphery to base area ratio.

However, the lack of the ability to fully integrate silicon power devices with other passive and active components on the same substrate material is a serious disadvantage of conventional silicon i.c.s at high frequencies caused by the substrate material having poor characteristics. Silicon-on-sapphire technology now under development in the USA may produce fully integrated circuits capable of operation up to S-band.

Gallium arsenide, however, has the unique ability at the present time of allowing the integration of both low-noise and power f.e.t. devices over a wide frequency range. The depletion mode f.e.t. is also capable of being used in a variety of other circuits.

2.1 Low-noise monolithic amplifiers

As has already been outlined, the noise performance of conventional hybrid amplifiers using GaAs f.e.ts has reached a level, particularly at frequencies below X-band, where further improvements will be difficult to implement. On the other hand, the noise figure of monolithic amplifiers still depends to a large extent on the circuti design concept. Results to date, using conventional lumped-element matching techniques indicate that a noise figure penalty of approximately 0·7 dB is paid at S-band for a 1·5 dB to 2 dB low-noise amplifier. This penalty is because of the loss of the matching components, particularly inductors. At X-band broadband amplifiers[5] have shown noise figures approximately 0·4 dB higher than their direct equivalents using microstrip techniques. Where techniques more closely akin to Si i.c. design are used, such as common-gate/common-source/source-follower feedback amplifiers, noise figures can be somewhat higher than the optimum for the f.e.t. device type used — by as much as 3 dB in S-band.

In order to achieve a good yield of working circuits, and achieve a low cost, certain performance figures such as the noise figure may well have to be compromised in a GaAs i.c. solution.

2.2 Mixers — f.e.t.s

Dual-gate field-effect transistors are now gaining popularity as low noise mixers where the device can give attractive conversion gains. Cripps et al.[6] have reported an image rejection mixer having a conversion gain of up to 15 dB with an image rejection of at least 20 dB at X-band and with a noise figure of approximately 8 dB. The basic circuit arrangement is shown in Fig. 1. An attractive alternative to the dual-gate mixer which can achieve much lower noise figures is to use a double-gate f.e.t. with a single-gate mixer f.e.t. as shown in Fig. 2.

The r.f. and l.o. frequencies are combined in the first f.e.t. where the r.f. port is tuned for minimum noise figure. Isolation between the l.o. and r.f. ports is very similar to the reverse isolation of the device and is typically -20 dB. The second single gate f.e.t. is biased to optimise the conversion gain and output compression point. For example a mixer circuit using such a technique at 4 GHz with a 200 MHz i.f. has a single-sideband noise figure of 3 dB with a conversion gain of 15 dB (using 0·5 μm f.e.t.s). Thus, for low-noise-receiver front ends, the use of a GaAs f.e.t. preamplifier and dual-gate f.e.t. mixer is capable of producing a low overall receiver noise figure with low-cost relatively-high-noise-figure i.f. amplifiers.

Lower-frequency f.e.t. mixers essentially employ a variable-gain amplifier stage where the gain modulating function is provided by f.e.t. variable resistors. Recently Van Tuyl[7] has reported a double-balanced mixer which operates up to approximately 3 GHz using such a technique (Fig. 3). Such a circuit can be extended to higher frequencies on GaAs than it can using similar approaches with Si bipolar transistors or m.e.s.f.e.t.s.

Monolithic balanced mixers at higher frequencies (where the f.e.ts are operating closer to their cut-off frequency) can be realised using circuit techniques such as that shown in Fig. 4. Here use is made of active power splitters and

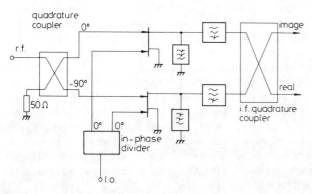

Fig. 1 *Dual-gate f.e.t. image rejection mixer*

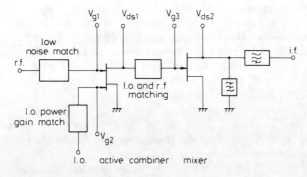

Fig. 2 *Double-gate/single-drain f.e.t. mixer*

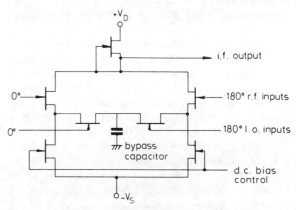

Fig. 3 *Double balanced monolithic mixer circuit*

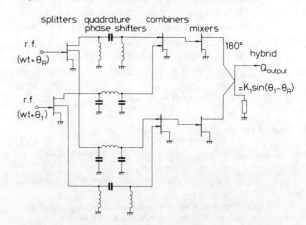

Fig. 4 *Balanced monolithic mixer circuit for 4 GHz and above*

combiners to produce either quadrature or antiphase r.f. and l.o. signals which are then applied to two single-gate mixers. The resulting two i.f. components can be combined in a conventional 180° i.f. hybrid coupler or by the use of an active circuit employing f.e.t.s, the latter being particularly convenient for monolithic circuits.

2.3 Diodes

Ristow et al.[8] have shown that planar GaAs Schottky-barrier diodes can be monolithically integrated to produce a mixer with an s.s.b. noise figure of 10 dB (including a 1·5 dB i.f. noise contribution) at 15 GHz (i.f. being 70 MHz). Further work by Courtney et al.[9] has produced mixers with noise figures of 11 to 12 dB at 30 GHz which have been monolithically integrated with a GaAs f.e.t. i.f. amplifier at 2 GHz.

2.4 Switches

The use of both single and dual-gate GaAs f.e.t.s has received considerable attention in the design of fast switches. Several configurations are possible to implement the f.e.t. as a switch (shown in Fig. 5). The series configuration makes use of the saturation and pinch-off conditions of the f.e.t. and provides an inherently broadband (untuned) response with zero d.c. bias power since the gate-source Schottky diode is always reverse biased and no external voltage is applied to the drain of the f.e.t. Gaspari and Yee[10] have reported an 8-way switch utilising a tuned series connected f.e.t. switch. The single or dual-gate f.e.t. can be used in a conventional amplifier configuration with tuning and/or attenuator pads on gate and drain to maintain low input and output v.s.w.r.s under on and off conditions (Fig. 5c).

An alternative configuration giving high broadband isolation is the π network of f.e.t.s as shown in Fig. 5d. This circuit may also be used as a matched attenuator and only relies on having known voltage-to-attenuation laws for the series and shunt f.e.t.s much as with *pin* diodes.

Vorhaus has reported a multithrow dual-gate f.e.t. switch[11] having isolations in excess of 25 dB in X-band using a novel 4-sided structure with a common-source connection to ground. This ground is supplied by introducing a via through the GaAs substrate. Most of the f.e.t. structures used to date as switching elements are small signal devices. Extensions to the dual-gate switch using power f.e.t.-type structures, i.e. long gate width devices, are currently under investigation. Garver[12] has also proposed a 'control f.e.t.' which, by controlling the doping density of the 'channel' region and employing an overlay dielectric, should enable the switching of, at least, several watts of r.f. power. Work to date using conventional 600 μm wide power f.e.t.s has shown that powers in excess of 200 mW can be switched at X-band with 20 dB of isolation using this method.

2.5 Phase-shifting circuits

Phase-shifting circuits associated with the receivers of phased arrays are usually *pin* diode controlled circuits while circuits for very high peak power applications depend on the use of ferrite materials.

Four types of phase-shifter configurations are generally employed: the switched line, reflection, loaded line and high-pass/low-pass. Of these the most promising for monolithic realisation using either planar Schottky diodes or m.e.s.f.e.t.s as the switching elements is the high-pass/low-pass. Of these the most promising for monolithic realisation using either planar Schottky diodes or m.e.s.f.e.t.s as the switching elements is the high-pass/low-pass structure.[12] The circuit is capable of being made small on GaAs by the use of lumped elements as shown in Fig. 6. The circuit uses f.e.t.s as switches so the insertion loss of the circuit is determined mainly by the 'on' resistance of the series switch. The 'off' capacitance of the f.e.t. is also important. Such a circuit provides a matched transmission behaviour providing up to 180° of phase shift over a 20% bandwidth. The dual-gate f.e.t. can itself be used to provide phase shift particularly for the smaller phase angles of a digital phase shifter.

From Fig. 7a it may be seen that, provided the designer sacrifices device maximum available gain, a multitude of

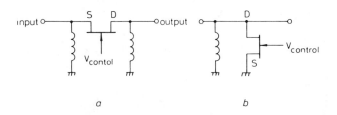

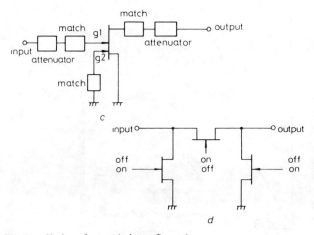

Fig. 5 *Various f.e.t. switch configurations*

a Series configuration
b Shunt configuration
c Dual-gate switch
d Matched attenuator

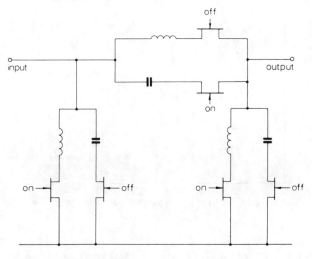

Fig. 6 *High-pass/low-pass phase shifter using m.e.s.f.e.t. switches*

voltage settings on the first and second gate are available for a specific gain G. By matching the device over a specified bandwidth somewhat greater than that required it is possible to produce a variable transmission phase shift with two different voltages on gate 1 and 2 as shown in Fig. 7b.

In X-band, phase shifts of over 50° have been obtained (Fig. 8). As the S_{11} of the dual-gate f.e.t. changes with the first gate voltage, amplitude changes are incurred at the band edges as shown in Fig. 9, and these are somewhat larger than deemed acceptable particularly for the higher

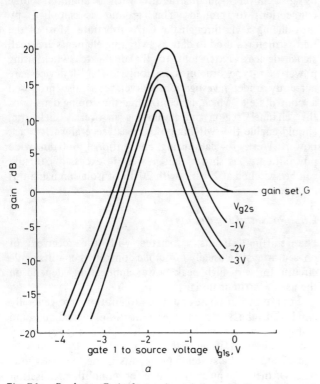

Fig. 7A *Dual-gate GaAs f.e.t. gain variation with first and second gate voltages*

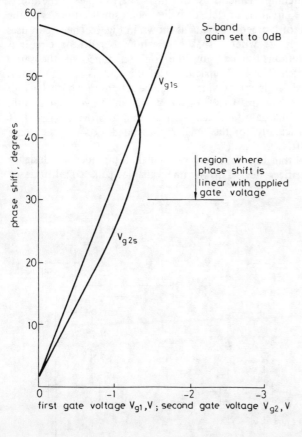

Fig. 7B *Voltages needed on gates 1 and 2 of dual-gate f.e.t. to achieve the phase shift on S21 shown*

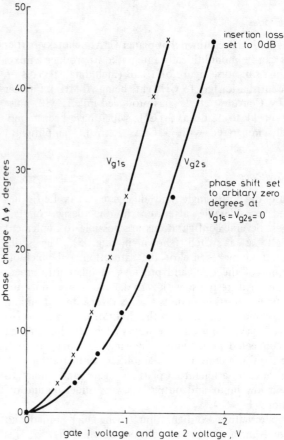

Fig. 8 *Dual-gate f.e.t. phase shifter*
X-band; 9·75–10·25 GHz

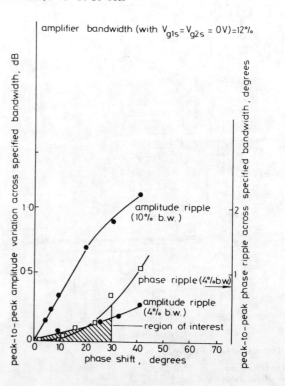

Fig. 9 *Amplitude and phase variations with bandwidth for dual-gate f.e.t. phase shifter*

phase shifts. However, by matching over a bandwidth typically 100% greater than is needed phase and amplitude ripples can be reduced substantially. For example (Fig. 9), if a phase shifter element giving 22.5 degrees needs to operate to an amplitude ripple of ±0.1 dB over the 2.7 to 3.2 GHz bandwidth then the matching circuits into the gates and drain of the device need to produce flat gain over the 2.2 to 3.7 GHz band.

A major advantage of this scheme is that the circuit is truly analogue allowing phase adjustments which are not quantised as in a digital phase shifter. Variations in transmission phase caused by temperature can be compensated by varying the gate bias. Phase variations between modules in a phased-array system can be adjusted by the use of such a technique thus improving sidelobe levels.

Amplitude adjustment (under constant phase conditions) can also be produced using the circuit arrangement as shown in Fig. 10A, where gate 1 and gate 2 bias are set to minimise transmission phase variations over the wanted level of attenuation. Fig. 10B compares the transmission phase variations over the attenuator settings with the second gate voltage variations and both the first and second gate voltage variations.

A circuit somewhat like that schematically shown in Fig. 11 can be adopted to produce a 5 bit phase-shifter

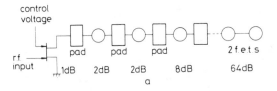

Fig. 10A *Single ended 6-bit attenuator chain using dual-gate f.e.t.s*

where the $\pi/2$ and π phase shifts can be produced using active combiners, the overall unit having insertion gain. Both the $\pi/2$ and π phase shift circuits can use lead/lag phase networks where each f.e.t. cell of the combiner is turned on or off in turn such that, in any one switch state, the impedance presented to the matching circuits is the same as in the other state. Alternatively, control f.e.t.s can be introduced into $\pi/2$ and π phase bits as switches.

GaAs f.e.t. active splitters or active combiners can be realised having good amplitude and phase equality as well as high port-to-port isolation. Such techniques are very useful in monolithic circuits for producing f.e.t. 'versions' of well known passive circuits such as 3 dB quadrature couplers. Such techniques lead to dramatic reductions in GaAs usage. For example, at S-band a reduction of at least 60% in GaAs is possible.

2.6 Monolithic GaAs f.e.t. oscillators

It is now well established that GaAs f.e.t. oscillators are efficient and have less stringent power-supply and heat-sink requirements than other solid-state oscillators. Being a planar device the f.e.t. is also ideally suited for monolithic microwave-oscillator functions and can be married to varactor-diode tuning capacitors if required. For a three terminal device, like the f.e.t., suitable feedback and impedance matching elements are needed to induce negative resistance and consequently provide useful power at microwave frequencies.

Monolithic GaAs f.e.t. oscillator circuits using grounded-gate arrangements have been realised. In these circuits the feedback element is a lumped inductor on the gate terminal. Overall oscillator size is less than 2×2 mm at X-band. Circuits providing 10 mW output power at 8.4 and

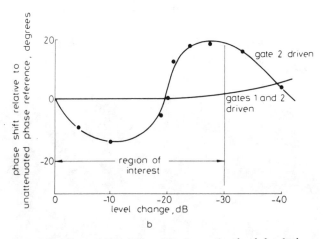

Fig. 10B *Phase shift of S_{21} with attenuation level for dual-gate GaAs f.e.t. switch over 20% b.w. centred at 5.68 GHz*

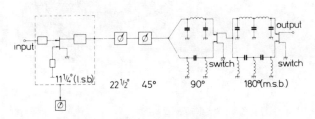

Fig. 11 *Dual gate f.e.t. active-phase shifter*

Matching circuits not shown

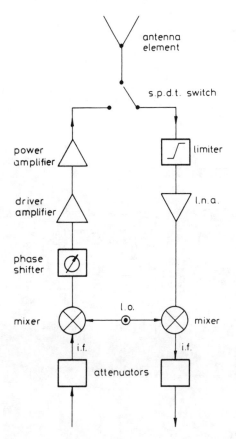

Fig. 12 *Active phased-array schematic diagram*

12·8 GHz have been realised at 6% and 4% d.c. to r.f. conversion efficiency, respectively.[14] Further work on frequency stabilisation and electronic tuning is in progress.

3 Active-array modules

Considering the circuit of Fig. 12 it may be appreciated that all the individual 'building blocks' to which reference has already been made may be fabricated using GaAs f.e.t. technology. Fig. 13 shows a low-noise receiver front-end realisation for an image rejection mixer. The r.f. signal is equally divided into two channels where a phase lead/lag network is employed to produce the 0 and π/2 signals for a 'single-ended' mixer. The l.o. buffer amplifier/splitter uses a common-gate/common-source combination to produce 18 to 20 dB gain at S-band whilst providing +10 dBm l.o. power to the dual-gate mixers. A lumped-element hybrid version of this latter circuit is shown in Fig. 14 together with its response.

The dual-gate mixer for this particular low-noise receiver module was designed using the large-signal S-parameters on gate 2 of the f.e.t. which differ markedly from the small-signal parameters as shown in Fig. 15. The r.f. is terminated in a short circuit at the drain of the device whilst the 50 Ω shunt resistor provides stabilisation at r.f. The i.f. is extracted via a low-pass filter at a 200 Ω impedance level. The image-rejection feature is introduced by combining the two i.f. outputs (Fig. 13) in a quadrature coupler which is a conventional lumped-component circuit.

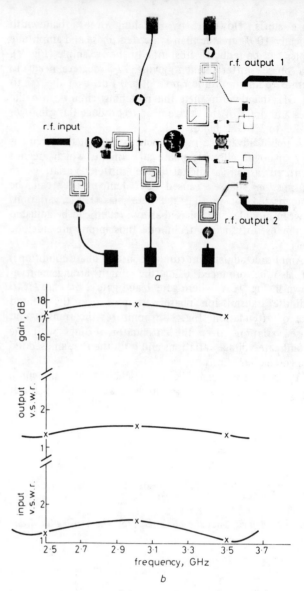

Fig. 14 *Common-gate/common-source f.e.t. amplifier/splitter*

a Photograph
b Response curves

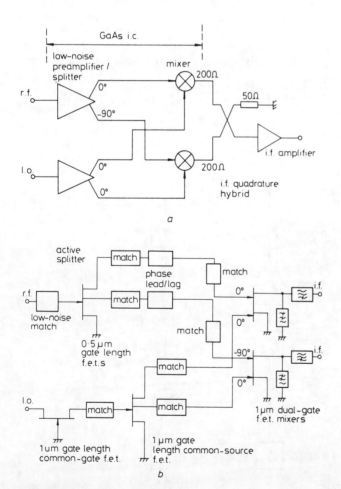

Fig. 13 *GaAs i.c. S-band receiver*

a Schematic diagram
b Detail of front-end

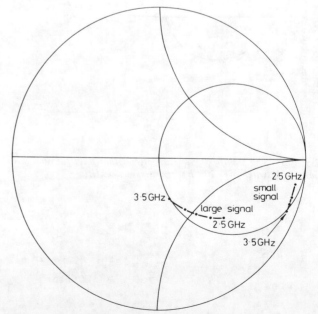

Fig. 15 *Small and large signal reflection coefficients of second gate of dual-gate f.e.t. mixer*

The r.f. preamplifier/splitter, l.o. buffer and mixers can be fabricated monolithically on GaAs. Fig. 16, for example, shows the chip layout for an S-band low-noise preamplifier/splitter designed to give 10 dB gain and a 2·0 dB noise figure in the 2·7 to 3·4 GHz frequency band.

Fig. 12 also shows an example of a transmit module for an array. Of interest here is the ability to be able to monolithically integrate the driver and power amplifier stages to produce a cost-effective circuit. Depending on radar function and frequency, circuits being designed at the present time include stages capable of delivering 0·5 to 1 W at X-band and up to 15 W at S-band.[15] Such circuits use a combination of microstrip and lumped-component matching on GaAs with substrate thicknesses of 100 μm. Use is made of both air-bridge connected f.e.t. source contacts and vias through the GaAs to produce very low-source reactance to ground. In many cases the use of pulsed d.c. drive to the f.e.t.s can be considered producing substantial increases in output power especially when the f.e.t.s are produced with a high power/area ratio.

4 GaAs monolithic-circuit technology

The potential performance advantages of GaAs over silicon as a high-frequency i.c. material have been recognised since the early 1960s, but early results with linear microwave circuits were disappointing. Silicon technology advanced at a tremendous pace because of the well-controlled diffusion processes available for the material. Diffusion techniques for GaAs, however, are virtually useless and early GaAs material varied in quality. The development of the GaAs Schottky-barrier gate f.e.t., using thin epitaxial n-type active layers on buffered semi-insulating material, was the first major breakthrough in producing monolithic microwave circuits.

Circuits produced on GaAs contain both active and passive components where the active components are fabricated on either an n-type epitaxial layer or an n-type implanted layer. Passive components are produced on the semi-insulating GaAs material which has a low microwave loss and a dielectric constant of 13. There are basically two techniques that can be used to produce microwave passive components on the material — either lumped elements or distributed transmission lines. For frequencies below X-band, and where power dissipation is not a problem, lumped-element matching is to be preferred since it enables the the optimum usage of GaAs area. For frequencies in J-band and above and where thermal dissipation is important (such as in power f.e.t. amplifiers) microstrip techniques are preferred.[16]

Lumped-element characterisation has received considerable attention over the years[17,18] but is still subject to

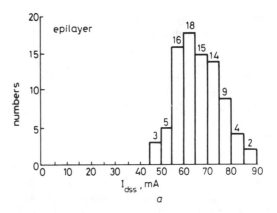

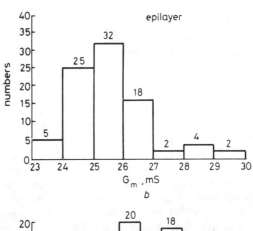

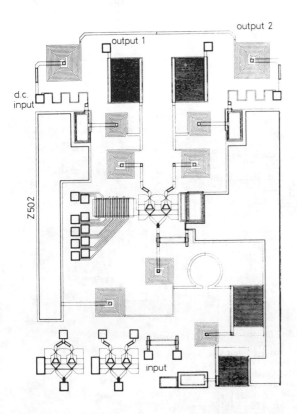

Fig. 16 *Chip layout for GaAs i.c. (S-band low-noise preamplifier/splitter)*

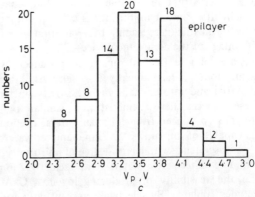

Fig. 17 *GaAs f.e.t. over wafer with v.p.e. grown layer*

a Saturated drain current
b Transconductance variation
c Pinch-off voltage
Number of samples = 86

investigation as new frequency bands and components are included. Results are now available for a considerable number of inductor, capacitor and resistor structures indicating the applicability of both classical and more modern analytical methods to their design. Obviously the i.c. design is as accurate as the determination of passive component values and the accuracy of both models and S-parameter characterisation of active devices. Thus, in many cases at the moment, it is necessary to either introduce some form of adjustment into the monolithic circuit or prepare initial mask sets covering a limited number of possible alternative solutions to a particular problem.

Qualification of GaAs material is also essential in enabling the circuit designer to guarantee a confidence limit on active-device performance variations. The uniformity of d.c. characteristics of f.e.t.s over a typical epitaxially grown layer is shown in Fig. 17 in terms of saturated drain current I_{dss}, transconductance g_m and pinch-off voltage V_p. Such parameters control the bias current of the f.e.t.s in a self-biased approach such as that adopted in Fig. 16 for example. It may be seen that the standard-deviation values are of the order of 10%.

Depletion-mode GaAs i.c.s at present use n^+ on n vapour phase epitaxy material. The n^+ region is $0.4\,\mu m$ thick and the doping level is 10^{18} carriers/cm³; the n region is similarly $0.4\,\mu m$ thick with 1.5×10^{17} carriers/cm.³ The n^+ region is used to lower the ohmic contact resistance of source and drain contacts for f.e.t.s and for low-resistance diodes. The epitaxial layer is grown on a high-resistivity buffer layer (10^{13} carriers/cm³) to isolate it from the GaAs substrate. This ensures long-term stability of the material parameters and, together with the uniformity of thickness and doping concentration in the n region, is an important factor in yield control.[19]

Where Schottky mixer diodes are fabricated on the same chip as f.e.t.s, an n on n^+ on n structure is used. This enables the diode to be placed between the upper n region and the n^+, and the f.e.t.s between the n^+ and the lower n region.

GaAs material technology has been progressing steadily since the early sixties. Vapour phase epitaxial processes have centred around the AsCl-Ga-H₂ process, developed by Knight et al.,[20] and the alkyl process.[21] The former process produces very high purity GaAs and together with the development of the buffer layer has lead to the introduction of state-of-the-art f.e.t.s. The technology has been extended to cope with 5 cm circular wafers upon which is grown a multilayer structure suitable for high-frequency monolithnic circuits. Present uniformity is ± 3% in doping density and ± 2% in thickness leading to a variation of 8% in pinch-off voltage of the completed devices.

GaAs i.c.s are being fabricated from layers produced by ion implanting donor species directly into semi-insulating substrate material and into epitaxial high-resistivity buffer layers. These devices have demonstrated similar performance to those produced from epitaxial wafers. The ion implanter is capable of implanting 90 5 cm diameter wafers in one processing sequence with a high degree of uniformity.

By combining large area wafers with selective ion implantation the feasibility of producing low-cost GaAs circuits becomes available. By selectively implanting n and n^+ regions a planar processing technique can be used to produce m.m.c.s. The conventional mesa isolation structure is no longer required since the intrinsic isolation afforded by the substrate is directly used. Thus overall process yield is improved since metalisation no longer has to go over mesa edges and some increase in packing density is also possible.

5 Circuit design of analogue monolithic circuits

5.1 Lumped elements

In order to design low-noise amplifiers, for example, with lumped-component matching it is necessary to accurately measure the scattering parameters of a large number of lumped inductors, capacitors and resistors. Fig. 18 shows the physical appearance of single-loop and multiturn spiral inductors and capacitors used at present. The single-loop inductors and interdigitated capacitors are used for small-value components and the multitrun square-spiral inductors and overlay capacitors for larger-value requirements. The equivalent circuits of each element including parasitic capacitances and resistive losses are used in the c.a.d. of monolithic circuits.

The Q values achieved for lumped inductors are shown to be around 50 with the correct geometry, and the interdigital capacitor Q can be optimised by the correct use of dimensions, metal thickness and aspect ratio.

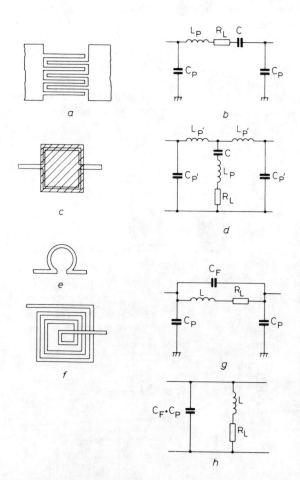

Fig. 18 *Lumped elements for GaAs i.c.s*

a Interdigital capacitor
b Series-capacitor equivalent circuit
c Overlay capacitor
d Shunt-capacitor equivalent circuit
e Single-loop inductor
f Multiturn square spiral inductor
g Series inductor equivalent circuit
h Shunt-inductor equivalent circuit

5.2 Circuit design

Figs. 19a, b and c show three realisations of broadband-matching circuits for the Plessey GAT 5 device, the first being a semidistributed/semilumped design suitable for a monolithic circuit, the second being a purely lumped design and the third being a purely distributed circuit.

Fig. 19d shows the same circuit as 19c but includes discontinuity parasitics caused by the transmission line T junctions, changes in line width etc., following Easter et al.[22] Fig. 20 shows the effects of these discontinuities on the gain of the circuit of Fig. 19c and also shows a reoptimised design including parasitics following c.a.d. It may be appreciated that the effect is dramatic and indeed this is usually the case throughout the frequency range S- to J-band where at the lower frequencies the impedances presented by GaAs f.e.t.s for example tend to be sensitive to loss and certain parasitic components. Fig. 21 is a similar graph resulting from the design of Fig. 19b with and without consideration of lumped-element loss and parasitics.

6 Yield and cost of GaAs monolithic circuits

Just as with the well established Si i.c. technologies of the 1970s and lower-frequency circuits it is envisaged that the GaAs i.c. will be able to make significant inroads into the reduction in the cost of microwave circuits particularly where high-volume requirements are concerned. This is because batch processing of the total microwave circuit becomes a small part of the cost of the units. In the case of phased-array radars where the number of elements can vary from a few hundred to many thousands, GaAs monolithic circuits become attractive in making such systems feasible.

In order to estimate the cost of a typical monolithic circuit it is necessary to assess accurately the overall yield of individual chips throughout the various production stages. In order to do this consider three examples of typical chips using different technological approaches:

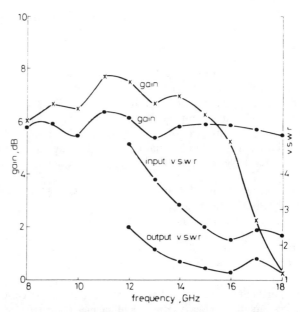

Fig. 20 *Theoretical gain and v.s.w.r. of microstrip distributed amplifier*

x Unoptimised for discontinuity parasitics
● Optimised to include discontinuity parasitics
Only optimised values of v.s.w.r. shown

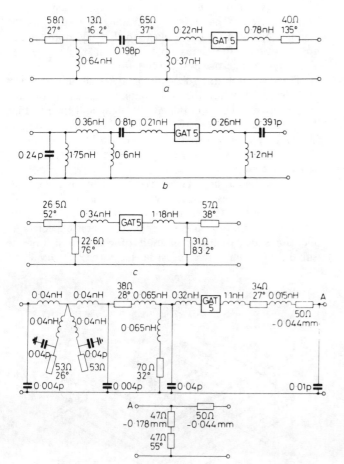

Fig. 19 *Broadband matching circuits for Plessey GAT5 device*

a Semidistributed/semilumped circuit
b Lumped circuit
c Distributed circuit
d Distributed circuit including discontinuities

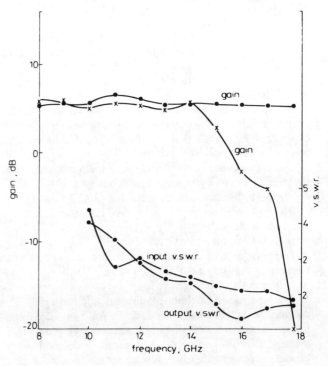

Fig. 21 *Theoretical gain and v.s.w.r. of monolithic amplifiers*

x Unoptimised for L.e. parasitics and loss
● Optimised to include parasitics and loss
Only optimised values for u.s.w.r. shown

Table 1: Yields of GaAs i.c.s on vapour phase epitaxial and ion implanted material

Circuit description	Present small area v.p.e.	5 cm v.p.e.	7·5 cm v.p.e.	5 cm ion implanted	7·5 cm ion implanted
0·3 μm e.b. preamplifier	15%	38%	42%	63%	70%
0·5 μm photolithography mixer	15%	38%	42%	63%	70%
1 μm photolithography i.f. amplifier	30%	46%	50%	65%	73%

(a) The realisation of a low-noise J-band amplifier using electron-beam exposed f.e.t.s with 0·3 μm gate lengths.

(b) The realisation of a low-noise single-ended mixer using dual-gate f.e.t.s with 0·5 μm gate lengths.

(c) The realisation of an i.f. amplifier with an a.g.c. facility using 1 μm gate length f.e.t.s.

Table 1 gives a comparison of the expected yields of the three circuits based on the use of present-day processing techniques. Many of the relatively simple monolithic circuits already realised in the USA and Europe have used small area v.p.e. material. Present yields are dominated by three factors:

(i) edge of wafer defects which account for 35% loss on typical v.p.e. wafer areas

(ii) gate metallisation faults

(iii) variations in d.c. parameters over the slice caused by active-layer thickness variations and material defects; surface defects usually account for a small percentage of failures

As indicated in Table 1 the use of 5 cm diameter GaAs wafers greatly increases the yield while going to ion-implanted material gives an overall yield of around 60%. On 5 cm material edge defects account for 12% of failures.

On this basis, assuming the chip sizes to be approximately 4 mm², 400 0·3 μm gate-length f.e.t. working circuits would be produced per 5 cm ion-implanted wafer while between 500 and 600 working chips for the mixer and i.f. circuits would be produced.

This review of the use of GaAs i.c.s in phased arrays has so far not considered the impact of cost — obviously most important.

In order to accurately assess the cost of a particular monolithically based module, consider the example of combining the three previously mentioned circuits into an overall receiver front end. For circuits presently being considered this involves the chips being put into a microwave package much as shown in Fig. 22 together with d.c. regulation circuits, temperature sensing circuits etc. Thus the overall cost of a module can be subdivided into:

(i) cost of basic materials including GaAs

(ii) cost of packaging — microwave chip package and overall module

(iii) cost of processing GaAs m.m.c.s

(iv) cost of d.c. regulators etc (using the lowest cost techniques available e.g. thick film or i.c.)

(v) assembly cost

(vi) testing cost (including individual chip testing)

As may be appreciated from Table 2, for 5 or 7·5 cm ion-implanted GaAs wafers, the ultimate cost of each module is dictated by the assembly and testing stages. Thus it becomes apparent that, in order to realise the lowest cost, assembly must be reduced by eventually integrating as many functions as possible onto chip and must be made as simple as possible to the point where automatic procedures can be adopted. This is also equally applicable to testing where both d.c. and r.f. testing must be achieved on an automatic basis. Presently automatic d.c. probe testing is achieved and some work on automatic r.f. testing at the wafer stage has also been demonstrated. However, this is an area where much concentration will be needed in the future to enable the lowest production costs to be achieved for the higher volume applications.

The cost of a circuit consists of more than the cost of the component parts — to consider i.c. costs in isolation from the systems they go into may be wrong. However, it has already been indicated that the cost of phased-array components based on GaAs i.c.s is made up of the cost of making the chip, packaging it and testing it, taking account of the yield at each stage. The cost of design and product engineering has to be recovered as well.

The cost of producing the chips depends on the cost of the process in man hours, materials, capital depreciation and the yield. The cost of packaging is related to the number of chips in that package so that the more circuits to be put onto the same chip by increasing packing density the lower the package cost per function.

The cost of testing depends on the test time, the cost and depreciation of the test equipment, man hours and yield. This applies to both wafer probing and final testing. One major advantage of GaAs i.c.s is that they will require the minimum of adjustment and select on test procedures. The cost of assembly depends on package type, cost, labour, yield etc.

A yield loss at any stage is significant but it is desirable to achieve the best possible yield towards the end of the process. Testing cost does not depend heavily on the process technology used. Slice costs are expected to increase substantially as a result of going from 1 μm to sub-0·5 μm geometries because the processes used are

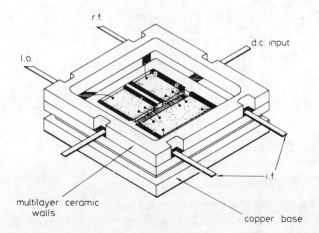

Fig. 22 *S-band GaAs receiver front-end package*

Table 2: Number of working Chips per slice for 5 cm ion implanted wafers

Cost breakdown for complete monolithic receiver front end for 500 off Process	Small area v.p.e. slice	5 cm ion implanted	5 cm ion implanted (2000 off)
preamplifier manufacture	31%	2%	3·5%
mixer manufacture	20%	1%	1·5%
i.f. manufacture	11%	1%	1·5%
packages	4%	10%	9·5%
assembly	25%	63%	56%
testing quality assurance	9%	23%	28%

Assumptions: (i) final number of working slices to 'start' slices is 1 in 3
(ii) assembly of chips into packages gives a 50% yield with 80% tested yield

Results:
- r.f. preamplifier chip — 80
- mixer/l.o. buffer amplifier chip — 100
- i.f. amplifier chip — 175

Figures given are percentage costs of each stage towards total cost

different in terms of mask making, yield and higher capital costs involved. The use of electronbeam technology also produces a relatively lower throughput. Taking into account slice sizes (7·5 cm ion-implanted wafers) and packing density going from 1 μm to 0·25 μm f.e.t. devices will probably increase the chip cost on 'typical' chip complexities by a factor of at least two or three times. Photolithography costs are reflected in the cost of equipment — a factor of 5:1 between conventional u.v. and electron beam. The advantages to be gained in going to sub-0·5 μm geometries on GaAs i.c.s are very uncertain both for analogue and digital circuits because of the high capital cost of electron-beam equipment and the slow rate of throughput. This situation could be transformed by technical advances in electron-beam (e.b.) machines and/or high speed resists which could increase work rate and reduce capital cost per slice. Thus it is seen that the use of 0·5 μm or greater f.e.t. geometries combined with large area GaAs slices, simple packaging techniques and multifunction chips will significantly reduce the overall cost to a point where the use of GaAs i.c.s in phased-arrays systems can become a realistic concept.

7 Conclusions

This paper has attempted to give a general overview of the use of GaAs monolithic circuits as related to phased-array radars. Some of the circuits specifically related to such applications have been described including low-noise receiver preamplifiers, mixers, switches and phase shifters as well as power amplifiers all using the GaAs f.e.t. A summary of yield and cost targets for r.f. monolithic circuits has also been given since this is the area in which system feasibility is often defined.

8 Acknowledgments

The author is grateful to the directors of Plessey Research (Caswell) Ltd. for permission to publish this paper. Some of the work described in this paper is being carried out with the support of the Procurement Executive, Ministry of Defence, sponsored by DCVD.

9 References

1 SUZUKI, T., KADAWAKI, Y., ITO, M., NAKATANI, M., and ISHII, T.: 'Super low noise packaged GaAs FETs for Ku band'. Proceedings of the 1980 IEEE/MTT-S International Microwave Symposium Washington DC, May 1980, Paper S3

2 FUKATA, M., MINURA, T., SUZUKI, H., and SUYAMA, K.: '4 GHz 15 W power GaAs MESFET', *IEEE Trans.*, 1978, **ED-25**, pp. 559–563

3 TSERNG, H.Q., and MACKSEY, H.M.: 'Ultra wideband medium power GaAs MESFET amplifiers'. Digest of technical papers, ISSCC, San Francisco, Feb. 1980, pp. 166–167

4 WADE, P.C., and DRUKIER, I.: 'A low X-band pulsed GaAs FET'. Digest of Technical Papers, ISSCC, San Francisco, Feb. 1980, pp. 158–159

5 PENGELLY, R.S., and TURNER, J.A.: 'Monolithic broadband GaAs f.e.t. amplifiers', Electron. Lett., 1976, **12**, pp. 251–252

6 CRIPPS, S.C., NIELSEN, O., and COCKRILL, J.: 'An X-band dual gate MESFET image rejection mixer'. 1978 IEEE MTT-S International Microwave Symposium Digest, Ottawa, Canada (78CH1355–7 MTT), pp. 300–302

7 VAN TUYL, R.: 'A monolithic GaAs FET signal generation chip'. Digest of Technical Papers, ISSCC, San Francisco, Feb. 1980, pp. 118–119

8 RISTOW, D., ENDERS, N., and KNIEPKAMP, H.: 'A monolithic GaAs Schottky barrier diode mixer for 15 GHz'. European Microwave Conference Digest, Paris, 1978, pp. 707–711

9 COURTENAY, W.: Memorandum, Lincoln Laboratories, MIT, Lexington, Mass. USA

10 GASPARI, R.A., and YEE, H.H.: 'Microwave GaAs FET switching'. IEEE MTT-S International Microwave Symposium Digest, Ottawa, Canada (78CH1355–7 MTT), 1978, pp. 58–60

11 FABIAN, W., VORHAUS, J.L., CURTIS, J.E., and NG, P.: 'Dual-gate GaAs FET switches'. GaAs IC Symposium, Research Abstracts, Lake Tahoe, Sept. 1979, Paper 28

12 GARVER, R.V.: 'Microwave semiconductor control devices', *IEEE Trans.*, 1979, **MTT-27**, pp. 523–529

13 GARVER, R.V.: 'Broadband diode phase shifters', *ibid.*, 1972, **MTT-20**, pp. 314-323

14 JOSHI, J., COCKRILL, J., and TURNER, J.A.: 'Monolithic microwave GaAs FET oscillator'. Digest of GaAs IC Symposium, Lake Tahoe, USA, Sept. 1979

15 TSERNG, H.Q., and MACKSEY, H.M.: 'Microwave GaAs power FET amplifiers with lumped element impedance matching networks'. IEEE MTT-S International Microwave Symposium Digest, (79CH-1355–7 MTT), June 1978, pp. 282–284

16 PUCEL, R.A.: 'Some design considerations for gallium arsenide monolithic circuits'. Active microwave semiconductor devices and circuits, Cornell University, 14th–17th Aug., 1979

17 PENGELLY, R.S., and RICKARD, D.C.: 'Design, measurement and application of lumped elements up to J band'. 7th European Microwave Conference Proceedings, Copenhagen, Sept. 1977, pp. 460–464

18 HOBDELL, J.L.: 'Optimization of interdigital capacitors', *IEEE Trans.*, 1979, **MTT-27**, pp. 788–791

19 SLAYMAKER, N.A., and TURNER, J.A.: 'Microwave FET amplifiers with centre frequencies between 1 and 11 GHz'. Proceedings of the European Microwave Conference, Brussels, 1973, Paper A.5.1

20 KNIGHT, J.R., EFFER, D., and EVANS, P.R.: 'The preparation of high purity Gallium Arsenide by vapour phase epitaxial growth', *Solid-State Electron.*, 1965, **8**, pp. 178–180

21 MANESEVIT, H.M., and SIMPSON, W.I.: 'The use of metal-organics in the preparation of semiconductor materials', *J. Electrochem. Soc.* 1973, **120**, pp. 135–137

Microwave Switching With GaAs FETs

Device and Circuit Design Theory and Applications

Yalcin Ayasli
Raytheon Research Division
Lexington, MA

Introduction

There is a growing need for high-performance, small, versatile, and inexpensive microwave switches in phased-array systems and electronic warfare applications. The use of GaAs FETs as switch elements to help meet these needs has been reported recently in various microwave applications[1-7]. These switches have already been demonstrated to provide subnanosecond switching speeds, multiwatt power-handling capability, virtually zero control power dissipation, and compatibility with monolithic applications.

This article discusses the design considerations for a switch FET, theoretical determination of its equivalent circuit under switching conditions, rf and dc circuit design requirements, and large-signal operation considerations for power switching applications. Design examples and experimental data from monolithic TR switches, phase shifters, and high-power switches will be given.

GaAs FET as a Switch

The FET switch is a three-terminal device with the gate voltage V_g controlling the switch states. In a typical switch mode, the high impedance state corresponds to a negative gate bias larger in magnitude than the pinchoff voltage ($|V_g| > |V_p|$), and the low impedance state corresponds to zero gate bias. These two linear operation regions of the FET are shown schematically in Figure 1.

Note that in either state virtually no dc bias power is required. Therefore the switches can practically be classified as passive as far as the overall power consumption is concerned: this leads to enor-

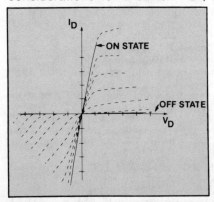

Fig. 1 Two linear operation regions of an FET switch.

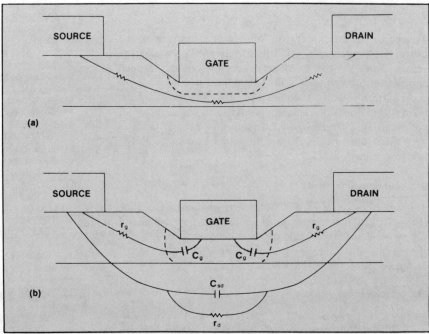

Fig. 2 Schematic cross-section of an FET showing various resistive and capacitive regions. a) No gate bias b) $|V_g| > |V_p|$.

Reprinted with permission from *Microwave J. Magazine*, vol. 25, pp. 61-64, 66, 68, 70-72, and 74, Nov. 1982.
Copyright © 1982 by Horizon House-Microwave, Inc.

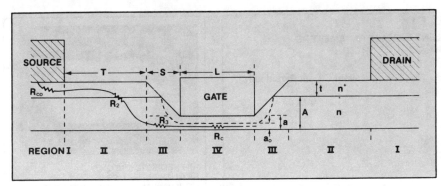

Fig. 3 Schematic cross-section of an FET showing various resistive sections contributing to the R_{ON} of the switch.

mous simplifications in driver requirements. Although the FET itself is a three-terminal device, the switch is bidirectional.

The cross-section of a simple FET structure appears in Figure 2. When no gate bias is applied, the channel is open except for the zero field depletion layer thickness. Hence, for current levels less than the saturated channel current I_{dss}, the FET can be modeled as a linear resistor.

When a negative gate voltage V_g is applied between gate and source so that $|V_g| > |V_{pinchoff}|$, the channel can be completely depleted of free charge carriers. Under this bias condition, the FET can be modeled by series and parallel combination of resistors and capacitors. The approximate region of the FET responsible for each element is shown in Figure 2b.

Assuming that the gate termination represents a high rf impedance at the frequency of operation, the OFF-state equivalent circuit can be expressed as a parallel combination of a resistor and a capacitor. For $1/\omega C_g \gg r_g$, the effective drain-to-source capacitor is simply $(C_{sd} + C_g/2)$ and the effective drain resistor is the parallel combination of R_d and $2/(\omega^2 C_g^2 r_g)$. The figure of merit for a switch FET can simply be expressed as the ratio of its effective OFF state resistance to its ON resistance.

Design Consideration for a Switch FET

Let us examine the effect of various device parameters on the device equivalent circuit. The important parameters are the channel geometry, gate length, the channel doping and the pinchoff voltage of the FET.

— R_{ON}

The ON resistance of a switch FET has contributions from four components, which are schematically illustrated in Figure 3. With the notation of Figure 3, it is clear that:

$$R_{ON} = R_c + 2(R_{co} + R_2 + R_3).$$

Each of these resistive components can be related to the channel parameters of the FET. Let us examine each one separately.

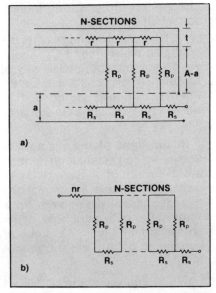

Fig. 4 Equivalent incremental resistance circuit for calculating the contribution of Region II. a) actual resistance distribution b) approximate resistance distribution for $R_s \gg r$.

R_{co}: The constant resistance R_{co} can be calculated as[8]:

$$R_{co} = \frac{2.1}{W\, t^{0.5}\, n_+^{0.66}}\ \Omega$$

where W is the gate periphery in mm, t the thickness of the n+ region under the source contact in μm. n_+ is the doping density of the contact layer expressed in units of 10^{16} cm^{-3}.

R_2: To estimate the value of R_2, region II can be modeled in terms of incremental resistances, as shown in Figure 4.

However, to obtain a simple formula, we can use the circuit model in Figure 4 when $R_s \gg r$. Making an analysis similar to Berger[9], and using Fukui's formula for the resistivity of GaAs[8], the resistance R_2 can be expressed in a closed-form equation as:

$$R_2 = \frac{1.1\, T}{W\, t\, n_+^{.82}} + \frac{1.1}{w\, n^{.82}} \left(\frac{A-a}{a}\right)^{0.5}\ \Omega.$$

In this paper, all the doping levels are in units of 10^{16} cm^{-3}, and all the lengths are expressed in μm except the gate periphery W, which is in mm.

R_3: This is the contribution of the bevelled section. Here it is difficult to define the current paths with any accuracy. We will assume that the conduction region is confined to a region of height a and length S. If we further assume that the slope of the bevel is 45°, then $S = t + A-a$. Hence, R_3 becomes:

$$R_3 = \frac{1.1\,(t + A-a)}{W\, a\, n^{.82}}\ \Omega.$$

R_c: The open channel resistance can be estimated as[8]:

$$R_c = \frac{1.1\, L}{W\, a_o\, n^{.82}}\ \Omega.$$

In this equation, a_o is the fraction of the channel which is open. For GaAs FETs with a given pinchoff voltage V_p and channel doping, it can be calculated as:

$$a_o = \frac{.378}{n^{0.5}}\left[(V_p + 0.85)^{0.5} - (0.85)^{0.5}\right]$$

The channel resistance calculated through the equations above is typically a lower limit to what one observes experimentally. This could be due to the proximity of the interface between the channel and buffer layer, which makes it difficult to precisely define either the mobility or the doping level under the gate.

TABLE I
CALCULATION OF EQUIVALENT CIRCUIT PARAMETERS FOR A TYPICAL 1 MM GaAs FET SWITCH

Design Parameters	Equivalent Circuit Parameters
W = 1 mm	R_{ON} = 2.7 Ω
L = 1 μ	C_{sd} = 0.14 pF
V_p = -4 V	R_{OFF} = 3 KΩ
n = 10^{17} cm^{-3}	R_g = 1.4 Ω
n^+ = 2 x 10^{18} cm^{-3}	C_g = 0.22 pF
A = .37 μm	
T = 2 μm	
S = 0.3 μm	
t = 0.2 μm	
a_o = 0.15 μm	
a = 0.26 μm	

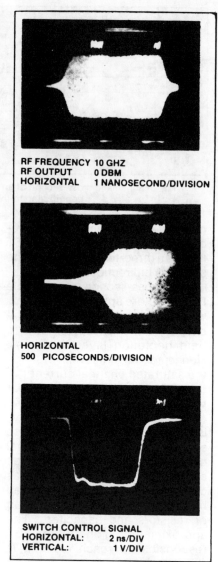

RF FREQUENCY 10 GHZ
RF OUTPUT 0 DBM
HORIZONTAL 1 NANOSECOND/DIVISION

HORIZONTAL
500 PICOSECONDS/DIVISION

SWITCH CONTROL SIGNAL
HORIZONTAL: 2 ns/DIV
VERTICAL: 1 V/DIV

Fig. 5 Monolithic TR switch switching data.

— OFF-State Equivalent Circuit Parameters

C_{sd}: The source-drain capacitance C_{sd} represents the fringing capacitance between the source and drain electrodes.

A good estimate of these capacitances can be obtained by considering the electrostatic coupling between two parallel conductors on a surface of a semi-infinite dielectric medium, representing the GaAs chip. Using this model, the capacitance expression becomes[10]:

$$C_{sd} = (\varepsilon_r + 1)\varepsilon_o W \frac{K\left[(1-k^2)^{1/2}\right]}{K(k)}$$

where K(k) is the complete elliptic integral of the first kind. The argument k is related to the geometry of electrodes as:

$$k = \left[\frac{(2L_s + L_{sd})L_{sd}}{(L_s + L_{sd})^2}\right]^{1/2}$$

where L_{sd} is the interelectrode spacing between the drain and source electrodes. In these expressions, it is assumed that $L_s = L_d$ and $L_s \gg L_{sd}$.

Source-drain capacitance is typically in the 0.14 pF range for 1 mm FETs and it is independent of device parameters such as channel doping and pinchoff voltage.

r_d: The resistor in parallel with C_{sd} represents the rf losses associated with C_{sd}. Our experience indicates that 3 KΩ for 1 mm gate devices representing a Q of around 25 at X-band frequencies is a reasonable choice.

C_g: C_g represents the drain-to-gate and gate-to-source capacitances. These capacitances are equal because of symmetry. If there is a gate pad of significant area, its capacitance to ground needs to be added to the equivalent circuit.

There is no precise way of estimating C_g. A simple-minded approach that gives C_g values in good agreement with experimental results is to use half the gate capacitance with the channel fully depleted so that:

$$C_g = 0.06 \frac{WL}{a} \text{ pF}.$$

Note that a is related to both channel doping and the pinchoff voltage of the device.

r_g: The resistor r_g represents the changing resistance of C_g. Again, we do not know a precise way of calculating this resistance. Experimental evidence suggests that one-half of R_{ON} is a reasonable value for r_g.

In the light of the discussion above, we can estimate the equivalent circuit of a typical GaAs switch FET. Table 1 summarizes typical channel parameters for a 1 mm total gate periphery, 1 μm gate length FET with -4 V pinchoff voltage and 10^{17} cm^{-3} channel doping.

These equivalent circuit values translate to equivalent off-to-on resistance ratios of 800 at 10 GHz. This ratio decreases to 420 at 20 GHz, which is probably the end of the useful range for 1 μm gate length switch FETs. To push the operating frequency to higher frequencies, one needs to go to submicron gate lengths to decrease C_g and R_{ON} and thus increase the switching equivalent resistance ratio.

We can also estimate the switching speed from the equivalent circuit. Assuming that the gate bias circuit is fed at 50 ohms impedance level and a total of 6 pFs are used in the bias circuit low pass filter, one reaches the conclusion that charging time constants are in the 0.3 nsec range. Experimental evidence also indicates that switching times of 1 nsec are quite possible, as shown in Figure 5. By optimizing the gate bias circuit design, switching times which are significantly smaller than 1 nsec are feasible.

Circuit Design Considerations

Switching circuits with FETs can be designed in essentially the same way as PIN diodes; this is done using on and off state equivalent circuits as required in the overall circuit configuration.

Although in this article we will discuss the monolithic circuit applications of switching FETs, it should be pointed out that it is quite possible also to use discrete FETs in hybrid form. Similar to PIN diodes, the FETs can be in series or shunt mode with respect to the transmission lines. There are, however, some salient features of FET switch circuit design that need to be mentioned.

The FET is a three-terminal device; the switching occurs only through the gate control voltages and no other bias is required for the operation of the phase shifter. The rf transmission lines do not carry any dc voltage and therefore there is no need for dc blocking capacitors between various switch elements: a significant design advantage.

Gate control voltages corresponding to the two switch states are $V_{g1} = 0V$ and $V_{g2} = -V$ where $V > V_p$. In either state, the gate junction is reverse biased and the gate current is either zero or negligible. Hence, the fact that switching control voltages need to be applied at negligible currents simplifies the requirements of the control circuit design.

In the off-state of the FET switch, note that gate-drain and gate-source capacitances are equal because both source and drain terminals are at ground potential. As a consequence of this, the drain terminal is not isolated from the gate terminal: the rf impedance of the gate bias circuit very much affects the equivalent drain-source impedance. In our designs, the gate bias circuit is configured as a two-section low pass filter providing an effective rf open to the FET at the gate terminal. When gate terminating impedance is very high, the equivalent drain-source capacitance can simply be approximated as $C_d + C_g/2$.

Note that the total drain capacitance shunting r_d represents a reactance of the order of 50 ohms at X-band frequencies. Therefore, to realize the switching action, this capacitance must be either resonated or its effect must be included in the design of the impedance matching sections. This is an important design consideration for FET switches, as it directly relates to the operation bandwidth.

Tuning out the effective drain-source capacitance can simply be accomplished by connecting an inductive reactance between the drain and source terminals. Monolithic circuit technology also allows distributing the switch FET and its associated drain capacitance along a transmission line structure. Consider the distributed switch approach shown in Figure 6[5]. In this structure, the drain-source capacitance and the overlay inductance are treated as the per-unit-length capacitance and per-unit-length inductance of an artificial transmission line. The source pad is grounded by a via hole.

The configuration shown in Figure 6 has two major design advantages: First, by integrating the FET into the rf transmission line, the intrinsic switch element of the FET is placed at the point where it is most effective. This eliminates the contribution of undesirable FET parasitic elements such as the drain pad capacitance and the extra transmission line sections that would have been required to connect a discrete FET to the rf line. Second, by including the effect of the overlay inductance itself or by adding extra inductive elements between sections, the drain-source capacitance of the FET in the high impedance state can be effectively tuned out

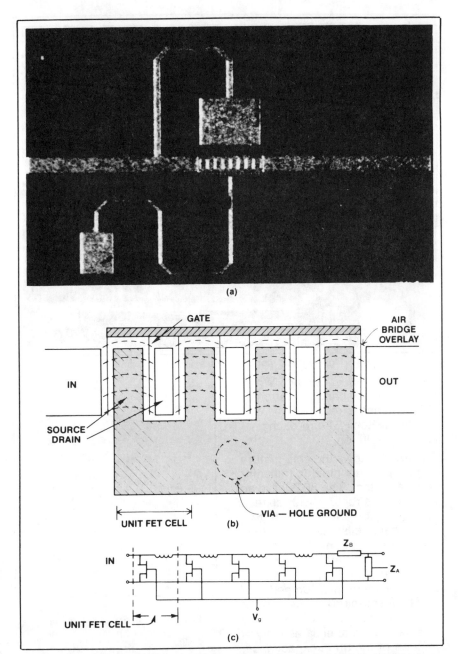

Fig. 6 Monolithic distributed switch approach. a) 1 × 1 switch in monolithic form b) FET structure c) equivalent circuit.

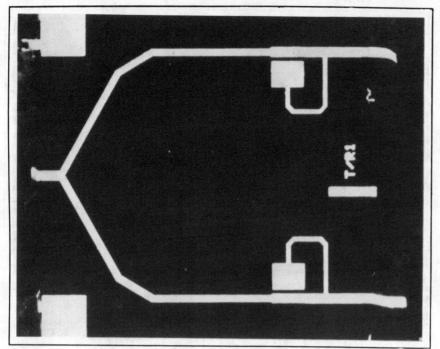

Fig. 7 X-band monolithic transmit/receive switch.

over a wider frequency range. When wide frequency band operation of FET switches is required, the distributed switch FET approach provides the solution.

Figure 7 shows the chip picture of an X-band monolithic transmit/receive switch[3]. The FETs have an interdigital structure with sixteen 100 μm wide channels. If two single-gate cells sharing a single drain finger are considered to be a unit cell, then there are eight unit cells connected by an overlay structure to form a lumped element transmission line. Because in this circuit the overlay inductance is small, it compensates only a fraction of the drain capacitance. The U-shaped shorted stub acts as the main tuning element. The experimental results for the switch indicate a 1 dB insertion loss bandwidth of 8-12 GHz with minimum insertion loss around 0.5 dB. Isolation between the transmit and receive arms is better than 30 dB across the band. The chip dimensions are 3 x 3 x 0.1 mm.

For microwave power switching using FETs, there are additional considerations. One of these considerations is the maximum allowable rf voltage swing across the device. The variation of the rf voltage on the drain and gate terminals with respect to the grounded source is shown in Figure 8 for one period[7].

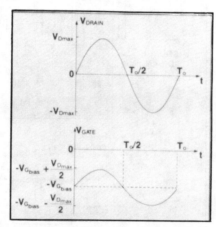

Fig. 8 Variation of the large signal rf voltages on drain and gate terminals over one period.

In this figure, it is assumed that the gate terminal of the switch FET is rf open. This condition should be realized in the design of the gate bias circuitry.

Under the above assumption and because gate-to-drain impedance and gate-to-source impedance are equal, half of the drain voltage swing appears in the gate terminal, as illustrated in Figure 8. Constraints on the terminal voltages can be summarized as follows. During the first half of the period, the total gate voltage should not fall below the pinchoff voltage V_p. During the entire cycle, the difference between the drain and gate voltages should not exceed the gate-drain breakdown voltage.

These constraints can be expressed mathematically as:

$$-V_{G_{bias}} + \frac{V_{Dmax}}{2} = -V_p$$

and

$$V_{Dmax} + V_{G_{bias}} - \frac{V_{Dmax}}{2} = V_B.$$

From these two equations, the maximum allowable drain voltage and the required gate bias condition can be solved as:

$$V_{Dmax} = V_B - V_p$$

and

$$V_{G_{bias}} = \frac{V_B + V_p}{2}.$$

Hence, if the impedance level that the switch FET sees in its high impedance state is Z_o, then the maximum power that can be transmitted in the switch closed position can be calculated as:

$$P_{max} = \frac{1}{2} \frac{(V_B - V_p)^2}{Z_o}.$$

Using the notation of the schematic TR switch shown in Figure 9, Z_o can approximately be calculated by transferring the 50 ohm antenna terminal impedance through the quarter wavelength MS line of characteristic impedance Z_1, as:

$$Z_o \simeq \frac{Z_1^2}{50}.$$

Hence,

$$P_{max} = 25 \frac{(V_B - V_p)^2}{Z_1^2}.$$

The maximum power calculation under switch closed condition is relevant to the transmitter arm of a transmit/receive (TR) switch. For the receiver arm, the constraints are different. During the time when the transmitter power is on, the switch on the receiver arm is in the open state. In this low impedance condition, it should be able to sustain essentially the short circuited current in the receive arm due to the transmitter pulse. Again using the notation of Figure 9, the peak value of this current can be calculated as:

$$I_{max_{switch\ open}} = \frac{\sqrt{100\ P_{transmitted_{max}}}}{Z_2}$$

Equations for P_{max} and I_{max} fully specify the constraints on the switch FETs under power-switching conditions. First note that constraints on the transmitter and the receiver arm switches are completely independent. Also note that in the equation for P_{max}, periphery of the device is not a parameter. Thus, once Z_1 is chosen from power requirements, the device periphery on the transmitter arm can be determined purely from small signal insertion loss analysis. On the other hand, the equation for I_{max} does bring in the device periphery as a design parameter for the receiver arm switch, since the maximum current that a device can support in its linear region before it reaches the saturation is directly proportional to its gate periphery. The characteristic impedance of the receiver arm, Z_2, also enters into the equation. It is easy to see that, by increasing Z_2, one can meet the requirement for I_{max} readily.

The switch described in Figure 9 is fabricated on 0.1 mm GaAs substrate[7]. A photograph of the finished chip is shown in Figure 10. The chip dimensions are 4.5 x 3.7 mm. In the transmit arm, a single-gate, interdigitated 3.2 mm total gate periphery FET is used. The receive arm uses a single-gate FET of 1.6 mm total periphery.

The dc gate circuitry is provided monolithically on-chip. It is essentially a low pass filter which provides high impedance to the gate of the device and isolates the bias pad at the edge of the chip from rf leakage, at X-band frequencies.

The experimental data indicates insertion losses in the 0.8-1.6 dB range with higher than 25 dB isolation. No degradation from small signal performance is observed up to 10 W of CW microwave power[7].

The FET switches, with their fast switching times and negligible dc power consumption, are ideal for passive phase-shifting circuits. As an example for this application, consider the X-band four-bit phase shifter chip shown in Figure 11 with 22.5°, 45°, 90° and 180° phase bits[6]. Starting in the upper right with the 180° bit, the microwave signal travels counterclockwise through the 45° (upper left), 22.5° (lower left) and 90° (lower right) bits, exiting on the right edge of the chip. The circuit is passive and reciprocal so that the signal can equally well traverse this path in the opposite direction. The chip size is 6.4 x 7.9 x 0.1 mm.

The schematic circuit diagram of the four-bit phase shifter is shown in Figure 12. The 22.5 and 45 degree bits are designed to provide constant phase shifts over the frequency bandwidth using the loaded line technique. Each loading stub is composed of a suitably designed three section transforming and matching network which is terminated by a 1200 μm switch FET. The principle of operation of this circuit and of the ones using PIN diodes is the same.

The 90° and 180° bits are designed using the switched-line technique. Note, however, that instead of the conventional four switching elements, only three 800 μm switch FETs are used in these circuits. Equal insertion loss between two phase states is maintained by designing the short and long arms of the phase shifter at different impedance levels.

For all bits, the switching is performed only through the gate control voltages and no other bias is required for the operation of the phase shifter. Thus, rf microstrip lines do not carry any dc voltage (in fact, they are dc grounded) and therefore there is no need for dc blocking capacitors between individual phase bit circuits.

Experimental performance of the phase shifter indicates 5.1 ± 0.6 dB

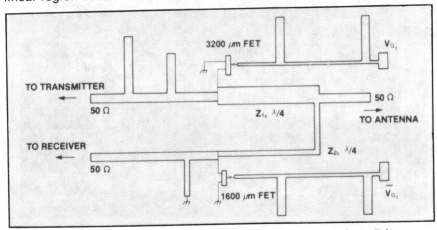

Fig. 9 The schematic circuit diagram of the 10 W transmit-receive switch.

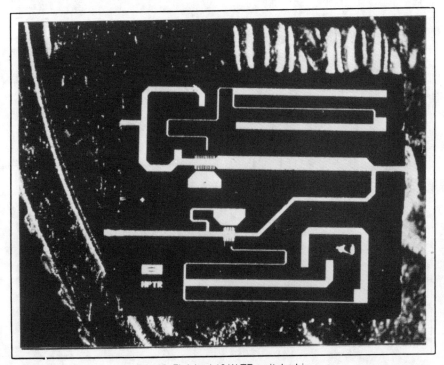

Fig. 10 Finished 10 W TR switch chip.

insertion loss with 16 distinct phase states between 0° and 360°[6]. The performance of the phse shifter is satisfactory for typical phase array applications. Their small size, negligible dc power requirements and subnanosecond switching times make the monolithic phase shifters good candidates for future frequency-agile airborne phased-array systems.

Conclusions

The GaAs FET is examined in detail as a microwave frequency switching element. Its equivalent circuit as a switch is related to device geometry and channel parameters. The rf circuit design considerations characteristic of FET switching circuits are discussed with examples from recent monolithic circuit designs. These examples demonstrate the versatility of GaAs FETs as switches.

When circuit size, dc power consumption, switching speed and producibility in large quantities are of prime importance, GaAs monolithic FET switch circuits will lead the way for future systems applications.

REFERENCES

1. Gaspari, R.A., and H.H. Yee, "*Microwave GaAs FET Switching,*" 1978 IEEE International Microwave Symposium Digest, pp. 58-60.
2. McLevige, W.V., and V. Sokolov, "*Microwave Switching with Parallel-Resonated GaAs FETs,*" IEEE Electron Device Lett. (8), (Aug. 1980).
3. Ayasli, Y., R.A. Pucel, J.L. Vorhaus, and W. Fabian, "*A Monolithic X-Band Single-Pole, Double-Throw Bidirectional GaAs FET Switch,*" IEEE GaAs Integrated Circuits Symposium, Nov. 4-6, 1980.
4. McLevige, W.V., and V. Sokolov, "*Resonated GaAs FET Devices for Microwave Switching,*" IEEE Trans. Electron Devices ED-28, pp. 196-204 (Feb. 1981).
5. Ayasli, Y., J.L. Vorhaus, R.A. Pucel, and L.D. Reynolds, "*Monolithic GaAs Distributed FET Switch Circuits,*" IEEE GaAs Integrated Circuit Symposium, Oct. 27, 1981, San Diego, California.
6. Ayasli, Y., A. Platzker, J.L. Vorhaus, and L.D. Reynolds, "*A Monolithic Single-Chip*

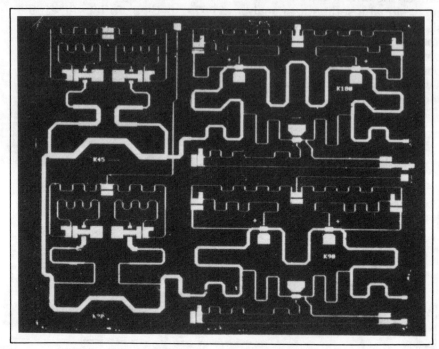

Fig. 11 Four-bit passive phase shifter chip.

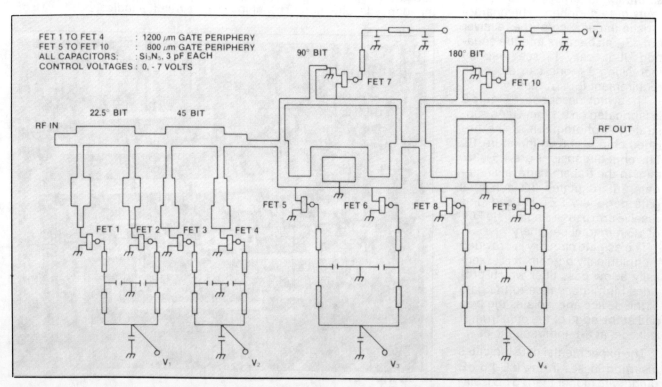

Fig. 12 The schematic circuit diagram of the four-bit phase shifter.

X-Band Four-Bit Phase Shifter," IEEE MTT-S International Microwave Symposium, Dallas, Texas, June 15-17, 1982.

7. Ayasli, Y., R. Mozzi, L. Hanes, and L.D. Reynolds, *"An X-Band 10-W Monolithic Transmit-Receive GaAs FET Switch,"* IEEE Microwave and Millimeter-wave Monolithic Circuits Symposium, Dallas, Texas, June 18, 1982.

8. Fukui, H., *"Determination of the Basic Device Parameters of a GaAs MESFET,"* Bell Syst. Tech. J. 58 (3), p. 771 (March 1979).

9. Berger, H.H., *"Contact Resistance on Diffused Resistors,"* 1969 IEEE ISSCC Digest of Technical Papers, Feb. 1969, pp. 160-161.

10. Pucel, R.A., H.A. Haus, and H. Statz, *"Signal and Noise Properties of Gallium Arsenide Microwave Field-Effect Transistors,"* Advances in Electronics and Electron Physics, Vol. 38, p. 195, 1975.

A Monolithic Single-Chip X-Band Four-Bit Phase Shifter

YALCIN AYASLI, MEMBER, IEEE, ARYEH PLATZKER, MEMBER IEEE, JAMES VORHAUS, AND LEONARD D. REYNOLDS, JR.

Abstract — X-band GaAs monolithic passive phase shifter with 22.5°, 45°, 90°, and 180° phase bits are developed using FET switches. By cascading all four bits, a four-bit digital phase shifter with 5.1 ± 0.6-dB insertion loss is realized on a single $6.4 \times 7.9 \times 0.1$-mm chip.

I. INTRODUCTION

THE USE OF FET's as microwave switches has been reported earlier [1]–[4]. In this paper we will describe the use of such FET switches in two different types of GaAs monolithic passive phase shifter bits. We will also describe the fabrication and performance of a single-chip MMIC X-band four-bit passive phase shifter formed by cascading these switch FET circuits.

The basic switch element, a single-pole single-throw circuit in shunt mode of operation is shown in Fig. 1(a). The FET switch is a three-terminal device with the gate voltage V_G controlling the switch states. In a typical switch mode, the high impedance state (switch closed) corresponds to a negative gate bias larger in magnitude than the pinchoff voltage ($|V_g| > |V_p|$), and the low-impedance state (switch open) corresponds to zero gate bias. These two linear operation regions of the FET are shown schematically in Fig. 1(b). Note that in either state virtually no dc bias power is required. Therefore, the switches can practically be classified as passive as far as the overall power consumption is concerned.

Although the FET itself is a three-terminal device, the switch is bidirectional. The equivalent circuit for the two states of the switch can be represented as shown in Fig. 2, for a typical 1000-μm switch FET. These values are dependent on the channel geometry, channel doping, and pinchoff voltage of the device. The important point to note about this equivalent circuit is the following. Gate-drain and gate-source capacitances are equal because both source and drain terminals are at ground potential. As a consequence of this, the drain terminal is not isolated from the gate terminal: the RF impedance of the gate bias circuit very much affects the equivalent drain-source impedance. In the present design, the gate bias circuit is configured as a two-section low-pass filter providing an effective RF open to the FET at the gate terminal. With this approach, the

Manuscript received March 23, 1982; revised July 24, 1982.
Y. Ayasli, J. Vorhaus, and L. D. Reynolds are with the Raytheon Company Research Division, Lexington, Ma 02173.
A. Platzker is with the Raytheon Company, Missile Systems Division, Bedford, MA 01730.

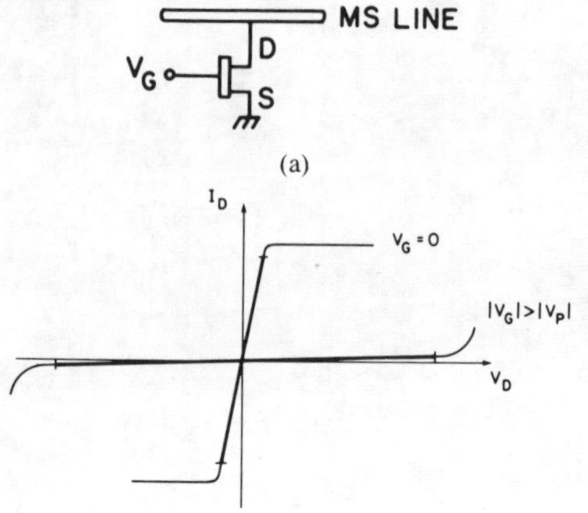

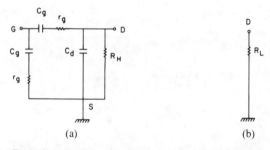

Fig. 1. (a) Basic GaAs switch in shunt mode of operation. (b) Switch FET linear operating regions.

Fig. 2. Equivalent circuit used in the designs for a switch FET. (a) High resistance state; (b) low resistance state. $R_L = 3$ Ω, $R_H = 3$ kΩ, $C_d = 0.2$ pF, $r_g = 1.6$ Ω, $C_g = 0.2$ pF.

equivalent drain–source capacitance can simply be approximated as $C_d + C_g/2$.

Note that, unlike the p-i-n diode, the total capacitor shunting of high impedance R_H represents a reactance of the order of 50 Ω at X-band frequencies. Therefore, to realize the switching action, thus capacitance must be either resonated or its effect must be included in the design of the impedance-matching sections. This represents an important design consideration for FET switches as it directly relates to the frequency bandwidth of operation.

II. FOUR-BIT PHASE-SHIFTER CIRCUIT

The schematic circuit diagram of the four-bit phase shifter is shown in Fig. 3. The 22.5° and 45° bits are designed to provide constant phase shifts over the frequency bandwidth using the loaded-line technique. Each loading

Reprinted from *IEEE Trans. Microwave Theory Tech.*, vol. MTT-30, pp. 2201–2205, Dec. 1982.

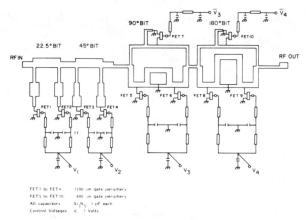

Fig. 3. The schematic circuit diagram of the four-bit phase shifter.

Fig. 4. Picture of individual phase bit circuits.

stub is composed of a suitably designed three-section transforming and matching network which is terminated by a 1200-μm switch FET. The principle of operation of our circuits and of the ones using p-i-n diodes is the same. Both utilize the fact that the phase of a signal passing through a loaded transmission line is a function of that load. A phase shift is obtained when the load is altered between two states. If the main line is symmetrically loaded, it is possible to obtain, over appreciable bandwidths, phase shifts at approximately constant insertion losses and low input and output VSWR. In order to do so, the loads seen by the main RF line at the two phase states must assume two distinct values. In general, the impedances of the switching elements, whether they are p-i-n diodes or MESFET's, do not present the proper impedances to the main line and thus require transformation. The transformation is bilinear and its physical realization constitutes the design of the circuits.

The 90° and 180° bits are designed using the switched-line technique. Switching between lines of different electrical lengths is accomplished by two single-pole double-throw (SPDT) switches similar in principle of operation to the X-band bidirectional switch reported earlier [4]. Note, however, that instead of the conventional four switching elements, only three 800-μm switch FET's are used in these circuits. Equal insertion loss between two phase states is maintained by designing the short and long arms of the phase shifter at different impedance levels.

The insertion loss in the long arm of the phase shifter has contributions from three sources. These are the microstrip line losses, the OFF-state losses of the two FET's in the long arm, and the ON-state loss of the FET in the short arm. The insertion loss of the short arm has contributions from the microstrip line losses, the OFF-state loss of the FET in the short arm, and the ON-stage losses of the two FET's in the long arm. If the line impedances are chosen equal in both arms, then the insertion loss of the long arm is always greater than the insertion loss of the short arm. By increasing the impedance level of the short arm, however, its insertion loss can be increased, while the contribution of the ON-state loss of the FET in the short arm to the

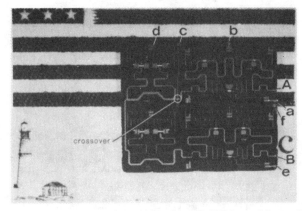

Fig. 5. Four-bit passive phase shifter chip. A = RF input/output, B = RF output/input, a = 180° bit reference arm control signal, b = 180° bit phase delay arm control signal, c = 22.5° bit control signal, d = 45° bit control signal, e = 90° bit reference arm control signal, f = 90° bit phase delay arm control signal.

TABLE I
FOUR-BIT PASSIVE PHASE SHIFTER STATISTICS

```
Chip Size:  6.4 x 7.9 x 0.1 mm
Number of Ports (all have integral beam leads):
    rf:  2
    Control Signal:  6
Number of FETs:  10
Total Gate Periphery (Gate Length = 1 μm):  9.6 mm
Number of Capacitors (3 pF each):  16
Number of Air Bridges:  77
Number of Via Holes:  26
Total Transmission Line Length:  13.9 cm (5.5 in.)
```

long arm insertion loss is decreased. Thus, it is possible to use only three switching FET's for the switched line phase shifter and still equalize the insertion losses of the two phase states.

In the present design, the long arms of both 90° and 180° phase bits are 58-Ω MS lines, whereas the 90°-bit short arm is a 74-Ω line, and the 180°-bit short arm is an 86-Ω line. The increase from 74 Ω to 86 Ω reflects the need for additional loss compensation for the 180° bit. To minimize the total gate periphery used, the 90° and 180°

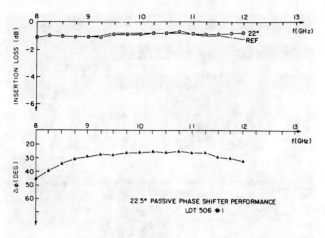

Fig. 6. 22.5° passive phase shifter performance (not corrected for jig losses).

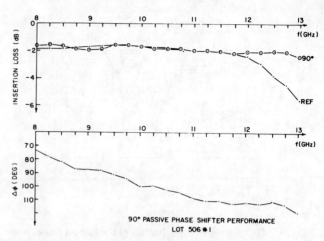

Fig. 8. 90° passive phase shifter performance (not corrected for jig losses).

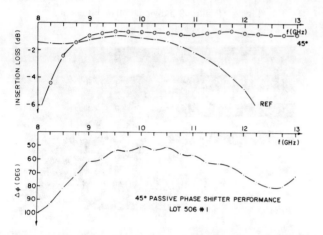

Fig. 7. 45° passive phase shifter performance (not corrected for jig losses).

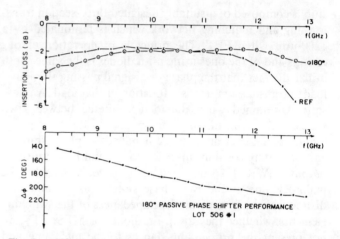

Fig. 9. 180° passive phase shifter performance (not corrected for jig losses).

phase bits are designed at the 65-Ω impedance level. The transformation back to 50-Ω is achieved by one-section impedance transformation.

For all bits, the switching is performed only through the gate control voltages and no other bias is required for the operation of the phase shifter. Thus, RF microstrip lines do not carry any dc voltage (in fact they are dc grounded) and therefore there is no need for dc blocking capacitors between individual phase bit circuits.

The realization of the individual phase shifter bits are shown in Fig. 4. The chip size is $3.15 \times 4.95 \times 0.1$ mm for the larger bits and $3.15 \times 3.15 \times 0.1$ mm for the smaller bits.

The four bits are cascaded to form a complete phase shifter on a single $6.4 \times 7.9 \times 0.1$-mm GaAs chip (Fig. 5). Starting in the upper right with the 180° bit, the microwave signal travels counterclockwise through the 45° (upper left), 22.5° (lower left), and 90° (lower right) bits, exiting on the right edge of the chip. The circuit is passive and reciprocal so that the signal can equally well traverse this path in the opposite direction.

As identified in Fig. 5, there are two control lines for each of the larger bits and one control line for each of the smaller bits. Also seen in the figure is the one crossover necessary in the circuit where the RF line between the 180° and 45° bits crosses the line connecting the control port of the 22.5° bit with the chip edge. All the necessary gate bias circuitry including RF bypass capacitors is provided monolithically on the GaAs chip, with integral beam leads for RF and control lines at the edge of the chip. The FET switches in each phase bit have a total gate periphery of 2.4 mm. The four-bit passive phase shifter chip statistics are summarized in Table I.

III. Circuit Fabrication

Circuits are processed on vapor phase epitaxy layers grown by the $AsCl_3$ system on semi-insulating GaAs substrates. The three layer structures consist of a high-doped contact layer ($n > 2 \times 10^{18} cm^{-3}$, $t = 0.2$ μm), an active layer of moderate doping ($n = 9 \times 10^{16} cm^{-3}$, $t = 0.4$ μm), and an undoped buffer region ($n < 5 \times 10^{13} cm^{-3}$, $t = 2.0$ μm). Device isolation is achieved with a combination of a shallow mesa etch and a damaging $^{16}O^+$ implant.

Ohmic contacts are formed by alloying the standard Ni/AuGe metallization into the surface. The ohmic metal also forms the bottom plates of the thin-film capacitors. The gates, which are recessed, consist of a Ti/Pt/Au (1000/1000/3000 Å) metallization and are nominally 1 μm long.

The capacitor dielectric is a plasma-assisted CVD silicon nitride layer with a nominal thickness of 5000 Å and a relative dielectric constant of 6.8. The final frontside processing steps define the transmission line structures, capacitor top plates and air-bridge interconnects. All of these are fabricated out of plated gold about 3 to 4 μm thick. The air-bridges are used to connect from the GaAs surface to the top plates of the MIM capacitors without having to cross the dielectric step and risk shorting of the structure.

After plating, the wafer is lapped to its final thickness of 100 μm by first mounting it upside down on an alumina substrate. Via-holes are etched through the wafer to ground points on the frontside. The via-holes are aligned by looking through the slice with infrared optics to see the frontside pattern. Finally, a chip dicing grid is defined in the back by alignment to the via-hole pattern and the region between the grid lines (the chip back) is plated to a thickness of 12 to 15 μm with gold. The grid lines are etched through to the frontside, the wafer dismounted, and the chips allowed to simply fall apart.

IV. THE EXPERIMENTAL PERFORMANCE

The insertion loss and relative phase shift for each of the four bits are presented in Figs. 6 to 9. None of the insertion-loss data is corrected for the approximately 0.5-dB test fixture loss. The differential phase shift follows the linear phase versus frequency behavior for the switched line 90° and 180° phase bits and the constant phase versus frequency behavior for the 22.5° and 45° loaded-line phase bits. The phase error at around 10 GHz is less than 10° for any bit.

The four individual phase shifter bits are cascaded to form a complete phase shifter. The text fixture with the circuits is shown in Fig. 10. The overall insertion loss for each of the sixteen phase states is shown in Fig. 11. There is about 0.5 dB of test fixture loss which has not been subtracted from this data. The cascaded phase shifter thus has slightly less than 5 dB of total insertion loss with a variation over all states of less than ±0.3 dB across the design band. Fig. 12 shows the relative phase shift of each of the phase shifter states.

Encouraged by the performance of the cascaded circuit, we have fabricated the first single-chip four-bit passive phase shifter circuit, described earlier.

This single-chip four-bit phase shifter is characterized in the measurement jig shown in Fig. 13.

Fig. 14 shows the insertion loss (not corrected for approximately 1 dB of jig losses) for all 16 states. At 9.5 GHz, the insertion loss is 5.1 ± 0.6 dB. Fig. 15 shows the differential phase shift in the 8.5- to 10.5-GHz frequency band. The circuit is matched to 50-Ω input and output system with better than 10-dB return losses for all 16 states over the 2.5-GHz frequency band.

The phase linearity of the reference state is also measured. Deviation of phase around a defined zero is less than ±10° for larger than 3-GHz frequency band.

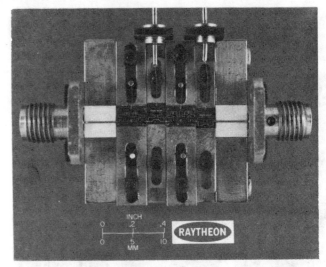

Fig. 10. Cascaded passive phase shift bit chips in test fixture.

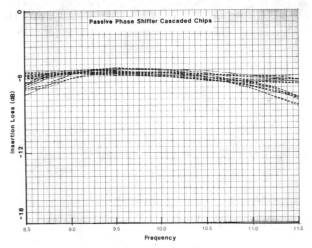

Fig. 11. Passive phase shifter cascaded chips insertion loss performance (not corrected for jig losses).

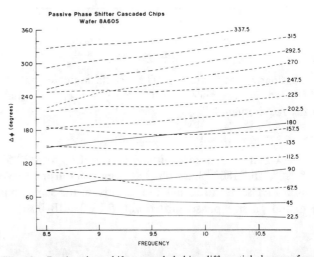

Fig. 12. Passive phase shifter cascaded chips differential phase performance.

Total RMS phase error of the four-bit phase shifter and the predicted performance of a typical airborne phased-array antenna system utilizing this monolithic phase shifter

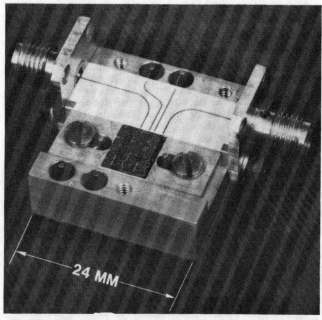

Fig. 13. Measurement jig for the single-chip four-bit phase shifter.

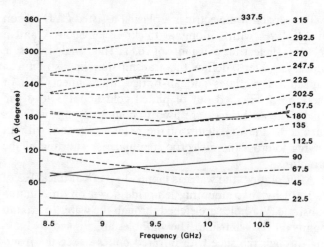

Fig. 15. Phase shift versus frequency for all 15 nonzero states of a four-bit passive phase shifter.

TABLE II
ARRAY RANDOM ERROR EFFECTS BASED ON FOUR-BIT PASSIVE MONOLITHIC PHASE SHIFTER DATA (100 ARRAY ELEMENTS ARE ASSUMED)

Frequency (GHz)	8.0	8.5	9.0	9.5	10.0	10.5	11.0	11.5	12.0
Total rms Phase Error (σ_T)	39.4°	25.4°	17.3°	11.4°	8.7°	11.1°	14.7°	17.6°	13.1°
Gain Reduction (dB) (G/G_0)	-1.7	-0.8	-0.4	-0.2	-0.1	-0.2	-0.3	-0.4	-0.5
rms Sidelobe Level (dB) (SLL)	-23.3	-27.1	-30.4	-34.0	-36.4	-34.3	-31.8	-33.9	-29.5
Beam Pointing Error (%) $\Delta\theta/\theta$ 3 dB	6.9	4.4	3.0	2.0	1.5	1.9	2.6	3.1	3.3

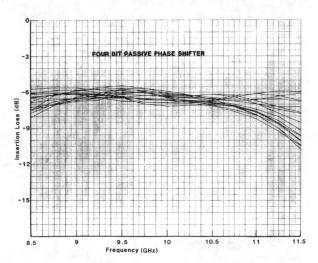

Fig. 14. Insertion loss for all 16 states of a four-bit passive phase shifter (no correction for jig losses has been made).

is summarized in Table II. The results indicate that the predicted performance is satisfactory for typical phase array requirements.

V. CONCLUSION

An X-band GaAs monolithic phase shifter with 22.5°, 45°, 90°, and 180° phase bits has been made using GaAs switch FET's. By cascading all four bits on the same chip, a digital passive phase shifter with 5.1 ± 0.6-dB insertion loss and 16 distinct phase states between 0° and 360° is realized. The performance of the phase shifter is satisfactory for typical phased array applications. Their small size, negligible dc power requirement, and subnanosecond switching times make these monolithic phase shifters prime candidates for future frequency agile airborne phase array systems.

REFERENCES

[1] R. A. Gaspari and H. H. Yee, "Microwave GaAs FET switching," in 1978 IEEE Int. Microwave Symp. Dig., pp. 58–60.
[2] W. V. McLevige and V. Sokolov, "Microwave switching with parallel-resonated GaAs FETs," IEEE Electron Devices, vol. ED-1, Aug. 1980.
[3] Y. Ayasli, R. A. Pucel, J. L. Vorhaus, and W. Fabian, "A monolithic X-band single-pole, double-throw bidirectional GaAs FET switch," in IEEE GaAs Integrated Circuits Symp., (Las Vegas, NV.), Nov. 4–6, 1980.
[4] Y. Ayasli, J. L. Vorhaus, R. A. Pucel, and L. D. Reynolds, "Monolithic GaAs distributed FET switch circuits," in IEEE GaAs Integrated Circuit Symp., (San Diego, CA), Oct. 27, 1981.

AN ANALOG X-BAND PHASE SHIFTER

D. E. Dawson, A. L. Conti, S. H. Lee, G. F. Shade*, and L. E. Dickens*

Westinghouse Electric Company
Baltimore, Maryland 21203

ABSTRACT

A hybrid-coupled phase shifter has been fabricated monolithically using reverse-biased Schottky varactor diodes to continuously vary phase with an analog control voltage. A phase shift of 105° is obtained at X-Band, and with design improvements a phase shift of 180° over the full 8-12.4 GHz is expected. Phase shift variation with power level is reduced by using back-to-back varactors on the direct and coupled ports of the coupler. A six-to-one reduction in size compared to a four-bit switched-line phase shifter (also at X-Band) results from this approach.

INTRODUCTION

The capacitance vs. voltage characteristic of a reverse-biased Schottky diode was used to make a monolithic analog phase shifter at X-Band. A Lange coupler was fabricated on 0.010" thick GaAs along with selectively ion implanted varactors and bias resistors. The circuit (Fig. 1) consists of a back-to-back varactor pair on each port of a 3-dB coupler. The input signal splits and is incident upon the two varactor pairs. Signal reflected from the varactor pairs adds at the output and is varied in phase relative to the input as the diode capacitance is varied by the external bias. By adjusting the tuning inductance in series with the diodes, 180° of phase shift over an 8-12.4 GHz bandwidth is possible. The layout and fabricated chip of Figures 2 and 3 show the airbridged Lange coupler formed in a U-shape with the varactors and bias/isolation resistors located in the middle. The varactors are grounded around the edge of the substrate. The resistors are small enough in value to provide a return path for the diode leakage currents, but large enough not to load the RF signal path.

Fig. 1. Hybrid-Coupled Phase Shifter Circuit Uses Back-to-Back Varactors (similiar to VCO circuits).

---

*Current address: Ford Microelectronics, Colorado Springs, CO 80908.

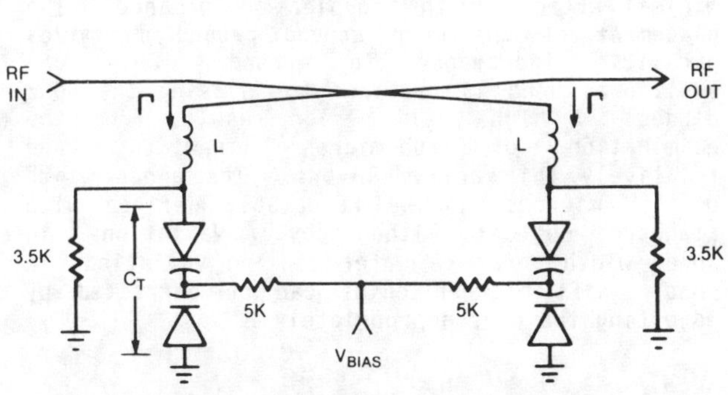

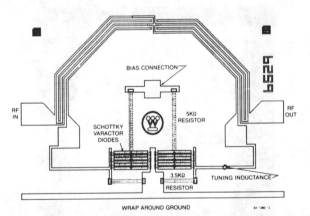

Fig. 2. CAD Chip Layout.

Fig. 3. Chip Size Is 0.077 x 0.100 Inches.

Reprinted from *IEEE Microwave and Millimeter-Wave Monolithic Circuits Symp. Dig. Papers*, 1984, pp. 6-10.

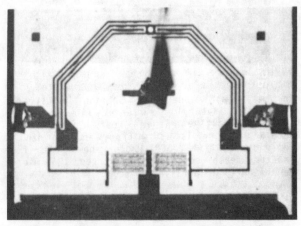

DIODE GEOMETRY AND FABRICATION

Surface oriented diodes made on the same active layer as FETs punchthrough at 2-4 V. as dictated by the FET pinchoff voltage. This gives the maximum range of capacitance. However, as the depletion layer nears pinchoff the diode series resistance increases because lateral current has less active layer through which to flow [1]. A buried N^+ layer can be used [2,3] but it involves more complicated processing. Instead, the approach here has been to make the active layer implantation as deep as possible so that breakdown occurs before punchthrough to allow the conduction layer to completely surround the cathode side of the depletion region and maintain low series resistance (Fig. 4). The doping concentration was chosen to be as high as possible for low series resistance but low enough to insure 80% depletion of the active layer without breakdown in order to achieve the greatest change in capacitance. N^+ contacts implanted deeper than the active layer were desired to provide low resistance from the ohmic contact to the depletion region at all bias levels. However, substrate overheating and photoresist carbonation during implantation limit the deepest practical N^+ implant to 0.6 microns, and therefore an active layer thickness of 0.5 microns was chosen.

The active and N^+ layers have been selectively implanted into LEC semi-insulating substrates (4000 - 5000 cm^2/V-sec mobility) with the implant schedule of Table I and capless annealed at 750°C. A concentration of 9×10^{16} cm^{-3} and a mobility of 3000 cm^2/V-sec are obtained. The active layer depth (LSS theory) of 0.4 microns increased to 0.5 microns after annealing. The diode (edge) breakdown is 10 V.; bulk breakdown for 10^{17} cm^{-3} material is 15 V. The depletion depth at 10 V. is 0.4 microns and punchthrough does not occur.

The circuit layout has been carefully designed for optimum use of chip area while minimizing cross-coupling effects. A value of 300 ohms/square was used for the implanted resistor design and gave a feasible and easy to implement

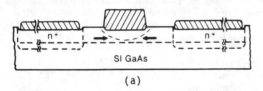

(a)

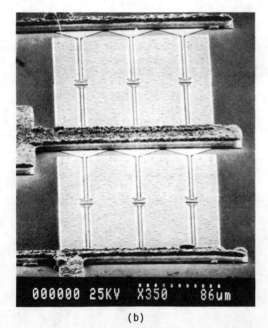

(b)

Fig. 4. (a) Diode Cross-Section illustrates restricted lateral current flow through undepleted active layer near punchthrough. Anode-N^+ spacing is 1.0 micron; anode width is 1.5 microns. (b) SEM Photo shows the four varactors each consisting of three 135 micron anode fingers in parallel.

Table I. Implant Schedule.

N^+ Implant Schedule

$^{28}Si^{++}$	2.42 E13/cm^2	500 KeV
$^{28}Si^{+}$	1.50 E13/cm^2	240 KeV
$^{28}Si^{+}$	7.00 E12/cm^2	100 KeV
$^{28}Si^{+}$	2.57 E12/cm^2	60 KeV

(The GaAs is etched 700 Angstroms prior to implant for a deeper N+ layer and for later aligning).

Active Layer Implant Schedule

$^{28}Si^{+}$	3.51 E12/cm^2	300 KeV
$^{28}Si^{+}$	1.79 E12/cm^2	120 KeV
$^{28}Si^{+}$	5.10 E11/cm^2	60 KeV

resistor layout for the values needed. The Lange coupler uses six airbridges and is folded to reduce overall chip size. The input/output VSWRs are less than 1.5:1 indicating that the bends had minimal affect on the coupler performance. The placement of the wrap around ground minimizes parasitic inductance to ground. Only one additional bond is required for biasing the four diodes. A high yield design results from the elimination of sub-micron linewidths and relatively thin active layers. The anode width of 1.5 microns can be repeatably defined with standard optical lithography. Variations in anode width across a wafer causing variations in phase shift between chips can be corrected by adjusting the bias appropriately.

Fabrication of the phase shifter utilizes all positive photoresist with a total of seven masking levels. Fig. 5 shows the first five mask levels. The remaining two are for the airbridge formation. All circuit paths are plated to 4 microns of gold to reduce RF losses.

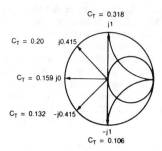

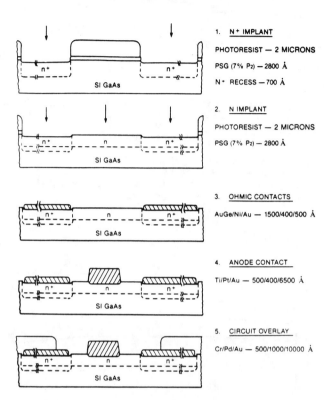

Fig. 5. Selective ion implant, ohmic, anode, and overlay fabrication sequence.

Fig. 6. Phase Shift Varies 180° as C_T varies from $C_T(0) = 0.318$ pF ($C_j(0) = 0.636$ pF) to 1/3 the zero-bias value. A frequency of 10 GHz and series inductance of 1.59 nH are assumed.

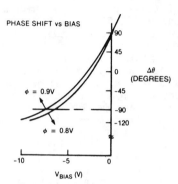

Fig. 7. Phase Shift Varies Smoothly With Bias. Frequency is 10 GHz.

DESCRIPTION OF CIRCUIT DETAILS

The phase shifter configuration makes use of a 3 dB quadrature (Lange) coupler to provide a matched input and output (low VSWR) for the phase shifting elements. Figure 1 shows the hybrid coupler type of phase shifter with series L-C loading elements, where the capacitive element is a series pair of voltage variable (varactor) diodes. Figure 6 is a Smith chart diagram showing the reflection coefficient for various varactor capacitances. A full 180 degree phase change is obtained for a 3:1 capacitance variation. Figure 7 is a plot of the relative phase shift as derived from the Smith chart of Figure 6 where the capacitance values have been related to the bias voltage according to the equation for Schottky barrier (abrupt) junctions; i.e., $C(V) = C(0)/(1-V/\phi)^{1/2}$, where ϕ has been given the two values of 0.8 and 0.9 volts. Notice that this form of phase shifter has a very smooth, well behaved variation of phase as a function of bias voltage. This makes the job of microprocessor control easier, including temperature stabilization.

Figure 8 shows the calculated transmission phase characteristics of a single section of the hybrid-coupled phase shifter. The fact that this type of phase shifter does yield a phase change which is relatively independent of frequency can be seen from this figure. Examination of a broader frequency range shows that at least 180° of phase shift can be obtained (for a 0 to 10 volt swing) over the frequency band of 8 to 12.4 GHz.

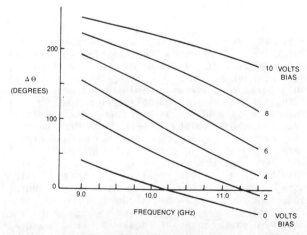

Fig. 8. Calculated Transmission Phase vs. Frequency is smooth, and the step size with bias is also uniform over the frequency band.

The phase shift variation with signal level is greatest for the analog phase shifter at zero bias where the capacitance vs. voltage slope is steepest. The series back-to-back configuration has a compensating effect compared to the single varactor. As the RF swing increases the capacitance of one varactor, the capacitance of the other is decreased. The variation of the total capacitance, C_T, given by $1/C_T = 1/C_1 + 1/C_2$ is of second order. However, if the onset of forward conduction is reached on one of the varactors, the compensating effect becomes out of balance. By being d-c biased in parallel the 3:1 capacitance variation is maintained, but the variation caused by the RF swing is reduced by the back-to-back series connection at RF. A 0.4 V. peak swing applied to a 0.318 pF single varactor causes a capacitance swing with a maximum of 0.449 pF and a minimum of 0.258 pF. The average is 0.354 pF, a 10% increase. With 0.4 V. peak across each of two 0.636 pF varactors back-to-back the maximum capacitance is 0.328 pF, and the minimum is 0.318 pF. The average (allowing for a capacitance swing twice each cycle) is 0.325 pF, a 2% change. Following the Smith chart of Fig. 6, the phase shift variation at zero bias for the 10% change of the single varactor is six degrees, and the back-to-back configuration shifts one degree. In addition, the back-to-back configuration is operating at a 6 dB greater power level because the signal is across two varactors (each of half the impedance) instead of one. Therefore, for a given tolerated phase error, the back-to-back varactors will handle more signal swing or bias closer to zero, or slightly forward, allowing greater phase shift.

The series resistance of the varactor string is measured to be about 3.0 ohms. At series resonance, this gives about 1.0 dB return loss (in a 50 ohm system). If we assume that the Lange coupler plus transmission lines have about 0.6 dB insertion loss per pass, then the phase shifter is expected to have about 2.2 dB insertion loss at resonance. The loss should be slightly less at the +90 & -90 degree and other off-resonance points.

PERFORMANCE

Measured series resistance of a single varactor junction is 1.5 ohms. The zero-bias capacitance was predicted to be 0.636 pF, and the resulting devices have 0.9 pF. The design value was based on a parallel-plate model and did not account for fringing. The C-V characteristics have a 3:1 change in capacitance as expected. The resulting measured phase shift is less than 180° because the range of capacitance and tuning inductance were incorrect. A phase shift of 105° at 11 GHz was measured (Fig. 9). The calculated results of a revised model of the phase shifter with $C_j(0) = 0.9$ pF and a series tuning inductance of 0.6 nH are shown in Fig. 10. The calculations are in good agreement with the measured phase shift of Figure 9. The measured insertion loss is 2.5 dB and varies ±0.5 dB as the phase is varied from 0° to 105°. The transmission phase versus frequency for different

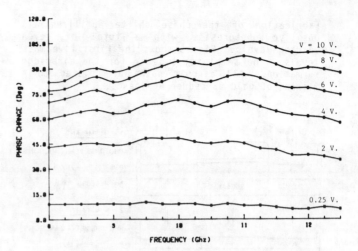

Fig. 9. Measured Phase Shift is flat with frequency. Varactors could be slightly forward biased to get more phase shift. Phase change is measured from zero bias.

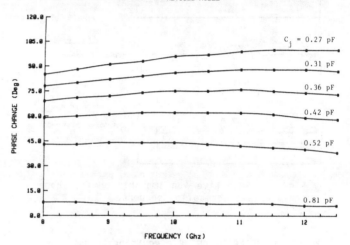

Fig. 10. To compare with the measured phase shift $C(0) = 0.9$ pF was substituted for expected value of 0.6 pF. Comparison is within 10°. Phase change is calculated using zero bias as the reference state.

power levels and bias levels is shown in Figures 11 and 12. Deviation from small-signal performance at zero-bias (worse case) occurs at +12 dBm. The phase measurement is phase locked but not vector error corrected and has 5° of error. The measured device, because of the inductance of 0.6 nH and $C_j(0)$ of 0.9 pF, is resonant at 9.6 GHz where the circuit is most sensitive to change in capacitance. A 10% change in capacitance causes a 10° phase shift which is larger than that discussed above. A 0.4 V. peak RF voltage across each varactor corresponds to 8 dBm of input power. The phase shift variation at 5-10 dBm is about 1°. With bias applied the power handling increases to 15 dBm (Fig. 12).

CONCLUSION

A 180° phase shifter has been designed and fabricated using Schottky varactor diodes hybrid-coupled and controlled by an analog voltage. The insertion loss has been minimized by choosing a 0.5-micron deep active layer for the surface oriented varactors. Phase shift variation with signal level is reduced by the use of series connected back-to-back varactors instead of single varactors. 105° of phase shift has been achieved experimentally with 0-10V. bias. If the measured diode capacitance is utilized in the circuit model, the measured and calculated phase shift agree within 10° and indicates the circuit works as expected. With further work 180° of phase shift can be obtained by reducing the anode area. The chip size is 0.077 x 0.100 inches (2 are required for 360°) and is 1/6 that of a 360° 4-bit switched-line phase shifter [4] (Fig. 13). The analog phase shifter offers a significant improvement for T/R module applications where the phase shifter chip is among the largest sized chips.

REFERENCES

1. G. E. Brehm, B. N. Scott, and F.H. Doerbeck, "Fabrication Techniques for X-Band Monolithic VCOs," IEEE 1982 Microwave and Millimeter Wave Monolithic Circuits Symposium, Digest of Papers, pp. 57-60.

2. A. Chu, et al., "Monolithic Frequency Doublers," IEEE 1983 Microwave and Millimeter Wave Monolithic Circuits Symposium, Digest of Papers, pp. 45-49.

3. C. Chao, et al., "94 GHz Monolithic GaAs Balanced Mixers," IEEE 1983 Microwave and Millimeter Wave Monolithic Circuits Symposium, Digest of Papers, pp. 50-53.

4. G.F. Shade, et al., "Monolithic X-Band Phase Shifter", IEEE 1981 GaAs IC Symposium Research Abstracts, p.37.

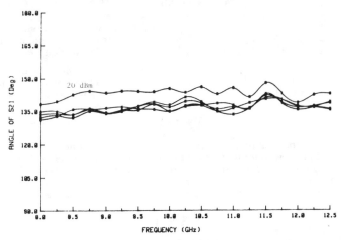

Fig. 12. Transmission Phase at 1.0V. Bias vs. Frequency. Deviation from small signal occurs at 20 dBm.

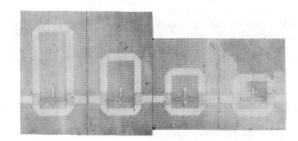

Fig. 13. The analog phase shifter (2 sections for 360°) is 1/6 the size of a switched-line phase shifter.

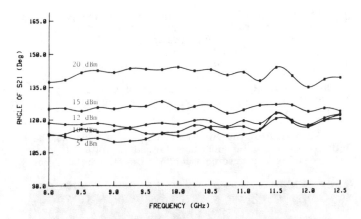

Fig. 11. Transmission Phase at Zero-Bias vs. Frequency. Deviation from small-signal occurs at 12 dBm.

A MULTI-CHIP GaAs MONOLITHIC TRANSMIT/RECEIVE MODULE FOR X-BAND

R. A. Pucel, Y. Ayasli, D. Wandrei, J. L. Vorhaus
Research Division, Raytheon Company
Lexington, Massachusetts 02173

S. Temple, R. Waterman, A. Platzker, C. Cavicchio
Missile Systems Division, Raytheon Company
Bedford, Massachusetts 01730

Abstract

The design, construction, and performance of an X-band multi-chip GaAs monolithic transmit/receive module is described. The module consists of a four-bit FET phase-shifter, two-stage low-noise amplifier, four-stage power amplifier, and associated FET switches.

Introduction

Monolithic microwave integrated circuits (MMICs) fabricated on GaAs substrates offer the potential for significant reductions in size, weight, and cost with increased reliability as compared with conventional hybrid integrated circuits. By monolithic we mean an approach wherein all active and passive components and interconnections are formed into or onto a semi-insulating substrate by some deposition scheme.

The advantages accruing from this approach are especially attractive in applications where electronically scanned active phased-array antennas are used. Such antennas consist of a two-dimensional array of closely-packed (about a half-wavelength in separation) transmit-receive modules. These modules individually phase-shift a transmitted or received signal in accordance with a sequence of command signals. By an appropriate control of the phase-shift of each module, the direction and shape of the transmitted or received antenna radiation pattern is determined. By introducing gain into the module, an active array is obtained.

A block diagram of an active transmit/receive module is shown in Fig. 1. As can be seen the module consists of a power amplifier (marked "transmitter"), low-noise small-signal amplifier (marked "receiver") and four single-pole, double-throw switches which control the direction of signal flow. A key element in the module is the phase shifter which, in addition to introducing a differential phase shift in the receive/transmit signal flow path, may also have gain. By an appropriate sequence of the settings of the SPDT switches s_1 to s_4 the module can be used to either amplify a source signal received from the "left" or a received signal coming from the "right". In either mode of operation, the active phase-shifter is introduced into the signal path in the proper direction of signal flow by switches s_2 and s_3.

We wish to describe in this paper, a preliminary (brassboard) module of eleven individual monolithic GaAs chips consisting of the four switches, four phase-shift bits, a two-stage low-noise amplifier, and a cascaded pair of two-stage power amplifiers. All of the monolithic chips are interconnected by appropriate microstrip lines printed on alumina substrates which are coplanar with the GaAs chips.

It should be pointed out that the alumina substrates, which contain 50-ohm microstrip lines only, are merely used to interconnect the GaAs chips. The GaAs chips contain, not only the appropriate rf circuitry and active elements (FETs), but in addition all of the bias circuitry.

Although this module is considered to be a multi-chip monolithic circuit, it is a convenient, intermediate stage leading eventually to a higher level of integration involving fewer chips. This transition to a higher level of integration is being carried out at our laboratory. It is appropriate to report on this multi-chip approach, since, to our knowledge, this complexity of monolithic integration has not been reported for any microwave circuit to date.

The Transmit/Receive Module

Figure 2 is a photograph of the transmit/receive module. The operating band is 9-10 GHz. Starting from the left, one may note the three monolithic switches, s_1, s_2, s_3, followed by four chips in a rectangular array which are the four bits of an active phase-shifter. Further to the right are two cascaded chips, each of which is a two-stage power amplifier, followed by another switch, s_4. The low-noise amplifier chip is at the lower right.

All eleven of the functional chips use FETs as the active element. The FETs are formed in epitaxial layers grown by the arsenic trichloride system on 0.1 mm

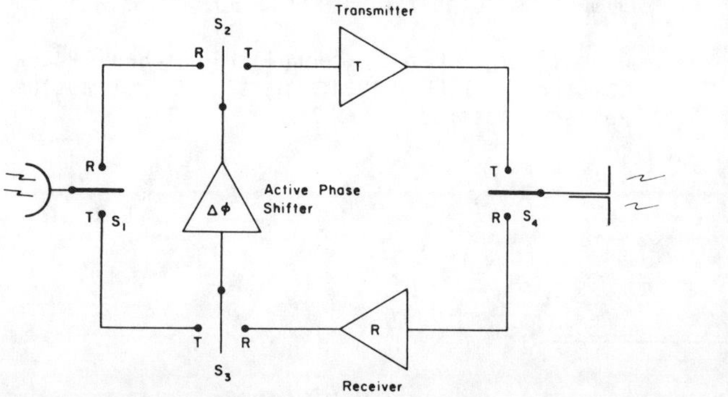

Fig. 1 Block diagram of an active transmit/receive module.

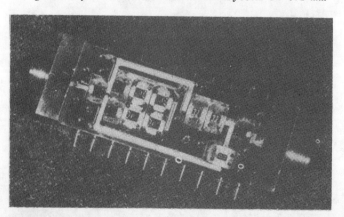

Fig. 2 Photograph of an assembled transmit/receive module.

thick semi-insulating GaAs substrates. These active elements are electrically separated by a combination of a partial etching and isolation implanting of the unused active epi-layer external to the device areas. All conductor metallization is gold, approximately 3 microns thick. The gate lengths of the FETs are all, nominally, one micron, except for the low-noise amplifier, for which gates 0.7 microns long were used. Both via hole grounding and air-bridge technologies were used.

Brief Description of Individual Chips

The low-noise and power amplifier circuits are of a conventional narrow-band design and their performances have been reported elsewhere.[1,2] Photographs of these two circuits in monolithic chip form are shown in Fig. 3 and 4, respectively.

The low-noise amplifier consists of two 150 μm periphery FETs biased for minimum noise, with 50-ohm input and output beam-leaded rf ports. The transmitter

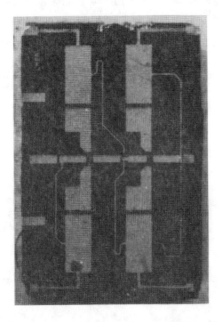

Fig. 3 Photograph of low-noise amplifier monolithic chip.

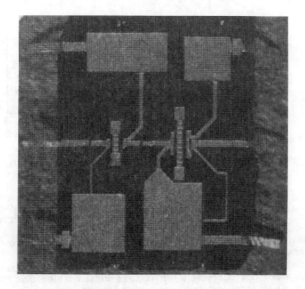

Fig. 4 Photograph of two-stage power amplifier monolithic chip.

also consists of two stages, the first a 600 μm periphery FET driver and a 1600 μm periphery "final" or power stage. The chip face dimensions are 2.5 and 3.2 mm. Here again, beam-leaded 50-ohm input and output ports are used for ease in testing. The module actually employs two identical transmitter circuits in cascade for more gain. In subsequent designs, all four stages will be redesigned and integrated into one chip. The phase-shifter consists of four cascaded "bits". Each bit contains a dual-gate FET with a "split" second gate and drain, and a common first gate. By appropriate control of the bias voltages on the two second gates, the amplified input signal can be made to exit one of two channels corresponding to either segment of the split drain. Each of the two channels is terminated on one of the two 3 dB ports of a Wilkinson combiner printed on the chip. The output of the bit appears at the remaining port the combiner. By introduction of an appropriate excess phase shift in one channel (produced by an additional microstrip line segment in one of the dual gate channels of each of the four chips), a differential phase shift can be obtained from each of the four phase-shifter chips. Because of the dual-gate FETs, each bit has some gain associated with it, that is, the phase-shifter is "active". The details of the phase-shifter design have been given elsewhere.[3]

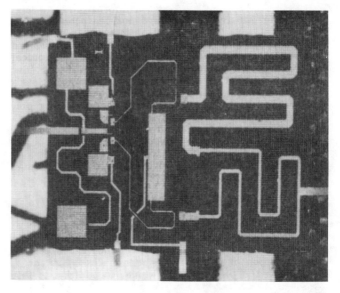

Fig. 5 Photograph of an active phase-shifter bit on monolithic form.

Figure 5 is a photograph of one of the four phase-shifter bits. Although the 100-ohm balancing resistor for the combiner is not visible, the two 90°-lines of the combiner, terminating in the common 50-ohm output port are clearly visible. The gate periphery of each FET channel is 140 μm.

All of the chips described above use on-chip bias filtering consisting of thin-film capacitors which employ Si_3N_4 dielectric films. Depending on the circuit, the capacitors range from 10 pF to 40 pF.

The switches use FETs in their resistive mode.[4] Their design and operation have been described earlier. Because the FETs are operated either at pinchoff, or at zero drain bias, no dc "hold" power is required in either state of the switch. This is a distinct advantage over PIN-type switches when minimization of dc prime power is important. The gate periphery of the FETs used is 1600 μm. The chip face dimensions are 3.2 and 3.2 mm. Figure 6 is a photograph of a monolithic switch.

Experimental Results Obtained with T/R Module

Figure 7 displays the differential phase shift and gain in the transmit mode at the center band frequency $f_0 = 9.5$ GHz. The median gain is 20 dB with a spread of approximately ± 1.5 dB. Actually, because the center band of the active phase-shifter, itself, was somewhat higher than the center-band frequency, a considerable gain slope occurred in the pass band. The phase-shift data exhibit an approximately linear variation with bit state, on the average, though the average slope is approximately 5% higher than the design specification. The saturated output power in the passband was approximately 250 mW.

Figure 8 illustrates the measured gain and phase response in the receive mode. The average gain is 13.6 dB with a variation of ± 2 dB over the bit states. The gain, of course, is lower than in the transmit mode because the low-noise amplifier has only two stages rather than four as in the transmitter. The phase-shift response is similar to the transmit mode as one would expect.

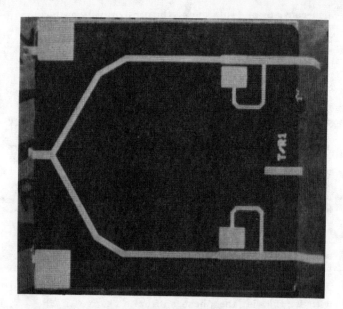

Fig. 6 Photograph of a monolithic FET switch.

Each of the monolithic chips was characterized individually after it was mounted on its carrier prior to installation in the T/R module housing. The assembled module was operated with gain over all 16 phase states, in both the receive and transmit modes.

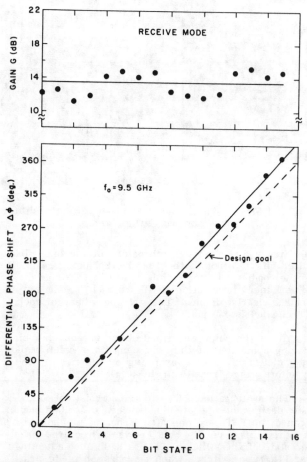

Fig. 7 Measured gain and differential phase shift at center band as a function of bit state (transmit mode).

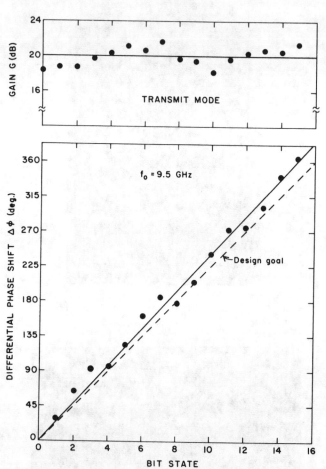

Fig. 8 Measured gain and differential phase shift at center band as a function of bit state (receive mode).

The noise figure in the receive mode was measured at selected points in the passband. The results are shown in Fig. 9. The spread in the data at each frequency (indicated by the vertical bars) reflects the variation in noise figure over the 16 bit states. These data were not corrected for fixture losses which are estimated, from independent measurements on the fixture itself, to be of the order of 0.25 dB. Contributing also to the noise figure is the antenna switch loss,

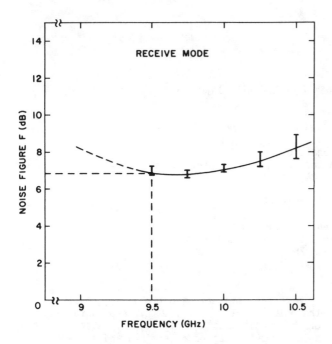

Fig. 9 Measured noise figure in receive mode as a function of frequency.

estimated to be 1.2 dB, and a microstrip loss on the alumina substrates of an amount equal to 0.15 dB. This leaves 5.2 dB at 9.5 GHz. Tests on the low-noise amplifier chip itself yield a noise figure of 5.0 dB (corrected for fixture losses). The difference in these two values is attributed to the noise figure of the active phase-shifter following the low-noise amplifier, inferred to be about 9 dB.

In this presentation, the philosophy underlying the module design will be covered in some detail. In addition, a companion design, now under test, using a passive phase-shifter will be described, if results are available, and comparisons will be made between the two approaches.

Conclusions

We have shown that a transmit/receive module for phased-array antennas, based entirely on FET circuits, is technically feasible. Furthermore, a monolithic version of such a module (in multi-chip form) has exhibited promising performance for X-band applications. A salient feature of this design is the realization of an active phase-shifter, that is, one with gain.

Acknowledgments

The authors wish to acknowledge the members of the materials and device technology laboratories for their expert assistance and to Mr. Robert Bierig for his encouragement.

References

1. J. L. Vorhaus, R. A. Pucel, Y. Tajima, and W. Fabian, "A two-stage all-monolithic X-band power amplifier." ISSCC Digest of Technical Papers, pp. 74-75 (1980).

2. R. A. Pucel, Y. Tajima, J. Vorhaus, and A. Platzker," An all-monolithic GaAs X-band low-noise amplifier," GaAs IC Symposium Research Abstracts, 1981, paper no. 29.

3. J. L. Vorhaus, R. A. Pucel, and Y. Tajima, "Monolithic dual-gate GaAs FET digital phase shifter," GaAs IC Symposium Research Abstracts, 1981, paper no. 35.

4. Y. Ayasli, R. A. Pucel, J. L. Vorhaus, and W. Fabian, "A monolithic X-band single pole, double-throw bidirectional GaAs FET switch," GaAs IC Symposium Research Abstracts, 1980, paper no. 21.

A 20 GHz 5-BIT PHASE SHIFT TRANSMIT MODULE WITH 16 dB GAIN*

A. Gupta, G. Kaelin, R. Stein, S. Huston, K. Ip and W. Petersen**

Rockwell International Corporation
Microelectronics Research and Development Center
Thousand Oaks, CA 91360

M. Mikasa and I. Petroff

Rockwell International Corporation
Strategic Defense and Electro-Optical Systems Division
Anaheim, CA 92803

ABSTRACT

A fully monolithic transmit module chip operating over the 17.7-20.2 GHz band is described. This 6.3 x 4.7 mm MMIC contains five phase shifter bits, five gain stages and digital interface circuitry. An overall gain of 15 dB has been achieved at band center, with only ± 0.5 dB gain variation with changes in phase shifter state. Phase shifts are within 5° of the desired values. This MMIC represents the highest level of integration yet reported for circuits operating at 20 GHz.

1.0 INTRODUCTION

Rockwell International is currently developing a fully monolithic 20 GHz transmit module for the NASA Lewis Research Center. The module is required to have 5-bit phase control and gain of 16 dB over the 17.7-20.2 GHz frequency band and control of the phase shift state via a 5-bit, TTL compatible, digital input signal. Detailed design goals are summarized in Table 1. The design of this circuit has been implemented in terms of four functional building blocks. These are: 1) a 5-bit passive phase shifter, 2) a two-stage buffer amplifier to overcome the loss through the phase shifters, 3) a three-stage power amplifier, and 4) a digital interface to convert the control signal to appropriate voltages for the phase shifters. The 5-bit phase shifter is implemented as five distinct phase shifter circuits of 180°, 90°, 45°, 22.5° and 11.25°. The interconnection of these building blocks is shown in Fig. 1.

The circuit design approach taken in the development of the fully monolithic transmit module was to implement each of the above building blocks and subcircuits as an individual circuit for test and design verification. Next, the individual circuit designs were integrated into a monolithic IC with only minimal layout changes, resulting in the MMIC described herein. By keeping the layout of the subcircuits unmodified, any interaction between circuits can be determined and analyzed. The final step will be, in addition to making any corrections, to compress the layout for minimum area and highest yield.

Table 1

Design Goals of 20 GHz Monolithic Transmit Module Chip

RF Band:	17.7-20.2 GHz
RF Gain:	> 16 dB
Gain Variation:	± 0.5 dB over entire band and < ± 0.2 dB over any 500 MHz band
Power Output:	> 0.2 W at 1 dB gain compression
Power Added Efficiency:	> 15%
Phase Shifter:	5-bits with the following phase shift at band center (18.9 GHz): 180°, 90°, 45°, 22.5° and 11.25° (± 3° each). Total module phase shift should be proportional to frequency with a phase error < ± 6° within the RF band
Phase Control:	5-bit TTL signal

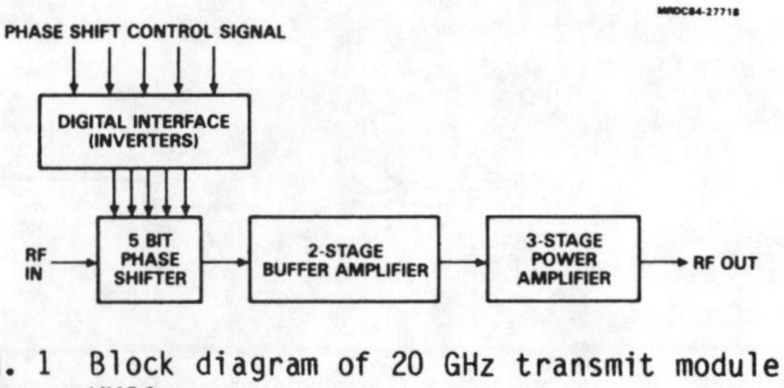

Fig. 1 Block diagram of 20 GHz transmit module MMIC.

* Supported by the NASA Lewis Research Center under Contract No. NAS3-23247
** W. Petersen is currently with Microwave Monolithic Inc., Simi Valley, CA.

Reprinted from *IEEE Gallium Arsenide Integrated Circuit Symp. Tech. Dig.*, 1984, pp. 197-200.

A summary of the performance of the functional building blocks is given in Sec. 2, and their monolithic integration into a 6.3 x 4.7 mm MMIC is described in Sec. 3. Circuit fabrication is discussed in Sec. 4, and concluding remarks are presented in Sec. 5 of this paper.

2.0 FUNCTIONAL BUILDING BLOCKS

The first step in the development of the fully monolithic transmit module was to design, fabricate and test the subcircuits of the four building blocks shown in Fig. 1. These were designed as individual circuits on two different mask sets; the standard MMIC fabrication steps were used, assuring compatibility for subsequent integration. A summary of the results obtained is presented below. Experimental results for each of the amplifier circuits have been reported previously.

2.1 Two-Stage Buffer Amplifier (1)

The buffer amplifier was designed to have a gain of at least 12 dB from 17.7-20.2 GHz to overcome the expected loss of 12.5 dB (2.5 dB/bit) through the 5-bit phase shifter chain. Measured performance was 13 ± 0.75 dB over the band.

2.2 Three-Stage Medium Power Amplifier (2)

Design goals for the power amplifier were 16 dB of gain and 200 mW (23 dBm) of output power with power-added efficiency of 15% from 17.7-20.2 GHz. The starting point in the design of this high efficiency power amplifier is the selection of the output FET characteristics; the power density and gain can be traded off under control of the device doping profile. Although localized ion implantation can be used to fabricate FETs with different doping profiles on the same chip, the design of this amplifier is based on a single, standard FET profile, the same as used in the passive phase shifter switches and small signal amplifiers to avoid processing complexity and thus increase yield. With this constraint, the design goals of 16 dB gain, 200 mW power and 15% efficiency are quite challenging. Measured performance of the 1.5 x 3.1 mm, three-stage amplifier chip was 15 dB gain from 16.5-20.2 GHz and a saturated output power of 21 dBm, or 120 mw. Discrete FETs with the same doping profile and size as the output stage of the amplifier were evaluated for power performance, and were found to be adequate for the present requirements. It is therefore expected that only minor changes in the matching networks will be required to meet the output power goal.

2.3 5-Bit Phase Shifter

The phase shifter submodule consists of a cascade of five binary phase shifters with phase increments of 11.25°, 22.5°, 45°, 90° and 180°, respectively, at band center (18.9 GHz). Each bit is implemented as a switched line phase shifter (Fig. 2). This circuit approach was chosen as being the most appropriate when issues such as phase shift accuracy, amplitude variation with phase change, power consumption, and sensitivity to element values were considered. The phase shifter consists of two single-pole, double-throw (SPDT) switches capable of connecting either line A or line B (Fig. 2) to the RF signal. The difference in the line lengths, of A and B provide the desired phase shift. FETs with a 1 μm gate length are employed in a series-shunt configuration to implement the SPDT switch, as shown in Fig. 3. For a series FET width of 300 μm and shunt FET swidth of 120 μm, insertion loss of 0.7 dB and isolation of 15 dB are obtained in the RF band. Two such switches in each SPDT switch provide an isolation of 30 dB, which is adequate for this application.

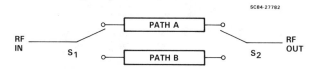

Fig. 2 Outline of a switched line phase shifter.

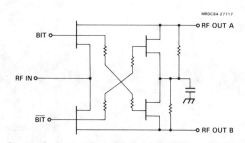

Fig. 3 Implementation of an RF SPDT switch using FETs in a series-shunt configuration.

The five phase shifter bits can be clearly identified in Fig. 5 which is a photograph of the fully monolithic transmit module. Measured phase shift and insertion loss data on an individual 90° phase shifter are presented in Fig. 4. Data for the other phase shifter bits are similar. The measurements were made on devices mounted in the Universal MMIC test fixture reported earlier (3). From Fig. 4, it is clear that the phase shift at band center is not exactly 90°. Such small errors in phase shift can be corrected easily by changing the lengths of the straight sections of delay line in the phase shifter in the final design. The insertion loss of this phase shifter is also higher than the expected 2.5 dB at band center. Typical measured values range from 2.5 to 4 dB. Although measurement uncertainties of ± 0.5 dB contribute to this variation, other reasons for the higher insertion loss are being determined as the chips are being evaluated. These include a larger FET "on" resistance than used in the model, mismatch losses, and a parasitic leakage current (through the semi-insulating substrate) between negatively biased ion implanted resistors too closely spaced

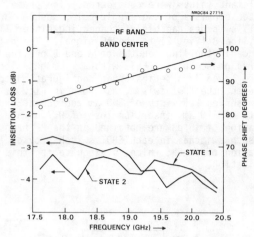

Fig. 4 Measured phase shift and insertion loss on an individual 90° phase shifter bit.

conditioned to provide ~ +1 V to turn a FET "on" and -5 V to turn a FET switch fully "off" (assuming a FET threshold of -3 V). The terminals of the phase shifter FET switch are maintained at +5 V and the inverter provides 0 or +7 V to the gate terminal (through a 20K resistor). Five such inverters are included on the fully monolithic chip.

3.0 MONOLITHIC TRANSMIT MODULE

A photograph of the fully monolithic, 6.3 x 4.7 mm, 20 GHz transmit module chip is shown in Fig. 5. Except for the inclusion of some low frequency (< 50 MHz) stabilization circuitry, the circuits are essentially unchanged from their discrete form. Many bonding pads and test circuits are scattered throughout the chip for diagnostic purposes. In the final version, the layout will be compressed and these will be eliminated in making the chip smaller. Gain and total phase shift vs frequency for the reference state is shown in Fig. 6. Gain rolloff at the high end occurs at a lower frequency than predicted. This has been consistently observed on the five-stage amplifier chains from this mask set, but was not observed in the discrete amplifiers fabricated earlier (1,2).

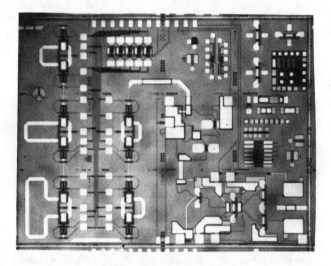

Fig. 5 Photograph of the fully monolithic 6.3 x 4.7 mm, 20 GHz transmit module chip.

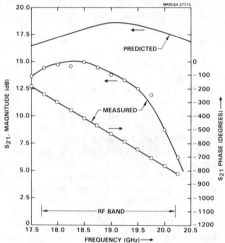

Fig. 6 Gain and total phase shift vs frequency for the reference state of the monolithic transmit module chip.

in the layout. This leakage current causes an uncontrolled voltage drop across the gate bias resistors, so that the switch FET is only partially turned off. Reasons for the higher than expected insertion loss have not yet been fully resolved, and work in this area is continuing; there appears to be no fundamental obstacle to achieving the low value predicted.

2.4 Digital Interface

Since the compound SPDT switch of Fig. 3 requires both true and complement control signals, an inverter circuit is required for each of the five input control lines to minimize interconnections. Furthermore, the TTL input signal must be

Overall gain is lower than expected by ~ 2.5 dB due to the higher phase shifter insertion loss. Another problem encountered here is an in-band oscillation for large drain biases which has precluded output power measurements on these chips. The cause of this oscillation has been determined; some of the intermediate stages are potentially unstable, even though the entire amplifier was designed to be stable. Differential insertion loss (ΔIL) and phase shift ($\Delta \phi$) data are presented in Figs. 7 and 8, respectively. At band center, ΔIL is ≤ ±0.5 dB, although this degrades at the high end due to the premature gain rolloff. $\Delta \phi$ is within a few degrees of the desired value. These data were obtained in some of the earliest functional chips, from which minor processing improvements have been determined. It is expected that chips from lots in progress will deliver improved MMIC performance. The next version of this mask set will include minor circuit changes to solve the problems indicated above, such as the oscillation problem and those associated with the phase shifter switches, as well as small corrections in the delay line lengths to set precisely the desired phase shift.

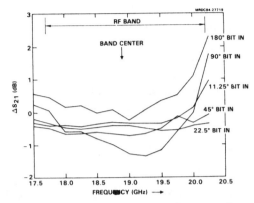

Fig. 7 Differential insertion loss vs frequency for five different phase shifter states.

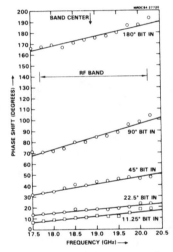

Fig. 8 Phase shift vs frequency for five different phase shifter states.

4.0 MMIC FABRICATION

The MMICs reported in this paper were fabricated by Rockwell's standard ion implantation based technology which has been describe earlier (4). However, due to the significantly increased complexity of the fully monolithic chip, as evidenced by the statistics in Table 2, fabrication yields had to improve considerably before completely functional chips could be obtained. A topic of interest is the overall fabrication yield of the monolithic transmit module; sufficient data on this design are not yet available to make a meaningful prediction of this number, although individual process step yields have been used in an analysis that predicts an RF yield of 20%. Clearly, the yield also depends on the design rules used for the circuits. The design of the final mask set will incorporate the detailed experience gained thus far in fabricting this circuit in adjusting the layout for maximum yield.

Table 2

Salient Statistics of Monolithic Transmit Module Chip

1.	Chip Size	6.3 x 4.7 mm^2
2.	Total Gate Periphery	11.5 mm x 0.65 μm
3.	Number of Metal-Insulator-Metal (MIM) Capacitors	32
4.	Total MIM Capacitance	180 pf
5.	Number of Resistors	106
6.	Number of Via Holes	28

5.0 CONCLUSIONS

This paper has describe a fully monolithic 20 GHz transmit module chip. The 6.3 x 4.7 mm MMIC contains five phase shifter bits, five gain stages and digital interface circuitry. An overall gain of 15 dB has been achieved at band center, with only ±0.5 dB gain variation with changes in phase shifter state. Phase shifts are within 5° of the desired values. Prior to the fully monolithic version, four functional building blocks, which together comprise the chip, were designed and tested. A summary of the performance of these circuits is also included. Amplifier gains are in accordance with the modeling, but the phase shifter insertion loss is ~ 1 dB/bit higher than expected. Further work is being done to reduce the loss, including minor changes in switch design and the design rules used for circuit layout. The monolithic chip shows gain rolloff at a lower frequency than was measured for the discrete circuits; also observed was an in-band oscillation. The causes of these problems are being determined and will be corrected in the final version.

The MMIC chip described here represents the highest level of integration yet reported for circuits operating at 20 GHz. With the experience gained here, highly functional monolithic ICs for phased array antenna systems should be available soon.

6.0 REFERENCES

(1) W.C. Petersen and A.K. Gupta, "A Two-Stage Monolithic Buffer Amplifier for 20 GHz Satellite Communications," Proc. 1983 IEEE Microwave and Millimeter Wave Monolithic Circuits Symp., pp. 37-39, May 1983.

(2) W.C. Petersen and A.K. Gupta, "A Three-Stage Power Amplifier for a 20 GHz Monolithic Transmit Module," 1983 GaAs IC Symp. Tech. Digest, pp. 119-122, Oct. 1983.

(3) J.A. Benet, "The Design and Calibration of a Universal MMIC Test Fixture," Proc. 1983 IEEE Microwave and Millimeter Wave Monolithic Circuits Symp., pp. 36-41, June 1982.

(4) A.K. Gupta, W.C. Petersen and D.R. Decker, "Yield Considerations for Ion Implanted GaAs MMICs," IEEE Trans. Elect. Dev., ED-30 No. 1.

Section III-E
Millimeter-Wave Circuits

MILLIMETER-WAVE circuit applications are enjoying a renaissance after numerous false starts. This rebirth is due, in no small measure, to the identification of potential military applications in the frequency range above 30 GHz.

Coincidental with this re-emergence of interest in millimeter waves is the serious attention being given to the extension of MMIC's to these higher frequency bands. The advantage of MMIC's, here, is based, not so much on possible cost savings, although this is important, but rather on the superior performance expected because of the diminution of circuit parasitics normally associated with other circuit techniques, such as the hybrid approach.

Receiver technology for the millimeter and submillimeter wave regions

Brian J. Clifton

Lincoln Laboratory, Massachusetts Institute of Technology
Lexington, Massachusetts 02173

Abstract

Increased interest in the millimeter and submillimeter wavelength regions during the past decade has stimulated the development of sensitive receivers for a wide range of applications. The extension of waveguide mixers into the submillimeter region and the development of various types of quasi-optical receivers are reviewed. The development of novel GaAs integrated circuit mixers for the millimeter and submillimeter regions is discussed. Receiver performance is reviewed and the potential impact of monolithic receiver technology on systems applications is considered.

Introduction

The past decade has seen an increased interest in the millimeter and submillimeter regions of the spectrum and in the related receiver technology.[1] The development of heterodyne receivers with improved sensitivity has resulted in a wide range of applications in radio astronomy, plasma physics, frequency standards and spectroscopy, radar, aeronomy and in satellite-based radiometry. The diversity of these applications with vastly different requirements generates the need for continuing development of heterodyne receivers with lower noise, improved reliability and at lower cost.

Although there are a number of different devices that are used as the non-linear mixing elements in heterodyne receivers, the most common device is the Schottky-barrier diode. Excellent results have been obtained with whisker-contacted diodes in waveguide and in quasi-optical mixer mounts, but it is considerably more difficult to make contact to the smaller-diameter diodes used at the higher frequencies and the reliability of the resulting diodes is inferior to that obtained at lower frequencies. The assembly of whisker-contacted diodes in mixer mounts is very labor intensive, time consuming and expensive. Naturally, there has been considerable interest in replacing the whisker-contact to the Schottky diode with a photolithographically-fabricated contact on the surface of the device. At microwave frequencies, beam-lead diodes have been integrated in a hybrid fashion with microstrip, stripline or dielectric-guide circuit elements. Similar techniques have been extended into the millimeter wave region with some success. The recent renewed interest in GaAs monolithic integrated circuits for the microwave region of the spectrum and the development of monolithic integrated circuit mixers for the millimeter and submillimeter regions offer the systems designer the possibility of totally integrated monolithic receivers with improved performance and reliability, and the potential for low-cost mass production. More difficult, but very exciting, is the possibility of building monolithic imaging arrays in the millimeter and submillimeter.

Receiver design concepts

The purpose of a heterodyne receiver is to convert a high-frequency signal to a lower frequency signal, preserving the phase information and adding a minimum of noise or degradation to the signal in the conversion process. A block diagram of a typical millimeter or submillimeter-wave heterodyne receiver is shown in Figure 1. The heart of any heterodyne receiver is a non-linear mixing element that combines a low-level radio frequency (RF) input signal with a local oscillator (LO) signal to produce a signal at the difference or intermediate freqeuncy (IF). The receiver elements after the IF amplifier are essentially independent of the frequency of the input signal. They depend mainly on the signal processing performed on the information in the IF signal and are not dealt with here.

The LO signal is usually many orders of magnitude larger than the RF signal and gives rise to the non-linearity of the mixer by driving or switching the mixer diode over some portion of its non-linear range. A typical millimeter or submillimeter mixer requires 1 to 10 mW of LO power, which is close to the maximum output power of typical present-day LO sources in the submillimeter region. Thus, in contrast to the microwave frequency range where there is usually abundant LO power available, the millimeter and submillimeter region requires that nearly all the available LO power be delivered to the mixer element.

Reprinted with permission from *Integrated Optics and Millimeter and Microwave Integrated Circuits, Proc. SPIE,* W. C. Pittman and B. D. Guenther, Eds. vol. 317, pp. 339–347, 1981.
Copyright © 1981, the Society of Photo-Optical Instrumentation Engineers.

The function of the diplexer is to combine the LO and RF signals and to send both signals to the mixer with minimum loss. In the microwave or lower-millimeter wave range, the diplexer could be a hybrid, magic T or even a directional coupler with, in all cases, the RF and LO energy propagating as guided waves confined within the component. At higher frequencies where quasi-optical techniques are more appropriate, the diplexer will be a quasi-optical component and the RF and LO energy will propagate in free-space beam modes within the diplexer.

The primary antenna receives the RF signal as a free-space plane wave and focuses the energy into a beam of appropriate diameter to send to the diplexer with minimum reflection and loss. The RF and LO energy exiting from the diplexer in overlapping beams must be coupled to a secondary antenna whose purpose is to convert the free-space energy propagating in beam modes into guided or confined wave energy that can be coupled to the Schottky-barrier junction where it appears as current through and voltages across the junction. Martin & Lesurf[2] provide a comprehensive discussion of the problems of applying classical geometrical optics to the design of submillimeter-wave optics and indicate how beam-mode concepts can be applied.

The function of all the elements preceding the mixer is to collect, focus and convert the energy of the RF and LO waves, which are propagating as free-space electromagnetic waves, into electrical current flowing in the non-linear mixing element. In the case of a Schottky-barrier diode, the actual diode has a diameter of the order of a few micrometers whereas the RF and LO free-space waves can have beam diameters of the order of centimeters. This implies considerable concentration of beam energy density in coupling the free-space energy to the diode junction. The coupling of electromagnetic energy must be achieved efficiently with minimum loss and little or no reflection. In general the impedance of the Schottky-barrier junction is quite different from that of free space so that the coupling function involves matching between impedance levels.

At microwave frequencies the RF and LO waves are guided and coupled to the diode by coaxial, stripline, microstrip, waveguide or other transmission-line media. Since the dimensions of the guiding media must be of the order of a wavelength to prevent energy propagation in undesirable modes, an upper frequency limit is imposed by problems in fabricating the guiding structures, by difficulties in mounting signal processing components within the guide structures, and by guide attenuation that increases rapidly as dimensions are reduced and frequency is increased.

For most practical purposes, mechanical fabrication is limited to components with dimensions greater than about 0.1 mm, although tolerances and surface finish can be held to much smaller dimensions. Typical matching and coupling structures in the microwave region have dimensions from a quarter to one-tenth wavelength. Thus, if similar structures are to be scaled to operate in the millimeter and submillimeter regions, present-day mechanical fabrication techniques will restrict fabrication of coupling and matching components to frequencies below 400 GHz. However, photolithographic techniques used in semiconductor device fabriction allow dimension control down to about one micrometer, and x-ray or electron beam lithography can extend dimension control down to below 0.1 μm. Clearly, it is attractive to consider using these techniques to build coupling and matching elements for the frequency range above 400 GHz where mechanical fabrication techniques are extremely difficult to apply. Thus, there is considerable interest in the fabrication of monolithic integrated circuits in which the elements of Figure 1, from the secondary antenna up to and possibly including the IF amplifier, are fabricated on a single piece of GaAs. Critical circuit elements can be located close to the diode junction with the possibility of much lower coupling loss.

<u>Discrete diode mixers</u>

The fundamental waveguide mixer is used extensively throughout the millimeter region with considerable success (Figure 2). A whisker-contacted axial diode is usually mounted in reduced-height waveguide in order to provide an optimum impedance environment to the diode. A discrete diode chip mounted flush with or on one broad wall of the waveguide, and a contact whisker centered across the waveguide and contacting one of many Schottky-barrier junctions on the diode chip comprise the non-linear mixing element. An impedance transformer is required to match the full-height waveguide impedance to the reduced-height mixer mount. At the higher millimeter-wave frequencies, the additional fabrication complexity and loss introduced by the transformer and reduced height waveguide usually favor mounting the diode in full-height guide. A movable or fixed backshort located approximately a quarter wavelength behind the diode presents a high impedance at the whisker-diode location and provides reactive tuning of the whisker and chip parasitics so that the diode junction is presented with a real impedance at the signal frequency. The backshort is a source of loss and both contacting and non-contacting shorts have been used. Non-contacting shorts give more repeatable results but are difficult to fabricate at the higher frequencies and have slightly more loss than the best contacting shorts. The coaxial RF choke in the IF output port can be difficult to fabricate at higher frequencies and is replaced in many designs by a photo-lithographically formed filter on suspended stripline.[3]

A number of workers[3-10] have reported excellent results with fundamental waveguide mixers operating both at room temperature and at cryogenic temperatures. Particularly noteworthy are papers by Vizard et al.[6] and by Keen et al.[7-9] which report results obtained with Mott barriers in which the n layer is fully depleted at zero bias. Keen et al.[8] reported a single side-band (SSB) mixer noise temperature of 98K and conversion loss of 5.0 dB at 115 GHz for a fundamental waveguide mixer cooled to 42 K. At higher frequencies, Erickson[10] reported a fundamental-mode waveguide mixer at 318 GHz having a SSB mixer noise temperature of 3100K and a conversion loss of 9.3 dB.

At lower millimeter wave frequencies discrete beam lead diodes have been integrated successfully with printed circuit mixer structures fabricated in various transmission line media.[11,12] The mixer reported by Cardiasmenos and Parrish[12] uses advanced GaAs beam-lead diodes mounted onto a fused silica suspended stripline mixer circuit. This is an impressive example of production-type mixer at 94 GHz with an uncooled SSB mixer noise temperature of 760 K and conversion loss of 6.0 dB. Unfortunately, the parasitics associated with the discrete beam-lead diodes will probably limit their applications to frequencies below 200 GHz.

Above 400 GHz it becomes exceedingly difficult to fabricate fundamental waveguide mixers. As a result mixers have been built in a quasi-optical free-space environment in which the propagating energy is not constrained to a single mode. In the simplest mounts a Schottky diode is located at the focus of a spherical reflector with a whisker contact used as a simple wire antenna. Gustincic[13-15] developed a quasi-optical biconical mixer mount, shown in Figure 3, in which a diode and whisker contact wire are mounted between the apexes of the cone-shaped pieces forming the biconical antenna. An interesting feature of this mount is that the characteristic impedance is essentially independent of frequency and is determined by the cone angle. Thus optimum RF matching to the diode should be possible simply by selecting the appropriate cone angle. GaAs Schottky-barrier diodes 2.5 µm in diameter were used in several versions of this mixer at frequencies up to 671 GHz.[16]

Lincoln Laboratory[17] adopted a long-wire antenna approach in a number of corner-reflector mixer mounts covering the 120 µm to 1mm wavelength range. The corner-reflector diode mount (Figure 4) uses a 4-wavelength long antenna wire located 1.2 wavelengths from the corner of a 90° corner reflector. The diode chip is mounted flush with the surface of the ground plane. Diodes having diameters in the range 1 µm to 2 µm are contacted by lowering the antenna whisker into contact with a diode, while observing the contacting operation through a high power optical microscope. The corner reflector mixer has an almost circular beam cross-section with a 14° beamwidth at the 3 dB points. The corner reflector mixers have been used successfully to detect interstellar carbon monoxide at 434 µm[17,18] in the Orion molecular cloud, and have also been used for fusion plasma diagnostics at 400 µm and 120 µm. The results of blackbody radiometric measurements using a standard Y-factor method interpretation of the data are shown in Table I.

TABLE I

QUASI-OPTICAL CORNER REFLECTOR MIXER PERFORMANCE

λ (µm)	FREQUENCY (GHz)	TOTAL SYSTEMS NOISE TEMPERATURE (DSB) K	MIXER TEMPERATURE (K)	CONVERSION LOSS (dB)
434	692	3,000	2,600	8.6
184	1630	19,000	≈17,000	15.6
119	2521	32,000	≈28,000	17.8

Monolithic Integrated Circuit Receivers

The rapid growth of interest in monolithic microwave integrated circuits is evidenced by the recent special issue of IEEE Transactions on Electron Devices[19] devoted to this area. At microwave and low millimeter wave frequencies, a considerable amount of GaAs surface area will be occupied by passive circuit elements, and the active device area to overall chip area ratio will be very small. Thus material cost, processing cost and active device yield tend to make a monolithic approach less cost effective at these lower frequencies than a hybrid approach in which active device and passive circuit technology are separated. Overall chip area can be reduced in some cases by replacing and simulating passive circuit functions with active devices. However, the penalty is usually increased noise and power dissipation.

The earliest reported work on millimeter wave integrated circuits on GaAs was by Texas Instruments[20,21] in 1968 in which single-ended and balanced mixers were fabricated for use at 94 GHz. The passive integrated circuit components were microstrip lines deposited on 100 μm thick semi-insulating GaAs substrates. More recently Chu et al.[22] have fabricated a 31 GHz monolithic receiver front end, as shown in Figure 5, using microstrip circuits on 175 μm thick semi-insulating GaAs. The balanced mixer-preamplifier combinations typically exhibit a conversion gain of 4 dB and a single-sideband noise figure of 11.5 dB.

Clifton et al.[23,24] have reported the development of a novel GaAs monolithic integrated circuit mixer (Figure 6) which is impedance matched to fundamental waveguide. It consists of a slot coupler, coplanar transmission line, surface-oriented Schottky-barrier diode, and RF bypass capacitor monolithically integrated on the GaAs surface. In this mixer the radiation propagates through the GaAs to a slot coupler fabricated photolithographically in a metallic ground plane on the surface of the GaAs, in contrast to monolithic mixer structures using microstrip circuits in which the radiation is guided along the surface of the GaAs dielectric. The bulk dielectric is actually part of the impedance matching circuit. The slot coupler is connected to a diode by an appropriate section of coplanar line, and an integrated bypass capacitor completes the mixer circuit providing a short circuit to millimeter wave frequencies and an open circuit at the IF. At 110 GHz, a monolithic mixer module mounted in the end of a waveguide horn has an uncooled double-side-band (DSB) mixer noise temperature of 339 K and a conversion loss of 3.8 dB. The same mixer module cooled to 77 K has a DSB mixer noise temperature of 50 K and a conversion loss of 4.5 dB. A similar monolithic mixer module designed to operate at 350 GHz has an uncooled DSB mixer noise temperature of 6500 K.

Monolithic Receiver Fabrication Considerations

At microwave and lower millimeter wave frequencies microstrip techniques can be applied very effectively to build passive circuit components on semi-insulating GaAs. However, in the submillimeter regime it is more appropriate to use waveguide and quasi-optical techniques. Thus, beam waveguide, focusing optics, antenna structures and dielectric waveguides are concepts that can be applied to the design of receiver circuits in the millimeter and submillimeter. The monolithic integrated circuit mixer (Figure 6) developed at Lincoln Laboratory combines antenna concepts with properties of the bulk GaAs dielectric to produce a very efficient coupling structure at 110 GHz. We believe that these concepts can be applied throughout the millimeter and a large portion of the submillimeter region.

In considering the Schottky-barrier diode, similar design rules apply to monolithic mixers as to conventional whisker-contacted diode mixers. However, since the millimeter wave circuit is connected directly to the Schottky diode, diode lead parasitics can be eliminated in a correctly designed monolithic mixer circuit. Diode junction capacitance and series resistance are parasitic elements that degrade mixer performance and should be minimized. A Schottky-barrier diode designed for use at frequencies near 100 GHz would have an equivalent circular diameter of about 2 μm on GaAs material with a carrier concentration in the low 10^{17} cm^{-3} range. Figure 7, which is a detailed drawing of the surface-oriented diode portion of the Lincoln Laboratory monolithic mixer, illustrates the main fabrication and topological details of the device. The n layer should be thin to minimize the contribution of the undepleted epitaxial layer to diode series resistance, and will typically be 0.1 to 0.2 μm thick with a carrier concentration of $1-2 \times 10^{17}$ cm^{-3}. If the n layer is fully depleted at zero bias, the resulting Mott barrier not only has lower series resistance than a Schottky barrier but also requires lower local oscillator drive power and gives lower noise performance in cooled mixer applications. Such devices require very uniform epitaxial layers with a very sharp transition between the n$^+$ and n layers and an n layer thickness of less than 0.1 μm. The Schottky barrier can be formed using standard metallization techniques. The particular choice of Schottky-barrier metal may be influenced, however, by the temperature stability of the barrier during subsequent device fabrication. The most critical and difficult aspect of device fabrication is the location of the Schottky-barrier junction at one edge of the conducting pocket of GaAs. This must be accomplished with a registration accuracy of better than ± 0.5 µm since this directly determines the resulting junction capacitance. A stripe geometry with the long axis perpendicular to the pocket edge is superior to a circular geometry since registration error has less effect on junction capacitance and, in addition, spreading resistance is lower. Reduction of spreading resistance requires an ohmic contact of low resistivity in close proximity to the Schottky barrier and an n^+ epitaxial layer with a thickness of several skin depths at the operating frequency. However, if proton bombardment is used to isolate the conducting pocket of GaAs, the total epitaxial layer thickness that can be converted to semi-insulating is determined by the proton energy available. For example, 400 keV protons will reliably convert up to 3 µm of n^+ GaAs with a carrier concentration of 2×10^{18} cm-3.

The diode ohmic contact region can be defined on the GaAs surface by a variety of means, such as by ion implantation or by etching away the n layer and alloying an evaporated contact metallization into the exposed n^+ layer. Specific contact resistance as low as 2×10^{-8} ohm-cm^2 has been achieved in the Lincoln Laboratory diode by alloying an evaporated Ni/Au-Ge/W/Au metallization into the n^+ layer. Other workers have reported extremely low specific contact resistance for non-alloyed ohmic contacts formed by direct metallization onto an epitaxial Ge layer grown on top of the n^+ layer. Similar results might be expected for non-alloyed metallizations onto n^+ layers with carrier concentration in the 10^{19} cm^{-3} range.

The technique selected to isolate and define the conducting area of the diode depends on many factors. Although proton bombardment has proven to be a convenient and successful technique, it has certain limitations. Thicker epitaxial layers, higher n^+ carrier concentration, or dopant tails extending into the semi-insulating substrate may not be reliably converted by the available proton energy. This can result in an undesirable buried conducting layer which attenuates electromagnetic radiation and degrades mixer performance. In addition, subsequent processing steps must be limited to temperatures below 300°C to prevent annealing of the proton damage. This could place severe restrictions on the fabrication sequence for more complex monolithic integration. Mesa etching might be an alternative, but it results in a non-planar surface which in most cases precludes subsequent high-resolution photolithography.

As we move from the integration of single components such as mixers or amplifiers to a completely monolithic receiver front end, we find that material that is optimum for mixer diodes is not suited to MESFET devices. Selective epitaxy might prove to be the most cost effective technique to provide such starting material, since this technique offers the possibility of tailoring the material characteristics within a particular pocket to suit a specific active device. Of course, ion implantation and proton bombardment will still be valuable techniques to modify material characteristics during fabrication. Although selective epitaxy into etched pockets in a wafer of semi-insulating GaAs has been used by many workers to produce isolated regions of conducting GaAs, problems of non-planar surfaces, moat-like voids at the periphery of the pockets, and quality of the epitaxial layers within the pockets have delayed the use of these techniques for monolithic mixers in the millimeter and submillimeter. The recent development[25,26] of ion beam assisted etching of semi-insulating GaAs to create pockets with vertical walls and smooth surfaces appears to offer an ideal pocket structure into which high-quality n on n^+ layers can be grown by MBE. The resulting void-free planar wafers should be ideal for the fabrication of monolithic mixers and receivers.

Monolithic mixer circuits can be scaled to operate at higher frequencies simply by scaling the circuit dimensions smaller in the ratio of the higher to lower frequency. However, contact bonding pads on the millimeter monolithic mixers must have minimum linear dimensions of about 50 µm if conventional ribbon or wire bonding techniques are used to connect to the device. As the devices are scaled into the submillimeter wave regime, contact pads will inevitably become too small and alternative techniques must be considered to interface to the device. In addition, the module thickness is reduced by the same scale factor. Techniques have been developed at Lincoln Laboratory that allow controlled thinning and separation of individual modules using front-to-back alignment and a cantilevered extended ground plane and coplanar IF lead as shown in Figure 8. Such devices are easy to integrate with subsequent low-frequency circuits and provide an integrated package that can be conveniently handled and interfaced with high-frequency quasi-optical components.

Conclusions

The state-of-the-art in low noise receiver technology in the 100-1000 GHz frequency range is summarized in Figure 9. Monolithic integrated circuit mixers appear to be competitive with the best waveguide mixers at 110 GHz. Above 400 GHz quasi-optical discrete diode mixers still dominate and significant improvement in performance has been obtained with corner reflector mixer mounts used for radio astronomy at 700 GHz.

Considerable progress has been made in the past few years in the development of monolithic integrated circuits for the microwave, millimeter and submillimeter wave regions of the spectrum. Material quality has improved and work is underway to provide high-quality large-area semi-insulating substrates. More work is still required to provide high-quality selective epitaxy for advanced monolithic integrated circuits. Device fabrication technology is well developed and given low-cost high-quality GaAs wafers there is no reason why monolithic receivers should not dominate the millimeter and submillimeter wave spectrum during the next decade. Integration of a low-noise submicrometer-gate FET IF-amplifier with the monolithic mixer offers the intriguing possibility of a cooled monolithic receiver at 3mm with a total system DSB noise temperature of 100 K or less. The monolithic mixers with appropriate integrated packaging concepts should be scalable to 3 THz. More difficult, but very exciting, is the possibility of building monolithic imaging arrays in the millimeter and submillimeter.

Acknowledgment

The author would like to acknowledge the contributions of many workers who have helped develop receiver technology to its present state. In particular, I wish to thank my colleagues in the Solid State Division at Lincoln Laboratory whose work, comments, and encouragement have contributed significantly to the preparation of this paper. The preparation of this paper was supported by the Department of the Air Force and the U.S. Army Research Office.

References

1. Clifton, B. J., "Schottky Diode Receivers for Operation in the 100-1000 GHz Region," Radio Electron Eng., Vol. 49, pp. 333-346, 1979.
2. Martin, D. H., and Lesurf, J., "Submillimeter Wave Optics", Infrared Physics, Vol. 18, pp. 405-412, 1978.
3. Wrixon, G. T., "Low Noise Diodes and Mixers for the 1-2 mm Wavelength Region," IEEE Trans. Microwave Theory Tech., Vol. MTT-22, pp. 1159-1165, 1974.
4. Kerr, A. R., "Low-noise Room-temperature and Cryogenic Mixers for 80-120 GHz, IEEE Trans. Microwave Theory Tech., Vol. MTT-23, pp. 781-787, 1975.
5. Kerr, A. R., Mattauch, R. J., and Grange, J. A., "A New Mixer Design for 140-220 GHz", IEEE Trans. Microwave Theory Tech. Vol. MTT-25, pp. 399-401, 1977.
6. Vizard, D. R., Keen, N. J., Kelly, W. M., and Wrixon, G. T., "Low Noise Millimeter Wave Schottky Barrier Diodes with Extremely Low Local Oscillator Power Requirements," 1979 IEEE-MTT-S International Microwave Symposium Digest, IEEE Cat. No. 79CH1439-9MTT-S, pp. 81-83, 1979.
7. Keen, N. J., Haas, R. W., and Perchtold, E., "A Very Low Noise Mixer at 115 GHz, using a Mott Diode Cooled to 20K", Electron Lett., Vol. 14, pp. 825-826, 1978.
8. Keen, N. J., Kelly, W. M., and Wrixon, G. T., "Pumped Schottky Diodes with Noise Temperatures of less than 100K at 115 GHz," Electron Lett., Vol. 15, pp. 689-690, 1979.
9. Keen, N. J., "Low-Noise Millimeter-Wave Mixer Diodes: Results and Evaluation of a Test Program," IEE Proc., Vol. 127, Pt.I, pp. 188-198, 1980.
10. Erickson, N. R., "A 0.9 mm Heterodyne Receiver for Astronomical Observations," 1978 IEEE-MTT-S International Microwave Symposium Digest, IEEE Cat. No. 78CH1355-7 MTT, pp. 438-439, 1978.
11. Meier, P. J., "Printed-Circuit Balanced Mixer for the 4- and 5-mm Bands," 1979 IEEE-MTT-S International Microwave Sysmposium Digest, IEEE Cat No. 79CH1439-9MTT-S, pp. 84-86, 1979.
12. Cardiasmenos, A. G., and Parrish, P. T., "A 94 GHz Balanced Mixer using Suspended Substrate Technology", 1979 IEEE-MTT-S International Microwave Symposium Digest, IEEE Cat. No. 79CH1439-9MTT-S, pp. 22-24, 1979.
13. Gustincic, J. J., "A Quasi-optical Radiometer", Second International Conference and Winter School on Submillimeter Waves and their Applications Digest, IEEE Cat. No. 76CH1152-8MTT, pp. 106-107, 1976.
14. Gustincic, J. J., Receiver Design Principles", Proc. Soc. Phot-Opt. Instrum. Engrs., Vol. 105, pp. 40-43, 1977.
15. Gustincic, J. J. "A Quasi-optical Receiver Design", 1977 IEEE-MTT-S International Microwave Symposium Digest, IEEE Cat. No. 77CH1219-5MTT, pp. 99-101, 1977.

16. Gustincic, J. J., DeGraauw, T. R., Hodges, D. T., and Luhmann, Jr. N. C., "Extension of Schottky Diode Receivers into the Submillimeter Region," Unpublished report 1977.

17. Goldsmith, P. F., Erickson, N. R., Fetterman, H. R., Clifton, B. J., Peck, D. D., Tannenwald, P. E., Koepf, G. A., Buhl, D., and McAvoy, N., "Detection of the J=6→5 Transition of Carbon Monoxide," Astrophys. J., Vol. 243, pp. L79-L82, 1981.

18. Fetterman, H. R., Koepf, G. A., Goldsmith, P. F., Clifton, B. J., Buhl, D., Erickson, N. R., Peck, D. D., McAvoy, N., and Tannenwald, P. E., "Submillimeter Heterodyne Detection of Interstellar Carbon Monoxide at 434 Micrometers", Science, Vol. 221, pp. 580-582, 1981.

19. IEEE Tans. Electron Devices, Vol. ED-28, No. 2, 1981.

20. Mehal, E. W., and Wacker, R. W., "GaAs Integrated Microwave Circuits," IEEE Trans. Electron Devices, Vol. ED-15, pp. 513-516, 1968.

21. Mao, S., Jones, S., and Vendelin, G. D., "Millimeter-Wave Integrated Circuits," IEEE Trans. Electron Devices, Vol. ED-15, pp. 517-523, 1968.

22. Chu, A., Courtney, W. E., and Sudbury R. W., "A 31 GHz Monolithic GaAs Mixer/Preamplifier Circuit for Receiver Applications," IEEE Trans. Electron Devices, Vol. ED 28, pp. 149-154, 1981.

23. Clifton, B. J., Alley, G. D., Murphy, R. A., and Mroczkowski, I. H., "High Performance Quasi-Optical GaAs Monolithic Mixer at 110 GHz." IEEE Trans. Electron Devices, Vol. ED-28, p. 155-157, 1981.

24. Clifton, B. J., Alley, G. D., Murphy, R. A., Piacentini, W. J., Mroczkowski, I. H., and Macropoulos, W., "Cooled Low Noise GaAs Monolithic Mixers at 110 GHz," 1981 IEEE-MTT-S International Microwave Symposium Digest, IEEE Cat. No. 81CH1592-5MTT, pp. 444-446, 1981.

25. Geis, M. W., Efremow, N. N., and Lincoln, G. A., "A Novel Dry Etching Technique." (Submitted to the Journal of Vacuum Science & Technology).

26. Geis, M. W. Lincoln, G. A., and Efremow, N. N., "A Novel Dry Etching Technique," Proceedings of the 16th symposium on Electron, Ion, and Photon Beam Technology, 1981.

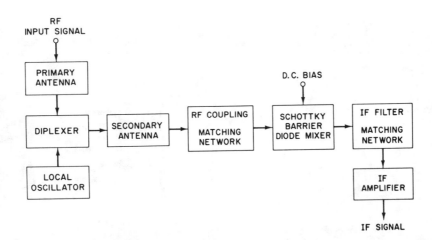

Figure 1. Block diagram of a typical millimeter or submillimeter-wave heterodyne receiver.

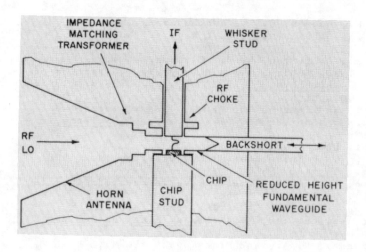

Figure 2. Cross-sectional detail of a typical millimeter wave fundamental waveguide mixer.

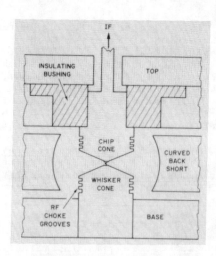

Figure 3. Quasi-optical biconical mixer mount. (Gustincic)

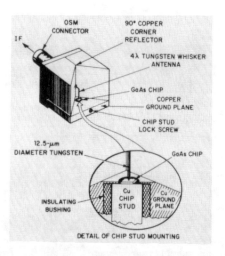

Figure 4. Corner-reflector mixer mount.

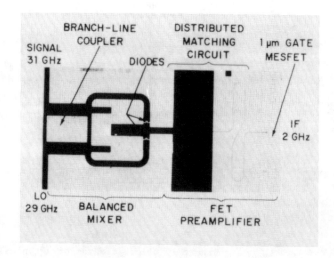

Figure 5. 31-GHz GaAs monolithic receiver showing a balanced mixer and MESFET amplifier. (Chu et al.)

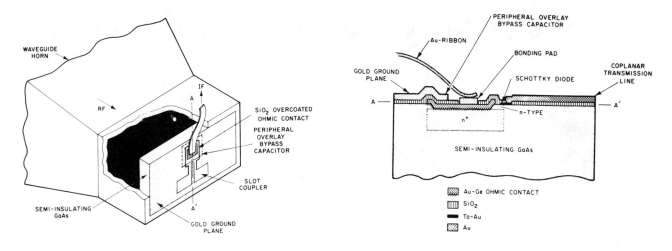

Figure 6. Monolithic integrated-circuit mixer mounted in TE_{10} waveguide horn.

Figure 7. Cross-section through A-A' showing fabrication details.

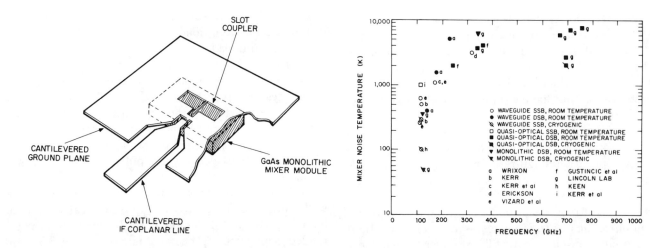

Figure 8. Integrated mixer module for the submillimeter with cantilevered ground plane and IF coplanar line.

Figure 9. State-of-the-art- in low noise receiver technology in the 100-1000 GHz range.

GaAs Integrated Microwave Circuits

EDWARD W. MEHAL AND ROBERT W. WACKER, MEMBER, IEEE

Abstract—GaAs has many desirable features that make it most useful for microwave and millimeter-wave integrated circuits. The process of selective epitaxial depositions of high purity single-crystal GaAs with various doping concentrations into semi-insulating GaAs substrates has been developed. These high-resistivity substrates ($>10^6$ ohm·cm) provide the electrical isolation between devices, eliminating the difficulties and deficiencies normally encountered in trying to obtain isolation with dielectrics, back-etching, p-n junctions, etc. This monolithic approach to integrated circuits thus allows for improved microwave performance from the devices since parasitics are reduced to a minimum.

Planar Gunn oscillators and Schottky barrier diodes have been fabricated for use in a completely monolithic integrated millimeter wave (94 GHz) receiving front end. The Gunn oscillators are made in a sandwich-type structure of three selective deposits whose carrier concentrations are approximately 10^{18}–10^{15}–10^{18} cm^{-3}. The Schottky diodes consist of two deposits with concentrations of 10^{18} and 10^{17} cm^{-3}. The Schottky contact is formed by evaporating Mo-Au onto the 10^{17} cm^{-3} deposits; all ohmic contacts are on the surface and are alloyed to the N^+ regions.

INTRODUCTION

VARIOUS APPROACHES have been investigated to fabricate integrated microwave circuits. Hybrid integrated circuits have been fabricated to perform mixing, harmonic generation, and signal generation functions. Silicon monolithic integrated microwave circuits have been extensively investigated for solid-state radar applications. GaAs has also been investigated for integrated microwave circuits. The unique properties of GaAs make it well suited for this type of application. High-resistivity single-crystal GaAs is available to serve as a substrate material. A high-resistivity substrate is required both to provide the necessary electrical isolation between components and to minimize the transmission losses in the circuit. Microwave oscillators can be readily produced from GaAs which can be incorporated into these circuits as signal sources.

Varactor and mixer functions can be accomplished using GaAs Schottky barrier diodes. The combination of desirable materials, properties, and available microwave devices makes GaAs a valuable material for integrated microwave circuits.

Manuscript received February 1, 1968. This work was partly supported by the Air Force Avionics Laboratory, Wright-Patterson Air Force Base, under Contracts AF 33(615)-1275 and AF 33(615)-5102. This special issue on Microwave Integrated Circuits is published jointly with IEEE TRANSACTIONS ON ELECTRON DEVICES, July 1968, and the IEEE JOURNAL OF SOLID-STATE CIRCUITS, June 1968.
The authors are with Texas Instruments Incorporated, Dallas, Tex. 75222

SEMI-INSULATING GaAs

A high-resistivity substrate is desirable for integrated microwave circuits to minimize transmission losses and to provide electrical isolation between devices in the circuit. Single-crystal GaAs can be produced with resistivities $>10^6$ ohm·cm which serve ideally as substrates. This material is produced by either oxygen or chromium doping. Oxygen-doped semi-insulating GaAs is prepared by boat growth in the presence of a moderate amount of Ga_2O (50 to 100 Torr).[1] The high resistivity of this material is the result of deep donor centers caused by oxygen. High resistivity results from having the concentration of shallow acceptors larger than shallow donors, and the concentration of deep donors larger than the net shallow acceptor concentration. Normally the boat grown material would be n-type because of the shallow donors produced by silicon. However, in oxygen-doped material the silicon concentration is suppressed by having Ga_2O present during growth. The oxygen also introduces the deep donor level.

Chromium-doped semi-insulating GaAs is prepared by introducing chromium at a concentration above the normal background donor concentration.[2] Chromium introduces a deep acceptor level at 0.79 eV. There is no necessity for reducing the concentration of normally produced background donor and acceptor impurities; therefore this material can be readily produced in sealed and semisealed crystal pullers in single-crystal form.

Both of these forms of semi-insulating GaAs would be well suited for integrated circuit use because of their high resistivity $>10^7$ ohm·cm. However, if any extended temperature processing is required such as epitaxial growth or high temperature contacts, the chromium-doped material is superior. Oxygen-doped GaAs is reported to undergo a heat treating effect and lower resistivity GaAs results.[3] Chromium-doped material does not appear to undergo any similar heat treating effect.

SELECTIVE GROWTH

The selective growth process was developed to produce device structures for monolithic integrated circuits.[4] The main process steps are shown in Fig. 1. The $(\overline{111})$ oriented substrate is chemically polished to remove surface damage. A masking layer is then deposited. SiO_2 or Si_3N_4 with thickness of 3000 Å to 5000 Å are used as masks. The desired deposit pattern

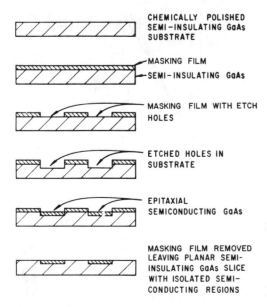

Fig. 1. Main process steps in epitaxial selective growth.

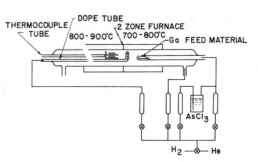

Fig. 2. Ga-AsCl₃ reactor system.

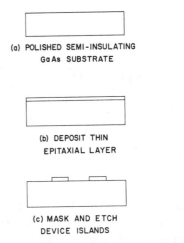

Fig. 3. Mesa-etching process for monolithic circuits.

is etched into this masking material using photolithographic methods and etches. Holes are then etched into the GaAs using the masking film to protect the remaining surface of the slice. The masked slice is then processed through an epitaxial deposition step.

Good geometrical control of selectively grown regions is required for predictable device performance. The etches most commonly used for GaAs (H_2SO_4:H_2O_2:H_2O, Br_2:CH_3OH, NaOCl) produce holes which have a nonplanar hole bottom.[5] This situation is eliminated by using 0.7-MH_2O_2–1.0-MNaOH. This etch is reaction-rate limited rather than diffusion-rate limited as are the other etches.

The selective depositions are performed utilizing the Ga-AsCl₃ reactor system, Fig. 2. Run conditions are selected which produce a minimum of deposition on the walls of the reactor tube in the deposition zone. Typical run conditions are: feed temperature, −825°C; substrate temperature, −750°C; H_2 flow through AsCl₃, −100 cc/min; AsCl₃ temperature, −25°C; H_2 dilution flow, −100 cc/min; and 50 cc/min additional H_2 flow through the auxiliary input to insure continuous flow through all parts of the reactor during subsequent operations. These deposition conditions produce a deposition rate of ∼5 µ/h on nonmasked ($\overline{111}$) oriented substrates.

The feed materials used are commercially available 99.9999 percent gallium from AIAG and distilled reagent grade AsCl₃. The hydrogen used is purified by the palladium diffusion process. The GaAs deposits produced with no intentional doping have carrier concentrations $\leq 10^{15}$ cm^{-3} which are below the requirements for most microwave devices. Doping is accomplished by introducing S_2Cl_2 through an auxiliary doping tube. Deposits with carrier concentrations up to 2×10^{18} cm^{-3} can be produced with this technique.

An alternate method for producing device structures for monolithic circuits which does not utilize selective growth is a mesa etching process. This process is illustrated in Fig. 3.[6] A semi-insulating substrate is processed through a normal epitaxial deposition procedure producing a thin layer (1–5 µm). The desired device geometry is then defined using photolithographic methods. The slice is then etched leaving islands electrically isolated from one another. Contacts and circuit patterns can then be applied. The mesa step is small enough with thin layers so that contact continuity is maintained.

Devices

The principal devices produced for integrated microwave circuits are tunnel diodes, varactor diodes, Schottky barrier mixer diodes, and Gunn oscillators. These have all been prepared in planar form by using either the "mesa" method or selective growth. The Schottky barrier mixer diode is perhaps the least complex of these devices to fabricate. Fig. 4 illustrates the design used in conjunction with selective epitaxial growth. This structure results in a minimum of parasitics, particularly expanded contact capacitance, to produce improved microwave performance.

Two separate deposition steps are used, a heavily doped n^+ pocket 10 µm in thickness and 1 by 2 mils in area, followed on top by a thin n-type layer of ∼0.5

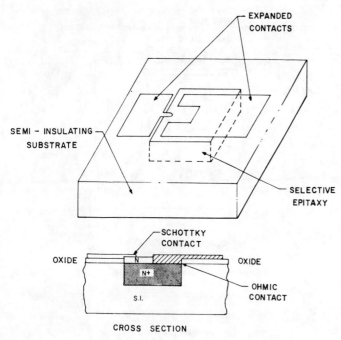

Fig. 4. Design structure of planar Schottky barrier diode.

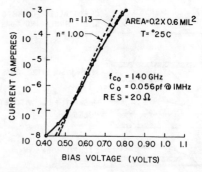

Fig. 5. I–V characteristic of planar Schottky barrier diode.

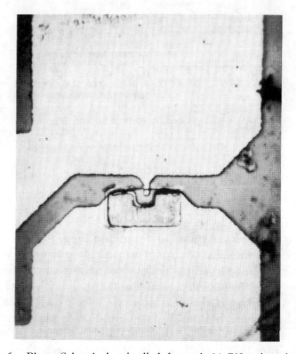

Fig. 6. Planar Schottky barrier diode for use in 94-GHz mixer circuit.

μm. A phosphorous-doped silicon oxide film is used to prevent any substrate conversion to lower resistivity material at the surface. The n^+ deposit is typically doped to 2×10^{18} cm^{-3}, while the n-layer is 10^{17} cm^{-3} with a measured room temperature Hall mobility of 5000 cm^2/V·s.

Ohmic contact is then made to the n^+ region by etching the contact pattern through the n-layer and then evaporating 90 wt.% Ag–5 wt.% In–5 wt.% Ge onto the slice. After the excess metal has been removed, the contact is alloyed at 600°C. The diode's active area is then defined by etching the oxide film with the appropriate photomask. Before evaporating the Mo–Au Schottky contact, which also serves as the expanded contact, the defined n-area must be adequately cleaned. This is accomplished by a short etch with a sulphuric acid–peroxide–water mixture, then a rinse in EDTA (ethylene diamine tetracetic acid), followed by a final rinse in a buffered HF solution. Immediately afterwards, the Mo–Au evaporation is made and the excess is etched away.

A typical I–V characteristic curve is presented in Fig. 5, to be compared with a reference curve whose exponent n has a value of 1.00. Improvements in the microwave performance for use as a mixer in a 94-GHz integrated microwave receiver were obtained by reducing the contact area to 0.2 by 0.2 mils, as shown in Fig. 6. This resulted in lower diode capacitance and cutoff frequencies as high as 500 GHz. Theoretical calculations for this diode structure, based on the work of Cox and Strack[7] on contact resistance, predict an f_{co} of 3300 GHz. However, the resistance due to the undepleted portion of the n-layer limits this value to ~1200 GHz since selective deposits thinner than 0.5 μm cannot as yet be reproduced.

These planar diodes can also be used as varactors rather than mixers for microwave frequency multipliers. The fabrication is the same as that for the mixer diodes, with the exception of the active device area. Using the same n-type deposit, which has a reverse breakdown of 10 to 15 volts, the device area was increased to 0.2 by 0.4 mils for use in a tripler circuit from 31.3 to 95 GHz. For lower frequency applications, the device area would be correspondingly increased.

The planar Gunn oscillator has been prepared in a similar manner using selective growth technology. This device structure requires three separate deposits, as seen in Fig. 7. The fabrication procedures using the phosphorous-doped oxide as a mask, the NaOH etchant to cut the holes in the substrate, the Ag–In–Ge ohmic contacts are identical to those of the planar Schottky diodes. The first n^+ deposit provides the "bottom" ohmic contact. Although its thickness or depth is 15 μm for operation at 30 GHz, this varies according to the specific

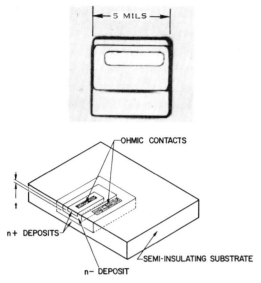

Fig. 7. Structure and deposits of selectively grown planar Gunn oscillator.

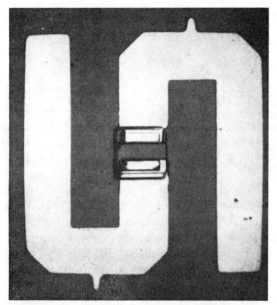

Fig. 8. Gunn oscillator in full wavelength cavity circuit.

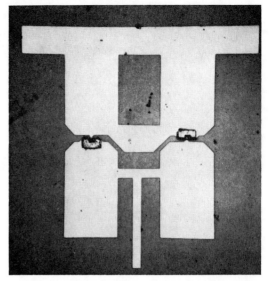

Fig. 9. Monolithic balanced mixer circuit.

frequency requirement as determined by the next, or n-layer thickness.

After remasking this 5 by 5 mils deposit, a smaller, 3 by 5 mils, hole is etched into it to a specific depth—5 μm for 30-GHz operation. The doping level for this deposit is $\sim 3 \times 10^{15}$ cm^{-3}. The top ohmic contact is provided for by the third selective growth, a 1 by 4 mil^2 n^+ pocket, deposited in a 2-μm hole etched into the n-deposit. As with the Schottky diodes, the Ag–In–Ge alloy contacts are applied to the two n^+ regions after the surface cleanup. A final Mo–Au film is evaporated onto the slice and, by using the appropriate photomask, yields either simple expanded contacts for probing the devices or an actual strip-line circuit. A complete oscillator in a full-wavelength cavity circuit can be seen in Fig. 8.

Integrated Circuits

These planar microwave devices can be readily incorporated into simple integrated microwave circuits. An example of this is the balanced mixer circuit shown in Fig. 9, for operation at 94 GHz, which incorporates two of the Schottky barrier diodes described earlier. It consists of two RF inputs to the diodes in a branch-line coupler circuit, and an IF output terminal along with two quarter-wavelength open circuited stubs providing an RF short to ground. The total chip size is 30 by 30 mils with a 4 mil thick substrate. A larger chip size is necessary at lower frequencies because of the increased mixer circuit area. At 10 GHz, the chip would be 300 by 300 mils. Although this size requirement is still practical using GaAs, a hybrid approach using ceramic substrates and planar beam-lead devices becomes more practical at lower frequencies.

Summary

GaAs has many desirable features which make it useful for integrated microwave circuits. It is desirable as a substrate material because of its availability in single-crystal high-resistivity form. A process for selective growth of device quality material is developed and has been used to produce Gunn oscillators and Schottky barrier diodes. Simple circuits have been fabricated utilizing these devices.

References

[1] J. F. Woods and N. G. Ainslie, *J. Appl. Phys.*, vol. 34, p. 1469, 1963.
[2] G. R. Cronin and R. W. Haisty, *J. Electrochem. Soc.*, vol. 111, p. 874, 1964.
[3] J. M. Woodall and J. F. Woods, *Solid State Comm.*, vol. 4, p. 33, 1966.
[4] E. W. Mehal, R. W. Haisty, and D. W. Shaw, *Trans. Metallurgical Soc.*, vol. 236, p. 263, 1966.
[5] D. W. Shaw, *J. Electrochem. Soc.*, vol. 113, p. 958, 1966.
[6] D. P. Holmes and P. L. Baynton, *Proc. Internat'l Symp. on GaAs*, p. 236, 1966.
[7] R. H. Cox and H. Strack, *Solid State Electronics*, vol. 10, p. 1213, 1967.

A 31-GHz Monolithic GaAs Mixer/Preamplifier Circuit for Receiver Applications

ALEJANDRO CHU, WILLIAM E. COURTNEY, AND ROGER W. SUDBURY, MEMBER, IEEE

Abstract—The portion of a monolithic receiver containing integrated Schottky mixer diodes and MESFET's with microstrip circuitry has been developed and tested at 31 GHz. This work is part of a program to establish the feasibility of monolithic receivers and transmitters at microwave and millimeter-wave frequencies.

Receiver designs using high-cutoff frequency diodes in a mixer configuration followed by a MESFET amplifier are capable of operating from microwave through millimeter-wave frequencies. However, the fabrication of monolithic receiver designs requires the integration on the same wafer of devices with different material requirements. We have developed a compatible integration scheme which is fundamental to the fabrication of monolithic receivers at millimeter-wave frequencies. Fabrication and design considerations for the 31-GHz balanced mixer and IF preamplifier are described. Completed monolithic units typically exhibit a conversion gain of 4 dB from the signal frequency of 31 GHz to the IF frequency of 2 GHz. The associated noise figure is typically 11.5 dB.

I. INTRODUCTION

THERE ARE a number of potential systems such as phased array radars requiring large numbers of monolithic analog GaAs circuits at microwave and millimeter-wave frequencies. Obvious reductions in size and weight can be realized by integration. However, the greatest benefit may be achieved in microwave component fabrication by implementing the mass-production techniques of silicon integrated-circuit technology which early workers [1] recognized could lead to significant decreases in the cost of microwave and millimeter-wave components.

Highly innovative efforts by Mao *et al.* [2], Mehal and Wacker [3], Vendelin [4], in the 1960's utilized a variety of GaAs devices in millimeter-wave components. Their work demonstrated flexibility in the implementation of circuit functions using different types of devices. Although circuits with active devices such as Gunn diodes or Schottky diodes were fabricated individually, attempts to combine different device types in monolithic circuits failed. There are many references to microwave monolithic circuits in the literature, but quite often the work described pertains to hybrid technology. To date, GaAs monolithic analog circuits have been primarily MESFET amplifiers [5]-[7] and Schottky-diode mixers [4], [8]-[10]. Combinations of devices have been reported [11] where the epitaxial layer was optimized for MESFET's.

Feasibility of fabricating different types of microwave device on the same GaAs wafer remains to be demonstrated. Development of the 31-GHz receiver section required the integration of MESFET's and high-frequency diodes. This paper reports the successful combining of high-cutoff-frequency Schottky diodes and a MESFET on the same wafer in order to fabricate modules consisting of a balanced mixer and a low-noise IF amplifier. This capability provides an expanded fabrication flexibility for a variety of monolithic circuits beyond the present receiver component.

Circuit design considerations for the balanced mixer and the MESFET amplifier are presented in Section II. A brief discrip-

Manuscript received July 9, 1980; revised November 12, 1980. This work was supported by the Department of the Army. The views and conclusions contained in this document are those of the contractor and should not be interpreted as necessarily representing the official policies, either expressed or implied, of the United States Government.

The authors are with Lincoln Laboratory, Massachusetts Institute of Technology, Lexington, MA 02173.

tion of the fabrication technique used in the monolithic integration of mixer diodes and MESFET's is given in Section III. Results for the receiver components are presented in Section IV. Data are presented for a unit constructed of separately tested mixer and amplifier chips and for the monolithic version.

II. CIRCUIT DESIGN

Fabrication of a complete millimeter-wave transceiver in a monolithic format requires considerable flexibility in the circuit layout which can be provided by microstrip lines. In general, microstrip also has a lower attenuation than coplanar waveguide and allows for a rugged, and lower thermal resistance, mount as the substrate need not be suspended.

The choice of a mixer followed by amplification provides flexibility in the choice of the final operating frequency as the mixer design can be scaled over a large frequency range. The balanced mixer offers advantages for monolithic implementation as it requires low local-oscillator power and, for limited bandwidth requirements, the circuitry required can be fabricated on a reasonably sized area.

With low-noise amplification at the IF frequency, submicrometer-gate MESFET fabrication is avoided. For the initial amplifier design only a single stage of amplification has been fabricated, with a second stage planned. To avoid two-level metallization, distributed elements are used in the matching circuit. Although the equivalent circuits for lumped components are complex [12], their use at the IF frequencies will conceivably result in reducing further the substrate area required.

A. Microstrip Properties

Measurements of the dielectric properties of semi-insulating GaAs have previously been reported [13]. These measurements gave a relative dielectric constant of 12.95 and a loss tangent of less than 6×10^{-4} up to 36 GHz. A substrate thickness of 0.178 mm was chosen as a balance between low loss (~0.3 dB/cm at 30 GHz for a 50-Ω line), dispersion (~3-percent change in effective dielectric constant from 0-30 GHz for a 50-Ω line), and coupling (to provide at least two linewidths of separation of the arms in the branchline coupler). This is also a convenient thickness for handling and for fabrication.

Fig. 1 shows the effective dielectric constant as a function of frequency for a 0.2-mm-thick substrate of GaAs, and also for a 0.254-mm-thick substrate of alumina for comparison. Measurements were made using half-wavelength open-circuited resonators. In both cases, the formula of Getsinger [14]

$$\epsilon_{eff} = \epsilon_s - \frac{(\epsilon_s - \epsilon_{e0})}{1 + G(f/f_p)^2} \quad (1)$$

for the effective dielectric constant agrees very well with experiment when the empirical parameter G in the formula is given by

$$G = 0.5 + 0.01 Z_0. \quad (2)$$

ϵ_s is the substrate relative dielectric constant, ϵ_{e0} is the microstrip effective dielectric constant at zero frequency, Z_0 is the microstrip characteristic impedance at zero frequency, and

$$f_p = \frac{Z_0}{2\mu_0 h} \quad (3)$$

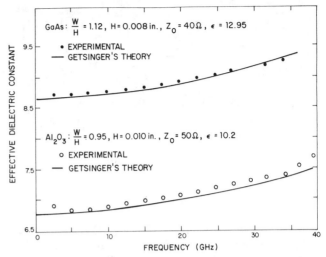

Fig. 1. Effective dielectric constant of microstrip on GaAs and Al_2O_3 substrates as a function of frequency.

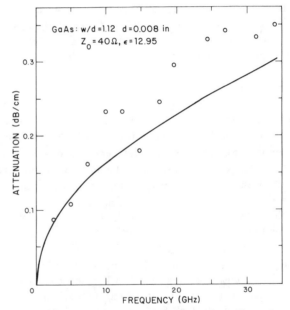

Fig. 2. Attenuation as a function of frequency for a 40-Ω microstrip line on a semi-insulating GaAs substrate.

where μ_0 is the permeability of free space and h is the substrate thickness.

Attenuation as a function of frequency is shown in Fig. 2 for an open-line resonator of 40-Ω impedance on a 0.2-mm-thick semi-insulating GaAs substrate. The solid line corresponds to the conductor loss predicted by the theory of Pucel et al. [15]. As can be seen, the measured attenuation is approximately 0.35 dB/cm at 30 GHz. The difference between the measured and theoretical attenuation is attributed to radiation from the unshielded resonators. The dielectric loss tangent is low enough that attenuation due to dielectric loss is negligible.

B. Balanced Mixer

The configuration of the balanced mixer shown in Fig. 3 is similar to that of Vendelin [4] and Leighton [8]. The 90° branch hybrid coupler has been modified from a conventional design to provide slightly greater bandwidth and isolation. The normal impedance ratio between the main and side arms of a

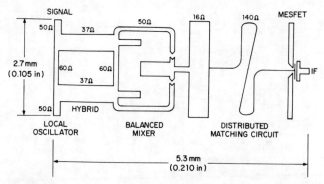

Fig. 3. Layout and parameters of the balanced mixer plus MESFET amplifier.

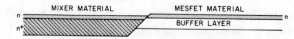

Fig. 4. Epitaxial material requirements for MESFET and Schottky diodes.

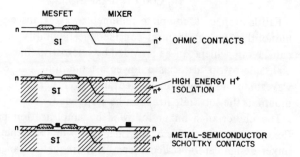

Fig. 5. Fabrication process for the Schottky diode and MESFET on semi-insulating GaAs.

conventional hybrid coupler is 1.414 ($\sqrt{2}$). In the modified design, this impedance ratio has been changed to 1.7. The effect of this change is to cause the amplitude versus frequency response of the output arms to cross above and below the design frequency rather than meet tangentially at the design frequency. This results in a bandwidth of 20 percent over which the amplitudes of the two output arms match to within 0.25 dB. The 90° phase relationship is also maintained to within ±6° over this bandwidth. Input match, however, is increased from less than 1.1:1 at the design frequency to 1.2:1. The stubs on the output arms of the hybrid are used to peak the isolation and to shift the isolation peak with respect to the overall hybrid design frequency. The design has been adjusted by use of the stubs so that the local oscillator frequency is closer to the peak of the isolation than the signal frequency. The value of the IF frequency (2.0–2.5 GHz) and the bandwidth desired resulted in the local oscillator and the signal frequencies being positioned symmetrically about the design frequency at either edge of the design band and away from the isolation peak of the conventional design. The quarter-wave stub between the Schottky diodes provides an RF short. The passive circuits were designed and tested on 0.127-mm-thick D-13 substrates[1] at 4.4 GHz corresponding to a 7 to 1 frequency scaling. Scaling experiments were also carried out on 0.427-mm-thick Emerson and Cumming high-K material with a relative permitivity of 13, corresponding to a 24:1 frequency scaling. The latter is useful since the passive circuits can be constructed of adhesive-backed copper tape without requiring photolithography, etching, or plating.

C. MESFET Amplifier

The structure of the input matching circuit for the MESFET amplifier is shown in Fig. 3. The equivalent input circuit at the MESFET gate is a 25-Ω resistor in series with a 0.6-pF capacitor. The goal for the amplifier was to achieve a bandwidth of the order of 300 MHz in the frequency range 2.0–3.0 GHz. As mentioned in the previous section, matching structures and filters consume significant substrate area at these lower frequencies which limits design flexibility. The matching circuit consists of a shunt open-circuited stub followed by a length of high-impedance line, the design parameters of which are shown in Fig. 3. This design was constrained to fabrication of distributed elements consisting of open-circuited stubs and lengths of transmission line that would fit the 2.7-mm by 2.7-mm chip. This results in a simple one-level metallization, but the long (5.6-mm) narrow high-impedance line approaches fabrication constraints. An impedance value of 130 Ω for the high-impedance line was obtained by de-embedding the line impedance from microwave measurements of the monolithic amplifier. The measured dc resistance of this line was 6 Ω. No matching circuit has been added to the output side of the amplifier since this will eventually become the interstage matching for a two-stage amplifier.

III. Monolithic Fabrication

A. Device Considerations

GaAs Schottky diodes can be used throughout the millimeter-wave frequency range, and MESFET's have many advantages in designs where low-noise performance, power capability, and versatility are needed. Both devices are used in a large variety of active circuits. Consequently, integrating mixer diodes and MESFET's on the same chip has implications beyond the present circuit configuration. Difficulty in integrating these devices arises from the dissimilar epitaxial requirement for the Schottky diodes and the MESFET as illustrated in Fig. 4.

The right-hand side of the bottom illustration in Fig. 5 shows a cross section of a planar diode with a junction area of 30 μm^2. The n^+ layer in the Schottky diode provides a low-resistance path for the diode current. Hence, the series resistance of the diode approaches the resistance of the n layer which is proportional to the layer thickness. The low series resistance is responsible for the high cutoff frequency of the diodes. If the n^+ layer were not present, the series resistance would increase by a factor of 10.

The MESFET's on the left-hand portion of Fig. 5 are fabricated on an n-type epitaxial layer. The gate is 1 μm long by 500 μm wide, and located in a drain-source spacing of 5 μm. To reduce the source resistance of the device, n^+ regions are usually implanted selectively into the drain and source areas. In the present fabrication, n^+ regions for the MESFET were not used to simplify processing. This results in a higher noise figure for the amplifier.

---

[1] Trans-Tech., Inc., Gaithersburg, MD.

B. Fabrication Techniques

Wafers of semi-insulating GaAs with epitaxial layers suitable for MESFET fabrication are used as the starting material. The GaAs surface is etched through a mask to a depth equal to the combined thickness of n and n^+ layers for the mixer diodes. The n^+ and n layers are then selectively grown in the etched area. The extent to which the surface can be made planar after epitaxial growth affects the yield in subsequent photolithographic steps.

The fabrication of devices combines the processing operations for conventional MESFET's and high-frequency Schottky diodes. The primary fabrication steps are shown in Fig. 5. Ohmic contacts are fabricated using an AuGe based system. First, the top n-type material is removed by etching from the ohmic contact areas of the planar diodes to expose the n^+ layer. The contact materials are evaporated through photoresist openings onto the n-type GaAs for the MESFET's and onto n^+-type GaAs for the mixer diode. The ohmic contacts are then formed by alloying.

High-energy proton bombardment is used to isolate devices as depicted in Fig. 5. The devices are completed by evaporation of the Schottky metal through photoresist openings to form the gates of the MESFET's and the anodes of the mixer diodes. Monolithic integration is completed by patterning the interconnecting circuit elements with photoresist and plating gold to a thickness of several micrometers.

A photograph of a wafer produced using this process is shown in Fig. 6. The overall dimensions of the wafer are 18.0 mm by 14.0 mm, and it contains six balanced mixers, six MESFET amplifiers, and six combined receiver modules. A photograph of the 2.5-mm × 5-mm receiver module is illustrated in Fig. 7. Individual components from the wafer, i.e., balanced mixers and amplifiers, are used to evaluate the noise match between the balanced mixer and the amplifier. Since the interaction of functions within the monolithic circuit cannot be evaluated directly, the test units become a necessity. Typical parameters of the 500-μm-wide-gate MESFET are a transconductance of 50 mmhos, a saturation current of 100 mA, and a pinchoff voltage of 2.5 V. The mixer diodes with a junction area of 30 μm^2 typically exhibit a zero-bias junction capacitance of 0.06 pF and a series resistance of 12 Ω.

Key features of this fabrication sequence are the possibility of applying the technology without further process developments to the fabrication of receivers at higher frequencies, and the adaptability of the fabrication sequence to a conventional production facility. It is apparent that these factors will contribute to the rapid development of monolithic microwave and millimeter-wave components.

IV. TEST RESULTS

A. Balanced Mixer

The conversion loss of the monolithic mixer as a function of IF frequency is shown in Fig. 8. The local-oscillator (LO) frequency was 29 GHz at a power level of 30 mW and the signal frequency varied from 30.5 to 31.8 GHz. From 1.5- to 2.8-GHz IF frequency, the conversion loss is 6.0 ± 0.5 dB at the chip level. This is consistent with a 4.0–4.5-dB diode conversion loss plus 1.0–1.5-dB circuit losses. The peak LO–RF isolation

Fig. 6. Dark-field view of the GaAs wafer showing the separate mixers, amplifiers, and receiver modules.

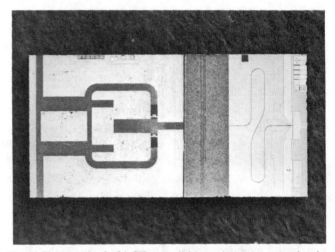

Fig. 7. Photograph of a 31-GHz monolithic receiver component showing the balanced mixer and MESFET amplifier.

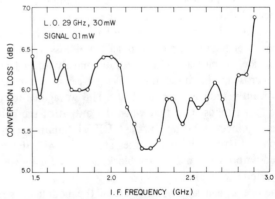

Fig. 8. Conversion loss as a function of frequency for the monolithic balanced mixer.

is 12.0 dB and is 7 dB over most of the frequency range. The low isolation is due to the fact that in the present design the LO is not matched into the Schottky diodes, and hence, the reflections are seen at the RF port. When bias structure is added to the diodes to reduce LO power requirements, the circuit will have to be modified to provide for better match into the diodes.

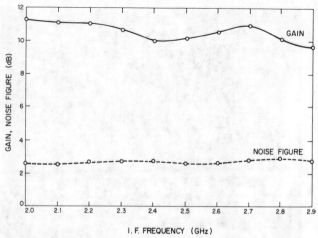

Fig. 9. Gain and noise figure of the monolithic MESFET amplifier as a function of frequency.

B. MESFET Amplifier

The gain of the MESFET amplifier varies from 12 dB at 2.0 GHz to 10 dB at 2.6 GHz with a noise figure that is greater than 4 dB. Noise-figure measurements indicated that if the open-circuit stub length is reduced to one-half the length shown in Fig. 3, the amplifier will be close to a minimum noise-figure design. Gain and noise figure for the redesigned amplifier are shown in Fig. 9. The noise figure is less than 3 dB with an associated gain of 10 dB from 2.0–3.0 GHz. In all cases, no matching structure has been added at the output, which is terminated in 50 Ω.

C. Combined Balanced Mixer and MESFET Amplifier Test Results

A circuit combining the balanced mixer and MESFET amplifier chips is mounted in a test structure as shown in the photograph of Fig. 10. The test structure consists of waveguide to coaxial to microstrip transitions for the LO and RF signal ports. On all ports, the inputs and outputs from the GaAs chips are connected to 0.25-mm-thick alumina substrates with 50-Ω microstrip lines to simplify substrate handling. This test unit is much larger than required and has a 1.0-dB loss from the input waveguide to the GaAs chip level, due mostly to the length of the microstrip lines on alumina. Designs can be constructed to significantly reduce insertion loss. The waveguide-to-coaxial transition is chosen so that when the output of the balanced mixer is joined to the gate of the MESFET amplifier, the mixer circuit is floating and hence, the gate of the MESFET can be biased without the need for a blocking capacitor between the mixer and the amplifier.

The overall gain and noise figure of the module constructed from a mixer chip followed by an amplifier chip, are shown in Fig. 11. With an LO frequency of 29 GHz, the conversion gain from RF to IF varies from 5 to 3 dB in the signal frequency range of 31–31.5 GHz, corresponding to an IF frequency of 2.0–2.5 GHz. The noise figure is 11.2 dB at 31.0 GHz and increases to 13.0 dB at 31.6 GHz. The present experimental arrangement provides for 38 mW of LO power and the noise figure is still decreasing at this value. The asymptotic value appears to be approximately 0.5 dB below the 11.2-dB figure

Fig. 10. Photograph of the Ka-band test structure and view of the module in position.

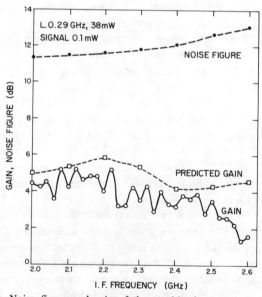

Fig. 11. Noise figure and gain of the combined mixer and amplifier chips as a function of frequency.

quoted previously. Also shown in Fig. 11 is the value of conversion gain from RF to IF predicted from the separate measurements of the mixer and amplifier.

The 31-GHz monolithic receiver component containing the balanced mixer and MESFET amplifier on the same 2.5-mm by 5-mm chip is shown in Fig. 7. This is mounted in the test structure of Fig. 10, as described earlier. The noise figure and overall gain for the monolithic unit as a function of IF frequency is shown in Fig. 12. As in the case of the two-chip version,

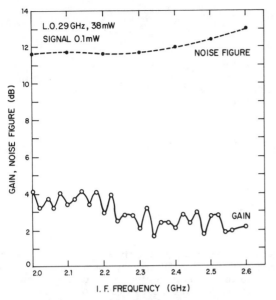

Fig. 12. Noise figure and gain of the 31-GHz monolithic receiver component as a function of frequency.

there was not sufficient 29-GHz LO power to achieve minimum noise figure. The measured single-sideband noise figure was 11.5 dB at an IF frequency of 2.0 GHz and increased to 13.3 dB at 2.6 GHz. The gain was approximately 4 dB from 2.0 to 2.25 GHz and 2.5 dB from 2.3 to 2.6 GHz.

V. Conclusions

A monolithic 31-GHz section of a receiver front end combining a balanced mixer and MESFET amplifier has been successfully fabricated on a semi-insulating GaAs substrate. Two different device types, Schottky-mixer diodes and MESFET's, were fabricated on the same wafer. Initial results of overall gain of 4 dB and 11.5-dB noise figure are reported. Performance of the devices can be improved and additional amplification added. It is anticipated that a 10-dB noise figure is easily achieved, and that ultimately a 6- to 7-dB noise figure could be achieved.

Acknowledgment

The authors wish to thank L. Mahoney, G. Lincoln, W. Macropoulos, and M. Pierce for help in the monolithic fabrication; C. Bozler and R. McClelland for assistance in the material growth; J. Lambert and P. Kaylan for help in the mask generation; J. Donnelly and R. Brooks for assistance in the implantation; W. Macropoulos for assistance in the packaging; J. Taylor, R. Aucoin, and W. Piacentini for help in the testing; and W. Lindley and R. Murphy for many helpful discussions.

References

[1] T. M. Hyltin, "Microstrip transmission on semiconductor transmission lines," *Proc. Inst. Elect. Eng.*, vol. 115, pp. 43–48, Jan. 1968.
[2] S. Mao, S. Jones, and G. D. Vendelin, "Millimeter-wave integrated circuits," *IEEE Trans. Microwave Theory Tech.*, vol. MTT-16, no. 7, pp. 455–461, July 1968.
[3] E. W. Mehal and R. W. Wacker, "GaAs integrated microwave circuits," *IEEE Trans. Microwave Theory Tech.*, vol. MTT-16, pp. 451–454, July 1968.
[4] G. D. Vendelin, "A Ku-band integrated receiver front end," *IEEE J. Solid-State Circuits*, vol. SC-3, no. 3, pp. 255–257, Sept. 1968.
[5] R. S. Pengelly and J. A. Turner, "Monolithic broad-band GaAs FET amplifiers," *Electron. Lett.*, vol. 12, no. 10, pp. 251–252, May 1976.
[6] R. S. Pengelly, J. Arnold, J. Cockrill, and M. G. Stubbs, "Prematched and monolithic amplifiers covering 8-18 GHz," in *9th European Microwave Conf. Dig.* (Brighton, England, 1979), pp. 293–297.
[7] R. A. Pucel, P. Ng, and J. Vorhaus, "An X-band GaAs FET monolithic power amplifier," in *Microwave Theory and Tech. Symp. Dig.* (Orlando, FL, 1979), pp. 387–389.
[8] W. Leighton, "Monolithic X-band microstrip-line mixers on semi-insulating gallium arsenide," Ph.D. dissertation, Carnegie-Mellon University, Pittsburgh, PA, 1970.
[9] D. Ristow, N. Enders, and H. Kniepkamp, "A monolithic GaAs Schottky barrier diode mixer for 15 GHz," in *8th European Microwave Conf. Dig.* (Paris, France, 1978), pp. 707–711.
[10] R. P. G. Allen and G. R. Antell, "Monolithic mixers for 60-80 GHz," in *3rd European Microwave Conf. Dig.* (Brussels, Belgium, 1973), paper A.15.4.
[11] R. L. Van Tuyl, "A monolithic GaAs FET RF signal generation chip," *IEEE ISSC Dig. Tech. Papers* (San Francisco, CA 1980), vol. 23, pp. 118–119.
[12] R. S. Pengelly and D. C. Rickard, "Design, measurement and application of lumped elements up to J-band," in *7th European Microwave Conf. Dig.* (Copenhagen, Denmark 1977), pp. 460–464.
[13] W. E. Courtney, "Complex permittivity of GaAs and CdTe at microwave frequencies," *IEEE Trans. Microwave Theory Tech.*, vol. MTT-25, no. 8, pp. 697–701, Aug. 1977.
[14] W. J. Getsinger, "Microstrip dispersion model," *IEEE Trans. Microwave Theory Tech.*, vol. MTT-21, no. 1, pp. 34–39, Jan. 1973.
[15] R. A. Pucel, D. J. Masse, and C. P. Hartwig, "Losses in microstrip," *IEEE Trans. Microwave Theory Tech.*, vol. MTT-16, no. 6, pp. 342–350, June 1968.

A 69 GHz MONOLITHIC FET OSCILLATOR*

D. W. Maki, J. M. Schellenberg, H. Yamasaki, and L. C. T. Liu

Hughes Aircraft Company
Electron Dynamics Division
3100 West Lomita Boulevard, Torrance, California 90509
(213) 517-6422

ABSTRACT

A monolithic oscillator was fabricated using conventional planar FET technology. The active device used was a 0.35x60 micron FET fabricated on an active layer formed by ion implantation into an undoped VPE buffer layer. Frequency stability is achieved using either an on-chip microstrip resonant circuit or by adding a 30 mil diameter dielectric resonator directly onto the 50 mil square GaAs chip. With no external tuning the oscillator delivered 0.45 milliwatts at 64 GHz. By using an external E-H waveguide tuner, 0.7 milliwatts of power at 65.7 GHz was achieved. The oscillator was tunable from 55 to 75 GHz by adjusting the source-gate tuning inductor and the drain tuning.

INTRODUCTION

Monolithic GaAs integrated circuits have been demonstrated for a wide variety of applications at frequencies from DC to 20 GHz. At millimeter wave frequencies, however, monolithic IC applications have been limited primarily to diode based mixers due to a lack of available active devices. Recently, quarter micron gate length discrete FETs have demonstrated themselves to be useful, active devices at frequencies up to 60 GHz and monolithic FET based amplifiers and oscillators are becoming attractive components for millimeter wave systems. This paper describes the development of a simple V-band monolithic oscillator using a planar quarter micron FET as its active device.

DEVICE DESCRIPTION

The oscillator design was based on the properties of a planar GaAs FET with a 0.25x60 micron gate reported previously.[1] The device structure consists of a pair of 0.25x30 micron gate fingers fabricated using a direct write E-beam lithography system with the gate offset towards the source in the channel to reduce the source resistance. The active layers were formed using both VPE and ion implantation. The discrete devices were DC characterized and their S-parameters were measured from 2-18 GHz using a conventional network analyzer. This data was used to construct a lumped device equivalent circuit shown in Figure 1. This model was then used to project the performance of the device up to 100 GHz. This is, admittedly, an approximation but it was felt that this would provide better design data than trying to measure S-parameters directly at 60-70 GHz. Figure 2 shows a plot of the calculated maximum available gain of FETs from four separate wafers plotted versus frequency and based on measured 2-18 GHz S-parameters. The devices show an f_{max} of 88 to 105 GHz and should make useful amplifiers up to 60 GHz and oscillators to 80 GHz. This calculated data is roughly confirmed by the fact that 60 GHz hybrid amplifiers fabricated from these devices provide 6-7 dB gain.[1]

OSCILLATOR DESIGN

Unlike two-terminal Gunn or IMPATT diodes, typically used to fabricate millimeter wave oscillators, an FET does not possess an inherent negative resistance and external feedback elements must be provided to the device to obtain oscillation.

Microwave oscillators generally can be reduced to the basic Colpitts configuration shown in Figure 3a. Here, the frequency of oscillation is set by the resonant frequency of the inductance-capacitance loop, with the FET acting as the active source sustaining the loop current. Actually, the capacitances (C_1 and C_2 in Figure 3) often are internal device interelectrode capacitances which are simply incorporated into the oscillator circuit. Any of the three device terminals can be connected to ground. The circuit still can be analyzed as a Colpitts oscillator. Selection of the terminal to connect to ground should be based on ease of coupling to the load, effects of parasitics and thermal considerations.

Two popular oscillator configurations, the common gate and the common drain, also are shown in Figure 3. In

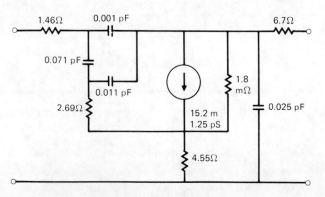

Fig. 1 Discrete FET equivalent circuit.

---

*This work was supported by the Naval Research Laboratory on Contract Number N00014-82-C-2498.

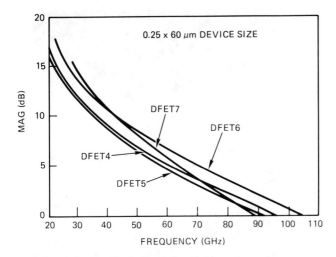

Fig. 2 Calculated maximum available gain vs frequency.

both configurations, C_1 usually is the internal gate-to-source interelectrode capacitance. Often, C_2 also is internal. In the common gate and common drain configurations, the load usually is coupled to the oscillator at the drain and source terminals, respectively. To oscillate, the Colpitts oscillator simply requires a net inductive reactance connected between the gate and drain.

Using the equivalent circuit of DFET 6 and calculating the input impedance looking into the drain of a grounded gate FET with an inductance in the source lead, we form a Colpitts oscillator and obtain impedance versus frequency as shown in Figure 4. A small signal negative resistance is shown from 61 to more than 100 GHz, but its magnitude at 90 is only -6 ohms with a Q of 9. A simple series inductance can be used to match the drain circuit and a microstrip resonator or a dielectric resonator as shown in Figure 5 can be used to stabilize the output. Abe et al[2] have described the technique of using

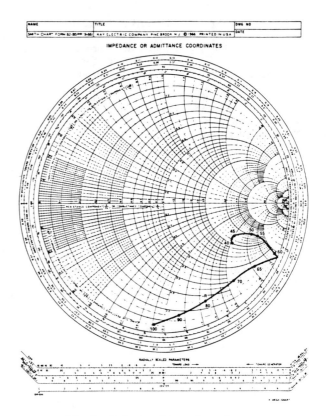

Fig. 4 Calculated input impedance.

a dielectric resonator between drain and load to stabilize an FET oscillator. Basically, the dielectric puck is coupled to the microstrip transmission line forming a single pole bandstop filter. This is placed approximately a half wavelength from the active device. The oscillator can be then mechanically tuned by optimizing both the distance from the device and the separation from the microstrip line. This violates some of the basic principles of monolithic IC's in that we are adding a chip component and are physically tuning the circuit, but it provides twice the Q of a microstrip cavity and allows us to compensate for design uncertainties.

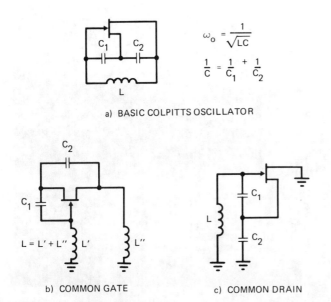

Fig. 3 FET oscillator circuits.

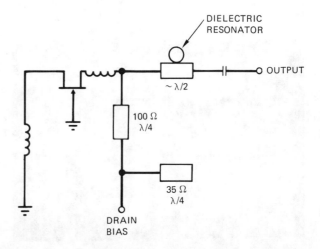

Fig. 5 Equivalent circuit of dielectric resonated oscillator.

OSCILLATOR FABRICATION

The monolithic oscillator is shown in Figure 6. It consists of a 0.35×60 μm single strip gate. The gate is directly grounded, and the source is grounded through a pair of high impedance microstrip transmission lines which can be adjusted easily using a wire bonder to vary the length. A simple, low-Q matching circuit is used to achieve oscillation over a range of frequencies with a minimum of tuning on chip. The chip size, 50×50×4 mils, allows additon of an on-chip dielectric resonator.

The high reactance at the gate terminal mandates that a high impedance inductor be used for matching, to avoid seriously degrading the small negative resistance. Conventional microstrip on 100 μm GaAs is limited by photolithographic and loss considerations to a maximum impedance of about 85 ohms. This impedance is inadequate for the oscillator circuit. For this application, a new, high-impedance transmission line was developed using a 12 micron high airbridge. The large airgap reduces the capacitance to ground by a factor of two and can, therefore, increase the impedance of the structure by as much as 40 percent.

The active device and the end of the high airbridge are shown in Figure 7.

The oscillators were fabricated on an active layer formed by ion implanting singly charged silicon at an energy of 100 KeV for a total dose of 6E12 atoms/cm^2 into a VPE undoped buffer layer. Isolation was accomplished by mesa etching through the active layer. The circuits were processed using optical masks generated on an E-beam pattern generator, and the gates were direct written on the same machine. The gate uses a Ti-Pt-Au metallization system and was designed to be 0.25 μm long but slight overexposure yielded 0.35 μm gates.

The wafer contained 161 devices, of which 53 or 33 percent were visual and dc good. The FET has an average I_{DSS} of 17.7 mA, with a standard deviation of 15 percent. The pinch-off voltage is 3 V, and the transfer conductance (gm) at zero bias is 8 ms. This gm is somewhat lower than seen on devices from wafers DFET 4 through 7 on which the design was based.

The oscillators were tested by coupling them to a waveguide, as shown in Figure 8. A microstrip probe on a 7-mil-thick quartz substrate protrudes into the broad wall

Fig. 7 Details of FET.

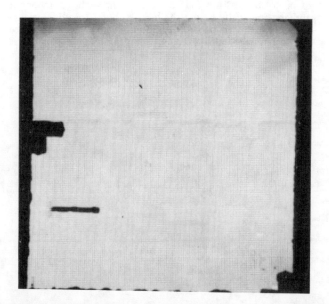

Fig. 6 Monolithic oscillator.

Fig. 8 Oscillator coupled to V band waveguide.

of the V-band waveguide, forming an E-plane probe. A pair of these probes including an additional 0.3 inch of microstrip line has approximately 1 dB of insertion loss at 60 GHz. A low impedance transmission line formed on K-100 material forms an off-chip bias decoupling network. The complete waveguide test fixture is shown in Figure 9.

With no external tuning the oscillator delivered 0.45 mW at 64 GHz. An external E-H tuner and gold foil tuning on the microstrip transmission lines were used to optimize performance. Figure 10 shows the optimum power output versus drain voltage. We achieved 0.7 milliwatt at 65.7 GHz with 4 volts on the drain and 1 percent efficiency. As shown in the Figure, dropping the drain voltage did reduce the power output but varied the output frequency less than 50 MHz. The oscillator frequency was continuously adjustable up to 75 GHz, with the output power tapering off to zero.

As mentioned earlier, it is possible to adjust the negative resistance of the oscillator by tuning the gate-source inductance and retuning the drain circuit. Figure 11 shows the result of tuning this inductor from 0.06 to 0.14 nH by means of bonding straps across the source transmission lines. A maximum frequency of 78.2 GHz was obtained with 0.2 mW of power delivered. This result was obtained with the aid of an EH tuner on the output.

To improve the stability of the oscillator we machined dielectric pucks, fabricated from Barium tetratitanate which has a dielectric constant of 37. The resonators were tested on a microstrip on 7 mil quartz test fixture shown in Figure 12. An unloaded Q of 247 was obtained at 76 GHz. The oscillators have been measured with several sizes of resonators and preliminary results show 1 to 1.2 mW of power delivered at 57.2 GHz and 0.5 mW at 67.9 GHz.

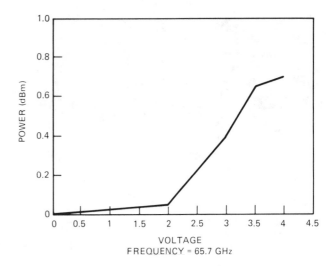

Fig. 10 Power versus voltage for the monolithic oscillator.

SUMMARY

A monolithic FET oscillator has been developed with 0.5-1.0 mW available from 60-70 GHz with an efficiency of 1 percent or less. The oscillators can be stabilized using conventional dielectric cavities but are difficult to work with due to the small size. Microstrip cavities are somewhat easier to work with, but have unloaded Q's of half that available from dielectric pucks. The conversion efficiency is somewhat lower than expected and work is continuing.

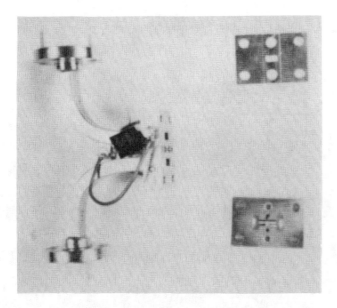

Fig. 9 Waveguide test hardware.

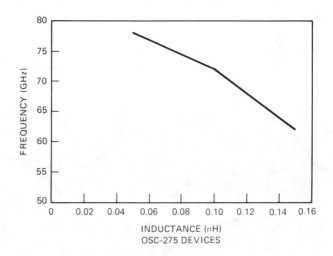

Fig. 11 Maximum frequency of oscillation as a function of G-S inductance.

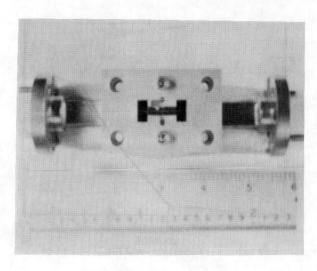

Fig. 12 Resonator test fixture.

ACKNOWLEDGMENTS

The authors would like to acknowledge the contributions and support of the management and staff of the Torrance Research Center. In particular we would like to acknowledge the efforts of Pam Busted for processing, Anneliese Grohs for the E-beam lithography, Sarah Rodriquez for assembly, and Robert Lipman for RF assembly and testing.

REFERENCES

1. "A 60 GHz GaAs FET Amplifier," E. T. Watkins, J. M. Schellenberg, L. H. Hackett, H. Yamasaki and M. Feng, 1983 IEEE MTT-S Digest.

2. "A Highly Stabilized Low-Noise GaAs FET Integrated Oscillator . . .," H. Abe, Y. Takayama, A. Higashisaka and H. Takamizawa, IEEE Transactions on MTT, Vol. MTT-26, No. 3, March 1978, pp. 156-162.

Ka-Band Monolithic GaAs Balanced Mixers

CHENTE CHAO, MEMBER, IEEE, A. CONTOLATIS, STEPHEN A. JAMISON,
AND PAUL E. BAUHAHN, MEMBER, IEEE

Abstract—Monolithic integrated circuits have been developed on semi-insulating GaAs substrates for millimeter-wave balanced mixers. The GaAs chip is used as a suspended stripline in a cross-bar mixer circuit. A double sideband noise figure of 4.5 dB has been achieved with a monolithic GaAs balanced mixer filter chip over a 30- to 32-GHz frequency range. A monolithic GaAs balanced mixer chip has also been optimized and combined with a hybrid MIC IF preamplifier in a planar package with significant improvement in RF bandwidth and reduction in chip size. A double sideband noise figure of less than 6 dB has been achieved over a 31- to 39-GHz frequency range with a GaAs chip size of only 0.5×0.43 in. This includes the contribution of a 1.5-dB noise figure due to IF preamplifier (5–500 MHz).

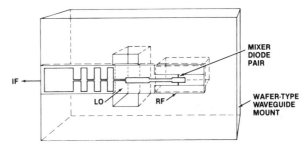

Fig. 1. Millimeter-wave integrated mixer chip design containing two planar mixer diodes and matching circuits for RF waveguide, LO coupling, and IF filter.

I. INTRODUCTION

THE PERFORMANCE of GaAs devices has been steadily improved with recent advances in material, process, and device technology. Monolithic integration of passive elements and active devices on GaAs substrates becomes increasingly attractive for use at millimeter-wave frequencies as opposed to the more conventional MIC hybrid approach where the effects of parasitics are difficult to control. Considerable attention has been given to the development of millimeter-wave monolithic balanced mixers which serve as an important building block for a number of potential systems. Recently, a monolithic mixer IF preamplifier using a microstrip circuit approach has been demonstrated with good performance at millimeter-wave frequencies. [1] The circuit requires an on-chip coupler to combine the local oscillator and signal frequencies increasing the chip size, but has the advantage of simple interfaces. The present paper describes a monolithic balanced mixer using a suspended stripline circuit approach in a cross-bar mixer circuit [2] which does not require an on-chip coupler. The monolithic mixer has also been integrated with a hybrid MIC IF amplifier in a wafer-type waveguide package. In this approach the only passive elements on the monolithic GaAs chip are the coupling circuits for millimeter-wave frequencies. The low-pass filter is incorporated with the bipolar IF preamplifier on a hybrid MIC.

II. CIRCUIT DESIGN

The basic configuration of the monolithic balanced mixer is illustrated in Fig. 1. The GaAs chip is used as a suspended stripline and is coupled to local oscillator power and RF signal via two full-height waveguide ports. It consists of two planar mixer diodes and matching circuits for RF waveguide, LO coupling, and IF filter. There are several unique features of the mixer that significantly reduce the complexity and take advantage of the monolithic circuit technique. The diodes are electrically in series with respect to the RF signal. The RF impedance of a single mixer is usually much lower than that of waveguide. With two diodes in series, the RF input impedance of the chip at the signal port can be matched easily with straight full-height waveguide. Furthermore, if a reduced height or ridged waveguide is used to match the impedances of circuit and chip over the desired frequency range, then the RF bandwidth of this mixer could be increased to nearly full waveguide bandwidth.

The diodes are electrically in parallel with respect to the local oscillator and are in a direction such that the induced LO currents in the diodes are out of phase. This eliminates the need for a magic tee to cancel the noise contributed by the local oscillator. The conventional waveguide magic tee is inherently a narrow-band component and is very expensive to make, particularly at millimeter-wave frequencies. Furthermore, the two diodes are physically close to each other and can have nearly identical parameters because they are monolithically fabricated. Such a well-matched diode pair gives excellent LO noise suppression.

The isolation between the local oscillator and signal ports can be very high over a wide range of frequencies since the dominant TE_{10} mode in the RF waveguide is orthogonal to the quasi-TEM LO input.

III. CHIP LAYOUT AND FABRICATION

The monolithic mixer described above has been fabricated with advanced semiconductor technology available in our laboratory. The monolithic chip design and the key

Manuscript received May 1, 1982.
The authors are with Honeywell Corporate Technology Center, Bloomington, MN 55420.

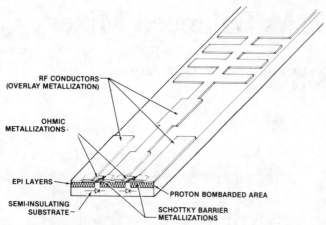

Fig. 2. Monolithic mixer chip shown with two planar Schottky-barrier mixer diodes isolated parasitically by a proton ion-implantation technique and RF matched by overlay metallization patterns.

Fig. 4. Monolithic mixer chip in a wafer-type mount.

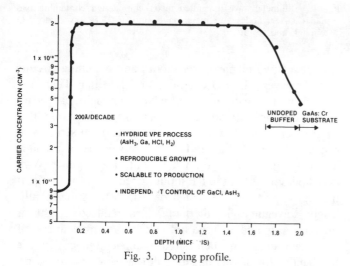

Fig. 3. Doping profile.

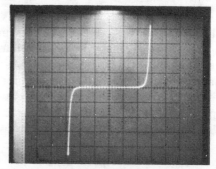

Fig. 5. I–V characteristics of the monolithic diode pair. Horizontal: 0.2 V/div; Vertical: 1 μA/div.

technology needed for its fabrication are illustrated in Fig. 2. The diodes are fabricated with VPE-grown n-n$^+$ layers on a semi-insulating substrate isolated by proton bombardment [3]. High-quality n-n$^+$ layers are grown on 10-mil semi-insulating GaAs:Cr substrates which have been qualified by our established qualification test procedure. The n$^+$ layer has a doping density of 2×10^{18} cm^{-3} and is 2 μm thick and the n layer which is on top of n$^+$ layer has a doping density of 9×10^{16} cm^{-3} and is 0.1 μm thick. These layers have been grown by our vapor-phase epitaxial reactor with the hydride process (AsH$_3$, HCl, Ga, H$_2$). Carrier concentration as a function of depth from the surface of the n-n$^+$ epi-layer is shown in Fig. 3. Schottky-barrier (TiW/Au) and contact (AuGe/Ni/TiW/Au) metallizations are then deposited for the planar mixer diodes. The chip is bombarded by high energy protons everywhere except at the diodes which are protected by a thick layer of photoresist and gold metal. This proton bombardment process isolates the diodes from the circuit parasitics. After the protective layers are removed, overlay metallization is deposited for the RF matching circuits. Individual chips are cut from the large wafer and mounted in the wafer-type waveguide package as shown in Fig. 4.

Diodes with a zero biased cutoff frequency of better than 600 GHz have been achieved. Fig. 5 shows the I–V characteristics of a monolithic diode pair.

IV. Monolithic Balanced Mixer-Filter

Fig. 6 shows the double sideband noise figure of the mixer filter chip as a function of the local oscillator power for the three different LO frequencies. A double sideband noise figure of 4.5 dB has been achieved with the monolithic GaAs balanced mixer chip over a 30- to 32-GHz frequency range at a LO power of approximately 10 mW. This includes the contribution of a 1.5-dB noise figure due to an IF preamplifier which has a bandwidth of 5–500 MHz.

This performance is quite competitive with the best conventional mixers at this frequency. Measurements of noise figure as a function of LO frequency for three different mechanical tuning positions are also plotted in Fig. 7. This indicates that some tuning is possible with the monolithic balanced mixer but the current chip configuration works best around 31 GHz with a low noise IF bandwidth of about 500 MHz.

The isolation between the local oscillator and signal

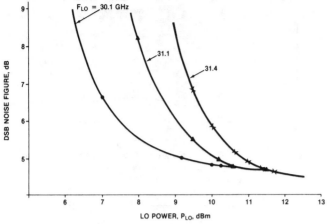

Fig. 6. Noise figure versus P_{LO} of the monolithic mixer.

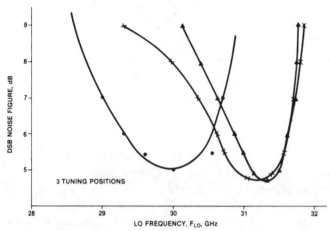

Fig. 7. Noise figure versus F_{LO} of the monolithic mixer.

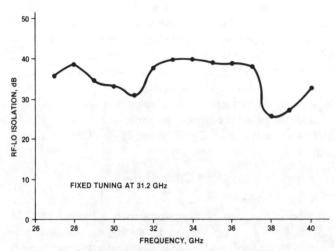

Fig. 8. Isolation versus frequency of the monolithic mixer.

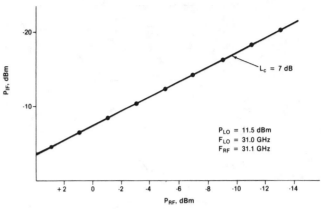

Fig. 9. Single sideband conversion loss of the monolithic mixer.

ports of the mixer is excellent as indicated in Fig. 8. An isolation of better than 30 dB has been achieved over a frequency range of 27–37 GHz when the mixer is tuned at 31 GHz. This excellent isolation property is a direct consequence of the mixer circuit design approach, i.e., the LO and RF ports are decoupled because the E-fields of the dominant modes are orthogonal to each other.

The single-sideband conversion loss of the mixer is about 7 dB as shown in Fig. 9. The corresponding single sideband noise figure is 7.6 dB. This means that the noise due to the diode is very low and the diodes are well matched. This confirmed our belief that the monolithic diode pair fabricated in close proximity to each other would have nearly identical parameters and therefore the LO noise suppression property of the mixer is excellent.

Excellent correlation between the experimental data and theoretically calculated results was obtained for the monolithic mixer filter chip. The theoretical model has also been used to aid the optimization of the monolithic mixer. Theoretical performance of the mixer was determined with the aid of a very complete mixer analysis program [4] which was obtained from Kerr and Siegel at NASA Goddard Space Flight Center. This program has been significantly modified to increase its speed and convenience for mixer design without degrading its accuracy. An extensive series of calculations with the program indicates that the order of importance of the parameters affecting the performance of a mixer diode are the series resistance, capacitances, and then the series inductance. The optimum inductance value is a function of the circuit impedance level and other factors.

Using realistic parameters for the various diode characteristics, the performance of the monolithic mixer indicated in Fig. 10(a) was obtained. The equivalent circuit is given in Fig. 10(b). The performance predicted for the series inductance estimated from the diode lead geometry for the monolithic mixer is indicated. The measured performance is essentially identical. It is clear that a slight adjustment of the circuit could give considerably better noise performance by improved impedance matching between the diode and waveguide.

V. Integration of Monolithic Mixer and Hybrid IF Preamplifier

A monolithic GaAs balanced mixer has been combined with a hybrid IF preamplifier in a planar waveguide package. Fig. 11 is a schematic of the circuit configuration and Fig. 12 is a photograph of the package. The size of the monolithic balanced mixer has been minimized by incorporating only those circuit elements which are critical for matching of devices at millimeter-wave frequencies on the

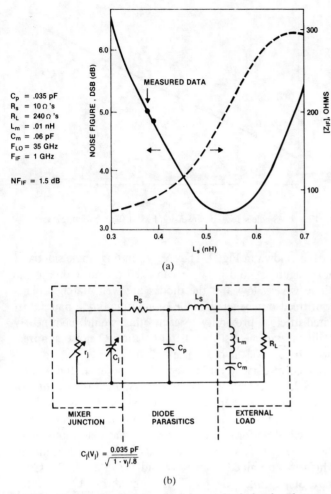

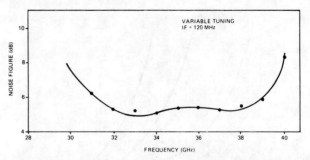

Fig. 13. Monolithic mixer noise figure versus frequency demonstrates wide tunable operating frequency range for flexible system applications (includes 1.5-dB IF preamplifier noise).

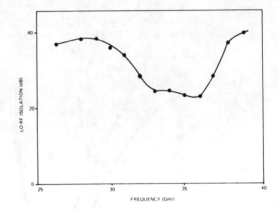

Fig. 14. Monolithic mixer LO-RF isolation versus frequency.

Fig. 10. (a) Comparison of measured data and calculated performance of the monolithic mixer filter chip. (b) Equivalent circuit used for mixer junction embedding impedance calculation.

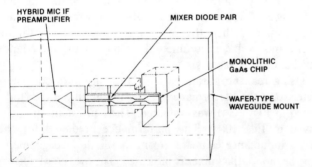

Fig. 11. Packaging technique for combining a monolithic balanced mixer and a hybrid IF preamplifier.

Fig. 12. Ka-band monolithic GaAs balanced mixer chip integrated with a hybrid MIC IF preamplifier on a planar package.

GaAs substrate. Significant reduction in chip size and improvement in RF bandwidth have been achieved.

Fig. 13 shows noise figure versus frequency of a monolithic chip from 30 to 40 GHz. A double sideband noise figure of less than 6 dB has been achieved over an 8-GHz bandwidth with a GaAs chip size of only 0.05×0.43 in. This includes the contribution of a 1.5-dB noise figure from IF preamplifier which has a bandwidth of 5–500 MHz. This performance gives an improvement of RF bandwidth by a factor of 4 and a reduction of chip size by a factor of 5 with respect to a result reported previously [2]. The isolation between the local oscillator and signal ports of the mixer is very good as indicated in Fig. 14. An isolation better than 20 dB over 26–40 GHz has been achieved.

VI. CONCLUSIONS

A monolithic GaAs balanced mixer chip with a minimum chip size can be combined with a hybrid MIC IF preamplifier in a unique circuit configuration to achieve high-performance and potentially cost-effective components for millimeter-wave receiver applications.

ACKNOWLEDGMENT

The authors would like to express their appreciation to T. Peck and J. Abrokwah for supplying VPE material and to D. Hickman for technical assistance.

REFERENCES

[1] A. Chu, W. E. Courtney, and R. W. Sudbury, "A 31-GHz monolithic GaAs mixer/preamplifier circuit for receiver applications," *IEEE*

Trans. Electron Devices, vol. ED-28, no. 2, pp. 149–154, Feb. 1981.

[2] C. Chao, A. Contolatis, S. A. Jamison, and E. S. Johnson, "Millimeter-wave monolithic GaAs balanced mixers," presented at 1980 Gallium Arsenide Integrated Circuit Symp., Las Vegas, NV, paper 32.

[3] R. A. Murphy, C. O. Bozler, C. D. Parker, H. R. Fetterman, P. E. Tannenwald, B. J. Clifton, J. P. Donnelly, and W. T. Lindley, "Submillimeter heterodyne detection with planar GaAs Schottky-barrier diodes," *IEEE Trans. Microwave Theory Tech.*, vol. MTT-25, pp. 494–495, June 1977.

[4] D. N. Held and A. R. Kerr, "Conversion loss and noise of microwave and millimeter-wave mixers: Part I—Theory," and "Part II—Experiment," *IEEE Trans. Microwave Theory Tech.*, vol. MTT-26, pp. 49–61, Feb. 1978.

94 GHz Planar GaAs Monolithic Balanced Mixer

P. Bauhahn, T. Contolatis, J. Abrokwah, C. Chao
Honeywell Physical Science Center
and
C. Seashore

Honeywell Defense Systems Division

ABSTRACT

A 94 GHz GaAs monolithic balanced mixer in a planar microstrip integrated circuit configuration has been demonstrated. A double sideband noise figure of 5.6 dB has been achieved at 94.5 GHz. This includes a 1.5 dB noise contribution of the IF preamplifier. The chip size is 0.076 x .034 inch with integral beam leads.

I. INTRODUCTION

While several monolithic mixers have been constructed [1-3], this is believed to be the first monolithic rat race mixer at 94 GHz. A double sideband noise figure of 5.6 dB was obtained at 94.5 GHz. This includes a 1.5 dB noise contribution of the IF preamplifier but not transmission line and mismatch losses which are about 5 dB. The monolithic mixer consists of a pair of GaAs planar Schottky barrier mixer diodes, the matching circuitry and the LO and RF transmission lines. The chip of .076 x .034 in. size is connected between two quartz microstrip probe transitions. Computer aided design and low frequency models were used for the design of the mixer circuit. The microstrip circuit approach will permit the integration of multiple functions on a single chip for low cost receiver applications at mm-wave frequencies.

II. CIRCUIT DESIGN

The configuration of the planar monolithic balanced mixer is illustrated in Figure 1. It consists of a rat race hybrid, two monolithically integrated diodes, matching circuits and beam lead RF interconnects. The IF output filter, which is off chip, is connected to the local oscillator input line. This helps reduce the size of the chip but has little effect on mixer performance. The mixer is relatively insensitive to local oscillator noise since the Schottky diodes are antiphase with respect to this input. On the right and left of the rat race hybrid are filters to suppress the response of the mixer at the sum of the local oscillator and signal frequencies. These filters, which are formed by two shunt stubs, a short high impedance transmission line and the diode capacitance, are needed to optimize the mixer conversion loss. A ground for the local oscillator and signal frequencies is provided by an open-circuited quarter wave stub. The IF ground is connected to the pads at the top and bottom of the chip.

III. CHIP FABRICATION

The monolithic mixer has been fabricated using the process outlined in Figure 2. The circuits are fabricated on n-n^+ layers grown by VPE on a semi-insulating substrate. The n^+ layer has a doping density of approximately $2 \times 10^{18} cm^{-3}$ and is 3 microns thick. The n layer on top of the n^+ layer has a doping density near 5×10^{16} and is about 0.1 microns thick.

The first step in fabricating the mixers is (a) formation of the ohmic contacts (Ni/Au-Ge/Ni) by a liftoff process. After alloying these contacts (b) they are covered by SiON using plasma assisted CVD. The Schottky contacts (Mo-Au) are fabricated by dielectric assisted liftoff and the diodes isolated by deep mesa etching using a combination of ion milling and wet etching. The surface of the wafer is planarized (c) using spin-on polyimide and wet etching. Formation of an air bridge contact to the Schottky and beam leads for interconnects finishes topside processing. After thinning the wafers to 6 mils and backside metallization the mixers were separated by sawing. A photograph of one of the mixer diodes in a completed mixer is given in Figure 3.

IV. MONOLITHIC 94 GHz MIXER PERFORMANCE

The diode parameters obtained are listed below. Note that C_t includes transmission line capacitance. The junction capacitance itself should be about 15 fF.

Ideality n   1.06
Series Resistance R_s 15.   ohms
Total Capacitance C_t .125 pF
Reverse Current   1   ma at 6V

The noise figure calculated for a similar mixer diode and its matching circuit is given in Figure 4.[4] This does not include the effect of losses and mismatch in either the circuit or test fixture. The measured noise figure for a monolithic mixer excluding 5 dB transition losses is shown in Figure 6. This includes the effect of a 1.5 dB NF IF preamplifier. The noise figure versus local oscillator power is plotted for this same mixer (#1) in Figure 7.

Another mixer, with slightly better performance, is also indicated. It is clear from these results that further optimization is possible but monolithic mixers appear to be promising devices for use at these frequencies.

V. CONCLUSION

A monolithic GaAs balanced mixer with minimum chip size is potentially useful for 94 GHz applications.

ACKNOWLEDGEMENT

The authors would like to express their appreciation to D. Hickman for technical assistance.

REFERENCES

1. A. Chu, W.E. Courtney, and R.W. Sudbury, "A 31 GHz monolithic GaAs mixer/preamplifier circuit for receiver applications," IEEE Trans. Electron Devices, vol. ED-28, no. 2, pp. 149-154, Feb. 1981.

2. C. Chao, A. Contolatis, S.A. Jamison, and E.S. Johnson, "Millimeter-wave monolithic GaAs balanced mixers," presented at 1980 Gallium Arsenide Integrated Circuit Symp., Las Vegas, NV, paper 32.

3. R.A. Murphy, C.O. Bozler, C.D. Parker, H.R. Fetterman, P.E. Tannenwald, B.J. Clifton, J.P. Donnelly, and W.T. Lindley, "Submillimeter heterodyne detection with planar GaAs Schottky barrier diodes," IEEE Trans. Microwave Theory Tech., vol. MTT-25, pp. 494-495, June 1977.

4. D.N. Held and A.R. Kerr, "Conversion loss and noise of microwave and millimeter-wave mixers: Part I-Theory," and "Part II-Experiment," IEEE Trans. Microwave Theory Tech., vol. MTT-26, pp. 49-61, Feb. 1978.

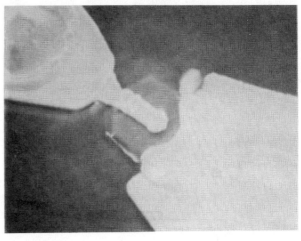

Figure 3. Schottky Diode Used in Monolithic Mixer.

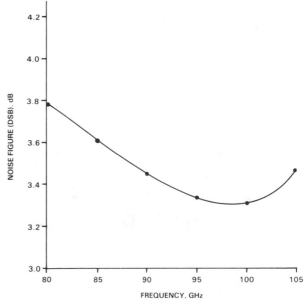

Figure 4. GHz Monolithic Mixer Calculated Noise Figure Including 1.5 dB IF Preamplifier Noise but Excluding Rat-Race and Transition

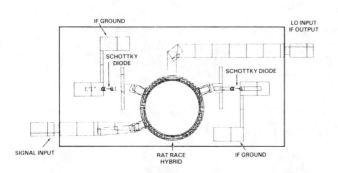

FIGURE 1. 94 GHz PLANAR MONOLITHIC BALANCED MIXER LAYOUT

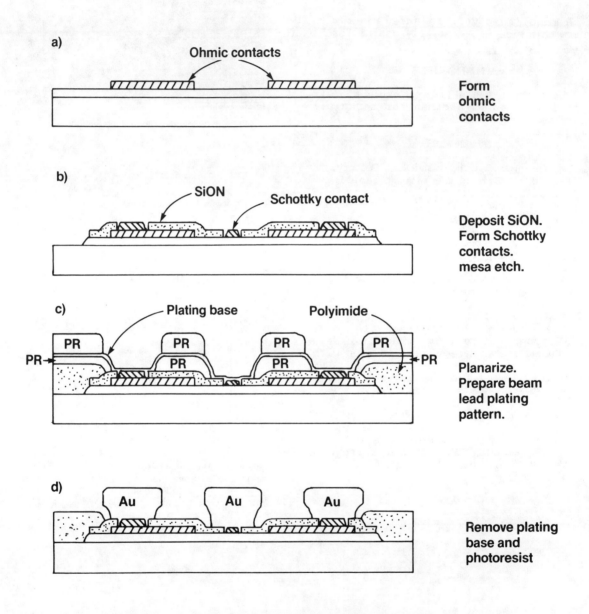

Figure 2. 94 GHz Monolithic Mixer Fabrication

Figure 5. Packaged Monolithic 94 GHz Mixer

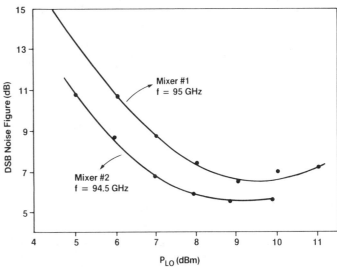

Figure 7. Noise Figure vs. Local Oscillator Power Excluding 5 dB Transition Loss

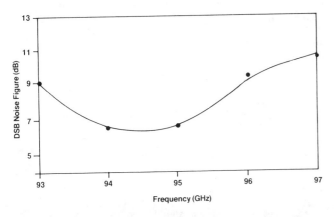

Figure 6. Noise Figure versus Frequency Excluding 5 dB Transition Losses

A W-BAND MONOLITHIC GaAs CROSSBAR MIXER

Lloyd T. Yuan

Hughes Aircraft Company, Electron Dynamics Division

3100 West Lomita Boulevard

Torrance, California 90509

ABSTRACT

A fully monolithic crossbar mixer has been designed and fabricated for operation at W-band (75 to 110 GHz). A typical conversion loss of 9 dB with a variation of less than ±0.5 dB has been measured, and a minimum conversion loss of 7.5 dB has been measured at 76 GHz.

INTRODUCTION

This paper describes the design, fabrication and performance of a W-band monolithic crossbar mixer. The monolithic crossbar mixer was fabricated with two in situ GaAs Schottky barrier diodes on a 4-mil thick semi-insulating GaAs substrate. The crossbar structure of the mixer was fabricated in a suspended stripline circuit. A special probe coupling was provided for the LO port which substantially reduces the susceptive loading of the IF port, resulting in extremely flat response over an IF of 8.5 GHz. The typical conversion loss measured was 9 dB for broadband operation, while for a narrow band case, a minimum conversion loss of 7.5 dB was measured at 76 GHz.

MIXER DESIGN

The heart of the monolithic crossbar mixer is a circuit substrate board comprising a pair of in situ GaAs Schottky barrier diodes connected across the stripline on the substrate, as shown in Figure 1. In operation, the LO signal is injected through a waveguide-to-suspended stripline transition and applied to the Schottky barrier diodes with the polarity shown in Figure 2. LO coupling to the diodes is through a specially designed E-probe with appropriate matching circuits as shown in Figure 3. This new probe design not only substantially reduces the size of the mixer housing but also the susceptive loading of the IF port due to excessive line length. As a result, a broader bandwidth and flatter response is achieved than with a conventional crossbar mixer. The RF signal is applied directly from the waveguide to the diode pair. The IF signal is extracted via a microstrip line etched on the same substrate.

Electrically, the mixer is well matched at the RF and IF ports by virtue of the series connection of the diodes as seen from the RF port, and the parallel connection as seen from the IF. This provides a higher impedance level to the RF signal and a lower impedance level to the IF signal than that of a single diode mixer. Therefore, an inherent impedance match condition for both the RF and IF signals is achieved for broadband performance. Basically, the planar crossbar mixer consists of a circuit board sandwiched between a split-block metal housing. One half of the housing consists of a built-in back-short for the RF signal, and the other half is a waveguide port for coupling the RF signal to the mixer diodes. As the LO signal is fed through the center conductor of the circuit board, it provides the right phase difference to the two diodes for a single balanced mixer operation. In addition, because of the orthogonality of the LO and RF E-fields, it also provides excellent RF to LO isolation. This design does not require any hybrid

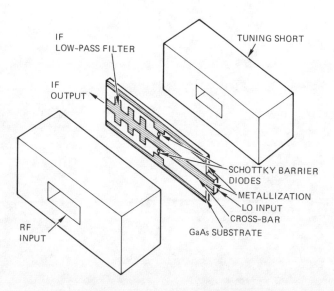

Figure 1. Schematic Diagram of Planar Crossbar Mixer

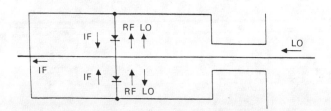

Figure 2. Diode Polarity of the Crossbar Mixer

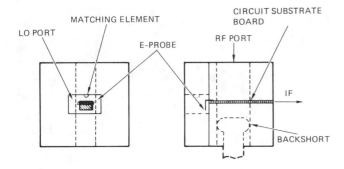

Figure 3. LO Probe Coupling Arrangement of the Mixer

coupler or RF diplexer, and therefore greatly simplifies the hardware design.

MONOLITHIC MIXER FABRICATION

In the fabrication of the monolithic crossbar mixer, an epitaxial layer-structure of n-n^+ grown in a semi-insulating substrate has been used. The typical characteristics of the epitaxial layers are:

Layers	Carrier Concentration (cm^{-3})	Thickness (μm)
n	$1 \sim 2 \times 10^{17}$	$0.1 \sim 0.15$
n$^+$	$>1 \times 10^{18}$	$2 \sim 3$

The GaAs substrate with the epi-layers serve not only as the dielectric substrate material for the crossbar mixer circuit but also as the starting material for device fabrication. The processing procedures for the fabrication of the monolithic mixers follows in general the conventional integrated circuit techniques. The two Schottky barrier diodes are fabricated in situ on the substrate utilizing the GaAs beam lead diode fabrication techniques developed at Hughes. Figure 4 shows the processing sequence for the fabrication of the monolithic mixer. First, a two layer epitaxial structure of n on n$^+$ is initially grown on a semi-insulating GaAs substrate by VPE, followed then by a proton bombardment process to convert the n and n$^+$ layers outside of the diodes to high resistivity material for isolation. In the ohmic contact area, a Se$^+$ implanted process is used to form a high conductivity channel to the n$^+$ layer, followed then by depositing and alloying the ohmic contact metallization, AuGe/Ni/Au, in the implanted area. Figure 5 shows the completed monolithic crossbar mixer.

RF PERFORMANCE

The mechanical configuration of the W-band monolithic crossbar mixer with a built-in LO is shown on the right hand side of Figure 6. On the left hand side is the detached LO housing built with the Hughes high power Gunn diode. RF performance of the monolithic crossbar mixer is encouraging. The conversion loss measured over an RF range of 84.6 to 93.1 GHz displays an extremely uniform response as shown in Figure 7. The typical conversion loss is 9 dB with a variation of less than ± 0.5 dB over an IF range of 0.5 to 8.5 GHz. For a narrow band operation, a minimum conversion loss of 7.5 dB was measured at an RF of 76 GHz and an IF of 15.5 GHz.

CONCLUSIONS

In conclusion, a W-band fully monolithic crossbar mixer with an extremely uniform response was successfully fabricated on a 4-mil thick semi-insulating GaAs substrate using planar integrated circuit techniques. RF performance of the monolithic mixer is compatible to

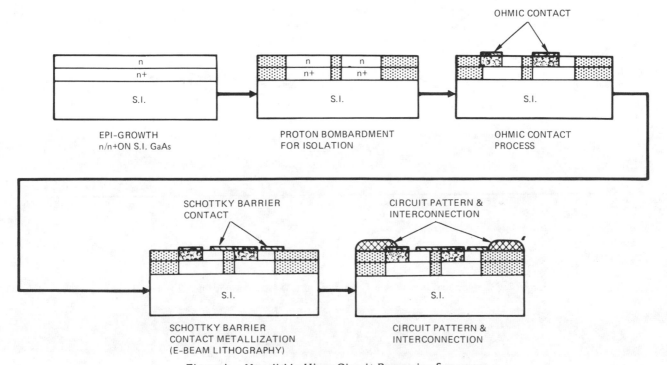

Figure 4. Monolithic Mixer Circuit Processing Sequence

Figure 5. W-band Monolithic Crossbar Mixer

Figure 6. Monolithic Crossbar Mixer/LO Assembly

that of its waveguide counterpart. The success of this development offers potentially low cost batch-processing of millimeter-wave integrated circuits and future integration of millimeter-wave components, all on the same common GaAs substrate.

ACKNOWLEDGEMENTS

The author wishes to thank Edward Roth for the RF measurements and to Danny Wong for processing of the monolithic mixer chips.

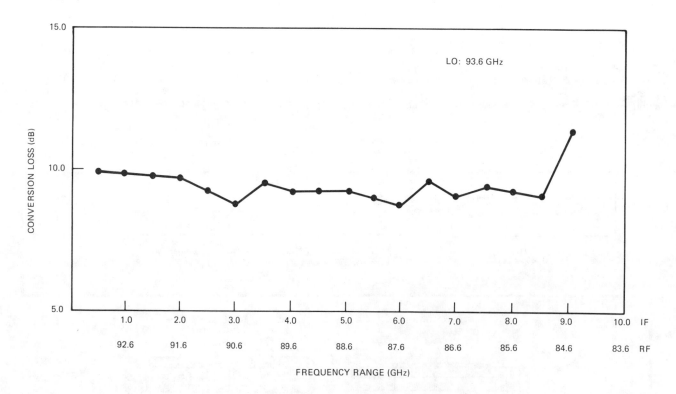

Figure 7. Conversion Loss Versus Frequency of a W-band Monolithic Crossbar Mixer

Section III-F
Special Components and Circuits

THIS section contains papers which describe the design and performance of various passive components for MMIC's, as well as specialized passive and active circuits such as couplers and wide-band gain-controlling elements.

GaAs Monolithic Lange and Wilkinson Couplers

RAYMOND C. WATERMAN, JR., MEMBER, IEEE, WALTER FABIAN, ROBERT A. PUCEL, FELLOW, IEEE, YUSUKE TAJIMA, MEMBER, IEEE, AND JAMES L. VORHAUS

Abstract—An important part of monolithic microwave integrated circuits are the passive distributed circuit elements which divide and combine RF signals. This paper describes the design and performance of a monolithic GaAs X-band three-port Wilkinson coupler [1] and a monolithic GaAs X-band four-port interdigitated Lange coupler [2], [3]. These couplers are useful both as power dividers and power combiners. In the Wilkinson, the input power is split equally between the output ports with zero differential phase shift. For the Lange, there is a phase difference of 90° between the outputs. All ports are well matched and the output ports are highly isolated.

The transmission mode chosen for these couplers is microstrip on semi-insulating GaAs with conductor-to-ground plane spacing of 0.1 mm. This low-loss, high-permittivity medium is compatible with the present monolithic GaAs FET technology which thus allows combining more complicated monolithic microwave integrated circuits on a single chip.

The fabrication process takes advantage of the technology developed for processing GaAs FET's. For example, connecting the coupling conductors for the Lange coupler requires RF crossovers. To minimize crossover capacitance, an air-bridge interconnection technique is used.

Calculated coupling, isolation, and VSWR data for the Wilkinson and Lange couplers are compared with actual measured performance showing good agreement with expected results. Measured loss, minus fixture contributions, shows 0.25 dB for the Wilkinson and 0.75 dB for the Lange.

I. INTRODUCTION

AS MICROWAVE monolithic integrated circuit (MMIC) technology advances, the need for passive distributed circuit elements which can divide and combine RF signals becomes evident. Such circuits would be useful in monolithic balanced amplifiers, mixers, and power combiners to name just a few examples. To be of maximum utility, the designs need to have high port-to-port isolation, low loss and, above all, be compatible with existing MMIC fabrication technologies.

This paper describes the design, fabrication, and performance of a monolithic three-port Wilkinson coupler [1] and a monolithic four-port interdigitated Lange coupler [2], [3]. Both circuits are fabricated on GaAs and are designed for operation at X-band frequencies.

These couplers have been used both as power dividers and power combiners. For example, the Wilkinson circuit when operated as a divider splits the input power equally between its two output ports with zero phase differential. The Lange also provides an equal power split with the additional advantage of a relative phase shift between the output ports of 90°. Because of this phase shift, the circuit is known as a quadrature hybrid coupler. Common uses of the quadrature hybrid include balanced amplifiers, mixers, and discriminators.

The transmission mode chosen for these couplers is microstrip on semi-insulating GaAs with conductor-to-ground plane spacing of 0.1 mm. The fabrication process takes advantage of the technology developed for processing GaAs FET's. This makes the couplers themselves completely compatible with existing MMIC technology where FET's are used as the active elements.

Section II describes the designs for the Wilkinson and Lange couplers. The fabrication technology is described in Section III. Finally, the experimental results for both circuits are given in Section IV and are compared to the model predictions.

II. COUPLER DESIGN

A. Wilkinson Coupler

A standard design approach was used for the Wilkinson divider. It was modeled as a three-port in a 50-Ω system (Z_0 = 50 Ω) implying 70.7-Ω quarter wavelength transmission lines in the branch arms (Fig. 1(a)).

The 100-Ω isolation resistor between the output ports was modeled both as a lumped element and a distributed element. In the distributed model, the resistor was simulated by a cascade connection of multiple small segments consisting of L's, C's, and R's (Fig. 1(b)). The values used for the lumped L, C, and R components are the equivalent distributed values per unit length divided by 20 (the number of segments used) for a distributed line with dimensions equal to those used to fabricate the thin-film resistor (see Section III for dimensions).

The design center frequency of the Wilkinson was 9.5 GHz. Thus the transmission lines in the branch arms were each 2.85 mm long and 0.03 mm wide, assuming a 0.1-mm-thick GaAs substrate with a dielectric constant of 12.5.

The coupler was modeled using COMPACT,[1] a computer-aided microwave design program. It was found that with the distributed model, the center frequency shifted upwards by 16 percent from the initially designed value. The predicted loss was about 0.25 dB at center frequency. The loss is mostly due to conductor losses and was calculated assuming 0.56 dB/wavelength.

B. Lange Coupler

The particular Lange coupler design chosen was to have a center-band frequency of 12 GHz and a bandwidth of at least

---

Manuscript received August 22, 1980; revised October 6, 1980. This work was partially supported by the U.S. Army Electronics Research and Development Command under Contract DAAK20-79-C-0269.

R. C. Waterman, Jr., is with Raytheon Company, Missile Systems Division, Bedford, MA 01730.

W. Fabian, R. A. Pucel, Y. Tajima, and J. L. Vorhaus are with Raytheon Research Division, Waltham, MA 02154.

[1] Compact Engineering, Inc., 1088 Valley View Court, Los Altos, CA 94022.

Reprinted from *IEEE Trans. Electron Devices*, vol. ED-28, pp. 212-216, Feb. 1981.

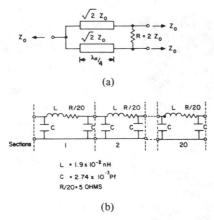

Fig. 1. (a) Circuit model of Wilkinson coupler showing isolation resistor as a lumped element. (b) Distributed isolation resistor model used in place of lumped resistor in circuit model shown in Fig. 1(a).

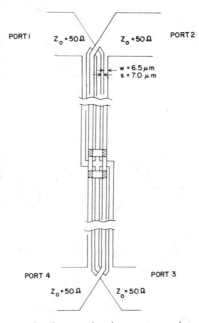

Fig. 2. Lange coupler layout showing crossover interconnections at either end and in the center.

4 GHz. The coupling section thus consisted of 2.2 mm of 6.5-μm-wide interdigitated conductors with 7-μm spaces. The structure was fed at all four ports by 50-Ω transmission lines. The design that was used showing the crossover interconnections at either end and the center of the structure is shown in Fig. 2. This four-port hybrid ideally couples half of the power incident on port 1 to port 2 (the coupled port) and directs the remaining power to port 3 (the direct port) with no power going to port 4 (the isolated port).

This design which was modeled on COMPACT has a calculated total loss at 12 GHz of 0.75 dB. The major contributor to this attenuation is conductor loss due to odd-mode currents. Even-odd mode analysis for coupled lines yields an odd-mode impedance (Z_{0o}) of 19.9 Ω and an even mode impedance (Z_{0e}) of 121.5 Ω for the 3-dB Lange coupler. The low odd-mode impedance implies that the level for the odd-mode currents will be high. In addition, the current distribution on the coupling conductors is concentrated at the conductor edges further increasing the $i^2 R$ losses. This loss can be lowered by increasing the conductor-to-ground plane spacing. For example, increasing this dimension from 0.1 to 0.2 mm reduces the loss by more than a factor of two by permitting the conductor width to increase while maintaining the same impedances. However, a number of conflicting considerations involving both electrical performance and fabrication technology have dictated the choice of 0.1-mm-thick substrates for all of our MMIC development [4]. In order to maintain compatibility with this technology, we have chosen to use the thinner substrates.

III. Coupler Fabrications

All aspects of the fabrication technology used to produce the couplers were compatible with or derived directly from our basic GaAs FET process technology. This was done to insure that once developed, the coupler circuit could be integrated with the FET's (and other passive circuitry) into a workable MMIC process.

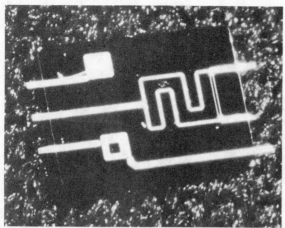

Fig. 3. Wilkinson structure showing coupler, integral beam leads, and two spiral inductor test structures.

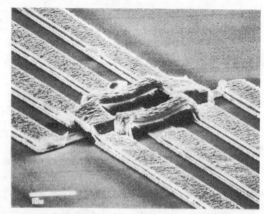

Fig. 4. SEM micrograph of air bridge in Lange coupler.

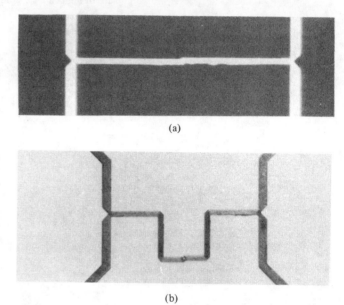

Fig. 5. (a) Lange coupler structure, showing straight coupling section. (b) Lange coupler structure showing folded coupling section.

The couplers were fabricated variously on semi-insulating (SI) GaAs substrates or undoped buffer layers epitaxially grown on the SI material. Some of the structures were also put down on a layer of intentionally doped material grown on the substrate. This layer had been previously bombarded with $^{16}O^+$ ions to convert it from conducting to insulating. This was done to establish the compatibility of using the couplers with our standard device isolation process. In all cases, the performance was independent of the method used to obtain a high-resistivity GaAs substrate beneath the coupler.

The Wilkinson structure is shown in Fig. 3. It consists of two 50-Ω transmission lines coming in from the top which connect to the thin-film isolation resistor and the two branch lines. The branch lines themselves are meandered somewhat to give a more compact overall structure. They join at the other end of the coupler in a T-junction with another 50-Ω transmission line. There are also two spiral inductor test structures on this chip but they do not affect the performance of the Wilkinson in any way.

Each of the input and output lines of the coupler ends in a gold beam lead which is fabricated as an integral part of the circuit processing. These so-called integral beam leads extend about 0.25 mm beyond the chip edge and make it easier to connect the chip into a test fixture for measurement since they eliminate the need for any on-chip bonding.

The chip is 3.1 × 2.3 mm. The coupler itself occupies an area of only about 1.3 × 1.0 mm excluding the input and output lines. Within the coupler, all transmission lines are separated from each other and from the thin-film resistor by a distance equal to at least two substrate thicknesses (0.2 mm) to provide isolation of the microwave signals. This criterion dictates just how compact the coupler can be made. Special care was taken in the design to assure complete symmetry between the branch lines with respect to line length and number of bends.

The 100-Ω thin-film resistor is formed from a 3000-Å-thick film of Ti which has a resistivity of 132 $\mu\Omega \cdot$ cm. The resistor is 622.8 μm long and 27.4 μm wide. The interconnections at the ends consist of a gold-topped underlayer (either AuGe/Ni/Au or Ti/Pt/Au), then the Ti film, and finally a 4-6-μm-thick plated gold-top connection which is part of the transmission lines. The purpose of the gold underlayer is to provide a good, low-resistance contact between the Ti film and the transmission lines. The titanium tends to form an oxide where it is exposed to the air which can introduce a significant resistance between the plated gold top layer and the isolation resistor itself. The underlayer guarantees intimate contact between gold and the Ti deposited on top of it. In an actual MMIC process, this underlayer would be deposited with either the source-drain or the gate metal layers and would thus not cost any additional process steps.

The Lange couplers were fabricated in essentially the same way as the Wilkinson couplers. No thin-film resistors were used on these test circuits although, in an actual MMIC application, a 50-Ω load resistor would be needed on the isolation port. The only difference between the two couplers was the need for interconnections in the Lange structure. These were made using the same technique that was developed for fabricating low-inductance, plated gold, air-bridge source interconnections in our interdigitated power FET devices. Fig. 4 shows an SEM micrograph side view of the center intercon-

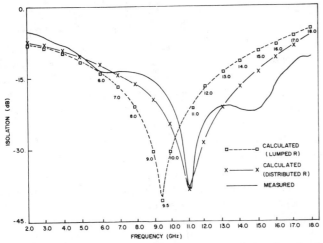

Fig. 6. Comparison of calculated and measured isolation data for the Wilkinson coupler.

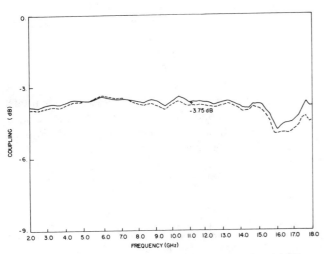

Fig. 7. Measured amplitude balance of Wilkinson coupler with all ports matched into 50 Ω, including fixture losses.

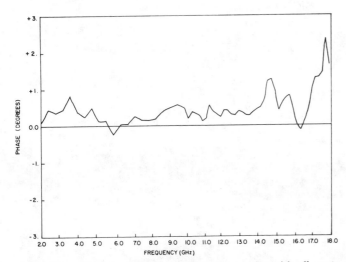

Fig. 8. Measured phase balance of Wilkinson coupler with all ports matched into 50 Ω.

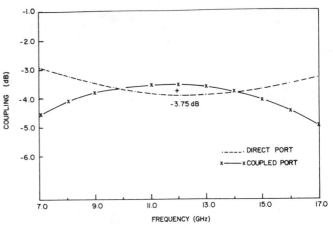

Fig. 9. Calculated amplitude balance for Lange coupler.

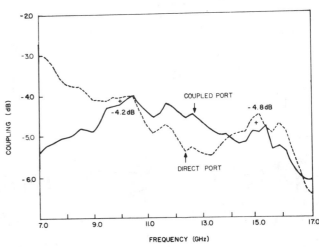

Fig. 10. Measured amplitude balance for Lange coupler, including fixture losses.

nection portion of the coupler. The cross shaped structure running under the air bridge is formed in the same way as the underlayer in the thin-film resistor. The plated air-bridge structure is formed at the same time as the plated coupler conductors (and all other transmission lines in the circuit) and thus requires no additional process steps.

Two different coupling layouts were designed and evaluated for the Lange coupler. They are shown in Fig. 5(a) and (b). The layout shown in Fig. 5(b) permits compression of the coupling section for compactness and alteration of the input/ output port positions for layout flexibility. No difference in the performance of these two structures was observed.

IV. Experimental Results

A. Wilkinson Coupler

Network analyzer measurements of isolation for the Wilkinson demonstrates a center frequency of 11.5 GHz. This shows excellent agreement with predictions when the thin-film isolation resistor is treated as a distributed element (see Fig. 6).

Figs. 7 and 8 show measured amplitude balance and phase balance with all ports matched into 50 Ω. Over the frequency range from 2 to 18 GHz, the amplitude balance is within ± 0.2 dB. The total measured loss in the coupler at 11.5 GHz

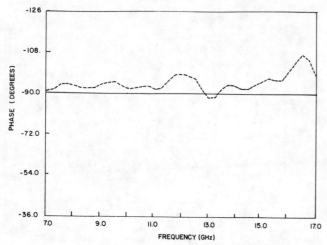

Fig. 11. Measured coupled port phase relative to direct port for the Lange coupler.

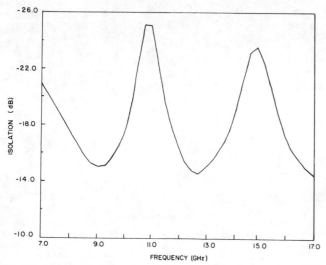

Fig. 12. Measured isolation for the Lange coupler showing a maximum of 14 dB over frequency range of 7 to 17 GHz.

is 0.75 dB, assuming a perfect power split. When fixture losses of 0.45 dB are taken into account, the coupler loss is 0.30 dB. This is in exact agreement with calculated losses. The phase balance is within 2.3° and the input reflection coefficient is below 0.2 from 6 to 18 GHz.

B. Lange Coupler

The Lange coupler calculated amplitude balance is compared with measured amplitude balance in Figs. 9 and 10. The average measured coupling from the input to the coupled port and from the input to the direct port is −4.2 dB at 10 GHz and −4.8 dB at 15 GHz. In addition, a dip in the direct port coupling is shown at 13 GHz. These rather high losses are due, in part, to the test fixture. Measured loss in the text fixture alone showed 0.45 dB at 10 GHz and 0.6 dB at 15 GHz. The test fixture has a 10-mil alumina substrate on the input and output side of the GaAs chip which launch into SMA connectors. The alumina substrate has 0.45 in of 50-Ω transmission line on it. When these fixture losses are included, the measured data are in exact agreement with calculated losses at 10 GHz and within 0.5 dB at 15 GHz.

Fig. 11 shows relative phase difference between the coupled and the direct port. The measured variation from 90° is 10° maximum from 7 to 16 GHz. Fig. 12 shows that the coupler has at least 14-dB isolation over this frequency range.

The Lange coupler was tested with up to 6 W of average input power at 10 GHz with no observed degradation in performance.

V. CONCLUSION

The results reported on in this paper demonstrate the realization and performance predictability of two different microwave couplers for monolithic circuits on GaAs. Measured loss, minus fixture contributions, shows 0.25 dB for the Wilkinson and 0.75 dB for the Lange. In addition, fabrication of these couplers is compatible with present MMIC technologies.

REFERENCES

[1] E. J. Wilkinson, "An N-way hybrid power divider," *IRE Trans. Microwave Theory Tech.*, vol. MTT-8, pp. 116–118, Jan. 1960.

[2] J. Lange, "Interdigitated stripline quadrature hybrid," *IEEE Trans. Microwave Theory Tech.*, vol. MTT-17, pp. 1150–1151, Dec. 1969.

[3] Y. Tajima and S. Kamihashi, "Multiconductor couplers," *IEEE Trans. Microwave Theory Tech.*, vol. MTT-26, pp. 795–801, Oct. 1978.

[4] R. A. Pucel, "Some design considerations for gallium arsenide monolithic circuits," in *1979 Proc. 7th Bien. Cornell Electrical Engineering Conf. on Active Microwave Semiconductor Devices and Circuits*, pp. 17–31.

A MICROWAVE PHASE AND GAIN CONTROLLER WITH SEGMENTED-DUAL-GATE MESFETs IN GaAs MMICs

Y.C. Hwang, Y.K. Chen, R.J. Naster
Electronics Laboratory, General Electric Co.
Syracuse, New York 13221

D. Temme
Lincoln Laboratory, Massachusetts Institute of Technology
Lexington, Massachusetts 02173

ABSTRACT

A novel segmented-dual-gate MESFET device which provides precise gain control over broad microwave bandwidth by using prescribed gate-width-ratio is presented. The digitally-controlled precision microwave gain scaler has potential application as an ultra-wide band microwave attenuator or active microwave phase shifter. The design and test results of GaAs MMIC active attenuator and hybrid phase shifter are described.

Introduction

Gallium Arsenide Monolithic Microwave Integrated Circuit (MMIC) technology has matured rapidly during the past few years. Functions usually implemented with passive elements and transmission lines may be replaced by active elements resulting in reduction of chip sizes.

In this paper, a circuit approach utilizing segmented-dual-gate-MESFETs (SDGFETs) to obtain precise gain control is presented. The gain control is achieved by properly scaled gate-width-ratios among the dual-gate MESFETs (DGFETs). Since the gate-width is the least sensitive processing variable, great process resilience is achieved.

Segmented-dual-gate MESFETs

Conventionally, dual-gate MESFETs have been used for gain control [1] by applying an analog voltage on gate #2. The control is non-linear and it is difficult to provide a control voltage with precision and repeatability. While the voltage and impedance applied to the gate #2 are changed, the transfer gain and phase also change depending on the biasing, geometry and process-dependent characteristics of the device. Thus precision cannot be maintained.

In the scheme presented here, we operate the dual gate devices only in the ON/OFF mode by switching the gate #2 bias between two fixed voltage levels, namely saturation and pinch-off, while the bias on gate #1 is kept constant. Several segments of dual-gate MESFETs are integrated together by connecting the gate #1's, sources and drains of each section together as shown in Figure 1. Each dual-gate segment is only operated either fully on or fully off with the input gate bias remaining the same. The layout geometry among dual-gate segments is the same to preserve linear scaling of electrical parameters. The gain of this device is proportional to the collective width of the turned-on segments. By selectively controlling the gate #2's, the gain of this device is programmed exactly. This "scaled-by-width-ratio" can also preserve the relative gain settings against bias and temperature fluctuations as well as process variations.

Modeling of dual-gate FET

It is somewhat difficult to model a dual-gate GaAs MESFET not only because of the physics of its complicated geometry and terminal impedance at gate #2, but also of interacting bias condition [2]. A dual-gate MESFET can be imagined as two single gate MESFET connected in series with the source of the second FET merged with the drain of the first FET. Each FET has two operation regions; the linear and saturated region. The equivalent circuit of the dual-gate MESFET then reflects the state (one out of four) generated from the two serial FETs. That is, different DC bias conditions generate different equivalent circuits for a single DG-MESFET. Since our dual-gate MESFET is only operated in

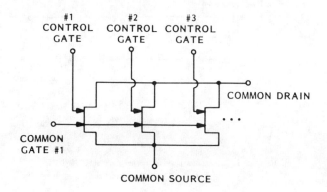

Figure 1. Implementation of a Segmented-Dual-Gate-MESFET Device

---

This work is supported in part by the Department of the Air Force. The U.S. Government assumes no responsibility for the information presented.

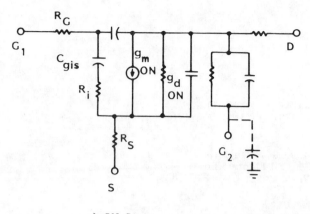
a) ON State

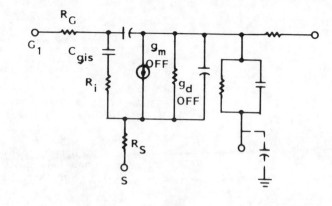
b) OFF State

Figure 2. Simplified Equivalent Circuit Model of a Dual-Gate MESFET Operated in ON/OFF States

either the ON or OFF state, two equivalent circuits for these two biases will be enough.

Figure 2(a) shows the simplified equivalent RF small signal model for MESFET in the ON-state with the second gate represented by a diode (parallel R and C). Figure 2(b) shows the OFF-state equivalent circuit model of the same FET with the same input impedance but different output transconductance and impedance to reflect the pinched-off DG-FET state. The model can be linearly scaled with gate width provided that the geometry and terminal bias voltages and impedances remain the same.

Binary Weighted Scaler

To verify this concept, a binary weighted scaler in the ratio of 1-2-4-8 has been designed and processed on GaAs (Figures 3, 4, and 5). The gate widths of the segments are 50, 100, 200 and 400 µM respectively. The control voltages are fed through 2K ohm resistors on chips. The gates are 1 µM long and separated with 2 µM spacing. The measured performance is shown in Figure 6. The phase and amplitude flatness and tracking over the wide bandwidth is excellent.

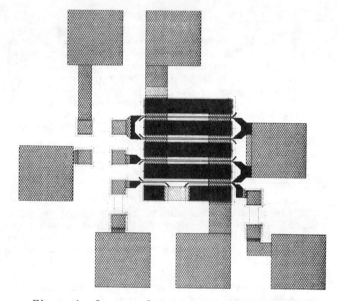

Figure 4. Layout of the 1-2-4-8 Binary Scaler

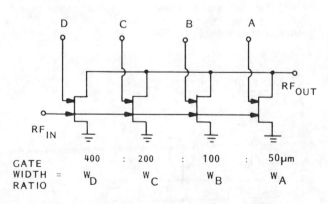

Figure 3. Segmented Dual-Gate MESFETs with Binary Weighting

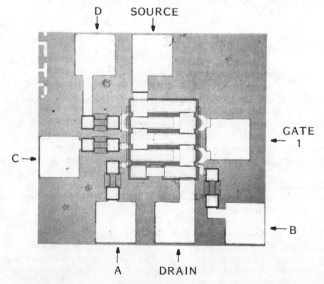

Figure 5. 1-2-4-8 Binary Scaler on Chip

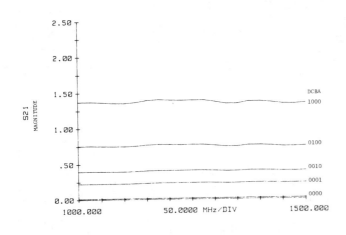
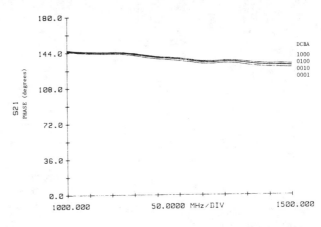

a) Magnitude vs. Frequency   b) Phase vs. Frequency

Figure 6. Magnitude and Phase Characteristics of the 1-2-4-8 Binary Scaler

Microwave sine/cosine function scaler

Another device has been designed to adjust the amplitude for sine function states with settings. The sine function scaler modulates its output signal magnitude by a factor of the sine function value of the designated phasor angle. The cosine function scaler utilizes an identical structure as the sine function scalar except that it is driven by the 90-degree complementary angle of the phasor. Table 1 shows the list of sine and cosine function values for a 4-bit phase shifter. We have adopted this broad band sine function scaler based on the segmented-dual-gate MESFET devices to design a phase shifter which is less sensitive to process variations and whose phase shift is solely determined by the gate width ratios.

Table 1 also shows a sine function for 22.5-degree resolution which is implemented with the "divided-by-13" scheme. From a segmented-dual-gate-MESFET device with three channels of 1:4:8 gate width ratio, values of sine or cosine function at 0, 22.5, 45, 67.5, and 90 degrees can be synthesized by combining the right on and off dual-gate FET segments. Table 2 shows a "divide-by-50" scheme to synthesize a 5-bit phase shifter of 11.25 degrees resolution. Figure 7 shows the photomicrograph of a finished GaAs sine scaler.

Digitally controlled active phase shifter

A phase shifter which consists of a vector modulation scheme and sine/cosine function scalers built on segmented-dual-gate MESFETs is presented. Figure 8 shows the block diagram of a vector modulator. A pair of balanced in-phase (I) and 90-degree (Q) phase vectors are generated from the input RF signal by a 90-degree phase splitter. Since both the sine and cosine scaler will be fabricated on the same MMIC chip, the same insertion gain and phase will be contributed to both the I and Q vectors by the two scalers. The I and Q vectors presented at the output summing point are modulated by the cosine and sine value of the designated phase respectively. The final phase shift is determined solely by the designed width ratios from the two scalers.

TABLE 1

SINE SCALER OF 22.5-DEGREE RESOLUTION FOR 4-BIT PHASE SHIFTER

Angle (degree)	Sine Value	1/13 Approximation	Simulated by control the width of a segmented-dual-gate MESFET device		
			W_{8x}	W_{4x}	W_{1x}
0	0	0/13 (.000)	OFF	OFF	OFF
22.5	.383	5/13 (.384)	OFF	ON	ON
45.0	.707	9/13 (.692)	ON	OFF	ON
67.5	.924	12/13 (.923)	ON	ON	OFF
90.0	1	13/13 (1.000)	ON	ON	ON

TABLE 2

SINE FUNCTION SCALER WITH 11.25° RESOLUTION FOR A 5-BIT PHASE SHIFTER

Angle (degree)	Sine Value	1/50 Approximation	Width control of a segment-dual-gate-MESFET device				
			W_{20x}	W_{14x}	W_{11x}	W_{10x}	W_{5x}
0	0	0/50 (.000)	0	0	0	0	0
11.25	.195	10/50 (.200)	0	0	0	1	0
22.5	.383	19/50 (.380)	0	1	0	0	1
33.75	.556	29/50 (.580)	0	1	0	1	1
45.0	.707	35/50 (.700)	1	0	0	1	1
56.25	.831	41/50 (.820)	1	0	1	1	0
67.50	.924	46.50 (.920)	1	0	1	1	1
78.75	.981	49/50 (.980)	1	1	0	1	1
90.00	1	50/50 (1.000)	1	1	1	0	1

0: segment FET OFF
1: segment FET ON

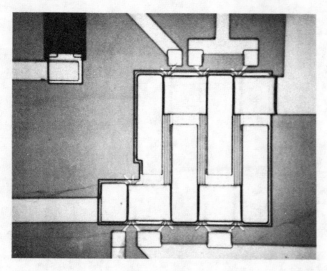

Figure 7. Segmented Dual-Gate MESFETs with 1-4-8 Weighting for Sine Function Scaler

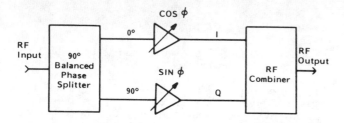

Figure 8. Block Diagram of a Vector Modulator Approach of Microwave Phase Shifter

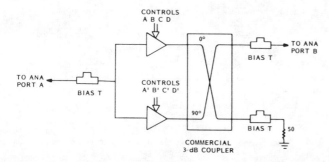

Figure 9. Phase Shifter with the Sine Function Scaler Chip

The phase shifter bandwidth is only limited by the response of the 90-degree phase splitter and the input and output matching circuits connected to it. Since the resultant phase shifts are determined by the turned-on gate width ratios, they are less sensitive to processing variations and the operating environment. The amplitude of the phase shifter is also constant over different phase states and some trimming capability is easy to achieve by fine-tuning the bias on gate #2's.

Figure 9 shows the hybrid test schematic for a 0-90 degree phase shifter that utilizes the vector modulator scheme with the segmented-dual-gate MESFETs

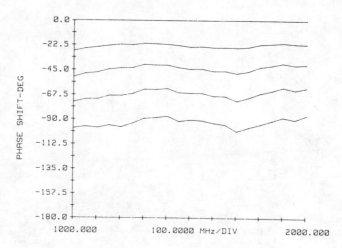

Figure 10. Relative Phase Shift Variations over L-Band

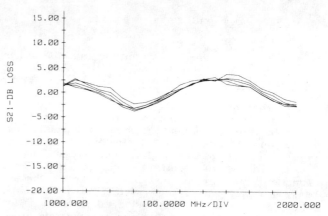

Figure 11. Relative Magnitude Variations over L-Band

shown in Figure 5. Figure 10 shows the phase characteristic of the circuit over the 1000 to 2000 MHz frequency range. The magnitude fluctuation shown in Figure 11 is due mainly to the coupler characteristic.

Active wideband attenuator

If the reference state of a segmented-dual-gate-MESFET device is set such that all segments are biased ON, a specific attenuation can be achieved by turning off a portion of the dual-gate segments. In this case the relative attenuation depends on the width ratio. The "relative" attenuation is independent of any process variations as long as the segments are of the same geometry and they are integrated in such a small area that the fabrication process treats them equally. However, the "insertion" gain of this attenuator will be set by different processing and biasing conditions.

Figure 12 shows the circuit diagram for such an active attenuator. The linear gain S21 of this device can be programmed from 0 to prescribed steps if the gain of a fully turned on device is normalized as unity. In the MMIC format, all contributing channel segments are fabricated in a very close proximity.

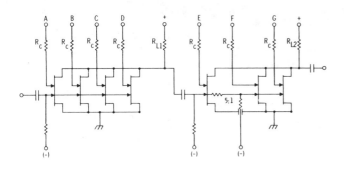

Figure 12. Attenuator with Segmented-Dual-Gate MESFETs

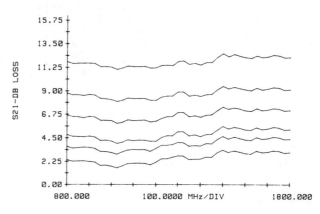

Figure 14. Measurement Result of the Primary Portion of the MMIC Active Attenuator Chip

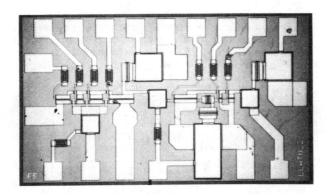

Figure 13. MMIC of the L-Band Attenuator

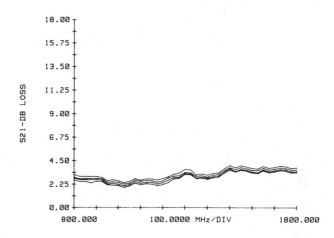

Figure 15. Measurement Result of the Trimming Portion of the MMIC Active Attenuator Chip

Transistor parameters among the dual-gate segments are scaled linearly by the ratio of their width, because they are fabricated under identical processing environments.

However, three major factors need to be considered in order to obtain intended RF gains by linear scaling of the combined width ratios. They are: (1) the loading effect from paralleling DGFETs of finite output impedance; (2) the back-gating effect from the non-ideal substrate, and (3) fringing fields from using DFGETs of short gate width. These non-ideal factors are circumvented by using good material, a set of carefully constructed design rules, and accurate simulations.

The performance for the monolithic attenuator shown in Figure 13 is plotted in Figures 14 and 15. It has a broadband capability from UHF up to 2000 MHz with maximum error within 0.5 dB from the designed primary attenuation settings and 0.2 dB from the designed trimmer settings.

Conclusions

A microwave scaler using a segmented-dual-gate-MESFET with GaAs technology has been presented. It has applications in broadband attenuators and phase shifters. Accurate amplitude control and phase shift control when implemented in a phase vector modulator are possible. By segmenting the gate #2, the need to accurately set in an analog bias on the gate #2 of a continuous dual gate FET is eliminated.

References

[1] Charles A. Liechti, "Performance of Dual-Gate GaAs MESFETs as Gain-Controlled Low Noise Amplifiers and High-Speed Modulators," IEEE, T-MTT, Vol. MTT-23, No. 6, June 1975, pp. 461-469.

[2] C. Tsironis, R. Meierer, "Microwave Wideband Model of GaAs Dual-Gate MESFET," IEEE T-MTT, Vol. MTT-30, No. 3, Mar. 1982.

GaAs MONOLITHIC WIDEBAND (2-18 GHz) VARIABLE ATTENUATORS

Y. Tajima, T. Tsukii,* R. Mozzi, E. Tong, L. Hanes, B. Wrona
Raytheon Company, Research Division
Lexington, Massachusetts 02173

*Electromagnetic Systems Division
Goleta, California 93017

Summary

GaAs monolithic variable attenuators have been developed. They operate in a very wide frequency band, are very small, and are controlled by one voltage. Insertion loss of 2-3 dB and a dynamic range of attenuation of 10 dB were obtained in the 2-18 GHz frequency range.

Introduction

Voltage-controlled variable attenuators have been widely used for automatic gain control circuits as well as for various switches. In broadband microwave amplifiers, this circuit is indispensable for temperature compensation of gain variation. It is usually realized by PIN diodes and a pair of hybrid couplers. However, the frequency performance is limited by the bandwidth of the hybrid couplers and, more significantly, it is not compatible with the GaAs monolithic circuits because of the PIN diodes. There has been a need to come up with a new type of variable attenuator as the requirements of GaAs monolithic circuits have become more sophisticated.

We will introduce, in this paper, a GaAs monolithic variable attenuator which operates in a very wide frequency band (2-18 GHz), is very small in size (1.5 × 1.2 mm), and is comparable in performance to the PIN attenuators.

Attenuator Design

The basic mechanism of the circuit is the change in the low field resistance of a zero-biased FET controlled by gate voltage. Three FETs are connected in T or Π shape, as shown in Figs. 1a and 1b. The electrical characteristics of each FET are expressed as a parallel combination of R and C, as shown in Figs. 1c and 1d, where R is a varying value as a function of gate voltage. The value of resistance R varies from the open-gate resistance ($\sim R_S$) to infinite when the gate voltage is changed from the built-in voltage (positive) of the gate barrier to the pinchoff voltage (negative). On the other hand, capacitance C is considered to be fairly constant with the gate voltage.

At a low frequency when the effect of capacitors can be neglected, resistance R_1, in a series arm, and resistance R_2, in a shunt arm, as in Fig. 1c, have to have a certain combination in order to meet the matching conditions and to obtain a given attenuation. This combination is specified by the following equations:

$$Z_0^2 = R_1^2 + 2 R_1 R_2,$$
$$A = \frac{V_2}{V_1} = \frac{R_2}{R_1 + R_2 + Z_0},$$

where Z_0 is the matching impedance. Similar equations can be derived for the circuit in Fig. 1d.

Figure 2 shows the combination of R_1 and R_2 as a function of the attenuation when Z_0 is 50 Ω. The minimum attenuation, or insertion loss, is mostly determined by the minimum achievable value for R_1.

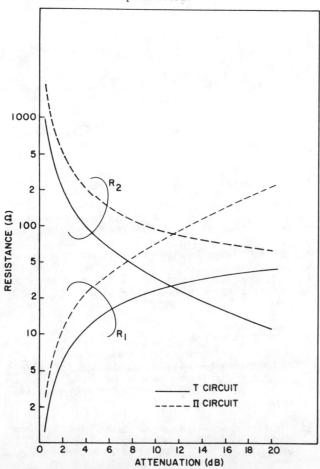

Fig. 2 Values of R_1 and R_2 as a function of attenuation.

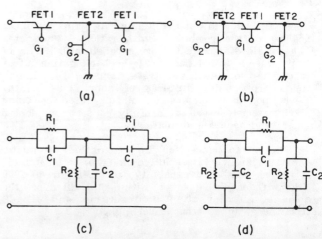

Fig. 1 Basic circuit topologies (a,b) of variable attenuators and their equivalent circuits (c,d).

Reprinted from *IEEE MTT-S Int. Microwave Symp. Dig.*, 1982, pp. 479-481.

For the same R_1, Π circuits have less insertion loss than T circuits. The value of R_1 can be reduced by increasing the number of gate fingers, although parasitic capacitors, C_1, will also increase. Larger capacitors limit the dynamic range of attenuation at high frequency. In terms of the dynamic range, T circuits become advantageous over Π circuits.

An attenuator with T circuit topology was designed using the FET parameters shown in Table 1. Gate widths of 600 and 200 µm were chosen for the series and shunt element FETs, respectively. Simulated performance showed a 0.7 dB insertion loss and 13 dB dynamic range of attenuation over the frequency range of 2-18 GHz.

TABLE 1

VALUES OF R AND C OF A 100 µm WIDE CHANNEL

	R	C
ON	35 Ω	0.03 pF
OFF	10 KΩ	0.03 pF

Figure 3 shows the mask layout of an attenuator with its circuit diagram. Input and output striplines are connected to drain contacts of FETs a and c, while source contacts are connected by a common airbridge which serves as a connection to the drain contacts of FET b. Source contact of FET c is grounded by a via hole.

Isolation between the rf circuit and dc control circuit was achieved by thin-film resistors and FET resistors which have 25 µm wide gates. These elements were inserted between gate terminals and bias terminals in order to reduce the leakage of rf signal to dc terminals. The completed circuit is shown in Fig. 4. The chip size is 1.5 × 1.2 mm.

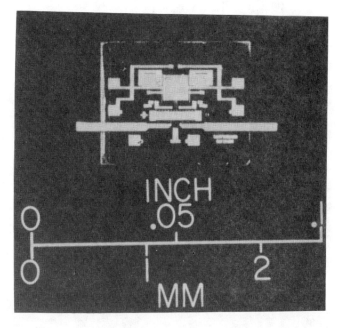

Fig. 4 Variable attenuator (photo).

Power-handling capability of this attenuator is limited by the performance of the shunt FET in two ways. If the rf swing becomes too large and exceeds the knee voltage of the IV curve, the resistance of the FET is no longer linear but shows increasing resistance with the voltage swing. Thus the attenuation will start to change with the input level. Also, if this rf voltage swings into the breakdown region, the attenuation will increase.

In order to increase the knee voltage and breakdown voltage of the FET, dual gates were used for the shunt FET in the high-power version of the attenuator. By use of a dual gate structure, the knee voltage as well as the breakdown voltage was increased considerably.

A Control Circuit Design

The attenuator chip requires two control voltages which do not change linearly with respect to rf attenuation in dB. A control circuit was designed and built in order to provide these voltages from a single control voltage and to establish a linear relationship between rf attenuation in dB and the control voltage.

The circuit developed for this purpose consists of a noninverting linear amplifier and an inverting nonlinear amplifier employing dual operational amplifiers, diodes and resistors. Figure 5 is a schematic representation of the control circuit. In this design, a potentiometer R_V provides a linear relationship with rf attenuation.

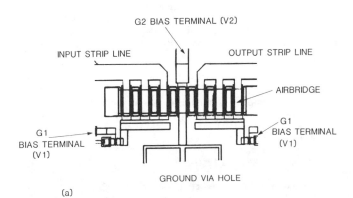

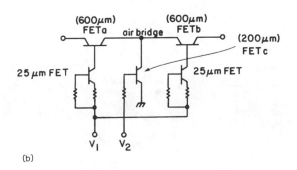

Fig. 3 Variable attenuator structure (a) and its equivalent circuit (b).

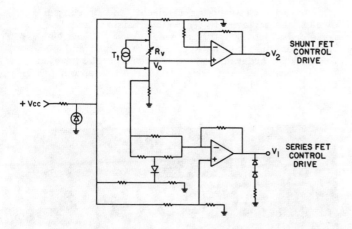

Fig. 5 Control circuit diagram.

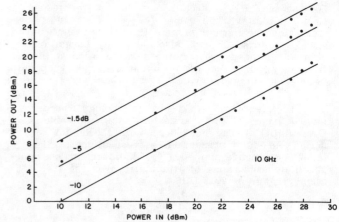

Fig. 7 Power performance of an attenuator at various attenuator levels.

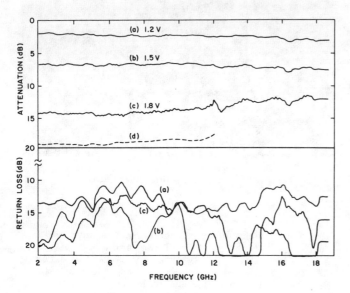

Fig. 6 Frequency performance of an attenuator at various control voltages.

Experiment

The control unit and the attenuator circuit were combined and tested. The rf performance is shown in Fig. 6. The minimum insertion loss was between 2 and 3 dB in the frequency range of 2-18 GHz. The amount of attenuation varied up to 12 dB uniformly when the control voltage varied from 1.2 to 1.8 V. The return loss was kept larger than 10 dB under all conditions. When the frequency band was limited to 2-12 GHz, the dynamic range of variable attenuation increased to 16-17 dB, as shown by curve d of Fig. 6.

These performances are already as good as or superior to attenuators of the PIN diode/Lange coupler combination in bandwidth, size, insertion loss, and dynamic range of attenuation. There is, however, some discrepancy between designed values and actual experimental data, such as higher insertion loss and smaller dynamic range. This discrepancy was found to be due to the difference in the capacitance values between drain and source terminals.

The high-power version showed similar small-signal performance except that it could take the input power as high as 600 mW and did not show any degradation (Fig. 7).

Conclusion

GaAs monolithic variable attenuators have been developed. By using a control circuit, a single control voltage could vary the attenuation electrically from 2 dB to 12 dB in the frequency region of 2-18 GHz. This dynamic range of attenuation was as large as 17 dB in the 2-12 GHz band.

Dual gate FETs were used for the high power version of the attenuator. The performance did not degrade up to 600 mW input power.

Design of Interdigitated Capacitors and Their Application to Gallium Arsenide Monolithic Filters

REZA ESFANDIARI, MEMBER, IEEE, DOUGLAS W. MAKI, AND MARIO SIRACUSA

Abstract —Theoretical expressions for the interelectrode capacitance and conductor losses for an array of microstrip transmission lines are presented. The effect of finite conductor thickness is included in the analysis by introducing equations for the effective width of the transmission lines. Good agreement between theory and experiment is observed up to 18 GHz. Experimental results obtained from a lumped-element GaAs monolithic bandpass filter are in excellent agreement with theory. The filter has 1.5-dB insertion loss at 11.95 GHz and greater than 22-dB loss in the stopband. The filter measures $0.58 \times 1.3 \times 0.203$ mm.

I. INTRODUCTION

A MAJOR TASK in the development of gallium arsenide (GaAs) monolithic circuits is the careful analysis and design of lumped-element microwave components. This paper presents design considerations and experimental results for lumped-element interdigital capacitors fabricated on GaAs semi-insulating substrates. The lumped capacitor is formed by the fringing field of an interdigital gap between fingers. Because of size limitations, the capacitance values obtained are typically less than 1 pF.

Interdigital capacitors have proven to be useful components in GaAs monolithic integrated circuits due to their simplicity of construction, relatively high Q, and repeatability. Their use affords a considerable size reduction when compared with equivalent distributed matching structures and they are higher yield, lower loss, and more repeatable than overlay capacitors.

In Section II, an analysis of interdigital capacitors is presented. In previous works [1], [2], the authors have used single-strip microstrip losses in their analysis. This approach significantly underestimates the losses of the odd-mode component. In the present technique, the loss components α_e and α_o are calculated for an array of microstrip lines by applying the incremental inductance rule [3]. The effect of finite strip thickness is also considered in the analysis which gives results that agree well with experiment. In Section III, a derivation of loss factors is presented. From these formulas the loss of a microstrip line in an array of lines is obtained for a given conductor thickness. The conductor loss equations are presented in terms of the fringing and parallel-plate capacitance of the microstrip lines, which make it very convenient for use in a computer-aided design approach. To support the theoretical analysis, several interdigital capacitors have been designed and fabricated on GaAs semi-insulating substrates. The fabricated devices have been tested on a semiautomatic network analyzer over the 2- to 18-GHz frequency range. In Section IV, the experimental results are presented and compared to the theory. A monolithic two-pole bandpass filter is presented in Section V. Finally, Section VI summarizes the results.

II. THEORETICAL ANALYSIS

Consider the interdigital capacitor depicted in Fig. 1. For the purpose of analysis, let us disregard the fingers and consider their effects later in the calculation. The two terminal strips can be regarded as a pair of coupled microstrip transmission lines and a four-port admittance matrix may be calculated using the theory of coupled transmission lines in an inhomogeneous media [4]. In the case of a center-tapped capacitor, where neither end of the terminal strips is grounded, if one assumes ports 1, 2, 3, and 4 are open circuited, then the impedance matrix for each half section is given by

$$Z_{11}'' = Z_{22}'' = \frac{1}{2}\left[Z_{Te}\coth(\gamma_{Te}l_T/2) + Z_{To}\coth(\gamma_{To}l_T/2)\right] \tag{1}$$

$$Z_{21}'' = Z_{12}'' = \frac{1}{2}\left[Z_{Te}\coth(\gamma_{Te}l_T/2) - Z_{To}\coth(\gamma_{To}l_T/2)\right] \tag{2}$$

$$y_{T11}'' = 2y_{11}'' \tag{3a}$$

$$y_{T12}'' = 2y_{12}''. \tag{3b}$$

y_{T11}, y_{T12} are the elements of the matrix where $[Y_T] = [Z_T]^{-1}$ and l_T is the length of the terminal strip. For the end-tapped configuration where ports 2 and 4 are open circuited, the impedance matrix is

$$Z_{T11} = Z_{T23} = \frac{1}{2}\left[Z_{Te}\coth(\gamma_{Te}l_T) + Z_{To}\coth(\gamma_{To}l_T)\right] \tag{3}$$

$$Z_{T31} = Z_{T13} = \frac{1}{2}\left[Z_{Te}\operatorname{csch}(\gamma_{Te}l_T) - Z_{To}\operatorname{csch}(\gamma_{To}l_T)\right]. \tag{4}$$

Manuscript received April 29, 1982; revised August 2, 1982.
The authors are with the Hughes Aircraft Company, Torrance Research Center, Torrance, CA 90509.

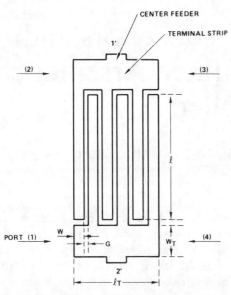

Fig. 1. Top view of an interdigital capacitor.

Now the task is to calculate the characteristic impedances Z_{Te}, Z_{To} and propagation constants γ_{Te}, γ_{To} for the terminal strip taking the effects of the fingers into consideration. We assume that the capacitor dimensions are much less than a quarter of a wavelength, and the fingers can be represented by an effective distributed shunt admittance across the terminal strip. Therefore, at a specified angular frequency ω, the characteristic admittance and propagation constant for the terminal strip are given by

$$Z_{Te} = \sqrt{\frac{R_T + j\omega L_{Te}}{j\omega C_{Te} + N_F(y_{11} + y_{21})/2l_T}} \quad (4)$$

$$Z_{To} = \sqrt{\frac{R_T + j\omega L_{To}}{j\omega C_{Te} + N_F(y_{11} - y_{21})/2l_T}} \quad (5)$$

$$\gamma_{Te} = \sqrt{(R_T + j\omega L_{Te})[j\omega C_{Te} + N_F(y_{11} + y_{21})/2l_T]} \quad (6)$$

$$\gamma_{To} = \sqrt{(R_T + j\omega L_{To})[j\omega C_{To} + N_F(y_{11} - y_{21})/2l_T]} \quad (7)$$

where C_{Te}, C_{To}, L_{Te}, and L_{To} represent the even- and odd-mode capacitances and inductances for the terminal strip, R_T is the resistance of the conductors, and N_F is the number of fingers. y_{11} and y_{21} are the elements of the admittance matrix for $N_F/2$ interdigital sections in parallel. These admittances are averaged over the terminal strip length in the above formulas. The values of y_{11} and y_{21} can be calculated as follows: Assuming each two-finger pair forms a system of coupled transmission lines, open circuited at opposite ends, one can deduce the matrix y from the theory of coupled transmission lines in an inhomogeneous media [4]. Therefore

$$Z_{11} = Z_{22} = \frac{1}{2}[Z_{oe}\coth\gamma_e l + Z_{oo}\coth\gamma_o l] \quad (8)$$

$$Z_{21} = Z_{12} = \frac{1}{2}[Z_{oe}\csc\gamma_e l - Z_{oo}\csc\gamma_o l] \quad (9)$$

$$[Y] = [Z]^{-1} \quad (10)$$

where Z_{oo}, Z_{oe}, γ_o, and γ_e are the odd- and even-mode impedances and propagation constants of the fingers, and l is the length of overlap of the fingers. The impedances Z_{oo}, Z_{oe} are calculated by obtaining the total capacitance of finger. The total capacitance is the summation of parallel-plate and fringing capacitances. The fringing capacitance is computed by applying a theoretical technique given by Smith [5]. The fringing capacitance for each finger is calculated by considering the effect of the immediately adjacent fingers. The loss factors α_o and α_e for the propagation constant γ_o, γ_e are calculated in the following section.

III. Conductor Losses

The even- and odd-mode attenuation constant due to ohmic losses for coupled microstrip lines can be determined using the incremental inductance rule of Wheeler [3]. Hence

$$\alpha = \frac{R_s}{2Z_0\mu_0}\frac{\delta L}{\delta n} \quad (10)$$

where R_s is the surface resistivity, δL is the change in the inductance that occurs when all cross-sectional conductors are perturbed by a distance δn inward, normal to the conductor surface (Fig. 2). In this case, it is more convenient to write (10) in terms of capacitance

$$\alpha = \frac{R_s}{2Z_0\eta c C^{a2}}\frac{\delta C^a}{\delta n} \quad (11)$$

where η is the free-space impedance, c is the velocity of light, and C^a is the total capacitance of microstrip with air as dielectric. C^a is a function of h, the distance between conductors and ground plane, G the gap spacing between the neighboring conductors, and t and w the thickness and width of the conductor, respectively (see Fig. 2). By considering to first order the variation in capacitance that occurs due to an inward normal perturbation δn at each surface, there results

$$\delta h = \delta G = +2\,\delta n$$
$$\delta w = \delta t = -2\,\delta n.$$

The variation in C^a is

$$\delta C^a = \frac{\partial C^a}{\partial h}\delta h + \frac{\partial C^a}{\partial t}\delta t + \frac{\partial C^a}{\partial w}\delta w + \frac{\partial C^a}{\partial G}\delta G \quad (12)$$

and furthermore

$$C^a = C_p^a + C_f^a$$

where $C_p^a = \epsilon_0 w/h$ is the total parallel-plate capacitance in picofarads per centimeter and C_f is the fringing capacitance. Then α_o for the odd mode can be written as

$$\alpha_o = \frac{R_s}{Z_{00}\eta c C^{a2}}\left[\frac{\epsilon_0 w}{h^2} + \frac{\epsilon_0}{h}\left(1 + \frac{\partial w}{\partial t}\right) - \frac{\partial C_{f0}^a}{\partial h}\right.$$
$$\left. - \frac{\partial C_{f0}^a}{\partial G} + \frac{\partial C_{f0}^a}{\partial w}\left(1 + \frac{\partial w}{\partial t}\right)\right]. \quad (13)$$

A similar expression holds for the even-mode attenuation constant.

When the microstrip conductor is of finite thickness t, the impedance can be evaluated by using the concept of

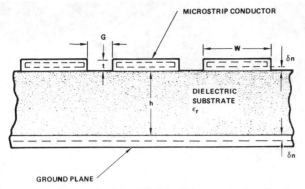

Fig. 2. An array of coupled microstrip line cross section showing an inward perturbation of an all metal surface.

effective width to obtain $\partial w/\partial t$. An expression for effective width W^t has been obtained by Jansen [6] for coupled microstrip lines. Similar expressions have been found to work well for an array of microstrip lines

$$W_e^t = W + \Delta W [1 - \exp(-A)] \qquad (14)$$

$$W_o^t = W_e^t + \Delta t \qquad (15)$$

where

$$\Delta t = \frac{2h}{\epsilon_r} \frac{t}{G} \qquad (16)$$

$$A = 0.35 \frac{\Delta w}{\Delta t} \qquad (17)$$

and Δt is the increase in effective width for the odd mode when compared with the even mode. This increase Δt has been approximated by modeling the excess capacitance, over the $t = 0$ case, by parallel-plate capacitance. The constants have been determined by a fit to the measured results. The error in the above approximations increases for $G > t$.

An expression for Δw the effective increase in microstrip width of single microstrip transmission line due to strip thickness is given approximately by Wheeler and Hughes et al., [3], [7], [8], and is repeated below

$$\Delta w = \frac{t}{\pi} \left(1 + \ln \frac{4\pi w}{t} \right), \qquad \frac{w}{h} \leq \frac{1}{2\pi}$$

$$\Delta w = \frac{t}{\pi} \left(1 + \ln \frac{2h}{t} \right), \qquad \frac{w}{h} \geq \frac{1}{2\pi}. \qquad (18)$$

The partial derivative $\partial w/\partial t$ is obtained by computing $\partial w^t/\partial t$ from (14) and (15). The result for the even mode is

$$\frac{\partial w^e}{\partial t} = \frac{\partial \Delta w}{\partial t} [1 - (1 + A)\exp(-A)] + \frac{2A^2 h}{\epsilon_r G} \exp(-A) \qquad (19)$$

and for the odd mode is

$$\frac{\partial w^o}{\partial t} = \frac{\partial w^e}{\partial t} + \frac{2h}{\epsilon_r G} \qquad (20)$$

where

$$\frac{\partial \Delta w}{\partial t} = \frac{1}{\pi} \ln \left(\frac{4\pi w}{t} \right), \qquad \frac{w}{h} \leq \frac{1}{2\pi}$$

$$= \frac{1}{\pi} \ln \left(\frac{2h}{t} \right), \qquad \frac{w}{h} \geq \frac{1}{2\pi}. \qquad (21)$$

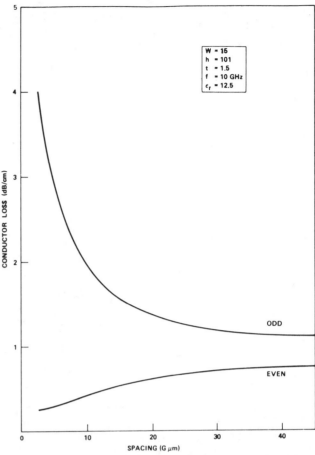

Fig. 3. Even- and odd-mode conductor losses in a microstrip array as a function of finger gap spacing.

Utilizing (19) and (20), the values of α_o and α_e are calculated from (13).

The derivatives in (13) were evaluated by applying a finite-difference approximation with the grid dimension taken to be several skin depths. In all of the following calculations, the resistivity of the conductor is assumed to be 3×10^{-6} $\Omega \cdot$cm and all dimensions are in micrometers. Results for a microstrip line in an array of microstrip lines are calculated by considering the effect of closest neighboring strip lines. Fig. 3 shows a plot of α_o and α_e as a function of conductor spacing for a microstrip in an array of microstrip lines. The conducting material is Cr-Au on a 0.1-mm-thick GaAs substrate. The even-mode attenuation constant is always less than the odd-mode value. The even-mode losses approach those of an infinite conducting sheet above a ground plane as the gap spacing approaches zero. Both even- and odd-mode losses approach those of a single isolated microstrip line for large spacing. Dielectric losses are neglected in these calculations since the loss tangent of semi-insulating GaAs is $2-3 \times 10^{-4}$ in the 2-12-GHz frequency range, which gives a dielectric loss of less than 0.015 dB/in. The variation of conductor attenuation for an array of lines as a function of finger width is shown in Fig. 4. Note that for finger width less than 10 μm, the odd-mode conductor losses increase sharply. The even-mode losses stay approximately constant for finger widths above 15 μm. This value is approximately equal to the loss of an infinite conducting sheet above a ground plane. For

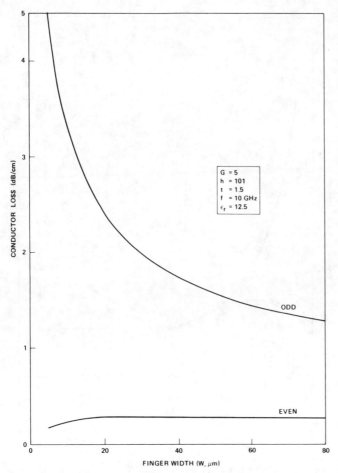

Fig. 4. Even- and odd-mode conductor losses in a microstrip array as a function of strip width.

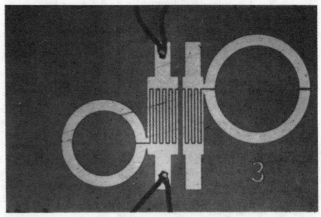

Fig. 5. A resonance test structure for interdigital capacitor testing.

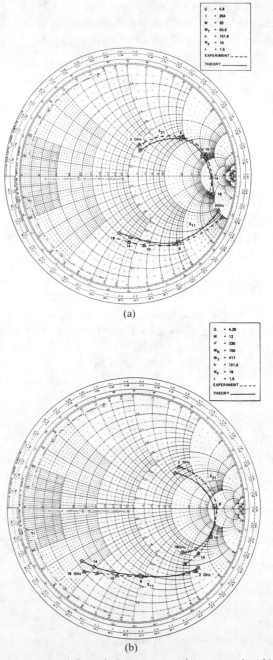

Fig. 6. Measured and theoretical S-parameters for two samples of interdigital capacitors. (a) CAP 1 and (b) CAP 2.

small finger width ($w < 10$ μm) even-mode losses decrease due to a decrease in even-mode capacitance. Also Fig. 3 indicates that conductor spacings of less than 5 μm have a substantial effect on odd-mode losses. Although larger finger width and wider spacing reduce the overall conductor losses, it increases the parasitic capacitance and increases the physical size of the device.

IV. THEORETICAL AND EXPERIMENTAL RESULTS

Several interdigital capacitors of different values were designed and processed on semi-insulating GaAs substrates using high-resolution photolithographic techniques. Several of these capacitors were resonated with a high-impedance microstrip line to facilitate high-frequency Q measurement. The S-parameters of the devices were measured using an HP 8409A semiautomatic network analyzer over the 2–18-GHz frequency range. Fig. 5 shows a picture of one of the devices assembled on a test fixture. Equations (1)–(10) with the aid of (13) were used to calculate the Y-parameters for a number of interdigital capacitors. The Y-parameters were then converted to S-parameters for comparison with experimental results. Fig. 6(a) and (b) shows the calculated and measured S-parameters for two interdigital capacitors which will be designated by CAP 1 and CAP 2, respectively. The dimensions are given on the figure. The parasitic inductances which are associated with the bond-

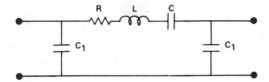

Fig. 7. Equivalent circuit model for interdigital capacitors.

TABLE I
EXPERIMENTAL AND THEORETICAL ELEMENTS VALUES FOR THE CIRCUIT MODEL OF FIG. 7 FOR TWO SAMPLES OF INTERDIGITAL CAPACITOR

CAP 1	R (Ω)	L (nH)	C (pF)	C_1 (pF)
1*			0.248	0.092
2	1	0.065	0.247	0.075
3	1.122	0.167	0.242	0.073
CAP 2				
1			0.609	0.100
2	1.019	0.076	0.607	0.149
3	1.062	0.173	0.590	0.177

*1—Calculated static capacitance. 2—From measured S-parameters. 3—From coupled line theory.

ing wires are extracted from the measured S-parameters data presented in Fig. 6(a) and (b). The measured S-parameters were fitted using a least squares error technique to the circuit model presented in Fig. 7. Fig. 6(a) and (b) indicates good correlation between experiment and theory. Table I presents the element values of the device equivalent circuit (Fig. 7) obtained by data-fitting techniques from the measured and calculated S-parameters over the frequency range of 2–18 GHz. These data are presented for the capacitors CAP 1 and CAP 2 for which the S-parameters are given in Fig. 6(a) and (b). The static capacitance is also presented. The values of static gap capacitance for both a 4- and an 8-mil-thick substrates with various line separation and finger widths are plotted in Fig. 8(a) and (b). These curves are obtained from coupled line theory (5) where calculation of even- and odd-mode fringing capacitances for coupled lines is made for finite line widths assuming a periodic array of lines. The effect of thickness is also considered in Fig. 8. The thickness effect makes a substantial difference for small finger spacing. From Fig. 8, the capacitance C can simply be calculated by

$$C = (N_F - 1)Cg \cdot l.$$

These plots are helpful for designing interdigital capacitors with optimum aspect ratios which will be discussed in the next section. The value of capacitance C_1 in Fig. 7 is the summation of parallel-plate capacitance and the even-mode fringing capacitance to ground. The Q of the interdigital capacitors was measured for several devices. The unloaded Q obtained by a resonance technique is between 35 and 45 for the combined LC microstrip circuit at 12–14 GHz. One can extract the measured Qc for the interdigital capacitor if an RLC circuit model is assumed for the combined resonance circuit shown in Fig. 5. By use

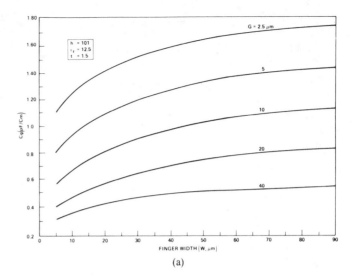

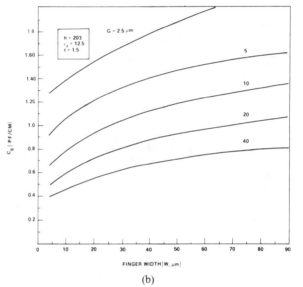

Fig. 8. Static gap capacitance values for different finger widths and gaps on GaAs semi-insulating substrates. (a) $h = 101$ μm and (b) $h = 203$ μm.

of this technique, Qc's in excess of 60 at 8 GHz were obtained for some of the devices which were tested.

Another technique for obtaining Q is from the circuit model of Fig. 7. This can be done after the measured and calculated S-parameters are fitted to the circuit model of Fig. 7. Fig. 9 presents the Q-values verses of frequency for an interdigital capacitor (CAP 2) obtained using such a technique. This method gives a better agreement with the theory which is also plotted in Fig. 9. Fig. 10 shows the effect of conductor thickness on capacitance values. For this particular example there is about an 11-percent difference between the capacitance at zero thickness and for a 2-μm-thick conductor. A design knowledge of the minimum acceptable thickness of strip would be an advantage in the fabrication of microstrip circuits, in that, the thinner the strip used, the less undercutting of line would be experienced during etching. Also, it would be easier to maintain the integrity of device dimensions and less costly. Horton et al. [9] and later Rizzoli [10] have shown theoreti-

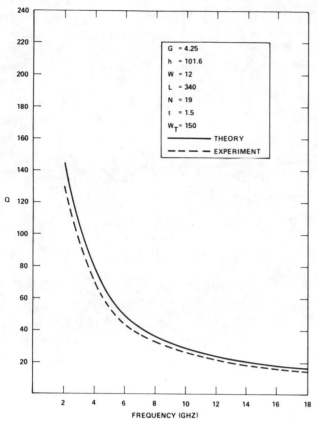

Fig. 9. Variation of Q as function of frequency for a 0.602-pF capacitance.

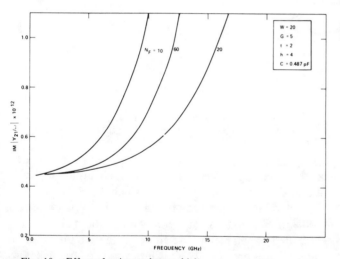

Fig. 10. Effect of strip conductor thickness on capacitance values.

cally that losses increase as $t \to 0$ for single microstrip lines and an absolute minimum exists for $t = \delta$, where δ is the skin depth. They argue that there exists an optimum strip thickness of approximately three times the skin depth for lowest loss. For a gold conductor this thickness is about 2.4 μm at 10 GHz. Equations (13)–(21) show a sharp increase for both even- and odd-mode losses for conductor thicknesses below 2 μm. An extensive experimental study on optimum conductor thickness has not been done but Q-measurements for two different metallization thicknesses

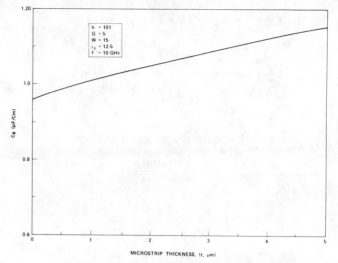

Fig. 11. Variation of the imaginary part of Y_{21} as a function of frequency for a 0.487-pF capacitor for various number of fingers.

were performed. A reduction in Q_u of more than 50 percent was measured when metallization thickness was reduced from 1.5 to 0.7 μm. The effect of the metal thickness on losses was also observed on the insertion-loss measurement done on the bandpass filter which will be discussed in the next section. Another important consideration is the dependence of capacitance values on frequency. The theoretical calculation indicates that the value of capacitance varies with frequency due to the change in finger length l and terminal strip length w_T. The optimum ratio of w_T/l for a specific capacitance corresponds to a minimum in $\partial C/\partial f$, where f is frequency. For series-connected interdigital capacitors, the capacitance slope is proportional to the slope of the imaginary part of y_{21}/ω. Fig. 11 is a plot of IM(y_{21}/ω) versus frequency for a 0.487-pF capacitance for different w_T/l ratios. It is apparent that a minimum slope exists for various aspect ratios. The search for minimum slope can be done efficiently with a computer-assisted design procedure.

V. Circuit Application

From the above analysis, the following design criteria for interdigitated capacitors on 0.202-mm GaAs substrates have been formulated. For minimal area, select capacitance values of less than 1.0 pF. To reduce losses, choose finger widths larger than or equal to 10 μm and gaps not less than 5 μm. Finally, for minimum dependence of capacitance value on frequency, choose the optimum aspect ratio w_T/l. Based on these criteria, a two-pole 0.5-dB ripple Chebyschev bandpass filter with a 5-percent bandwidth centered at 11.95 GHz was designed and fabricated on a semi-insulating GaAs substrate. Fig. 12(a) and (b) shows the initial circuit and the circuit model used for optimizing the final device layout. Fig. 12(b) shows the initial circuit plus all the parasitics from the lumped-element interdigitated capacitor, physical layout, and connections of each component. The capacitance values were kept to less than 1

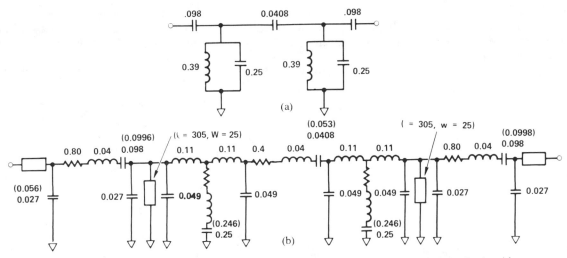

Fig. 12. (a) Equivalent circuit of an X-band two-pole bandpass filter. (b) Equivalent circuit with parasitics. Values in parentheses are from measured S-parameters. Capacitance and inductance are in picofarads and microhenrys, the dimensions are in micrometers.

pF by applying standard filter design procedures [11]. Before fabricating the final filter, three steps in optimizing the circuit were performed. Since no circuit tuning or tweaking is practical in GaAs monolithic integrated circuits, it is necessary to perform most of the circuit tuning by computer-aided analysis and optimization. The following optimization procedure proved to be effective for the present filter. First, the initial circuit (Fig. 12(a)) was optimized over the frequency band of interest. Next, we included all the parasitics due to interdigital capacitors which could be approximated by the above theory. After inclusion of all of these parasitics the circuit was optimized again. This procedure will adjust the most sensitive parameters to compensate for the effect of the parasitics. The third and final optimization or parameter adjustment comes after the physical layout of the device. Layout parasitics arise, for example, due to extension of sections of microstrip lines for the purpose of physically connecting components. The filter was fabricated using direct-write electron-beam lithography. The capacitors use 15-μm fingers and 5-μm gaps and are formed in 1.5-μm-thick gold on a 0.202-mm semi-insulating GaAs substrate. The filter size is 0.58 × 1.32 mm. This is an order of magnitude smaller in area than a distributed version of the same filter which was also fabricated. Fig. 13 shows an enlarged picture of the fabricated monolithic filter on GaAs semi-insulating substrate. Fig. 14 shows the response of this filter plotted directly from the output of a semiautomatic network analyzer. The in-band insertion loss is 1.5 dB and the image rejection at 9.5 GHz is greater than 22 dB. The losses were reduced about 0.5 dB by chemically polishing the back of the substrate using a sodium hypochlorite solution before depositing the back metallization. Also we have achieved about 0.5-dB reduction in losses by using thicker plated metallization (about 2.5 μm) over the microstrip portion of the filter. The filter response in Fig. 14 is achieved after polishing and plating thicker gold.

Fig. 13. An X-band 2-pole lumped-element GaAs monolithic bandpass filter.

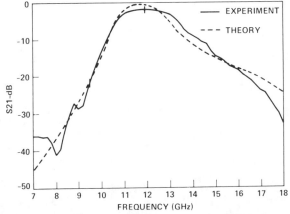

Fig. 14. Measured and theoretical insertion loss of the filter from 2–18 GHz.

VI. Conclusion and Summary

Theoretical expressions for the capacitance and conductor loss for an array of microstrip lines have been presented in a form suitable for interactive computer-aided design. Conductor losses have been calculated for an array of microstrip transmission lines using the incremental inductance rule. By including the conductor thickness in the analysis, a better correlation between theory and experiment can be achieved. To design an optimum interdigitated capacitor one should first select W and G for maximum Q-values and minimum parasitics. This selection can be aided with curves similar to Figs. 3, 4, and 8. If a minimum dependence of capacitance on frequency is important for an application, a search for minimum variation in capacitance slope may be easily performed. From these analyses, an iterated procedure for designing a monolithic filter on GaAs has been presented. Good correlation between theory and experiment suggests that the interdigital capacitors remain essentially lumped up to 18 GHz. The losses are highly dependent on gap width, finger width, and metallization thickness. Q-values are higher than those of MOM capacitors, however, not as high as reported in an earlier work [1]. Nevertheless, with careful design, reasonable Q's can be achieved for application in monolithic GaAs integrated circuits.

Acknowledgment

The authors wish to acknowledge the support of their colleagues in the Torrance Research Center, in particular L. Hackett for electron-beam fabrication, J. Schellenberg for critical discussions and RF measurements, and T. Midford for continuing support.

References

[1] G. D. Alley, "Interdigital capacitors and their application to lumped element microwave circuit," *IEEE Trans. Microwave Theory Tech.*, vol. MTT-18, pp. 1028–1033, Dec. 1970.
[2] J. L. Hobdell, "Optimization of interdigital capacitors," *IEEE Trans. Microwave Theory Tech.*, vol. MTT-27, pp. 788–791, Sept. 1979.
[3] H. A. Wheeler, "Transmission-line properties of a strip on a dielectric sheet on a plane," *IEEE Trans. Microwave Theory Tech.*, vol. MTT-25, pp. 631–641, Aug. 1977.
[4] G. I. Zysman and A. K. Johnson, "Coupled transmission lines networks in an inhomogeneous dielectric medium," *IEEE Trans. Microwave Theory Tech.*, vol. MTT-17, pp. 753–759, Oct. 1969.
[5] J. I. Smith, "The even- and odd-mode capacitance parameters for coupled lines in suspended substrate," *IEEE Trans. Microwave Theory Tech.*, vol. MTT-19, pp. 424–431, May 1971.
[6] R. H. Jansen, "High-speed computation of single and coupled microstrip parameters including dispersion, high order modes, loss and finite thickness," *IEEE Trans. Microwave Theory Tech.*, vol. MTT-26, pp. 75–82, Feb. 1978.
[7] H. A. Wheeler, "Formulas for skin effect," *Proc. IRE*, vol. 30, no. 9, pp. 412–424, Sept. 1942.
[8] M. Caulton, J. J. Hughes, and H. Sobol, "Measurement of the properties of microstrip transmission lines for microwave integrated circuit," *RCA Rev.*, vol. 27, no. 3, pp. 377–391, Sept. 1966.
[9] R. Horton, B. Easter, and A. Gopinath, "Variation of microstrip loss with thickness of strips," *Electron. Lett.*, vol. 7, no. 17, Aug. 1971.
[10] V. Rizzoli, "Losses in microstrip arrays," *Alta Freq.*, vol. XLVI, no. 2, pp. 86–94, Feb. 1975.
[11] G. L. Matthaei, L. Young, and E. M. T. Jones, *Microwave Filters, Impedance Matching Networks and Coupling Structure*. New York: McGraw-Hill, 1964.

Part IV
CAD, Measurement, and Packaging Techniques

AN intrinsic feature of the MMIC approach is that once a circuit is designed and fabricated, it cannot be "trimmed" or tuned for optimum performance in any convenient or cost-effective manner. Thus, the burden is passed to the designer to provide a design based on as complete a description of the actual circuit as is possible.

Essential to the achievement of this goal are the use of computer-aided design (CAD) techniques, accurate device and component models, and reliable measurement techniques for obtaining these accurate models. However, all of these measures are for naught if the predicted good circuit performance is compromised by a poorly designed package.

We cannot emphasize too strongly the central role played by CAD in the MMIC approach. Indeed, were it not for the availability of sophisticated CAD software, the extraordinary development of the MMIC field, as chronicled in this volume, would not have taken place, and this book would not exist.

The results of a CAD design procedure, however, are no more reliable than the quality of the input data, most of which is based on prior measurements of devices and circuits. These measurements, of course, must take into account all parasitics introduced by the package.

The following papers address these very important topics as they apply to good MMIC design practice.

COMPUTER-AIDED DESIGN OF HYBRID AND MONOLITHIC MICROWAVE INTEGRATED CIRCUITS - STATE OF THE ART, PROBLEMS AND TRENDS

Rolf H. Jansen *

ABSTRACT

A survey is given describing the status of computer-aided design techniques as they are used today in the development of microwave integrated circuits. Some still existing deficiencies and fundamental problems are outlined. Parallel to this, the special requirements for the design of monolithic microwave circuits are discussed, as well as the impact which comes from the growing availability of economic, decentralized computer power. Some considerations concerning future developments are presented and a few representative design examples will be shown.

INTRODUCTION

Thirty years after the introduction of stripline and microstrip as a microwave transmission and circuit medium computer-aided design, simulation and optimization techniques have finally found their way into most production-oriented industrial laboratories. In the competitive development of microwave integrated circuits (MICs), computer-aided design (CAD) has become a necessity due to the very restricted possibilities of tuning and trimming such circuits. In some respect and at least for hybrid MICs, microwave CAD can be said to have reached a certain state of maturity today /1/-/3/. However, as a closer look into the topic shall reveal, microwave CAD is just in a start position to reach a higher level of complexity and sophistication in the coming years.

For a description of the state of the art, first, a brief general sketch of the development of microwave CAD is given. The different overlapping methodical aspects of the CAD of MICs as visualized in Fig. 1 are considered separately, with stress put on characterization, modeling and analysis where many problems are still open today. Linkage between the different aspects is also shown in Fig. 1 which represents in some way a fictitious computer storage map. A first, fundamental point of view is the availability and generation of reliable and accurate design data. This field denoted characterization comprises the description of transmission lines, discontinuities, lumped and distributed passive, as well as active components. It includes the rigorous numerical methods and the computer-aided measurement techniques which today are employed to obtain such information in a systematic manner. A second aspect is the process of modeling by which available information on the behaviour of elementary components is transferred into mathematical mo-

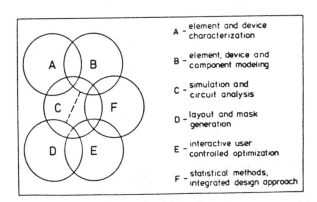

Fig. 1 Pictorial representation of the methodical aspects of microwave CAD and their linkage

* University of Duisburg, Dept. of El. Eng./ATE
Bismarckstr. 81, D-4100 Duisburg, W.-Germany

dels which are suited for CAD purposes. The validity and accuracy of existing models is discussed. A further important point concerns the methods of network analysis presently used for the simulation of MICs. It is outlined that one of the conceptual problems of the characterization of MICs is hidden here and is responsible for existing deficiencies, like for example model uncertaincies and design problems caused by spurious coupling. Additional items are the systematic variation of design parameters in CAD, i.e. optimization techniques, the automated generation of MIC mask layouts and the use of statistical methods in microwave CAD. These secondary level aspects of the computer-aided design of MICs are treated briefly from the view of a microwave engineer. They are not discussed very much in detail since, mathematically, they represent tools of relatively high maturity which can serve the needs of MIC design well. For a visualization of the state of the art some typical circuit examples emerging from the application of the described CAD techniques are shown.

The considerations presented here are made with emphasis on the realistic, layout-oriented design of MICs, i.e. with direct recourse to the relationship between layout geometry, physical construction and electrical behaviour. Also, the impact is considered which is presently inducing new developments in microwave CAD from two sides. One of these is the advent of truly monolithic MICs on gallium arsenide with improving yield figures, competitive technical specifications and with a steadily opening market /4/-/6/. The other one consists in a drastical reduction of the prices of computer power or, equivalently, the availability of high computational speed and large storage in economic, compact and decentralized configurations /7/. While the design and fabrication of monolithic MICs poses new problems to microwave CAD, the availability of cheep computer power will aid to overcome most of these. The special aspects coming along with the design of monolithic MICs are touched in this paper.

GENERAL SKETCH OF THE DEVELOPMENT OF MICROWAVE CAD

Most of the concepts which form the basis of todays microwave CAD have been developed and formulated already 8 - 15 years ago /8/-/39/, at least as linear circuits are concerned. This early phase of microwave CAD started concurrently with the introduction of useful microwave transistors and the associate explosive development of MIC technology by about the end of the 1960s. Only a few years before, Wheeler /40/ had published the first analytical formulas for the TEM characteristics of microstrip lines and, by this, provided the key equations to MIC design in the low-GHz region. Computer-aided network optimizaiton for low frequency applications is considerably older /41/ and has developed parallel to the appearance of electronic computers. For this reason, it was a natural way of proceeding of microwave CAD to modify and extend the existing low frequency concepts and to match them to the requirements of MIC design. The computer-aided design of MICs did not confront researchers with completely new problems, except in the field of element and device characterization. Nevertheless, microwave measurement and analysis is predominently performed in the frequency domain and in terms of scattering parameters. MIC topologies are comparatively simple, however, characterization is much more complicated than for low frequencies.

Under these byconditions, computer-aided MIC design developed in the years 1968 - 1974/75 with first attempts to combine the different methodical aspects into comprehensive CAD tools, however, mainly characterized by independent research in the separate branches. Already in this phase, the automatic network analyzer (ANA) and computer-corrected S-parameter measurement techniques became indispensable for element and device characterization. Resonance methods were employed extensively for the experimental investigation of elementary MIC structures and quasi-static numerical TEM-techniques were developed to provide further design data. In contrast to this, frequency dependent numerical hybrid mode techniques were just evolving at the end of the named period and were considered too laborious for the direct application in MIC design at that time. In the field of modeling, much progress had already been achieved by 1975 and a collection of analytical models was published in a microstrip handbook by Hammerstad and Bekkadal /39/. In addition, the magnetic wall waveguide model of microstrip was used to model microstrip discontinuities and junctions up to about 12 GHz. Lumped elements were employed in MIC design, however, their description was still relatively crude. For active devices, like bipolar and field-effect transistors, low-frequency small-signal equivalent circuits had been modified and extended into the microwave region, employing computer-matching on the base of measured S-parameters. Also, some results with sufficient usefulness for CAD were available for the modeling of nonlinear microwave behaviour.

For the methods of microwave circuit analysis used by about 1974 a first survey has been given by Monaco and Tiberio /32/. They include network representation in terms of S-parameters, admittances and impedances, with interconnections defined by topological matrices. Mixed formulations and the ABCD-matrix formalism were also used frequently. To speed up network algorithms, sparse matrix techniques were employed and first tendencies became visible in CAD programs, to avoid larger network matrices by the use of syntax-oriented step-by-step topology descriptions. The circuit elements implemented into the majority of analysis programs of that time were still highly idealized, and the consideration of parasitics in simulation and design was only in a rudimentary stage. On the other hand, computer-aided mask layout was available as part of several in-house design packages and was used for the automated production of microwave circuit masters. In addition, most of the larger CAD programs contained auxiliary graphical design aids for the interactive inspection of circuit functions.

As fas as the mathematical tool of optimization is concerned, reviews made by Bandler /9/,/16/ and a later one by Charalambous /33/ in 1974 show, that a high level of sophistication had already been achieved. The general concept of least pth optimization had been developed and many efforts had been invested to increase the efficiency of optimization algorithms. Both, direct search methods and conjugate-gradient algorithms were used in CAD packages. However, it appears that the high theoretical level achieved in optimization techniques was not the standard implemented in general purpose microwave CAD programs. The same prevailed for the use of statistical methods in the CAD packages of that time. The concept of wave sensitivities was available, including that of large-change sensitivities. Optimization considering tolerance assignment and final dircuit alignment had been reported in some cases. However, in most cases sensitivity, worst case and Monte Carlo analysis were used as isolated supplementary means of CAD, not integrated into the design algorithm.

In the years following 1974/75, the situation of microwave CAD has not changed fundamentally except in a few specialized fields. Most of the progress achieved in this second phase of CAD is of gradual character, either with respect to the extension to higher frequencies, to better design accuracies or to improved interaction between formerly separated design aspects and increased CAD program complexity. It has particularly been spurred since the first gallium-arsenide (GaAs) microwave field-effect transistors (FETs) became commercially available and opened up the region beyond 12 GHz for a broad range of MIC design applications /42/-/45/. Only recently the revival of monolithic microwave integrated circuits (MMICs) has further increased the need for improved CAD tools and is now likely to open a third phase of development. References /1/-/6/ give a good survey on the concurrent progress of CAD and FET technology up to about the year 1980/81. In this paper, particularly those items will be discussed and referenced which are considered characteristic or of fundamental importance for the present and future status of microwave CAD.

ELEMENT AND DEVICE CHARACTERIZATION

Accurate and reliable characterization is one of the fundamental prerequisits of microwave component modeling and CAD. The degree of accuracy to which the electrical performance of MICs can be predicted depends predominantly on the results coming from this field, though statistical design approaches may alleviate the requirements to some degree. The transmission and circuit media used for the realization of MICs and MMICs determine which kinds of elements and devices have to be characterized. Among these media, microstrip plays and will play a decisive role due to its quasi-TEM properties and its suitability for not only small scale integration. Consequently, most research efforts aimed at characterization for CAD purposes have been reported in the microstrip area and for the typical devices used therein. A considerable amount of results is available for the slot- and fin-line medium too, whereas coplanar and suspended substrate CAD techniques are still in an embryonic state. So, the discussion mainly has to concentrate to the microstrip area and Fig. 2 gives an according survey over the contents of what comprises the term characterization here.

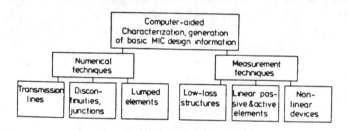

Fig. 2 Techniques and objects of MIC element and device characterization

Today, in contrast to the situation of 1974/75, efficient, user oriented computer program packages are available for the accurate frequency-dependent hybrid mode characterization of MIC transmission lines of all relevant types of cross-sectional geometry /46/-/52/ and, finally, are used in everyday design by many microwave engineers. With such computer programs, frequency dispersion, modal characteristic impedances and line loss can be calculated to any practical required degree of accuracy even on minicomputers. The numerical technique employed in most cases and with particular efficiency is Galerkin's method in the spectral domain in conjunction with properly chosen expansion functions. On the whole, the characterization problem can be considered solved for MIC transmission lines. Some work may still be necessary to account for the effects of finite metallization thickness in a better way, since these are not negligible in a variety of monolithic MIC structures.

The situation is different if the rigorous frequency dependent characterization of discontinuities and junctions is concerned which are always present in hybrid and monolithic MICs. The numerical treatment of these basic circuit constituents requires the solution of three-dimensional hybrid-mode boundary-value problems which even on modern computers can be time consuming and costly. Up to about 1980, the available results were either restricted to the static and stationary limit of capacitive and inductive behaviour, respectively, or too confined and laborious to be of great usefulness for microwave CAD /1/,/2/, except for some work specialized to transverse fin-line discontinuities /53/. Only recently, the first hybrid mode approaches to the general case of MIC discontinuities and junctions were published providing a larger amount of frequency dependent design data /54/-/57/. Much remains to be done in this field, however, if the improvement of numerical methods and the development of computers is extrapolated into the near future we can expect a solution to this problem within the 1980s. Such progress would be of particular importance for a predictable design of monolithic MICs.

The rigorous numerical characterization of MIC lumped element structures in terms of frequency is a very similar item. It is difficult due to the comparatively complicated geometries of lumped elements. Accurate characterization of these, therefore, is presently performed predominantly by successively improved approximate descriptions in conjunction with experimental data, see for example ref. /58/.

Computer-aided measurement techniques for MIC element and device characterization have become standard tools of microwave CAD. They are needed whenever rigorous and reliable theoretical descriptions do not exist, are too laborious or not applicable because the physical input parameters are unknown. For lossless, small microstrip structures, resonance measurements have been used extensively and are described in a recent overview by Edwards /59/. Similar approaches applied to finline discontinuities have been reported by Hoefer and coworkers /60/. These techniques use the fact that frequency can be measured very accurately even in the microwave and mm-wave region. On the other hand, they require great care in the deembedding of discontinuity data from the measurement results. Several improvements regarding the accuracy, generality and practicability of MIC resonance measurement methods have been achieved very recently /61/-/64/. Due to the relatively large substrate areas required, these methods are not well suited for monolithic MIC element characterization, except at mm-wave frequencies.

As the working horse in general linear MIC characterization, today the two-port automatic network analyzer is used together with accuracy enhancement software. Inspection of the technical literature and industrial development laboratories shows, that some standardization has already been achieved if S-parameter measurement with respect to coaxial reference planes (APC-7, APC-3.5, SMA) is concerned. Some of the papers which illuminate the state of the art and still existing difficulties are references /65/-/67/. The 6-term error model and the associated calibration procedure developed at Hewlett-Packard appear to be applied most frequently, the through-short-delay method is also used often. For calibration and error correction with respect to microstrip and other MIC reference planes, in-house computer programs are used exclusively. The same prevails for MIC test fixtures, for which standardized versions would be desirable but are not yet available commercially. At least, some progress has been made very recently by the development of a fairly universal transistor test fixture /68/. Also, the first broadband solutions to the construction of measurement probes for monolithic elements and devices have been published /69/. The characterization of nonlinear devices employing the ANA is feasible /70/-/72/ and offers attractive conceptual advantages for CAD.

MATHEMATICAL MODELING OF ELEMENTS, DEVICES AND COMPONENTS

Whereas the term characterization refers to the process of procuring reliable design information, modeling is concerned with the transformation of this into forms which are compatible with CAD requirements. The border line between these two aspects of microwave CAD cannot be drawn accurately, since any kind of description contains idealizations to some degree and is therefore a model. However, with the present status of computers, the distinction is characteristic, because most of the methods used for characterization are too laborious to be implemented into MIC optimization algorithms directly where mathematically brief and fast descriptions are needed. Fig. 3 shows a possible classification scheme for the kind of models which today are employed in microwave CAD.

Models characterized by explicit analytical expressions, or a straightforward sequence of these are denoted analytical. In contrast to this, numerical models typically are those for which differential equations or idealized field problems have to be solved by some algorithm in order to obtain the electrical response. The easiest and fastest way to transfer basic design information into a CAD program is by interpolation from suitable data files. This is frequently done in conjunction with ANA measurement systems, transistor data banks or computed tables when other descriptions are not suitable. Interpolation from component data files is a useful and unavoidable feature of modern general purpose CAD packages /1/,/3/, however, it has some disadvantages compared to explicit formulas and becomes impractical for more than a few variables.

Equivalent circuit models with elements described in explicit form as functions of the relevant design parameters play the most important role in the present status of computer-aided MIC design. The structure of such models is usually formed by qualitative physical considerations.

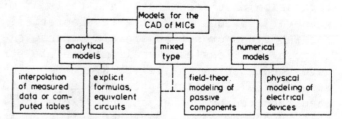

Fig. 3 Classification scheme for microwave CAD models

To obtain good accuracy over a sufficient wide range of applicability it is common practice to solve the modeling problem by multidimensional computer-assisted curve fitting procedures. This is done on the basis of the results of extensive theoretical and/or experimental characterization of the element or device under investigation. Despite computer assistance, the generation of accurate explicit formulas is a laborious and cumbersome procedure in many cases, still depending much on the microwave engineer's intuition. Nevertheless, some progress has been reported in the last years for the modeling of MIC transmission lines, discontinuities and junctions, particularly for the microstrip medium /63/, /64/,/73/-/79/ and also for fin-lines /60/,/80/,/81/. Explicit, very accurate formulas for single and coupled microstrip characteristics are now available with validity into the mm-wave region. Fairly accurate equations have been published for the elementary discontinuities like open end, gap, bend and impedance step. To the author's experience, the description of the symmetric T-junction still has to be improved for higher frequencies. Also, for the nonsymmetrical T-junction and both types of the double T-junction improvement is necessary due to the frequent application of these structures in MIC and MMIC design. Discrete linear components like capacitor and resistor chips can be modeled by standard procedures from measured data /63/. In a similar way the linear modeling of bipolar and single-gate field-effect transistors has emerged to a routine procedure performed by computer-optimization for the elements of predescribed small-signal equivalent circuits /63/,/82/-/85/. Such models can, for example, be used for the extrapolation of FET behaviour into the mm-wave region /86/. For the dual-gate FET, design-oriented refined linear circuit models are also available now /87/,/88/. Finally, as a result of recent MMIC activities, near analytical circuit models of nonlinear GaAs FET performance have been elaborated for use in CAD /89/,/90/.

Accurate, wide range modeling of microwave lumped passive elements, like interdigital and overlay capacitors as well as spiral inductors by explicit formulas still presents some problems. It is of particular technical importance since interest in lumped element techniques has strongly grown with the latest developments of monolithic MICs /91/. At the same time, this is particularly difficult due to the simultaneous effects of complicated geometry, finite thickness metallization, skin losses and coupling which are present in lumped circuit elements. Typical accuracies achieved with analytical formulas are in the order of 10 % compared to experimental results /5/. It is common practice to produce empirical formulas from measurements for a restricted range of geometries in order to describe parasitic effects and to better predict the first resonance frequency of an element. Another, more general approach which leads to better results consists in the use of multiple coupled line parameters together with empirical corrections /58/,/92/-/94/. However, the models created in this way are not truly analytical ones since the required line parameters have still to be computed numerically. Therefore, there is clearly need for further improvement and the generation of truly CAD-compatible analytical models for microwave lumped elements.

Where explicit formulas of sufficient quality do not exist, or where a more direct relationship to physical parameters is desired, as in the case of semiconductor devices, numerical modeling might be useful even if it often does not result in fast models of good CAD-applicability. An important, representative example for this is the derivation of approximate design information using the planar circuit or planar waveguide concept which dates back to 1972 /95/-/99/. By empirical adjustment of the relevant effective model parameters, good modeling accuracies have been demonstrated with this approach in the frequency region up to 12 GHz. For this reason, it is still used today for the analysis and the design of direct coupled filter and coupler configurations. Due to its construction, the planar waveguide model is not suited for the treatment of structures with multiply connected metallization. Numerical modeling is also frequently applied to the simulation of active microwave devices on the basis of physical and material parameters /100/-/101/. Usually, this involves the iterative solution of the transport equations of the device and is too laborious for direct use in CAD. Nevertheless, it is very useful for optimizing device performance technologically via insight into the physical dependencies.

SIMULATION AND CIRCUIT ANALYSIS

Most of the techniques which today are commonly used for analyzing microwave integrated circuits have already been discussed by Monaco and Tiberio in 1974 /32/. The discussion be restricted to scattering and/or transfer scattering parameters, since this applies to lumped elements and distributed components likewise, has the advantage of generality and conforms to the best state of the art. With this presumption, MIC analysis only seems to be a matter of processing suitably defined multiport scattering and topological matrices in order to evaluate circuit performance with respect to the exterior ports. Beyond this conventional method, syntax-oriented topology descriptions have been developed which do not require the manipulation of large matrices. They root back to the subnetwork growth algorithm proposed by Monaco and Tiberio in 1970 /15/. Some of them utilize the fact that a high percentage of the interconnections in MICs can be realized by cascading two-ports. Generalization is achieved by modification of the analysis algorithm to provide branching and three-port capability /23/,/102/. In a study performed under guidance by the author, the feasibility of a general syntax-oriented network reduction scheme based on S-parameters has been demonstrated and its computational efficiency has been proved /103/. A similar interconnection scheme has just been introduced into SUPERCOMPACT, a commercially available CAD program /3/. Since intelligent computer-aided MIC design is always an interactive procedure, syntax-oriented analysis algorithms will probably be preferred in the future. They seem best suited for the interaction between designer and computer. In addition, they are conformal with the trend to install CAD tools on economic, smaller computers due to their efficiency and low storage requirements.

Behind this, a fundamental, still unsolved problem of MIC analysis remains. It refers to the fact that any kind of network representation applied to MIC configurations can only be an approximation, even in the purely passive case. Due to the hybrid mode and open character of MIC media like microstrip, this becomes worse with increasing frequency. Some of the associated problems are mirrored in recent discussions on the so-called frequency dependent characteristic impedance of microstrip /77/,/104/-/107/. With a design-oriented point of view, it turns out that the power-current definition of this quantity promises to give the best results if used in a TEM-type analysis. This applies to virtually all of today's CAD packages and is also in accordance with experimental results. In addition, it takes care of the fact that characterization of devices by computer-corrected ANA measurements contains the TEM assumption implicitly. Furthermore, the unified use of this definition could overcome the uncertainties existing with respect to conversion between equivalent circuits and S-Parameters in MIC design.

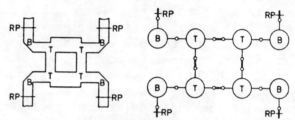

The best way to surmount the outlined fundamental difficulties appears to be the development of field-theoretical techniques which can be applied to MIC geometries more complicated than discontinuities and junctions. If, for example, a branch-line coupler like that of Fig. 4a has to be designed at high frequencies (15 GHz or more on standard alumina), spurious coupling and parasitics are so strong that the usefulness of a decomposition according to Fig. 4b is questionable.

Fig. 4 Illustration of the usual treatment of microstrip configurations in MIC analysis
(B = bend, T = T-junction, RP = reference plane)

That such a coupler can be built with practicable specifications is a matter of experience. However, for a treatment like that provided by existing CAD packages accurate performance simulation cannot be expected. The same applies for densely packed subareas in monolithic MICs. Therefore, only if the criteria of the classical waveguide concept are not invalidated, decomposition and subsequent TEM analysis can be expected to give reasonably good results.

Several available approximate approaches are at least able to take these considerations into account partially. Among them, the most general in terms of shape is the original planar circuit concept /95/,/97/ followed by the segmentation and desegmentation method /108/-/110/. Restricted geometries combined with a higher quality of approximation have been treated using the magnetic wall waveguide model of microstrip /96/,/99/,/111/,/112/ and also further improved models /113/-/116/. Especially, Sorrentino's and Pileri's work /116/ has demonstrated substantial improvement as it is able to predict experimentally verified effects which could not be described by former lossless theories. All these approximate approaches have in common that they can take into account not only the fundamental quasi-TEM mode but also higher order mode excitation and interaction. Their most severe limitation is that the modes used for expansion of the field do not always have physical justification and that coupling between dislocated structures cannot be described by them. Rigorous hybrid mode analyses up to the level of filter and matching structures have been reported for the fin-line medium /117/-/119/ and for specialized transverse geometries simpler than those commonly used in microstrip circuits. Full wave analyses for MIC structures of higher complexity than that of discontinuities and junctions have not yet been published, however, some trends for possible solutions have emerged already /120/,/54/-/57/. It is believed that the impact coming from MMIC design requirements and modern computer technology will speed up these developments in the coming years. This would not only enhance the quality of prediction in the design process but also increase design flexibility by the ability to characterize more complicated circuit structures.

NEW DEVELOPMENTS IN MIC DESIGN PHILOSOPHIES

Whereas early CAD approaches used highly idealized circuit elements and optimization in terms of electrical component values, modern MIC design programs take into account parasitics and discontinuities as far as possible /1/-/3/,/36/,/63/,/74/,/102/,/121/-/123/. There is a clear trend to optimize MIC designs in terms of geometrical layout parameters with a direct view to realizability and automated mask generation, for which novel solutions have also been shown recently /124/,/125/. From many papers in the field of monolithic MIC design, it becomes obvious that the larger microwave industry firms are developing inhouse CAD instruments of high complexity. It will be characteristic for coming CAD tools that they have a high level of integration of all the software and hardware aspects that have to be controlled for predictable and economic MIC design.

The same trend to higher design complexity is visible in a variety of CAD approaches which in addition include tolerance aspects /126/-/132/. The integrated approach to microwave design developed by Bandler et al. /126/ is a milestone in the CAD of MICs as it combines several formerly separated concepts into a single instrument. It is particularly aimed at realistic microwave design with a unified treatment of tolerances and model uncertainties. It involves parasitics, design centering, tolerancing and tuning as well as the problem of mismatch and uncertain reference planes. Appropriate modeling and approximation techniques have been proposed for statistical MIC design /128/ and the efficiency of computational methods for statistical design optimization has been further improved /128/,/129/,/132/,/133/. First systematic attempts to derive insensitive circuit topologies have been reported /5/,/134/. Statistical concepts like the few ones, that can be referenced here will become indispensable means if monolithic MIC mass production becomes reality. In restricted form and with dropping computer prices they may even be of interest for hybrid MIC manufacturing.

In the field of nonlinear microwave CAD, numerous results have been published for spealized circuit applications, however, only recently the first unified approaches to computer-aided nonlinear MIC design seem to be emerging /135/-/140/. They are formulated in the frequency domain which is an essential theoretical and practical bycondition of realistic microwave CAD. For everyday design problems, modifications of the harmonic balance method /136/,/137/,/139/,/140/ will probably have the best chance to become general purpose algorithms. Problems incorporating even strong nonlinearities have been treated and the application to class-C amplifiers, oscillators and frequency dividers has been demonstrated. A particularly elegant design approach by Lipparini et al. /140/ eliminates the need for a repeated complete analysis of the circuit by a special optimization strategy. In fact, this might even eliminate the now existing strict distinction between linear and nonlinear design in coming CAD packages.

REFERENCES

/1/   Gupta et al., Artech House, Dedham, Mass., 1981.
/2/   Edwards, J. Wiley and Sons, Chichester, 1981.
/3/   Besser et al., IEEE MTT-Symp. Dig., Los Angeles, 1981, 51-53.
/4/   Pucel, IEEE Trans., MTT-20, 1981, 513-534.
/5/   Pengelly, J. Wiley and Sons, Chichester, 1982, ch. 10.
/6/   Magarshack, Proc. 12th Eur. Microw. Conf., Helsinki, 1982, 5-15.
/7/   HP-Novum, Hewlett-Packard GmbH, Frankfurt, Nr. 3, 1982, 4-5.
/8/   Gelnovatch, Burke, IEEE Trans., MTT-16, 1968, 429-439.
/9/   Bandler, IEEE Trans., MTT-17, 1969, 533-552.
/10/ Bandler, Macdonald, IEEE Trans., MTT-17, 1969, 552-562.

/11/ Houston, Read, IEEE MTT-Symp. Dig., Dallas, 1969, 392-396.
/12/ Lump, IEEE Int. Conf. Rec., 1970, 320-321.
/13/ Gelnovatch, Chase, IEEE Journal, SC-5, 1970, 303-309.
/14/ Trick, Vlach, IEEE Trans., MTT-18, 1970, 541-547.
/15/ Monaco, Tiberio, Alta Frequenza, Vol. 39, 1970, 165-170.
/16/ Bandler, Proc. 2nd Eur. Microw. Conf., Stockholm, 1971, B 8/S, 1-8.
/17/ Wright, Gasoriek, Proc. 2nd Eur. Microw. Conf., 1971, B 8/2, 1-4.
/18/ Iuculano et al., Proc. 2nd Eur. Microw. Conf., 1971, B9/1, 1-4.
/19/ Andreasciani, De Leo, Proc. 2nd Eur. Microw. Conf., 1971, B9/3, 1-4.
/20/ Bandler, Seviora, IEEE Trans., MTT-20, 1972, 138-147.
/21/ Charalambous, Bandler, IEEE Trans., MTT-21, 1973, 815-818.
/22/ Judd, AGARD Conf. Proc., 130, Lyngby, 1973, 14/1-8.
/23/ Tauritz, AGARD Conf. Proc., 130, Lyngby, 1973, 15/1-23.
/24/ Toussaint, Hoffmann, AGARD Conf. Proc., 130, Lyngby, 1973, 12/1-14.
/25/ Altmäe, Proc. 3rd Eur. Microw. Conf., Brussels, 1973, A.12.2.
/26/ Bandler, Proc. 3rd Eur. Microw. Conf., Brussels, 1973, A.13.1.
/27/ Marazzi et al., Proc. 3rd Eur. Microw. Conf., Brussels, 1973, A.13.2.
/28/ Perlman, Gelnovatch, Advances in Microwaves, Vol. 8, Academic Press, New York, 1974, 321-399.
/29/ Howe, Jr., Proc. 4th Eur. Microw. Conf., Montreux, 1974, 301-307.
/30/ Bandler et al., IEEE MTT-Symp. Dig., Atlanta, 1974, 275-277.
/31/ Bennett, Fynn, Marconi Review, 37, 1974, 150-163.
/32/ Monaco, Tiberio, IEEE Trans., MTT-22, 1974, 249 - 263.
/33/ Charalambous, IEEE Trans., MTT-22, 1974, 289-300.
/34/ Sanchez-Sinencio, Trick, IEEE Trans., MTT-22, 1974, 309-316.
/35/ Marchent, Computer Aided Design, Vol. 7, 1975, 179-189.
/36/ Madsen, IEEE Trans., MTT-23, 1975, 803-809.
/37/ Charalambous, Conn, IEEE Trans., MTT-23, 1975, 834-838.
/38/ Rauscher, AGEN Comm., Zürich, 1975, 53-88.
/39/ Hammerstad, Bekkadal, Microstrip Handbook, ELAB Report, STF44 A74169, Univ. of Trondheim, Norway, 1975.
/40/ Wheeler, IEEE Trans., MTT-13, 1965, 172-185.
/41/ Temes, Calahan, Proc. IEEE Vol. 55, 1967, 1832-1863.
/42/ Liechti, Tillmann, IEEE Trans., MTT-22, 1974, 510-517.
/43/ Luxton, Proc. 4th Eur. Microw. Conf., Montreux, 1974, 92-96.
/44/ Pengelli, Proc. 5th Eur. Microw. Conf., Hamburg, 1975, 301-305.
/45/ Liechti, IEEE Trans., MTT-24, 1976, 279-300, 250 references.
/46/ Mittra, Itoh, Advances in Microwaves, Vol. 8, Academic Press, New York, 1974, 67-141.
/47/ Jansen, Proc. 7th Eur. Microw. Conf., Copenhagen, 1977, 135-139.
/48/ Jansen, IEEE Trans., MTT-26, 1978, 75-82.
/49/ Mirshekar-Syakhal, Davies, IEEE Trans., MTT-27, 1979, 694-699.
/50/ Jansen, Proc. IEE, MOA-3, 1979, 14-22.
/51/ Itoh, Trans. IEEE, MTT-28, 1980, 733-736.
/52/ Schulz, Pregla, Arch. Elektr. Übertr., AEÜ-34, 1980, 169-173.
/53/ El Hennawy, Schünemann, Proc. 9th Eur. Microw. Conf., 1979, 448-452.
/54/ Jansen, IEE Proc. Vol. 128, Pt. H, 1981, 77-86.
/55/ Jansen, Koster, Proc. 11th Eur. Microw. Conf., 1981, 682-687.
/56/ Koster, Jansen, IEEE Trans., MTT-30, 1982, 1273-1279.
/57/ Worm, Ph.D. Thesis, Fernuniversität Hagen, 1983.
/58/ Esfandiari et al., IEEE Trans., MTT-31, 1983, 57-64.
/59/ Edwards, IEEE MTT-Symp. Dig., Dallas, 1982, 338-341.
/60/ Pic, Hoefer, IEEE MTT-Symp. Dig., Los Angeles, 1981, 108-110.
/61/ Gruner, IEEE Trans., IM-30, 1981, 198-201.
/62/ Rizzoli, Lipparini, IEEE Trans., MTT-29, 1981, 655-660.
/63/ Jansen, Mikrowellen Magazin, Vol. 8, 1982, 433-437.
/64/ Kirschning et al., IEEE MTT-Symp. Dig., Boston, 1983, 495-497.
/65/ Da Silva, McPhun, Microwave Journal, June 1978, 97-100.

/66/ Speciale, Proc. 9th Eur. Microw. Conf., Brighton, 1979, 350-354.
/67/ Perlman et al., Microwave Journal, April 1982, 73-80.
/68/ Pollard, Lane, IEEE MTT-Symp. Dig., Boston, 1983, 498-500.
/69/ Strid, Gleason, IEEE Trans., MTT-30, 1982, 969-975.
/70/ Mazumder, van der Puije, IEEE Trans., MTT-26, 1978, 417-420.
/71/ Rauscher, Willing, IEEE Trans., MTT-27, 1979, 834-840.
/72/ Yang, Peterson, IEEE MTT-Symp. Dig., Dallas, 1982, 345-347.
/73/ Hammerstad, Jensen, IEEE MTT-Symp. Dig., Washington D.C., 1980, 407-409,
/74/ Hammerstad, IEEE MTT-Symp. Dig., Los Angeles, 1981, 54-56.
/75/ Kirschning et al., Electronics Letters, Vol. 17, 1981, 123-125.
/76/ Kirschning, Jansen, Electronics Letters, Vol. 18, 1982, 272-273.
/77/ Jansen, Kirschning, Arch. Elektr. Übertr., AEÜ-37, 1983, 108-112.
/78/ Kirschning et al., Electronics Letters, Vol. 19, 1983, 377-379.
/79/ Kirschning, Jansen, Accurate, wide range design equations for the frequency dependent characteristics of parallel coupled microstrip lines, IEEE Trans., MTT-31, 1983, subm. for publ.
/80/ Sharma, Hoefer, IEEE MTT-Symp. Dig., Los Angeles, 1981, 102-104.
/81/ Hoefer, Burton, IEEE MTT-Symp. Dig., Dallas, 1982, 311-313.
/82/ Hartmann, Strutt, IEEE Trans., MTT-22, 1974, 178-183.
/83/ Dawson, IEEE Trans., MTT-23, 1975, 499-501.
/84/ Vendelin, Omori, Microwaves, June 1975, 58-70.
/85/ Jansen, Koster, Arch. Elektr. Übertr., AEÜ-31, 1977, 475-477.
/86/ Watkins et al., IEEE MTT-Symp. Dig., Boston, 1983, 145-147.
/87/ Mau, IEEE MTT-Symp. Dig., Los Angeles, 1981, 43-45.
/88/ Tsironis, Meierer, IEEE Trans., MTT-30, 1982, 243-251.
/89/ Shur, IEEE Trans., ED-25, 1978, 612-618.
/90/ Madjar, Rosenbaum, IEEE Trans., MTT-29, 1981, 781-788.
/91/ Aitchison, Proc. 12th Eur. Microw. Conf., Helsinki, 1982, 37-46.
/92/ Camp et al., IEEE MTT-Symp. Dig., Boston, 1983, 46-49.
/93/ Duême et al., IEEE MTT-Symp. Dig., Boston, 1983, 65-68.
/94/ Cahana, IEEE MTT-Symp. Dig., Boston, 1983, 245-247.
/95/ Okoshi, Miyoshi, IEEE Trans., MTT-20, 1972, 245-252.
/96/ Wolff et al., Electronics Letters, Vol. 8, 1972, 177-179.
/97/ Silvester, IEEE Trans., MTT-21, 1973, 104-108.
/98/ Silvester, Csendes, IEEE Trans., MTT-22, 1974, 190-201.
/99/ Wolff, Menzel, Proc. 5th Eur. Microw. Conf., Hamburg, 1975, 263-266.
/100/ Shur, Eastman, IEEE Trans., ED-25, 1978, 606-611.
/101/ Moglestue, Comp. Methods in Applied Mechanics and Eng., Vol. 30, 1982, 173 - 208, plus references.
/102/ Jansen, Arch. Elektr. Übertr., AEÜ-32, 1978, 145-152.
/103/ Boukamp, M.E.E. Theses, University of Aachen, 1978.
/104/ Kuester et al., Proc. Int. URSI Symp., München, 1980, paper 335 B.
/105/ Jansen, Koster, IEEE MTT-Symp. Dig., Dallas, 1982, 305-307.
/106/ Getsinger, IEEE MTT-Symp. Dig., Dallas, 1982, 342-344.
/107/ Getsinger, Measurement and modeling of the apparent characteristic impedance of microstrip, IEEE Trans., MTT-31, 1983, to be publ.
/108/ Okoshi et al., IEEE Trans., MTT-24, 1976, 662-668.
/109/ Chadha, Gupta, IEEE Trans., MTT-29, 1981, 71-74.
/110/ Sharma, Gupta, IEEE Trans., MTT-29, 1981, 1094-1098.
/111/ Kompa, Mehran, Electronics Letters, Vol. 11, 1975, 459-460.
/112/ D'Inzeo et al., IEEE Trans., MTT-26, 1978, 462-471.
/113/ Jansen, Arch. Elektr. Übertr., AEÜ-30, 1976, 502-504.
/114/ Bonetti, Tissi, IEEE Trans., MTT-26, 1978, 471-477.
/115/ D'Inzeo et al., IEEE Trans., MTT-28, 1980, 1107-1113.
/116/ Sorrentino, Pileri, IEEE Trans., MTT-29, 1981, 942-948.
/117/ Arndt et al., Proc. 11th Eur. Microw. Conf., Amsterdam, 1981, 309-314.
/118/ Shih et al., IEEE MTT-Symp. Dig., Dallas, 1982, 471-473.

/119/ El Hennawy, Schünemann, IEE Proc., Vol. 129, Pt. H, 1982, 342-350.
/120/ Jansen, Arch. Elektr. Übertr., AEÜ-30, 1976, 71-79.
/121/ Hosseïni et al., Electronics Letters, Vol. 12, 1976, 190-192.
/122/ Kirchhoff, Frequenz, Vol. 34, 1980, 218-223.
/123/ Saviani, Giarola, IEEE MTT-Symp. Dig., Dallas, 1982, 462-464.
/124/ Dowling et al., IEEE MTT-Symp. Dig., Dallas, 1982, 465-467.
/125/ Childs, McGregor, IEEE MTT-Symp. Dig., Dallas, 1982, 468-470.
/126/ Bandler et al., IEEE Trans., MTT-24, 1976, 585-591.
/127/ Bandler et al., IEEE MTT-Symp. Dig., Ottawa, 1978, 79-81.
/128/ Bandler, Abdel-Malek, Proc. 7th Eur. Microw. Conf., Copenhagen, 1977, 153-157.
/129/ Glesner, Haubrichs, IEEE Conf. on Computer-aided Design of Electronic Circuits, Hull (England), 1977.
/130/ Tromp, IEEE Trans., MTT-26, 1978, 973-978.
/131/ Tromp, Hoffman, Revue HF (Belgium), Vol. 10, 1978, 50-65.
/132/ Jansen, Arch. Elektr. Übertr., AEÜ-33, 1979, 262-264.
/133/ Singhal, Pinel, IEEE Trans., CAS-28, 1981, 692-702.
/134/ Riddle, Trew, IEEE MTT-Symp. Dig., Boston, 1983, 521-523.
/135/ Chua, Ng, Electr. Circuits and Systems, Vol. 3, 1979, 165-185.
/136/ Filicori et al., IEEE Trans., MTT-27, 1979, 1043-1051.
/137/ Naldi et al., Proc. 10th Eur. Microw. Conf., Warsaw, 1980, 485-489.
/138/ Chua, Ushida, IEEE Trans., CAS-28, 1981, 953-971.
/139/ Rizzoli, Lipparini, IEEE MTT-Symp. Dig., Dallas, 1982, 453-455.
/140/ Lipparini et al., IEEE Trans., MTT-30, 1982, 1050-1058.

COMPUTER-AIDED DESIGN FOR THE 1980's

Les Besser, Charles Holmes, Mike Ball, Max Medley, and Steven March
Compact Engineering
1131 San Antonio Road
Palo Alto, California 94303

ABSTRACT

This paper describes a new, third generation computer program called SUPER-COMPACT, that opens a new era in automated microwave circuit design. Although the program is moderately large (approximately 50,000 FORTRAN statements), it can be run efficiently on midi-computers, as well as on most of the large-scale computer systems. The program combines analysis, optimization, and synthesis with interactive graphics for maximum user convenience and efficiency. Databanks, which provide scattering and noise parameters for transistors, dielectric information for substrate materials, and a library of available circuit topologies are integral to the program. While SUPER-COMPACT retains all of the previous capabilities of its first and second generation predecessors,[1,2,3] it uses a completely new approach to analyze and optimize microwave circuits. The program utilizes a novel interconnection scheme based upon scattering parameters without requiring the manipulation of large matrices. Both the inputting and outputting of data are handled through interactive graphic terminals.

Program Description

SUPER-COMPACT was created as the outcome of nearly ten years of experimentation to provide a single, general-purpose program to fulfill most needs of a microwave circuit designer. Several years of research, including a survey of about 5,000 designers, was spent to determine the input/output format and list of capabilities that satisfy the needs of the experts but be simple enough to be used by beginners. The result is a combination of numerous programs and subprograms to provide convenient data file creation and editing, circuit analysis and optimization, matching network synthesis, coupler and transmission line analysis, as well as several large databanks containing vital information on transistors and dielectric substrate materials.

Some of the capabilities of the program will be illustrated through the design of a 2-4 GHz low-noise amplifier. Specifications include 10 ±1 dB gain and noise figure less than 2 dB within the octave passband. The design procedure is the following:

Selection of the Active Device

The transistor should have optimum noise figure (NF_{opt}) of less than 1.8 dB (allowing .2 dB for input circuit losses) with associated gain in excess of 10 dB between 2-4 GHz. These specifications are submitted to the Transistor Databank[4] resulting in a list of devices, both FET and bipolar, that are capable of providing the desired gain. Any parameter of a specific transistor, including stability information, can be examined in either tabulated or graphical form. After a review of the available devices, a Hewlett-Packard HFET-1101 GaAs field-effect transistor was selected.

Device Modeling

The program models the input and output impedances of the transistor using the negative image device modeling[5] technique which provides the actual impedances presented by the device input and output under the desired gain and noise figure conditions. The modeling was done in terms of distributed elements since both input and output networks will be synthesized in distributed form. Lumped element modeling of the transistor input and output impedances could have been used if lumped-element matching had been desired. The resultant model is shown in Figure 1.

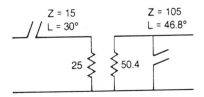

FIGURE 1: Transistor Equivalent Circuit

Topology Selection

Using the <u>internally derived</u> transistor impedance models, the program provides a list of alternative circuit topologies which may be synthesized for the given task. The user chooses a set of acceptable matching circuits by specifying the number of high-pass, low-pass and unit elements.

Matching Network Synthesis

The selected topology is submitted internally for synthesis; the program automatically selects the desired gain slope compensation by examining the transistor's gain versus frequency characteristic. The input equivalent circuit of the device includes a parasitic series open stub that presents problems to physical realization if not fully absorbed in the synthesis. The SUPER-COMPACT synthesis typically begins from the termination where the parasitic exists, and in this case the gain and ripple specifications are interactively adjusted until an "exact absorption" is reached -- meaning that the series stub will not have to be realized in the circuit.

Combining Data Files

At the completion of the synthesis, both networks are stored automatically in forms of disk files. Through the internal program editor, the user next combines the synthesized networks with the actual transistor. The synthesized circuit is shown on Figure 2.

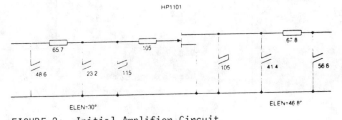

FIGURE 2: Initial Amplifier Circuit

Initial Analysis and Optimization

The overall circuit is then analyzed and submitted for optimization. Due to the accurate device modeling, the synthesized circuit provides an excellent starting point for the optimization -- the initial gain variation was less than 1.5 dB. SUPER-COMPACT includes two different techniques: adaptive random and a gradient type. The combination of the two techniques assures a high degree of confidence in finding the global optimum of the circuit. In order to examine the interim performance, the optimization may be halted interactively at any time. The targeted gain of 10 dB was achieved with an overall flatness of ±0.20 dB. The optimized response is shown in Figure 3.

Statistical Analysis

At the conclusion of the optimization, the circuit is submitted for a Monte Carlo analysis to examine the effects of active and passive component tolerances. Subsequent yield analysis shows the expected distribution of an actual production run, as shown on Figure 4.

Conversion to Physical Dimensions

Finally, the program offers a transmission line synthesis to convert the electrical parameters of the microstrip lines to physical dimensions. The microstrip models include the effects of dispersion, discontinuities, dielectric and conductor losses, and finite conductor thickness with multiple metallizations.[6] Although layout is not offered at this time, a future update to the program will also include the artwork generation.

```
Circuit: STAGE
S-matrix, ZS =    50. +j    0.   ZL =    50. +j    0.

                 S11              S21              S12              S22           |S21|
 FREQ(Hz)    Mag    Ang       Mag    Ang       Mag    Ang       Mag    Ang        dB
 2.000E+09  0.930  -74.1     3.134  117.6     0.065   43.9     0.823  -19.6      9.922
 2.500E+09  0.831  171.9     3.231   37.6     0.083  -31.6     0.695  -67.0     10.186
 3.000E+09  0.762   89.5     3.098  -25.5     0.092  -90.2     0.616 -111.5      9.821
 3.500E+09  0.652    3.6     3.186  -89.7     0.105 -149.7     0.532 -161.4     10.064
 4.000E+09  0.536 -127.0     3.169 -169.9     0.113  135.0     0.411  132.8     10.019

                        NOISE FIGURE DATA
 FREQ.    MIN. NOISE FIG.   OPT. NOISE SOURCE      ACTUAL NF      NORM
   HZ           DB            MAGN      ANG            DB          RN
 2.000E+09      1.25         0.352      86.8          1.39        0.104
 2.500E+09      1.33         0.384    -164.6          1.54        0.047
 3.000E+09      1.42         0.380     -90.6          1.69        0.174
 3.500E+09      1.51         0.280      -9.3          1.69        0.311
 4.000E+09      1.60         0.212     150.3          1.73        0.166
```

FIGURE 3: Optimized Data for the 2-4 GHz Amplifier

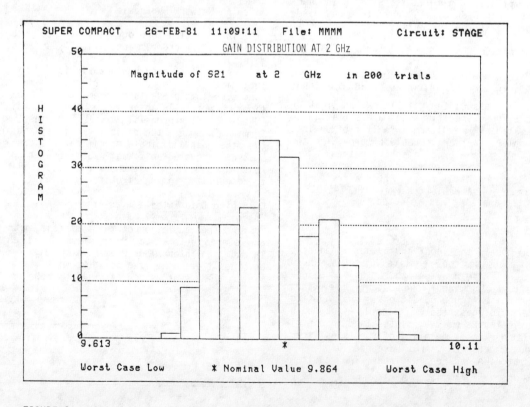

FIGURE 4: Histogram for 2 GHz Gain Distribution

Circuit Layout

At this point the design is completed. The final circuit with physical transmission line dimensions is shown in Figure 5. The corresponding wide band gain response is shown in Figure 6.

The 1980's will see computer-aided design of microwave circuits combined with automated layout and artwork generation, manufacturing, and test. SUPER-COMPACT will be expanded to offer interactive layout where the effects of significant parasitics and discontinuities will be accounted for during analysis and optimization.

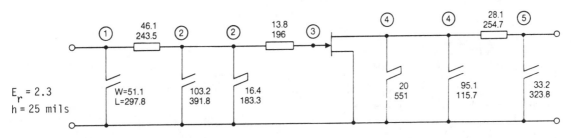

FIGURE 5: Optimized Circuit for 2-4 GHz Amplifier

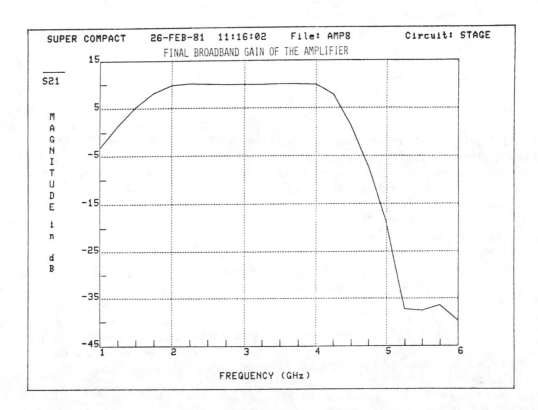

FIGURE 6: Broadband Gain for 2-4 GHz Amplifier

References

1. L. Besser, "A Fast Scattering Matrix Program," presented at the IEEE International Conference on Communications, San Francisco, California, June 1970.

2. P. Bodharamick, L. Besser, R. W. Newcomb, "Two Scattering Matrix Programs for Active Circuit Analysis," IEEE Transactions, Vol. CT-18, November 1971, p. 610-618.

3. L. Besser, C. Hsieh, S. Ghoul, "Computerized Optimization of Transistor Amplifiers and Oscillators using COMPACT," presented at the 1973 European Microwave Conference, Session A:12.3.

4. L. Besser, "Synthesize Amplifiers Exactly," Microwave Systems News, Vol. 9, Oct. 1979, p. 28-40.

5. M. W. Medley, Jr., J.L. Allen, "Broadband GaAs FET Design Using Negative-Image Device Models," IEEE Transactions, Vol. MTT-29, Sept. 1979, p. 784-787.

6. S. March, "Empirical Formulas for Single and Coupled Covered Microstrip for use in the Computer-Aided Design of MICs," submitted to the 1981 European Microwave Conf., Sept. 1981, Eindhaven, The Netherlands.

GaAs FET Large-Signal Model and its Application to Circuit Designs

YUSUKE TAJIMA, MEMBER, IEEE, BEVERLY WRONA, AND KATSUHIKO MISHIMA

Abstract—A large-signal GaAs FET model is derived based on dc characteristics of the device. Analytical expressions of modeled nonlinear elements are presented in a form convenient for circuit design. Power saturation and gain characteristics of a GaAs FET are studied theoretically and experimentally. An oscillator design employing the large-signal model is demonstrated.

I. INTRODUCTION

INFORMATION on the large-signal behavior of GaAs FET's is very limited. Often GaAs FET power amplifiers and oscillators are designed using small-signal S parameters, necessitating either tweaking of the circuits or cut and trys later to improve the circuit. In power amplifier design, the load pull method [1] is frequently used to obtain the optimum load condition at large-signal operation. However, this cannot predict gain performance of the designed circuit for small- or large-signal levels.

Willings et al. [2], [3] proposed a large-signal GaAs FET model by measuring bias dependence of small-signal S parameters and established expressions for an instantaneous equivalent circuit in terms of terminal voltages. They made it possible to predict large-signal performance such as power saturation and distortion at arbitrary input levels. Their use of time-domain analysis, however, is not always convenient for circuit design where frequency-domain elements, such as Ls and Cs, are commonly used. Also their expressions for equivalent circuit elements are based on S parameters measured at all the possible bias conditions the RF signal can sweep. This is not easy, especially when RF swing is large; for instance, when it swings into large drain voltage region or forward gate bias condition.

In this paper, we will first discuss the derivation of RF equivalent-circuit elements in terms of signal voltages, based on static characteristics of an FET, such as dc drain current–voltage curves, gate current–voltage curves, etc., all of which are parameters easy to obtain experimentally. Assuming sinusoidal variation to all the terminal voltages, nonlinear element values are obtained analytically with these curves built in as analytical functions. Thus nonlinear expressions for arbitrary size FET's can be calculated once these curves are determined by dc measurements.

Manuscript received July 21, 1980; revised October 6, 1980. This work was supported in part by the U.S. Army Electronics Research and Development Command under Contract DAAK20-79-C-0269.
Y. Tajima and B. Wrona are with Raytheon Research Division, Waltham, MA 02154.
K. Mishima is with Toshiba Corporation, 1-Komukai Toshibacho, Saiwaiku, Kawasaki, Japan.

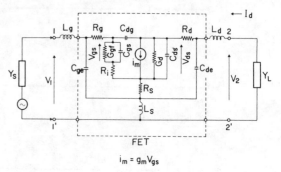

Fig. 1. FET equivalent circuit.

By connecting outside circuits to this nonlinear model, FET performance can be obtained consistently at any level from small signal to large signal. With this we studied power saturation and variation of the optimum load conditions, comparing our results with experimental data. An application of the nonlinear model to an FET oscillator design is also demonstrated. Available output power is mapped on a load impedance chart which gives a full description of an oscillating FET.

II. LARGE-SIGNAL EQUIVALENT CIRCUIT

Under large-signal operation, element values of the FET equivalent circuit, shown in Fig. 1, vary with time because at large driving levels they become dependent on terminal voltages. We may consider two of the terminal voltages to be independent and choose the set V_{gs} and V_{ds}, V_{gs} being the voltage across the gate capacitance and V_{ds} across the drain conductance. If we restrict our interest to the signal frequency and ignore the effects due to higher harmonic components, these voltages can be written as

$$V_{gs} = V_{gs0} + v_{gs} \cos(\omega t + \phi) \quad (1)$$

$$V_{ds} = V_{ds0} + v_{ds} \cos \omega t \quad (2)$$

where V_{gs0} and V_{ds0} are the dc bias voltages, v_{gs} and v_{ds} amplitudes of signal frequency components, and ϕ the phase difference between gate and drain voltages. The equivalent circuit for the signal frequency can now be expressed as a function of the following parameters which are independent of time: $V_{gs0}, V_{ds0}, v_{gs}, v_{ds}, \omega$, and ϕ.

In order to avoid unnecessary complexity of calculations, we limit the nonlinear behavior to five elements, gate forward conductance G_{gf}, gate capacitance C_{gs}, gate charging resistance R_i, transconductance g_m, and drain conductance G_d. Here, G_{gf} represents the effect of the forward rectified current across the gate junction under large-signal operation. No voltage dependence was assumed for parasitic elements, i.e., lead inductances (L_g, L_d, L_s), contact resistances (R_g, R_d, R_s), and pad capacitances (C_{ge}, C_{de}). We also ignored the voltage dependence of drain channel capacitance C_{ds} and feedback capacitance C_{dg} due to their small values.

A. Expressions for g_m and G_d

Transconductance g_m and drain conductance G_d are defined as

$$g_m = \left(\frac{i_{ds}}{v_{gs}}\right)_{v_{ds}=0} \quad G_d = \left(\frac{i_{ds}}{v_{ds}}\right)_{v_{gs}=0} \quad (3)$$

where i_{ds} is the RF drain current amplitude. The instantaneous drain current can be written using g_m and G_d as

$$I_{ds}(t) = I_{ds0} + g_m v_{gs} \cos(\omega t + \phi) + G_d v_{ds} \cos \omega t \quad (4)$$

where I_{ds0} is the dc drain current. In this expression linear superposition of two terms is assumed.

Now, if we have a function which can simulate drain current I_{ds} as a function of V_{gs} and V_{ds}, as

$$I_{ds} = I_{ds}(V_{gs}, V_{ds}) \quad (5)$$

instantaneous current $I_{ds}(t)$ can be obtained by inserting (1) and (2) into (5). By multiplying $\sin \omega t$ to (4) and integrating over a complete period, g_m is obtained as

$$g_m = -\frac{\omega/\pi}{v_{gs} \sin \phi} \int_0^{(2\pi/\omega)} I_{ds} \sin \omega t \, dt. \quad (6)$$

Similarly, G_d is obtained as

$$G_d = \frac{\omega/\pi}{v_{ds} \sin \phi} \int_0^{(2\pi/\omega)} I_{ds} \sin(\omega t + \phi) \, dt. \quad (7)$$

Equations (6) and (7) are now functions of RF amplitudes v_{gs} and v_{ds}, as well as bias voltages V_{gs0} and V_{ds0}.

The equation for I_{ds} was arrived at empirically to simulate the typical dc transistor curves. Any other function that reproduced the I-V behavior could have been used.

$$I_{ds}(V_{ds}, V_{gs}) = I_{d1} \cdot I_{d2}$$

$$I_{d1} = \frac{1}{k}\left[1 + \frac{V'_{gs}}{V_p} - \frac{1}{m} + \frac{1}{m}\exp\left\{-m\left(1 + \frac{V'_{gs}}{V_p}\right)\right\}\right]$$

$$I_{d2} = I_{dsp}\left[1 - \exp\left\{\frac{-V_{ds}}{V_{dss}} - a\left(\frac{V_{ds}}{V_{dss}}\right)^2 - b\left(\frac{V_{ds}}{V_{dss}}\right)^3\right\}\right]$$

$$k = 1 - \frac{1}{m}\{1 - \exp(-m)\}$$

$$V_p = V_{p0} + p V_{ds} + V_\phi$$

$$V'_{gs} = V_{gs} - V_\phi \quad (8)$$

where V_{p0} (>0) is the pinchoff voltage at $V_{ds} \sim 0$, V_{dss} is the drain current saturation voltage, V_ϕ is the built-in potential of the Schottky barrier, I_{dsp} is the drain current when $V_{gs} = V_\phi$, and a, b, m, and p are fitting factors that can be varied from device to device.

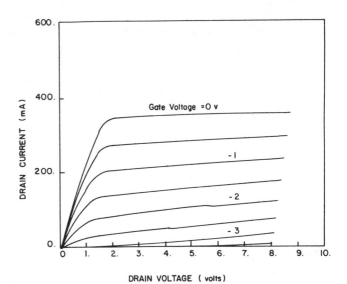

Fig. 2. I_D-V_D characteristics calculated from (8) with parameter values of Table I, column a.

TABLE I
ELEMENT VALUES OF THE FET EQUIVALENT CIRCUIT AND BIAS CONDITIONS
(Values in column a are for the amplifier, in column b for the oscillator)

	a	b		a	b
i_s (pA)	3	0.15	α (1/V)	30	30
C_{gso} (pF)	0.3	0.4	τ_i (ps)	3	2
I_{dsp} (mA)	70	55	V_{po} (V)	2	2.0
V_{dss} (V)	0.7	1			
R_g (Ω)	3	4	R_d (Ω)	3	4
R_s (Ω)	3	3	L_s (nH)	0.02	0.05
L_g (nH)	0.15	0.3	L_d (nH)	0.15	0.3
C_{ge} (pF)	0.01	0.02	C_{de} (pF)	0.01	0.02
C_{dg} (pF)	0.015	0.01	C_{ds} (pF)	0.02	0.02
a	-0.2	-0.2	m	3	3
b	0.6	0.6	p	0.2	0.2
C_F (pF)	---	10	L_F (nH)	---	1.5
ℓ_g (mm)	---	0.7	ℓ_d (mm)	---	0.7
V_{dso} (V)	7	5	V_{gso} (V)	-1	-0.4

Drain current versus voltage curves shown in Fig. 2 were obtained from (8) when these parameters were given as in Table I (column a).

B. Nonlinear Expressions for C_{gs}, G_{gf}, and R_i

Although the gate junction is also a function of V_{gs} and V_{ds}, we assumed here that it can be approximated by a Schottky-barrier diode between gate and source, with V_{gs} as the sole voltage parameter. Gate capacitance C'_{gs} and forward gate current i_{gf} can be found from Schottky-barrier theory as

$$C'_{gs} = C'_{gso}/\sqrt{1 - V_{gs}/V_\phi} \quad (-V_p \leq V_{gs}) \quad (9)$$

or

$$C'_{gs} = C'_{gs0}/\sqrt{1 + V_p/V_\phi} \quad (-V_p \geq V_{gs})$$

$$i_{gf} = i_s (\exp \alpha V_{gs} - 1) \tag{10}$$

where C'_{gs0} is the zero bias gate capacitance, i_s the saturation current of the Schottky barrier, and $\alpha = q/nkT$.

When V_{gs} varies according to (1), effective gate capacitance C_{gs} and gate forward conductance G_{gf} for the signal frequency are obtained from (1), (2), (9), and (10) as

$$C_{gs} = \frac{1}{\pi v_{gs}} \int_0^{2\pi} \left(\int^{V_{gs}} C'_{gs} \, dV \right) \cos \omega t \, d(\omega t) \tag{11}$$

$$G_{gf} = 2i_s \exp (\alpha V_{gs0}) I_1(\alpha v_{gs})/v_{gs} \tag{12}$$

where $I_1(x)$ is the first-order modified Bessel function.

Gate charging resistance R_i was assumed to vary in such a way that the charging time constant did not change [4]

$$R_i \cdot C_{gs} = \tau_i \text{(constant)}. \tag{13}$$

III. Amplifier Design

The device under analysis is based on an arbitrarily selected GaAs FET which had a 0.7-μm gate with 250-μm width. Nonlinear element parameters and parasitic element values are listed in Table I(column a).

Available output power of the device was calculated as a function of input power following the procedure described below.

By giving a driving RF voltage for v_{gs}, output voltage v_{ds} and phase shift ϕ are calculated assuming some initial values for nonlinear elements and signal and load admittances, Y_S and Y_L. Knowing v_{gs}, v_{ds}, and ϕ, nonlinear element values are revised according to (6), (7), (11), (12), and (13). The revised equivalent circuit with new element values determines a new value for v_{ds} and ϕ, which will again change the nonlinear element values. When this process converges, the equivalent circuit is obtained for the driving level of v_{gs}. From terminal voltages, v_1 and v_2, input and output powers P_1 and P_2 are calculated using admittances Y_S and Y_L.

The operating condition is dependent on terminating admittances Y_S and Y_L. Fig. 3 shows calculated saturation output power contours plotted on the load admittance chart. Source admittance Y_S was kept matched to the device input admittance as it varied with input power and Y_L. The figure shows the maximum power condition and saturation power rolloff as the load mismatches to the optimum load. Also shown in the figure are experimental data taken with known admittances while the input circuit was tuned for best output power. Considering the difficulty of measuring admittances, the agreement is very good.

Fig. 4 shows theoretical and experimental power gain performance for three different terminations. Although a small discrepancy in saturation power is found, the general dependence of small-signal gain and saturation power on termination admittance is successfully simulated. The discrepancy in saturation power could be due to gate-drain breakdown, which is not modeled in the calculation. The saturation mechanism of the model can be seen in Fig. 5, which shows variation of

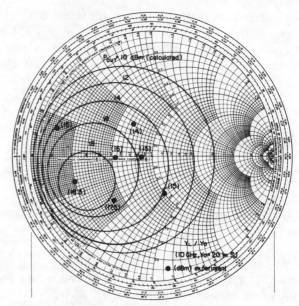

Fig. 3. Saturation power contours on load admittance chart.

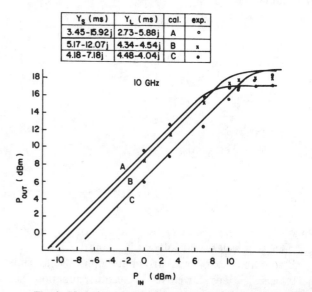

Fig. 4. Variation of gain with terminating impedance.

the nonlinear element values as a function of input power. Although C_{gs}, R_i, g_m, and G_d change a small amount, the increase of G_{gf} is significant in the saturation region. In other words, saturation of output power is caused by rectified current across the gate junction. This limits input power by increasing loss in the circuit resistances as power increases.

Using the model, output power performance of GaAs FET amplifiers can be simulated at arbitrary input levels and terminating impedances. The calculations provide predictions of both small-signal gain and saturation power necessary to circuit designers.

IV. Oscillator Design

Although we discuss here an oscillator circuit using a common source FET with a feedback circuit embedded between gate and drain, the calculation method is feasible for any other circuit topologies of oscillators.

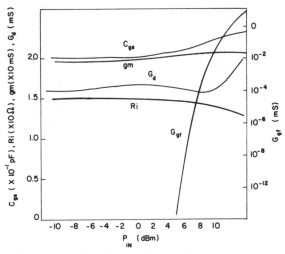

Fig. 5. Variation of nonlinear element values with input power.

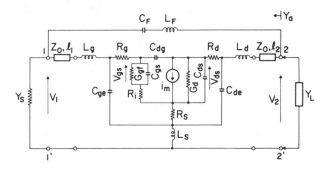

$i_m = g_m V_{gs}$

Fig. 6. FET oscillator equivalent circuit.

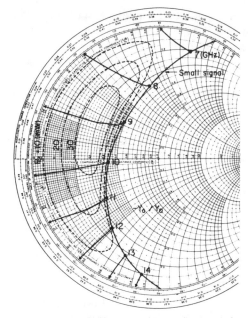

Fig. 7. Calculated oscillating device admittance for increasing values of available output power.

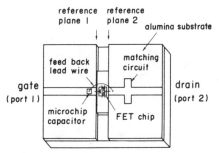

Fig. 8. FET oscillator circuit.

The equivalent circuit of an FET in an oscillating circuit is expressed by the same equivalent circuit as an amplifier except that a feedback circuit, composed of C_F and L_F, is now placed between the gate and drain ports (Fig. 6). Distances between the actual ports and feedback points are expressed by lengths l_1 and l_2 of transmission lines of impedance Z_0. The gate port is terminated by 20 mS(Y_s) for the purpose of stabilizing the circuit [5].

Large-signal behavior of the device was calculated following a procedure similar to that described in the previous section. Parameters used to model a 400 × 1 μm gate FET are listed in Table I(column b) for a bias condition of −0.4-V gate, 5-V drain. Variation of oscillating device admittance Y_a was obtained as a function of available output power P_2 which was calculated from terminal voltage v_2 and the real part of Y_a. Loci of Y_a, shown in Fig. 7, revealed a complete map of the available power and frequency within the unstable region of the device. This particular circuit has an optimum oscillating frequency around 9 GHz with a maximum output power of 35 mW and efficiency of 25 percent when load Y_L is $(6 + 4j)$ mS. With this knowledge, an oscillator can be designed at its optimum.

This calculation was confirmed by the circuit shown in Fig. 8. A GaAs FET was mounted between two microstrip

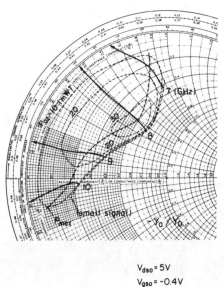

$V_{dso} = 5V$
$V_{gso} = -0.4V$
$Y_0 = 20mS$

Fig. 9. Measured oscillating device admittance.

circuits with a feedback circuit between drain and gate constructed by a lead wire and a microchip capacitor. The gate port was terminated by 50 Ω.

Device admittance Y_a was measured at the drain port (reference plane 2) and is shown in Fig. 9 which has the same pattern as the calculated results in Fig. 7. With a matching circuit designed to convert the optimum admittance to the 50-Ω load, the circuit operated as an oscillator. Its output power and oscillation frequency at the optimum bias condition were 30 mW and 8.8 GHz, very close to predicted data.

V. Conclusion

A technique to model the large-signal behavior of GaAs FET's was proposed. Because the model was derived based on dc current–voltage characteristics, it was easy to apply to any arbitrary size FET once its dc curves were given. By limiting our interest to the fundamental frequency of the signal, it was possible to analyze the circuit in the frequency domain rather than the time domain. Device analysis and circuit simulation have thus become simple enough for use in circuit design.

The model was used to study power performance of FET's. Output-power saturation values were predicted and confirmed by experiment. Because the terminating impedances of the circuit can be varied, the model is useful for predicting power performance of GaAs FET amplifiers.

By applying a feedback element to the nonlinear model of an FET, oscillatory behavior of the FET was studied. A complete admittance chart with available power and frequency in the unstable region of the device was calculated and confirmed by experimental data. The chart provided complete data for designing an oscillator at maximum efficiency.

References

[1] Y. Takayama, "A new load-pull characterization method for microwave power transistors," in *1976 MTT-S Int. Microwave Symp.*, *Dig. Tech. Papers*, pp. 218–220.

[2] H. A. Willing, C. Rauscher, and P. deSantis, "A technique for predicting large-signal performance of a GaAs MESFET," *IEEE Trans. Microwave Theory Tech.*, vol. MTT-26, pp. 1017–1023, Dec. 1978.

[3] C. Rauscher and H. A. Willing, "Quasi-static approach to simulating nonlinear GaAs FET behavior," in *1979 MTT-S Int. Microwave Symp.*, *Dig. Tech. Papers*, pp. 402–404.

[4] H. Statz, H. Haus, and R. Pucel, "Noise characteristics of gallium arsenide field effect transistors," *IEEE Trans. Electron Devices*, vol. ED-21, pp. 549–562, Sept. 1974.

[5] Y. Tajima, "GaAs FET applications for injection-locked oscillators and self-oscillating mixers," *IEEE Trans. Microwave Theory Tech.*, vol. MTT-27, no. 7, pp. 629–632, July 1979.

LARGE-SIGNAL GaAs FET AMPLIFIER CAD PROGRAM

A. Platzker and Y. Tajima*
Raytheon Missile Systems Division
Bedford, Massachusetts 01730

*Raytheon Research Division
Lexington, Massachusetts 02173

Abstract

A CAD program for the design of multistage power GaAs FET amplifiers has been developed. The program is capable of analyzing the circuit performance of power amplifiers as a function of their input powers and frequency. Either graphic or printed output is available.

In an optimizing mode, the program returns the load and source admittance values for optimum power performance of FET devices.

The small-signal performance of multistage FET amplifiers (hybrid or monolithic) can be predicted to a high degree of accuracy by using the measured S-parameters of the devices and the various matching circuit elements. Many CAD programs are available for performing the necessary calculations. Power amplifiers, however, are more difficult to design since the circuit parameters of the devices become dependent on the power level. A number of approaches to the design problem have been reported in the literature. Most of these approaches, namely the ones based on the load-pull method,[1] require a large number of measurements and are therefore very difficult to use in a one-power-stage circuit design and next to impossible in multi-power-stage designs. Tajima et al.[2] have successfully used a FET circuit model to which power-dependent elements have been added, to predict the power performance of a power amplification stage.

We have utilized their power-dependent FET model (Fig. 1) and constructed a multimoded CAD program, LSFET. The program can operate in three modes, one of optimization and two of analysis. In the optimization mode, the program calculates, at each frequency and input power level, the source and load impedances necessary for optimum power performance of a power FET device. In the analysis modes, the program can analyze the nonlinear performance of either one or two power amplifier stages. The amplifiers in either case, however, can contain any number of driving stages provided they operate in the linear regime.

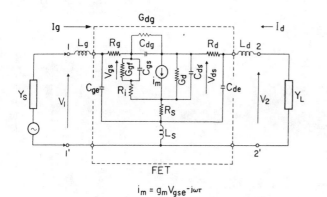

Fig. 1 GaAs FET large-signal equivalent circuit model.

The program accepts circuits whose block diagram is shown in Fig. 2. The input, interstage, and output matching networks can contain active or passive linear circuit elements whose parameters (S or Y) are calculated by other commercially available CAD programs (COMPACT for instance) and are written to files available to the nonlinear large-signal program.

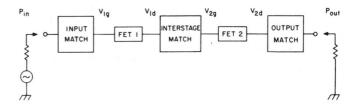

Fig. 2 Block diagram of FET amplifier with two power stages.

The FET parameters are stored in a data file. The file contains both the circuit elements shown in Fig. 1 and also the dc I-V curves, including breakdown characteristics, in parameter form. From the latter, the nonlinear g_m and g_d are calculated and their fundamental harmonic content used for the calculations. Our data files currently contain only FETs fabricated by our laboratory, for which we are able to obtain the complete necessary data.

The power calculation is performed by an iterative self-consistent method. Initially, a voltage v_{g2} is assumed. With this value, the nonlinear FET circuit elements and thereafter the voltage v_{d2} and current i_{d2} are calculated. Since the load across the FET is determined by the output matching network and the amplifier load, the ratio v_{d2}/i_{d2} is externally fixed. If this ratio does not agree with the ratio calculated from the nonlinear circuit model, v_{d2} is varied until a self-consistency is achieved. The whole process is repeated for the first power stage. Typically the convergence is fast.

The large-signal program, LSFET, is very easy to use. Prior to an analysis session, the parameters of the devices have to be resident in a library file, and the circuit descriptions of the various matching networks, as well as the frequency information, written into separate files. LSFET can interact with any program that is capable of calculating the necessary linear parameters. For our internal use, we have standardized on using the Y-parameters as calculated by the commercially available COMPACT program.

A sample analysis of a two-stage amplifier is shown in Figs. 3 and 4. In Fig. 3 the user's responses are underlined. From the presence of three file names on the first line, the program understands that an analysis of a two-stage design is called for. Two file names or none would have called for an analysis of a one-stage design or for an optimization.

Reprinted from *IEEE MTT-S International Microwave Symp. Dig.*, 1982, pp. 450-452.

```
BEGIN,,LSFET  ,DRI5IN,DRI5TR,DRI5OT
  82/03/18.  09.12.24.

  ENTER DEVICES CODES:     FIRST STAGE FOLLOWED BY CR, THEN
                           SECOND STAGE FOLLOWED BY CR.
  CR ALONE IF YOU WANT TO TYPE IN THE DEVICES PARAMETERS
? MM
? NN
  PRINT COMPACT SOURCE,INTERSTAGE AND LOAD FILES ?
  PRINT Y PARAMETERS GENERATED BY ABOVE FILES? (YY,YN,NY,NN=CR)?   NN




  82/03/18.  09.14.44.

LARGE SIGNAL ANALYSIS, TWO STAGE AMPLIFIER

FET PARAMETERS                                                 FIRST STAGE FET      SECOND STAGE FET
  (I) LINEAR ELEMENTS
       GATE,DRAIN,SOURCE RESISTANCES(OHM)                      4.00  4.00  3.00      2.50  2.50  2.60
       GATE,DRAIN,SOURCE INDUCTANCES(NH)                       0.00  0.00   .02      0.00  0.00   .02
       DRAIN-SOURCE,DRAIN-GATE,GATE PAD,DRAIN PAD CAPACITANCES"(PF)   .04   .02  0.00  0.00      .07   .02  0.00  0.00
       TRANSCONDUCTANCE TIME CONSTANT(PS)                       .36                   .50
  (II) DRAIN CURRENT PARAMETERS
       MAX DRAIN CURRENT(MA),KNEE AND PINCHOFF VOLTAGE(V)       60.    .50  2.80     100.    .50  2.80
       PARAMETERS A,B,N,W                                      -.20    .10  2.00  .50  -.20   .10  2.00  .50
       RESIDUAL DRAIN CONDUCTANCE(MS),PINCHOFF INCREMENT FACTOR 1.75   .12             3.05   .12
  (III) GATE PARAMETERS (VPHI=0.8V)
       SCHOTTKY SATURATION CURRENT(PA),ZERO BIAS GATE CAPACITANCE(PF)  5.00   .40    10.00   .55
       ALP(G/NKT), GATE CHARGING TIME CONSTANT(PS,RI*CGS)     30.00  1.80          30.00  1.17
  (IV) BREAKDOWN PARAMETERS
       BREAKDOWN VOLTAGE(V), INCREMENT RESISTANCE(OHM)         15.00  75.00          15.00  37.00
       BREAKDOWN RESISTANCE, INCREMENT FACTOR(OHM)             400.   1600.          200.    800.

  "FIRST STAGE BIAS VGS,VDS(V)= "   ?  -1.5 7
  "SECOND STAGE BIAS VGS2,VDS2(V)= "  ?  -1.5 7
  II: NUMBER OF GATE RF VOLTAGE INCREMENTS ?  40
```

Fig. 3 Sample LSFET input instructions and device parameter listings. User's responses are underlined.

```
FREQUENCY(GHZ)=12.00

                                                SMALL SIGNAL PERFORMANCE
                                                                                                MAG
                   S11(MAG,DEG)      S12(MAG,DEG)      S21(MAG,DEG)      S22(MAG,DEG)      MSG(DB)    K       S21(DB)

FIRST STAGE FET   .759E+00 -.860E+02  .693E-01  .478E+02  .110E+01  .113E+03  .797E+00 -.231E+02  10.547  1.056    .799
SECOND STAGE FET  .787E+00 -.106E+03  .800E-01  .385E+02  .138E+01  .994E+02  .680E+00 -.423E+02  12.357   .831   2.780
AMPLIFIER         .357E+00 -.171E+03  .109E-01 -.261E+02  .296E+01  .999E+02  .246E+00 -.718E+02  10.317 12.673   9.433

                                                LARGE SIGNAL PERFORMANCE

 PIN   POUT  GAIN  PAE   GGF     CGS    RI      GM     GD     GDG    IDS   IDG   IGS      V1            V2
(DBM) (DBM) (DB) (PCENT) (S)     (PF)  (OHM)   (S)    (S)    (S)    (MA)  (MA)  (MA)     (V,DEG)      (V,DEG)

 0.00  0.00  0.00  0.00 .562E-29 .236E+00 .762E+01 .197E-01 .175E-02 0.    26.  0.00  .00   .05  16.   .05  164.
                        .112E-28 .325E+00 .360E+01 .325E-01 .305E-02 0.    43.  0.00  .00   .05  13.   .07  156.
-8.65  1.34  9.99   .26 .120E-27 .236E+00 .762E+01 .201E-01 .175E-02 0.    26.  0.00  .00   .23  15.   .16 -131.
                        .176E-27 .325E+00 .360E+01 .342E-01 .305E-02 0.    43.  0.00  .00   .22  11.   .65 -173.
-2.61  7.26  9.87   .99 .273E-25 .237E+00 .760E+01 .201E-01 .175E-02 0.    26.  0.00  .00   .46  15.   .32 -131.
                        .260E-25 .325E+00 .360E+01 .338E-01 .305E-02 0.    43.  0.00  .00   .43  11.  1.29 -173.
  .97 10.75  9.79  2.18 .101E-22 .238E+00 .757E+01 .200E-01 .175E-02 0.    26.  0.00  .00   .69  16.   .49 -131.
                        .577E-23 .327E+00 .358E+01 .337E-01 .305E-02 0.    44.  0.00  .00   .65  11.  1.91 -173.
 3.55 13.24  9.69  3.78 .496E-20 .239E+00 .752E+01 .199E-01 .175E-02 0.    27.  0.00  .00   .93  16.   .66 -131.
                        .152E-20 .328E+00 .356E+01 .336E-01 .305E-02 0.    45.  0.00  .00   .86  12.  2.55 -173.
 5.60 15.18  9.58  5.75 .320E-17 .241E+00 .746E+01 .198E-01 .175E-02 0.    27.  0.00  .00  1.18  16.   .83 -130.
                        .440E-18 .331E+00 .354E+01 .336E-01 .305E-02 0.    46.  0.00  .00  1.08  12.  3.19 -172.
 7.33 16.76  9.43  8.01 .280E-14 .244E+00 .737E+01 .196E-01 .175E-02 0.    28.  0.00  .00  1.43  16.  1.01 -130.
                        .135E-15 .334E+00 .351E+01 .336E-01 .305E-02 0.    47.  0.00  .00  1.29  12.  3.82 -173.
 8.88 18.11  9.23 10.50 .357E-11 .249E+00 .724E+01 .194E-01 .175E-02 0.    29.  0.00  .00  1.70  16.  1.21 -130.
                        .434E-13 .338E+00 .347E+01 .336E-01 .305E-02 0.    49.  0.00  .00  1.51  12.  4.46 -173.
10.32 19.28  8.96 13.11 .754E-08 .255E+00 .705E+01 .191E-01 .175E-02 0.    30.  0.00  .00  1.98  16.  1.42 -129.
                        .143E-10 .343E+00 .341E+01 .337E-01 .305E-02 0.    50.  0.00  .00  1.73  12.  5.11 -173.
10.71 19.54  8.83 13.72 .656E-07 .258E+00 .699E+01 .191E-01 .175E-02 0.    31.  0.00  .00  2.06  16.  1.48 -129.
                        .614E-10 .344E+00 .340E+01 .337E-01 .305E-02 .96E-05 51.  .03  .00  1.78  12.  5.26 -173.
11.08 19.78  8.70 14.30 .530E-06 .260E+00 .692E+01 .190E-01 .175E-02 0.    31.  0.00  .00  2.13  16.  1.54 -129.
                        .263E-09 .346E+00 .338E+01 .337E-01 .305E-02 .28E-04 51.  .10  .00  1.83  12.  5.41 -173.
11.45 20.00  8.54 14.85 .451E-05 .263E+00 .683E+01 .189E-01 .175E-02 0.    32.  0.00  .00  2.21  16.  1.61 -129.
                        .113E-08 .347E+00 .337E+01 .336E-01 .305E-02 .51E-04 51.  .19  .00  1.89  12.  5.55 -173.
11.86 20.21  8.34 15.36 .407E-04 .267E+00 .674E+01 .188E-01 .175E-02 0.    32.  0.00  .04  2.29  16.  1.68 -129.
                        .485E-08 .349E+00 .335E+01 .336E-01 .305E-02 .76E-04 51.  .30  .00  1.94  12.  5.68 -173.
12.44 20.40  7.96 15.73 .392E-03 .271E+00 .663E+01 .187E-01 .175E-02 0.    33.  0.00  .44  2.39  16.  1.75 -130.
                        .209E-07 .351E+00 .333E+01 .335E-01 .305E-02 .10E-03 51.  .41  .00  2.00  12.  5.81 -173.
12.92 20.48  7.57 15.68 .995E-03 .273E+00 .660E+01 .187E-01 .175E-02 0.    33.  0.00 1.12  2.44  16.  1.77 -130.
                        .374E-07 .352E+00 .332E+01 .335E-01 .305E-02 .11E-03 51.  .43  .00  2.02  12.  5.87 -173.
13.81 20.55  6.75 15.24 .248E-02 .274E+00 .656E+01 .187E-01 .175E-02 0.    33.  0.00 2.84  2.52  16.  1.80 -130.
                        .671E-07 .353E+00 .331E+01 .335E-01 .305E-02 .12E-03 51.  .50  .00  2.04  12.  5.91 -173.
14.54 20.63  6.09 14.80 .389E-02 .275E+00 .655E+01 .186E-01 .175E-02 0.    33.  0.00 4.48  2.58  16.  1.82 -130.
                        .120E-06 .354E+00 .330E+01 .335E-01 .305E-02 .13E-03 51.  .53  .00  2.06  12.  5.97 -173.
14.54 20.64  6.10 14.90 .389E-02 .275E+00 .655E+01 .186E-01 .175E-02 0.    33.  0.00 4.48  2.58  16.  1.82 -130.
                        .139E-06 .354E+00 .330E+01 .334E-01 .305E-02 .14E-03 51.  .58  .00  2.07  12.  5.97 -173.
```

Fig. 4 Sample LSFET printed output.

Following the three self-explanatory opening questions, the program prints the parameters and equivalent circuit values of the selected devices. Bias information and the number of desired input power increments at each frequency are then supplied. A sample analysis, repeated at each frequency, is shown in Fig. 4. The program calculates the small-signal performance and follows by the large-signal analysis. As can be seen in the figure, all the power-dependent elements of the equivalent circuits of the two devices are printed at each power level (columns 5 through 13), the elements of FET1 in the first line, of FET2 in the second. In the first four columns, the input and output powers, the gain and power-added efficiency of the amplifier are printed.

The drain and gate currents, I_D and I_G are defined as positive when they flow into the terminals of the FET.

$$I_D = I_{DS} + I_{DG}$$

$$I_G = I_{GS} - I_{DG}$$

where I_{DS}, I_{DG}, and I_{GS} are given in columns 11, 12 and 13. The presence of gate current indicates the onset of power saturation. The direction of I_G identifies the power-limiting factor; positive I_G indicates the Schottky junction breakdown, while negative I_G is caused by forward conduction. The voltages V_1 and V_2 are defined in Figs. 1 and 2. Graphic representation of each parameter, as a function of frequency, is also available. Since the program uses nonlinear analysis, the input power increments are not regular. An interpolating routine is therefore also available.

A sample graphic representation is shown in Fig. 5. The figure shows the predicted power performance of a Ku-band monolithic amplifier fabricated at our laboratory, at various power levels as a function of frequency, and the measured performance at several points. The measured data is shown as measured, with jig losses included. The agreement between the predicted and actual performances is quite good.

Typically, our circuit design philosophy follows the steps shown in Fig. 6. We use our large-signal program only after a small-signal design has been performed. In the small-signal design, however, we constrain the impedance levels presented to the FETs to the values dictated by the large-signal requirements. To obtain these values, our large-signal program is run in its optimizing mode in which case it prints the optimum loads required for the maximum designed power. Once these loads are known, the design of the input, interstage, and output matching network can be done in a straightforward manner using available CAD programs. Numerous network topologies will provide suitable small-signal performance. From our design experience, however, we found that several of them will provide very poor high-power performance; several design iterations are typically necessary until both a small-signal and a large-signal design can be simultaneously achieved. To the latter end, our large-signal analysis has proven to be extremely valuable.

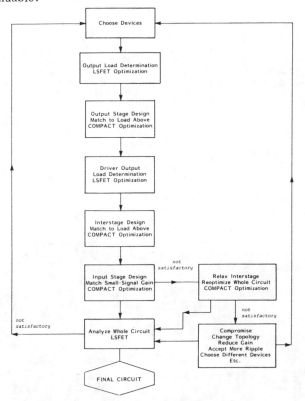

Fig. 6 Multistage power amplifier design steps.

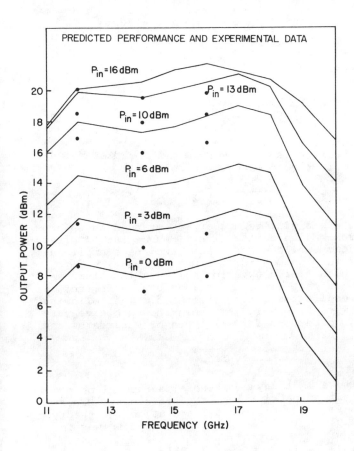

Fig. 5 Ku-band amplifier. Experimental data for P_{in} = 0, 3, 10, 13, 16 dBm.

References

1. J. E. Degenford et al., "A study of optimal matching circuit topologies for broadband monolithic power amplifiers," 1981 IEEE MTT-S International Microwave Symposium Digest, p. 351.

2. Y. Tajima, et al., "GaAs FET large-signal model and its application to circuit designs," IEEE Trans. ED-28, pp. 171-175 (1981).

ACCURATE MODELS FOR MICROSTRIP COMPUTER-AIDED DESIGN

E. Hammerstad and Ø. Jensen
ELAB
7034 Trondheim-NTH, Norway

Summary

Very accurate and simple equations are presented for both single and coupled microstrip lines' electrical parameters, i.e. impedances, effective dielectric constants, and attenuation including the effect of anisotropy in the substrate. For the single microstrip the effects of dispersion and non-zero strip thickness are also included.

Introduction

In microstrip design it is often observed that the physical circuit performance differs significantly from that theoretically calculated. This is due to factors such as attenuation, dispersion, and discontinuity effects, which again are functions of physical parameters and the actual circuit lay-out. The conclusion which may be drawn from this observation, is that optimum microstrip design should be based directly upon physical dimensions. This approach requires accurate models and for computer-aided design the models must also be in an easily calculable form.

The models presented in this paper represent partly completely new models, but also revisions and expansions of some which have been given earlier [1,2]. It has been a goal to obtain an accuracy which gives errors at least less than those caused by physical tolerances. The line models are based on theoretical data [1,2], while the discontinuity models also are based on experimental data.

Single microstrip

The model of the single microstrip line is based upon an equation for the impedance of microstrip in an homogeneous medium, Z_{01}, and an equation for the microstrip effective dielectric constant, ε_e.

$$Z_{01}(u) = \frac{\eta_0}{2\pi} \ln\left[\frac{f(u)}{u} + \sqrt{1+\left(\frac{2}{u}\right)^2}\right] \quad (1)$$

$$f(u) = 6 + (2\pi - 6)\exp\left[-\left(\frac{30.666}{u}\right)^{0.7528}\right] \quad (2)$$

Here η_0 is wave impedance of the medium (376.73Ω in vacuum) and u is the strip width normalized with respect to substrate height (w/h). The accuracy of this model is better then 0.01% for $u \le 1$ and 0.03% for $u \le 1000$.

$$\varepsilon_e(u,\varepsilon_r) = \frac{\varepsilon_r+1}{2} + \frac{\varepsilon_r-1}{2}\left(1+\frac{10}{u}\right)^{-a(u)\,b(\varepsilon_r)} \quad (3)$$

$$a(u) = 1 + \frac{1}{49}\ln\frac{u^4 + (u/52)^2}{u^4 + 0.432} + \frac{1}{18.7}\ln\left[1 + \left(\frac{u}{18.1}\right)^3\right] \quad (4)$$

$$b(\varepsilon_r) = 0.564\left(\frac{\varepsilon_r-0.9}{\varepsilon_r+3}\right)^{0.053} \quad (5)$$

The accuracy of this model is better than 0.2% at least for $\varepsilon_r \le 128$ and $0.01 \le u \le 100$.

Compared to earlier equations, these give much better accuracy, in the order of 0.1% for both impedance and wavelength. They are also complete with respect to range of strip width, while earlier ones where given in two sets.

Strip thickness correction

In correcting the above results for non-zero strip thickness, a method described by Wheeler[3] is used. However, some modifications in his equations have been made, which give better accuracy for narrow strips and for substrates with low dielectric constant.
For homogeneous media the correction is

$$\Delta u_1 = \frac{t}{\pi}\ln\left(1 + \frac{4\exp(1)}{t\coth^2\sqrt{6.517\,u}}\right) \quad (6)$$

where t is the normalized strip thickness. For mixed media the correction is

$$\Delta u_r = \frac{1}{2}(1 + 1/\cosh\sqrt{\varepsilon_r - 1})\,\Delta u_1 \quad (7)$$

By defining corrected strip widths, $u_1 = u + \Delta u_1$ and $u_r = u + \Delta u_r$, the effect of strip thickness may be included in the above equations:

$$Z_0(u,t,\varepsilon_r) = Z_{01}(u_r)/\sqrt{\varepsilon_e(u_r,\varepsilon_r)} \quad (8)$$

$$\varepsilon_{eff}(u,t,\varepsilon_r) = \varepsilon_e(u_r,\varepsilon_r) \cdot [Z_{01}(u_1)/Z_{01}(u_r)]^2 \quad (9)$$

Dispersion

As microstrip propagation is not purely TEM, both impedance and effective dielectric constant vary with frequency. Getsinger[4] has propsed the following model for dispersion in effective dielectric constant

$$\varepsilon_{eff} = \varepsilon_r - \frac{\varepsilon_r - \varepsilon_{eff}(0)}{1 + G(f/f_p)^2} \quad (10)$$

where $f_p = Z_0/(2\mu_0 h)$ may be regarded as an approximation to the first TE-mode cut-off frequency, while G is a factor which is empirically determined. Getsinger gave an expression for G which fitted experimental data for alumina substrates, but this did not give correct results for other substrates. The following expression for G has been shown to give very good results for all types of substrates now in use:

$$G = \frac{\pi^2}{12}\frac{\varepsilon_r - 1}{\varepsilon_{eff}(0)}\sqrt{\frac{2\pi Z_0}{\eta_0}} \quad (11)$$

There is today not general agreement on a model for microstrip impedance dispersion. Based on a parallel-plate model and using the theory of dielectrics, the following equation was found.

$$Z_0(f) = Z_0(o)\sqrt{\frac{\varepsilon_{eff}(o)}{\varepsilon_{eff}(f)}} \cdot \frac{\varepsilon_{eff}(f) - 1}{\varepsilon_{eff}(o) - 1} \quad (12)$$

While we would accept theoretical objections to this model on the grounds that microstrip impedance is rather arbitrarily defined, it may be pointed out that it agrees very well with the calculations of Krage & Haddad [5] and that in the limit $f \to \infty$ it is agreement with three of the definitions discussed by Bianco & al. [6]. It may also be noted that the model predicts a rather small but positive increase in impedance with frequency and that this would seem to fit experimental observations.

Coupled microstrips

The equations given below represent the first generally valid model of coupled microstrips with an acceptable accuracy. The model is based upon perturbations of the homogeneous microstrip impedance equation and the equations for effective dielectric constant. The equations have been validated against theoretically calculated data in the range $0.1 \le u \le 10$ and $g > 0.01$ (g is the normalized gap width (s/h)), a range which should cover that used in practise.

The homogeneous mode impedances are

$$Z_{01m}(u,g) = Z_{01}(u)/[1 - Z_{01}(u)\phi_m(u,g)/\eta_0] \quad (13)$$

where the index m is set to e for the even mode and o for the odd. The effective dielectric constants are

$$\varepsilon_{em}(u,g,\varepsilon_r) = \frac{\varepsilon_r+1}{2} + \frac{\varepsilon_r-1}{2} F_m(u,g,\varepsilon_r) \quad (14)$$

$$F_e(u,g,\varepsilon_r) = [1 + \frac{10}{\mu(u,g)}]^{-a(\mu)b(\varepsilon_r)} \quad (15)$$

$$F_o(u,g,\varepsilon_r) = f_o(u,g,\varepsilon_r)(1+\frac{10}{u})^{-a(u)b(\varepsilon_r)} \quad (16)$$

The modifying equations are as follows

$$\phi_e(u,g) = \varphi(u)/\{\psi(g)\cdot[\alpha(g)u^{m(g)}+[1-\alpha(g)]u^{-m(g)}]\} \quad (17)$$

$$\varphi(u) = 0.8645 u^{0.172} \quad (18)$$

$$\psi(g) = 1 + \frac{g}{1.45} + \frac{g^{2.09}}{3.95} \quad (19)$$

$$\alpha(g) = 0.5 \exp(-g) \quad (20)$$

$$m(g) = 0.2175 + [4.113 + (\frac{20.36}{g})^6]^{-0.251} + \frac{1}{323} \ln \frac{g^{10}}{1+(\frac{g}{13.8})^{10}} \quad (21)$$

$$\phi_o(u,g) = \phi_e(u,g) - \frac{\theta(g)}{\psi(g)} \exp[\beta(g)u^{-n(g)} \ln u] \quad (22)$$

$$\theta(g) = 1.729 + 1.175 \ln(1+\frac{0.627}{g+0.327 g^{2.17}}) \quad (23)$$

$$\beta(g) = 0.2306 + \frac{1}{301.8} \ln \frac{g^{10}}{1+(\frac{g}{3.73})^{10}} + \frac{1}{5.3} \ln(1+0.646 g^{1.175}) \quad (24)$$

$$n(g) = \{\frac{1}{17.7} + \exp[-6.424 - 0.76 \ln g - (\frac{g}{0.23})^5]\} \cdot \ln \frac{10+68.3 g^2}{1+32.5 g^{3.093}} \quad (25)$$

$$\mu(u,g) = g \exp(-g) + u \cdot \frac{20+g^2}{10+g^2} \quad (26)$$

$$f_o(u,g,\varepsilon_r) = f_{o1}(g,\varepsilon_r) \cdot \exp[p(g) \ln u + q(g) \cdot \sin(\pi \frac{\ln u}{\ln 10})] \quad (27)$$

$$p(g) = \exp(-0.745 g^{0.295})/\cosh(g^{0.68}) \quad (28)$$

$$q(g) = \exp(-1.366 - g) \quad (29)$$

$$f_{o1}(g,\varepsilon_r) = 1 - \exp\{-0.179 g^{0.15} - \frac{0.328 g^{r(g,\varepsilon_r)}}{\ln[\exp(1)+(g/7)^{2.8}]}\} \quad (30)$$

$$r(g,\varepsilon_r) = 1 + 0.15[1 - \frac{\exp(1-(\varepsilon_r-1)^2/8.2)}{1+g^{-6}}] \quad (31)$$

When the development of the above equations was done (1) and (2) were not available. Instead a simpler equation for homogeneous microstrip impedance with a maximum error of 0.3% for $u > 0.06$ was used:

$$Z_{01}(u) = \eta_0/(u+1.98 u^{0.172}) \quad (32)$$

The errors in the even and odd mode impedances where then found to be less than 0.8% and less than 0.3% for the wavelengths. The above model does not include the effect of non-zero strip thickness or assymetry. It is a goal at ELAB to include these factors, although the parameter range probably will have to be restricted to the strictly practically useful in regard of the complexity of the above equations.

Dispersion [7] is also not included presently. Getsinger has proposed modifications to his single strip dispersion model, but unfortunately it is easily shown that the results are asymptotically wrong for extreme values of gap width. However, it is possible to modify the Getsinger dispersion model so that the asymptotic values are satisfied for coupled lines, but the details remain to be worked out.

Attenuation

The strip capacitive quality factor is

$$Q_d = \frac{(1-q)+q\varepsilon_r}{(1-q)/Q_A + q\varepsilon_r/Q_s} \quad (33)$$

where q is the mixed dielectric filling fraction, Q_A is the dielectric quality factor of the upper half space medium, and Q_s is the substrate quality factor.

The strip inductive quality factor is approximated by [8]:

$$Q_c = \frac{\pi Z_{01} h f}{R_s c} \cdot \frac{u}{K} \quad (34)$$

where R_s is the skin resistance and K is the current distribution factor. Surface roughness will increase the skin resistance [1].

$$R_s(\Delta) = R_s(0)[1 + \frac{2}{\pi} \arctan 1.4 (\frac{\Delta}{\delta})^2] \quad (35)$$

where δ is the skin depth and Δ is the rms surface roughness.

For the single microstrip current distribution factor we have found

$$K = \exp(-1.2(\frac{Z_{01}(u)}{\eta_0})^{0.7}). \qquad (36)$$

to be a very good approximation to the current distribution factor due to Pucel et al.[9] provided that the strip thickness exceeds three skin depths.

The microstrip quasi-TEM mode quality factor, Q_o, is now given by

$$\frac{1}{Q_o} = \frac{1}{Q_d} + \frac{1}{Q_c} \qquad (37)$$

and the attenuation factor becomes

$$\alpha[\frac{dB}{m}] = \frac{20\pi}{\ln 10} \frac{c}{Q_o f \sqrt{\epsilon_{eff}}} \qquad (38)$$

The above loss equations are also valid for coupled microstrips[8], provided that the dielectric filling factor, homogenious impedance, and current distribution factor of the actual mode are used. Presently, no equations are available for the odd and even mode current distribution factors. Except for very tight coupling, however, the following approximation gives good results

$$K_e = K_o = \exp[-1.2(\frac{Z_{ole}+Z_{olo}}{2\eta_o})^{0.7}] \qquad (39)$$

Anistropy

From recent work [10,11] it may be shown that the above models are also valid with anisotropic substrates by defining an isotropyized substrate where

$$h_{eq} = h\sqrt{\epsilon_x/\epsilon_y} \qquad (40)$$

$$\epsilon_{eq} = \sqrt{\epsilon_x \epsilon_y} \qquad (41)$$

This method requires that one of the substrate's principal axes is parallel to the substrate (x-axis) and the other normal to the substrate (y-axis). The strip dimensions remain unchanged, but are normalized with respect to h_{eq}. The procedure for utilizing the above models then proceed through calculation of the mode capacitances, C_m.

$$C_m = 1/(v_{eqm} Z_{eqm}) \qquad (42)$$

Here v_{eqm} is the mode phase velocity and Z_{eqm} the mode impedance on the isotropyized substrate. Calculating the homogeneous mode capacitance, C_{m1}, the impedance and effective dielectric constant on the anistropic substrate are then

$$Z_{0m} = 1/(c\sqrt{C_m C_{m1}}) \qquad (43)$$

$$\epsilon_{effm} = C_m/C_{m1} \qquad (44)$$

The inductive quality factor of the anisotropic substrate, may be directly calculated with the above equations, while the substrate quality factor has to be modified:

$$Q_s = 2(\frac{1}{Q_{sx}} + \frac{1}{Q_{sy}})^{-1} \qquad (45)$$

Discontinuities

Due to the illnes, modelling of microstrip discontinuities could unfortunately not be finished to meet the digest deadline. The final paper presented at the symposium will give models for the microstrip open end, the gap, compensated right-angle bend, and the T-junction.

Acknowledgements

This work has been supported by the European Space Agency under contract no. 3745/78.

References

1. E.O.Hammerstad & F.Bekkadal:"Microstrip Handbook" ELAB-report, STF44 A74169, Feb. 1975, Trondheim.

2. E.O.Hammerstad:"Equations for Microstrip Circuit Design". Conference Proceedings, 5th. Eu.M.C., Sept. 75, Hamburg.

3. H.A.Weeler:"Transmission-Line Properties of a Strip on a Dielectric Sheet on a Plane". IEEE-trans. Vol. MTT-25, No. 8, Aug. 1977, p 631.

4. W.J.Getsinger:"Microstrip Dispersion Model" IEEE-trans., Vol. MTT-21, No.1, Jan. 1973, p. 34.

5. M.K.Krage & G.I.Haddad:"Frequency-Dependent Characteristics of Microstrip Transmission Lines" IEEE-trans, Vol. MTT-20, No. 10, Oct. 1972, P. 678.

6. B.Bianco et al.:"Some Considerations About the Frequency Dependence of the Characteristic Impedance of Uniform Microstrip". IEEE-trans, Vol. MTT-26, No. 3, March 1978, p. 182.

7. W.J.Getsinger: "Dispersion of Parallell-Coupled Microstrip" IEEE-trans. Vol. MTT-21, No. 3, March 1973, p. 144.

8. M.Mæsel:"A Theoretical and Experimental Investigation of Coupled Microstrip Lines". ELAB-report TE-168, April 1971, Trondheim.

9. R.A.Pucel et al.:"Losses in Microstrip". IEEE-trans., Vol. MTT-16, No. 6, June 1968, p. 342.

10. Ø.Jensen:"Single and Coupled Microstrip Lines on Anistropic Substrates" ELAB Project Memo No. 3/79, Project no.441408.04 June 1979, Trondheim.

11. M.Kobayashi & R.Terakado:"Accuratly Approximate Formula of Effective Filling Fraction for Microstrip Line with Isotropic Substrate and Its Application to the Case with Anistropic Substrate". IEEE-trans.,Vol.MTT-27,No 9,Sept.1979, p.776.

Design of Microwave GaAs MESFET's for Broad-Band Low-Noise Amplifiers

HATSUAKI FUKUI, SENIOR MEMBER, IEEE

Abstract—As a basis for designing GaAs MESFET's for broad-band low-noise amplifiers, the fundamental relationships between basic device parameters, and two-port noise parameters are investigated in a semiempirical manner. A set of four noise parameters are shown as simple functions of equivalent circuit elements of a GaAs MESFET. Each element is then expressed in a simple analytical form with the geometrical and material parameters of this device. Thus practical expressions for the four noise parameters are developed in terms of the geometrical and material parameters.

Among the four noise parameters, the minimum noise figure $F_{\min}$, and equivalent noise resistance R_n, are considered crucial for broad-band low-noise amplifiers. A low R_n corresponds to less sensitivity to input mismatch, and can be obtained with a short heavily doped thin active channel. Such a high channel doping-to-thickness (N/a) ratio has a potential of producing high power gain, but is contradictory to obtaining a low $F_{\min}$. Therefore, a compromise in choosing N and a is necessary for best overall amplifier performance. Four numerical examples are given to show optimization processes.

I. INTRODUCTION

THE GaAs Schottky-barrier gate field effect transistors (GaAs MESFET's) have demonstrated excellent noise and gain performance at microwave frequencies through K band [1]. The excellent microwave performance of GaAs MESFET's is certainly related to their channel properties. GaAs MESFET's to be used for broad-band low-noise amplifier applications, must have special requirements on their channel properties for optimum performance. The purpose of this paper is to investigate the fundamental relationships between the noise and small-signal properties, and the basic channel parameters of GaAs MESFET's. This information should be useful as a basis for device design.

II. REPRESENTATION OF NOISE PROPERTIES

A. Noise Parameters

From the circuit point of view, the GaAs MESFET can be treated as a black box of noisy two port. The noise properties of such a black box are then characterized by a set of four noise parameters in the binomial form [2]. A derivation of this form can be written as

$$F = F_{\min} + \frac{R_n}{R_{ss}} \left[\frac{(R_{ss} - R_{op})^2 + (X_{ss} - X_{op})^2}{R_{op}^2 + X_{op}^2} \right] \quad (1)$$

Manuscript received August 14, 1978; revised January 15, 1979.
The author is with Bell Laboratories, Murray Hill, NJ 07974.

where

F noise figure,
$F_{\min}$ minimum (or optimum) noise figure,
R_n equivalent noise resistance,
R_{ss} signal source resistance,
R_{op} optimum signal source resistance,
X_{ss} signal source reactance,
X_{op} optimum signal source reactance.

In this expression, $F_{\min}$, R_n, R_{op}, and X_{op} are the characteristic noise parameters of the device. Since the noise figure F is a function of its driving source impedance, the minimum noise figure $F_{\min}$ is achieved only when the driving source impedance is exactly at the optimum signal source impedance.

As has been well known, (1) can be represented on the source impedance Smith chart as a family of circles, each of which corresponds to a constant F value [3]. The spatial distance between two circles is related to R_n. The greater R_n corresponds to the shorter distance. In other words, the noise figure of a device with a small value of R_n is relatively insensitive to the variation in the signal source impedance. Thus small R_n is essential for a device to be used in a broad-band amplifier where a large tolerance is desirable in the input match. Furthermore, as will be seen later, R_n has a close relationship with power gain. Usually, the smaller R_n value corresponds to the higher gain in a given gate structure. In the design of a low-noise MESFET, therefore, obtaining a low R_n should be considered to be just as crucial as achieving a low $F_{\min}$.

It may be noted that an equivalent expression of (1), in terms of the source-reflection coefficient, is found elsewhere [4]. This expression is based on the circuit analysis using the s parameters. The s-parameter representation would be convenient to use in conjunction with the test facilities available for two-port investigation these days. However, the impedance parameter would provide us with direct insight into the operation of a device under consideration. Therefore, the noise parameters in the impedance form are adopted in this paper. If necessary, the noise parameters of this form can be transformed into other forms in a straightforward manner.

B. Noise Equivalent Circuit

It has been well accepted that the noise properties of a GaAs MESFET would be described by an equivalent circuit as shown in Fig. 1 [5], if the effect of reactive parasitic elements on the noise performance could be ignored. The major reactive parasitic elements are lead inductances and header stray capacitances. As the operating frequency increases, the impedance of such external elements may become comparable to those of the corresponding internal elements, and then the reactive parasitic effects are no longer negligible. Such a critical frequency, that the reactive parasitic effect begins to participate in the determination of the noise performance parameters, may vary from one noise parameter to another for a given

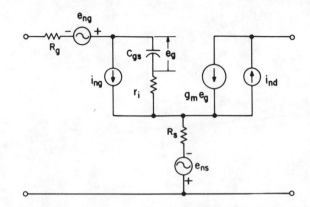

Fig. 1. Noise equivalent circuit of GaAs MESFET's. g_m is the transconductance which is assumed to be constant over the frequency range of interest. Element C_{gs} is the gate-source capacitance, r_i, the associated charging resistance, R_g, the ac gate metallization resistance, and R_s, the total source resistance series. Noise sources i_{ng}, i_{nd}, e_{ng}, and e_{ns} represent the induced gate noise, drain-circuit noise, thermal noise of R_g, and R_s, respectively.

device. For example, the critical frequency for $F_{\min}$ may be much higher than those for R_n, R_{op}, and X_{op}, analogous to microwave bipolar transistors [6].

III. Formulation of Noise Parameters in Terms of Device Geometrical and Material Parameters

A. Experimental Procedure and Results

Relationships between the noise parameters and equivalent circuit elements have been given in rigorous but complicated forms [5]. For practical purposes, however, it would be much more convenient if simple analytical forms of such relationships were available with reasonable accuracy.

In order to carry out this search, GaAs MESFET's with nominal gate length of 2 μm were used in the experiments described below. The MESFET's were mounted in the package which had a total input lead inductance of approximately 1 nH, a total output lead inductance of approximately 1 nH, a common-source lead inductance of approximately 0.12 nH, and a header stray capacitance of approximately 0.08 pF at both input and output [7]. As a result of such parasitic element values, these devices nearly satisfied the aforementioned requirement for possible elimination of the reactive parasitic effects on any of the four noise parameters at a test frequency of 1.8 GHz. Therefore, only the equivalent circuit elements shown in Fig. 1 will be referred to in the analyses described later on.

Table I shows the geometrical and material parameters of the six GaAs MESFET's used in the experiments. In Table I, N is the free carrier concentration in the active channel in units of 10^{16} cm^{-3}, L is the gate length in micrometers, a is the active channel thickness in micrometers, Z is the total device width in millimeters, and z is the unit gate width in millimeters. Since z was 0.25 mm for all devices, each device had either two or six paralleled unit

TABLE I
Geometrical and Material Parameters of Sample GaAs MESFET's

	DEVICE	N (10^{16}cm^{-3})	L (μm)	a (μm)	Z (mm)	z (mm)
a :	K864-1-02	11.0	1.85	0.174	0.5	0.25
b :	K976-1-04	7.5	1.85	0.40	0.5	0.25
c :	K949-1-12	5.5	1.85	0.44	0.5	0.25
d :	K949-3-02	5.5	2.3	0.35	1.5	0.25
e :	C75B-1-05	5.0	2.3	0.46	0.5	0.25
f :	C75B-3-04	5.0	2.3	0.38	1.5	0.25

TABLE II
Measured Values of Noise Parameters and Equivalent Circuit Elements of Sample GaAs MESFET's

DEVICE	F_{min} (dB)	R_n (Ω)	R_{op} (Ω)	X_{op} (Ω)	g_m (℧)	C_{gs} (pF)	R_g (Ω)	R_s (Ω)
a	1.80	31	40	85	0.031	1.0	2	8.5
b	1.29	-	45	145	0.021	0.65	2	6
c	1.56	-	45	140	0.019	0.62	3	6
d	1.70	13.3	23	47	0.047	2.0	1	3.5
e	1.73	94	50	125	0.018	0.68	2	8
f	2.04	15.3	35	45	0.044	2.2	4	3.5

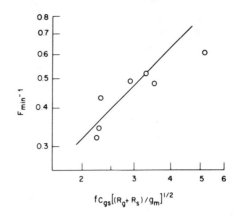

Fig. 2. Correlation between the minimum noise figure F_{min} and equivalent circuit elements, C_{gs}, g_m, R_g, and R_s.

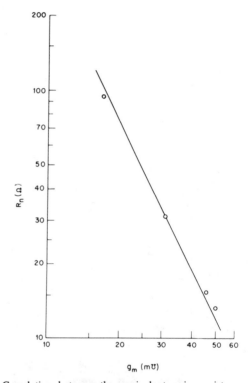

Fig. 3. Correlation between the equivalent noise resistance R_n and transconductance g_m.

gates. These devices were an early version of the low-noise GaAs MESFET reported elsewhere [8]. An epitaxial film, as the active channel, was grown directly on a semi-insulating substrate in the first four devices. In the last two devices, however, there was an additional undoped buffer layer between the substrate and active layer.

Four noise parameters F_{min}, R_n, R_{op}, and X_{op} were measured at 1.8 GHz under optimum gate-bias condition for each device, using the standard technique [9]. Equivalent circuit elements g_m, C_{gs}, R_g, and R_s were evaluated from the s-parameter measurement taken as a function of frequency under the zero gate-bias condition, in a similar way to that described in [10]. All the above parameters were measured at room temperature. The results are shown in Table II.

B. Derivation of Expressions for Noise Parameters in Terms of Equivalent Circuit Elements

It was assumed that the four noise parameters could be expressed in terms of equivalent circuit elements as follows:

$$F_{min} = 1 + k_1 f C_{gs} \sqrt{\frac{R_g + R_s}{g_m}} \quad (2)$$

$$R_n = \frac{k_2}{g_m^2} \quad (3)$$

$$R_{op} = k_3 \left[\frac{1}{4g_m} + R_g + R_s \right] \quad (4)$$

$$X_{op} = \frac{k_4}{f C_{gs}} \quad (5)$$

where k_1, k_2, k_3, and k_4 are fitting factors, and f is frequency.

Comparing these expressions with the experimental data shown in Table II would yield determination of the fitting factors. As seen in Figs. 2–5, it has been found that the expressions would well represent the measured values of the noise parameters if the fitting factors were chosen as follows:

$$k_1 = 0.016$$
$$k_2 = 0.030$$

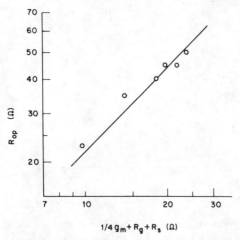

Fig. 4. Correlation between the optimum source resistance R_{op} and equivalent circuit elements g_m, R_g, and R_s.

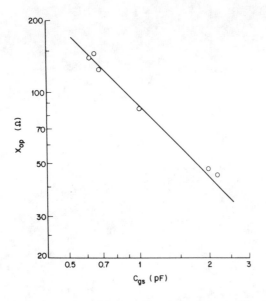

Fig. 5. Correlation between the optimum source reactance X_{op} and gate-source capacitance C_{gs}.

$$k_3 = 2.2$$
$$k_4 = 160$$

provided that R_n, R_{op}, X_{op}, R_g, and R_s are all in ohms, g_m in mhos, C_{gs} in picofarads, and f in gigahertz.

It may be remarked that the equivalent circuit elements as obtained with null gate bias are used in the above expressions for the noise parameters, although the noise parameters are provided with a certain gate bias. In spite of such a difference in the gate-bias conditions, the above relationships were empirically found to be present. If the equivalent circuit elements were measured under a gate-bias condition different from the null gate bias, the fitting factors would have to be modified.

A deviation of X_{op}, with increasing C_{gs}, from the linear relationship as seen in Fig. 5 was probably caused by an increased participation of the input lead inductance in X_{op}.

C. Semiempirical Expressions for Transconductance, Gate-Source Capacitance, and Cutoff Frequency

For design purposes, g_m and C_{gs} must be expressed in terms of the geometrical and material parameters of a device. Approximate expressions were then derived on a semiempirical basis as follows:

$$g_m = k_5 Z \left[\frac{N}{aL} \right]^{1/3} \mho \quad (6)$$

$$C_{gs} = k_6 Z \left[\frac{NL^2}{a} \right]^{1/3} \text{pF} \quad (7)$$

$$f_T = \frac{10^3 g_m}{2\pi C_{gs}} = \frac{9.4}{L} \text{ GHz} \quad (8)$$

in which fitting factors k_5 and k_6 were found to be 0.020 and 0.34, respectively, for the sample devices under the zero gate-bias condition.

Figs. 6 and 7 show comparisons of g_m and C_{gs}, respectively, between the calculated values using L, N, and a given in Table I and the measured values shown in Table II. They are in good agreement in both cases.

D. Simplified Expressions for Parasitic Resistances

Simplified expressions for R_g and R_s of the sample MESFET's would be given by [11]

$$R_g = \frac{17z^2}{hLZ} \; \Omega \quad (9)$$

$$R_s = \frac{1}{Z}\left[\frac{2.1}{a^{0.5}N^{0.66}} + \frac{1.1 L_{sg}}{(a-a_s)N^{0.82}} \right] \Omega \quad (10)$$

where h is the gate metallization height in micrometers, L_{sg} is the distance between the source and gate electrodes in micrometers, and a_s is the depletion layer thickness in micrometers at the surface in the source-gate space.

E. Practical Expressions for Noise Parameters

The substitution of (6)–(10) into (2)–(5) in association with the practical values of the fitting factors yields

$$F_{min} = 1 + 0.038 f \left[\frac{NL^5}{a}\right]^{1/6}$$
$$\cdot \left[\frac{17z^2}{hL} + \frac{2.1}{a^{0.5}N^{0.66}} + \frac{1.1 L_{sg}}{(a-a_s)N^{0.82}} \right]^{1/2} \quad (11)$$

$$R_n = 75 Z^{-2} \left[\frac{aL}{N}\right]^{2/3} \Omega \quad (12)$$

$$R_{op} = 2.2 Z^{-1} \left[12.5 \left(\frac{aL}{N}\right)^{1/3} \right.$$
$$\left. + \frac{17z^2}{hL} + \frac{2.1}{a^{0.5}N^{0.66}} + \frac{1.1 L_{sg}}{(a-a_s)N^{0.82}} \right] \Omega \quad (13)$$

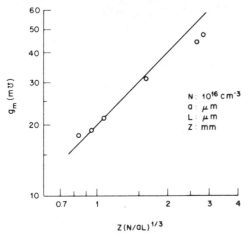

Fig. 6. Transconductance g_m as a function of channel parameters Z, L, a, and N.

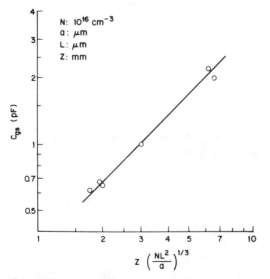

Fig. 7. Gate-source capacitance C_{gs} as a function of channel parameters Z, L, a, and N.

$$X_{op} = \frac{450}{fZ}\left[\frac{a}{NL^2}\right]^{1/3} \Omega. \quad (14)$$

Again, units are gigahertz (GHz) for f, millimeter (mm) for z and Z, micrometer (μm) for a, a_s, h, L, and L_{sg}, and 10^{16} cm^{-3} for N.

Since F_{min} is dominated by the parasitic resistances outside the gate region, F_{min} is structure sensitive [12]. Remember that (11) is suitable for the simplest structure of MESFET's. As sophistication increases in the structure, the proper expression for F_{min} may be obtained after the corresponding modification primarily to (10), and hence to (11). Applications of the gate recess structure [8], [12] and n$^+$-GaAs epitaxial layer [13] are the major examples of structural variations so far reported.

In (12), it is seen that *a small R_n value can be obtained with a short gate device having a heavily doped thin active channel.* The expression for R_n given in (12) may hold, regardless of the structural modification applied to a section of the channel outside the gate region.

IV. Discussions on Minimum Noise Figure, Associated Power Gain, and Input Impedance Match

A. Example 1

In order to see the characteristic variations of F_{min} and R_n as functions of N and a, the following conditions are assumed:

$$L = L_{sg} = 1.0 \ \mu\text{m}$$
$$R_g = 4.0 \ \Omega$$
$$f = 0.4 f_T = 3.76 \text{ GHz}.$$

Furthermore, a_s is assumed to be approximately equal to the gate depletion layer thickness under null gate-bias condition a_0 in numerical value. This parameter has been given by [11]

$$a_0 = \left[\frac{0.706 + 0.06 \log N}{7.23 N}\right]^{1/2} \mu\text{m}$$

for aluminum gates, as far as $a > a_0$. Under such conditions, (11) reduces to

$$F_{min}|_{f=0.4f_T} = 1 + 0.15[N/a]^{1/6}$$
$$\cdot \left[1.82 + \frac{1.9}{a^{0.5} N^{0.66}} + \frac{1}{(a-a_0) N^{0.82}}\right]^{1/2}. \quad (15)$$

The calculated values of F_{min} by (15) and of R_n by (12) are shown in Fig. 8, both as the contour mapping on the a-N plane. It can be seen that F_{min} is a weak function of N for a given value of a. Also, F_{min} takes a low value in the region where the a/N ratio is high. This is the contradictory condition to obtaining the low value of R_n. The high a/N ratio tends to provide not only the critical noise tuning but also the low power gain. Therefore, there is a compromise in choosing a and N for the best overall performance as an amplifying device.

In Fig. 8, there are two other curves, as shown with the dash–dotted line, which indicate practical limits for the selection of a and N. The upper curve corresponds to the a and N values which provide a drain–source breakdown voltage V_B, of 10 V if the gate is biased at the pinchoff voltage, $-V_p$. The lower curve indicates the thickness of the gate depletion layer at zero bias for aluminum gates.

B. Example 2

In order to see a clear distinction between the low a/N device and high a/N device, the following two representatives were assumed under the same conditions, as used in the previous section.

Device A: $N = 12.5 \times 10^{16}$ cm^{-3} and $a = 0.2$ μm.
Device B: $N = 5.0 \times 10^{16}$ cm^{-3} and $a = 0.5$ μm.

The noise performance of these devices was calculated using (12)–(15). The results are then plotted on the signal

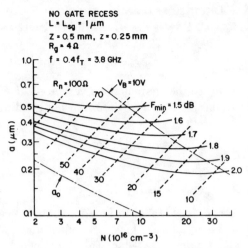

Fig. 8. Contours of equi-$F_{\min}$ and equi-R_n on the a-N plane for 1-μm gate devices.

	DEVICE	$N(10^{16}\,\text{cm}^{-3})$	$a\,(\mu\text{m})$
$L = L_{sg} = 1\,\mu\text{m}$	---- A	12.5	0.2
$Z = 0.5$ mm, $z = 0.25$ mm	——— B	5.0	0.5
$R_g = 4\,\Omega$, $f = 3.8$ GHz			
$Z_0 = 50\,\Omega$			

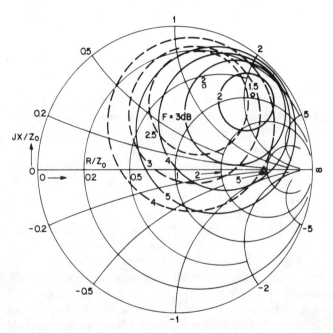

Fig. 9. The noise performance chart for two 1-μm gate devices.

source impedance chart [3], as shown in Fig. 9. It may be concluded that device B may reach the better noise performance in the case of a narrow-band application with the individually designed circuitry. However, device A may be more suitable in such a case that relatively large tolerances for the impedance variation are demanded. In addition, device A usually gives gain higher than that of device B, because of the higher value of g_m with device A.

C. Example 3

As seen in Fig. 9, device B, in spite of the smaller value of $F_{\min}$, would exhibit much worse noise figure than

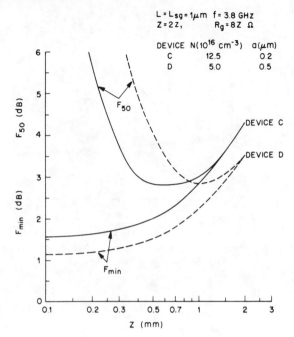

Fig. 10. Noise figures $F_{\min}$ and F_{50} as functions of total device width Z for two 1-μm gate devices.

device A when they are simply inserted into a 50-Ω coaxial system. Such a 50-Ω system insertion noise figure can be designated as F_{50}, which is

$$F_{50} = F_{\min} + \frac{R_n}{50}\left[\frac{(50 - R_{op})^2 + X_{op}^2}{R_{op}^2 + X_{op}^2}\right]. \quad (16)$$

Next shown are the dependence of $F_{\min}$ and F_{50} on Z, provided that $Z = 2z$. Two representative devices C and D were chosen, which were the same as devices A and B, respectively, except for an additional assumption that $R_g = 8Z$, in which R_g is in units of ohms and Z is in millimeters. Parameters $F_{\min}$ and F_{50} were then calculated as functions of Z, using (11)–(14) and (16). The results are shown in Fig. 10, in which $F_{\min}$ gradually increases with increasing Z for both devices. On the contrary, F_{50} varies as a strong function of Z and takes a minimum for moderate values of Z, as also shown in Fig. 10. For the overall performance as a general purpose low-noise MESFET with two paralleled unit-gates 1 μm long, an optimum value of the total gate width could be evaluated to be 0.4–0.5 mm for device C, and 0.8–1.0 mm for device D.

D. Example 4

In [14], the noise and gain performance as measured at 8 GHz for GaAs MESFET's with 0.5-μm-long gates was described. The device consists of four sections of 70-μm-wide units, i.e., $z = 0.07$ mm and $Z = 0.28$ mm. The height of aluminum as the gate metallization was 0.5 μm. The source-to-gate distance was 0.8 μm. The active layer thickness was claimed to be between 0.1 and 0.2 μm. The gate pinchoff voltage V_p of typical devices was 2.0 V. Since the representative value of N was 25×10^{16} cm^{-3}, the corre-

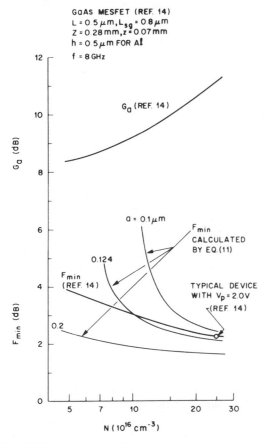

Fig. 11. Calculated values of F_{min} and measured values of F_{min} and associated power gain G_a for 0.5-μm gate devices.

sponding value of a was evaluated to be 0.124 μm. Thus F_{min} at 8 GHz was calculated using (11) as a function of N for the three values of a, i.e., 0.1, 0.124, and 0.2 μm.

As shown in Fig. 11, the measured F_{min} values happened to be between the two calculated curves of F_{min} for $a=0.1$ and 0.2 μm. Moreover, the measured F_{min} for $N \geqslant 10\times10^{16}$ cm^{-3} agreed well with the calculated F_{min} for $a=0.124$ μm. However, the measured F_{min} for $N<10 \times 10^{16}$ cm^{-3} appeared to be better than the calculated F_{min} for $a=0.124$ μm. This discrepancy could have been caused by an increased active layer thickness of the actual devices with $N<10\times10^{16}$ cm^{-3}, in order to maintain the zero gate-bias drain current to be finite in the positive direction. Such an increase in a should result in an improvement of F_{min} from that predicted for $a=0.124$ μm, as seen in Fig. 11. Therefore, it can be concluded that the experimental data on the noise performance, [14, fig. 2], is well explained by (11) of this paper.

In the same reference figure as mentioned above the associated power gain G_a was also shown as a function of N. The data are replotted in Fig. 11 in which G_a increases with increasing N and hence N/a, as has been predicted in the previous sections of this paper.

E. Remarks

In general, with increasing N/a, G_a increases and R_n decreases. If R_g is small enough as compared to R_s, F_{min} is improved with an increased value of N for a given value of a. This has been empirically known in the industry [15]. The reason is that R_s dominates F_{min} for a given value of L, and that R_s decreases with increasing the Na product. In the simple planar channel structure used here, F_{min} increases with decreasing a for a given N. However, in a sophisticated channel structure, this may no longer hold partly due to a possible decrease in the effective gate length [11], [13].

V. Conclusions

A set of four noise parameters for GaAs MESFET's of the planar channel structure were semiempirically found in terms of simple functions of its equivalent circuit elements. Each of the equivalent circuit elements was further expressed by the device geometrical and material parameters in a simple analytical form. Combining these two things together yielded the practical expressions for the four noise parameters in terms of the device geometrical and material parameters.

Among the four noise parameters, F_{min} and R_n were regarded as most crucial for a device to be used in a broad-band low-noise amplifier. Because a device with a small value of R_n behaves rather insensitively to the variation in the signal source impedance, the variation in the noise figure over the band can be expected to be small. In addition, as the inverse of R_n is closely related to power gain, smaller R_n can enhance the gain performance.

The aforementioned expression for R_n indicates that a small value of R_n can be obtained with a short gate device having a heavily doped thin active channel (i.e., a high N/a ratio). This is, in general, the contradictory condition to obtaining a low value of F_{min}. Although F_{min} is a weak function of N for a given value of a, F_{min} takes a low value in the region where the N/a ratio is low. However, the low N/a ratio tends to make the noise tuning critical and to degrade the power gain. Therefore, a compromise in choosing a and N is necessary for the best overall amplifier performance.

The proper selection of a and N may vary depending upon the particular purpose of an amplifier. Four examples were given to show the dependence of the noise, gain and input matching properties on a and N in a practical manner. If the gate metallization resistance were designed to be small enough to the total source series resistance, the minimum noise figure, associated power gain, and input matching sensitivity would all be improved with increasing N for a given a. This is a case which has been often observed in the industry.

Acknowledgment

The author is grateful to D. E. Iglesias and W. O. Schlosser for their help in measuring the noise properties and s parameters of the sample devices. He is also thankful to J. V. DiLorenzo, R. H. Knerr, R. Trambarulo, and H. Wang for their careful reading of the original manuscript.

References

[1] H. F. Cooke, "Microwave FET's—A status report," in *IEEE ISSCC Dig. Tech. Papers*, 1978, pp. 116–117.
[2] H. Rothe and W. Dahlke, "Theory of noisy fourpoles," *Proc. IRE*, vol. 44, pp. 811–818, June 1956.
[3] H. Fukui, "Available power gain, noise figure, and noise measure of twoports and their graphical representation," *IEEE Trans. Circuit Theory*, vol. CT-13, pp. 137–142, June 1966.
[4] J. A. Eisenberg, "Systematic design of low-noise, broad band microwave amplifiers using three terminal devices," in *Microwave Semiconductor Devices, Circuits and Applications, Proc. Fourth Cornell Conf.*, 1973, pp. 113–122.
[5] R. A. Pucel, H. A. Haus, and H. Statz, "Signal and noise properties of gallium arsenide microwave field-effect transistors," in *Advances in Electronics and Electron Physics*. New York: Academic, vol. 38, 1975, pp. 195–265.
[6] H. Fukui, "The noise performance of microwave transistors," *IEEE Trans. Electron Devices*, vol. ED-13, pp. 329–341, Mar. 1966.
[7] W. O. Schlosser, private communication.
[8] B. S. Hewitt *et al.*, "Low-noise GaAs M.E.S.F.E.T.S.," *Electron. Lett.*, vol. 12, pp. 309–310, June 10, 1976.
[9] "IRE Standards on Electron Tubes: Methods of Testing," 1962, 62 IRE 7 S1, pt. 9: Noise in linear twoports.
[10] J. Jahncke, "Höchstfrequenzeigenshaften eines GaAs MESFET's in Steifenleitungstechnik," *Nachrichtentech. Z.*, vol. 5, pp. 193–199, May 1973.
[11] H. Fukui, "Determination of the basic device parameters of a GaAs MESFET," *Bell Syst. Tech. J.*, vol. 58, pp. 771–797, Mar. 1979.
[12] B. S. Hewitt *et al.*, "Low-noise GaAs MESFET's: Fabrication and performance," in *Gallium Arsenide and Related Compounds (Edinburg) 1976, Conf. Series No. 33a*, The Inst. Physics, Bristol and London, 1977, pp. 246–254.
[13] H. Fukui, "Optimal noise figure of microwave GaAs MESFET's," *IEEE Trans. Electron Devices*, vol. ED-26, pp. 1032–1037, July 1979.
[14] M. Ogawa, K. Ohata, T. Furutsuka, and N. Kawamura, "Submicron single-gate and dual-gate GaAs MESFET's with improved low noise and high gain performance," *IEEE Trans. Microwave Theory Tech.*, vol. MTT-24, pp. 300–306, June 1976.
[15] H. F. Cooke, "Microwave field effect transistors in 1978," *Microwave J.*, vol. 21, no. 4, pp. 43–48, Apr. 1978.

Addendum to "Design of Microwave GaAs MESFET's for Broad-Band Low-Noise Amplifiers"

HATSUAKI FUKUI

It has been called to the author's attention that (3) in the above paper[1] appears to be inadequate [1], especially for scaling [2]. Considering this situation the expression should read

$$R_n = \frac{k_2}{g_m} \quad (3)$$

where $k_2 = 0.8$.

Manuscript received June 3, 1981.
The author is with Bell Laboratories, Murray Hill, NJ 07974.
[1]Hatsuaki Fukui, *IEEE Trans. Microwave Theory Tech.*, vol. MTT-27, pp. 643–650, July 1979.

This modification leads to rewriting (12) as follows:

$$R_n = \frac{40}{Z}\left[\frac{aL}{N}\right]^{1/3} \Omega. \quad (12)$$

Consequently, the numerical values for R_n in Fig. 8 should be, in descending order, 46, 39, 33, 29, 25, 21, 18, and 15. Figs. 9 and 10 are also slightly affected by the revised expression. However, the principal statement and conclusions remain unchanged.

The author wishes to thank Dr. R. A. Pucel for his encouragement concerning this amendment.

References

[1] S. Weinreb, "Low-noise cooled GASFET amplifiers," *IEEE Trans. Microwave Theory Tech.*, vol. MTT-28, pp. 1041–1054, Oct. 1980.
[2] A. F. Podell, "A functional GaAs FET noise model," *IEEE Trans. Electron Devices*, vol. ED-28, pp. 511–517, May 1981.

Addendum: *IEEE Trans. Microwave Theory Tech.*, vol. MTT-29, p. 1119, Oct. 1981.

ACCURATE COUPLING PREDICTIONS AND ASSESSMENTS IN MMIC NETWORKS

H.J. Finlay, J.A. Jenkins, R.S. Pengelly and J. Cockrill

Plessey Research (Caswell) Ltd., Caswell, Towcester,
Northants., U.K.

ABSTRACT

A major design requirement for GaAs MMIC networks is a knowledge of the coupling properties between transmission lines and within active switches. This is believed to be the first report on coupling effects between conductors spaced by more than a substrate thickness (h). Measurements and predictions are given on over 20 circuits for (i) coupled line pairs, with and without intervening shielding lines, (ii) transmission line crossover networks, and (iii) switching amplifier applications.

INTRODUCTION

With the present growth in MMIC technology and the need to design for optimum performance, it has become essential to obtain detailed characterisations of many transmission line networks in order to accurately model and predict their behaviour at microwave frequencies. One important aspect of these networks concerns the coupling properties between transmission lines and within active switches. One example includes high capacity satellite switched TDMA microwave switching matrices utilising coupled crossbar architectures which may soon be implemented using MMIC networks; a major design requirement is a knowledge of the coupling between the matrix transmission lines and crossover networks. Another example can be found in MMIC switching amplifiers where the 'off' state isolation can be degraded due to 'on-chip' coupling between metallised tracks. Coupling problems have also been observed between conductors on GaAs and bond wires during the development of a compact dual gate FET switch module (1). In order to predict these coupling effects for design purposes, it is necessary to make use of coupled line theories. Although some have existed for a considerable period of time, few have been verified as to their usefulness in GaAs ICs. A review of the theories has shown that some are very inaccurate while others give satisfactory results. Appropriate theories have been selected by comparison with measured results and measurement errors have been identified where possible. One of the few papers dealing with this subject has been reported by M. Le Brun et al (2) wherein closely spaced conductors have been characterised with and without intervening shielding strips for spacings smaller than a substrate thickness (h). In this paper we will address the coupling problems between various structures having separations greater than a substrate thickness (h). We believe this is the first report of coupling in GaAs under these conditions where coupling factors are more difficult to predict; more susceptable to error and typical of many practical cases. Predictions and measurements are presented for various coupled line pairs, with and without intervening shielding lines, transmission line crossover networks and switching amplifiers. Results on both forward (S_{13}) and backward waves (S_{12}) are given throughout from 2 to 18 GHz. Unlike reference (2), particular emphasis has been placed on the dominant S_{13} contributions.

CHARACTERISATION AND ASSESSMENTS

The characterisations were based on measurements from different circuits. A few selections are shown in Fig. 1,5,8. Firstly, various coupled line pairs (Fig. 1), having different lengths (L) and separations (S) were measured. Secondly, in Fig. 5 several cases having intervening terminated strips were characterised with outer line pair separations (S) and overall length L. Thirdly, a number of crossover transmission structures were evaluated. These were designed to minimise crossover capacitance to reduce coupling (Fig. 8).

Each circuit was placed on a precision carrier designed to give the best compromise between a number of constraints. The small GaAs substrates (.3 mms thick) were interfaced in most cases with .25 mms thick ceramic microstrip lines that were flared out to minimise extraneous coupling. These were then interfaced with .635 mms thick ceramic microstrip to interconnect with a precision microstrip launcher. Unfortunately, this requires a number of transitions but these have been previously characterised and have been found to give excellent results. For example, a .635 to .25 mm microstrip transition has a VSWR of less than 1.1 up to 18 GHz. The test fixtures were designed to be large enough to suit measurement equipment but sufficiently small in order to reduce the probability of box mode propagation. Box geometries were configured around the coupled sections to reduce waveguide modes at least up to 14 GHz.

The results of a number of coupled line pairs will now be presented. In Fig. 1,2 the measurements of two comparatively long 50 ohm lines on semi-insulating GaAs demonstrate good agreement with theory (L = 6.1 mms, S = .5, 2.0 mms). Track widths are 220 microns. Theoretical predictions using in house (3) and commercial programmes based on Compact/Super Compact are given (4). For shorter 50 ohm coupled line pairs, Fig. 3,4 (L = 2.0 mms) comparisons are again made for close and wide separations (S = .5, 4.0 mms). Predictions from M. Sobhy (5), Compact and Super Compact (4) for the GaAs coupled sections only are plotted. Predictions incorporating corrections for the microstrip feed lines are included based on in house (3) and Thomson CSF (2) programmes. From Figs. 5,6,7 the coupling results with an intervening terminated line are plotted. All lines are 50 ohm impedance, L = 2.0 mms, S = 1.2 and 4.2 mms respectively. In Fig. 5,7 the shielding lines are short circuited at both ends while in Fig. 6 they are loaded in 50 ohms. Theories from M. Le Brun (2) and A.J. Holden (3) again include corrections for feed lines. Single (uncorrected) coupled sections are plotted from Sobhy (5). In Fig. 8a, b, the four port couplings of a minimum capacitance matrix crossover network are presented using 6 micron air bridge structures. The degree of coupling is higher than predicted from a simple crossover capacitance model and is due to the proximity of the orthogonal input and output lines, Fig. 8a. Predictions taking into account line coupling were within 3 dB of measurement.

COMPARISON OF THEORIES WITH MEASUREMENTS

From these selected results, it is apparent that there are significant differences between the theories employed. The results from standard Compact (4) based on exact Bryant and Weiss analyses have given the largest discrepancies. In particular S_{13} is very inaccurate. However, the predictions of Super Compact which makes use of recent theories from Hammerstad and Jensen (6) have given good agreement in both S_{12} and S_{13} values. In-house analyses (3) based on multiconductor TEM approaches have been generated primarily to suit GaAs ICs. The programmes of Sobhy (5) make use of time domain analyses and converts to the frequency domain. As such they are quite general and higher order modes other than TEM can be included. Programmes STRIP and ESOPE from TCSF (2) use essentially a TEM approach and have been demonstrated previously for multiconductor and coplanar arrays in GaAs.

From the measurements it is apparent that the major contributions to the coupled powers is not S_{12} but S_{13}. This forward wave S_{13} can be up to 20 dB greater than S_{12} at higher frequencies. As the separation (S) increases the proportion of S_{13} power compared with S_{12} also increases particularly at higher frequencies. This is also true for the intervening shielded cases, Fig. 5,6,7. Therefore, unlike closely coupled cases, for S>h the forward wave may cause significant degradations in certain circuits. For example, parallel coupled couplers having S>h will have poor directivity in GaAs. Furthermore, unlike the closely coupled cases in (2) the shielding effect of a grounded or resistively terminated intervening strip gives only a minor (5 dB) reduction (S = .5) in coupling. This is in keeping with predictions (3). The intervening strip has a negligible shielding effect for S>1.0 mms. In order to validate the accuracy of these measurements a number of carefully controlled evaluations were recorded to determine the likely errors in measurements. For all the cases considered here, investigations were carried out to detect the presence of stray radiation or waveguide modes. Using profiled lossy material, no significant changes were detected in the measurements.

The extraneous coupling contributions due to the .25 and .635 mms ceramic feed lines, included in the in-house predictions were independently assessed using identical circuits by terminating the ceramic microstrip lines in precision chip resistor loads, Fig. 1,5. The spurious radiation in the test circuits is around -50 dB and -40 dB for the S_{12} and S_{13} respectively at 10 GHz - shown dotted Fig. 3(a),(b). Measurements above these curves will be less than 3 dB in error which is the case for most curves. All of these results were recorded on a calibrated automatic network analyser. For coupled lines having wide spacings S = 3, 4 mms, couplings reach an asymptotic level independent of distance and are not adequately described by TEM theory alone. Studies on surface wave theory have predicted levels of typically -35 dB at 12 GHz worst case and is independent of distance (S).

Finally, these findings have direct bearing on the performance of switching amplifier designs. Factors affecting switching ranges particularly in the 'off' state are coupling between conductors, crossovers, surface waves and between stages. An example of this is a two stage MMIC amplifier. Predicted gains were 15 dB while measured gains were 14 dB. In the 'off' state predicted isolation was -55 dB while measurements gave -25 dB. However, by redesigning this MMIC, improved gain (18 dB), 'off' isolation (-40 dB) and dynamic range (58 dB) was achieved over 7-11.0 GHz. Although only a preliminary result, this indicates that in order to optimise switching ranges in MMICs, gains should be increased and 'on-chip' coupling should be minimised by adopting the best trade offs in coupled powers.

CONCLUSIONS

The coupling characteristics of transmission line pairs and crossover networks on .3 mms thick GaAs have been presented for separations greater than .5 mms or 1.5h. Significant coupling values have been identified, particularly in the forward wave case, placing limitations on the directivity of GaAs side coupled couplers around the 20 dB level. The effects of intervening shielding lines between coupled pairs for S>1.5h are insignificant. The theories of Hammerstad (6), Le Brun (2), Holden (3) and Sobhy (5) have been shown to give accurate predictions in practice. This design information provides an essential groundwork for MMICs such as the design of switching amplifiers where good isolation and high dynamic ranges are required.

ACKNOWLEDGEMENTS

The authors are grateful to Mr. M. Gibson (ESTEC) and to the directors of Plessey Research (Caswell) Ltd. for permission to publish this paper. The assistance of R.S. Butlin, C. Suckling A.J. Holden and Zena Jackson is appreciated.

REFERENCES

1. H. Finlay, 'An optimised two Gigabit dual gate FET switch for satellite systems using QPSK modulators' 11th Eu.MC, p.219, 1981.
2. M. Le Brun et al 'Coupling and impedance between line and ground electrodes on GaAs, implications for MMIC design' 11th Eu.MC p.850 1981.
3. A.J. Holden 'Coupling in GaAs' Private Comm. Plessey Research.
4. Compact/Super Compact programmes, COMSAT Eng. Inc. CGIS.
5. M. Sobhy, Private Communication, Univ. of Kent, U.K.
6. E. Hammerstad et al 'Accurate models for microstrip computer aided design' IEEE Int. Microwave Symposium, p.407-409, 1980.

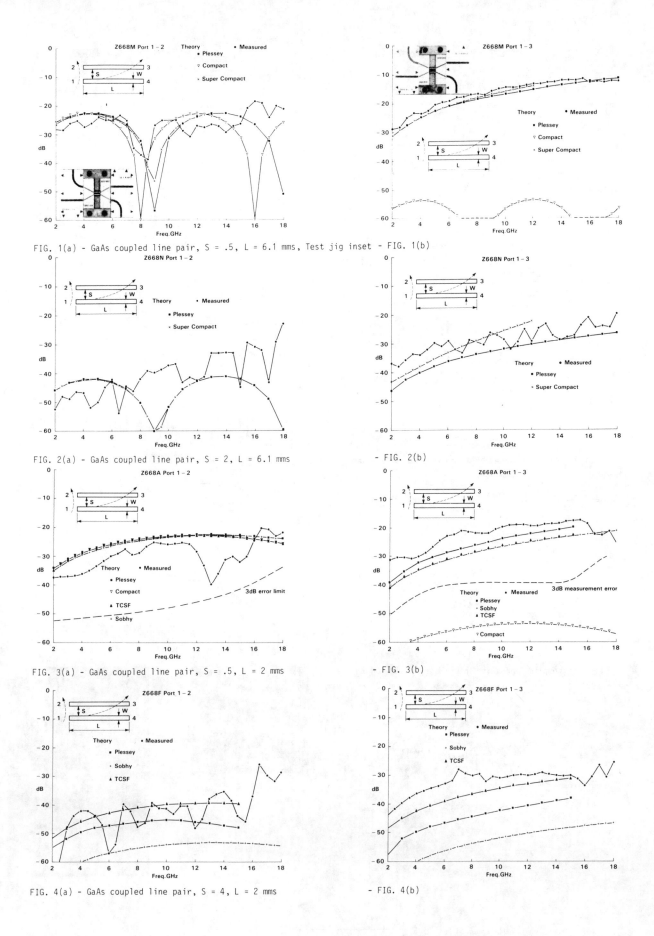

FIG. 1(a) - GaAs coupled line pair, S = .5, L = 6.1 mms, Test jig inset - FIG. 1(b)

FIG. 2(a) - GaAs coupled line pair, S = 2, L = 6.1 mms   - FIG. 2(b)

FIG. 3(a) - GaAs coupled line pair, S = .5, L = 2 mms   - FIG. 3(b)

FIG. 4(a) - GaAs coupled line pair, S = 4, L = 2 mms   - FIG. 4(b)

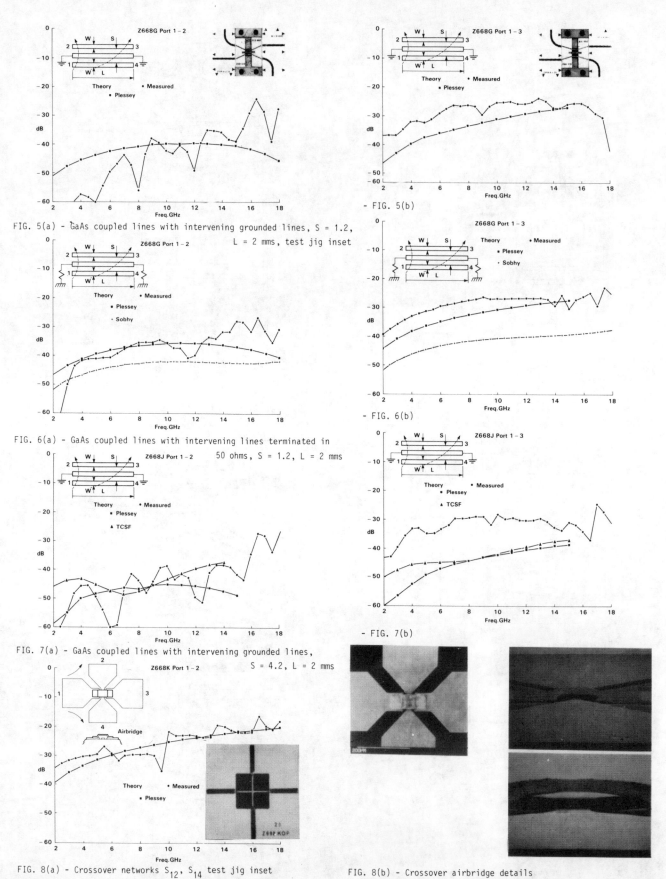

FIG. 5(a) - GaAs coupled lines with intervening grounded lines, S = 1.2, L = 2 mms, test jig inset

FIG. 5(b)

FIG. 6(a) - GaAs coupled lines with intervening lines terminated in 50 ohms, S = 1.2, L = 2 mms

FIG. 6(b)

FIG. 7(a) - GaAs coupled lines with intervening grounded lines, S = 4.2, L = 2 mms

FIG. 7(b)

FIG. 8(a) - Crossover networks S_{12}, S_{14} test jig inset

FIG. 8(b) - Crossover airbridge details

THE DESIGN AND CALIBRATION OF A UNIVERSAL MMIC TEST FIXTURE

James A. Benet

Rockwell International
Collins Communication Systems Division
Anaheim, California

ABSTRACT

A universal test fixture suitable for performing repeatable, nondestructive microwave tests for characterizing various sized monolithic microwave integrated circuit (MMIC) chips has been developed at Rockwell International. The fixture, which encloses the MMIC chip, is designed to accommodate multiple RF inputs and outputs as well as up to 36 independent isolated bias connections. A method for calibrating the fixture on an automatic network analyzer (ANA) without the use of known precision calibration standards was also developed. A description of the fixture and the calibration method is presented in this paper.

INTRODUCTION

Advances in microwave gallium arsenide (GaAs) technology has lead to the emergence of monolithic microwave integrated circuits (MMIC) to solve the problem of mass producing low cost, high reliable microwave circuits. The MMIC combines many active (GaAs) devices with appropriate passive circuit elements to produce a single chip under one square centimeter in size which performs the electrical tasks equivalent to several microstrip circuits occupying ten to twenty times the area.

A problem arises with regard to testing and characterizing these devices. A method or technique needed to be developed to accurately measure the microwave performance without destroying the chip (for reuse). To solve this problem a project was initiated to develop an appropriate MMIC test fixture.

DESIGN AND FABRICATION

Design Objectives

The fundamental design goal was to develop a fixture system to perform accurate, nondestructive RF tests on a wide variety of chips which differ both functionally and physically. The fixture must allow for quick connections and disconnections of the MMIC. The fixture also must accommodate a large number of bias inputs without necessitating bonding during the testing process. Multiple RF input/output connections, as well as the multiple bias connections, must also be provided. Provisions for monitoring bias voltages is another desirable feature. Metal walls completely enclosing the input/output circuits and the MMIC chip must be provided to maintain RF shielding and reduce external noise inputs to the device under test. Since a subcarrier would be required, it must be designed such that it could be produced inexpensively. Finally, since the overall objective is to obtain accurate RF measurements, the fixture must provide a means by which it can be calibrated on an automatic network analyzer (ANA). Hence, the development of calibration pieces and appropriate measurement software must be included in the design objectives.

MMIC Subcarrier

To achieve nondestructive testing of the fragile MMIC chips and to avoid the necessity of making bonding connections during tests, the MMIC chip is bonded to a .062 inch thick, copper subcarrier which is made from a copper-backed dielectric (or PC) board. The subcarrier is shown in Figure 1. The dielectric material, which is .010 inch thick, is machined away in the center section of the subcarrier where the MMIC device is mounted. Bias lines are etched on the PC board sections which extend out on each side of the center section. These microstrip lines fan out to a pattern of 18 metallized pads which are .040 inch square. Small tabs extend outward in the center section along the measurement axis to interface with the fixture. The entire subcarrier is gold-plated to prevent copper oxidation.

The subcarrier is designed to accommodate MMIC chips of various sizes from .5 x .5 x .125 mm up to 10 x 8 x .625 mm with up to 36 independent isolated bias input. Leader microstrip substrates can be placed in front and in back of the MMIC device to accommodate very small chips or to convert the MMIC to a beam lead device. Small SiO chip capacitors are bonded to the subcarrier between the MMIC chip and the PC board to provide RF isolation on the bias lines.

The overall size of the subcarrier is 1.8 x .84 inches; however, this size can be reduced considerably after the testing is completed by merely shearing off the leader tabs and most of the PC board sections. The final dimensions could be as small as 2 mm long by 5 mm wide depending on the size of the chip and the bias capacitors. In the testing configuration the length of subcarrier with respect to the RF path is .4 inch or less. Although the size of the PC board sections with the bias pads is always fixed, the center section length is "customized" to fit the length of the particular MMIC chip which is being tested.

To produce the subcarrier inexpensively, the bias line pattern is repeated sixty times on a 16 by 10 inch panel. The entire panel is then etched and the individual PC boards are cut from the panel using an automatic (computerized) routing machine. The tabs and grooves are machined in for each size by stacking and machining the boards in lots of ten or more. Produced in this manner, the subcarriers can be built inexpensively.

The design of the subcarrier does not severely restrict the configuration in which the MMIC chip is packaged. For example, the chip can be built on a thin metal base and bonded or epoxied to the subcarrier. Alternatively, the subcarrier could be made part of a ceramic package to house the MMIC chip. The chip could also be built on a post or pill package and inserted into the subcarrier through a hole drilled in the center of the subcarrier. Furthermore, the subcarrier itself could be modified to use a thicker metal base or to change the width of the center groove in the dielectric, if necessary.

Test Fixture

An assembly drawing depicting the various components of the test fixture is shown in Figure 2. The fixture is built on a dove-tail assembly with a fixed center block and two end blocks which move in and out in unison by rotating a right- and left-hand threaded drive rod. Microstrip housings, which are open on the inside end, are spring-mounted on the end blocks and overhang them enough to come together midway at the center block. At this end, the bottom floor of the microstrip housing is thin (.031 inch) and a notch in the floor is cut away to expose the bottom surface of a .015 inch thick alumina microstrip substrate. When the MMIC subcarrier is set in place, the tabs on the subcarrier slip into the notches in the microstrip housing and the tops of the tabs make pressure contact with the bottom surface of the microstrip substrates to achieve the ground continuity. Four springs, located in the spring-mounted block, are used in mounting each microstrip housing to the movable end blocks so as to exert a sufficient amount of downward pressure on the substrate to assure an adequate ground contact to the subcarrier tabs. The two inside springs exert a downward force while the two outside springs exert an upward force. This forces the ground contact on the tabs to be made at the very end of the microstrip substrate. A tapered wedge, which is located between the end block and the spring mount block, is used to tilt the microstrip housing up at the inside end to allow the subcarrier to be inserted easily. The foam spring cushion in the beam lead pressure lid is used to exert pressure to hold the MMIC beam lead down on the microstrip center conductor.

The microwave path through the test fixture is completely enclosed with the use of three kinds of lid covers. The microstrip housing lid covers most of the microstrip housing near the connector end. The beam lead pressure lid covers the remaining microstrip at the open (inside) end of the housing. A top lid cover fits over the MMIC chip and rests on top of the other two lids. This cover is built with side walls in the center to provide extra shielding around the test device. The side walls come down to .025 inch above the dielectric material on the subcarrier. Four spring-loaded lid clamp assemblies, which are attached to the microstrip housing assemblies, exert sufficient pressure on the covers to maintain proper shielding.

The fixture is designed for considerable flexibility to handle a variety of potential testing requirements. For example, circuits requiring multiple RF inputs and outputs can be readily tested by using the two auxiliary connectors on each microstrip housing. The microstrip substrate inside the housing can easily be changed to a three-conductor pattern for this application. In addition, microstrip couplers can be incorporated in the substrate to monitor power or to inject additional signals into the MMIC circuit. Special filters, diplexers, bias chokes, by-pass capacitors, or attenuators could also be made on the microstrip circuits. The response of these circuits would be removed during the calibration procedure. If desired, the entire microstrip housings could easily be replaced with a microstrip-to-waveguide transition for making measurements above 18 GHz.

Bias Interface

To provide bias to the MMIC chip, wire bonds are made from the MMIC chip to the SiO chip capacitors and then to the PC board on the subcarrier. The bias lines fan out to the .040 inch square bias pads on the sides of the subcarrier. A set of 18 spring-loaded pins, closely spaced in a phenolic block, produce a positive pressure contact to the bias pads on the subcarrier. Two dowel pins on each side of the center section of the fixture position the bias block assembly directly over the bias pads on the subcarrier and two clamps swing over to hold the bias block down. Each pin is soldered on the top of the bias block assembly to a wire in a nineteen lead, wire harness that has a connector on the other end. The extra wire goes to the fixture base to establish the ground reference. The connector attaches to a bias interface box which consists of 36 pairs of banana jacks and 36 SPDT switches. These switches provide a convenient way to change the voltage between two states on any bias pad, which is particularly useful in testing multibit phase shifters or devices requiring bias on/off switches. The bias voltages can be monitored at the bias interface box or directly on the subcarrier PC board by using a small probe.

FIXTURE CALIBRATION

Calibration Approach

Since the overall objective is to characterize MMIC chips (in particular with accurate S-parameter measurements), the calibration approach was derived for measurements on an automatic network analyzer (ANA). The software was written for measurements on a Hewlett Packard 8542B ANA, but it is applicable for any ANA model or system. The basic approach taken was to calibrate the fixture as part of the overall ANA system calibration, instead of characterizing the fixture separately from the ANA. The procedure used for calibration is described in the following paragraphs.

The twelve-term error model, outlined in Hewlett Packard Application Note 221A, forms the basis of the calibration. Since the equations for the forward and reverse parameters are identical in form, the notation used here will be given only for the forward direction. The equations for the reverse direction can be obtained by replacing S_{11} and S_{21} with S_{22} and S_{12}, respectively. The forward error terms become reverse terms. The equations for the measured reflection and transmission (Γ_m and T_m, respectively) are as follows:

$$\Gamma_m = E_D + \frac{E_R S_{11} - E_R E_L (S_{11}S_{22} - S_{12}S_{21})}{(1 - E_S S_{11})(1 - E_L S_{22}) - E_S E_L S_{12} S_{21}} \quad (1)$$

$$T_m = E_x + \frac{E_T S_{21}}{(1 - E_S S_{11})(1 - E_L S_{22}) - E_S E_L S_{12} S_{21}} \quad (2)$$

To determine the six error coefficients, reflection measurements are taken on a short and on an open, and both transmission and reflection measurements are taken on five offset transmission lines. Although five offsets are used, only three offsets are actually required to determine the error terms; however, the additional measurements are used to improve the accuracy by making use of a least error square fit of the data points. For the reflection measurements of the short and the open, S_{12} and S_{21} are assumed to be zero. For the short $S_{11} = -1$ and for the open $S_{11} = 1$. The measured reflection equation, (1), is reduced for the short and the open to:

$$\Gamma_{ms} = E_D - E_R/(1 + E_S) \quad (3)$$

$$\Gamma_{mo} = E_D + E_R/(1 - E_S) \quad (4)$$

For the reflection of the offset through transmission lines, $S_{11} = S_{22} = 0$ and $S_{12} = S_{21} = e^\gamma$, where γ is $\alpha l + j\beta l$. The reflection equations for the offsets reduce to:

$$\Gamma_{mT} = E_D + E_R E_L e^{2\gamma}/(1 - E_S E_L e^{2\gamma}) \quad (5)$$

The same assumptions are made for the transmission equation for offsets. Under these conditions equation (2) reduces to:

$$T_m = E_x + E_T e^\gamma/(1 - E_S E_L e^{2\gamma}) \quad (6)$$

Except for the open, the remaining calibration pieces are shown in Figure 4. These pieces are mounted on a gold-plated, .062 inch thick, copper subcarrier with interfacing tabs similar to the MMIC subcarriers. The short is a rectangular bar, .1 inch long with a cross-section of .015 by .015 inch. Beam leads are bonded to the top of the bar to interface with the microstrip center conductors in the microstrip housing. The offsets are made similarly, except that microstrip transmission lines on .015 inch alumina substrates replace the shorting bar. The offsets are made in incremental step sizes from .1 to .5 inch in length. No calibration piece is used to obtain the open. The housings are positioned slightly over the center section of the fixture with the wedge pushed back to raise the open end of the microstrip housing above the center section. Since both the center conductor and the ground plane of the microstrip are opened, the fringing capacitance is negligible.

The procedure used in the calibration is to make all the measurements first, store them, and then solve the equations to obtain the error coefficients. The effort term, E_D, is first determined from equation (5). The product of $E_S E_L$ is assumed to be very small compared to unity since these are source and load reflection error terms. If these terms are neglected initially, equation (5) will trace out a circle as a function of increasing offset lengths. The center of the circle is the E_D error term. The $E_S E_L$ product puts a small amount of distortion in the trace of the circle; but since the center of the circle is determined from a least-error square fit of all five points, the resulting error in determining E_D is negligible. Since E_D is now known, E_R and E_S can be determined from Eqs. (3) and (4) for the reflections of the short and open, respectively.

To determine the remaining error terms, it is necessary to first determine the complex propagation constant, γ, for the incremental offsets. This is accomplished by averaging the ratio of the transmission equations, Eq. (6), of offset number n + 1 to offset number n. This technique assumes E_X and $E_S E_L$ are mathematically negligible, which turns out to be a valid assumption. In any case, the averaging technique will diminish the error caused by this assumption. The E_L error term can now be determined from the reflection equations of the offset transmission lines using equation (5). Since this equation is repeated for each of the five offsets, the unknown error term, E_L, is determined using a least-error square fit of all the data points. The remaining two terms, E_X and E_T, are determined in a similar fashion by evaluating the transmission equations, Eq. (6), using the least-error square fit for the set of all five offsets.

The mathematics in the calibration approach may appear to be somewhat cumbersome; however, the technique is significant because it allows the error coefficients to be determined accurately from a few simple, insertable calibration pieces.

Measurement Software

The calibration and measurement software allow for the measurement of up to 101 frequency points. These points do not have to be spaced at equal intervals apart because a frequency file is stored in the software. Based on a single measurement run of the four S-parameters, up to twelve pages of corrected data can be printed or displayed at the discretion of the user. (The corrected S-parameters are obtained directly from the measured data using the equations given in Application Note 221A.) The first two pages of output data provide listings of the S-parameters in a magnitude and phase format for page one and in dB magnitude format in page two. Data pages three through six present the data in a Smith Chart or polar graphics form for each of the four parameters. The Smith Chart for the reflection measurements can be presented either as an impedance chart or can admittance chart. Pages seven through ten present a frequency plot of the magnitude in dB of each S-parameter. Automatic horizontal and vertical scaling is used in the plots. Plots of the forward group delay and maximum available gain are presented on the last two pages.

One convenient feature of the measurement software is that the input/output reference plane can be rotated by specifying the rotation length in inches of microstrip. Since the propagation constants of the microstrip lines were determined in the calibration software, this data is used for obtaining accurate line rotations independent of line loss or dispersion as these effects have already been included in the measured data.

TEST RESULTS

To obtain a proper evaluation of the fixture and the system calibration approach each must be evaluated separately. The fixture was first evaluated on the automatic network analyzer for reflections in the frequency domain and the results were convoluted into the time (or distance) domain to determine the source of the reflections. The system was then evaluated using the developed calibration software to determine if calibration routine was successful in removing the fixture reflection errors.

Fixture Results

The overall fixture VSWR is shown in Figure 5. A .4 inch microstrip transmission was measured. These reflections were convoluted into the time domain to determine their origins and the results are displayed in Figure 6. The results indicated that there was a problem associated with the output SMA connector interface. The interface was later repaired to correct the problem. The reflections around subcarrier were converted back into the frequency domain to produce the plot shown in Figure 7. This plot represents the interactions between the VSWR of the subcarrier interfaces between the two housings. The maximum VSWR is 1.6 to 1 which is equivalent to a maximum reflection coefficient of .117 at each interface.

The biggest contribution to this reflection was caused by misalignment of the center conductors of the microstrip transmission lines. Before these conductors were realigned, the fixture was calibrated using the correction software to determine if these errors could be removed by the software.

System Software

After calibrating the fixture with the ANA using the developed software, .4 inch offset was remeasured with the reference planes rotated in by .2 inch. The return loss data is shown in Figure 8. It can be seen that the reflections have been significantly reduced to demonstrate the validity of the calibration approach. Also, the transmission phase angle is near zero degrees which validates the rotation technique.

Further testing on actual MMIC chips is now in progress; however, insufficient data has been obtained at the time of this writing for inclusion in this paper.

CONCLUSIONS

A universal MMIC test fixture has been developed which has several unique features. It can accommodate various chip sizes, it can be calibrated on an automatic network analyzer using just microstrip type calibration pieces, it has provisions for multiple RF inputs and outputs, and it has provisions for use of up to

36 independent bias lines. Furthermore, no bonding is required during tests and the device under test can be enclosed on all sides. The fixture has the flexibility that it can be easily modified for special or unusual chip devices or testing requirements.

In addition to the hardware, software for calibrating and measuring MMIC chips in the fixture in an automatic network analyzer has also been developed. The software removes the measurement errors contributed by the fixture.

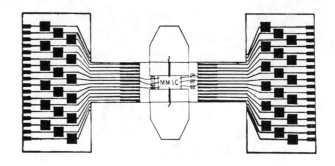

Figure 1. MMIC Subcarrier

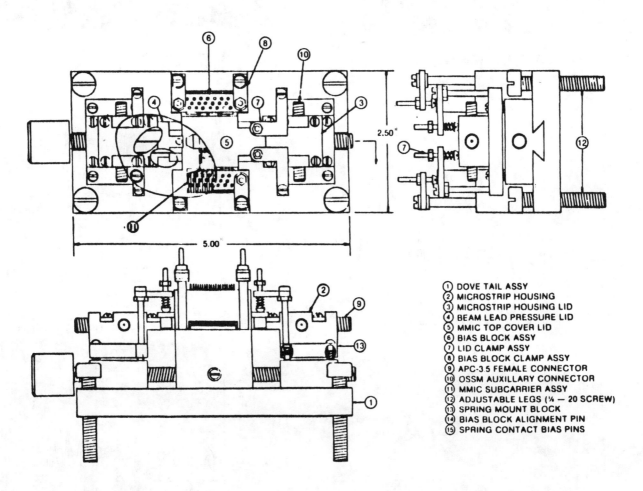

Figure 2. Test Fixture Assembly Drawing

Figure 3. MMIC Test Fixture

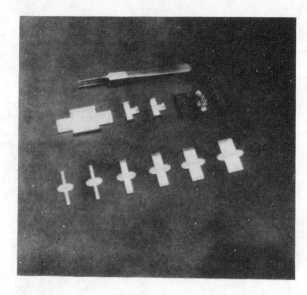

Figure 4. Calibration Pieces With Top Cover, BL Pressure Lid, and Subcarrier

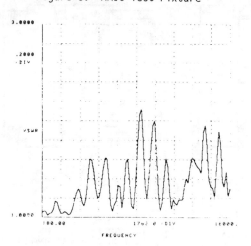

Figure 5. Overall Fixture VSWR

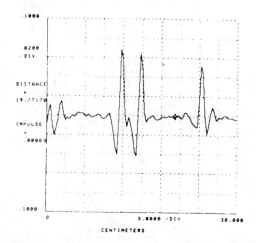

Figure 6. Fixture Time Domain Reflections

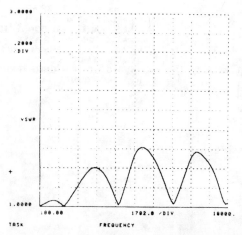

Figure 7. VSWR Without Connectors

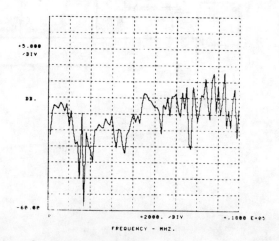

Figure 8. Return Loss Using Calibration Routine

Calibration Methods for Microwave Wafer Probing

Eric W. Strid and K. Reed Gleason

Tektronix, Inc.

Abstract: A new level of accuracy in the measurement of microwave parasitics has been achieved, due to the combined development of microwave wafer probes and on-wafer impedance standards. Repeatable losses and reflections in the probes can be readily removed from measured data, but radiation losses and crosstalk cannot be corrected and must be minimized. Oneport and twoport on-wafer standards for several probe footprints are shown, and their performance verified.

Introduction

Timely development of either monolithic microwave circuits or ultra-high-speed digital circuits requires precise knowledge of monolithic circuit element parameters and their variations and parasitics. The complex nature of some of these elements and their interactions has precluded accurate theoretical prediction or even scale modelling. Microwave wafer probes have been shown to be an accurate and convenient tool for the detailed network analysis of monolithic elements [1,2]. A wafer probe can be viewed as an adapter from coax to bonding pads, and as such will perturb microwave measurements in the same fashion that coaxial adapters affect measurements. Demonstrated microwave wafer probes allow uncorrected measurement accuracies similar to the accuracies achieved with SMA connectors. However, the combination of microwave probes with a corrected network analyzer and on-wafer impedance standards which are much smaller than a wavelength allows on-wafer S-parameter measurements with a new level of accuracy. In this paper we discuss the requirements on the probes to be usable with a corrected network analyzer, some typical oneport, twoport, and balanced calibration techniques and results, and a discussion of probes and calibrations for high-speed digital circuits.

Probe requirements for use with a corrected network analyzer

There are three classes of signal degradations which a wafer probe can cause: losses, reflections, and crosstalk. When using a probe (or any adapter) with a corrected network analyzer, the tolerable level of losses or reflections is relatively high; the only limit is maintaining sufficient signal level for good signal-to-noise ratio.

However, the losses and reflections must be as repeatable as the resolution desired. If significant signal power is radiated from the probe(s), the probe losses are normally not repeatable. This is because the wafer, wafer chuck, or other conductors are moved in relation to the probes, causing changes in the radiation impedance. Radiation from one line to another can also occur, creating crosstalk.

For twoport calibrations, the standard 12-element vector correction model [3] includes a leakage correction element for each direction, but the ability of this element to correct for crosstalk is very limited [1]. In practice, it has been found that limiting crosstalk between probe lines is simpler and more accurate than attempting to correct for it. Crosstalk can be caused either by coupling between transmission lines on the probe(s) or by common-lead inductances. Since the crosstalk is uncorrected, even in a corrected measurement, any crosstalk will appear in low-level transmission measurements. The allowable crosstalk level is approximately equal to the required transmission accuracy. Demonstrated pairs of single-line microwave probes achieve greater than 45 dB isolation through 18 GHz. Two-line probes with a signal-ground-signal contact configuration achieve a worst-case isolation of only about 20 dB through 18 GHz. This is due to the common-lead inductance of the ground contact (about 50 pH), the worst case being when all the contacts are shorted together.

On-wafer calibrations

A "two-tier" deembedding approach [4] is possible, wherein the probe parameters are measured and stored for removal from parameters measured from a coax calibration. However, since the probe contact to the standards on the impedance standard substrate (ISS) is faster and more repeatable than making coax connections, two-tier deembedding is a waste of time and accuracy. Therefore, the preferred approach in calibration is to use the on-wafer standards to calibrate directly at the probe tip(s) ("one-tier" deembedding).

Reprinted from *IEEE Microwave and Millimeter-Wave Monolithic Circuits Symp. Dig. Papers*, 1984, pp. 78-82.

The wafer probe adapts from a coaxial transmission medium to essentially twinstrip or coplanar waveguide or other coplanar lines on the wafer surface. However, since the dimensions of many monolithic structures for ICs and for impedance standards are small with respect to a wavelength, these on-wafer structures are lumped in nature. The impedance standards for both oneport and twoport calibrations are analyzed for accuracy using theoretical predictions of parasitics, measurement at low frequency, comparison to other standards, and scale modelling.

Impedance standards have been built on GaAs, Si, and alumina. The GaAs and alumina calibrations perform very similarly, while the Si calibrations show significantly more capacitance to the substrate, as expected.

The type of calibration standards used must correspond to the contact configuration of the probe(s). For simplicity, the standards for the probes shown in figure 1 [1,2] will be considered first. Generally, narrower contact spacings allow slightly more accurate calibration; bond pads as small as 50 um wide on 100 um centers are readily used.

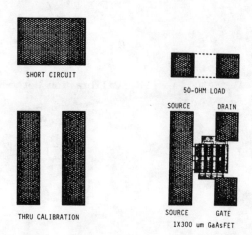

Figure 2. Minimum oneport and twoport standards for the probe footprint illustrated in figure 1, and a GaAsFET with a corresponding footprint.

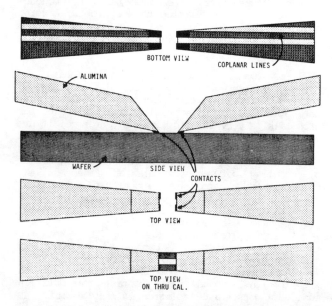

Figure 1. A simple coplanar probe configuration which has achieved accurate microwave results. One probe head has a ground-signal contact configuration and the other probe head has a signal-ground contact configuration. Note that the probe contact areas are visible from the top, since they extend just beyond the end of the probe board.

Oneport calibrations

Figure 2 shows the minimal set of twoport calibrations for the probes in figure 1, and a GaAsFET with the corresponding footprint. In Figure 2, the short standard is simply an area of metallization which creates a low inductance between the contacts. The 50-ohm load is a 50 um square resistor deposited on the GaAs. Its resistance can be measured at DC, its series inductance is calculated to be about 30 pH, and the parallel capacitance is calculated to be 4.9 fF on GaAs. The open-circuit standard is just the probe raised from the substrate. The stray capacitance can be empirically determined, as is done for coaxial calibrations. By ensuring that the corrected reflection coefficient magnitudes of high-Q coils and capacitors are less than one, the open-capacitance can be determined to within about 3 fF.

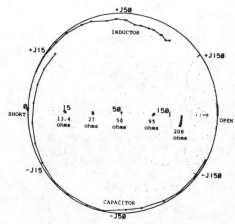

Figure 3. Corrected oneport measurements of various on-wafer impedance standards. The short, the 50-ohm resistor, and the open were used as the three standards for calibration.

Figure 3 shows oneport measurements of the calibration standards and other impedance standards after calibrating the system with a short, a 50-ohm termination, and an open circuit. The other standards are necessary to verify the accuracy of the calibration, since repetition of the calibration standards only proves that the system repeats its measurements. As can be seen in figure 3, the resulting measurements are extremely tightly grouped and demonstrate the lumped nature of these elements.

The measurement of a 50 X 150 um rectangle of metalization (just large enough to short the signal contact to the ground contact on the probe) is shown in Figure 4. About 30 pH of inductance is measured, comparing well with the expected inductance. Extra conductor under the end of the probe tip causes a small interaction between the conductor and the very end of the probe tip, resulting in an apparent negative inductance as large as -60 pH.

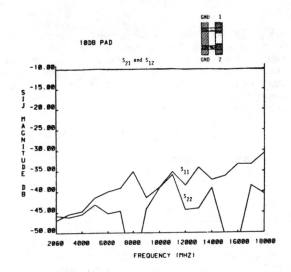

Figure 5. Corrected twoport measurement of a small 10-dB pad. The measured S21 magnitude typically varies ±0.1 dB over 2 to 18 GHz.

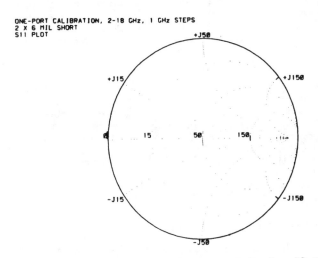

Figure 4. Corrected oneport measurement of a 50 X 150 um metallized rectangle, illustrating the ability to measure inductances below 50 pH.

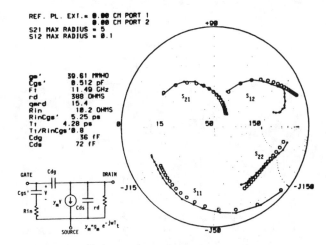

Figure 6. Measured 2-18 GHz S-parameters of a typical 1.0 X 300 um FET, using the microwave wafer probe. The circles are S-parameters of the simplified equivalent circuit calculated from the measured parameters.

Twoport calibrations

Twoport corrected S-parameter measurements with the standard 12-element error model use the above calibration for each port, plus a through connection and isolation calibration standards. The through standard connects the two ground contacts together and the two signal contacts together. For the isolation calibration, either the probes are open-circuited in air, minimizing any coupling between them, or the isolation error terms are simply set to zero.

Figure 5 shows the measurement of a 10-dB pad after the twoport calibration, verifying the basic accuracy of the standards. Figure 6 shows the measurement of a typical 1X300 um GaAsFET, along with its lumped equivalent circuit [5].

The above discussion illustrates the calibrations for a simple probe configuration. Calibrations for other types of probe footprints, including configurations for most commercially-available discrete FETs, are possible. Since most discrete devices have not been designed for RF probing, special ground connections between the gate probe tip and the drain probe tip are often necessary.

For the case of an MMIC, ground contacts should be provided next to the input and output pads to be probed, as well as next to any bias pads which require off-chip bypassing. Low-impedance bypass probes can be used for on-wafer testing, in place of the bypass capacitors to be used in the package. The through calibration standard for probing an MMIC with a fixed probe footprint is a 50-ohm transmission line with ends at positions corresponding to the input and output of the MMIC. However, greater accuracy can be achieved by using an adjustable probe footprint, since a minimum-length through calibration standard can be used.

Balanced calibrations

Oneport and twoport measurements with balanced probes have been performed. A 2-18 GHz 50-ohm balun (Cascade Microwave part no. 010-019) is connected between each port of the network analyzer and its corresponding balanced probe, as in figure 7. Neither probe head has a ground contact, but there is a virtual ground plane vertically through the center of each. In this fashion, most of the imperfections of the baluns are removed, as if they were just other adapters. The corrected oneport measurement results are very similar to those shown in figure 3. A corrected twoport measurement of a 10-dB pad is shown in figure 8.

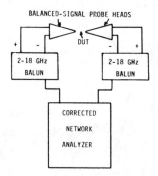

Figure 7. Connection of balanced probe heads with wideband baluns and a corrected network analyzer to achieve balanced on-wafer measurements.

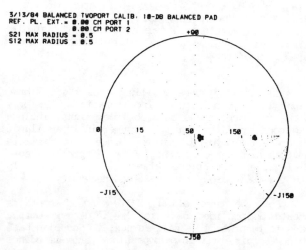

Figure 8. Corrected twoport measurement of a balanced 10-dB pad using the test setup shown in figure 7.

Calibrations for digital measurements

The measurement of a nonlinear circuit will not in general be correctable using linear techniques, such as the error models for a corrected network analyzer. Correction of output signals for frequency response of the output lines (in software) can be useful, and precorrection of input pulses (in hardware) to counteract the response of input lines may be practical in some cases. For the case of digital measurements, crosstalk and radiation again are not removable from the measurements. The main problems with ultrafast digital probing are the same problems with ultra-fast digital packaging: ground inductance and low-impedance power supply requirements. Except for die with balanced low-impedance outputs, wire probe tips are simply far too inductive. In addition to noise from common-lead inductance, crosstalk between transmission lines must also be minimized. For probe crosstalk measurements, a useful pattern is one which connects two signal line contacts and shorts them to a ground or power-supply contact. The transmission between two signal lines is a sensitive measure of the ground or power-supply impedance (usually a small inductance).

Digital circuit risetimes do not yet approach the risetimes of available cables and printable transmission lines, so the need for waveshape correction is relatively minor. More important to the digital designer is the accurate measurement of propagation delays. Through connections for accurate delay calibrations with multi-line probes can be built in the same style as for twoport ANA standards, with lines at enough different angles and positions to make throughs between each pair of signal lines.

Conclusions

The accuracy available with state-of-the-art microwave wafer probes exceeds the best accuracies possible in bonded-chip test fixtures. Wafer-probe measurements are so repeatable that the user can resolve which side of a bonding pad (about 50 pH) is being contacted by the probe tip. In contrast, the bondwires in a chip test fixture (at least 500 pH) often cannot be seperated from the device under test. In addition to accuracy improvements, wafer-probe measurements can be performed nondestructively, and eliminate processing steps to dice and bond up prototype chips. Balanced-signal probes have been demonstrated through 18 GHz, allowing testing of MMIC designs which make use of virtual grounds. During the design stage, MMICs should be laid out with RF on-wafer testing in mind.

References

1. E. W. Strid and K. R. Gleason, "A DC-12 GHz Monolithic GaAsFET Distributed Amplifier", IEEE Trans. Microwave Theory and Tech., Vol. MTT-30, No. 7, pp. 969-975, July 1982, and IEEE Trans. on Electron Devices, Vol. ED-29, No. 7, pp. 1065-1071, July 1982.

2. K. R. Gleason, et. al. "Precise MMIC Parameters Yielded by 18-GHz Wafer Probe", *Microwave System News*, pp. 55-65, May 1983.

3. J. Fitzpatrick, "Error Models for Systems Measurement," *Microwave J.*, pp. 63-66, May 1978.

4. D. Swanson,"Ferret Out Fixture Errors with Careful Calibration," *Microwaves*, pp. 79-84, Jan. 1980.

5. R. A. Minasian, "Simplified GaAs M.E.S.F.E.T. Model to 10 GHz," *Elect. Lett.*, Vol. 13, No. 8, pp. 549-551, Sept. 1, 1977.

WAFFLELINE - A Packaging Technique for Monolithic Microwave Integrated Circuits

Douglas E. Heckaman, Jeffrey A. Frisco,
Jerry B. Schappacher, Dawn A. Koopman

Harris Aerospace Systems Division
Melbourne, Florida 32901

ABSTRACT

WAFFLELINE, a high density packaging technique designed for monolithic microwave integrated circuits, is described. WAFFLELINE has been tested at frequencies up to 18 GHz, showing a standing wave ratio of 1.1:1 and signal isolation of crossover wires of greater than 30 dB. Adjacent signal wires, which are separated by 0.050 inches, have greater than 50 dB isolation. This low coupling, high density environment makes WAFFLELINE ideal for high speed monolithic applications.

INTRODUCTION

The field of monolithic microwave integrated circuits (MMIC's) is a rapidly growing one, bringing with it many advantages. However, in order to fully reap the benefits of MMIC's on the system and subsystem levels, a packaging scheme designed specifically for these circuits is necessary. Such a packaging scheme must address several problems currently associated with the interconnection of MMIC's. It must provide high circuit density, good, low loss RF connections at frequencies up to and beyond 18 GHz signal path crossovers with good isolation, high speed transmission, and a heat sink for thermal dissipation. Presently, stripline and microstrip printed circuit boards, twisted wire pairs, and coaxial lines launched via SMA connectors are used in the packaging of MMIC systems. All of these methods are limited, however, in that none provide all of the characteristics listed above.

WAFFLELINE is a packaging technique which was specifically designed to meet the stringent demands of monolithic subsystem assemblies. This structure is shown in Figure 1. The body of the structure is best described as a waffle iron-like grid. Dielectric coated signal wires lie in the channels defined by the grid. Hollowed out areas in the waffle iron structure allow packaged and unpackaged chips to be mounted and interconnected via these wires. The structure is completed with a top metal foil covering to isolate interior wires from outside RF signals and from each other. This unique structure allows higher density packaging than conventional methods, it permits crossovers, and it provides a heat sink for each monolithic chip in the subassembly. Electrically, WAFFLELINE is an extremely low dispersion, periodic transmission medium. Also, the speed of propagation is high, approximately 0.8 times the speed of light. Finally, WAFFLELINE has fast breadboard capability and is easily manufactured and repaired.

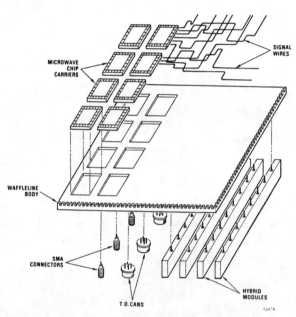

Figure 1. WAFFLELINE assembly - exploded view

DESIGN

WAFFLELINE design goals were to provide a high density interconnection technology with excellent electrical performance to 18 GHz. This interconnect would be used as a basis for subsystem integration of both MMIC's and hybrid components into a common assembly.

The concept of signal wires confined to an X-Y grid was initially adopted because of its suitability to fabrication using CAM techniques. Once this configuration was chosen it was necessary to design the grid such that a controlled characteristic impedance was obtained throughout the assembly, and transitions from the main body had a bandwidth of at least 18 GHz.

The WAFFLELINE transmission medium is a periodic structure with a physical period of 0.05 inches. The impedances of the two different cross sections of the structure are averaged to obtain the overall characteristic impedance. Approximate solutions of the characteristic impedance were obtained using equivalent coaxial circuits. Finally, WAFFLELINE dimensions were optimized through experimental verification. A cross section of the resulting design is shown in Figure 2. This structure supports a quasi-TEM mode of propagation in a near dispersionless medium. As shown, dielectric coated signal wires with inner conductor diameters of 0.008 inches and outer dielectric diameters of 0.019 inches were chosen. In order to permit wire crossovers, the channel depth must be greater than the outer diameter of the wire used, although some bulging of the aluminum foil covering is considered acceptable. Therefore, with crossovers and a desired Zo of 50 ohms in mind, a channel depth of 0.025 inches was selected.

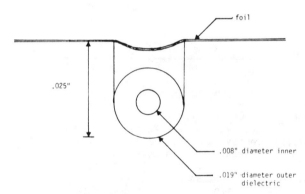

Figure 2. WAFFLELINE cross section

WAFFLELINE is not limited to the case of a 50 ohm characteristic impedance. Within the same assembly, geometries with impedances of 25 to 100 ohms are possible. Also, by using magnet wire, slow speed control wires and dc wires can share a common channel, in effect providing power bus channels. This minimizes crossovers while maximizing circuit density. Finally, high loss RF lines or lossy filtered lines are easily added by using wires constructed with lossy dielectric materials and nickel chromium conductors.

Another critical design parameter which had to be considered was the electrical and mechanical design of all transitions from the WAFFLELINE body. Our goal was to use WAFFLELINE for the interconnection of a variety of transmission line mediums. The first transition designed was from a SMA coaxial interface into the assembly. To allow a maximum number of I/O ports, a right angle launch was desired. During the design of this launch, several geometries were characterized. As would be expected, the smaller the transition geometry, the greater the inherent bandwidth. As a result of this study, two launch techniques were chosen.

The first technique was chosen due to the low cost of the connector and the ease of construction. In this case, a SMA feed through connector with a pin diameter of 0.020 inches and Teflon jacket diameter of 0.060 inches is inserted orthogonal to the WAFFLELINE body. The resulting intrusion of the pin into a given channel is 0.006 inches. The depth of penetration is critical to the transition performance and has been optimized for maximum bandwidth. While this transition provided excellent performance through 10 GHz, it was our belief that a further improvement was possible if the feed through dimensions could be reduced.

The second technique diminishes the transition geometry through use of a connector without the Teflon jacket, thereby reducing the outside diameter of the coaxial feed through. By making the feed through smaller, the resulting discontinuity is made electrically shorter, thus increasing the inherent matched bandwidth. Again, the connector pin depth is critical in providing a matched impedance transition. The final optimized dimensions are shown in Figure 3. The relative ease of matching this 90° transition over such a wide bandwidth can be directly related to the WAFFLELINE structure. This can be demonstrated through an analogy with a 90° launch into stripline. It is generally known that to provide a wideband matched transition it is necessary to insert pins through the stripline assembly around the launch, thereby providing a continuous path for ground currents. In the WAFFLELINE structure, these "pins" are replaced by the waffle grid. In effect, ground pins are spaced every 0.050 inch, or the period of the WAFFLELINE grid.

While this transition design was initially conceived as a SMA launch to the WAFFLELINE body, it can also be used for a transition into a variety of commercially available hybrid components, including TO cans and other hermetically sealed hybrid circuit modules.

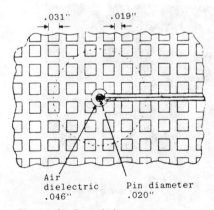

Figure 3. Transition geometry

Besides the transition from the external assembly into the WAFFLELINE structure, internal transitions were also needed. These transitions allow the ultimate goal of WAFFLELINE; the interconnection of MMIC chips. First, monolithic chips are packaged into Microwave Chip Carriers (MMC's). These MCC's permit decoupling and matching networks at the chip interface and allow the MMIC's to be hermetically sealed. Next, the MCC stripline I/O's are connected to the WAFFLELINE wires. Since WAFFLELINE is a quasi-coaxial structure, launch techniques from coax to strip and microstrip lines were used for interfacing MCC's.

TEST RESULTS

WAFFLELINE Through

For evaluation purposes, a two inch by two inch WAFFLELINE structure was built. Center-to-center channel spacings were 0.050 inches. Channels were 0.019 inches wide and 0.025 inches deep. As described in the design section, signal wires had a 0.008 inch center conductor diameter and a 0.019 inch outer dielectric diameter. These lines were connected to the WAFFLELINE via SMA connectors.

In order to examine various line configurations in WAFFLELINE, a Time Domain Reflectometer (TDR) with a 12 GHz bandwidth was used to view the lines at their interfaces and a network analyzer was used to determine S-parameters. Figure 4 gives a TDR reading of a through line in WAFFLELINE. The input (left side of picture) is from a precision 50Ω line and the output (right side) is to a 50Ω load. The scale on this display is 50 mρ per division. Clearly, all disontinuities are contained to within 20 mρ. The 90° launch into the WAFFLELINE is relatively smooth. As can be seen, the largest discrepancy occurs at the launch out of the WAFFLELINE into the 50Ω load, which is actually more like 52Ω. The associated return loss, consisting of the SMA launch into the WAFFLELINE, the through line and the output SMA, is shown in Figure 5. The reflection coefficient, or S11, is less than -20 dB through 18 GHz.

WAFFLELINE-To-MCC

Figure 5 gives the measured return loss from the SMA connector through a straight wire into a MCC containing a 65Ω resistor. By terminating the signal wire inside the WAFFLELINE structure in the MCC, a simulation of the return loss of integrated chip carriers in WAFFLELINE is provided. Originally, 50Ω thin film resistors packaged inside a MCC were used. However, the capacitive and ground parasitics associated with the alumina resistors in the miniature MCC's were large enough to mask essential return loss information. The 65Ω resistor used was ion implanted GaAs. Although 65Ω is a higher resistance than the desired termination, this MCC packaged GaAs resistor provided a more accurate evaluation of the WAFFLELINE-to-MCC launch.

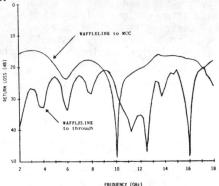

Figure 5. Return loss of WAFFLELINE to MCC and WAFFLELINE through

WAFFLELINE Cross-Talk

An important WAFFLELINE feature is that it allows crossovers. The measured return loss of a simple, unshielded crossover is given in Figure 6. As can be seen, the isolation between wires is better than 30 dB to near 18 GHz. Further isolation can be provided by thin metal foil shields placed at critical crossovers.

Another important aspect of WAFFLELINE is the decoupling between signal lines in adjacent channels. Measurements of the forward and back coupling of such a situation were made and in both cases, the isolation between the two wires was greater than 50 dB through 18 GHz.

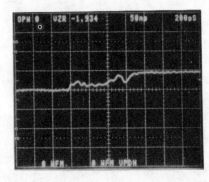

Figure 4. TDR view of WAFFLELINE through

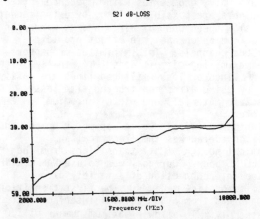

Figure 6. Isolation of signal wire crossovers

Subsystem Test

The test results given in the previous section have demonstrated that WAFFLELINE is a viable packaging scheme for the interconnection of MMIC's. Therefore, a WAFFLELINE test fixture has been built which can be used in the test of various integrated subsystems. This "universal" test fixture allows the demonstration of an integrated subsystem assembly with up to twenty MCC packaged circuits. There are 34 RF I/O ports and 25 dc or control lines. One subsystem which is presently under test in this fixture is a high speed digital test system. This digital formatter/deformatter which operates at 1.4 Gb/s will be used for testing other GaAs digital circuits. The ability to interconnect subsystems such as this digital circuit in this fixture makes breadboard fabrication a fast, simple process.

FABRICATION TECHNIQUES

WAFFLELINE Body

There are several techniques which can be used in the fabrication of WAFFLELINE. 1) Use of multi-blade circular saws mounted on a single arbor in an automatic feed milling machine. 2) Investment casting is a technique which yields a product that requires very little machining clean up. Using this technique, all the design features required for shape configuration and component mounting, as well as the waffle iron grid, can be formed at one time. 3) Injection moulding is a technique ideal for large production quantities. A possible drawback here, though, is that it requires a very high zinc alloy, and in some cases this may not meet the structural requirements of the intended application. All three of these fabrication techniques use an aluminum alloy. The WAFFLELINE fabrication technique used is best determined based on several tradeoffs, such as initial tooling costs, unit cost, desired WAFFLELINE material and the WAFFLELINE application.

Wire Construction

The tests reported in this paper have used a ruggedized Teflon* jacketed, silver-plated copper wire designed for Wire-Wrap+ interconnection technology. In the manufacture of this commercially available product, the center conductors are helically wrapped with layers of precision thickness Teflon PTFE resin tapes which are then sintered into a homogeneous mass conforming to MIL-W-16878D. This yields a highly concentric, uniform diameter microwave quality dielectric jacket. This outside vendor proprietary construction method, when compared with the usual extruded Teflon methods, also gives superior resistance to cold flow and rip propagation due to an improved molecular arrangement. The dielectric is stable to 260°C, rated to 300 volts and available in solid colors to MIL-S-140.

The wire vendor, in addition to the standard silver-plated copper conductor, can supply special conductors such as nickel-plated copper, pure nickel, silver, aluminum and gold alloys.

Assembly

In our MMIC test-bed we use a small elastomer ring around the MCC package to press the plated wires against the chip carrier lands to provide reconfigurable electrical connections. We found that silver-plated wires on gold lands provided repeatable microwave performance. Potentially, with gold-to-gold or other more optimum metal interfaces, this pressure connection method could be used in some commercial products. For high reliability or severe environment applications, these wire-to-land connections can be made using standard technology weld, solder or conductive epoxy bonding. We believe this wire bonding and wire placement can be machine automated if justified by quantity needs.

To date, timing, phasing and isolation requirements have been met with manual design techniques. Since the available circuit paths are on a standard X-Y grid pattern, the circuit design layout of WAFFLELINE integrated subsystem assemblies can make use of the available automatic layout optimization programs developed for the semiconductor and printed wiring board industries.

MECHANICAL ASPECTS

Mechanically, WAFFLELINE has several advantages over conventional packaging methods. First, WAFFLELINE can provide heat dissipation dependent on the system in which it is used, simply by varying the thickness of the structure. If necessary, cooling fluids can be run through the structure. Second, the various WAFFLELINE parts are not rigidly attached. Thus, for example, if the WAFFLELINE body and the chip packages had differing thermal coefficients of expansion, chips and their connections would not be subjected to undue stress. Finally, the assembly of WAFFLELINE, with the metal foil and elastometer cover, gives it superior resistance to vibration. These mechanical aspects contribute to the high reliability of the WAFFLELINE packaging scheme.

CONCLUSION

WAFFLELINE is proving to be an excellent packaging and interconnection technology for both digital and microwave monolithic circuits. It has demonstrated low reflections and high isolation through 18 GHz without many of the limitations of current interconnection techniques. It also offers flexibility to the subsystem designer by allowing a variety of package types and vendor parts to be integrated into a single assembly. Finally, the WAFFLELINE packaging concept provides the excellent RF performance and high packaging density required by monolithic microwave and digital circuits.

* Teflon is a DuPont trademark
+ Wire-Wrap is a Gardner-Denver trademark

A LOW COST MULTIPORT MICROWAVE PACKAGE
FOR GAAS ICS

D.A. Rowe,* B.Y. Lao, R.E. Dietterle, M.A. Moacanin+

Magnavox Advanced Products and Systems Co.

Torrance CA

ABSTRACT

This paper describes a multiport microwave chip carrier designed to solve the classical problems associated with packaging GaAs circuits with more than 2-4 I/O ports operating at GHz rates. The chip carrier contains nine copper/polyimide layers to provide 50-ohm stripline signal lines and a dedicated power plane. Chip capacitors located close to the IC serve to bypass two low impedance power lines. Heat sinking is achieved by thermal conduction through copper heat columns under the IC. Initial testing and evaluation has yielded excellent results in RF performance, manufacturing yield and cost, solder reflow capability, and low failure rate under thermal cycling. The design of the package and test results are presented.

INTRODUCTION

Recent progress in GaAs IC technology has made integrated cicuits at microwave frequencies a reality. In contrast, the technology to package and interconnect a system comprised of GaAs ICs has not kept up with device development. It has been pointed out [1] that currently no package with more than 6 RF I/O ports is commercially available that will provide performance at or beyond 4GHz clock rates.

The main difficulties encountered with existing IC packages are the uncontrolled transmission line characteristics of the signal lines, poor thermal performance, and inadequate power and ground connections to the IC. The signal lines typically have non-uniform or uncontrolled impedance, excessive cross-talk at high frequencies and large discontinuities associated with the abrupt change in geometry of the signal carrying structure, as an example, at the interface between the package and the motherboard. Current GaAs logic structures such as BFL and SDFL require 0.1 to 10 mW per gate. Because of this large power requirement (as compared to low power Silicon technologies), the thermal resistance of conventional packages can become the limiting item for even modest levels of chip complexity. Finally, in conventional packages, one is often forced to use high impedance (25-75 ohm) lines for power supply and ground connections to the chip. The large current transients in the output buffers when dropped across these high impedance lines can produce large voltages at the chip power supply nodes. This can cause circuit oscillation, time dependent shifts in logic thresholds and generally unreliable gate operation [2]. In this paper we present a multiport RF chip carrier which solves the above mentioned problems.

The next section briefly summarizes the process used to fabricate the package. The package design is then discussed and test results are given.

FABRICATION

The package is fabricated by the Augat-Pactel Corp. using a thin (2-4 mil) additive glass filled polyimide for the dielectric layers and plated copper for the electrical connections (vias) between the copper conducting layers (see Figure 1). All patterning, including the holes in the dielectric layers is accomplished by standard photolithographic techniques. The fine line resolution of this method allows the designer to form small transmission line structures while maintaining the neccessary metal line width accuracy for excellent control of the impedance. In addition, the packages are batch fabricated in a panel containing over 300 units making this technique very cost effective. Reference 3 contains more details on the manufacturing aspects of this technique.

PACKAGE DESIGN

To take advantage of the tooling and fixturing already available in the market, it was decided that the overall package would be a JEDEC standard size. Although this choice significantly increases the overall dimensions of the carrier and therefore the time delay in the signal path, it allows the inclusion of power supply bypass capacitors directly in the package. The 44-pin JEDEC standard was chosen to allow each of the 12 RF I/O ports to have a ground pin on either side, reducing both the ground inductance and the port to port coupling.

Figure 1 shows the cross-sectional structure of the package and Figure 2 shows a simplified version of the plan view. The basic design philosophy can be seen in Figure 1: Three ground planes are used to isolate the two fuctionally distinct regions of the package, namely, the signal plane which is stripline in nature and utilizes the top and intermediate ground planes, and the power plane which is also stripline and utilizes the intermediate and bottom ground planes.

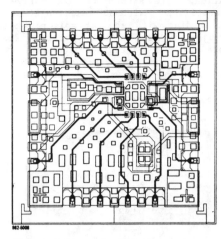

Figure 2 Plan view of the package

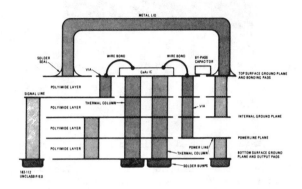

Figure 1 Cross-sectional view of the package

Four mil thick polyimide is used for the signal transmission line structure making the total ground to ground spacing 8 mils in this region. The glass filled polyimide has a relative dielectric constant of 3.6. The signal line conductor width is 4 mils which produces a characteristic line impedance of 50-ohms. Two mil thick polyimide was used for the power plane dielectric to reduce the impedance of the power supply lines. A characteristic impedance of 2.5-ohms was achieved for these conductors.

The top ground plane also serves to isolate the signal lines from the cavity formed by the metal lid, preventing undesirable coupling between ports and eliminating resonances associated with this cavity. Also, the three ground planes are periodically shorted together by vias to prevent the cavities formed between them from supporting unwanted resonances to beyond 10GHz. These vias also provide shielding between signal lines which reduces cross-talk. The overall configuration allows the signal and power lines to be surrounded by grounds in a well defined, controlled impedance environment.

The signal launching structure was designed to minimize the discontinuity created by the transition from microstrip on the motherboard to stripline on the package. This was accomplished by having short, high impedance launchers on the package. When these launchers are overlaid onto the 50-ohm lines on the motherboard, overall 50-ohm lines are achieved with only a small capacitive discontinuity. The geometry of the launcher is designed to dimensionally match that of the motherboard for both the signal line and ground plane.

The package is physically and electrically attached to the motherboard by reflow soldering. During fabrication, copper is selectively plated on the underside of the carrier to form bumps which are then coated with solder by screening.

Figure 3 is a photograph of the underside of the fabricated package showing the signal launching structures (small rectangles around the edge) and the two power connections (wide rectangles with three solder bumps each). The array of bumps in the interior of the package is a part of the thermal design discussed in the next paragraph. Figure 4 is a photograph of the top side of the package. The small rectangles are the signal pads for wire bonding to the chip. The large pads located away from the die attach area are for the power supply bypass capacitors. The crosses define the corners of the metal lid.

The copper thermal columns shown in Figure 1 form a continuous thermal path from the bottom of the chip to the underside of the package. An array of reflow solder bumps are then used to form a thermal connection to the substrate. At this point the heat is free to diffuse into the underlying material. This reduces the major heat transfer problem in conventional packages, i.e., the low thermal conductivity of the package substrate. Initial analysis suggests that a thermal resistance of 6 degrees centigrade per watt for the package is possible with this technique.

TEST RESULTS

The packages were DC tested for shorts and opens and the yield exceeded 85 percent. Both aluminum and gold wire bonds were used for electrical connections to the chip. In each case the minimum wire bond pull strength was greater than 5 grams. Even the small signal pads (5 X 8 mils) showed no signs of lifting. Vapor phase solder reflow was successful and subsequently sectioned packages showed no signs of delamination. Die attachment with silver epoxy was accomplished without difficulties, while eutectic bonding with special solder samples was also subjected to thermal cycling from -50 centigrade to 125 centigrade 100 times without problems. Each port was terminated with a 50-ohm load and TDR and return loss measurements were conducted before and after thermal cycling. No detectable changes were observed.

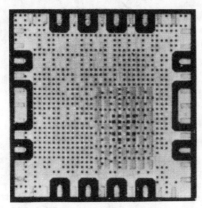

Figure 3 Photograph of the underside of the fabricated package

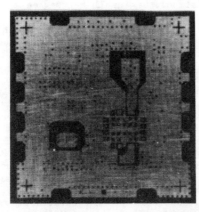

Figure 4 Photograph of the top side of the fabricated package

The TDR measurement of a typical terminated port is shown in Figure 5. The rise time of the pulse was 25 picoseconds. Position 1 denotes the SMA connector location while 2 and 3 represent the package to motherboard laucher and wire bond respectively. The average impedance of the signal line is approximately 52-ohms and the maximum deviation observed for the entire package, including the launcher but not the wire bond, was less than 20 millirho. The total time delay in the signal line is 180 picoseconds.

Return loss of the terminated line has been measured from 0.5 to 12GHZ. Greater than 10db return loss for the entire band was observed for all ports.

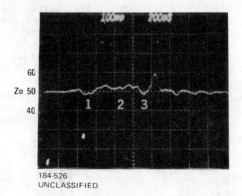

Figure 5 TDR measurement of a terminated signal line

CONCLUSIONS

We have developed a low-cost multiport chip carrier which solves the classical problems associated with packaging GaAs ICs. The package has several novel features including a separate power plane, fully shielded signal lines, copper thermal columns for excellent heat transfer and on-chip power supply bypassing capacitors. Test results show high DC and RF yield and excellent transmission properties for the signal structures including the launchers.

ACKNOWLEDGEMENT

The authors would like to thank S. Lebow of Augat-Pactel Corp. for fabricating the package and for his many helpful suggestions. We also wish to thank W. Oosterman, C. Price and S. Rintoul for their valuable contributions to this work.

REFERENCES

[1] D. Wilson, N. Fride, J. Kwiat, S. Lo, J. Churchill, J. Barrera, "Package Study for High Speed (GHz) Commercial GaAs Products," IEEE GaAs IC Symposium Technical Digest, p. 13 (1982).

[2] A. Frisch, "Reduction of Power Bias Noise in GaAs Digital Systems," ibid, p. 112.

[3] S. Lebow, "A Method of Manufacturing High Density Fine Line Printed Circuit Multilayer Substrates Which Can Be Thermally Conductive," Proceedings of 30th Electronic Components Conference, p.307 (1980).

* Current address: Aerospace Corp., El Segundo, CA
+ Current address: Gigabit Logic, Inc., Newbury Park, CA

Author Index

A

Abrokwah, J., 406
Akai, S., 79
Akazawa, Y., 332
Andrade, T., 172, 252, 313
Aono, K., 147
Asher, P. G., 249
Ayasli, Y., 244, 257, 350, 358, 368

B

Ball, M., 449
Barrett, D. L., 63
Bauhahn, P. E., 401, 406
Benet, J. A., 475
Besser, L., 449
Boire, D. C., 218
Brehm, G. E., 119, 186, 195

C

Cavicchio, C., 368
Cerretani, R. T., 116
Chang, C. D., 218
Chao, C., 327, 401, 406
Ch'en, D. R., 73
Chen, Y. K., 419
Chen, R. T., 73
Christou, A., 56
Chu, A., 116, 390
Clarke, R. C., 94
Clifton, B. J., 377
Cockrill, J. R., 190, 471
Conti, A. L., 363
Contolatis, A., 401
Contolatis, T., 406
Courtney, W. E., 116, 390

D

D'Avanzo, D. C., 87, 100
Dawson, D. E., 363
Decker, D. R., 128, 283
Degenford, J. E., 218
Dickens, L. E., 363
Dietterle, R. E., 490
Dormail, J., 223
Driver, M. C., 218
Durschlag, M. S., 112, 215, 223

E

Eldridge, G. W., 63, 94
Elta, M. E., 116
Esfandiari, R., 427
Estreich, D. B., 87, 319

F

Fabian, W., 414
Faguet, J., 166
Fairman, R. D., 73
Feng, M., 182
Finlay, H. J., 471
Fisher, R. A., 87
Freitag, R. G., 218
Frisco, J. A., 486

F (cont.)

Fujikawa, K., 147
Fukui, H., 463

G

Gleason, K. R., 481
Gold, R. B., 273
Goyal, R., 313
Gupta, A., 128, 283, 372

H

Hagio, M., 286
Hammerstad, E., 460
Hanes, L. K., 257, 270, 424
Harrop, P., 166
Hasegawa, H., 108
Heckaman, D. E., 486
Helix, M., 327
Hitchens, W. R., 273
Hobgood, H. M., 63
Holmes, C., 449
Honjo, K., 160, 175, 297
Hori, S., 153
Hornbuckle, D. P., 87, 305
Huang, C. C., 252, 313
Huston, S., 372
Hwang, Y. C., 419

I

Ip, K., 372
Ishihara, N., 332
Ishihara, O., 147
Itoh, H., 160

J

Jamison, S. A., 327, 401
Jansen, R. H., 437
Jenkins, J. A., 471
Jensen, O., 460
Joshi, J. S., 190

K

Kaelin, G., 372
Kamei, K., 153
Kano, G., 286
Katsu, S., 286
Kennan, W., 252
Kermarrec, C., 166
Kitagawa, T., 108
Klatskin, W., 179
Kocot, C., 132
Konaka, S., 332
Koopman, D. A., 486
Kukielka, J., 330
Kumar, V., 87

L

Lao, B. Y., 490
Lee, C. P., 138
Lee, S. H., 363
Lee, S. J., 138
Lehmann, R. E., 186

Liu, L. C. T., 179, 182, 396
Lokken, R., 327

M

Macksey, H. M., 119, 207
Mahoney, L. J., 116
Maki, D. W., 179, 182, 396, 427
March, S., 449
McCarter, S. D., 293
McOwen, S. A., 223
Medley, M., 449
Mehal, E. W., 386
Mikasa, M., 372
Mishima, K., 153, 452
Moacanin, M. A., 490
Moghe, S. B., 172, 313
Morris, A. M., 223
Mozzi, R. L., 223, 244, 257, 270, 424
Murai, S., 79

N

Nakatani, M., 147
Nambu, S., 286
Naster, R. J., 419
Nelson, S. R., 207
Ng, P., 327
Niclas, K. B., 231, 262, 273
Nishitani, K., 147
Nishiuma, M., 286

O

Odaka, T., 147
Ohara, M., 332
Ohtani, M., 147
Okano, S., 153
Oliver, J. R., 73

P

Palmer, C. D., 227
Pavio, A. M., 293
Pengelly, R. S., 35, 288, 339, 471
Petersen, W. C., 128, 283, 372
Peterson, V. E., 87
Petroff, I., 372
Platzker, A., 215, 358, 368, 457
Podell, A., 327
Pucel, R. A., 13, 244, 368, 414

R

Reynolds, L. D., Jr., 244, 257, 358
Rigby, P. N., 288
Rowe, D. A., 490

S

Saunier, P., 119, 227, 293
Sawada, T., 108
Schappacher, J. B., 486
Schellenberg, J. M., 249, 396
Scott, B. N., 195
Seashore, C., 406
Seymour, D. J., 186
Shade, G. F., 363
Shen, Y., 83, 138
Shibata, K., 153
Simons, M., 142
Siracusa, M., 182, 427
Snapp, C., 330
Sokolich, M., 179
Stein, R., 372
Stolte, C. A., 132
Storment, C., 179
Strid, E. W., 481
Sudbury, R. W., 390
Suffolk, J. R., 288
Sugiura, T., 160, 175
Sun, H.-J., 172, 313
Suzuki, T., 79

T

Tada, K., 79
Tajima, Y., 223, 270, 414, 424, 457
Takayama, Y., 297
Tatematsu, M., 153
Taylor, T. W., 87
Temme, D., 419
Temple, S., 368
Thomas, R. N., 63
Tong, E., 270, 424
Tserng, H. Q., 207
Tsironis, C., 166
Tsuji, T., 160, 175
Tsukii, T., 270, 424
Tucker, B. A., 262
Turner, J. A., 190

V

Vahrenkamp, R., 138
Van Tuyl, R. L., 87, 201, 305
Vorhaus, J. L., 112, 215, 244, 358, 368, 414

W

Wacker, R. W., 386
Wandrei, D., 368
Waterman, R. C., Jr., 368, 414
Webber, G. E., 327
Welch, B. M., 83, 138
Westphal, G. H., 186
Wickstrom, R. A., 218
Williams, R. E., 227
Wilser, W. T., 273
Wisseman, W. R., 119
Wrona, B., 270, 424, 452

Y

Yajima, Y., 452
Yamasaki, H., 249, 396
Yuan, L. T., 410

Subject Index

A

Active elements, 1, 35, 61
  amplifiers, 231
  arrays, 339
  switches, 471
Amplifiers, 13, 35
  AGC, 332
  AGC response, 313
  bipolar, 332
  broad-band, 215, 218, 223, 230, 286, 288, 297, 319, 330, 449, 463, 470
  CAD, 457
  cascaded, 252, 330
  CR coupling, 297
  direct-coupled, 230, 283, 305, 313
  distributed, 112, 230, 231, 249
  equalizing, 332
  feedback, 231, 273, 283, 286, 288, 293, 297, 305, 313, 330
  FET, 283, 286, 288, 297
  GaAs monolithic, 160, 172, 179, 182, 186, 207, 223, 227, 231, 244, 270, 283, 286
  K_u-band, 270, 457
  low distortion, 305
  low-noise, 151, 152, 160, 166, 172, 179, 182, 186, 286, 339, 463, 470
  medium power, 227
  MESFET, 231, 463
  multi-octave bandwidth, 151, 231
  multistage, 231, 293, 305, 457
  phase-splitting, 201
  power, 151, 206, 207, 218, 223, 257, 372, 457
  push-pull, 327
  resistive-feedback, 230
  switching, 471
  traveling-wave, 230, 244, 257
  ultrabroad-band, 297
  X-band, 112, 119, 182, 186, 215
  see also Balanced amplifiers; Bipolar monolithic amplifiers; Broadband amplifiers; Cascaded amplifiers; Direct-coupled amplifiers; Distributed amplifiers; Equalizing amplifiers; Feedback amplifiers; IF preamplifiers; Low-noise amplifiers; Microwave amplifiers; Matched feedback amplifiers; Multistage amplifiers; Power amplifiers; Preamplifiers; Push-pull amplifiers; Solid-state amplifiers; Traveling-wave amplifiers; Ultrawide-band amplifiers
Analog IC
  manufacture, 87
  monolithic, 339, 363
  phase shifters, 363
Annealing
  of ion-implanted GaAs, 94
  GaAs FET, 147
Attenuators, 339
  active, 419
  L-band, 419
  wide-band, 424
Automatic network analyzers
  calibration, 475

B

Backgating, 100
  in GaAs MESFET, 132
  substrates, 138

Balanced amplifiers
  MMIC, 327
  modules, 270
Bandpass filters, 35, 427
Bipolar monolithic amplifiers
  for gigabit optical repeaters, 332
Broad-band amplifiers, 230, 286
  CAD, 449
  feedback, 288, 330
  K_u-band, 270
  low-noise, 288, 463, 470
  monolithic, 179, 215, 218, 319
  use of Si bipolar technology, 330
  X-band, 215, 270
Buffers, 175, 182
  amplifiers, 372
Bulk GaAs
  semi-insulating, 73

C

CAD
  see Computer-aided design
Calibration
  for microwave wafer probing, 481
  of MMIC test fixture, 475
Capacitors
  interdigital, 166, 427
  MIM, 128, 147
  on GaAs, 112
  tantalum oxide, 116
  tantalum pentoxide, 223
Capless annealing
  transient, 94
Cascaded amplifiers
  performance, 252
Cell cluster matching, 218
Ceramics, 35
Chebyshev filters, 35
Chips
  amplifiers, 215, 244, 249, 257, 293, 313
  design, 56
  materials, 61
  mixers, 175
  monolithic mixer, 406
  multiport, 490
  oscillators, 190
  phase shifters, 363
  T/R modules, 338
Cluster matching
  use for wide-band amplifiers, 218
Computer-aided design, 435
  for 1980's, 449
  hybrid circuits, 437
  microstrip, 460
  monolithic MIC, 437
  multistage amplifier, 457
Computer-aided testing
  of MMIC's, 1
Computer programs
  SUPER-COMPACT, 449
Consumer market, 152
Control circuits
  for attenuator, 424

Converters
  low-noise, 175
Cost effectiveness, 11
Coupled microstrip
  CAD, 460
Couplers, 413
  GaAs monolithic, 414
  Lange-type, 414
  Wilkinson-type, 414
Coupling, accurate
  in MMIC networks, 471
Crossbar mixers
  GaAs monolithic, 410
  W-band, 410
Crossover networks, 471, 486
Crosstalk
  WAFFLELINE technique, 486
Crystals
  LEC, 63, 73
  mixed, 79

D

DBS
  see Direct broadcast satellites
Deembedding, 1
Depletion mode
  MESFET, 132
Dielectrics, 61
  surface passivation, 108
Digital IC
  manufacture, 87
Digital interfaces, 372
Digital measurements
  calibrations, 481
Diodes, 339
  mixers, 377, 406, 410
  Schottky, 363, 386, 390, 406, 410
Direct broadcast satellites
  receivers, 152, 153, 160, 166, 175
Direct-coupled amplifiers, 230
  feedback, 283, 313
  monolithic GaAs, 305, 313
Distributed amplifiers, 230
  four-stage, 231
  low-noise, 249
  monolithic, 249, 252
  noise figure, 262
Distributed switches
  monolithic, 350
Dose effects
  total, 142
Dual-gate FET, 252
  MESFET, 166, 175, 419

E

ECM systems
  use of power amplifiers, 151
Electrodes
  ohmic, 147
Electron beams
  submicron, 119
Electronic warfare, 227
Epitaxy, 61

F

Feedback amplifiers, 231, 297
  cascadable, 330
  dc, 283, 313
  gain bandwidth, 288
  low-noise, 288
  matched, 273
  monolithic, 283, 286, 288, 293, 313
  multistage, 293
  noise figure, 288
  two-stage, 313
FET's, 1, 61, 94
  common-gate, 186
  dual-gate, 166, 175, 252, 419
  GaAs, 119, 179, 223, 244, 252, 257, 339, 350, 457
  irradiation, 147
  K-band, 119
  light emission, 207
  models, 195, 252, 305, 419, 452
  oscillator, 396
  S-band, 119
  S-parameters, 223
  spaced electrode, 160
  switches, 350, 358, 368
  uniformity, 79
  X-band, 119, 339
Films
  surface passivation, 108
Filters
  GaAs monolithic, 427
  see also Bandpass filters; Chebyshev filters
Fixtures, tests
  universal, 475
Front ends, 13, 327

G

GaAs
  bulk, 73
  dislocation-free, 79
  FET, 119, 190
  high-purity, 63
  ion-implantation, 94
  MESFET, 132, 273, 463, 470
  microwave packaging, 490
GaAs MMIC's, 1, 13, 35
  manufacture, 56, 83
Gain-controlling elements
  microwave, 419
  wide-band, 413
Gain scalers, 419
Gamma rays
  radiation effects, 147
Gigabit/s data rate systems
  use of GaAs FET amplifiers, 297
Graphic terminals
  interactive, 449
Growth
  crystals, 73, 79
Gunn oscillators
  planar, 386

H

Heterodyne generation
  of RF signals, 201
Heterodyne receivers, 377
History
  of MMIC's, 1
Hybrid circuits, 1, 11
  CAD, 437
  versus MMIC, 35

I

IF preamplifiers, 406
  MIC, 401
Impedance inverters, 35

Implantation
 proton, 100
Inductances, 166
Integrated circuits
 analog, 83
 CAD, 435, 437
 digital, 83
 GaAs, 83, 87, 100, 108, 138, 142, 201, 313, 386
 linear, 56
 manufacture, 83, 87
 microwave, 327, 401, 437, 490
 monolithic, 179, 215, 283, 313, 319, 327, 332, 377
 packaging, 490
 proton isolation, 100, 138
 radiation effects, 142
 S-band, 94
 Si bipolar technology, 330
 substrates, 79
 testing, 87
Ion implantation, 61, 63, 218
 direct, 73
 GaAs, 94, 128
Interdigitated capacitors
 design, 427
Intermodulation
 measurement, 244

L

Lange couplers, 270, 414
Large-scale computer systems, 449
Large-signal models
 GaAs FET, 452
LEC
 see Liquid-encapsulated Czochralski growth
Limiters
 variable-threshold, 201
Liquid-encapsulated Czochralski growth
 of GaAs crystals, 63, 73
Lithography, 56
 optical, 249
LNA
 see Low-noise amplifiers
Local oscillators
 design, 201
Logic circuits, 142
Low-noise amplifiers, 151, 152
 broad-band, 463, 471
 distributed, 249
 low-cost, 172
 monolithic, 286, 288, 339
 three-stage, 153, 186
 two-stage, 179, 182, 368
 TVRO applications, 172
 with common-gate input, 186

M

Manufacturability
 of GaAs IC, 83
Matched feedback amplifiers
 gain, 273
Masks, 35
 computer-aided design, 1
 proton implantation, 100
Measurements, 435
 digital, 481
MESFET's, 13
 backgating, 132
 mixer, 166, 175
 ohmic contacts, 116
Metallization, 218

Microstrips, 390, 396, 437
 accurate models, 460
 CAD, 460
 couplers, 414
 planar, 406
 transmission lines, 427
Microwave amplifiers
 ultrawide-band, 273
Microwave circuits
 CAD, 437, 449
 GaAs monolithic, 339, 386
Microwave devices
 GaAs, 119
 switched, 338, 350, 358
 wafer probing, 481
Microwave engineers, ix
Military applications
 millimeter waves, 376
 T/R modules, 338
Millimeter-wave circuits, 151
 military applications, 376
 mixers, 401
 receivers, 377
MIM capacitors
 control, 128
 gamma rays, 147
Mixed crystals
 for IC substrates, 79
Mixers
 balanced, 390, 401, 406
 crossbar, 410
 discrete diodes, 377
 doubly balanced, 201
 dual gate MESFET, 166, 175
 GaAs monolithic, 190, 377, 390, 401, 406, 410
 millimeter waves, 401
 waveguides, 377
MMIC's (microwave monolithic integrated circuits)
 accurate coupling, 471
 active circuits, 413, 471
 applications, 151
 broad-band amplifiers, 230
 CAD, 435
 cascadable, 330
 computer-aided testing, 1
 cost effectiveness, 1, 11, 35
 couplers, 413, 414
 DBS receivers, 153
 definition, 1
 design, 1, 11, 13
 direct-coupled, 313
 fabrication, 11, 61, 128
 FET, 1, 119
 GaAs, 1, 13, 35, 56, 63, 116, 128, 147, 471
 history, 1
 hybrids, 1, 11, 35
 ion implantation, 128
 low-cost, 152, 327, 490
 materials, 61
 medium-power amplifier, 227
 military applications, 338
 millimeter waves, 376
 packaging, 486
 passive components, 413, 414
 radiation effects, 147
 reliability, 35
 repairability, 35
 reproducibility, 35
 SOS, 1, 13
 students, ix
 subcarrier, 475

S-X bands, ix
tantalum, 112
test fixtures, 475
T/R modules, 338, 372
X-band amplifiers, 112
Models
accuracy, 435
Modulators
on/off, 201
Monolithic MIC
CAD, 437
packaging, 486
Multichip T/R modules
for X-band, 368
Multistage amplifiers
feedback, 293
monolithic, 293, 305

N

Network analyzers
automatic, 475
corrected, 481
Neutron effects, 142
Noise
in distributed amplifier, 262
measurements, 190
Noise figure
feedback amplifiers, 286, 288, 293
for broad-band LNA, 463, 470
formulas, 262
Nondestructive testing
of MMIC, 475

O

Ohmic contacts, 56, 116
fabrication, 87
metallization, 119
Ohmic electrodes
degradation, 147
GaAs monolithic, 190, 195, 339
K_u-band, 195
local, 166, 201
microwave, 190
push-pull, 201
X-band, 195
On-chip components
thin films, 112
On-chip oscillators
tuning, 201
One-ports
calibration, 481
Optical repeaters
gigabit, 332
Oscillators
dielectric resonator, 153
FET, 396
large-signal model, 452
voltage-controlled, 119, 151, 195
see also Gunn oscillators; Local oscillators; On-chip oscillators

P

Packaging techniques, 435
GaAs IC, 490
high density, 486
WAFFLELINE, 486
Parasitic effects
feedback amplifiers, 293
measurement, 481

Passive components
for MMIC, 413, 414
Phased arrays, 338
active, 152, 206
microwave circuits, 339
Phase controllers
microwave, 419
Phase shifters, 338, 339
active, 419
hybrid, 419
five-bit, 372
four-bit, 358, 368
monolithic, 358, 368
X-band, 358, 363, 368
Phase splitters, 201
Photocurrent spectra, 108
Photoresists, 112
Planar mixers
crossbar, 410
Plasma CVD processes, 108, 128
Power amplifiers, 151
CAD, 457
microwave, 207, 223, 227
military applications, 206, 338
monolithic, 223, 227, 257
traveling-wave, 257
X-band, 112, 119, 368
Power combiners, 414
Power dividers, 414
Preamplifiers, 332
hybrid, 401
IC, 401, 406
monolithic, 390
two-stage, 116
Proton implantation, 138
Proton isolation
for GaAs IC, 100, 138
Push-pull amplifiers
inductively coupled, 327

Q

Quasi-optical receivers, 377

R

Radar systems
multifunction, 179
phased arrays, 339
Radiation effects
gamma rays, 147
in GaAs IC, 142
Receivers, 13, 35
baseband, 327
circuits, 152
DBS, 153, 160, 166
for millimeter waves, 377
for submillimeter waves, 377
home, 152
low-noise, 338
quasi-optical, 190, 377
use of FET, 339
Repeaters, optical
use of bipolar monolithic amplifiers, 332
Reproducibility
of MMIC, 35
Resonators
dielectric, 396
microstrip, 13

S

Sapphire
  low-loss properties, 13
Scalers
  binary weighted, 419
Schottky barrier diodes
  fabrication, 386
  mixer, 406, 410
Schottky gates
  formation, 56
  metallization, 119
Semi-insulating GaAs
  for MMIC's, 63, 73
  growth, 73, 79
Signal generation, 13
  heterodyne, 201
Silicon, 1
  bipolar technology, 330
  radiation effects, 142
Silicon dioxide
  films, 108
Silicon-on-sapphire, 1, 13
Single-chip phase shifters, 358
Single event upsets, 142
Solid-state amplifiers
  single-ended, 231
SOS
  see Silicon-on-sapphire
Space-borne systems
  use of power amplifiers, 206
Special components and circuits
  for MMIC, 413
Striplines, 490
Students at universities, ix
Subcarrier
  MMIC, 475
Submillimeter-wave region
  receiver technology, 377
Substrates, 61
  backgating, 138
  conduction, 138
  GaAs, 63
  growth, 166
  IC, 79
SUPER-COMPACT program, 449
Surface passivation dielectrics
  for GaAs IC, 108
Switches, active
  coupling, 471
Switching, microwave
  with GaAs FET, 350, 358

T

Tantalum-based process
  for MMIC, 112, 116
Tantalum oxide capacitors
  for GaAs monolithic IC, 116
Thin films, 13
  on-chip components, 112
T/R switches, 13
  X-band, 350
Transient capless annealing
  of GaAs, 94
Transmission lines, 230
  coupling, 471
  microstrip, 427
  slow-wave, 223
Transmitters
  use of FET, 339
Transmit module
  five-bit phase shift, 372
Transmit/receiver modules, 151, 338
  for X-band, 368
Traveling-wave amplifiers, 230
  monolithic, 244
  power, 257
TV broadcasting
  by DBS, 153
TVRO (TV receive only), 172
Two-ports
  calibration, 481

U

Ultrabroad-band amplifiers, 273, 297
Ultrawide-band amplifiers
  microwave, 273, 297
Universal text fixtures
  design and calibration, 475
  for MMIC, 475

V

Varactors
  monolithic, 195
  Schottky, 363
Variable attenuators
  GaAs monolithic, 424
VCO, 119, 151
  monolithic, 195
VLSI, GaAs, 108

W

Wafers
  fabrication, 87, 182
  implanted, 73
  microwaves, 481
  processing, 83
  substrates, 56
  thickness, 13
WAFFLELINE packaging technique, 486
Waveguides
  mixers, 377
  oscillators, 35
  tuners, 396
Wide-band amplifiers
  see Broad-band amplifiers
Wide-band attenuators
  GaAs monolithic, 424

Editor's Biography

Robert A. Pucel (S'48-A'52-M'56-SM'64-F'79) received the D.Sc. degree in electrical communications from M.I.T. in 1955. Both his Master's and Doctoral dissertations were performed in the field of network synthesis under the late Professor Guillemin.

In 1955, Dr. Pucel joined the Research Division of Raytheon in Waltham, MA, as a Staff Member and worked in the area of solid-state device research. From 1965 to 1970 he was Project Manager of the Microwave Semiconductor Devices and Integrated Circuits Program. He is now a Consulting Scientist in this group. His work has involved both theoretical and experimental studies of most microwave semiconductor devices, including their signal and noise properties. His activities have also included studies of propagation on dielectric and magnetic substrates, as well as research on miniature dielectric cavities. His most recent work is in the field of FET oscillator noise studies and monolithic microwave integrated circuits. He has published extensively on most of these topics.

Dr. Pucel is a co-recipient of the 1976 Microwave Prize granted by the IEEE Microwave Theory and Techniques Society. He also was the National Lecturer for the MTT Society for 1980-81 on the subject of GaAs microwave monolithic circuits. In this role he lectured in the United States, Canada, Europe, the Middle East, and Asia, including the People's Republic of China. He is a member of the Editorial Board of the MTT Society and is a Registered Professional Engineer in Massachusetts.